Print Reading for Architecture & Construction Technology

Print Reading for Architecture & Construction Technology

David Madsen Alan Jefferis

NOTICE TO THE READER

Publisher does not warrant or guarantee any of the products described herein or perform any independent analysis in connection with any of the product information contained herein. Publisher does not assume, and expressly disclaims, any obligation to obtain and include information other than that provided to it by the manufacturer.

The reader is expressly warned to consider and adopt all safety precautions that might be indicated by the activities herein and to avoid all potential hazards. By following the instructions contained herein, the reader willingly assumes all risks in connection with such instructions.

The Publisher makes no representation or warranties of any kind, including but not limited to, the warranties of fitness for particular purpose or merchantability, nor are any such representations implied with respect to the material set forth herein, and the publisher takes no responsibility with respect to such material. The publisher shall not be liable for any special, consequential, or exemplary damages resulting, in whole or part, from the readers' use of, or reliance upon, this material.

Delmar Staff
Associate Editor: Kimberly Davies
Project Editor: Elena M. Mauceri
Production Coordinator: Dianne Jensis
Art/Design Coordinator: Heather Brown

Printed in the United States of America
10 9 8 7 6 XXX 00

For more information, contact Delmar, 3 Columbia Circle, PO Box 15015, Albany, NY 12212-0515; or find us on the World Wide Web at http://www.delmar.com

Library of Congress Cataloging in Publication Data

ISBN: 0-8273-5429-0

Contents

Preface

Print Reading for Architecture and Construction Technology is a practical comprehensive workbook that is easy to use and understand. The text may be used as presented, following a logical sequence of learning activities for residential and light commercial architectural and construction technology print reading, or the chapters may be rearranged to accommodate alternate formats for traditional or individualized instruction. This is the only training and reference material students will need for architectural and construction technology print reading.

PREREQUISITES

An interest in print reading for architecture and construction technology, plus basic arithmetic, written communication, and reading skills are the only prerequisites required. Students who begin with an interest in architecture and construction technology will end with the knowledge and skills required to read complete sets of working drawings for residential and light commercial construction projects.

MAJOR FEATURES

Practical

Printing Reading for Architecture and Construction Technology provides a practical approach to reading prints as related to current common practices. One excellent and necessary foundation of print reading training is the emphasis on standardization and quality architectural and construction print examples. When students become professionals in the construction industry, this text will go along as a valuable desk reference.

Realistic

Chapters contain realistic examples, actual prints, illustrations, related tests, and print reading problems based on actual architectural prints. The examples demonstrate recommended presentation with actual architectural prints used for reinforcement.

Practical Approach to Problem Solving

The professional in the construction technology industry is responsible for accurately reading and interpreting architectural prints. This workbook explains how to read actual industry prints and interpret code requirements in a knowledge-building format; one concept is learned before the next is introduced. Print reading problem assignments are presented in order of difficulty and in a manner that provides students with a wide variety of print reading experiences. The concepts and skills learned from one chapter to the next allow students to read complete sets of working drawings in residential and light commercial architecture. The prints are presented as designs in a manner that is consistent with actual architectural office practices. Students must be able to think through the process of print reading with a foundation of how prints and related construction components are implemented. The goals and objectives of each problem assignment are consistent with recommended evaluation criterion based on the progression of learning activities.

Prints Prepared Using Computer-aided Drafting (CAD)

Computer-aided drafting is here to stay and is used in architectural drafting applications. This print reading workbook is written and presented with current CAD technology standards. This is an advantage as the students proceed into a construction career in the 21st century. All of the print reading examples and problems are prepared using CAD in a manner that displays the highest industrial standards.

Codes and Construction Techniques

National codes UBC, CABO, SBC, and BOCA are introduced and compared throughout the text as related to specific instruction and applications. Construction techniques differ throughout the country. This text clearly acknowledges the difference in construction methods and introduces the student to the format used to read complete sets of working prints for each method of construction. The print reading problem assignments are designed to provide drawings that involve a variety of construction alternatives.

Fundamental Through Advanced Coverage

This text may be used in the Architectural Drafting or Building Construction Technology curriculum that covers the basics of residential architecture in a one- or two-term sequence. In this application, students use the chapters directly associated with reading a complete set of working prints for a residence. The balance of the text may remain as reference for future study or as a valuable desk reference. The workbook may also be used in the comprehensive architectural construction technology program where a four- to five-term or semester sequence of residential and light commercial construction skills and theory is required. In this application, students may expand on the primary objective of reading a complete set of residential prints followed by light commercial projects. Additional coverage is also provided in the following areas:

1. Energy efficient construction techniques.
2. Heating and cooling thermal performance calculations.
3. Structural load calculations.

Course Plan

Architectural and construction technology print reading is the primary emphasis of many technical drafting and construction technology curriculums while other programs offer only an exploratory course in this field. This text is appropriate for either application. The content of this text reflects the common elements in a comprehensive architectural and construction technology curriculum and may be used in part or totally.

Section Length

Chapters are presented in individual learning segments that begin with elementary concepts and build until each chapter provides complete coverage of each topic. Instructors may choose to present lectures in short 15-minute discussions or divide each chapter into 40–50 minute lectures.

Applications

Special emphasis has been placed on providing realistic print reading examples and problems. The examples and problems have been supplied by architects, consulting engineers, and architectural designers. Each problem solution is based on the step-by-step layout procedures provided in the chapter discussions. Problems are given in the order of complexity so that students may be exposed to a variety of print reading experiences. Problems require the students to go through the same thinking process that a professional construction worker is faced with daily, including reading symbols, notes, finding and interpreting information, and many other activities. Chapter tests provide a complete coverage of each chapter and may be used for student evaluation or as study questions.

Instructor's Guide

The instructor's guide contains test and exercise solutions.

List of Reviewers

We would like to give special thanks and acknowledgement to the many professionals who reviewed the manuscript of this text in an effort to help us publish the best *Print Reading for Architecture and Construction Technology* text:

Gilbert Atkins — Mercer County Vocational Technical Center, Princeton, West Virginia

Charles Case — ITT Technical Institute, Indianapolis, Indiana

Victor Marshall — Washington-Holmes Vocational Technical Center, Chipley, Florida

Michael Price — Walters State Community College, Morristown, Tennessee

The quality of this text is also enhanced by the support and contributions from architects, designers, engineers, and vendors. The list of contributors is extensive and acknowledgment is given at each figure illustration; however, the following individuals and companies gave an extraordinary amount of support:

Dan Kovac — Piercy and Barclay Designers

Ken Smith — Structureform Masters, Inc.

Wally Grainer — Sunridge Designs

E. Henry Fitzgibbon, AIA — Soderstrom Architects, PC

TO THE STUDENT

This *Print Reading for Architecture and Construction Technology* workbook is designed for you, the student. The development and format presentation has been tested in actual conventional and individualized classroom instruction. The information presented is based on architectural and construction standards, drafting room practice, and trends in the building construction industry. This text is designed to be comprehensive coverage of architectural and construction technology print reading. Use the text as a learning tool while in school and take it along as a desk reference when you enter the profession. The amount of written text is complete but kept to a minimum. Examples and illustrations are used extensively. Many students learn best by studying examples. Here are a few helpful hints:

1. **Read the Text.** The text is intentionally designed to be easily read. It gives information in as few and easy to understand words as possible. Do not pass up the reading because the text helps you to understand clearly the prints that you will read.
2. **Look Carefully at the Examples.** The figure examples are presented in a manner that is consistent with architectural and construction technology standards. Look at the examples carefully and attempt to understand their specific applications. If you can understand why something is done a certain way, it is easier to read the actual prints. Print reading is often like doing a puzzle; you may need to search carefully the print to find the desired information.
3. **Use the Text as a Reference.** Few professionals know everything about standard practices, techniques, and concepts, so always be ready to use the reference if you need to verify how specific applications are handled. Become familiar with the use and definitions of technical terms. It will be difficult to memorize everything in this text, but after studying and using the concepts, applying print reading skills should become second nature.
4. **Learn Each Concept and Skill Before You Continue to the Next.** The text is presented in a logical learning sequence. Each chapter is designed for learning development, and chapters are sequenced so print reading knowledge develops from one chapter to the next. Print reading problem assignments are presented in the same learning sequence as the chapter content and also reflect progressive levels of difficulty.
5. **Do the Chapter Tests.** Each chapter has a test at the end. Answering these test questions gives you an opportunity to review the material that you just studied. This reinforcement helps you to learn the material fully.
6. **Do the Print Reading Problems.** There are several print reading problems after each chapter test. These problems require that you answer questions as you read actual architectural and construction technology prints. There is no substitute for reading actual prints. The practice of print reading becomes easier when you have had a chance to read several prints. By the time you complete the print reading problems, the content covered in the preceding chapter will be easier for you to identify on future prints.

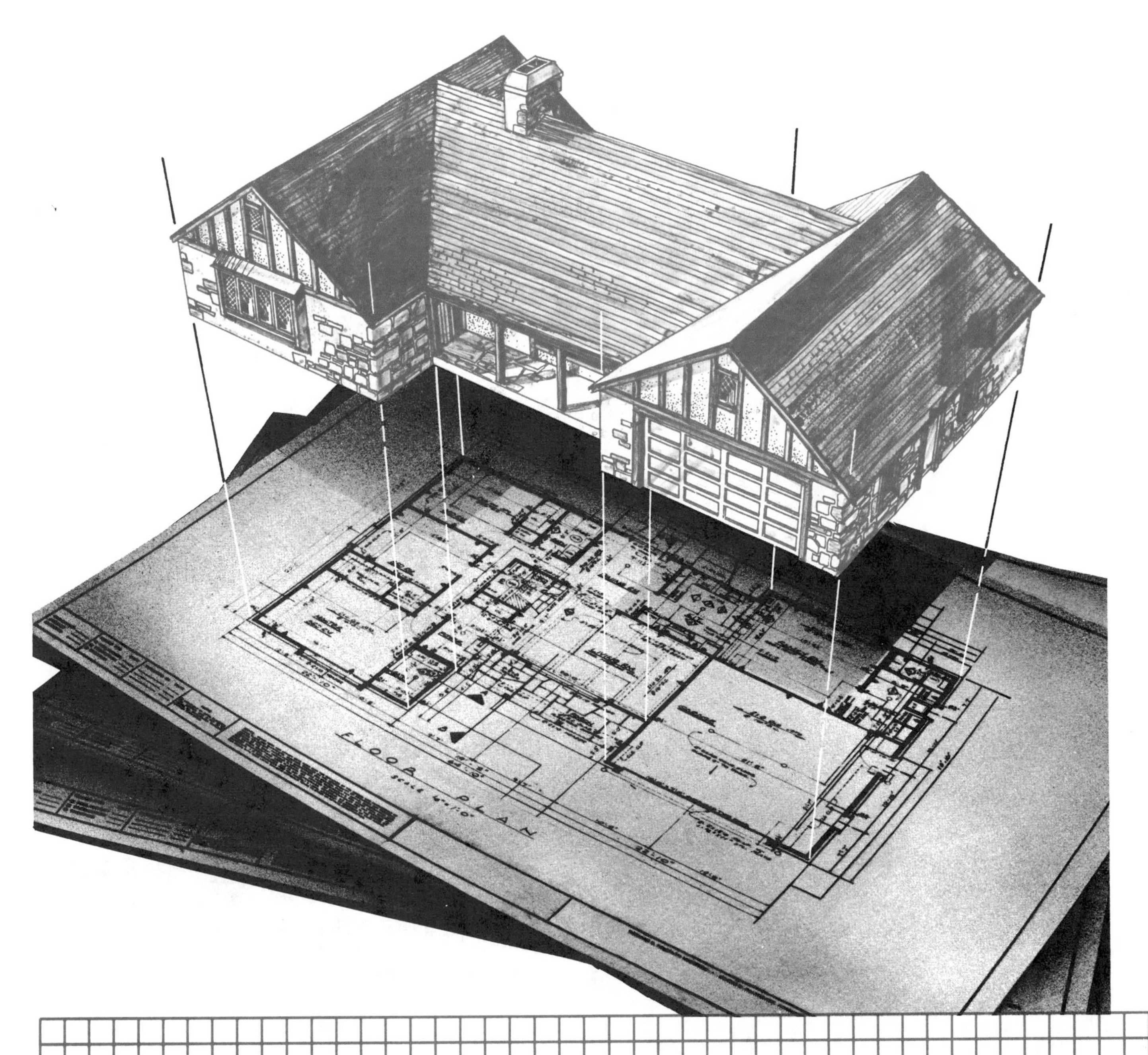

Part 1

Reading Residential Prints

Chapter 1

Introduction to Architectural and Construction Technology Print Reading

ARCHITECTURE IS the art and science of building construction. Architectural drawings, known as *plans*, are prepared by professional architects or designers for any home or commercial building before the actual construction can begin. Home building is referred to as *residential construction*, and *commercial construction* is any building for business or industrial purposes.

ABOUT PRINTS

This text refers to the reproduction of architectural drawings as *prints*. "Blueprint" is an old term now generally used in the construction business when referring to prints. Blueprinting is an old method that results in a print with a dark blue background and white lines. You may find some of these in a museum or in a company's drawing archives. The reproduction methods commonly used today are the diazo and photocopy processes.

The Diazo Print Process

Diazo prints are also known as blue-line prints because the resulting print is generally blue lines on a white background. The diazo process is an inexpensive way to make copies from any type of translucent material. The diazo print is made using a printing process that uses ultraviolet light passing through a translucent original drawing to expose a chemically coated paper or print material underneath. The light does not pass through the lines of the drawing. Thus the chemical coating beneath the lines remains unexposed. The print material is then exposed to ammonia vapor, which activates the remaining chemical coating to produce the blue lines. Diazo materials are also available that make black or brown lines. The diazo print process is demonstrated in Figure 1–1.

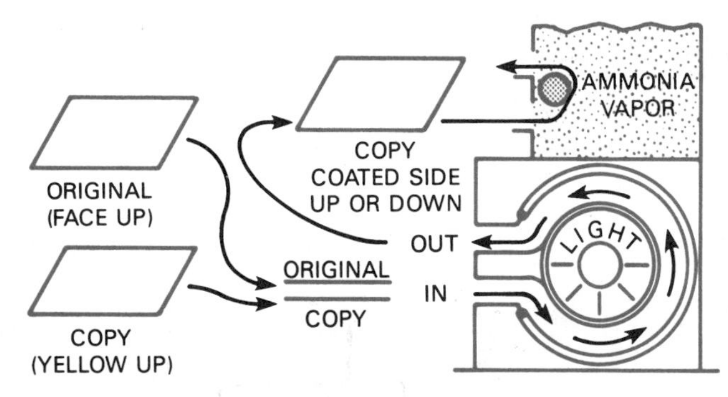

Figure 1–1 The diazo print process.

Photocopy Prints

The photocopy process is gaining wide use for copying engineering and architectural drawings. These copy machines are much like the copier found in the traditional office setting. The only difference is the capacity allows the copying of large drawings. The popularity of these machines is growing because there is no need for coated copy materials or the possible hazards of ammonia.

SHEET SIZES, TITLE BLOCKS, AND BORDERS

All professional drawings have title blocks. Standards have been developed for the information in the title block and on the sheet adjacent to the border so the drawing is easier to read and file than drawings that do not follow a standard format.

Sheet Sizes

Architectural drafting offices generally use sheet sizes of 8½ × 11 in., 12 × 18, 18 × 24, 22 × 34 in. (24 × 36), 28 × 42 in., 30 × 42 in., 30 × 48 in., or 36 × 49 in. Standard sheet sizes generally fold into 8½ × 11 in. format.

Zoning

Some companies use a system of numbers along the top and bottom margins and letters along the left and right margins, called *zoning*. Zoning allows the drawing to be read like a road map. For example, the reader can refer to the location of a specific item as D-4, which means that the item can be found at or near the intersection of D across and 4 up or down. Zoning may be found on architectural drawings for large commercial projects.

Architectural Drafting Title Blocks

Title blocks and borders are normally preprinted on drawing paper. Most architectural drafting firms use one basic sheet size with preprinted borders and title blocks.

Architectural drawing title blocks are generally placed along the right side of the sheet, although some companies place them across the bottom of the sheet. Each company uses a slightly different title block design, but the same general information is found in almost all blocks:

1. Drawing number. This may be a specific job or file number for the drawing.
2. Company name, address, and phone number.
3. Project or client. This is an identification of the project by company or client name, project title, or location.
4. Drawing name. This is where the title of the drawing may be placed. For example, MAIN FLOOR PLAN or ELEVATIONS. Most companies omit this information from the title block and place it on the face of the sheet below the drawing.
5. Scale. Some company title blocks provide a location for the general scale of the drawing. On any view or detail on the sheet that differs from the general scale, the scale of that view must be identified below the view title and both placed directly below the view. Most companies omit the scale from the title block and place it on the sheet directly below the title of each individual plan, view, or detail.
6. Drawing or sheet identification. Each sheet is numbered in relation to the entire set of drawings. For example, if the complete set of drawings has eight sheets, then each consecutive sheet is numbered 1 of 8, 2 of 8, 3 of 8, and so on, to 8 of 8. Commercial drawings are often divided into major divisions such as the A (architectural), S (structural), M (mechanical), and P (plumbing).

7. Date. The date noted is the date on which drawing or project is completed.
8. Drawn by. This is where the drafter, designer, or architect who prepared this drawing places his or her initials or name.
9. Checked by. This is the identification of the individual who approves the drawing for release.
10. Architect or designer. Most title blocks provide for the identification of the individual who designed the structure.
11. Revisions. Many companies provide a revision column in which drawing changes are identified and recorded. Where changes are made on the face of the drawing after it has been released for construction, a circle with a revision number or letter accompanies the change. This revision number is keyed to a place in the drawing title block where the revision number, revision date, initials of the individual making the change, and an optional brief description of the revision are located.

Figure 1–2 shows a typical architectural title block and its elements. The location of specific items may differ slightly and some companies require more detailed information.

LINES ON ARCHITECTURAL DRAWINGS

Drafting is a universal graphic language that uses lines, symbols, and notes to describe a structure to be built. Certain lines may be drawn thick so that they stand out clearly from other information on the drawing. Other lines are drawn thin. Thin lines are not necessarily less important than thick lines, but they may be subordinate for identification purposes.

Outlines

In architectural drafting outline lines are used to define the outline and characteristic features of architectural plan components, but the method of presentation may differ slightly from one office to another.

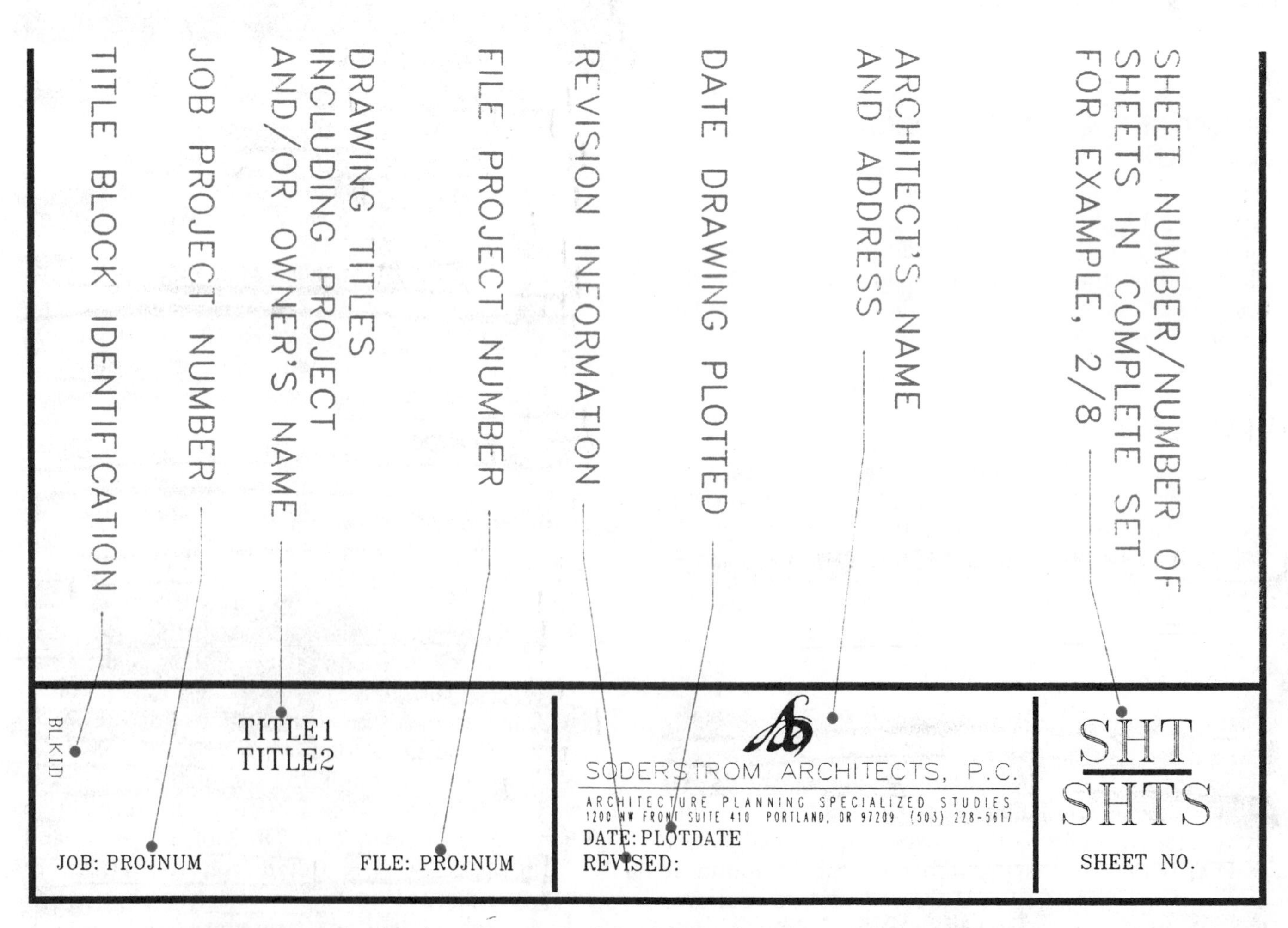

Figure 1–2 The sample architectural title block and its components. *Courtesy Soderstrom Architects.*

The following techniques may be alternatives for outline line presentation:

1. One popular technique is to enhance certain drawing features so they stand out clearly from other items on the drawing. For example, the outline of floor plan walls and partitions or beams in a cross section may be drawn thicker than other lines so that they are more apparent than the other lines on the drawing. See Figure 1–3.
2. Another technique is for all lines of the drawing to be the same thickness. This method does not differentiate one type of line from another. See Figure 1–4(a). This technique may use dark shading to accentuate features, such as walls in floor plans, as shown in Figure 1–4(b).

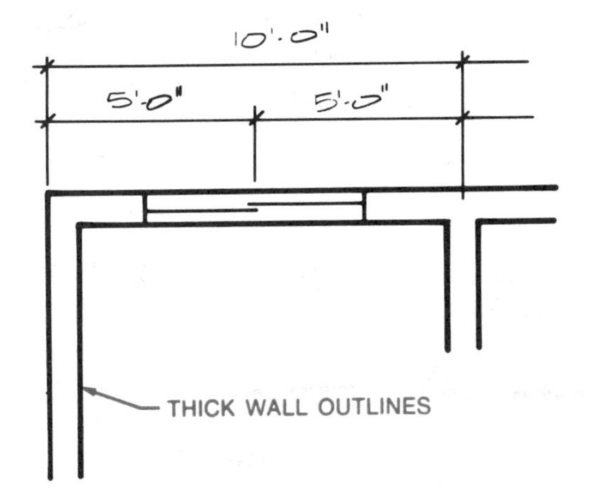

Figure 1–3 Thick outlines.

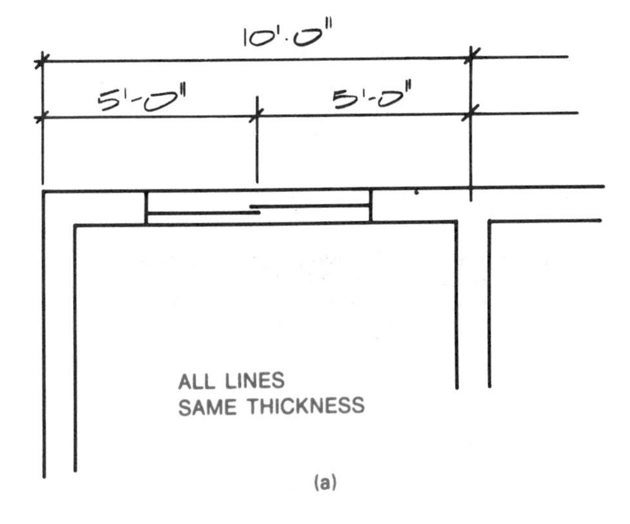

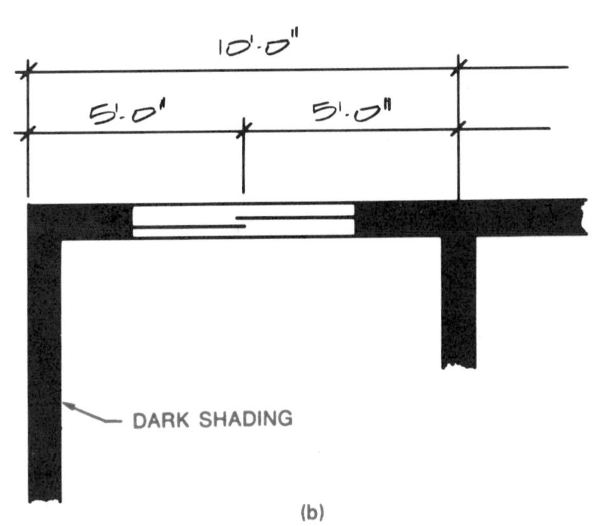

Figure 1–4 (a) All lines the same thickness; (b) accent with wall shading.

Dashed Lines

In architectural drafting dashed lines are used to show drawing features that are not visible in relationship to the view or plan. These dashed features may also be subordinate to the main emphasis of the drawing. Examples of dashed line representations include beams, Figure 1–5; and upper kitchen cabinets, undercounter appliances (dishwasher), or electrical circuit runs, Figure 1–6.

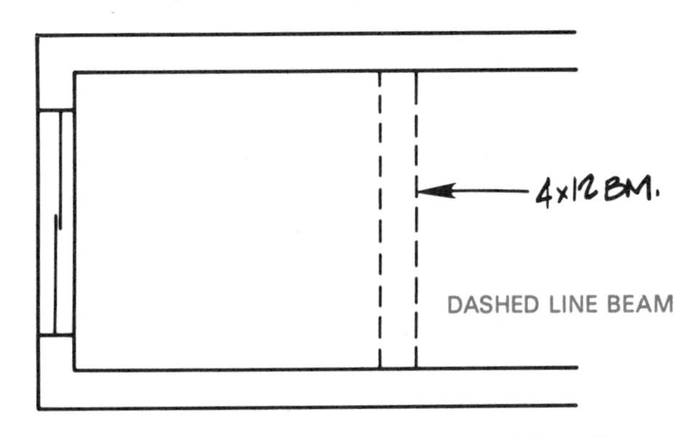

Figure 1–5 Dashed lines used to represent beam above.

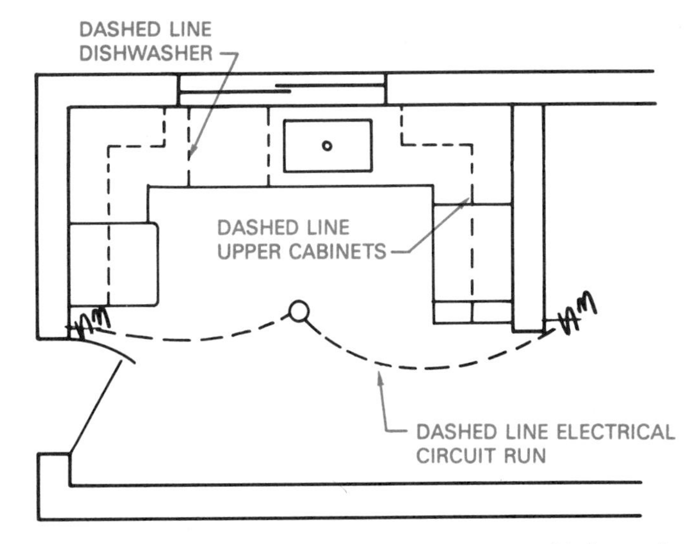

Figure 1–6 Dashed lines used to represent upper kitchen cabinets, dishwasher under the cabinet, and electrical circuit run.

Extension and Dimension Lines

Extension lines show the extent of a dimension, and dimension lines show the length of the dimension and terminate at the related extension lines with slashes, arrowheads, or dots. The dimension numeral in feet and inches is placed above and near the center of the solid dimension line. Figure 1–7 shows several dimension and extension line examples.

Leader Lines

Leader lines are used to connect notes to related features on a drawing. Figure 1–8 shows several examples.

Break Lines

The two types of break lines are the long break line and the short break line. The type of break line normally associated with architectural drafting is the long break line. Break lines are used to terminate features on a drawing when the extent of the feature has been clearly defined. Figure 1–9(a) shows several examples. The short break line may be found on some architectural drawings. This line, as shown in Figure 1–9(a), is an irregular line and may be used for a short area. Breaks in cylindrical objects, such as steel bars and pipes, are shown in Figure 1–9(b).

An architectural drawing showing a variety of typical lines is displayed in Figure 1–10 (page 8).

LETTERING ON AN ARCHITECTURAL DRAWING

Lettering is used on an architectural drawing to provide written information. Architectural lettering is generally done with all uppercase letters, and abbreviations are commonly used to save space and drafting

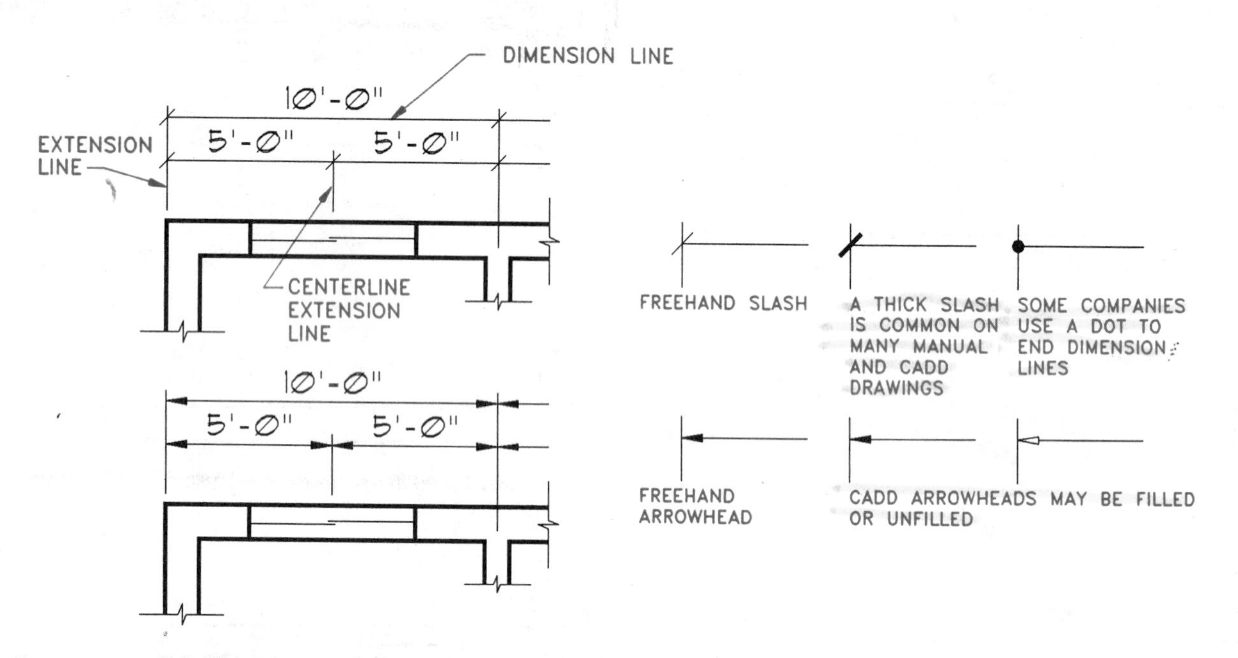

Figure 1–7 Dimension and extension lines.

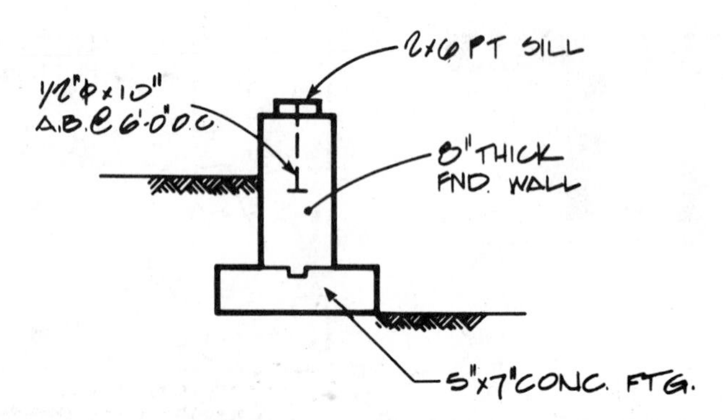

Figure 1–8 Sample leader lines.

time. Refer to the list of abbreviations in the appendix of this text for clarification of abbreviations. Lettering is shown in the form of title block information, drawing titles, room labels and related information, *specific notes*, *general notes*, and *schedules*.

Specific notes are also known as *local notes*. Specific notes may be placed close to the item or connected to the item being identified with a leader line. These are notes placed on the drawing that relate to a specific application or describe a specific feature, such as 3" CONC. FILLED GUARD POST. Specific notes are shown in Figure 1–11 (page 9).

A *general note* is information that applies to the entire drawing. General notes may be found individually anywhere on the drawing, or more commonly, they are grouped together in convenient areas of the drawing away from drawing components. General notes may be grouped together and have such titles as GENERAL NOTES, NOTES, FRAMING NOTES, CONCRETE NOTES, or MISCELLANEOUS NOTES, as shown in Figure 1–12 (page 10).

Schedules are charts of information used to describe such items as doors, windows, appliances, materials, fixtures, hardware, concrete reinforcing, and finishes. Schedules help keep drawings clear of unnecessary notes. Items in the schedules are keyed to the drawing with identification letters, numbers, and symbols. When using door and window schedules, for example, the key may label doors with a letter and windows with a number placed in different geometric shapes.

Look at Figure 1–13 (page 12) and notice the circles with letters by the floor plan, which are door symbols, and the hexagons with numbers by the floor plan, which are window symbols. These letters and numbers are then keyed to the door and window schedules in Figure 1–13.

HOW DO YOU FIND INFORMATION ON PRINTS?

Print reading is basically finding information on prints. You have seen that information may be displayed on a print in the form of lines, notes, and schedules. These items are located either in the title block or in the *field* of the drawing. The *field of the drawing* is anywhere within the border lines outside the title block. As you will learn later, many items on the drawings are made up of symbols. If you know the information you are looking for is usually displayed as a symbol, then you need to look for that symbol in a location where it is often found. Information in the form of symbols is discussed as related to specific applications throughout this book. For now, here are some general rules to follow on architectural prints:

- Scan the entire drawing while looking at the general layout. This allows you to become familiar with the major parts of the drawing and should quickly tell you if you are looking at the correct print. For example, if you want to look at the main floor plan you should not be looking at the sheet labeled FOUNDATION PLAN, unless they happen to be next to each other.

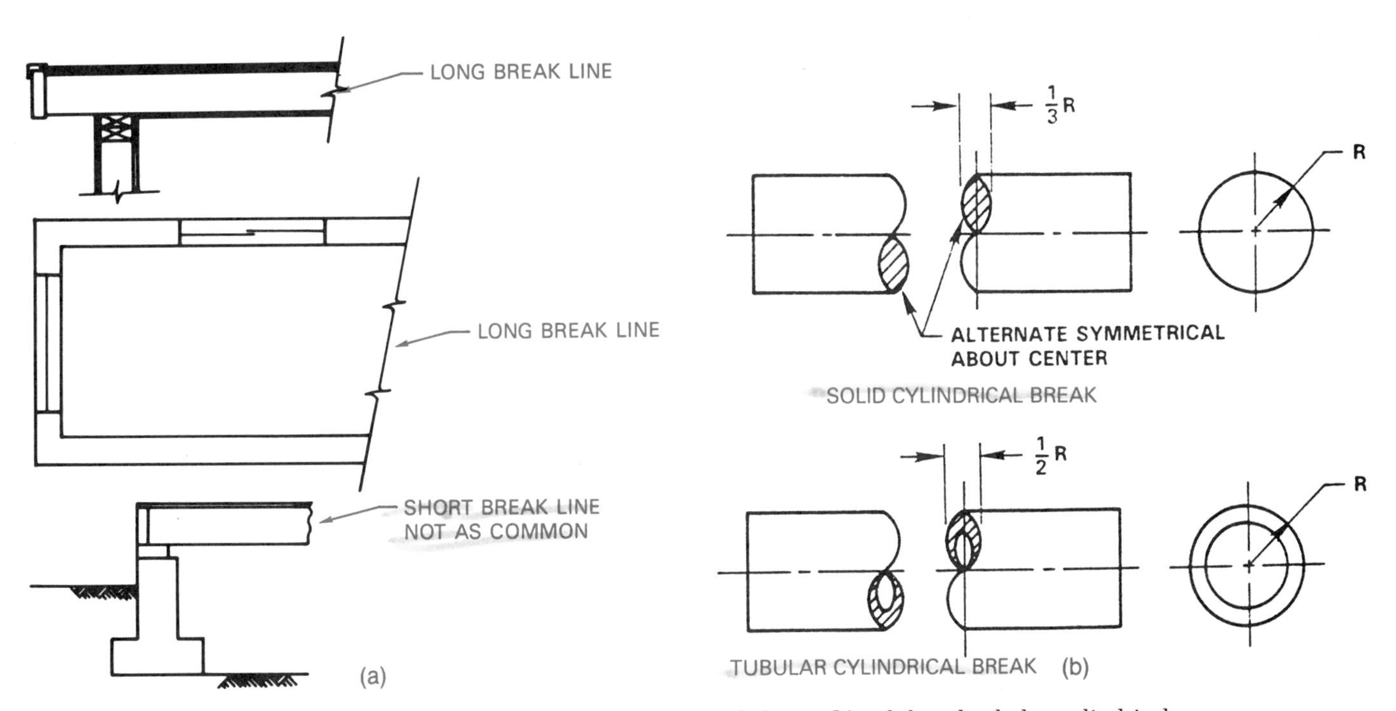

Figure 1–9 (a) Use of the long and short break lines; (b) solid and tubular cylindrical breaks.

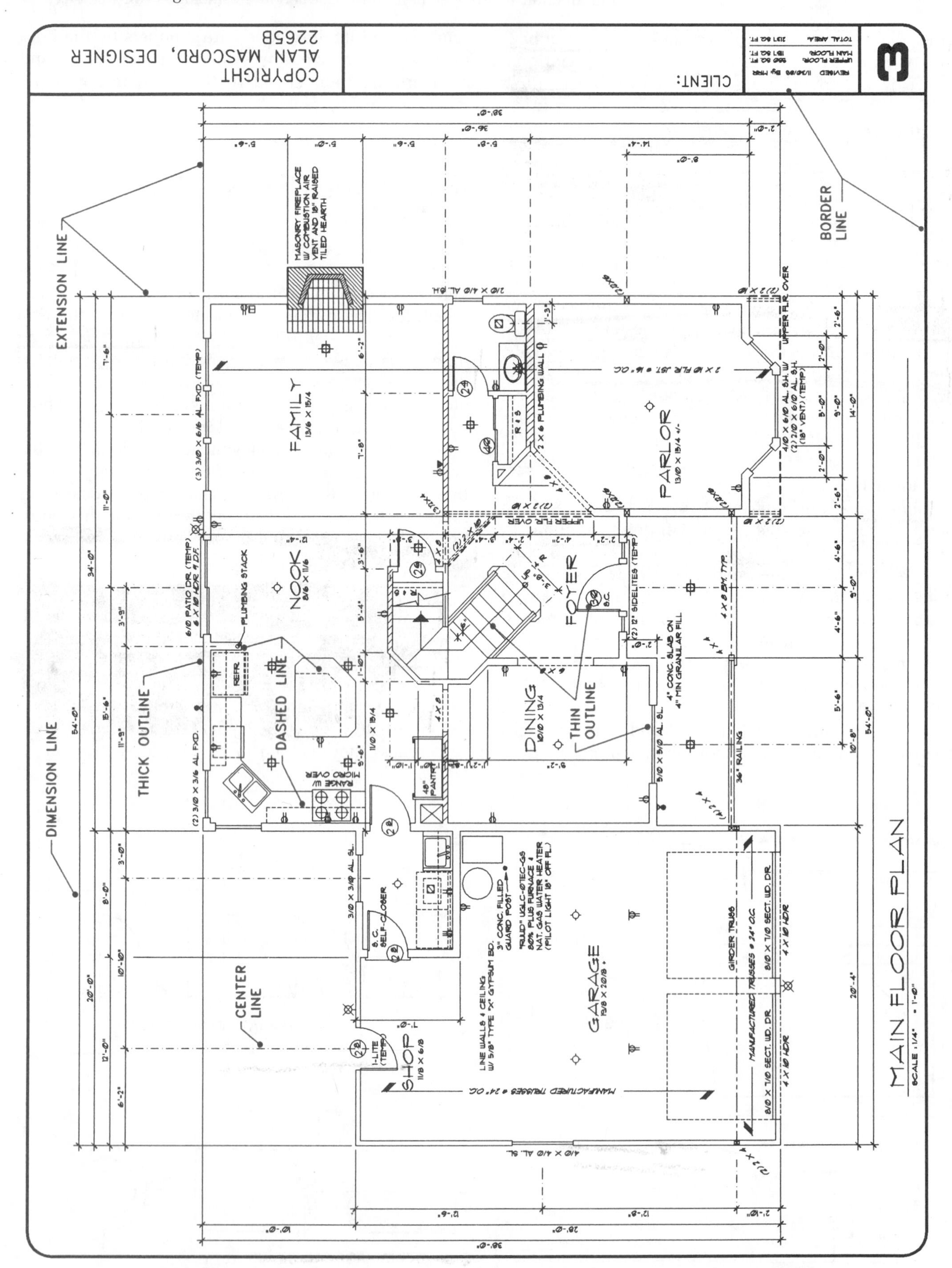

Figure 1–10 A floor plan showing a variety of different forms. *Courtesy Alan Mascord Design Assoc., Inc.*

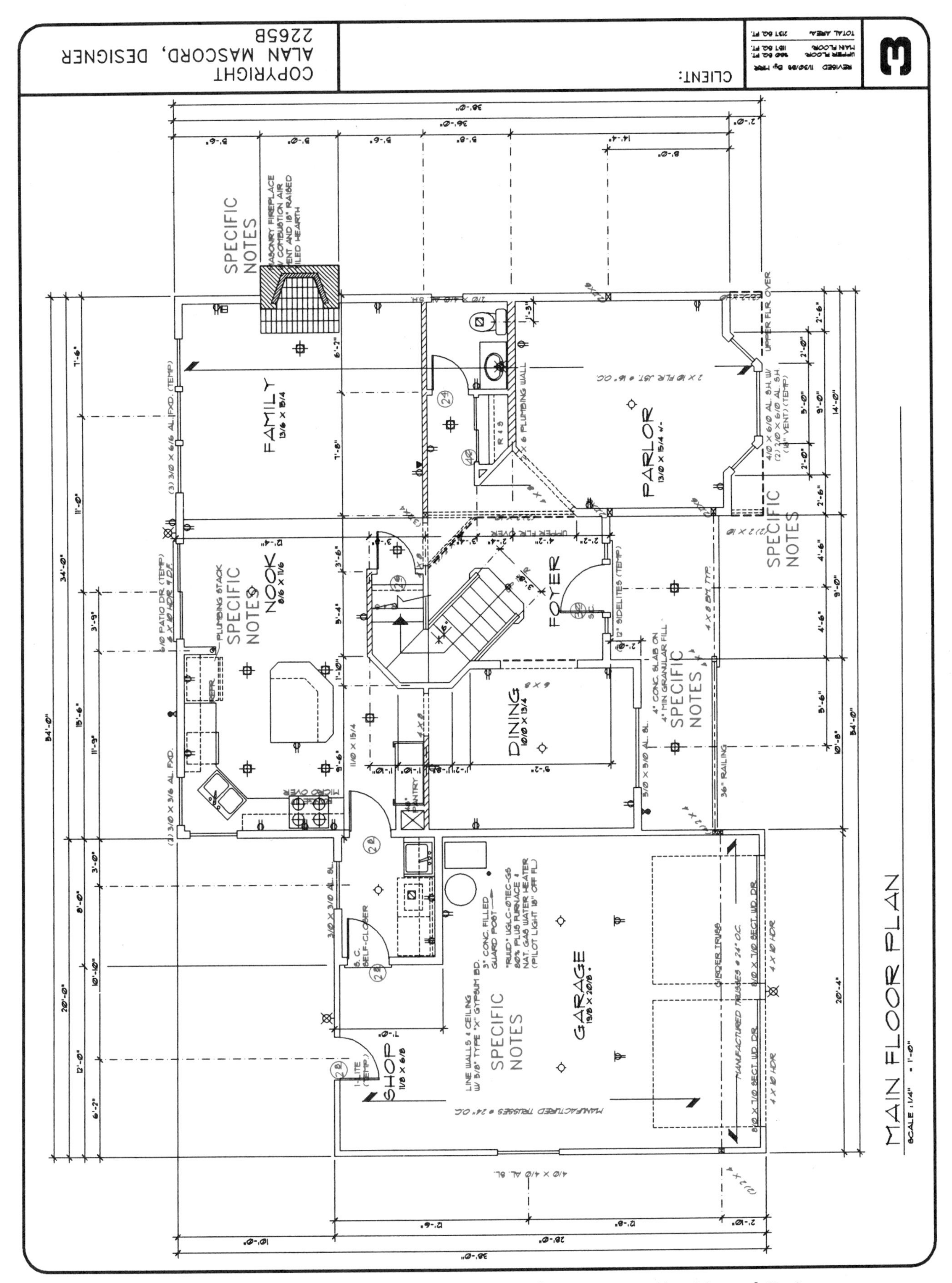

Figure 1–11 Specific notes shown on a floor plan. *Courtesy Alan Mascord Design Assoc., Inc.*

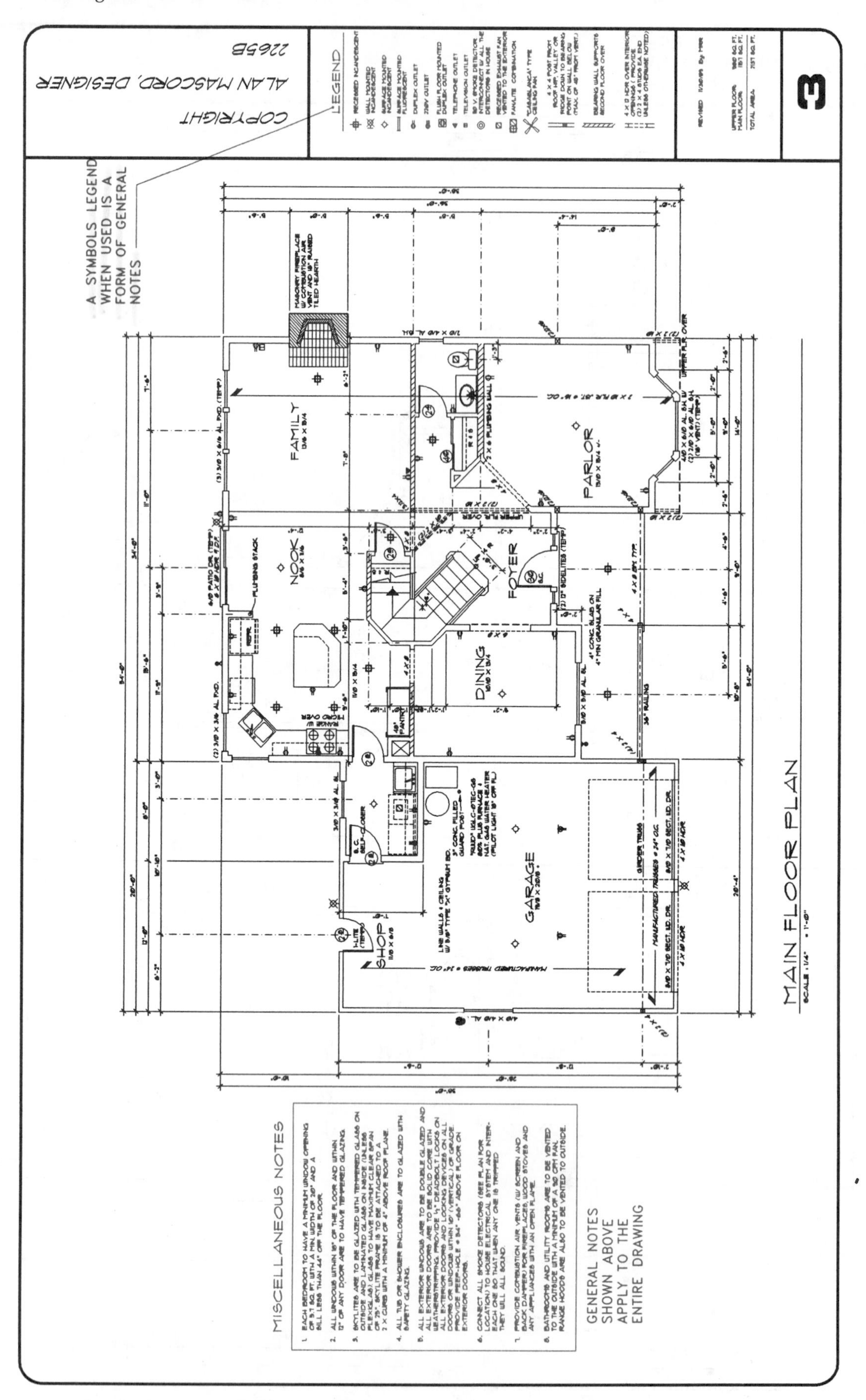

Figure 1–12 (a) A floor plan with general notes displayed; (b) typical general notes. *Courtesy Alan Mascord Design Assoc., Inc.*

MISCELLANEOUS NOTES

1. EACH BEDROOM TO HAVE A MINIMUM WINDOW OPENING OF 5.7 SQ. FT. WITH A MIN. WIDTH OF 20" AND A SILL LESS THAN 44" OFF THE FLOOR.
2. ALL WINDOWS WITHIN 18" OF THE FLOOR AND WITHIN 12" OF ANY DOOR ARE TO HAVE TEMPERED GLAZING.
3. SKYLITES ARE TO BE GLAZED WITH TEMPERED GLASS ON OUTSIDE AND LAMINATED GLASS ON INSIDE (UNLESS PLEXIGLAS). GLASS TO HAVE MAXIMUM CLEAR SPAN OF 25". SKYLITE FRAME IS TO BE ATTACHED TO A 2 X CURB WITH A MINIMUM OF 4" ABOVE ROOF PLANE.
4. ALL TUB OR SHOWER ENCLOSURES ARE TO GLAZED WITH SAFETY GLAZING.
5. ALL EXTERIOR WINDOWS ARE TO BE DOUBLE GLAZED AND ALL EXTERIOR DOORS ARE TO BE SOLID CORE WITH WEATHERSTRIPPING. PROVIDE ½" DEADBOLT LOCKS ON ALL EXTERIOR DOORS AND LOCKING DEVICES ON ALL DOORS OR WINDOWS WITHIN 10' (VERTICAL) OF GRADE. PROVIDE PEEP-HOLE @ 54" -66" ABOVE FLOOR ON EXTERIOR DOORS.
6. CONNECT ALL SMOKE DETECTORS (SEE PLAN FOR LOCATION) TO HOUSE ELECTRICAL SYSTEM AND INTER-EACH ONE SO THAT WHEN ANY ONE IS TRIPPED THEY WILL ALL SOUND.
7. PROVIDE COMBUSTION AIR VENTS (W/ SCREEN AND BACK DAMPER) FOR FIREPLACES, WOOD STOVES AND ANY APPLIANCES WITH AN OPEN FLAME.
8. BATHROOMS AND UTILITY ROOMS ARE TO BE VENTED TO THE OUTSIDE WITH A MINIMUM OF A 90 CFM FAN. RANGE HOODS ARE ALSO TO BE VENTED TO OUTSIDE.

Figure 1–12 (b)

- Look at the title block to find general information related to the project, such as the architect's name, client's name, project title, drawing number, and sheet number.
- Look at the drawings on the print for a quick understanding of what is included. For example, if you are looking at a floor plan, look at the title, the room layout, the major features, and the overall dimensioning format.
- Read the general notes to get a good understanding of the construction specifications and other information that relates to the entire drawing.
- Now that you have a good understanding of the major features found on the drawing, take more time to look for the specific information that you seek.

If you follow these basic guidelines you normally gain a quick understanding of what the print includes, and in many cases you are able to find the information needed. Sometimes, however, the drawing may be very complex and the information needed is difficult to find. When this happens, print reading can be time consuming. After you study this text and when you gain experience reading a variety of prints, you will normally be able to read a print quickly and interpret the information given.

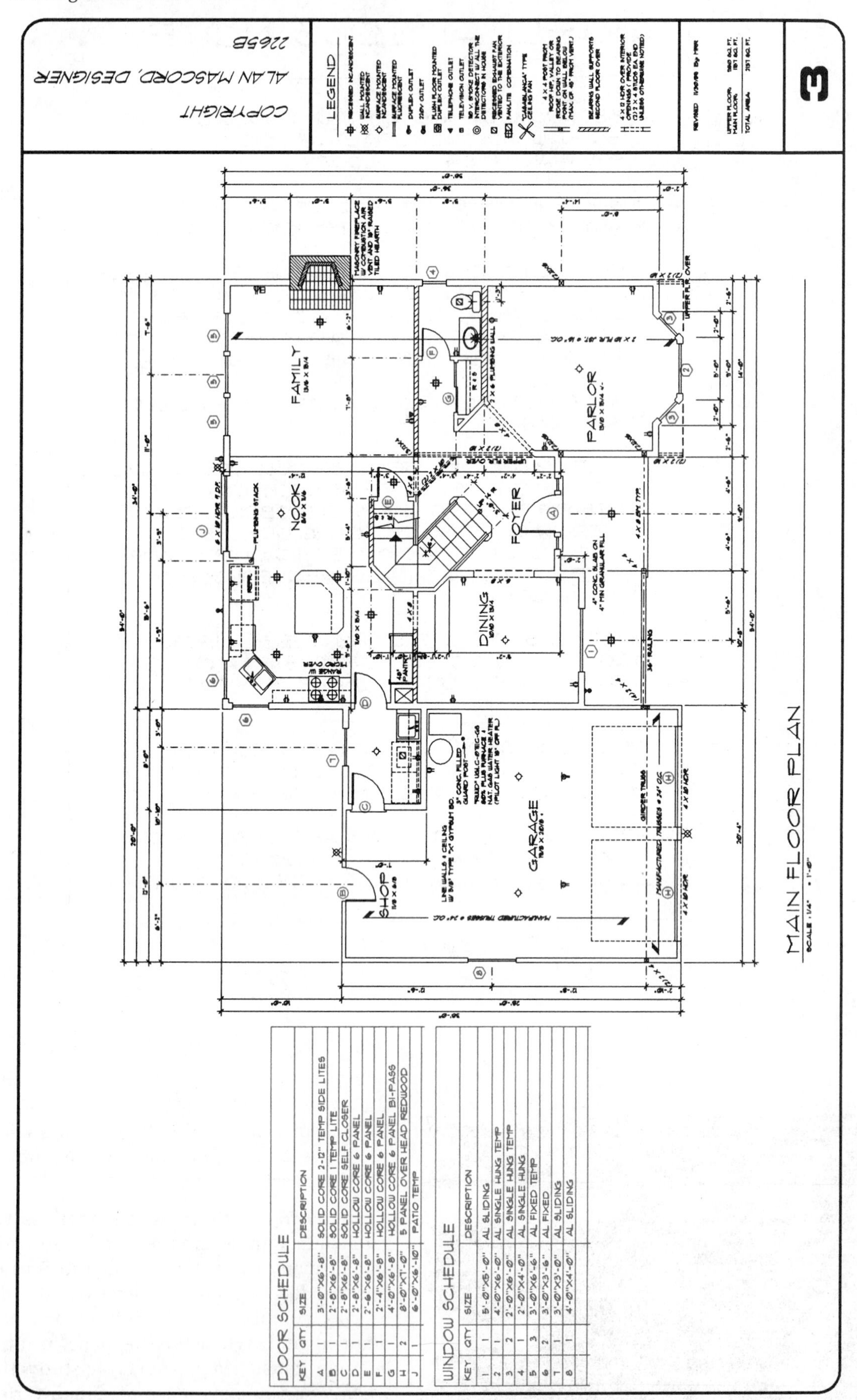

DOOR SCHEDULE

KEY	QTY	SIZE	DESCRIPTION
A	1	3'-0"X6'-8"	SOLID CORE 2-12" TEMP SIDE LITES
B	1	2'-8"X6'-8"	SOLID CORE 1 TEMP LITE
C	1	2'-8"X6'-8"	SOLID CORE SELF CLOSER
D	1	2'-8"X6'-8"	HOLLOW CORE 6 PANEL
E	1	2'-6"X6'-8"	HOLLOW CORE 6 PANEL
F	1	2'-4"X6'-8"	HOLLOW CORE 6 PANEL
G	1	4'-0"X6'-8"	HOLLOW CORE 6 PANEL BI-PASS
H	2	8'-0"X7'-0"	5 PANEL OVER HEAD REDWOOD
J	1	6'-0"X6'-10"	PATIO TEMP

WINDOW SCHEDULE

KEY	QTY	SIZE	DESCRIPTION
1	1	5'-0"X5'-0"	AL SLIDING
2	1	4'-0"X6'-0"	AL SINGLE HUNG TEMP
3	2	2'-0"X6'-0"	AL SINGLE HUNG TEMP
4	1	2'-0"X4'-0"	AL SINGLE HUNG
5	3	3'-0"X6'-6"	AL FIXED TEMP
6	2	3'-0"X3'-6"	AL FIXED
7	1	3'-0"X3'-0"	AL SLIDING
8	1	4'-0"X4'-0"	AL SLIDING

Figure 1–13 (a) This drawing shows how the door and window schedules are keyed to the floor plan; (b) schedules and reference symbols from enlarged. *Courtesy Alan Mascord Design Assoc., Inc.*

DOOR SCHEDULE			
KEY	QTY	SIZE	DESCRIPTION
A	1	3'-Ø"X6'-8"	SOLID CORE 2-12" TEMP SIDE LITES
B	1	2'-8"X6'-8"	SOLID CORE 1 TEMP LITE
C	1	2'-8"X6'-8"	SOLID CORE SELF CLOSER
D	1	2'-8"X6'-8"	HOLLOW CORE 6 PANEL
E	1	2'-6"X6'-8"	HOLLOW CORE 6 PANEL
F	1	2'-4"X6'-8"	HOLLOW CORE 6 PANEL
G	1	4'-Ø"X6'-8"	HOLLOW CORE 6 PANEL BI-PASS
H	2	8'-Ø"X7'-Ø"	5 PANEL OVER HEAD REDWOOD
J	1	6'-Ø"X6'-10"	PATIO TEMP

WINDOW SCHEDULE			
KEY	QTY	SIZE	DESCRIPTION
1	1	5'-Ø"X5'-Ø"	AL SLIDING
2	1	4'-Ø"X6'-Ø"	AL SINGLE HUNG TEMP
3	2	2'-Ø"X6'-Ø"	AL SINGLE HUNG TEMP
4	1	2'-Ø"X4'-Ø"	AL SINGLE HUNG
5	3	3'-Ø"X6'-6"	AL FIXED TEMP
6	2	3'-Ø"X3'-6"	AL FIXED
7	1	3'-Ø"X3'-Ø"	AL SLIDING
8	1	4'-Ø"X4'-Ø"	AL SLIDING

A B

1 2

Figure 1–13 (b)

NOTE: the circle w/ letter is for door locator; the stop sign shape w/ number is windows

CHAPTER 1 TEST

Fill in the blanks below with the proper word or short statement as needed to complete the sentence or answer the question correctly.

1. BLUEPRINT is an old term that is generally used in the construction business when referring to prints.
2. The reproduction methods commonly used today are the Diazo Print and Photocopy Print processes.
3. Diazo prints are also known as Blue line prints because the resulting print is generally blue lines on a white background.
4. When a diazo copy is made the ultraviolet light passes through the drawing paper and only the chemical behind the lines or letters remains. The ammonia then activates the remaining chemical to produce the blue lines.
5. State at least three reasons that the popularity of photocopying is increasing for copying architectural drawings. these copiers are much like standard copiers
   allows for copying large drawings
   doesn't have the ammonia health risks that Diazo Prints have.
6. List at least three standard drawing sheet sizes. 8½x11", 12x18, 36x49

7. A system of numbers along the top and bottom margins and letters along the left and right margins used to locate items on the drawing is called zoning ________.
8. List at least four items represented by dashed lines on an architectural drawing. ________
beams, upper kitchen cabinets, undercounter appliances, electrical circuit runs
9. Specific notes are also known as ________.
10. Define specific notes. ________
11. Define general notes. ________
12. Define schedules. ________
13. Briefly describe in your own words five general rules you can use to find information on an architectural drawing:
Rule 1: ________.
Rule 2: ________.
Rule 3: ________.
Rule 4: ________.

CHAPTER 1 EXERCISES

1. Given the title block on page 16, fill in the blanks identifying the type of information requested at each location.
2. Answer the following questions as you read the partial floor plan given on page 17:
   a. Give the sizes of the following rooms:
   Parlor ________
   Dining ________
   b. Give the complete word or words for each of the following abbreviations:
   AL ________
   BM ________
   CONC. ________
   FLR ________
   JST ________
   MIN ________
   OC ________
   SH ________
   TEMP ________
   W/ ________

c. Give the size and spacing of the joists over the parlor. ______________________________.
d. Give the specifications for the material to be used under the 4" concrete slab. ______________.
e. Is the note described in d a specific or a general note? ______________________________
f. Give the dimension from the front face of the bay window in the parlor to the center of the two 2×6's.
____

3. Answer the following questions as you read the partial floor plan on page 18:
a. Are the notes in the box at the left of the floor plan specific or general notes? ______________
b. What is the size of the garage? ____
c. Does the floor plan size given for the garage include the shop? ____________________
d. What is the dimension from the front face of the garage to the girder truss? ______________
e. Give the specifications for the construction members used in the ceiling of the garage between the front face of the garage and the girder truss. ______________________________________
f. Give the specifications for the construction members used in the ceiling of the garage from the girder truss to the back wall of the garage. ______________________________________
g. Give the specifications for all the exterior windows and doors. ______________________
____
h. Give the specifications for all bathroom and utility room fans. ______________________
____
i. Give the specifications provided for the furnace and hot-water heater. __________________
____
j. Give the complete word or words for each of the following abbreviations or symbols:
SQ ____________________
FT ____________________
CFM ____________________
@ ____________________
% ____________________

4. Answer the following questions as you read the partial floor plan on page 19:
a. What is the purpose of the letter inside the circles located by the door symbols? ______________
____
b. What is the purpose of the number inside the hexagons located by the window symbols? __________
____
c. Give the quantity, size, and description of the G doors. ______________________________
____
____
d. Give the quantity, size, and description of the 3 windows. ____________________________
____
____

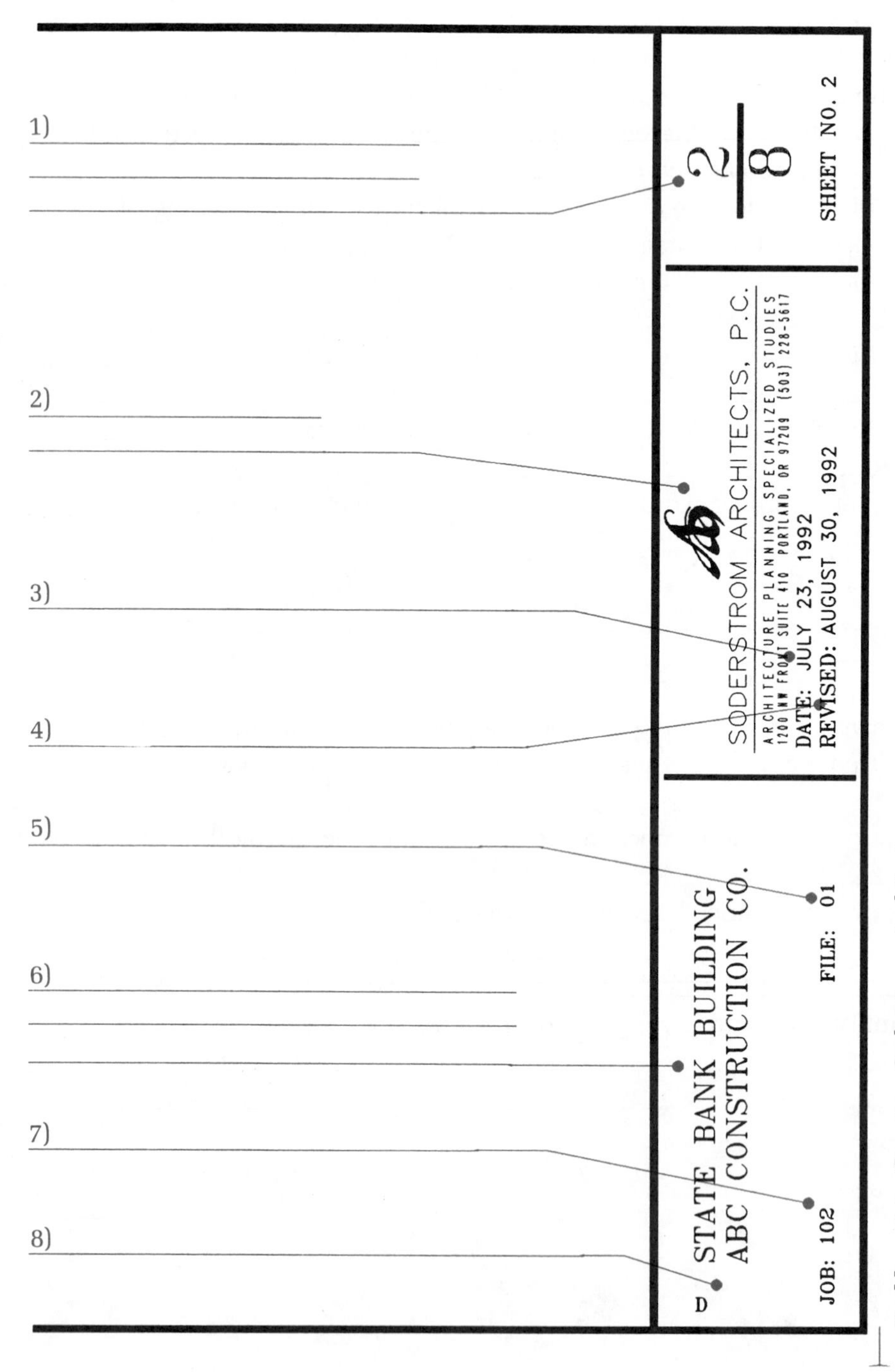

Problem 1–1. *Courtesy Soderstrom Architects.*

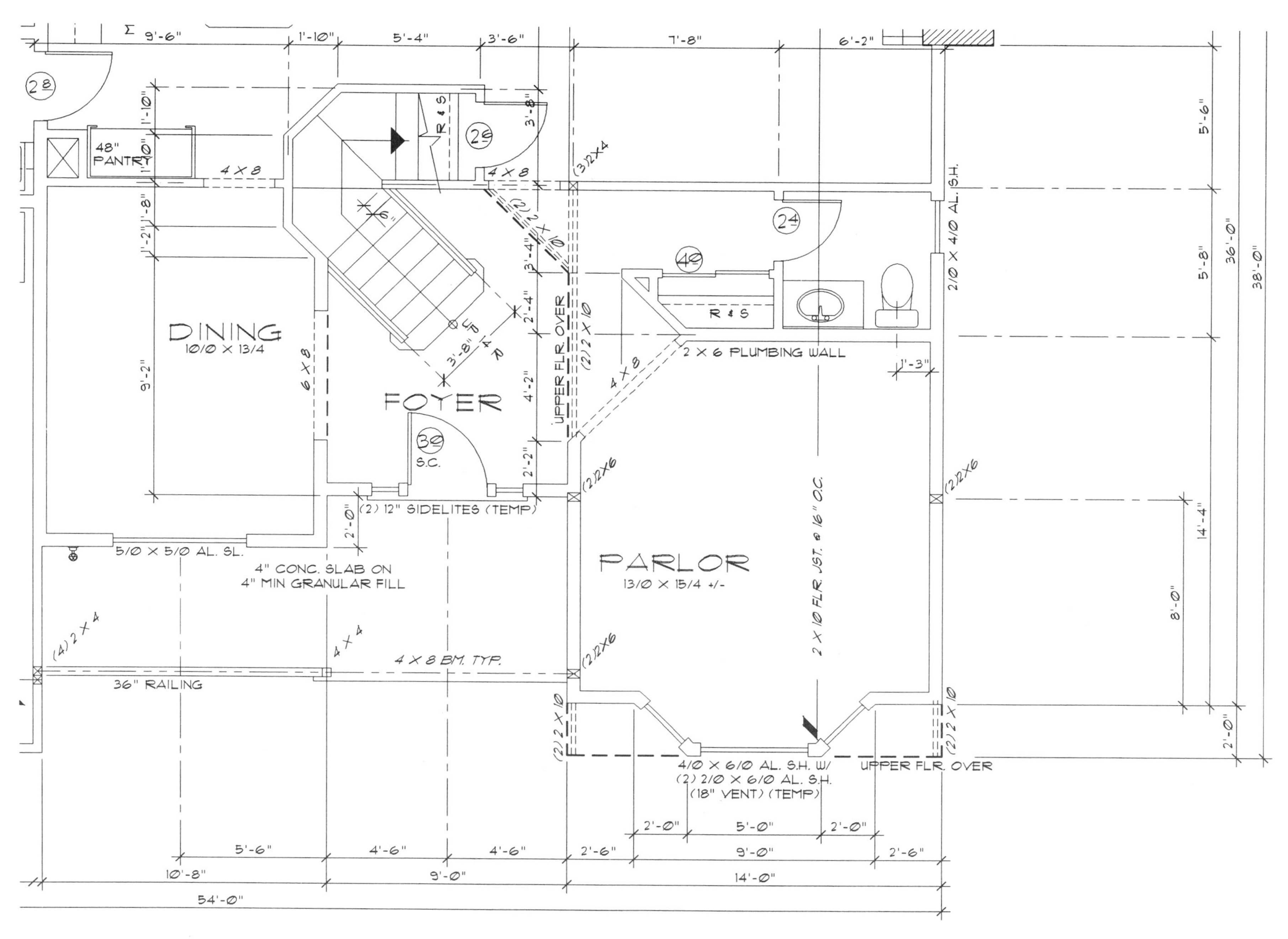

Problem 1–2. *Courtesy Alan Mascord Design Assoc., Inc.*

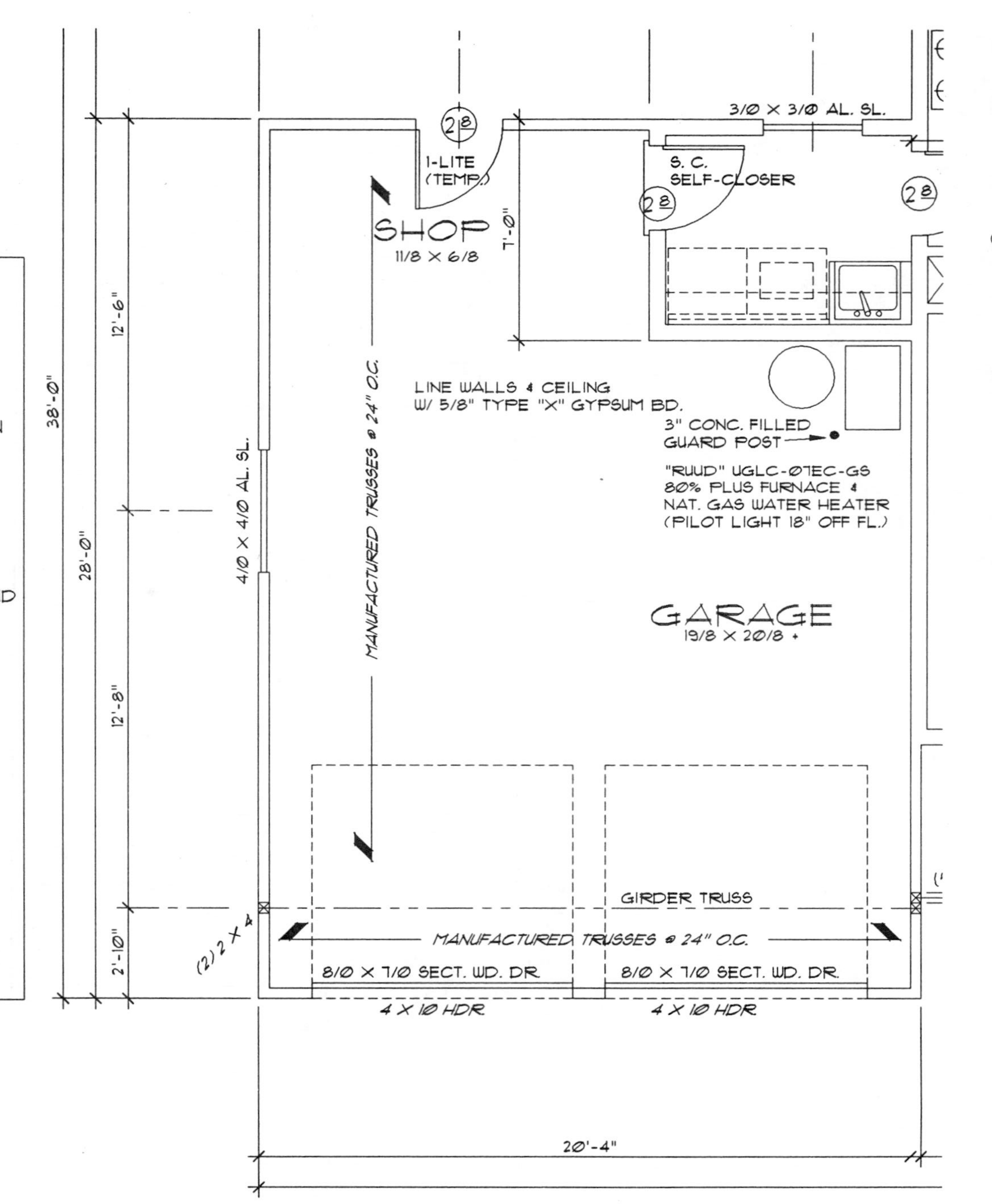

MISCELLANEOUS NOTES

1. EACH BEDROOM TO HAVE A MINIMUM WINDOW OPENING OF 5.7 SQ. FT. WITH A MIN. WIDTH OF 2Ø" AND A SILL LESS THAN 44" OFF THE FLOOR.
2. ALL WINDOWS WITHIN 18" OF THE FLOOR AND WITHIN 12" OF ANY DOOR ARE TO HAVE TEMPERED GLAZING.
3. SKYLITES ARE TO BE GLAZED WITH TEMPERED GLASS ON OUTSIDE AND LAMINATED GLASS ON INSIDE (UNLESS PLEXIGLAS). GLASS TO HAVE MAXIMUM CLEAR SPAN OF 25". SKYLITE FRAME IS TO BE ATTACHED TO A 2 X CURB WITH A MINIMUM OF 4" ABOVE ROOF PLANE.
4. ALL TUB OR SHOWER ENCLOSURES ARE TO GLAZED WITH SAFETY GLAZING.
5. ALL EXTERIOR WINDOWS ARE TO BE DOUBLE GLAZED AND ALL EXTERIOR DOORS ARE TO BE SOLID CORE WITH WEATHERSTRIPPING. PROVIDE ½" DEADBOLT LOCKS ON ALL EXTERIOR DOORS AND LOCKING DEVICES ON ALL DOORS OR WINDOWS WITHIN 1Ø' (VERTICAL) OF GRADE. PROVIDE PEEP-HOLE @ 54" -66" ABOVE FLOOR ON EXTERIOR DOORS.
6. CONNECT ALL SMOKE DETECTORS (SEE PLAN FOR LOCATION) TO HOUSE ELECTRICAL SYSTEM AND INTER-EACH ONE SO THAT WHEN ANY ONE IS TRIPPED THEY WILL ALL SOUND.
7. PROVIDE COMBUSTION AIR VENTS (W/ SCREEN AND BACK DAMPER) FOR FIREPLACES, WOOD STOVES AND ANY APPLIANCES WITH AN OPEN FLAME.
8. BATHROOMS AND UTILITY ROOMS ARE TO BE VENTED TO THE OUTSIDE WITH A MINIMUM OF A 9Ø CFM FAN. RANGE HOODS ARE ALSO TO BE VENTED TO OUTSIDE.

Problem 1–3. *Courtesy Alan Mascord Design Assoc., Inc.*

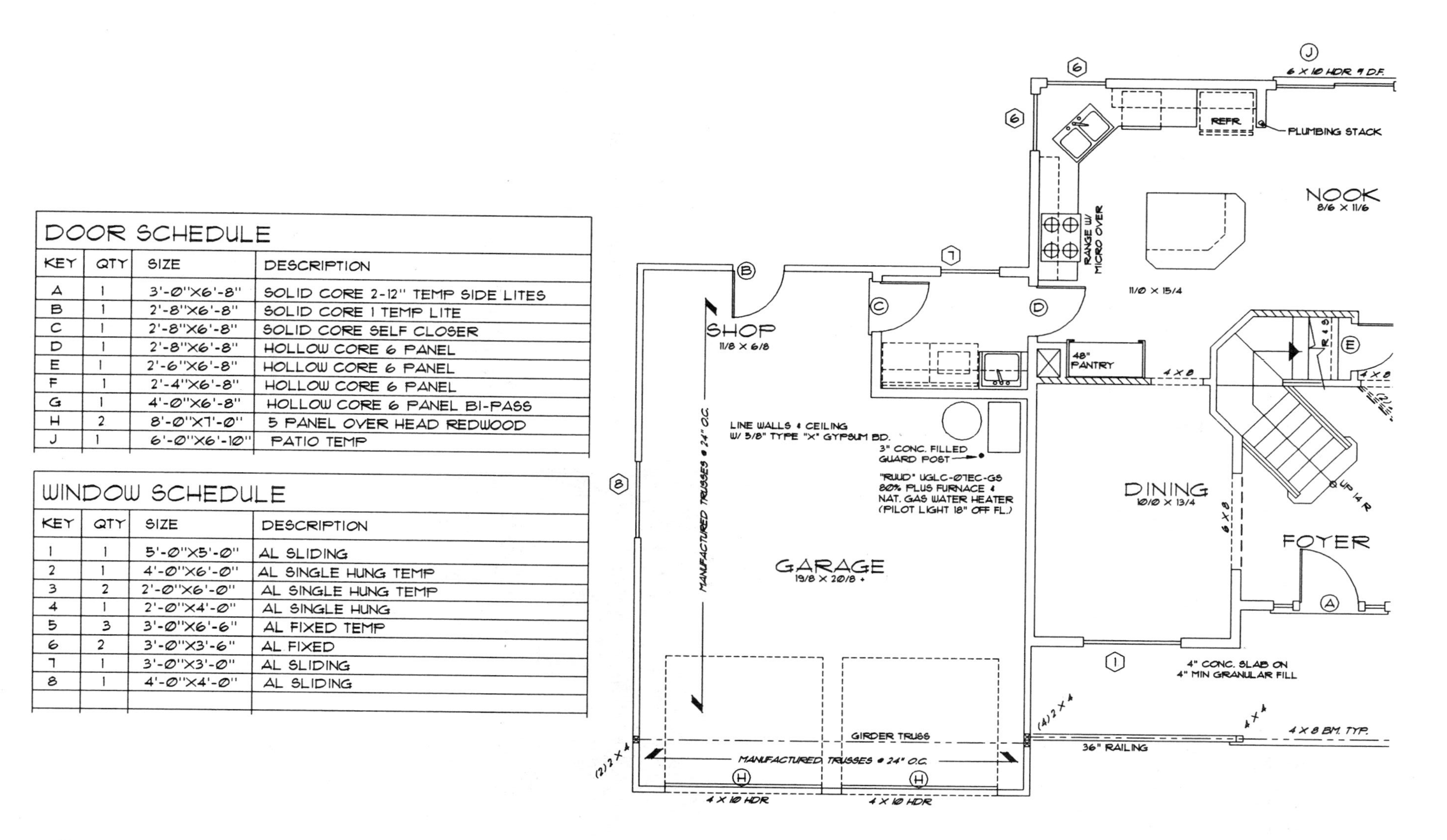

DOOR SCHEDULE

KEY	QTY	SIZE	DESCRIPTION
A	1	3'-0"X6'-8"	SOLID CORE 2-12" TEMP SIDE LITES
B	1	2'-8"X6'-8"	SOLID CORE 1 TEMP LITE
C	1	2'-8"X6'-8"	SOLID CORE SELF CLOSER
D	1	2'-8"X6'-8"	HOLLOW CORE 6 PANEL
E	1	2'-6"X6'-8"	HOLLOW CORE 6 PANEL
F	1	2'-4"X6'-8"	HOLLOW CORE 6 PANEL
G	1	4'-0"X6'-8"	HOLLOW CORE 6 PANEL BI-PASS
H	2	8'-0"X7'-0"	5 PANEL OVER HEAD REDWOOD
J	1	6'-0"X6'-10"	PATIO TEMP

WINDOW SCHEDULE

KEY	QTY	SIZE	DESCRIPTION
1	1	5'-0"X5'-0"	AL SLIDING
2	1	4'-0"X6'-0"	AL SINGLE HUNG TEMP
3	2	2'-0"X6'-0"	AL SINGLE HUNG TEMP
4	1	2'-0"X4'-0"	AL SINGLE HUNG
5	3	3'-0"X6'-6"	AL FIXED TEMP
6	2	3'-0"X3'-6"	AL FIXED
7	1	3'-0"X3'-0"	AL SLIDING
8	1	4'-0"X4'-0"	AL SLIDING

Problem 1–4. *Courtesy Alan Mascord Design Assoc., Inc.*

Chapter 2

Reading Floor Plans, Part 1: Floor Plan Symbols

THE FLOOR plans communicate the overall construction requirements to the builder. Symbols are used on floor plans to describe items associated with living in the home, such as doors, windows, cabinets, and plumbing fixtures. Other symbols that are more closely related to the construction of the home include electrical circuits and material sizes and spacing. Figure 2–1 shows a typical complete floor plan.

Figure 2–2 shows that floor plans are the representation of an imaginary horizontal cut made approximately 4 ft above the floor line. Residential floor plans are generally drawn at a scale of 1/4" = 1'–0".

In addition to knowing the proper symbols when reading plans, you should be familiar with the standard products the symbols represent. Products used in the structures, such as plumbing fixtures, appliances, windows, and doors, are usually available from local vendors. Occasionally special items must be ordered from a factory far enough in advance to ensure delivery to the job site at the time needed.

WALL SYMBOLS

Exterior Walls

Exterior wood frame walls are generally shown 6 in. thick at a 1/4" = 1'–0" scale, as seen in Figure 2–3. Exterior walls 6 in. thick are common when 2 × 4 studs are used. The 6 in. is approximately equal to the thickness of wall studs plus interior and exterior construction materials. When 2 × 6 studs are used, the exterior walls may be thicker. The wall thickness depends on the type of construction. If the exterior walls are to be concrete or masonry construction with wood framing to finish the inside surface, then they are drawn substantially thicker, as shown in Figure 2–4. Exterior frame walls with masonry veneer construction applied to the outside surface are drawn an additional 4 in. thick, with the masonry veneer represented as 4 in. thick over the 6 in. wood frame walls, as seen in Figure 2–5.

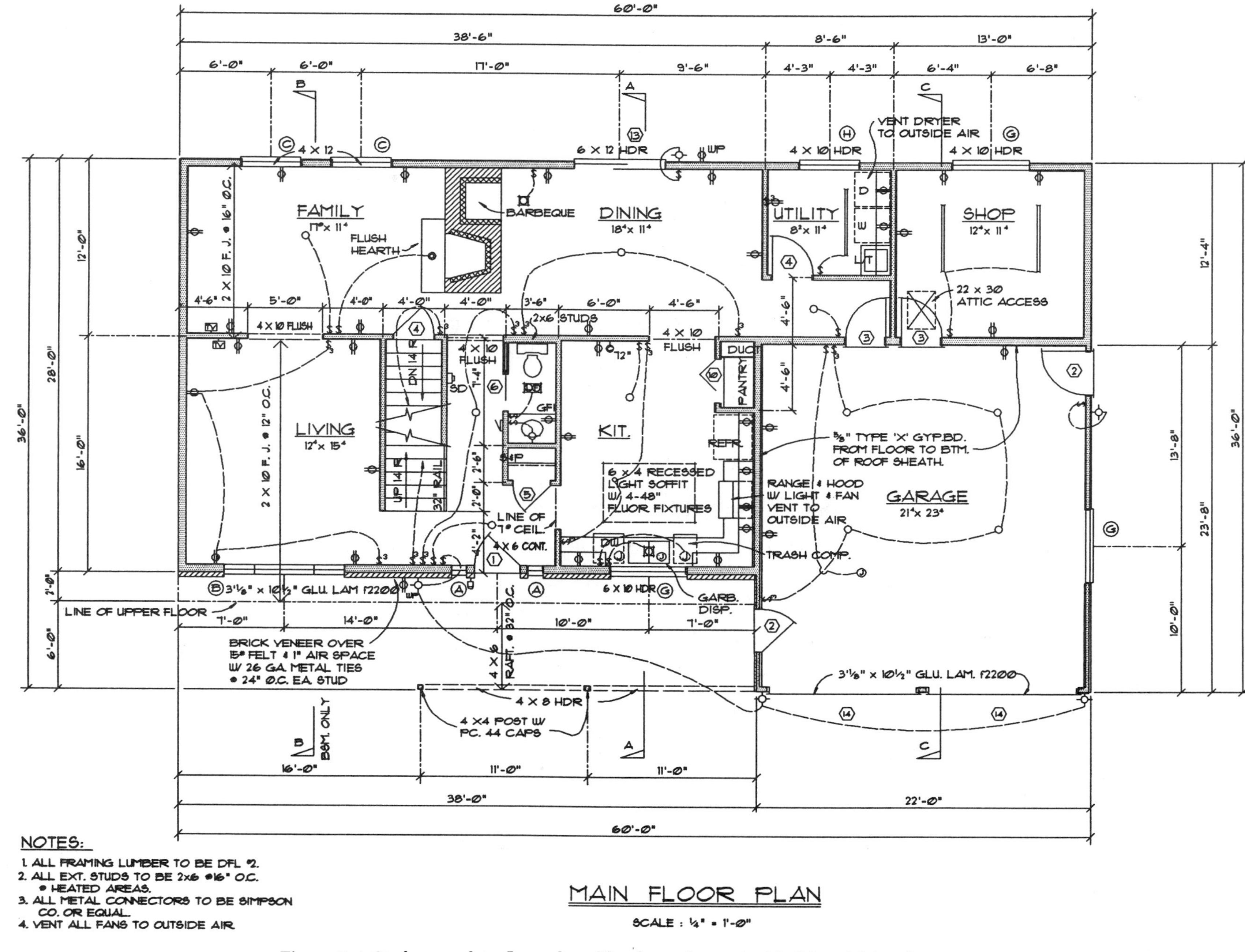

Figure 2–1 Study complete floor plans like those shown in (a), (b) and (c) to become familiar with reading prints. (a) Main floor plan.

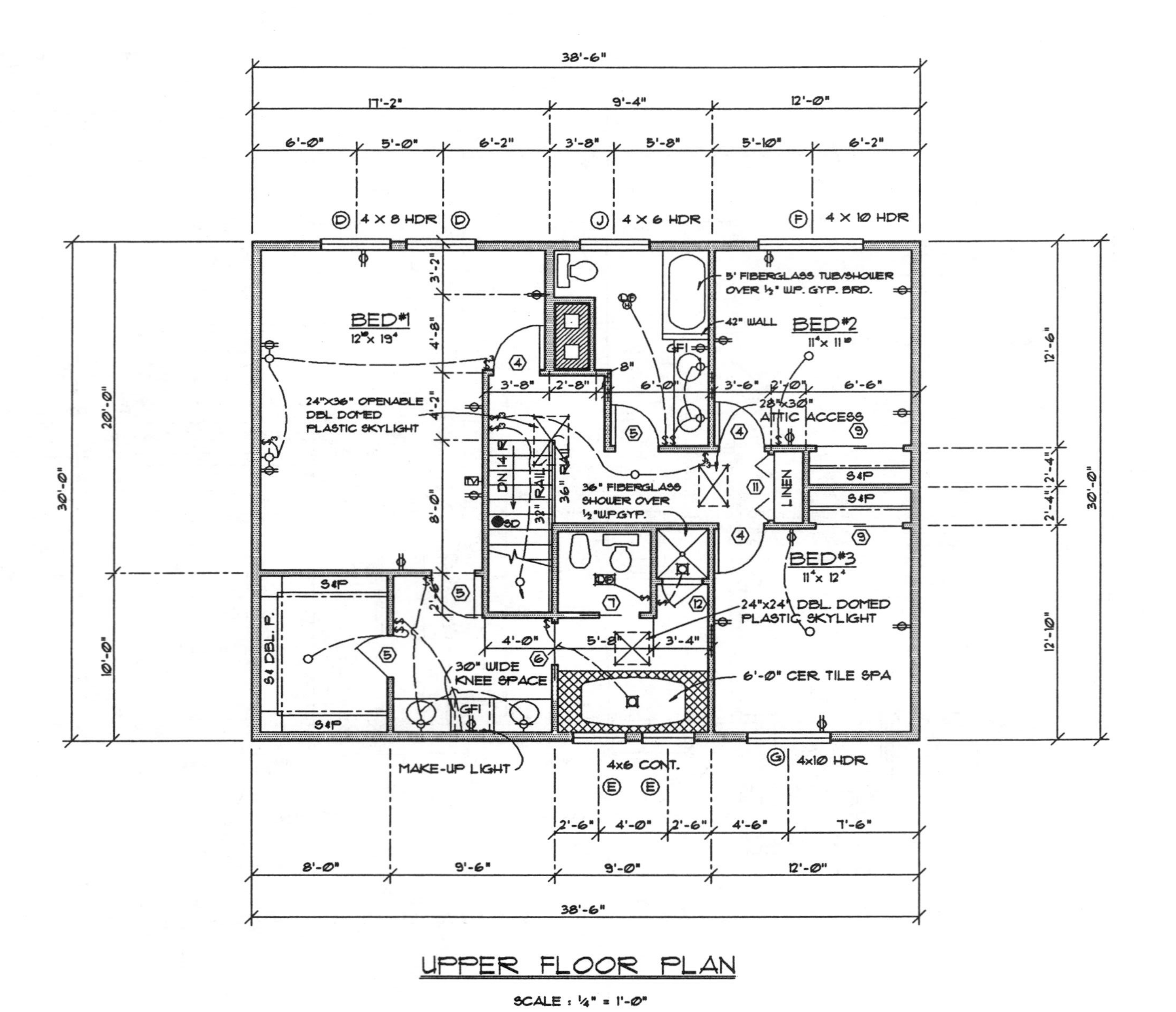

Figure 2–1(b) Second floor plan.

WINDOW SCHEDULE

SYM.	SIZE	MODEL	ROUGH OPEN	QUAN
A	$1^{0} \times 5^{0}$	JOB-BUILT		2
B	$8^{0} \times 5^{0}$	W4N5 CSM	8'-2¾"x5'-5⅞"	1
C	$4^{0} \times 5^{0}$	W2N5 CSM	4'-2¼"x5'-5⅞"	2
D	$4^{0} \times 3^{6}$	W2N3 CSM	4'-2¼"x3'-5½"	2
E	$3^{6} \times 3^{6}$	2N3 CSM	3'-5¼"x3'-5½"	2
F	$6^{0} \times 4^{0}$	G64 SLDG	6'-0½"x4'-0½"	1
G	$5^{0} \times 3^{6}$	G536 SLDG	5'-0½"x3'-6½"	4
H	$4^{0} \times 3^{6}$	G436 SLDG	4'-0½"x3'-6½"	1
J	$4^{0} \times 2^{0}$	A41 AWN	4-0½"x2'-0⅝"	3
K	$4^{0} \times 2^{0}$	G42 SLDG	4'-0½"x2'-0½"	3

DOOR SCHEDULE

SYM.	SIZE	TYPE	QUAN
1	$3^{0} \times 6^{8}$	S.C. R.P. METAL INSULATED	1
2	$3^{0} \times 6^{8}$	S.C.-FLUSH-METAL INSUL	2
3	$2^{8} \times 6^{8}$	S.C-SELF CLOSING	2
4	$2^{8} \times 6^{8}$	H.C.	5
5	$2^{6} \times 6^{8}$	H.C.	5
6	$2^{6} \times 6^{8}$	POCKET	2
7	$2^{4} \times 6^{8}$	POCKET	1
8	PR $2^{6} \times 6^{8}$	H.C.	1
9	$5^{0} \times 6^{8}$	BI-PASS	2
10	$3^{0} \times 6^{8}$	BI-FOLD	1
11	$4^{0} \times 6^{8}$	BI-FOLD	1
12	$2^{0} \times 6^{0}$	SHATTER PROOF	1
13	$6^{0} \times 6^{8}$	WOOD FRAME-TEMP. SLDG. GL.	1
14	$9^{0} \times 7^{0}$	OVERHEAD GARAGE	2

WALL AREAS *	2275 SQ. FT.
WINDOWS	250 SQ. FT.
SKYLITES	10 SQ. FT.
DOORS	77 SQ. FT.
TOTAL OPENINGS	337 SQ. FT.
% OPENINGS	15%

*BASEMENT EXCLUDED.

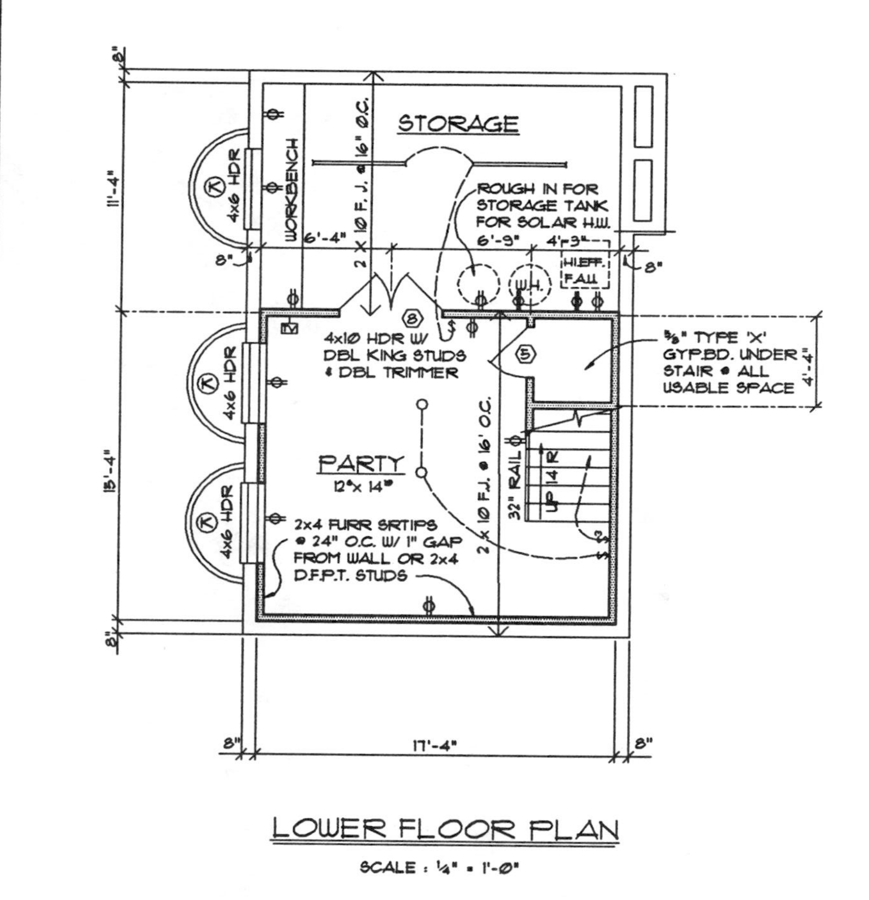

Figure 2–1(c) Basement floor plan with schedules.

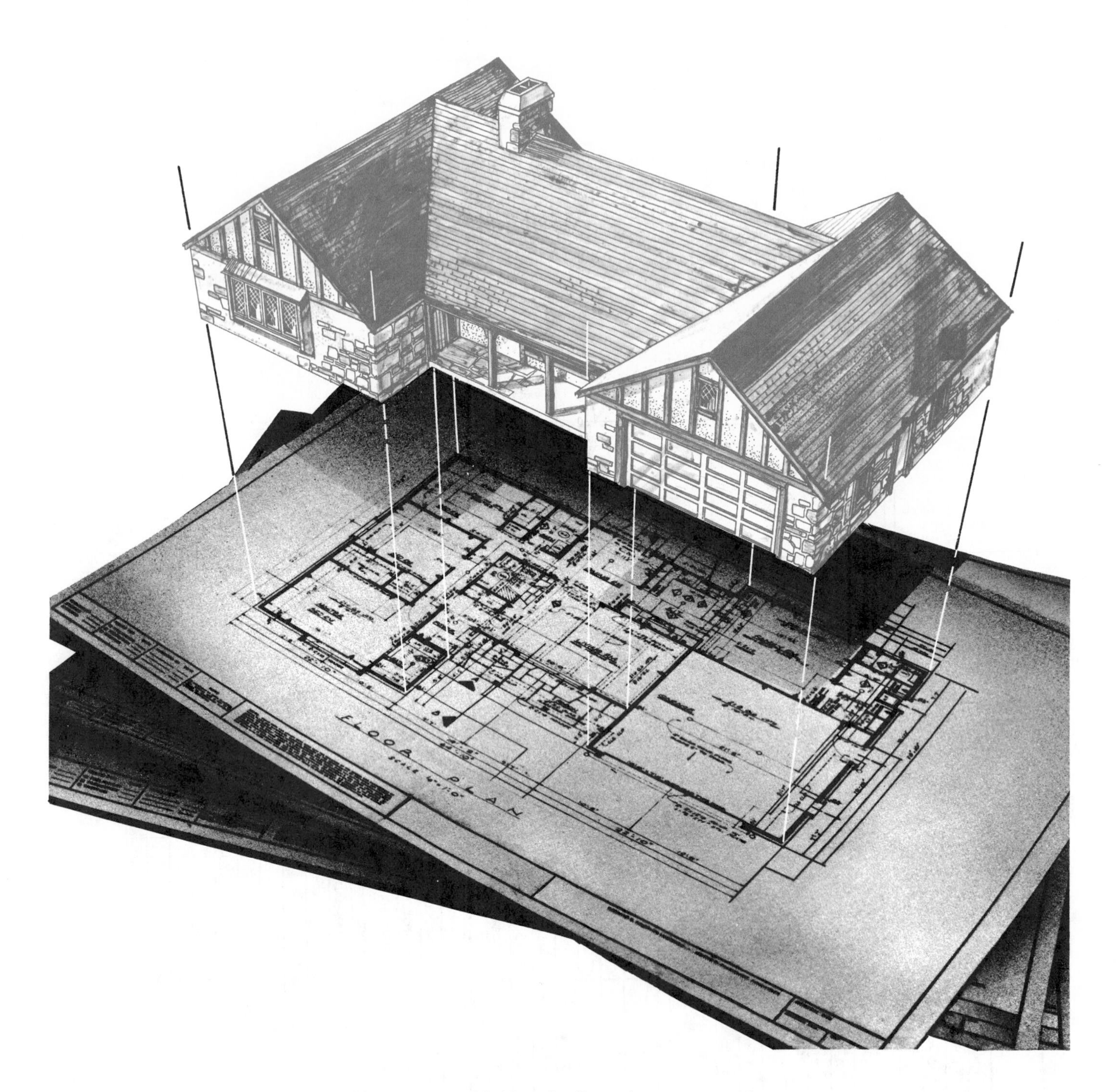

Figure 2–2 Establishing the floor plan representation.

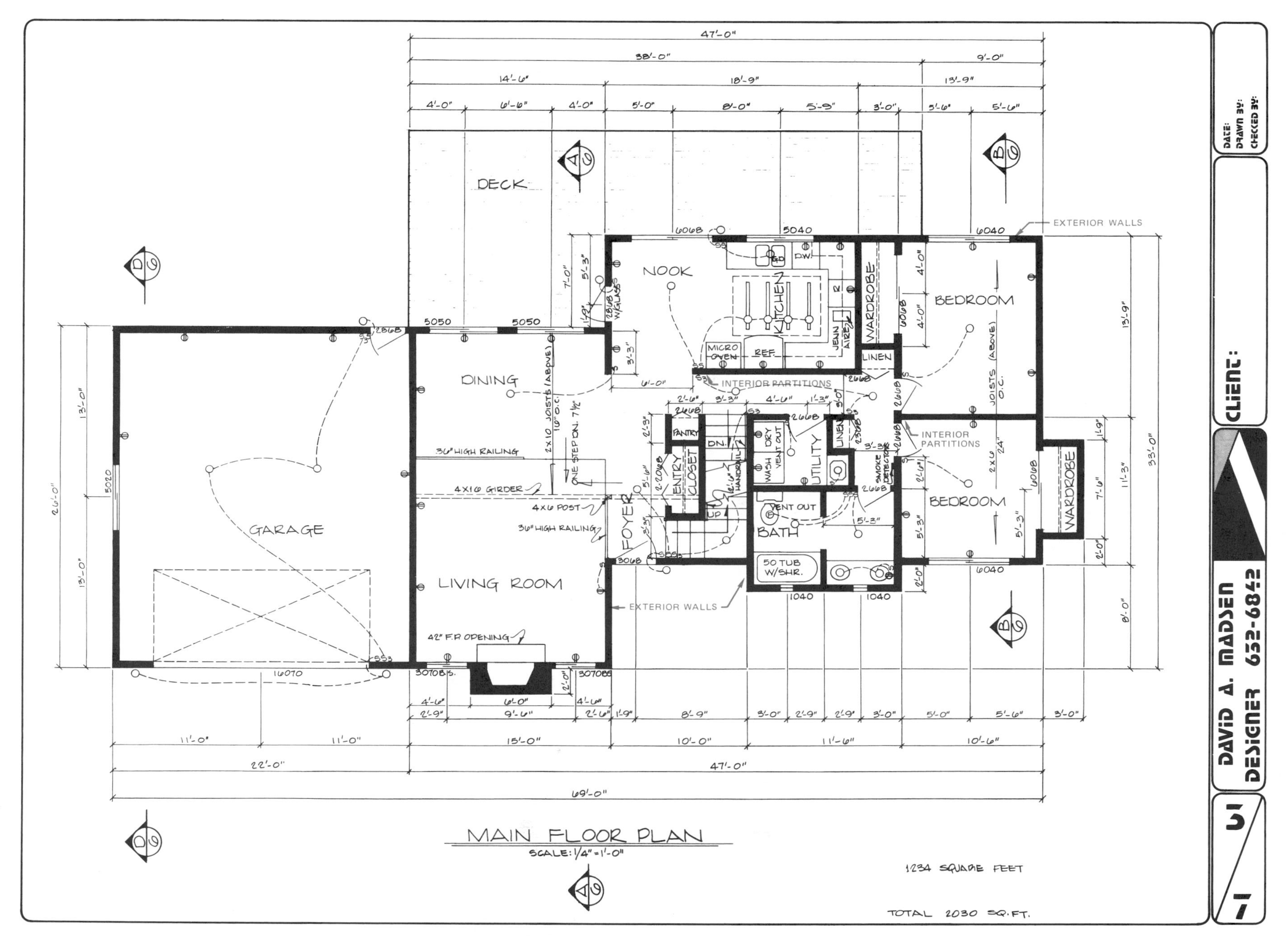

Figure 2–3 Typical floor plan. *Courtesy Madsen Designs.*

Interior Partitions

Interior walls, known as partitions, are frequently drawn 5 in. thick when 2 × 4 studs are used with drywall applied to each side. Walls with 2 × 6 studs are generally used behind a toilet to accommodate the soil pipe. Occasionally masonry veneer is used on interior walls and is drawn in a manner similar to the exterior application shown in Figure 2–5. In many architectural offices a wood frame exterior and interior walls are drawn the same thickness to save time.

Wall Shading

Several methods are used to shade walls, which is done so the walls stand out clearly from the balance of the drawing. Some walls are shaded very darkly for accent; others are shaded lightly for a more subtle effect, as shown in Figure 2–6. Other wall-shading techniques include closely spaced thin lines, wood grain effect, or the use of colored pencils.

Office practice that requires light wall shading may also use thick wall lines to help accent the walls and partitions so they stand out from other floor plan features. Figures 2–6(a) and 2–7 show how walls and partitions appear when outlined with thick lines.

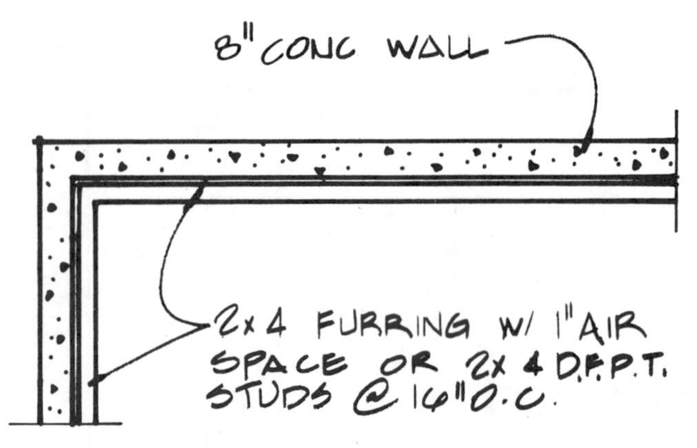

Figure 2–4 Concrete exterior wall.

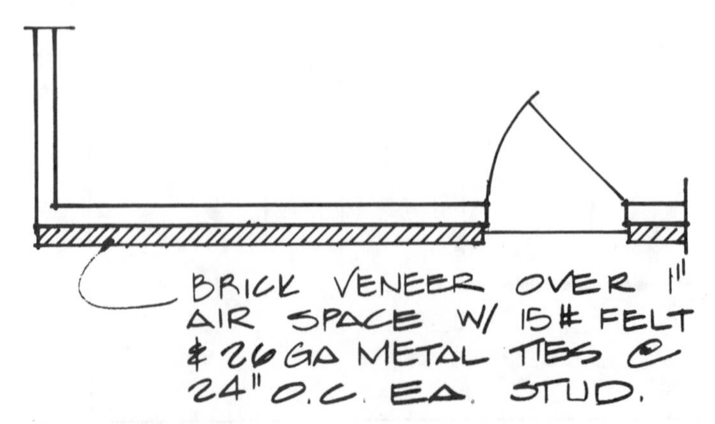

Figure 2–5 Exterior masonry veneer.

Partial Walls

Partial walls are used as room dividers where an open environment is desired, such as guard rails on balconies or adjacent to a flight of stairs. Partial walls require a minimum height above the floor of 36 in. and are often capped with wood or may have decorative spindles that connect to the ceiling. Partial walls are differentiated from other walls by wood grain or very light shading and should be defined with a note that specifies the height, as shown in Figure 2–8.

Guardrails

Guardrails are used for safety on balconies, lofts, stairs, and decks over 30 in. above the next lower

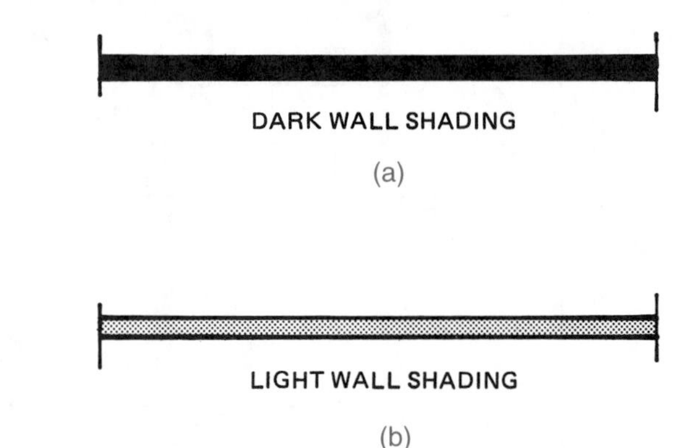

Figure 2–6 Wall shading.

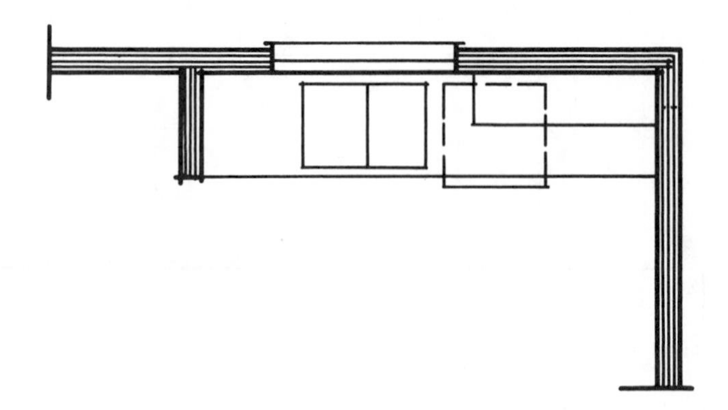

Figure 2–7 Thick lines used on walls and partitions.

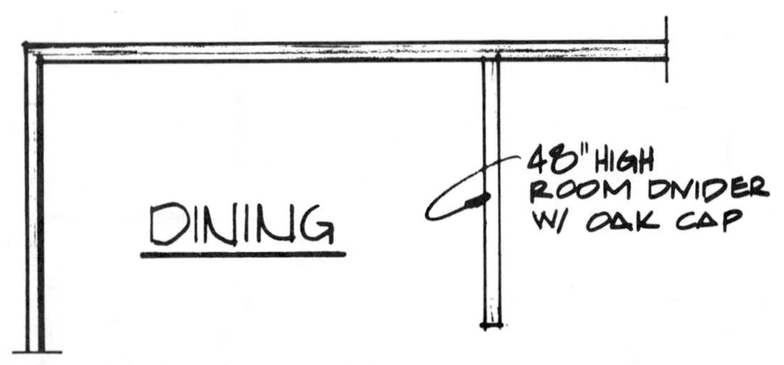

Figure 2–8 Partial wall used as a partition or room divider.

level. Residential guardrails are noted on floor plans as at least 36 in. above the floor and may include a note specifying that intermediate rails should not have more than a 6 in. open space. The minimum space helps ensure that small children do not fall through the rails. Decorative guardrails are also used as room dividers, especially in a sunken area or to create an open effect between two rooms. Figure 2–9 shows guardrail designs and how they are indicated on a floor plan.

DOOR SYMBOLS

Exterior doors are drawn on the floor plan with the sill shown on the outside of the house. The sill is commonly drawn as a projection. See Figure 2–10.

The interior door symbol, as shown in Figure 2–11, is drawn without a sill. Interior doors should swing into the room being entered and against a wall. Interior doors are usually slabs but may be the raised panel type or have glass panels.

Pocket doors are commonly used when space for a door swing is limited, as in a small room. Look at Figure 2–12. Pocket doors should not be placed where the pocket is in an exterior wall or where there is interference with plumbing or electrical wiring. Pocket doors are more expensive to purchase and install than standard interior doors because the pocket door frame must be built while the house is being framed.

A common economical wardrobe door is the bipass door, as shown in Figure 2–13. Bifold wardrobe or closet doors are used when complete access to the closet is required. Sometimes bifold doors are used on a utility closet that houses a washer and dryer or other utilities. See Figure 2–14. Bifold wardrobe doors often range in size from 4'–0" through 9'–0" wide, in 6 in. intervals.

Double-entry doors are common where a large, formal foyer design requires a more elaborate entry than can be achieved by one door. The floor plan symbol for double-entry and French doors is the same; therefore, the door schedule should clearly identify the type of doors to be installed.

Sliding glass doors are made with wood or metal frames and tempered glass for safety. Figure 2–15 shows the floor plan symbol for both a flush and a projected exterior sill representation. These doors are

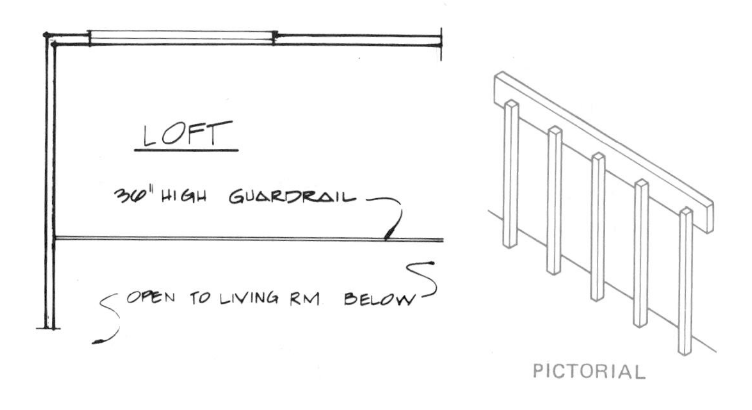

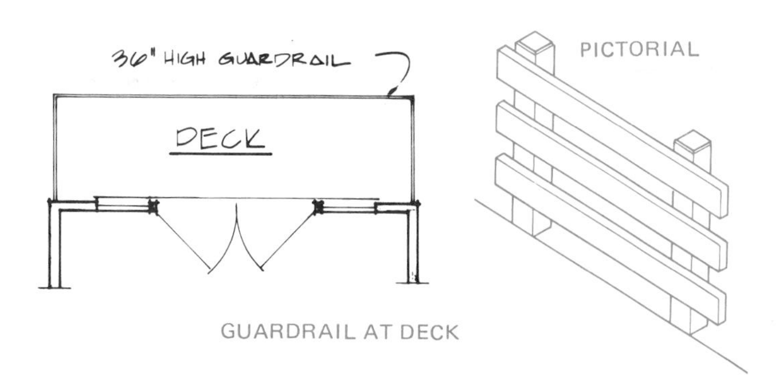

Figure 2–9 Guardrails.

used to provide glassed-in areas and are excellent for access to a patio or deck. Sliding glass doors typically range in size from 5'–0" through 12'–0" wide.

French doors are used in place of sliding glass doors when a more traditional door design is required. Sliding glass doors are associated with contemporary design and do not take up as much floor space as French doors. French doors may be purchased with wood *mullions* and *muntins* (the upright and bar partitions, respectively) between the glass panes or with one large glass pane and a removable grill for easy cleaning. See Figure 2–16. French doors range in size from 2'–4" through 3'–6" wide. Doors may be used individually, in pairs, or in groups of three or four. The three- and four-panel doors typically have one or more fixed panels, which is often specified on the plan.

Double-acting doors are often used between a kitchen and eating area so the doors swing in either direction for easy passage. See Figure 2–17. Common sizes for pairs of doors range from 2'–6" through 4'–0" wide in 2 in. increments.

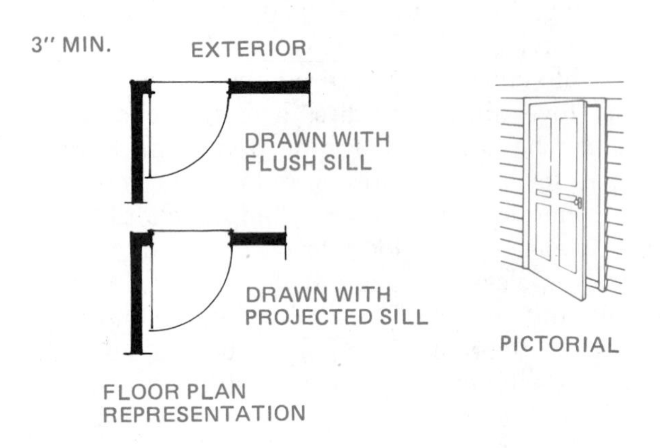

Figure 2–10 Exterior door.

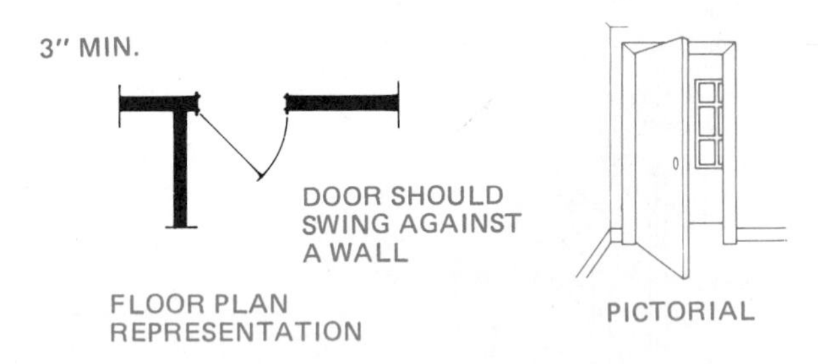

Figure 2–11 Interior door.

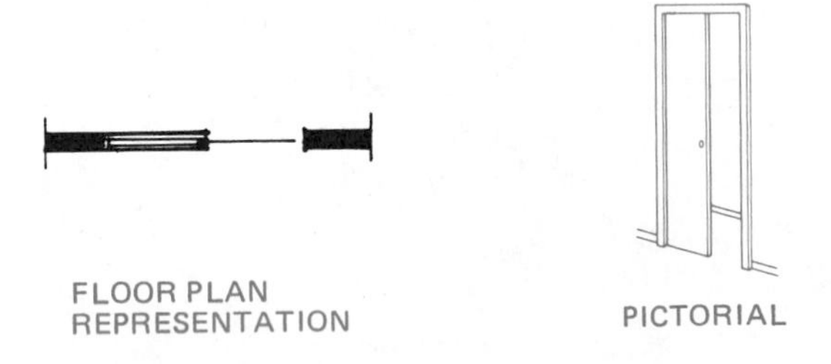

Figure 2–12 Pocket door.

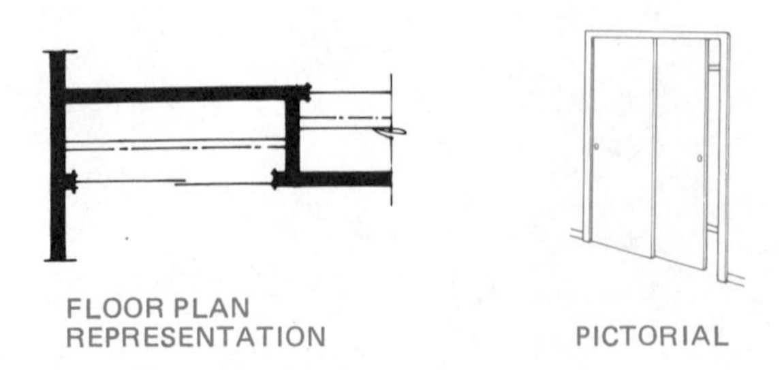

Figure 2–13 Bipass door.

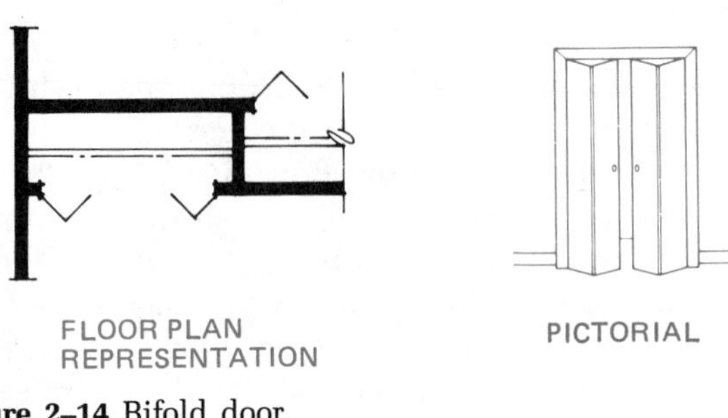

Figure 2–14 Bifold door.

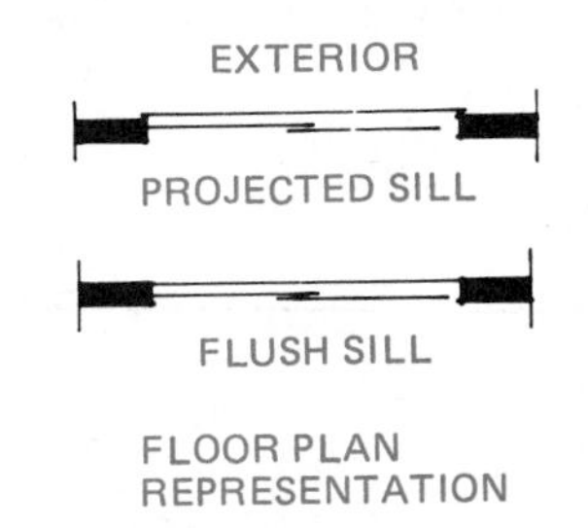

Figure 2–15 Glass sliding door.

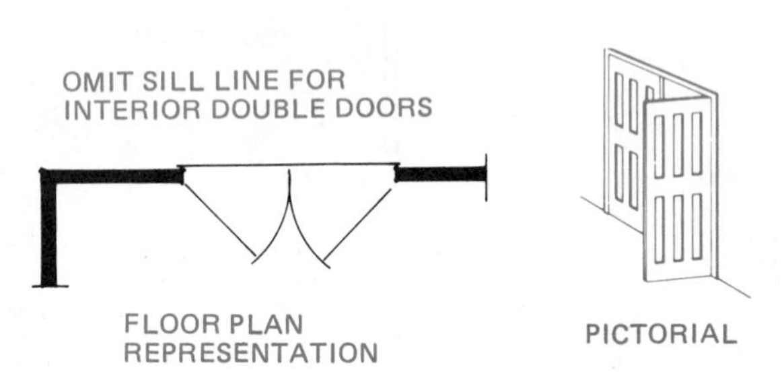

Figure 2–16 Double-entry or French doors.

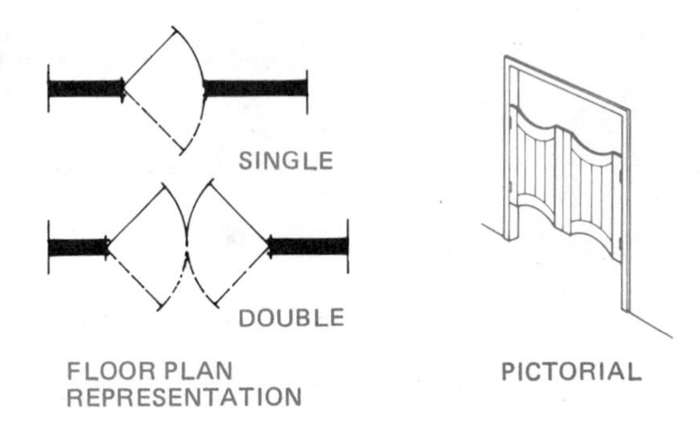

Figure 2–17 Double-acting doors.

Dutch doors are used when it is desirable to have a door that may be half-open and half-closed. The top portion may be opened and used as a pass-through. Look at Figure 2–18. Dutch doors range in size from 2'–6" through 3'–6" wide in 2 in. increments.

Accordion doors may be used for closets or wardrobes, or they are often used as room dividers where an openable partition is needed. Figure 2–19 shows the accordion door floor plan symbol. Accordion doors range in size from 4'–0" through 12'–0" in 1 ft increments.

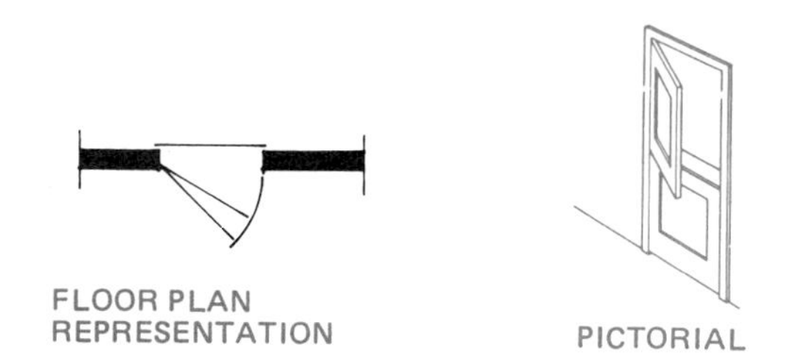

Figure 2–18 Dutch doors.

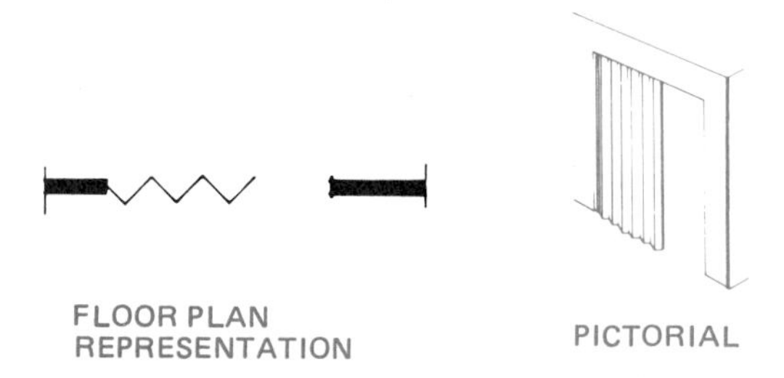

Figure 2–19 Accordion door.

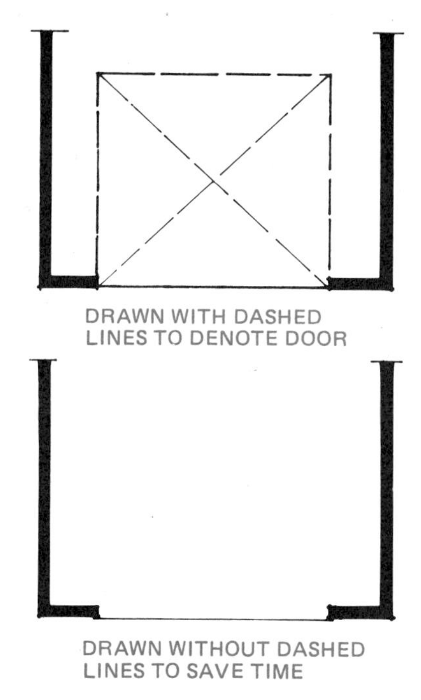

Figure 2–20 Overhead garage door.

The floor plan symbol for an overhead garage door is shown in Figure 2–20. The dashed lines show the size and extent of the garage door when open. The extent of the garage door may be shown when the door might interfere with something on the ceiling.

Garage doors range in size from 8'–0" through 16'–0" wide. An 8'–0" door is the minimum for a single car width. A 9'–0" door is common for a single door. A 16'–0" door is common for a double car door. Door heights are typically 7'–0" high, although 8'–0" to 10'–0" high doors are available.

WINDOW SYMBOLS

Windows are represented with a sill on the outside and inside. The sliding window, as shown in Figure 2–21 is a popular 50 percent openable window. Notice that windows may be drawn with exterior sills projected or flush, and the glass pane may be drawn with single or double lines by the preference of the specific architectural office. Some offices draw all windows with a projected sill and one line to represent the glass; they then specify the type of window in the window schedule.

Casement windows may be 100 percent openable and are best used where extreme weather conditions require a tight seal when the window is closed, although these windows are in common use everywhere. See Figure 2–22.

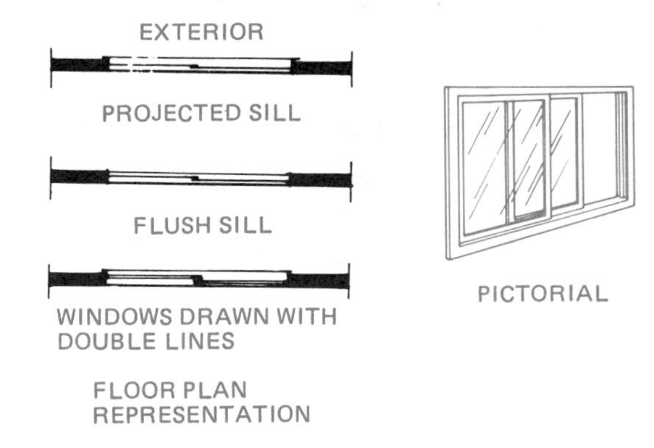

Figure 2–21 Horizontal sliding window.

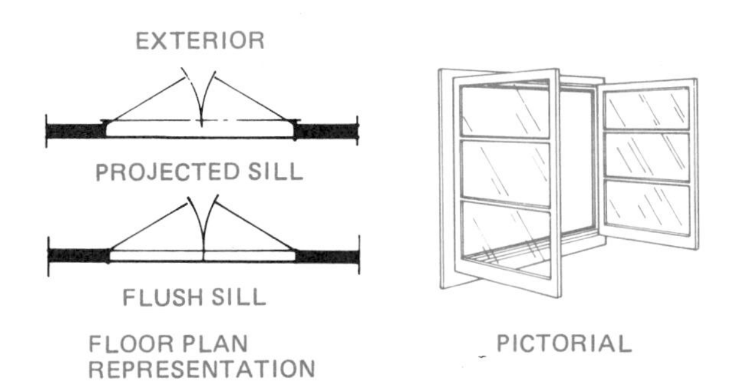

Figure 2–22 Casement window.

The traditional double-hung window has a bottom panel that slides upward, as shown in Figure 2–23. Double-hung wood frame windows are designed for energy efficiency and are commonly used in traditional as opposed to contemporary architectural designs. It is very common to group double-hung windows together in pairs of two or more.

Awning windows are often used below a fixed window to provide ventilation. Another common use places awning windows between two different roof levels; this provides additional ventilation in vaulted rooms. These windows are hinged at the top and swing outward, as shown in Figure 2–24. Hopper windows are drawn in the same manner; however, they hinge at the bottom and swing inward. Jalousie windows are used when a louvered effect is desired, as seen in Figure 2–25.

Fixed windows are popular when a large unobstructed area of glass is required to take advantage of a view or to allow solar heat gain. Figure 2–26 shows the floor plan symbol for a fixed window.

Bay windows are often used when a traditional style is desired. Figure 2–27 shows the representation of a bay. Usually the sides are built at 45° or 30°. The depth of the bay is usually between 18" and 24". The total width of a bay is limited by the size of the center window, which is typically a fixed panel, double-hung, or casement window. Bays can be either premanufactured or built at the job site.

A garden window, as shown in Figure 2–28, is a popular style for utility rooms or kitchens. Garden windows usually project between 12" and 18" from the residence. Depending on the manufacturer, either the side or top panels may open.

EXTERIOR
PROJECTED SILL
FLUSH SILL
FLOOR PLAN REPRESENTATION
PICTORIAL

Figure 2–23 Double-hung window.

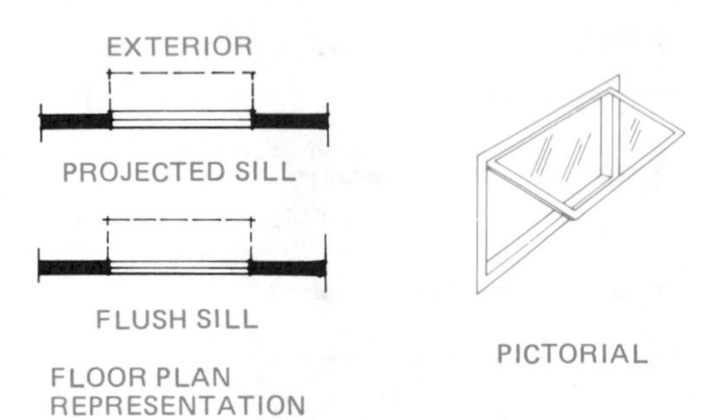

Figure 2–24 Awning window.

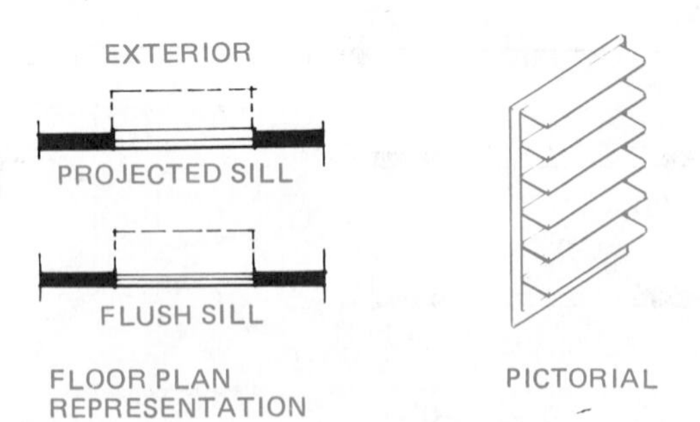

Figure 2–25 Jalousie window.

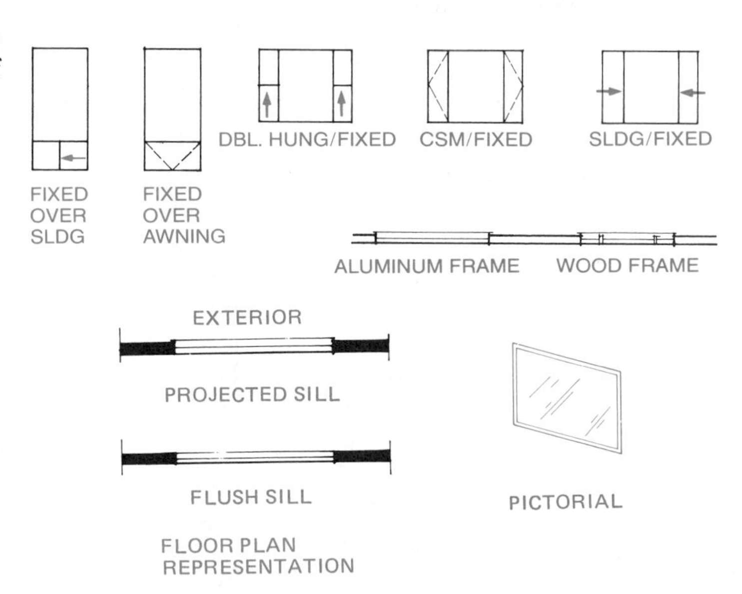

Figure 2–26 Fixed window.

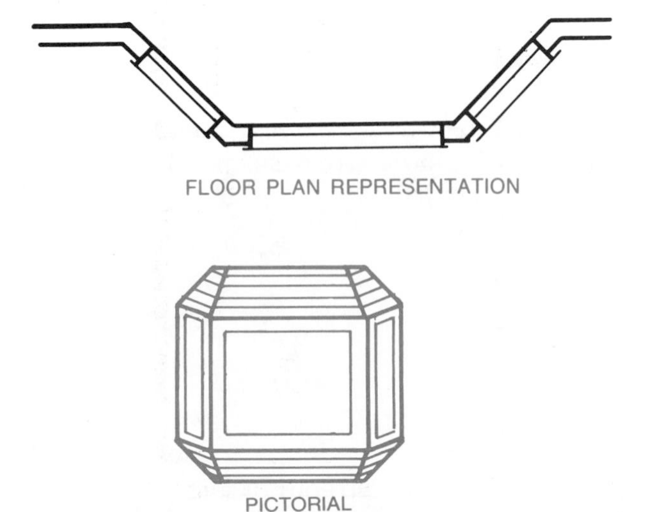

Figure 2–27 Bay window.

Skylights

When additional daylight is desirable in a room or for natural light to enter an interior room, consider using a skylight. Skylights are available as fixed or openable. They are made of plastic in a dome shape or are flat and made of tempered glass. Tempered doublepane insulated skylights are energy efficient, do not cause any distortion of view, and generally are not more expensive than plastic skylights. Figure 2–29 shows how a skylight is represented in the floor plan.

SCHEDULES

Numbered symbols used on the floor plan key specific items to charts known as *schedules.* Schedules are used to describe such items as doors, windows, appliances, materials, fixtures, hardware, and finishes. See Figure 2–30. Schedules help keep drawings clear of unnecessary notes since the details of the item are not on the drawing but on another part of the sheet or on another sheet. There are many different ways to set up a schedule, but it may include any or all of the following information about the product:

- Vendor's name
- Product name
- Model number
- Quantity
- Size
- Rough opening size
- Color

Schedule Key

When doors and windows are described in a schedule, the items must be keyed from the drawing to the schedule. The key may be to label doors with a number and windows with a letter and enclose the

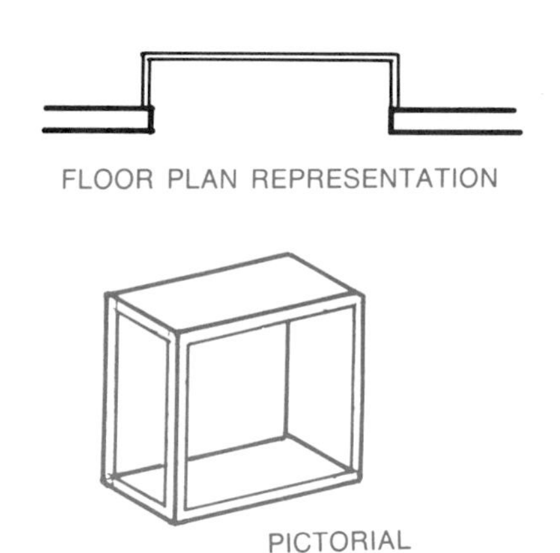

Figure 2–28 Garden window.

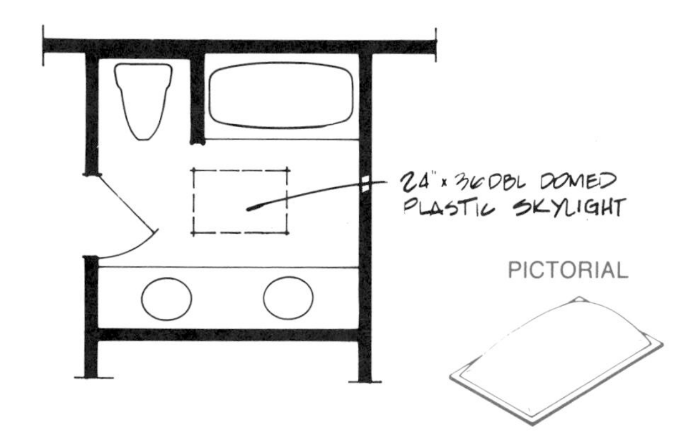

Figure 2–29 Skylight representation.

WINDOW SCHEDULE

SYM	SIZE	MODEL	ROUGH OPEN	QUAN.
A	1° x 5°	JOB BUILT	VERIFY	2
B	8° x 5°	W 4 N 5 CSM.	8'-Ø¾ x 5'-Ø⅛	1
C	4° x 5°	W 2 N 5 CSM.	4'-Ø¾ x 5'-Ø⅛	2
D	4° x 3^{6}	W 2 N 3 CSM	4'-Ø¾ x 3'-6½	2
E	3^{6} x 3^{6}	2 N 3 CSM	3'-6½ x 3'-6½	2
F	6° x 4°	G 64 SLDG.	6'-Ø½ x 4'-Ø½	1
G	5° x 3^{6}	G 536 SLDG.	5'-Ø½ x 3'-6½	4
H	4° x 3^{6}	G 436 SLDG.	4'-Ø½ x 3'-6½	1
J	4° x 2°	A 41 AWN.	4'-Ø½ x 2'-Ø⅛	3

(a)

DOOR SCHEDULE

SYM.	SIZE	TYPE	QUAN
1	3° x 6^{8}	S.C. R.P. METAL INSULATED	1
2	3° x 6^{8}	S.C. FLUSH METAL INSULATED	2
3	2^{8} x 6^{8}	S.C. SELF CLOSING	2
4	2^{8} x 6^{8}	HOLLOW CORE	5
5	2^{6} x 6^{8}	HOLLOW CORE	5
6	2^{6} x 6^{8}	POCKET SLDG.	2

(b)

Figure 2–30 Window and door schedule.

letters or numbers in different geometric figures, such as the following:

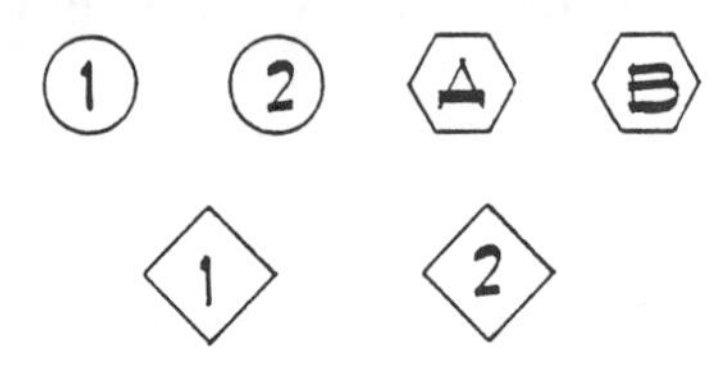

See also Figure 2–31. As an alternative method, you may see a divided circle for the key, the letter D for door or W for window above the dividing line and the number of the door or window, using consecutive numbers, below the line, as shown:

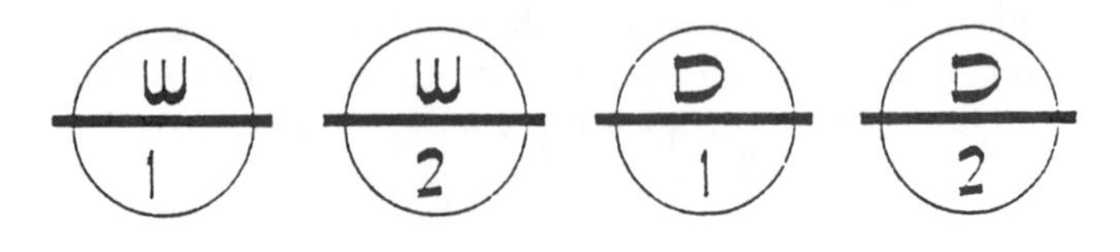

The exact method of representation depends upon individual company standards.

Schedules are also used to identify finish materials used in different areas of the structure. Figure 2–32 shows a typical interior finish schedule.

Door and window sizes may also be placed on the floor plan next to their symbols. This method, as shown in Figure 2–33 is easy but may not be used when specific data must be identified. The sizes are the numbers next to the windows and doors in Figure 2–33. Notice the first two numbers indicate the width. For example, 30, or 3/0, means 3'–0". The second two numbers indicate the height: 68, or 6/8, means 6'–8". A 6040, or 6/0 × 4/0 window means 6'–0" wide by 4'–0" high, and a 2868, or 2/8 × 6/8, door means 2'–8" wide by 6'–8" high. Standard door height is 6'–8". Some drafters use the same system in a slightly

INTERIOR FINISH SCHEDULE

ROOM	FLOOR					WALLS				CEIL.		
	VINYL	CARPET	TILE	HARDWOOD	CONCRETE	PAINT	PAPER	TEXTURE	SPRAY	SMOOTH	BROCADE	PAINT
ENTRY					●							
FOYER			●			●			●		●	●
KITCHEN			●				●			●		●
DINING				●		●			●		●	●
FAMILY		●				●			●		●	●
LIVING		●				●		●			●	●
MASTER BED.		●				●		●			●	●
MASTER BATH.			●				●			●		●
BATH #2			●			●			●	●		●
BED. #2		●				●			●		●	●
BED. #3		●				●			●		●	●
UTILITY	●					●			●	●		●

Figure 2–32 Finish schedules.

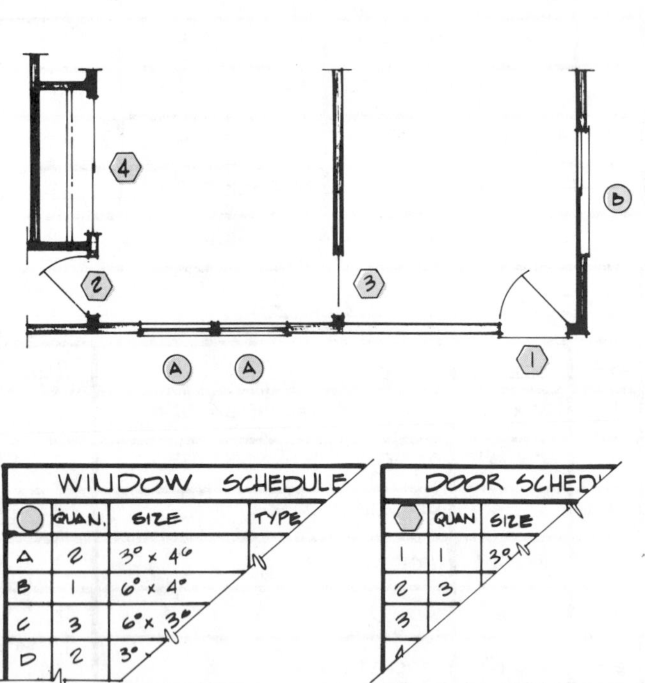

Figure 2–31 Method of keying windows and doors from the floor plan to the schedules.

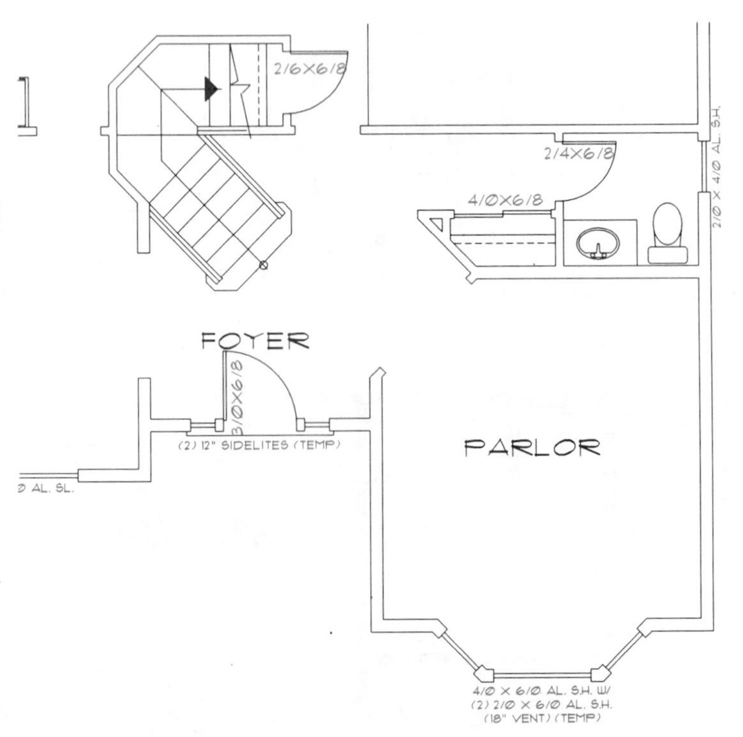

Figure 2–33 Simplified method of labeling doors and windows. *Courtesy Alan Mascord Design Assoc., Inc.*

different manner by presenting the sizes as $2^8 \times 6^8$, meaning 2'–8" wide by 6'–8" high, or with actual dimensions, such as 2'–8" × 6'–8". This simplified method of identification is better suited for development housing in which doors, windows, and other

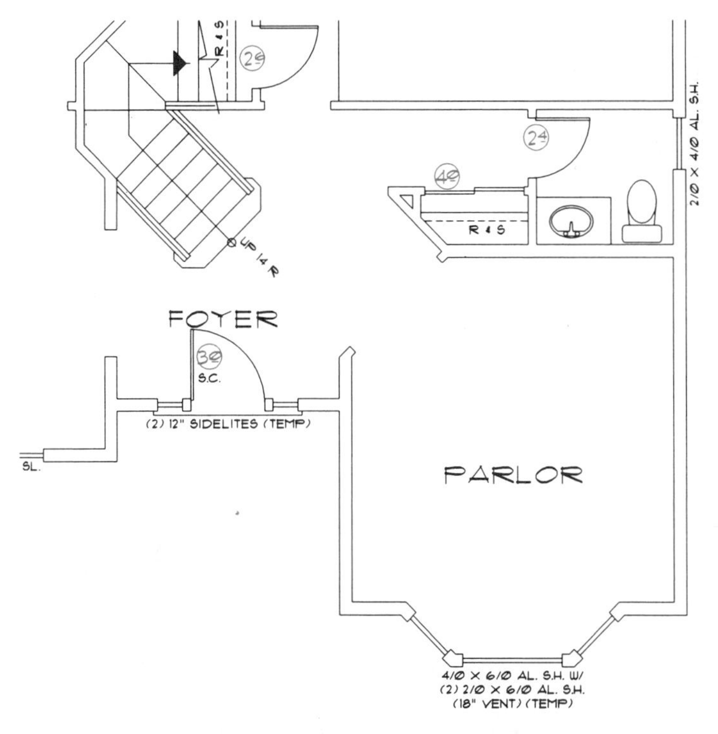

Figure 2–34a A simplified method of labeling door size. *Courtesy Alan Mascord Design Assoc., Inc.*

items are not specified on the plans as to a given manufacturer. These details are included in a description of materials or specification sheet for each individual house. The principal reason for using this system is to reduce drafting time by omitting schedules. The building contractor may be required to submit alternative specifications to the client so that actual items may be clarified.

Some plans show the window sizes given by the window symbol as previously discussed and show door sizes with a symbol like that in Figure 2–34a. The number 30 inside the circle designates the door width as 3'–0". The door height is a standard 6'–8".

Some architects and designers key the window specifications to a specific manufacturer's product by giving the catalog number next to the window symbol, as shown in Figure 2–34(b). Notice the general note that describes the manufacturer information. You must get a copy of the manufacturer's catalog to know the exact size and other related installation and construction information.

When reading a set of plans, you should know specific details about doors and windows: standard size, material, type of finish, energy efficiency, and

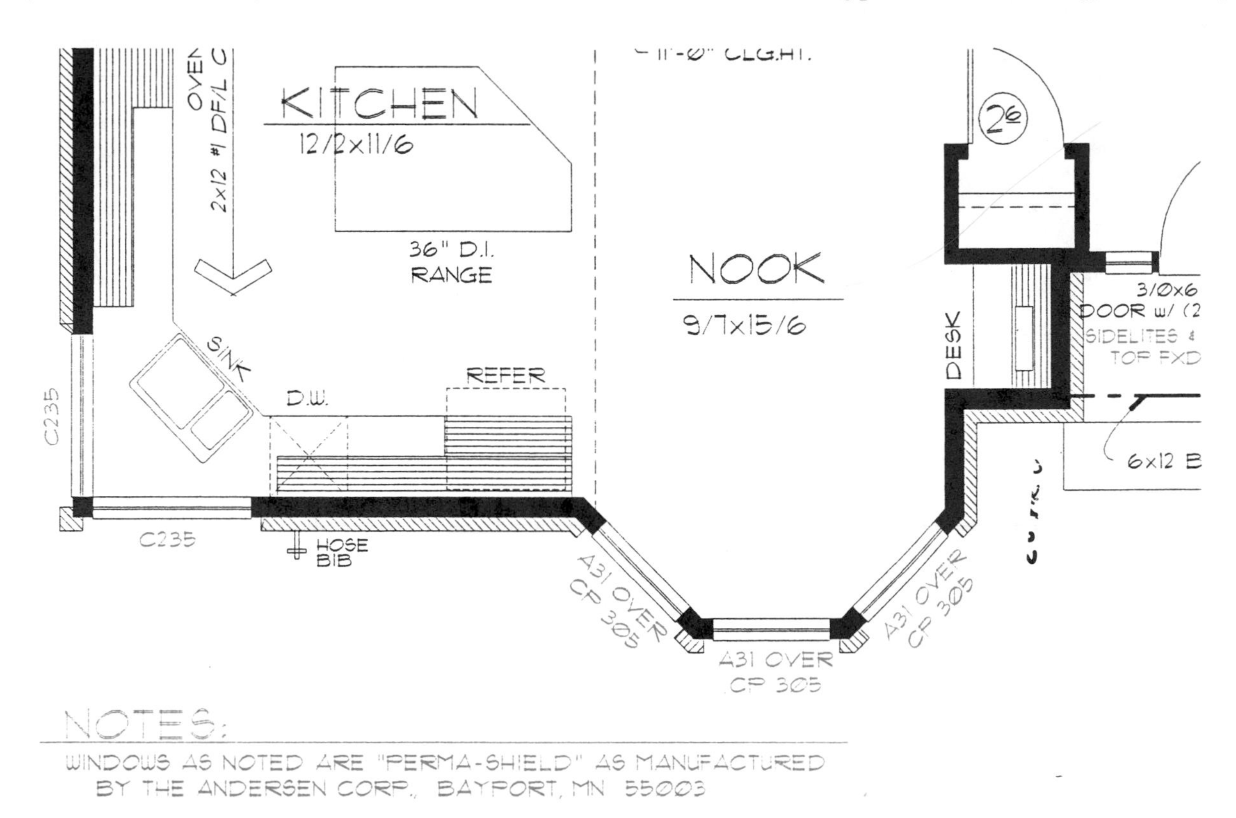

Figure 2–34b Labeling windows with manufacturer's catalog number. *Courtesy of Piercy and Barclay Designers, Inc.*

style. This information is available through vendors' specifications. In many cases you should obtain copies of vendors' specifications from the architect or supplier to ensure the proper product is installed. Figure 2–35 shows a typical page from a door catalog that shows styles and available sizes. Figure 2–36 is a page from a window catalog that describes the sizes of a particular product. Notice that specific information is given for the rough opening (framing size), finish size, and amount of glass. Vendors' catalogs also provide construction details and actual specifications for each product, as shown in Figure 2–37. This information may be important to the print reader to help ensure proper installation.

Doors with Beveled, Leaded Glass

Widths: 3'0"—single; 6'0"—double.
Heights: 6'8", 6'10", 7'0".

Embossed Doors

Widths: 2'8", 2'10", 3'0"—single; 5'4", 5'8", 6'0"—double.
Heights: 6'6", 6'8", 6'10",7'0".
NA: Certain styles not available in sizes as marked.

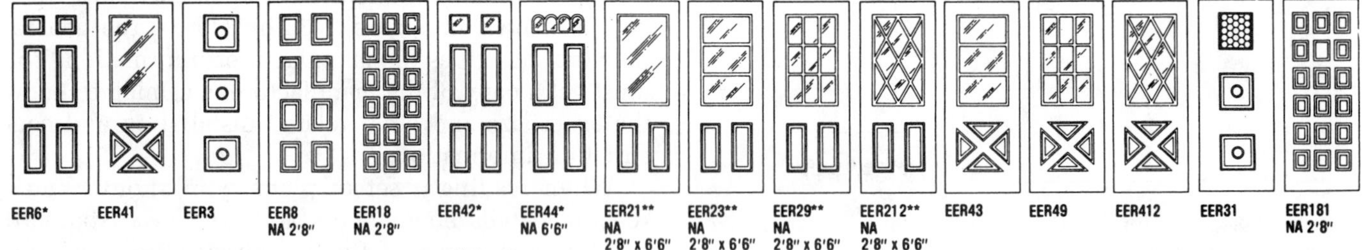

*Deadlocks on 2'8" 6-panel Embossed Doors must have a 2⅜" backset.
**Deadlocks on 2'8" 6-panel Embossed Doors with half-glass lites must have a 2⅜" backset and a 2" rose.

Figure 2–35 Sample from a door catalog page. *Courtesy Ceco Entry Systems, A United Dominion Company.*

Mas. Opg.	2-0½ (622)	2-4½ (724)	2-8½ (826)	2-10½ (873)	3-0½ (927)	3-2½ (978)	3-4½ (1029)	3-8½ (1130)	4-0½ (1232)
Rgh. Opg.	1-10⅜ (568)	2-2⅜ (670)	2-6⅜ (772)	2-8⅜ (822)	2-10⅜ (873)	3-0⅜ (924)	3-2⅜ (975)	3-6⅜ (1076)	3-10⅜ (1178)
Frame Size	1-9⅜ (543)	2-1⅜ (645)	2-5⅜ (746)	2-7⅜ (797)	2-9⅜ (843)	2-11⅜ (899)	3-1⅜ (949)	3-5⅜ (1051)	3-9⅜ (1153)
Sash Opg.	1-8 (508)	2-0 (610)	2-4 (711)	2-6 (762)	2-8 (813)	2-10 (864)	3-0 (914)	3-4 (1016)	3-8 (1118)
Glass Size	16" (406)	20" (508)	24" (610)	26" (660)	28" (711)	30" (762)	32" (813)	36" (914)	40" (1016)
2-10 9/16 (878) 2-9 9/16 (853) 2-9 1/16 (840) 2-6 (762) 12" (305)	WDH1612	WDH2012	WDH2412	WDH2612	WDH2812	WDH3012	WDH3212	WDH3612	WDH4012
3-2 9/16 (980) 3-1 9/16 (954) 3-1 1/16 (941) 2-10 (864) 14" (356)	WDH1614	WDH2014	WDH2414	WDH2614	WDH2814	WDH3014	WDH3214	WDH3614	WDH4014
3-6 9/16 (1081) 3-5 9/16 (1056) 3-5 1/16 (1043) 3-2 (965) 16" (406)	WDH1616	WDH2016	WDH2416	WDH2616	WDH2816	WDH3016	WDH3216	WDH3616	WDH4016

Figure 2–36 Sample from a window catalog page (additional sizes available). *Courtesy Marvin Windows.*

CABINETS, FIXTURES, AND APPLIANCES

Cabinets are found in kitchens, baths, dressing areas, utility rooms, bars, and workshops. Specialized cabinets, such as desks or built-in dressers, may be found in bedrooms. In general, the cabinets that are drawn on floor plans are built-in units.

Kitchens

In conjunction with cabinets, you will locate fixtures, such as sinks, butcher block cutting board coun tertops, lighting, and appliances, such as the range, refrigerator, dishwasher, trash compactor, and garbage disposer. Figure 2–38 shows the floor plan view of the kitchen cabinets, fixtures, and appliances. Sometimes upper cabinets are shown, as in Figure 2–39.

The range should have a hood with a light and fan and a note indicating how the fan will vent. Some ranges do not require a hood because they vent through the floor or wall. Vent direction should be specified on the plan. The garbage disposer is labeled with a note or with the abbreviation G.D. Houses with septic tanks should not have garbage disposers.

Pantries are popular and may be displayed as shown in Figure 2–40. A pantry may also be designed to be part broom closet and part shelves for storage.

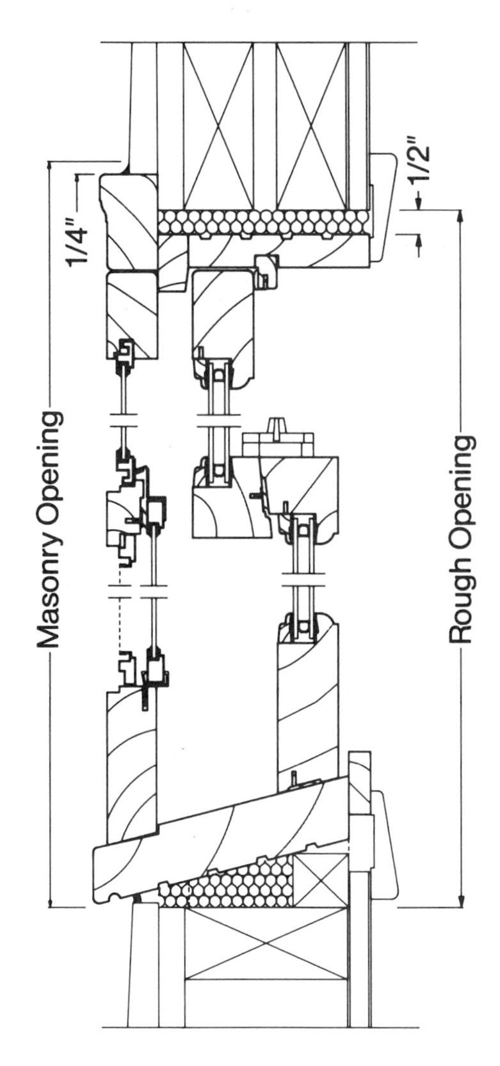

Figure 2–37 Vendor catalog details. *Courtesy Marvin Windows.*

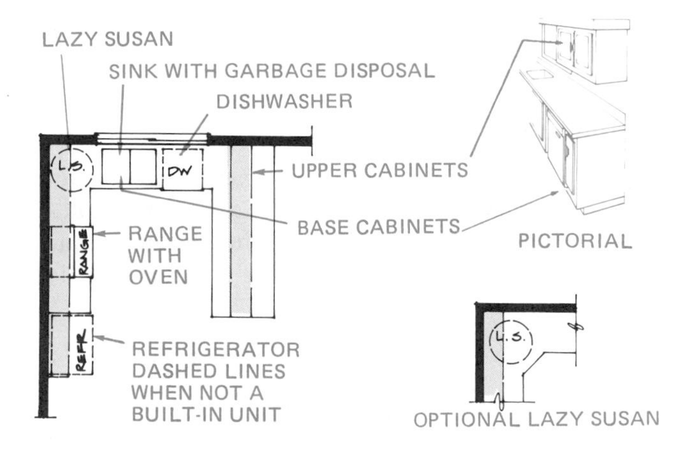

Figure 2–38 Kitchen cabinets, fixtures, and appliances.

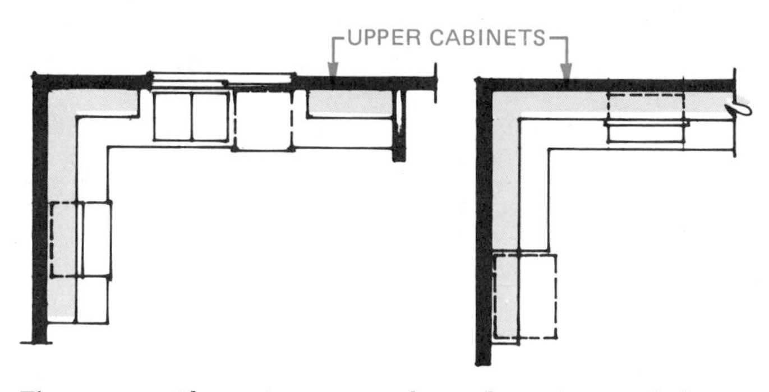

Figure 2–39 Alternative upper cabinet floor plan symbols.

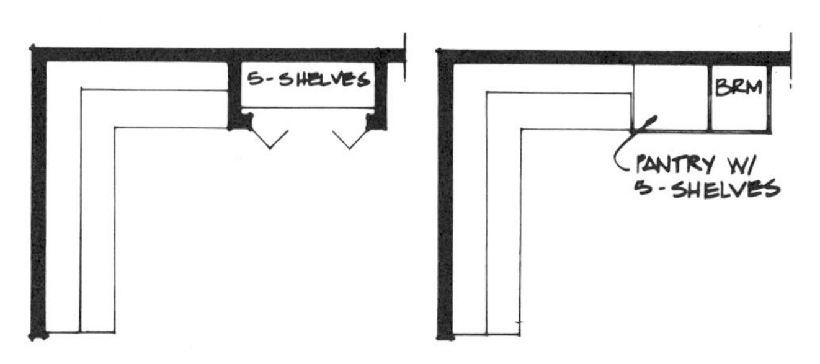

Figure 2–40 Kitchen pantry.

Figure 2–41 shows typical floor plan representation and cabinet sizes for a kitchen layout.

Bathrooms

Bathroom cabinets and fixtures are shown in several typical floor plan layouts in Figure 2–42. The vanity may be any length depending on the space available. The shower may be smaller or larger than the one shown depending on the vendor. Verify the size for a prefabricated shower unit before you install one. Showers built on the job may be any size or design. They are usually lined with tile, marble, or other materials. The common size of tubs is 30" × 60", but larger and smaller tubs are available. Sinks can be round, oval, or other shapes.

Often, because of cost considerations, the space available for bathrooms is minimal. Figure 2–43 shows some minimum sizes to consider.

Utility Rooms

The symbols for the clothes washer, dryer, and laundry tray are shown in Figure 2–44. The clothes washer and dryer symbols may be shown with dashed lines if these items are not part of the construction contract. Ironing boards may be built into the laundry room wall or attached to the wall surface, as shown in Figure 2–45. A note to the electrician is given if power is required to the unit or an adjacent outlet is provided.

The furnace and hot-water heater are sometimes placed together in a location central to the house. They may be placed in a closet, as shown in Figure 2–46. These utilities may also be placed in a separate room, the basement, or the garage.

Wardrobes and closets are utilitarian in nature because they are used for clothes and storage. Bedroom closets may be labeled WARDROBE. Other closets may be labeled ENTRY CLOSET, LINEN CLOSET, or BROOM CLOSET, for example. Wardrobe or guest closets should be provided with a shelf and pole. The shelf and pole may be shown with a thin line to represent the shelf and a dashed or centerline to show the pole, as in Figure 2–47. Sometimes two dashed lines are used to symbolize the shelf and pole.

When a laundry room is below the bedroom area, a chute may be provided from a convenient area near the bedrooms through the ceiling and into a cabinet directly in the utility room. The cabinet should be above or next to the clothes washer. Figure 2–48 shows a laundry chute noted in the floor plan.

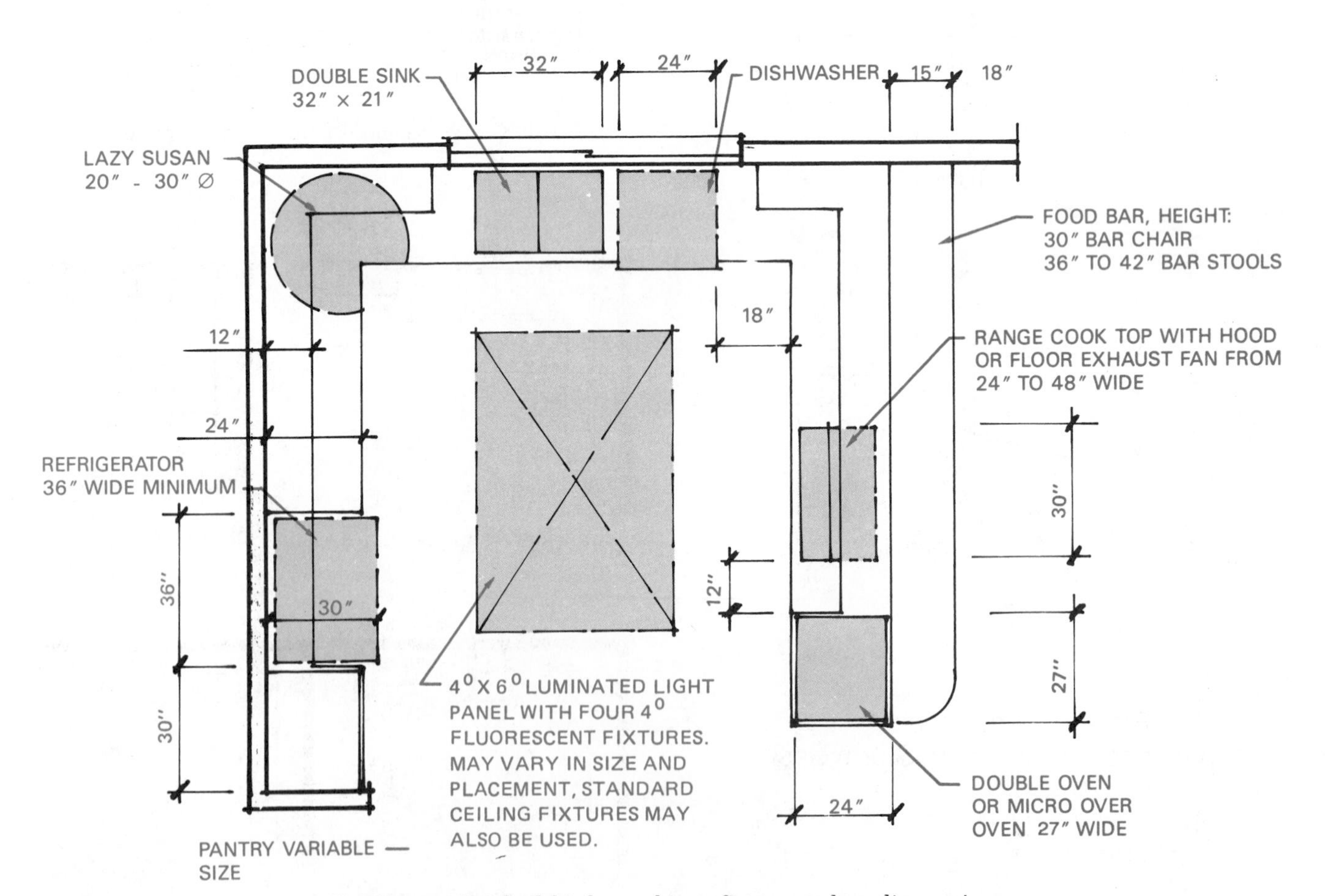

Figure 2–41 Standard kitchen cabinet, fixture, and appliance sizes.

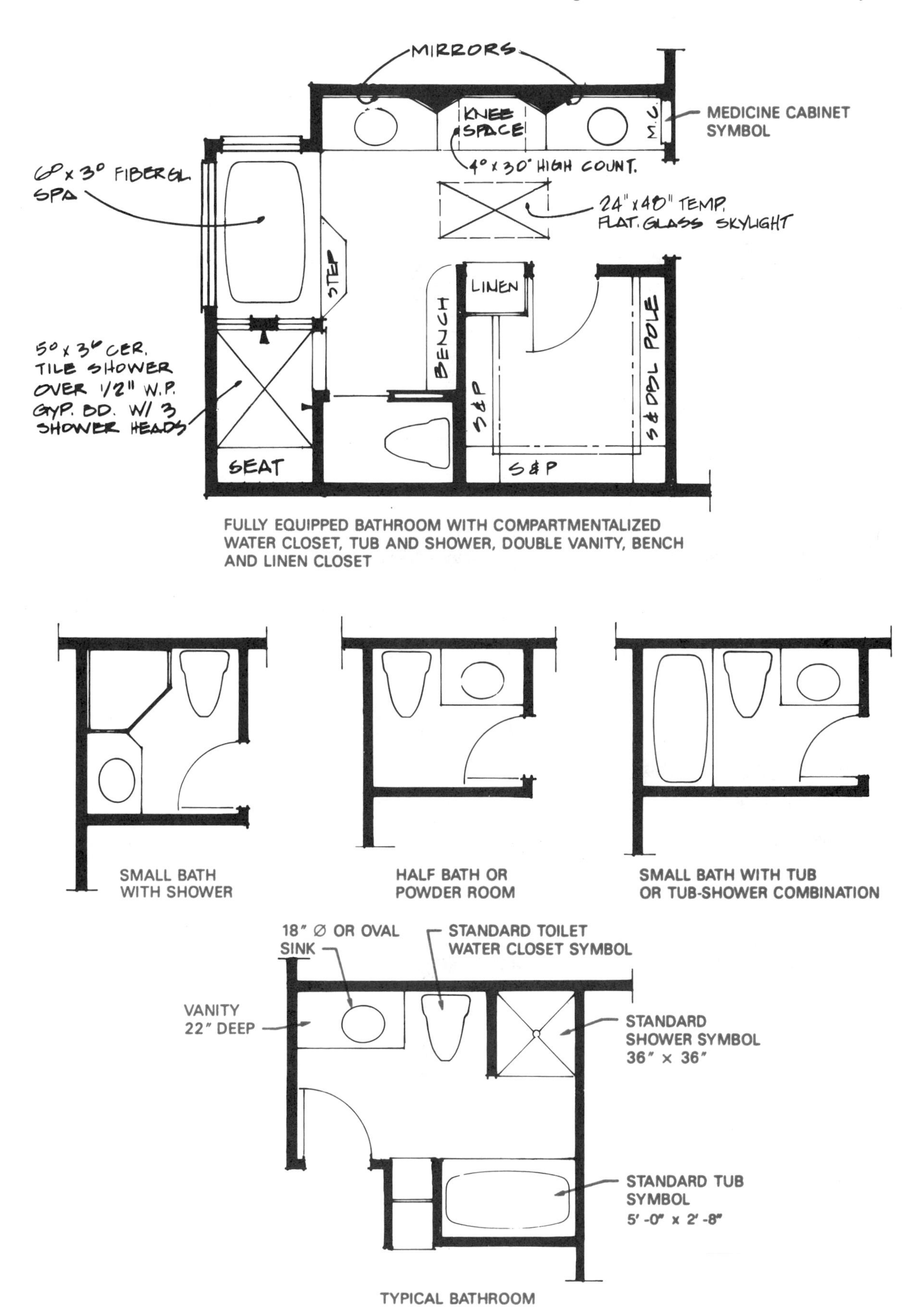

Figure 2–42 Layout of bathroom cabinet and fixture floor plan symbols.

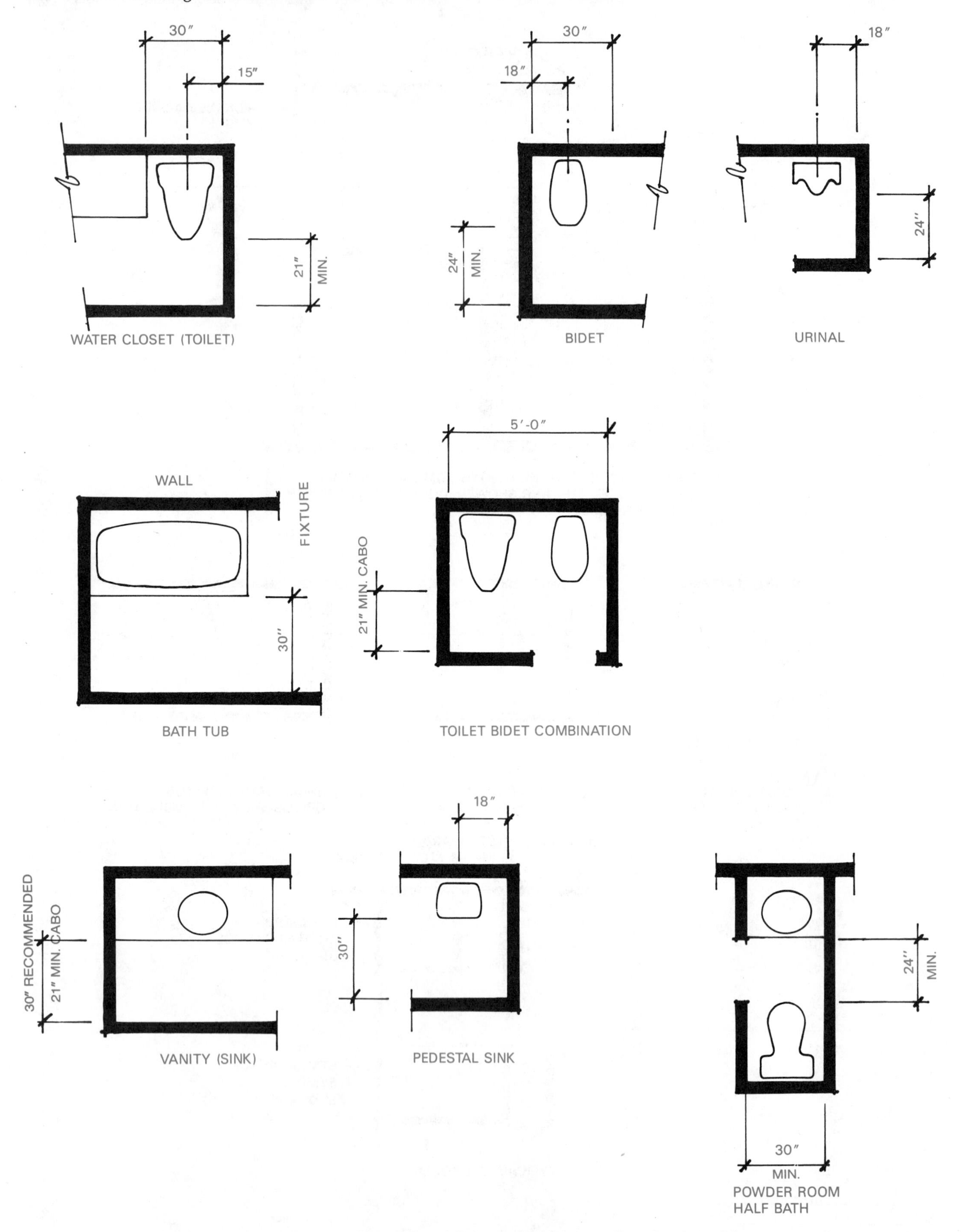

Figure 2–43 Minimum bath spaces. (CABO minimals)

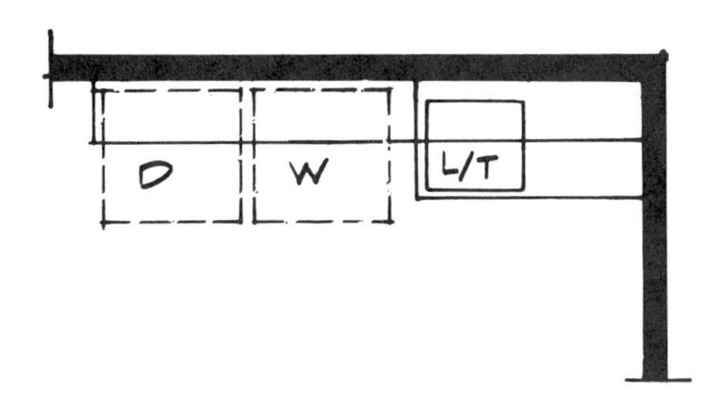

Figure 2–44 Washer, dryer, and laundry tray.

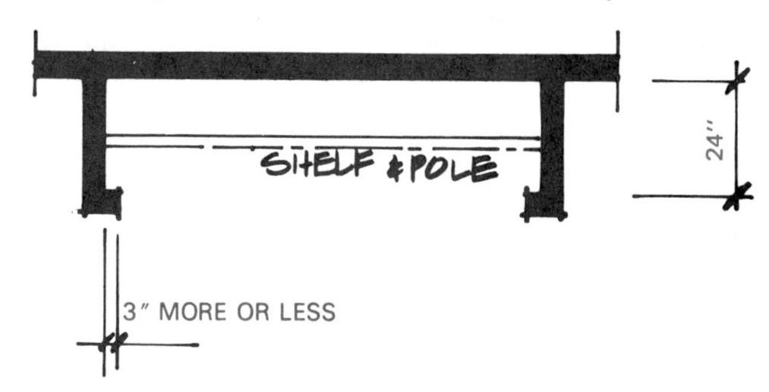

Figure 2–47 Standard wardrobe closet.

Figure 2–45 Ironing boards.

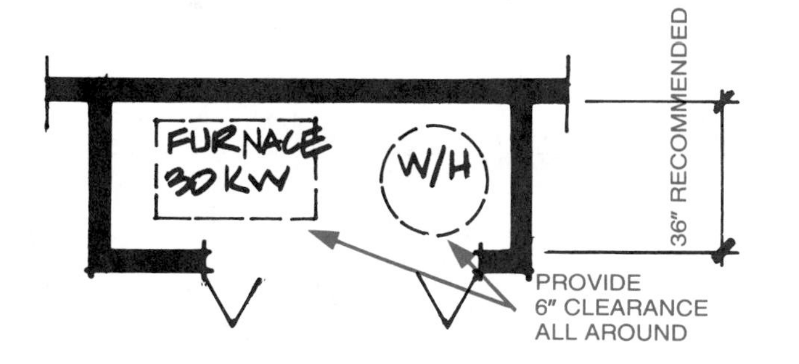

Figure 2–46 Furnace and water heater.

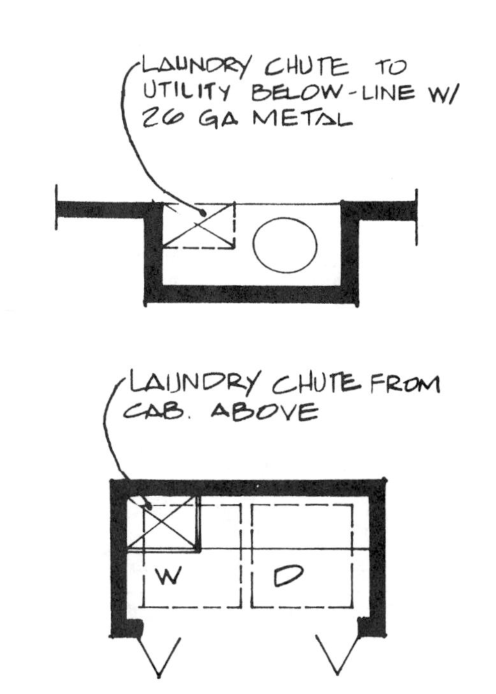

Figure 2–48 Laundry chute.

Figure 2–49 Room labels with floor finish material noted.

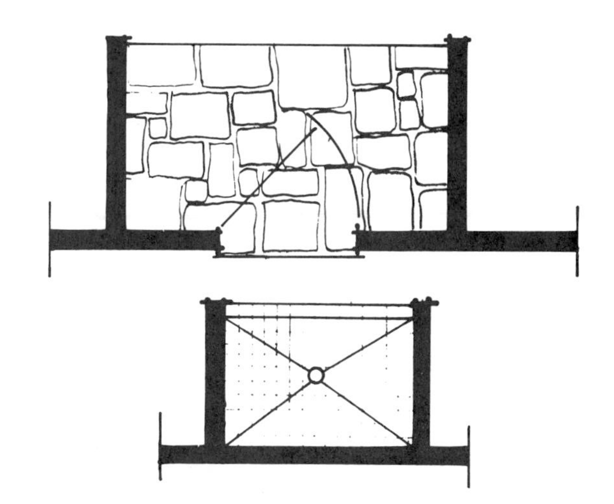

Figure 2–50 Floor finish material symbols.

FLOOR PLAN MATERIALS

Finish Materials

Finish materials used in construction may be identified on the floor plan with notes, characteristic symbols, or key symbols that relate to a finish schedule. The key symbol is also placed on the finish schedule next to the identification of the type of finish needed at the given location on the floor plan. Finish schedules may also be set up on a room-by-room basis. When flooring finish is identified, the easiest method is to label the material directly under the room designation, as shown in Figure 2–49. Other methods include using representative material symbols and a note to describe the finish materials, as shown in Figure 2–50.

Structural Materials

Structural materials are identified on floor plans with notes and symbols. The symbolic representation of a specific header size over a garage door open ing is a dashed line. See Figure 2–51. Other specific header sizes may be shown as a note over a window, as in Figure 2–51.

Ceiling beams are generally shown with dashed lines and labeled as shown in Figure 2–52. Construction members, such as beams, joists, rafters, or headers, that are labeled on the floor plan are considered at ceiling level unless otherwise specified. Construction members located in the floor are shown in the foundation plan or as part of the floor plan.

Ceiling joists, trusses, or roof joists for a vaulted ceiling are identified in the floor plan by an arrow that shows the direction and extent of the span. Ceilings are generally considered flat unless specified as *vaulted*, which means a sloping ceiling. The size and spacing of the members are noted along the arrow, as shown in Figure 2–53. Notice in the section at the beam in Figure 2–53 that the ceiling joists are over the beam, which leaves the beam exposed. The exposed beam may be made of quality material for a natural appearance, or it may be wrapped with gypsum wall board or other material for a finished appearance.

When it is not desirable to expose a beam in the room, the joists intersect flush with the bottom of the beam. To accomplish such an intersection, steel joist hangars are often used, as labeled in Figure 2–54. A number of structural connectors can be used in residential construction to connect joists to walls. The best thing to do is obtain a vendor's catalog from a company that manufactures structural connectors and become familiar with the many construction methods shown as examples in the catalog.

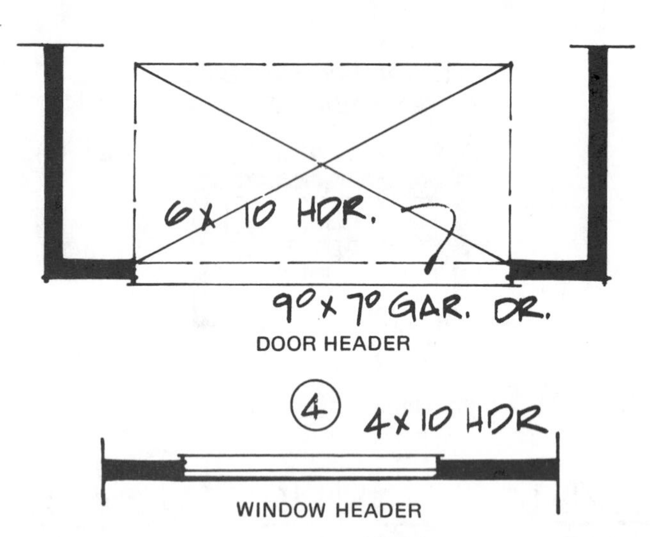

Figure 2–51 Header representation and notes.

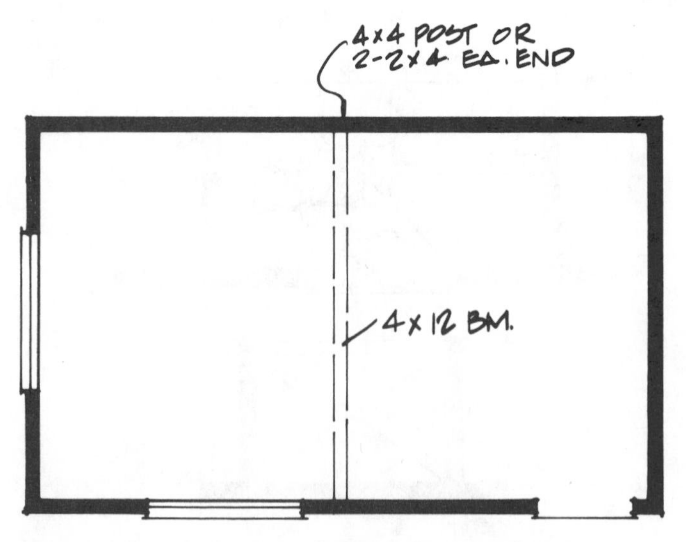

Figure 2–52 Beam and post representation and note.

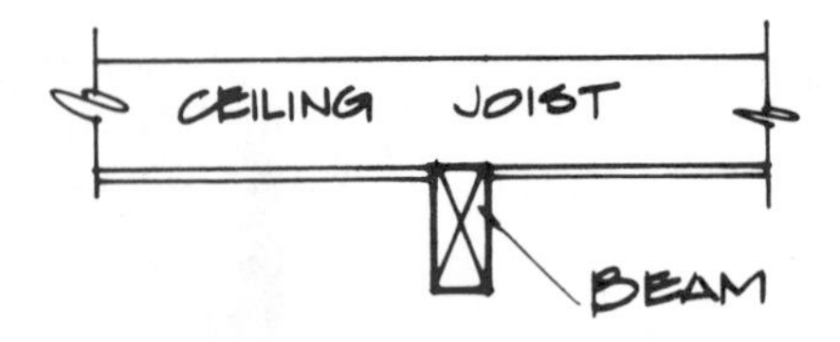

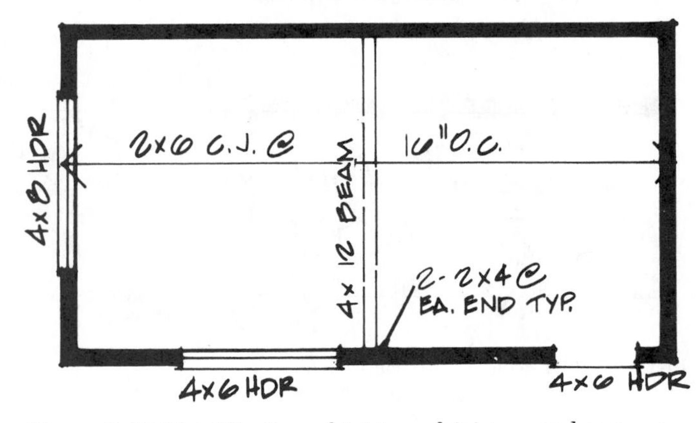

Figure 2–53 Identification of joists and joists over beam.

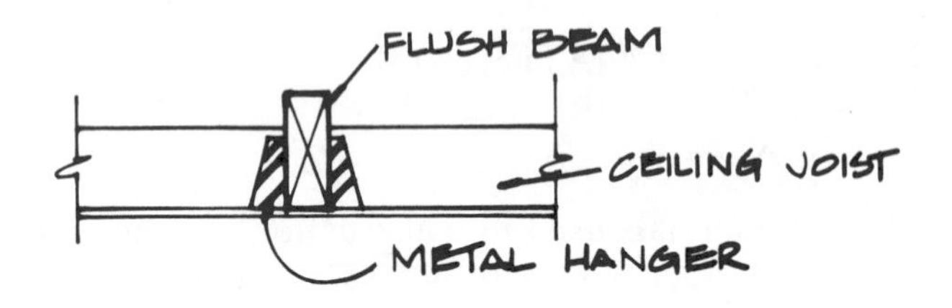

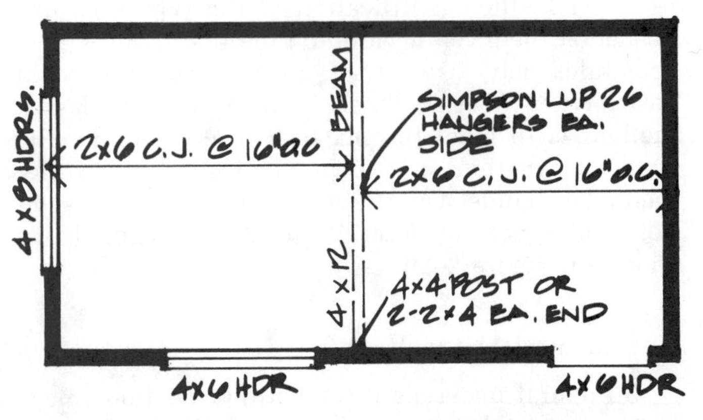

Figure 2–54 Identification of joists with joist hangers used at beam for a flush ceiling.

All construction members that relate directly to the floor plan but are not clearly identified in cross sections or details are shown and labeled on the floor plan. A great deal of information is included, and it is often difficult to include all the necessary data without crowding the drawing. This extensive information provides a challenge to the print reader.

STAIRS

Stairs on Floor Plans

Stairs are shown on floor plans by the width of the tread, the direction and number of risers, and the lengths of handrails or guardrails. Abbreviations used are DN for down, UP for up, and R for risers. The note 14R means there are 14 risers in the flight of stairs.

Figure 2–55 shows a common straight stair layout with a wall on one side and a guardrail on the other. Figure 2–56 shows a flight of stairs with guardrails all around at the top level and a handrail running down the stairs. Figure 2–57 shows stairs between two walls. Notice also that the stairs are drawn broken with a long break line at approximately midheight. Figure 2–58 shows parallel stairs with one flight going up and the other down. This situation is common when access from the main floor to both

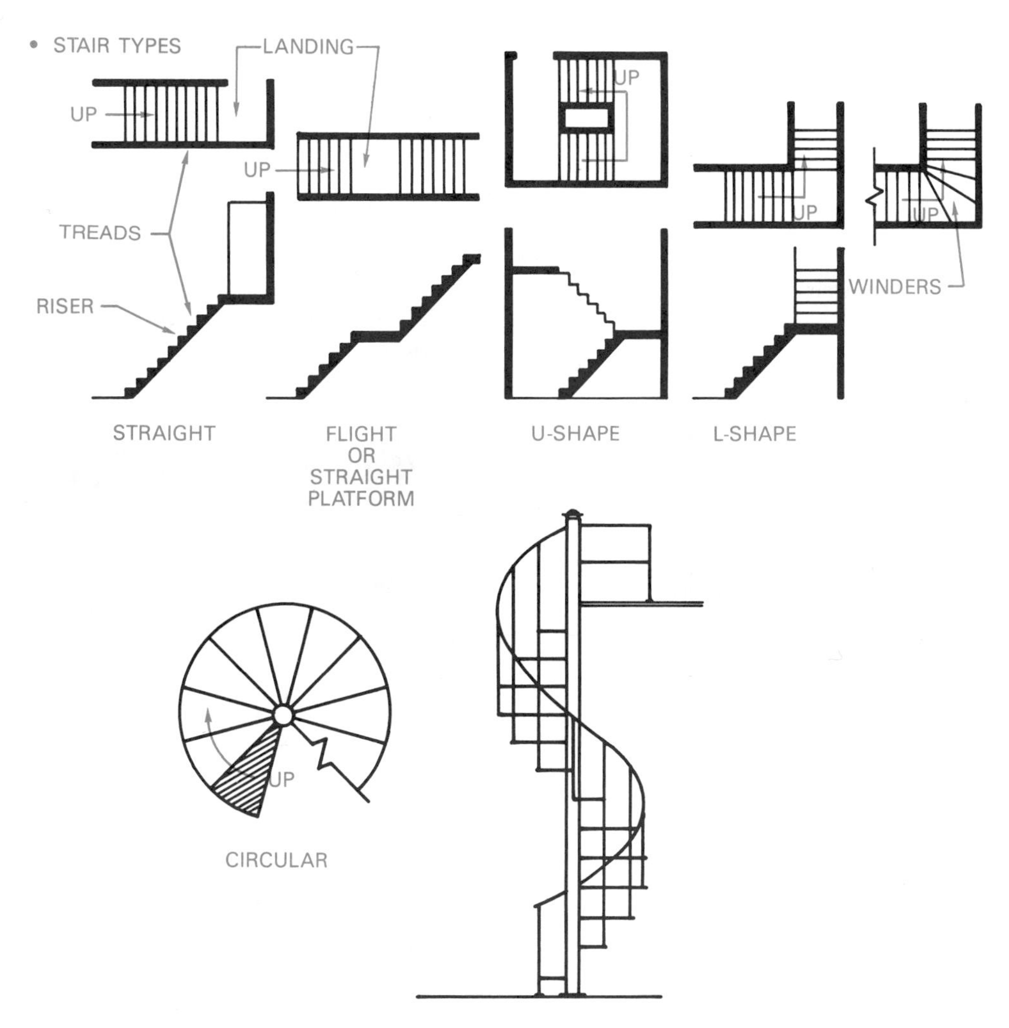

Figure 2–55 Stairs with a wall on one side and a guardrail on the other side.

Figure 2–56 Stairs with guardrails all around.

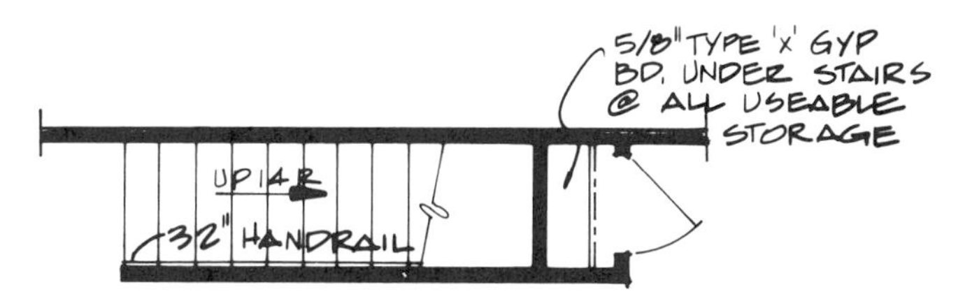

Figure 2–57 Stairs between two walls.

the second floor and basement is designed for the same area.

Stairs with winders and spiral stairs may be used to conserve space. Winders are used instead of a landing to turn a corner. The winder must be no smaller than 6 in. at the smallest dimension. Spiral stairs should be at least 30 in. wide from the center post to the outside, and the center post should have a 6 in. minimum diameter. Spiral stairs and custom winding stairs may be manufactured in several designs. Figure 2–59 shows plan views of winder and spiral stairs.

When a room is either sunken or raised, there is one step or more into the room. The steps are noted with an arrow, as shown in Figure 2–60. The few steps up or down do not require a handrail unless there are over three risers, but provide a handrail next to the steps when there are four or more. The guardrail shown in Figure 2–60 is for decoration only.

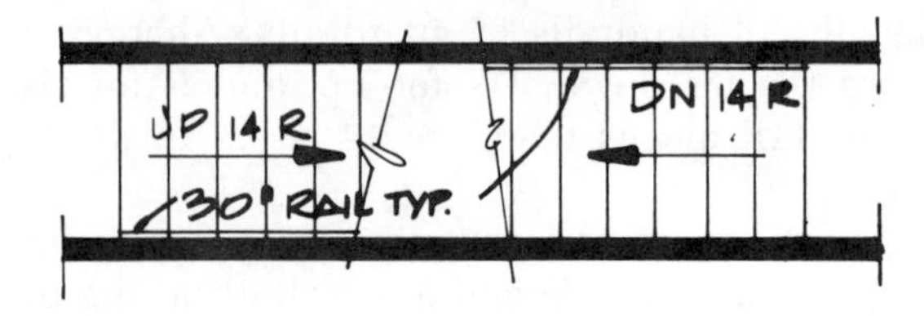

Figure 2–58 Stairs up and down in the same area.

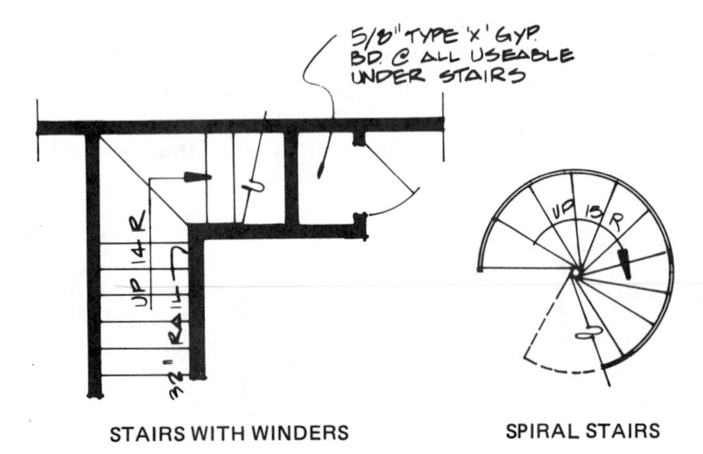

Figure 2–59 Winder and spiral stairs.

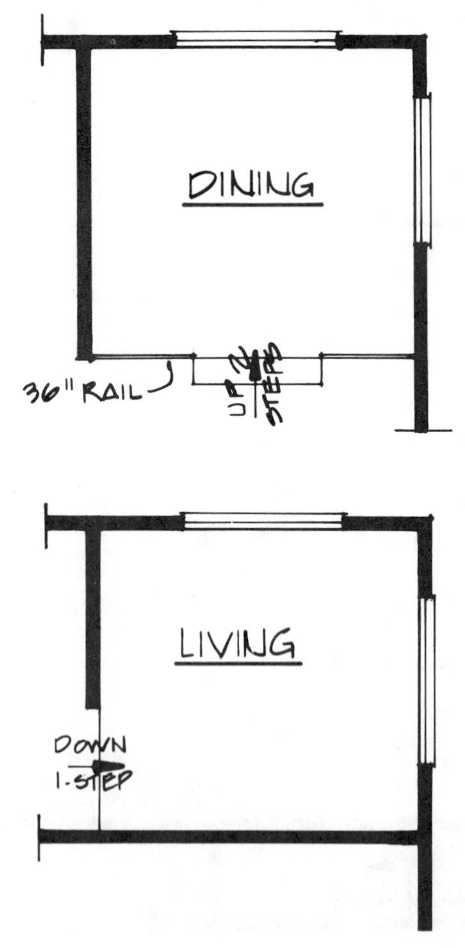

Figure 2–60 Sunken or raised rooms.

FIREPLACES

Floor Plan

Figure 2–61 shows several typical fireplace floor plan representations. Common fireplace opening sizes in inches are shown in the following table:

Opening Width	Opening Height	Unit Depth
36	24	22
40	27	22
48	30	25
60	33	25

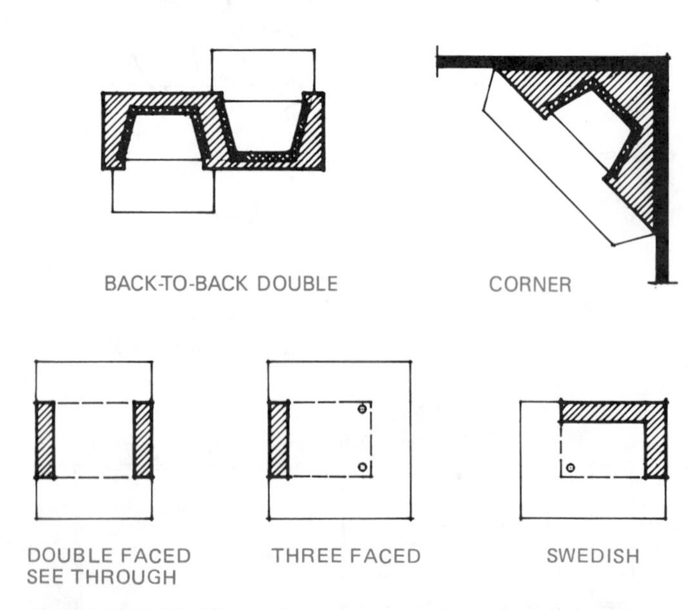

Figure 2–61 The floor plan representation of typical masonry fireplaces.

Steel Fireplaces

Fireplace fireboxes made of steel are available from various manufacturers. The fireplace shown on the floor plan may look the same whether the firebox is prefabricated of steel or constructed of masonry materials. See Figure 2–62.

There are also insulated fireplace units made of steel that can be used with wood-framed chimneys. These units are often referred to as zero clearance. *Zero clearance* means the insulated metal fireplace unit and flue can be placed close to a wood frame structure. Verify building codes and vendor's specifications before construction. See Figure 2–63. Figure 2–64 shows some common sizes of available fireplace units.

Wood Storage

Firewood may be stored near the fireplace. A special room or an area in a garage next to the fireplace may be ideal. A wood compartment built into the masonry next to a fireplace opening may also be provided for storing a small amount of wood. The floor plan representation of such a wood storage box is shown in Figure 2–65.

Fireplace Cleanout

When there is easy access to the base of the fireplace, you may see a cleanout (CO). The cleanout is small door in the floor of the fireplace firebox that allows ashes to be dumped into a hollow cavity built into the fireplace below the floor. Access to the fireplace cavity to remove the stored ashes is provided from the basement or outside the house. Figure 2–66 shows a fireplace cleanout noted in the plan view.

Figure 2–62 Manufactured firebox framed in masonry.

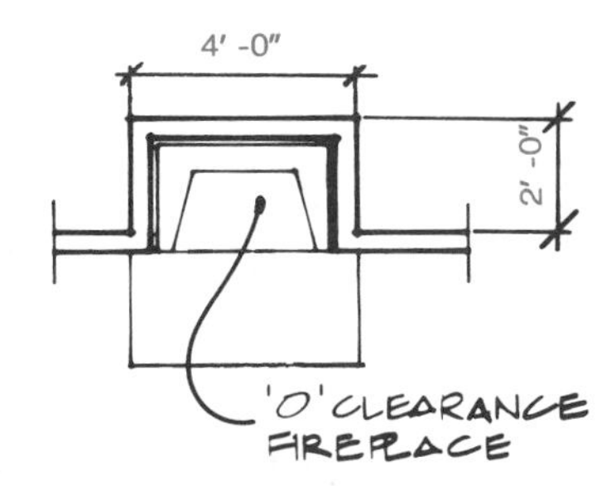

Figure 2–63 Manufactured firebox in wood frame.

Combustion Air

Building codes require that combustion air be provided to the fireplace. Combustion air is outside air supplied in sufficient quantity for fuel combustion. The air is supplied through a screened duct that is built by masons from the outside into the fireplace combustion chamber. By providing outside air, this venting device prevents the fireplace from using the heated air from the room, thus maintaining indoor oxygen levels and keeping heated air from going up the chimney.

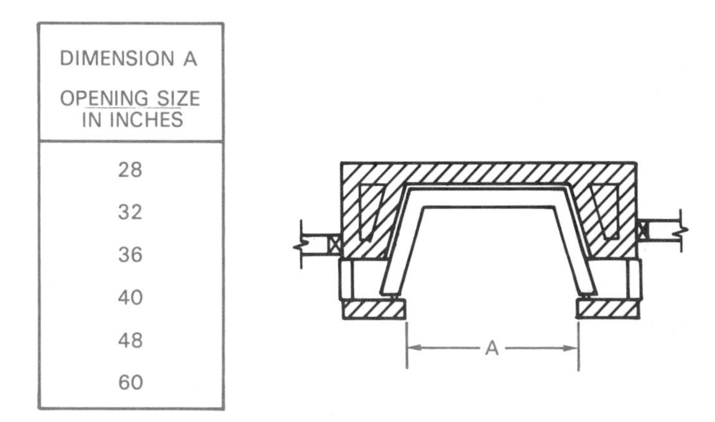

DIMENSION A OPENING SIZE IN INCHES
28
32
36
40
48
60

Figure 2–64 Prefabricated fireplace opening sizes. *Courtesy Heatilator, Inc.*®

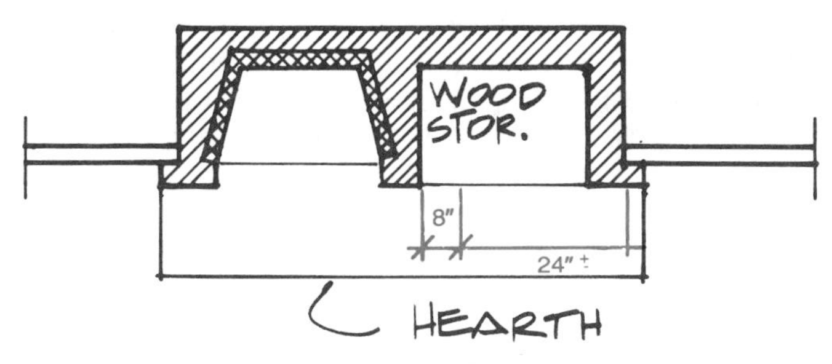

Figure 2–65 Wood storage built into the masonry structure.

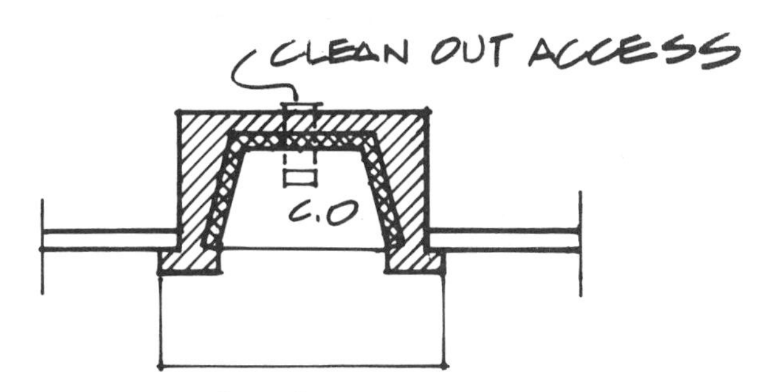

Figure 2–66 Fireplace cleanout.

Gas-Burning Fireplace

Natural gas may be provided to the fireplace either for starting the wood fire or for fuel to provide flames on artificial logs. The gas supply should be noted on the floor plan, as shown in Figure 2–67.

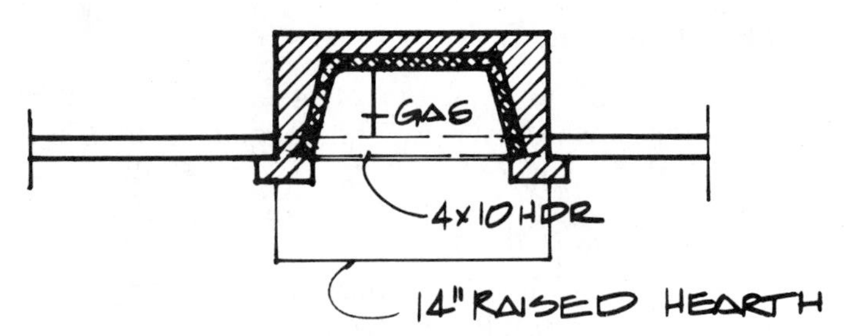

Figure 2–67 Gas supply to the fireplace.

Built-in Masonry Barbecue

The floor plan representation for a barbecue is shown in Figure 2–68. A built-in barbecue may be purchased as a prefabricated unit set into the masonry structure that surrounds the fireplace. There may be gas or electricity supply to the barbecue as a source of heat for cooking. As an alternative, the barbecue unit may be built into the exterior structure of a fireplace for outdoor cooking. The barbecue may also be installed separately from a fireplace, although there is some cost savings when the cooking unit is combined with the fireplace structure.

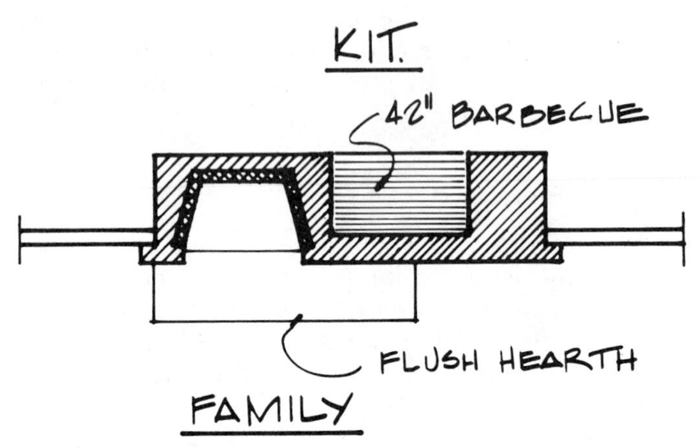

Figure 2–68 Barbecue built into the masonry fireplace structure.

SOLID FUEL-BURNING APPLIANCES

Solid fuel-burning appliances are such items as airtight stoves, freestanding fireplaces, fireplace stoves, room heaters, zero-clearance fireplaces, antique stoves, and fireplace inserts for existing masonry fireplaces. This discussion shows the floor plan representation and minimum distance requirements for the typical installations of some approved appliances.

Appliances that comply with nationally recognized safety standards may be noted on the floor plan with a note like the following:

ICBO APPROVED WOOD STOVE AND INSTALLATION REQUIRED [ICBO = International Conference of Building Officials. See Chapter 3.]

VERIFY ACTUAL INSTALLATION REQUIREMENTS WITH VENDORS' SPECIFICATIONS AND LOCAL FIRE MARSHAL OR BUILDING CODE GUIDELINES.

General Rules (Verify with Local Fire Marshal or Building Code Guidelines)

Floor Protection. Combustible floors must be protected. Floor protection material shall be noncombustible, with no cracks or holes, and strong enough not to crack, tear, or puncture with normal use. Materials commonly used are brick, stone, tile, and metal.

Wall Protection. Wall protection is critical whenever solid fuel-burning units are used in a structure. Direct application of noncombustible materials does not provide adequate protection. When solid fuel-burning appliances are installed at recommended distances from combustible walls, a 1 in. airspace is necessary between the wall and floor to the noncombustible material, plus a bottom opening for air intake and a top opening for air exhaust to provide positive air change behind the structure. This helps reduce superheated air adjacent to combustible material, Figure 2–69. Noncombustible materials include brick, stone, and tile over cement asbestos board. Minimum distances to walls should be verified with regard to vendors' specifications and local code requirements.

Combustion Air. Combustion air is generally required as a screened closable vent installed within 24" of the solid fuel-burning appliance.

Air Pollution. Some local areas have initiated guidelines that help control air pollution from solid fuel-burning appliances. The installation of a catalytic converter or other devices may be required. Check with local regulations.

Figure 2–70 shows floor plan representations of common wood stove installations.

A current trend in housing is to construct a masonry alcove within which a solid fuel unit is installed. The floor plan layout for a typical masonry alcove is shown in Figure 2–71. Strict code compliance is critical in these installations to avoid overheating of the space.

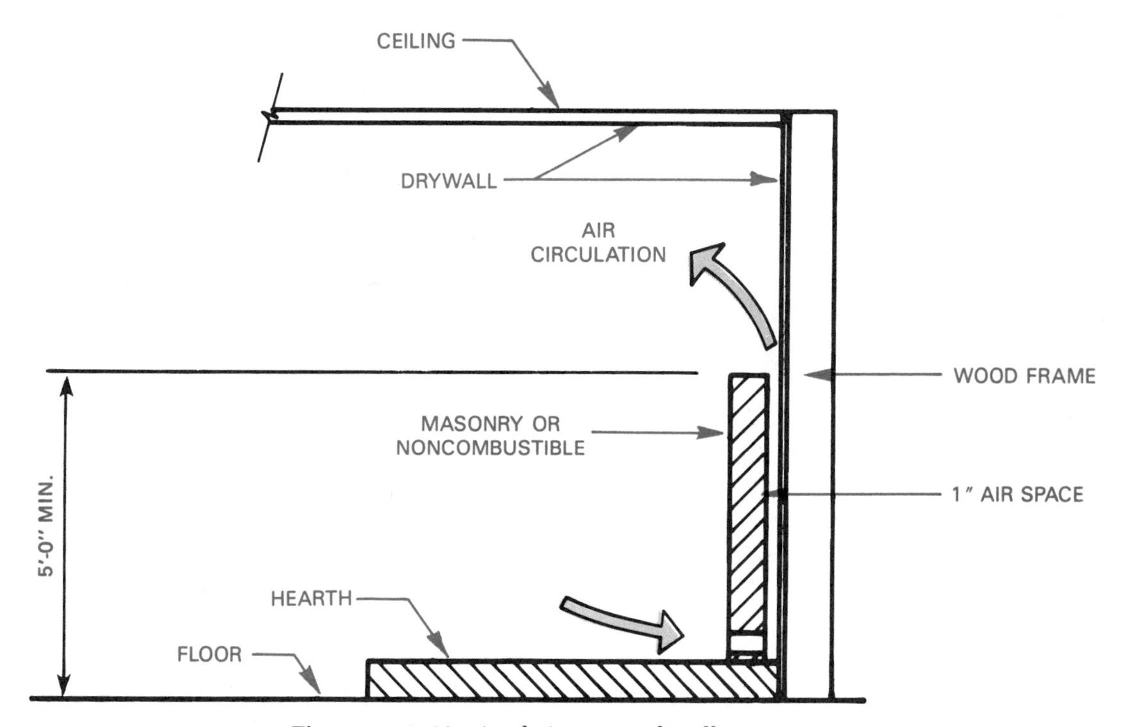

Figure 2–69 Air circulation around wall protection.

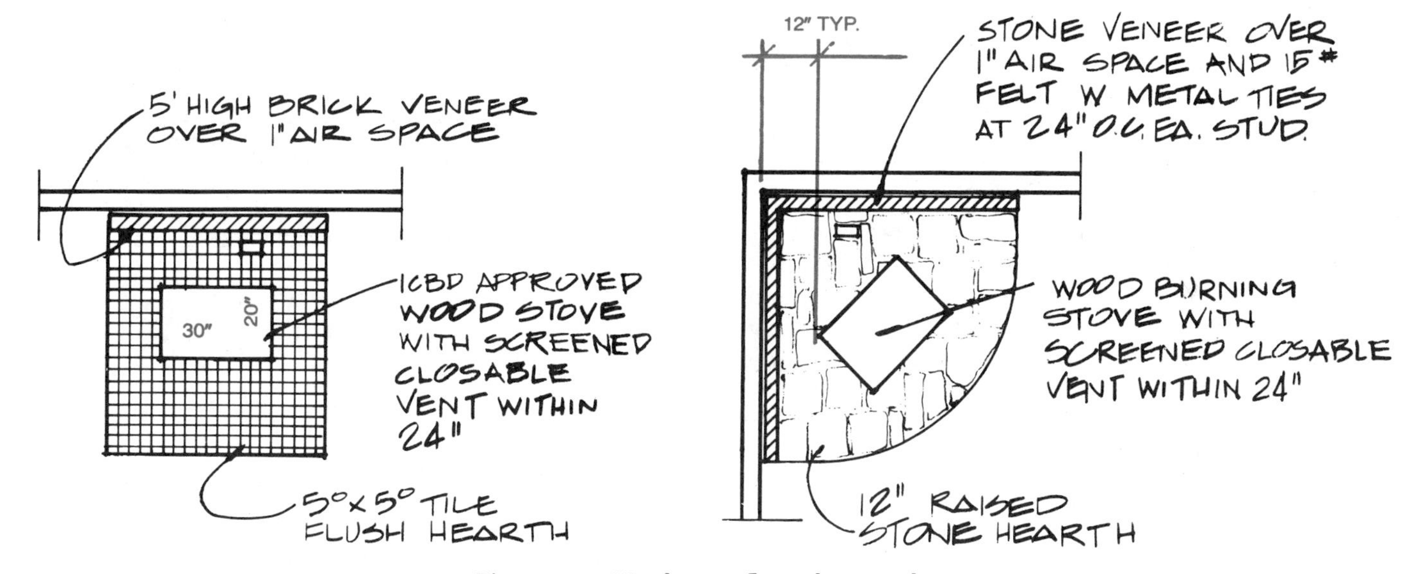

Figure 2–70 Wood stove floor plan samples.

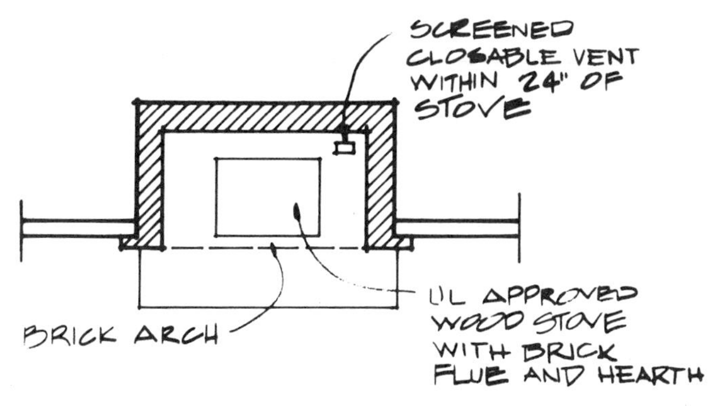

Figure 2–71 Masonry alcove for solid fuel-burning stove.

ROOM TITLES

Rooms are labeled with a name on the floor plan. Generally the lettering is larger than that used for other notes and dimensions. See Figure 2–72. The interior dimensions of the room are often lettered under the title as shown in Figure 2–73.

Figure 2–72 Room titles.

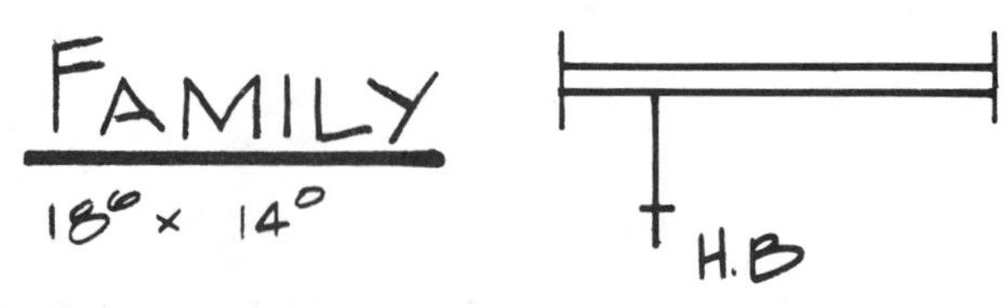

Figure 2–73 Room titles with room sizes noted.

Figure 2–74 Hose bibb symbol.

OTHER FLOOR-PLAN SYMBOLS

Hose Bibb

A hose bibb is an outdoor water faucet to which a garden hose may be attached. Hose bibbs are found at locations convenient for watering lawns or gardens and for washing a car. The floor-plan symbol for a hose bibb is shown in Figure 2–74.

Concrete Slab

Concrete slabs used for patio walks, garages, or driveways may be noted on the floor plan. A typical example is 4" THICK CONCRETE WALK. Concrete slabs used for the floor of a garage should slope toward the front to allow water to drain. The amount of slope is 1/8 in./ft minimum. Calling for a slight slope on patios and driveway aprons is also common.

Attic and Crawl Space Access

Access is necessary to attics and crawl spaces. The crawl access may be placed in any convenient location, such as a closet or hallway. The minimum size of the attic access must be 22" × 30". The crawl access should be 22" × 30" if located in the floor. The attic access may include a folding ladder (disappearing stairs) if the attic is to be used for storage. A minimum of 30" must be provided above the attic access. Crawl space access may be shown on the foundation plan when it is constructed through the foundation wall.

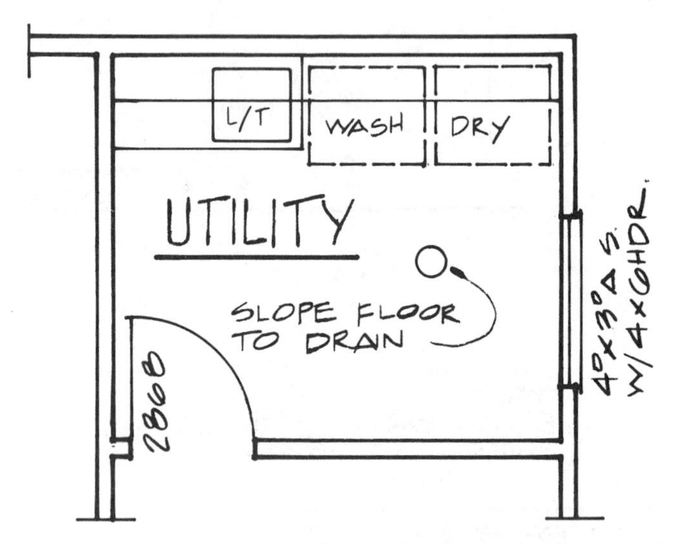

Figure 2–75 Floor drain symbol (symbol may also be square).

Floor Drains

Floor drains should be used in any location where water can accumulate on the floor, such as the laundry room, bathroom, or garage. The easiest application of a floor drain is in a concrete slab floor, although drains may be designed in any type of floor construction. Figure 2–75 shows a floor drain in a utility room.

Cross-Section Symbol

The location on the floor plan where a cross section is taken is identified with symbols known as cuttingplane lines. These symbols place an imaginary cutting plane through the structure at a particular location. The cross-sectional view relating to the cutting-plane line is usually found on another sheet. The arrows of the cutting plane are labeled with a letter, which is used to identify the cross-sectional view. Figure 2–76 shows a cross-section symbol (cutting-plane line). A number of methods are used in industry to label sections. The example in Figure 2–76 is a simple version. Other methods of identifying cutting planes and cross sections are more elaborate. Figure 2–77 shows three optional methods that are commonly used. The identification usually includes a letter designating the section and a number designating the drawing sheet where the letter is to be found. The identification letters and numbers may read from the right side or bottom of the sheet. The method used depends on company practice.

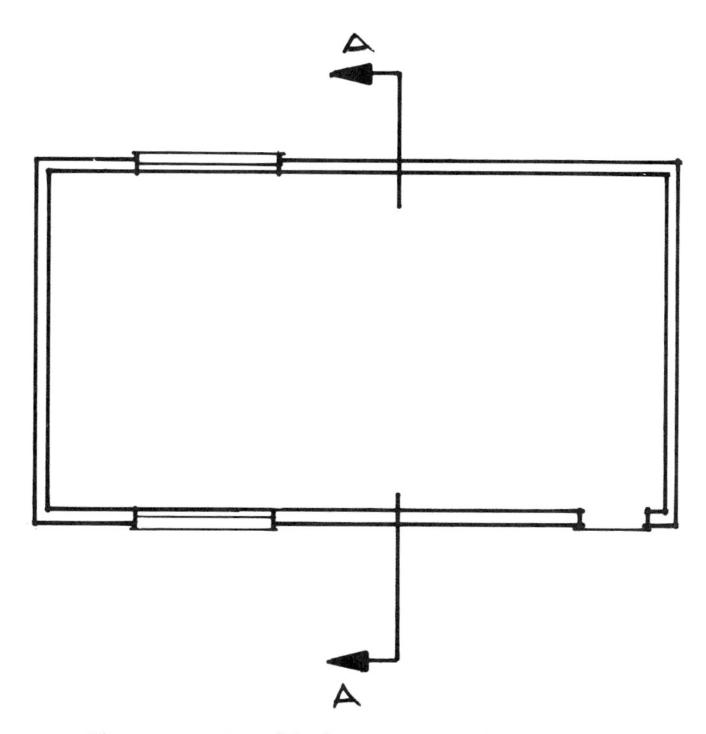

Figure 2–76 Simplified cutting-plane line symbol.

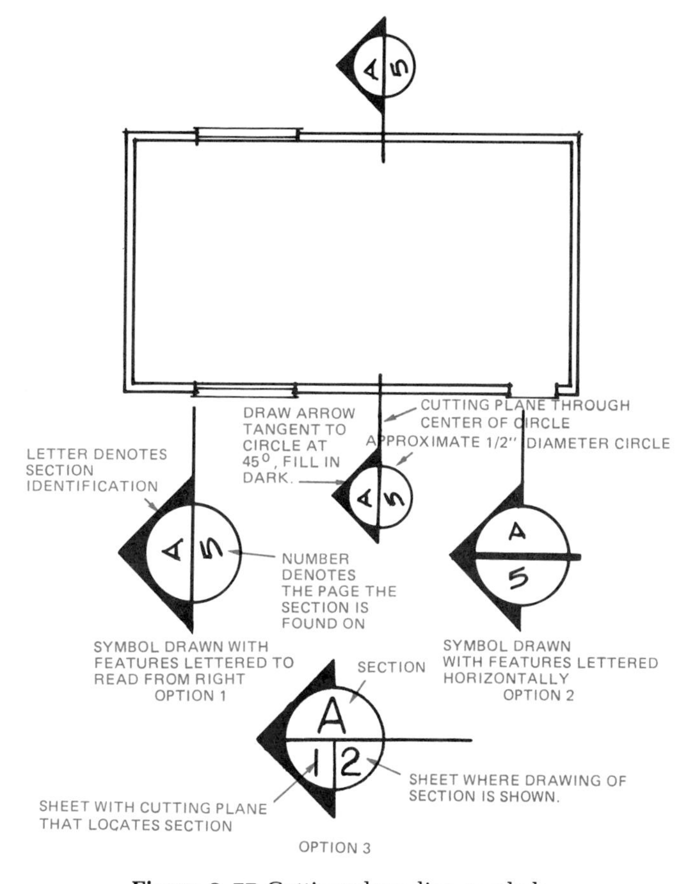

Figure 2–77 Cutting-plane line symbols.

CHAPTER 2 TEST

Fill in the blanks below with the proper word or short statement as needed to complete the sentence or answer the question correctly.

1. Residential floor plans are generally drawn at a scale of ______________________________.
2. Numbered symbols used on the floor plan key specific items to charts known as ______________________________.
3. Identify at least six items that may be found in schedules. ______________________________
______________________________.
4. What is the meaning of the numbers 3068 placed on the floor plan next to a door symbol? ______________________________
______________________________.
5. Describe the meaning of this set of numbers placed next to a window symbol on the floor plan: 6'–0" × 3'–6".
______________________________.
6. Why is it important to have the vendor's catalog or specifications available when reading prints that specify specific products in the window and door schedules? ______________________________
____
____
______________________________.
7. What does the abbreviation G D represent? ______________________________
8. Why do you think it would be a good idea to verify the exact size of a prefabricated shower before installation?
____
____
____
______________________________.
9. List a reason the clothes washer and dryer might be shown on the floor plan with dashed lines. ______________________________
____
______________________________.
10. List at least three names that may be given to closets on a floor plan. ______________________________
____
____
______________________________.
11. List at least four ways finish materials may be represented on a floor plan. ______________________________
____
____
______________________________.
12. Describe how a header is shown and specified on the floor plan. ______________________________
______________________________.
13. What does the abbreviation HDR represent? ______________________________
______________________________.
14. Explain how ceiling joists or trusses are represented and labeled on a floor plan. ______________________________
______________________________.
15. Identify two ways discussed in this chapter that represent how ceiling joists might intersect with a beam.
____
____
______________________________.
16. Discuss one good way to learn about the construction methods used when building with structural connectors. ______________________________
____
______________________________.

17. The minimum stair tread length is ______________________________.
18. Individual stair risers may range between ______________ in. in height, and risers should not vary more than ______________ in.
19. A clear height of __________ is the minimum amount of headroom required for the length of the stairs.
20. Stairs with over ______________ risers require handrails.
21. If you have a stairs with 14 risers, then how many runs are there?

____.

22. If each individual run is 10 in. in the stairs from question 21, then what is the total run? Give your answer in feet and inches and show your calculations:

____.

23. What does the term "zero clearance" mean? ______________________________

____

____

____.

24. What does the abbreviation CO mean as related to fireplace construction? ______________
25. What is combustion air? ______________________________

____.

26. Give four examples of solid fuel-burning appliances. ______________________________

____

____.

27. What does the abbreviation ICBO mean? ______________________________

____.

28. Why is strict code compliance so important when dealing with the construction of fireplaces or the installation of solid fuel-burning appliances? ______________________________

____.

29. What does the abbreviation HB represent? ______________________________
30. The location on the floor plan where a cross section is taken is identified with symbols known as ________

CHAPTER 2 EXERCISES

PROBLEM 2–1. Answer the following questions as you read the floor plan drawing on page 52:

1. Identify the type of wall shading used in this floor plan. ______________________________
2. Describe the brick veneer. (Change any abbreviations and symbols to whole words.) ______________

____

____.

3. Describe the type of windows and the manufacturer used in this plan. ______________________________

____.

4. Describe the entry door, sidelights, and window above. (Change any abbreviations and symbols to whole words and sizes to feet and inches.) ______________________________

____

____.

5. What type and size of door is used on the pantry? ______________________________
6. What is the kitchen ceiling height? ______________________________

7. What are the inside kitchen dimensions? ________________
8. Describe the type and size of the doors located at the master suite. ________________
________________.
9. What size and spacing of exterior studs are used in the construction of this house? (Change any abbreviations and symbols to whole words.) ________________
________________.
10. What size and spacing of interior studs are used in the construction of this house? (Change any abbreviations and symbols to whole words.) ________________
________________.
11. What size of beam is used between the kitchen and family room? ________________
12. Give the specifications of the headers that are used over all openings in bearing walls, except where otherwise noted. (Change any abbreviations and symbols to whole words.) ________________
____
________________.
13. Give the specifications for the ceiling joists over the kitchen and family room. (Change any abbreviations and symbols to whole words.) ________________
________________.
14. Give the specification provided for the crawl space access and identify its location. ________________
________________.
15. How many hose bibbs can you count? ________________
16. What is the size of the header between the entry and the living room? ________________

PROBLEM 2–2. Answer the following questions as you read the floor plan drawing on page 53:

1. Give the location, quantity, size, and description of door A. ________________
________________.
2. Give the location, quantity, size, and description of door G. ________________
________________.
3. Give the location, quantity, size, and description of door H. ________________
________________.
4. Give the location, quantity, size, and description of window 3. ________________
________________.
5. Give the location, quantity, size, and description of window 8. ________________
________________.
6. Describe the posts holding up the ends of the girder truss in the garage. ________________
________________.
7. Describe the ceiling construction members in the garage from the girder truss to the black wall. ________________
________________.
8. Describe the joists over the parlor that extend toward the back of the house. ________________
________________.
9. Give the specifications of the material used on the walls and ceiling in the garage. ________________
________________.
10. What is placed in front of the water heater and furnace to protect from automobile damage? ________________
________________.
11. Describe the floor used at the entry before you go into the foyer. ________________
________________.

12. What is the purpose of the 2 × 6 plumbing wall? ____________________

13. How many risers are there in the stairs? ____________________
14. What size are the garage door headers? ____________________

PROBLEM 2–3. Answer the following questions as you read the floor plan drawing on page 54:

1. Describe the cooking unit in the kitchen. ____________________
2. Give the door size, type, and header specifications as you exit the nook. ____________________
3. Give the complete fireplace specifications. ____________________
4. What does the symbol used for the fireplace hearth represent? ____________________
5. What is the size of the door that leads to the area under the stairs? ____________________
6. Give the specifications provided for the windows in the family room. ____________________

PROBLEM 2–4. Answer the following questions as you read the floor plan drawing on page 55:

1. Describe the type and size of doors located at the washer/dryer space. ____________________
2. Describe the type and size of the door entering bath 1. ____________________
3. Give the size and location of the attic access. ____________________
4. From what direction in this floor plan do the stairs run? ____________________
5. How many risers are there in the stairs? ____________________
6. How many runs are there in the stairs? ____________________
7. If each individual stair tread (run) is 10 in., what is the length of the total run? Show your calculations. ____________________
8. Explain what the note 4 × 10 BM IN CLG means. ____________________

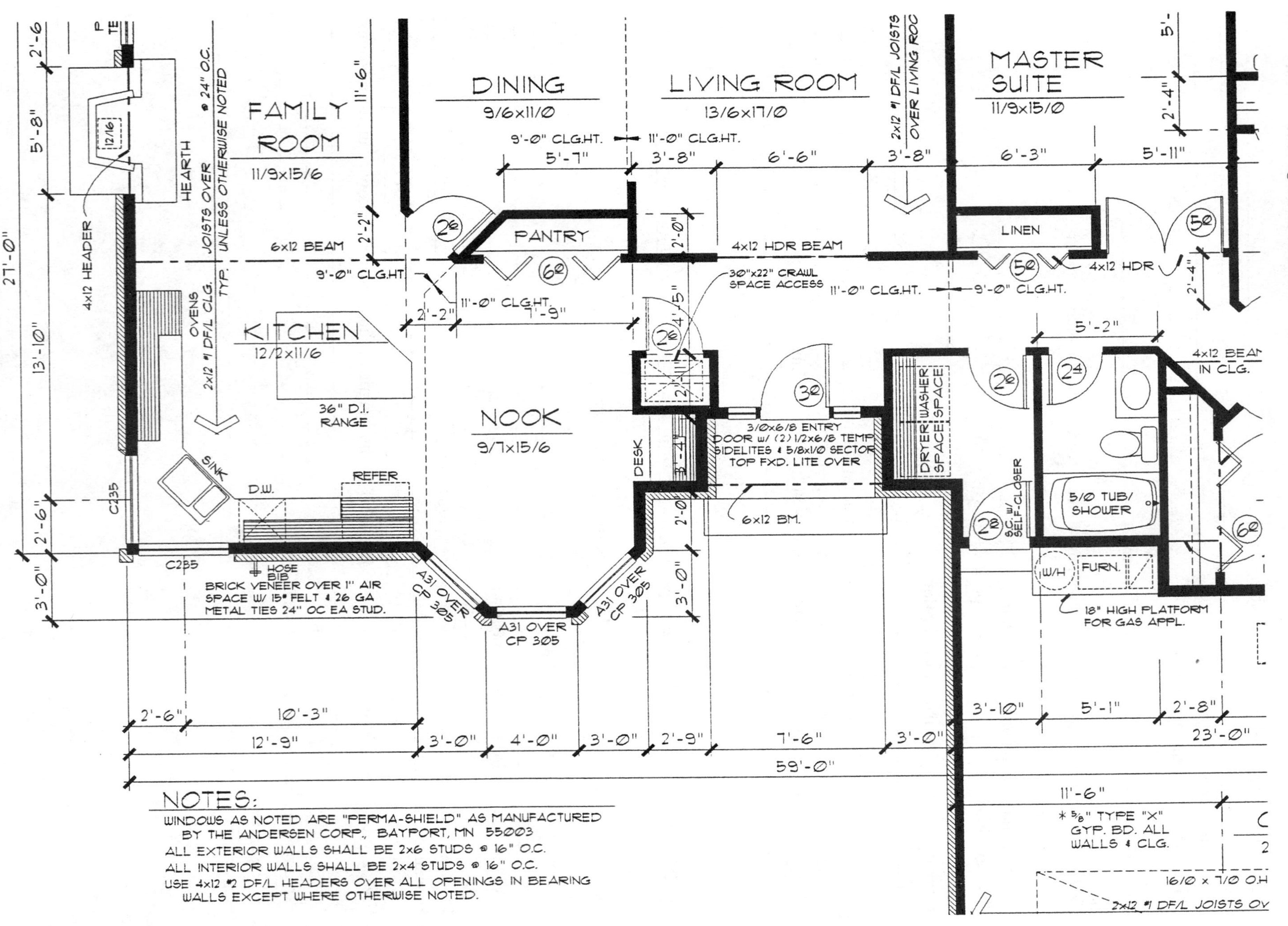

Problem 2–1. *Courtesy Piercy and Barclay Designers, Inc.*

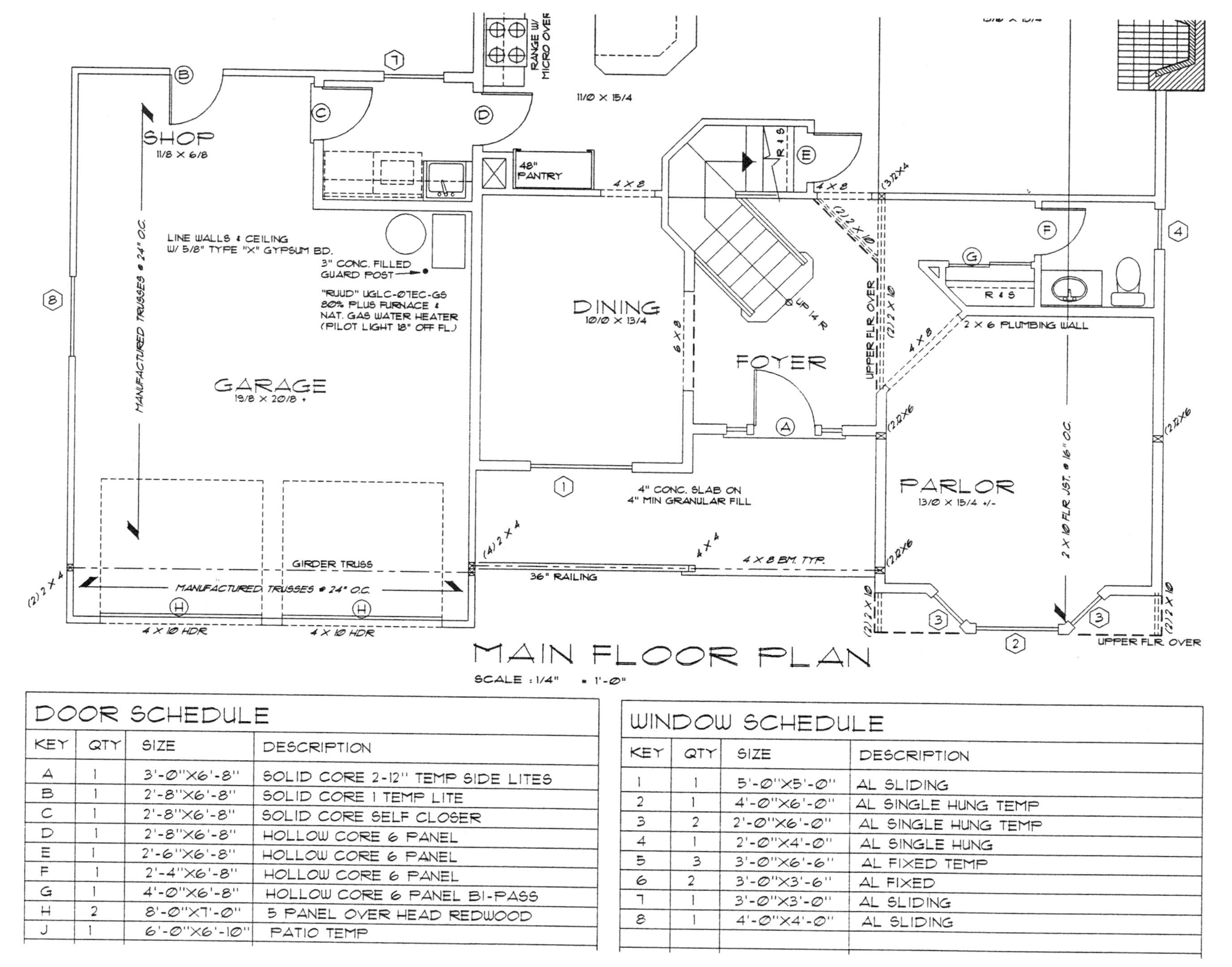

DOOR SCHEDULE

KEY	QTY	SIZE	DESCRIPTION
A	1	3'-0"X6'-8"	SOLID CORE 2-12" TEMP SIDE LITES
B	1	2'-8"X6'-8"	SOLID CORE 1 TEMP LITE
C	1	2'-8"X6'-8"	SOLID CORE SELF CLOSER
D	1	2'-8"X6'-8"	HOLLOW CORE 6 PANEL
E	1	2'-6"X6'-8"	HOLLOW CORE 6 PANEL
F	1	2'-4"X6'-8"	HOLLOW CORE 6 PANEL
G	1	4'-0"X6'-8"	HOLLOW CORE 6 PANEL BI-PASS
H	2	8'-0"X7'-0"	5 PANEL OVER HEAD REDWOOD
J	1	6'-0"X6'-10"	PATIO TEMP

WINDOW SCHEDULE

KEY	QTY	SIZE	DESCRIPTION
1	1	5'-0"X5'-0"	AL SLIDING
2	1	4'-0"X6'-0"	AL SINGLE HUNG TEMP
3	2	2'-0"X6'-0"	AL SINGLE HUNG TEMP
4	1	2'-0"X4'-0"	AL SINGLE HUNG
5	3	3'-0"X6'-6"	AL FIXED TEMP
6	2	3'-0"X3'-6"	AL FIXED
7	1	3'-0"X3'-0"	AL SLIDING
8	1	4'-0"X4'-0"	AL SLIDING

Problem 2–2. *Courtesy Alan Mascord Design Assoc., Inc.*

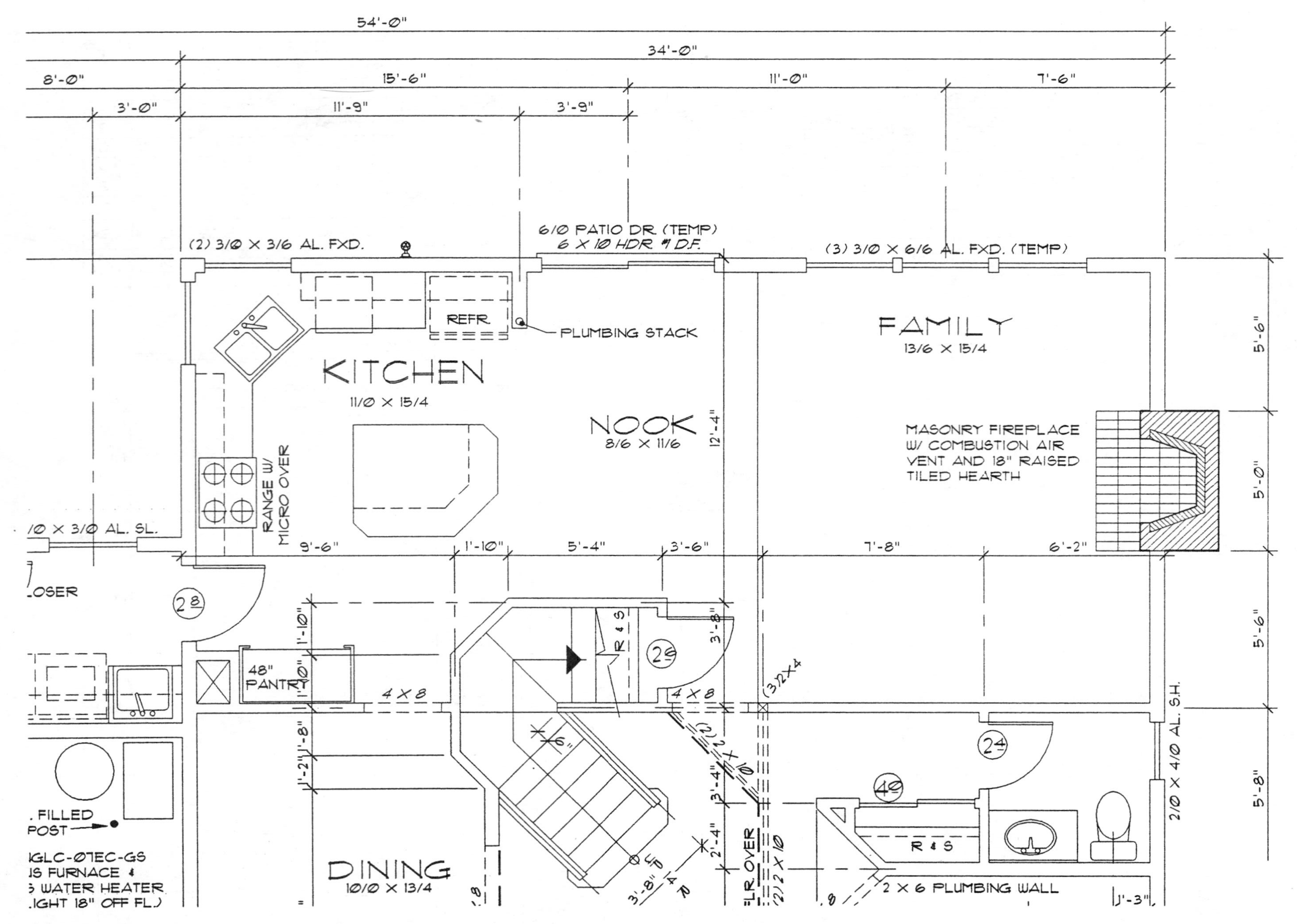

Problem 2-3. *Courtesy Alan Mascord Design Assoc., Inc.*

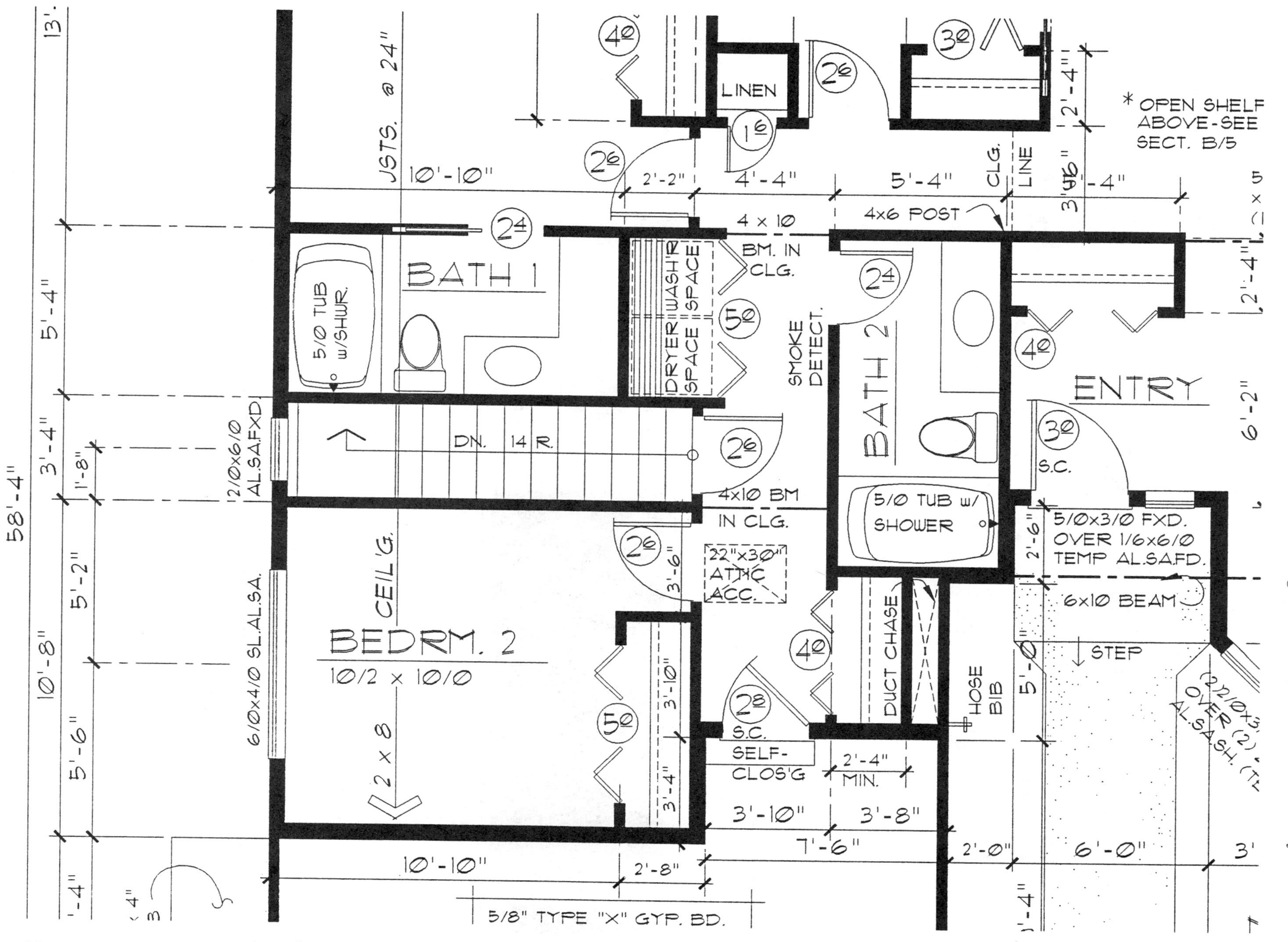

Problem 2–4. *Courtesy Piercy and Barclay Designers, Inc.*

Chapter 3

Reading Floor Plans, Part 2: Floor Plan Dimensions

ALIGNED DIMENSIONS

THE DIMENSIONING system most commonly used in architectural drafting is known as aligned dimensioning. With this system dimensions are placed in line with the dimension lines and read from the bottom or right side of the sheet. Dimension numerals are centered on and placed above the solid dimension lines. Figure 3–1 shows a floor plan that has been dimensioned using the aligned dimensioning system.

FLOOR PLAN DIMENSIONS

Dimensions on the drawing are made up of three parts: the extension lines, dimension lines, and dimension numerals. The extension lines project from the plan and continue slightly beyond the last dimension line to show the extent of the dimension. The dimension lines are used to display the length of the dimension and for placement of the dimension numeral. See Figure 3–2.

Dimension lines terminate at extension lines with dots, arrowheads, or slash marks, depending on the specific office practice. See Figure 3–3. Dimension numerals are displayed in feet and inches for all lengths over 12 in. Inches and fractions are used for units less than 12 in.

Exterior Dimensions

The overall dimensions on frame construction are understood to be given to the outside of the stud frame of the exterior walls. The first line of dimensions on the plan is the smallest distance from the exterior wall to the center of windows, doors, and partition walls. The second line of dimensions generally gives the distance from the outside walls to partition centers. The third line of dimensions is usually the overall distance between two exterior walls. See Figure 3–4. This method of applying dimensions eliminates the need for you to add dimensions at the job site and reduces the possibility of making an error.

Interior Dimensions

Interior dimensions locate all interior partitions

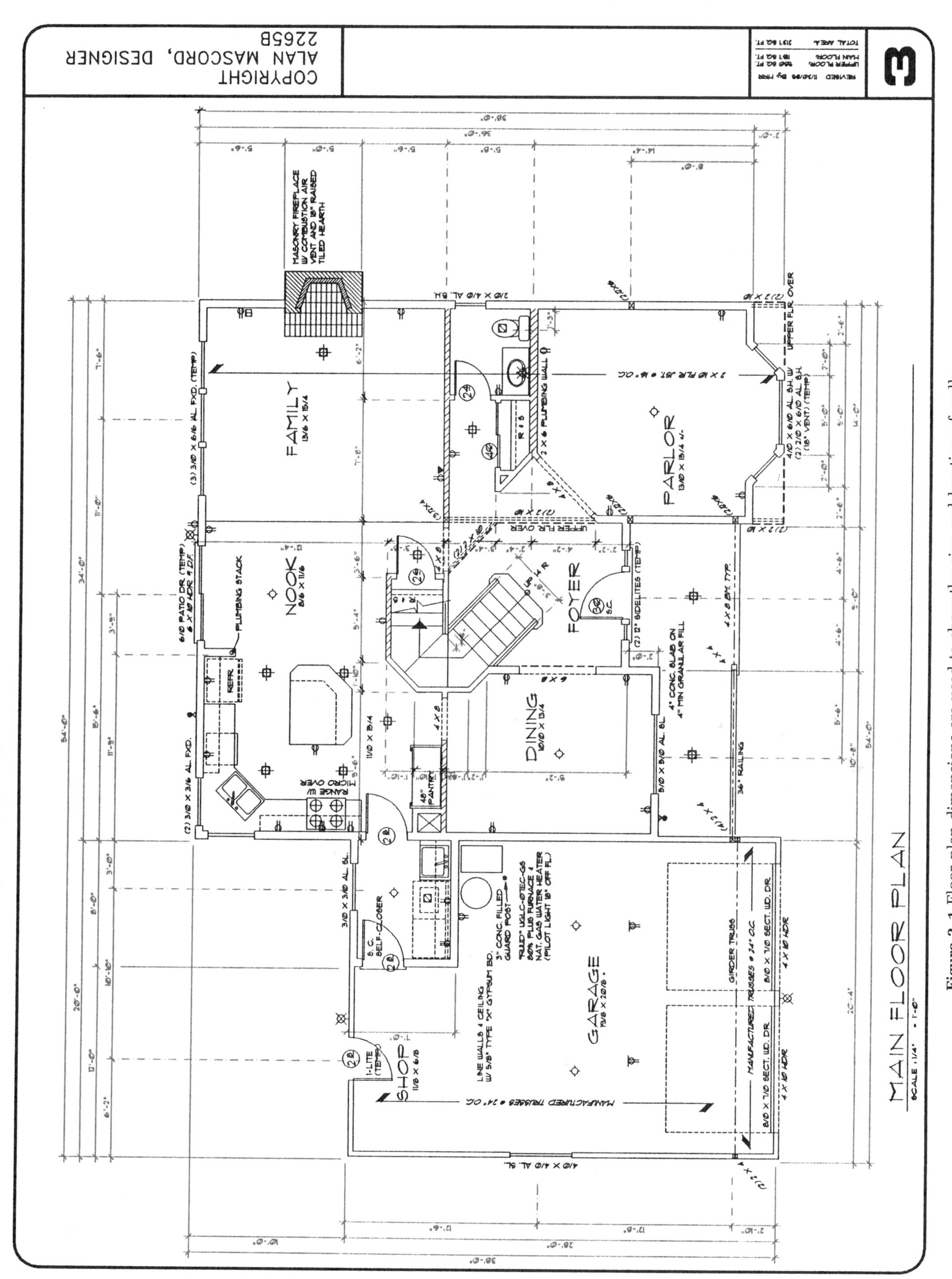

Figure 3–1 Floor plan dimensions are used to show the size and location of walls, partitions, doors, windows, and other construction items. *Courtesy Alan Mascord Design Assoc., Inc.*

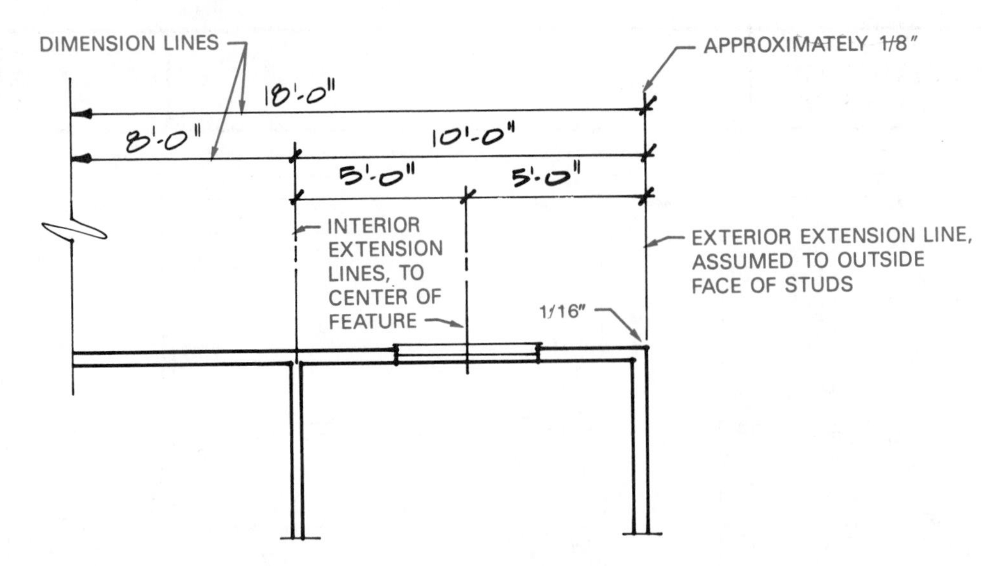

Figure 3–2 An example of extension lines and dimensions.

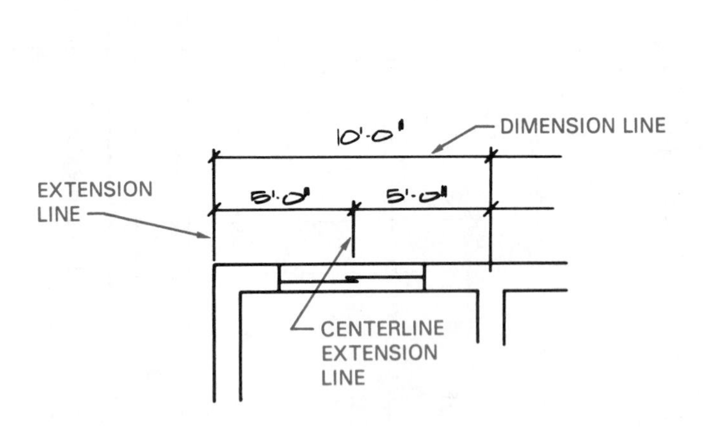

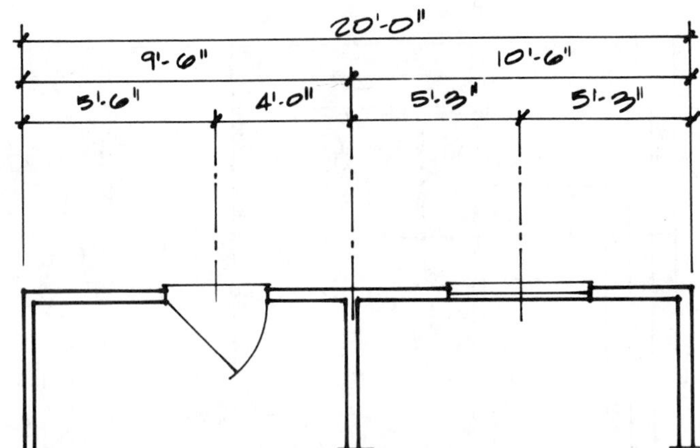

Figure 3–4 Typical exterior dimensions.

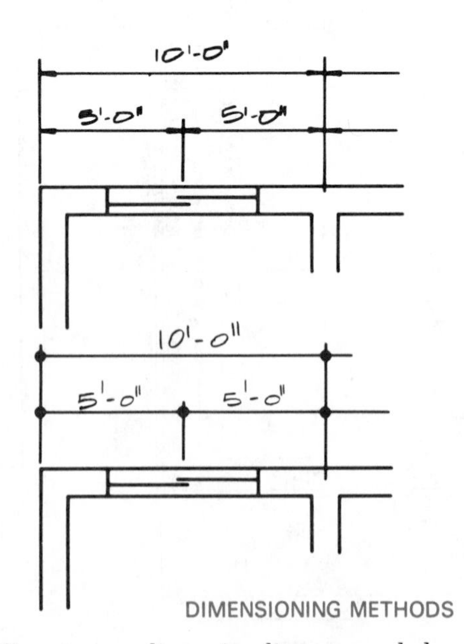

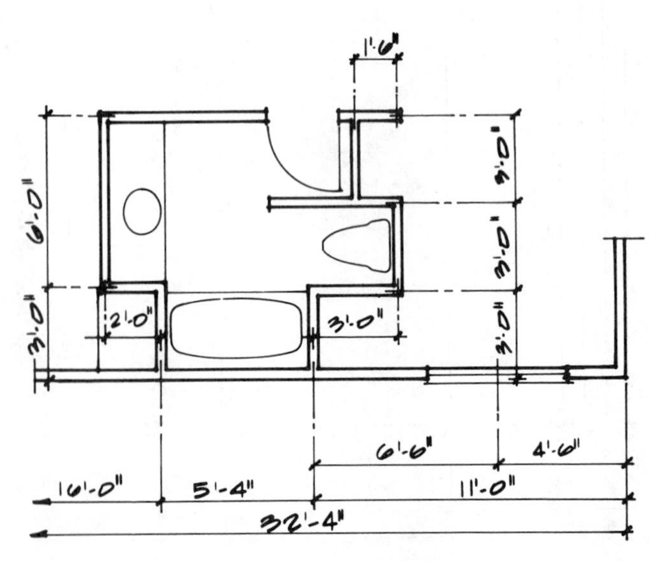

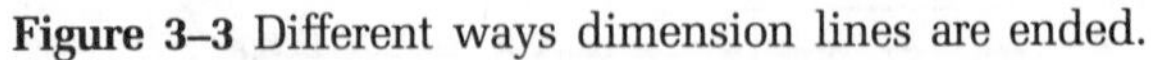

Figure 3–3 Different ways dimension lines are ended.

Figure 3–5 Typical interior dimensions.

and features in relationship to exterior walls. Figure 3–5 shows some common interior dimensions. Notice how they relate to the outside walls.

Standard Features

Some interior features that are considered standard sizes may not require dimensions. Figure 3–6 shows a situation in which the drafter elected not to dimension the depth of the pantry since it is directly adjacent to a refrigerator that has an assumed depth of 30 in. Notice that the refrigerator width is given because many different widths are available. The base cabinet is not dimensioned because cabinets are typically 24 in. deep. When there is any doubt, check with the architect or designer.

Other situations in which dimensions may be assumed are when a door is centered between two walls, as at the end of a hallway, or when a door enters a room and the minimum distance from the wall to the door is assumed. See the examples in Figure 3–7. Some dimensions may be provided in the form of a note for standard features, as seen in Figure 3–8. The walls around a shower need not be dimensioned when the note, 36" SQUARE SHOWER, defines the inside dimensions. The shower must be located, however. Be sure to verify the exact rough (framed) opening dimensions for a product of this type before the walls are framed. Different manufacturers often have different specifications for a standard 36 in. shower, for example.

One of the best ways to learn how to read dimensions is to study and evaluate existing plans.

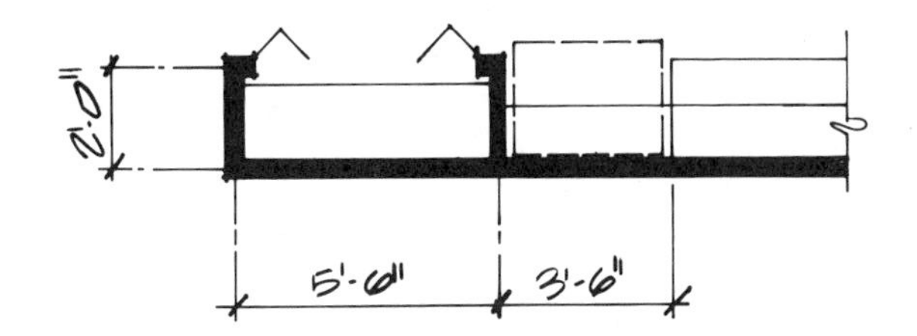

Figure 3–6 Assumed dimensions.

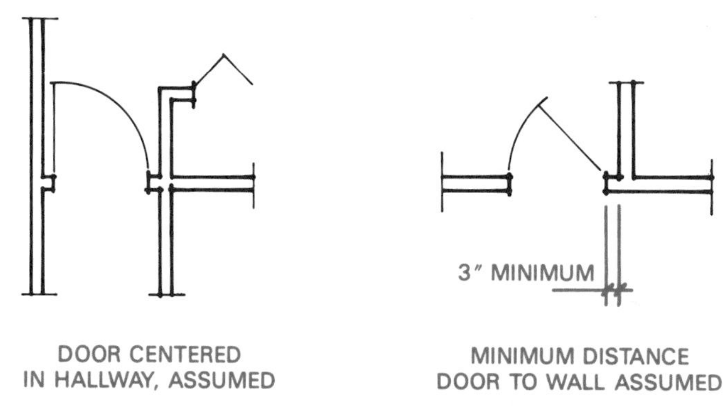

Figure 3–7 Assumed location of features without dimensions given.

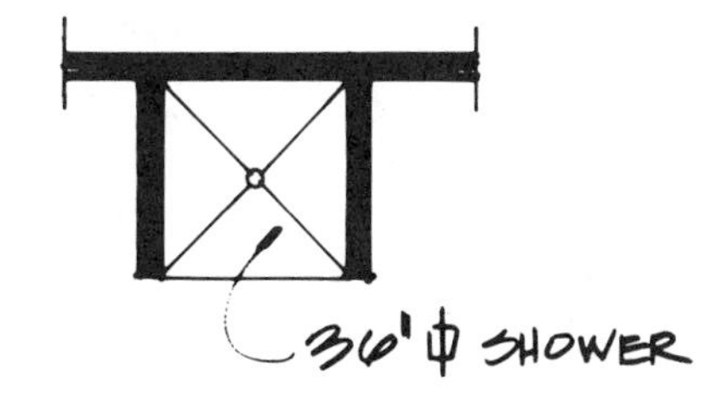

Figure 3–8 Standard items dimensioned with a specific note.

DIMENSIONING FLOOR PLAN FEATURES

Masonry Veneer

The dimensioning discussion so far has provided examples of floor plan dimensioning for wood-framed construction. Other method of residential construction include masonry veneer, concrete block, and solid concrete. Masonry veneer construction is the application of thin masonry, such as stone or brick, to the exterior of a wood-framed structure. Masonry veneer construction is dimensioned on the floor plan in the same way as wood framing, except that the veneer is dimensioned and labeled with a note describing the product.

Concrete Block Construction

Concrete blocks are made in standard sizes and may be used to construct exterior or interior walls of residential or commercial structures. Concrete block construction may be covered, for example by masonry veneer. Some structures use concrete block for the exterior bearing walls and wood-framed construction for interior partitions.

Dimensioning concrete block construction is different from that for wood-framed construction in that each wall, partition, and window and door opening is dimensioned to the edge of the feature, as shown in Figure 3–9.

Solid Concrete Construction

Solid concrete construction is used in residential and commercial structures. In residences it is mostly limited to basements and subterranean homes. Masonry veneer may be placed on either side of concrete walls for appearance. Wood framing and typical interior finish materials may also be added to the inside of concrete walls. Solid concrete construction

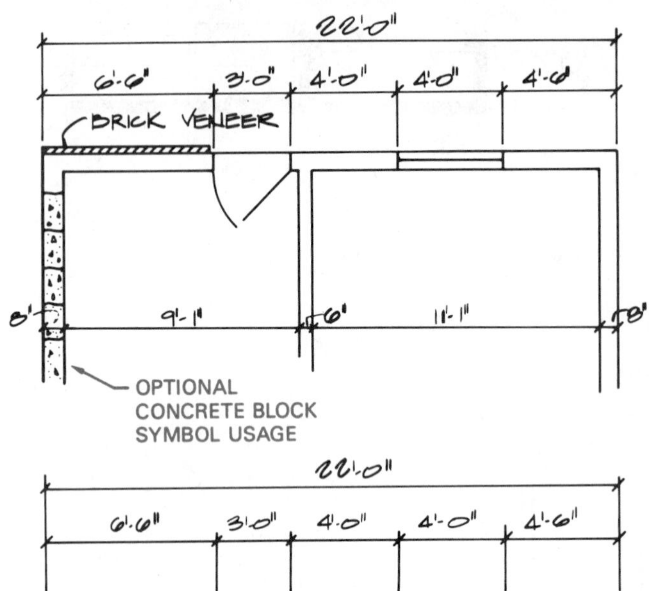

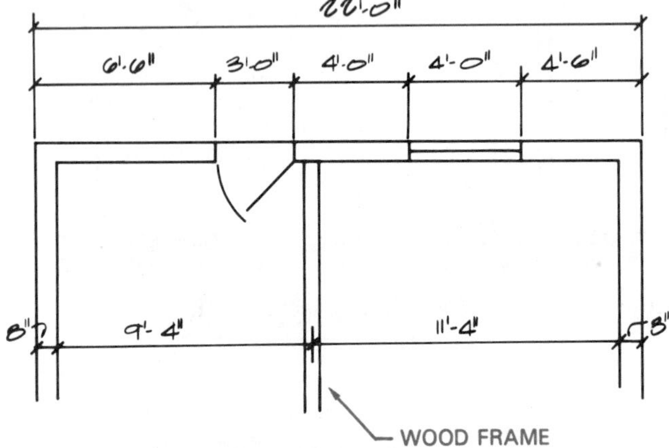

Figure 3–9 Dimensioning for concrete block construction.

is typically dimensioned in the same way as concrete block construction. See Figure 3–10.

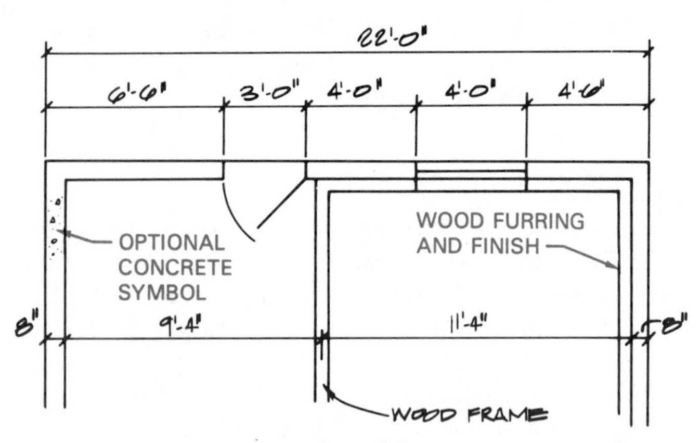

Figure 3–10 Dimensioning for solid concrete construction.

COMMON FRAMING NOTES:

1. ALL FRAMING LUMBER TO BE D.F. L. #2 OR BETTER.
2. ALL EXTERIOR WALLS @ HEATED LIVING AREAS TO BE 2 x 6 @ 24" O.C.
3. ALL EXTERIOR HEADERS TO BE 2 2x12 UNLESS NOTED, W/ 2" RIGID INSULATION BACKING UNLESS NOTED.
4. ALL SHEAR PANELS TO BE 3/8 STD. GRADE 32/16 PLY W/ 8d @ 4" o.c.@ EDGE, HDRS, & BLOCKING AND 8d @ 8" o.c. @ FIELD UNLESS NOTED.
5. PROVIDE 5/8" TYPE 'X' GYP. BD. UNDER STAIRS @ ALL USEABLE STORAGE.
6. PROVIDE 1/2" W.P.GYP. BD. AROUND ALL TUBS, SHOWERS & SPAS.
7. VENT DRYER AND ALL FANS TO OUTSIDE AIR THRU VENT W/ DAMPER.
8 ALL METAL CONNECTORS TO BE BY SIMPSON CO. OR EQUAL.
9. INSULATE W.H. TO R-11 AND PLACE ON 18" HIGH PLATFORM (GARAGE LOCATIONS ONLY)
10. BRICK VENEER TO BE OVER 1" AIR SPACE W/ 15# FELT AND METAL TIES @ 24" o.c. @ EA. STUD.
11. ALL TRUSSES TO BE @ 24" o.c. (DIRECTLY OVER STUDS). SUBMIT TRUSS CALCS. TO BUILDING DEPT. PRIOR TO ERECTION.
12. ALL BEDROOM WINDOWS TO BE WITHIN 44" OF FIN. FLOOR.
13. ALL DOORS TO BE 6'8" HIGH UNLESS NOTED. GARAGE DOOR TO BE 7'-0' HIGH.
14. ENTRY DOOR TO BE RAISED PANEL, METAL INSULATED.
15. ALL WINDOWS AND GLASS DOORS TO BE THERMAL PANE W/ BRONZE ANODIZED FRAMES
16. ALL GLASS WITHIN 18" OF DOORS TO BE TEMPERED.
17 ALL SKYLITES TO BE__" x__" DBL DOMED PLASTIC SKYLITES BY VELOX OR EQUAL.
   (OR)
   ALL SKYLITES TO BE __" x__"TEMPERED FLAT GLASS BY VELOX OR EQUAL.

Figure 3–11 Some typical general notes.

NOTES AND SPECIFICATIONS

Notes on plans are either specific or general. *Specific notes* relate to specific features within the floor plan, such as the header size over a window opening. The specific note is often connected to the feature with a leader line. Specific notes are also called local notes since they identify isolated features. *General notes* apply to the drawing overall rather than to specific items. General notes are commonly lettered in the field of the drawing. Some typical general notes are seen in Figure 3–11.

Some specific notes that are too complex or take up too much space may be lettered with the general notes and keyed to the floor plan with a short identification, such as with the phrase. SEE NOTE 1, or with a number with a symbol, such as ①.

Written specifications are separate notes that specifically identify the quality, quantity, or type of materials and fixtures used in the entire project. Specifications for construction are prepared in a format different from drawing sheets. Specifications may be printed in a format that categorizes each phase of the construction and indicates the precise methods and materials to be used. Architects and designers may publish specifications for a house so the client knows exactly what the home contains, including even the color and type of paint. This information also sets a standard that allows contractors to prepare construction estimates on an equal basis.

Lending institutions require material specifications to be submitted with plans when builders apply for financing. Generally, each lender has a form to be used that supplies the description of all construction materials and methods along with a cost analysis of the structure.

USING METRIC

The unit of measure commonly used is the millimeter (mm). Canada is one country that uses metric dimensioning.

When materials are purchased from the United States, it is often necessary to make a *hard conversion* to metric units. This means that the typical inch units are converted directly to metric. For example, a 2 × 4 that is planed to $1\frac{1}{2} \times 3\frac{1}{2}$ in. converts to 38 × 89 mm. When making the conversion from the imperial inch to millimeters, use the formula 25.4 × inch = millimeters.

The preferred method of metric dimensioning is called a *soft conversion*. This means that the lumber is milled directly to metric units. The 2 × 4 lumber is 40 × 90 mm using soft conversion. This method is much more convenient when drawing plans and measuring in construction. When plywood thickness is measured in metric units, $\frac{5}{8}$ in. thick equals 17 mm and $\frac{3}{4}$ in. thick equals 20 mm. The length and width of plywood also changes from 48 × 96 in. to 1200 × 2400 mm. Modules for architectural design and construction in the United States are typically 12 or 16 in. In countries using metric measurement, the dimensioning module is 100 mm. For example, construction members may be spaced 24 in. on center (O.C.) in the United States, but the spacing in Canada is 600 mm O.C. The spacing between studs at 16 in. O.C. in the United States is 400 mm O.C. in Canada. These metric modules allow the 1200 × 2400 mm plywood to fit exactly on center with construction members. Interior dimensions are also designed in 100 mm increments. For example, the kitchen base cabinet measures 600 mm deep.

Reading Metric Units on a Drawing

When reading metric dimensions on a drawing, all dimensions within dimension lines are in millimeters, and the millimeter symbol (mm) is omitted. When more than one dimension is quoted, the millimeter symbol (mm) is found only after the last dimension. For example, the size of a plywood sheet reads 1200 × 2400 mm, or the size and length of a wood stud reads 38 × 89 × 600 mm. The millimeter symbol is omitted in the notes associated with a drawing except when referring to a single dimension, such as the thickness of material or the spacing of members. For example, a note might read 90 × 1200 BEAM, the reference to a material thickness, 12 mm gypsum, or the spacing of joists, 400 mm O.C.

Metric Scales

When drawings are produced in metric, the floor plans, elevations, and foundation plans are generally drawn at a scale of 1:50 rather than the $\frac{1}{4}$" = 1'–0" scale used in the imperial system. Larger scale drawings, such as construction details, are often drawn at a scale of 1:5. Small-scale drawings, such as plot plans, may be drawn at a scale of 1:500. Figure 3–12 shows a floor plan drawn completely using metric dimensioning. The preferred method of soft conversion is used, and the metric scale is 1: 50.

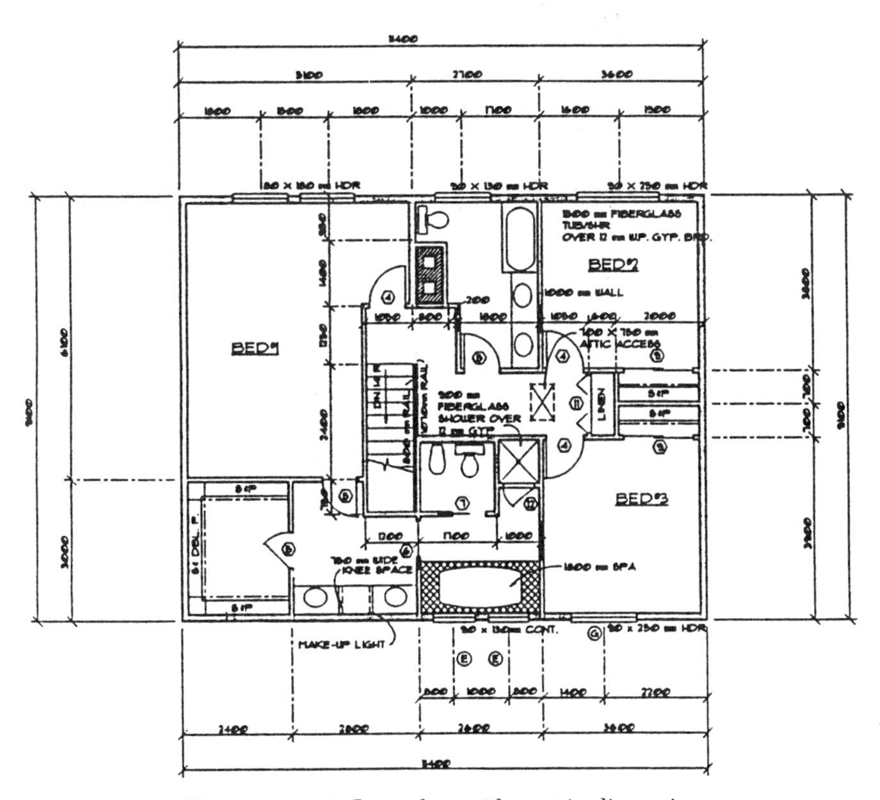

Figure 3–12 A floor plan with metric dimensions.

CHAPTER 3 TEST

Fill in the blanks below with the proper word or short statement as needed to complete the sentence or answer the question correctly.

1. Fully explain how dimension numerals are displayed on an architectural drawing. ____.
2. Show the foot symbol. ______________________________
3. Show the inch symbol. ______________________________
4. Why is it possible to have a drawing that does not show the foot and inch symbols on dimension numerals? ____.
5. The overall dimensions on frame construction are understood to be given to the ______________ stud frame of the exterior walls.
6. Dimensions are normally given to the ______________ of windows, doors, and interior partitions.
7. Identify at least two situations in which items may not be dimensioned on a drawing. ____.
8. Why is it important to verify the exact rough (framed) opening dimensions for a product before the walls are framed? ____.
9. The application of thin masonry, such as stone or brick, to the exterior of a wood-framed structure is called ______________________________.
10. Give an example of a specific note. ______________________________
11. Give an example of a typical general note. ______________________________
12. Describe construction specifications. ____.

CHAPTER 3 EXERCISES

PROBLEM 3–1. Answer the following questions as you read the floor plan drawing on page 64:

1. Describe how the brick veneer is shown, and give the specifications provided. ____.
2. Give the dimensions of the portion of the master bath that projects out from the rest of the house. ______________________________.
3. What is the size of the whirlpool tub? ______________________________
4. What is the size of the shower? ______________________________
5. Describe the material used to finish the shower. ______________________________
6. Describe the construction materials used as a shower door and between the shower and tub. ____.

7. Give the total dimension from the outside wall of the master bath to the center of the wall behind the water closet. (Show your calculations.) ____.
8. Give the specifications of all construction components of the decor column. (Change all abbreviations and symbols to whole words.) ____.
9. What size beam runs from the decor column to the outside wall of the master bath? ____________
10. Give the dimensions of the water closet compartment. ____________

PROBLEM 3–2. Answer the following questions as your read the floor plan drawing on page 65:

1. Give the exterior dimensions of the fireplace. ____________
2. How far does the living room project out past the dining room? ____________
3. What is the width of the living room as measured from the center of the master suite wall to the outside of the upper left corner of the living room? (Show your calculations.) ____________.
4. What is the length of the living room as measured from the outside wall to the center of the opposite wall? (Show your calculations.) ____________.
5. Give the complete specifications of the joists over the living room. ____________.
6. How far is the fireplace from the upper left outside corner of the living room? ____________
7. What are the interior dimensions of the living room? ____________
8. How does the builder know where to frame in the 2'–6" door that goes into the dining room? ____________.

PROBLEM 3–3. Answer the following questions as your read the floor plan drawing on page 66:

1. What are the exterior dimensions of the garage (overall width × depth measured from wall at bedroom 3)? ____________.
2. Give the size and specifications of the garage door. ____________.
3. What is the size of the garage door header? ____________
4. Describe the type and location of the attic access. ____________.
5. Give the dimensions of bedroom 3 from the outside of the garage wall to the center of the wardrobe closet and from the garage side of the wall to the center of the wall at bedroom 2. ____________.
6. Describe the construction material used on the walls and ceiling of the garage. ____________.
7. Give the location dimension for the 2"–8" door with one tempered light found in the garage. ____________
8. Fully describe the joist over the garage. ____________.

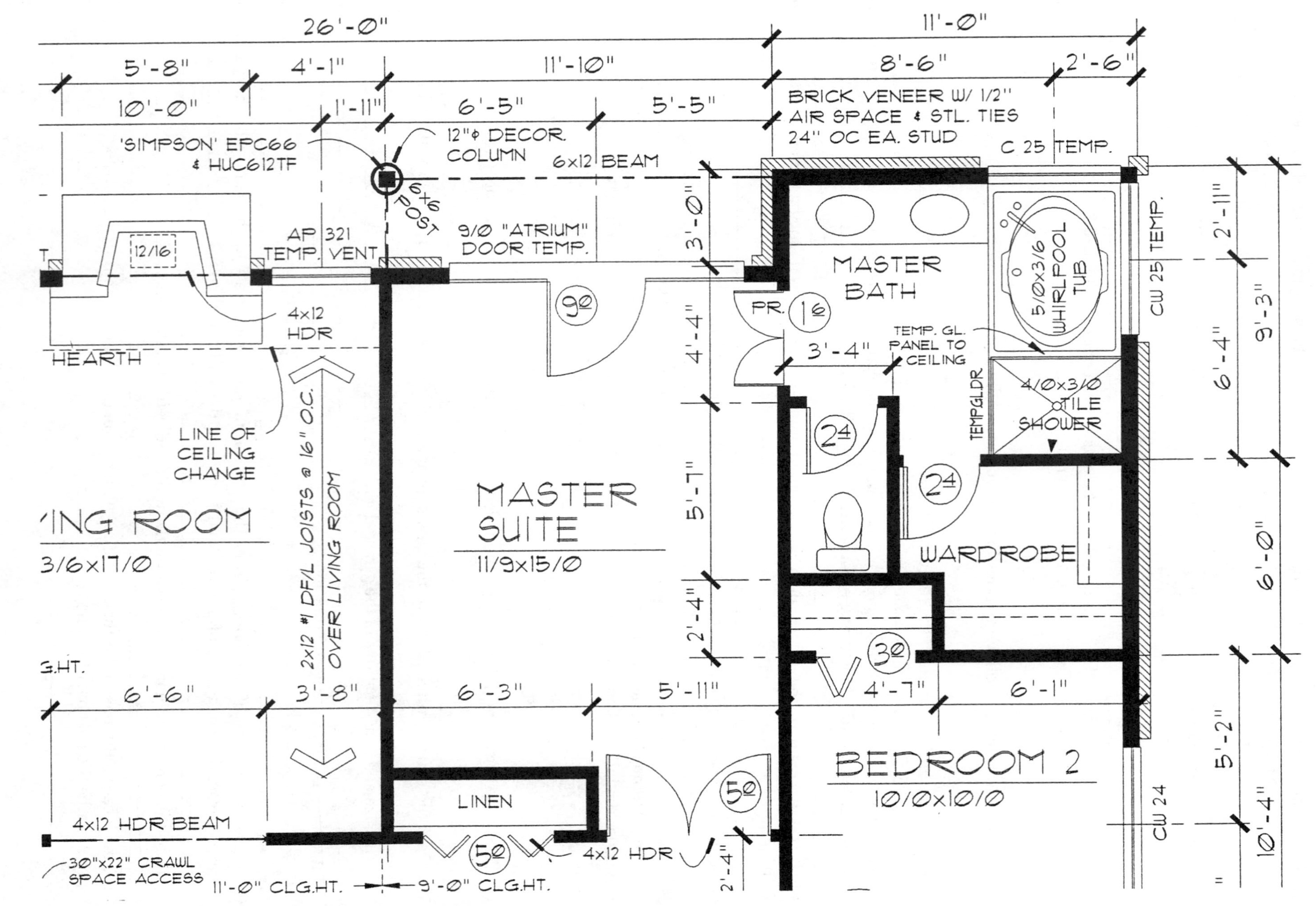

Problem 3–1. *Courtesy Piercy and Barclay Designers, Inc.*

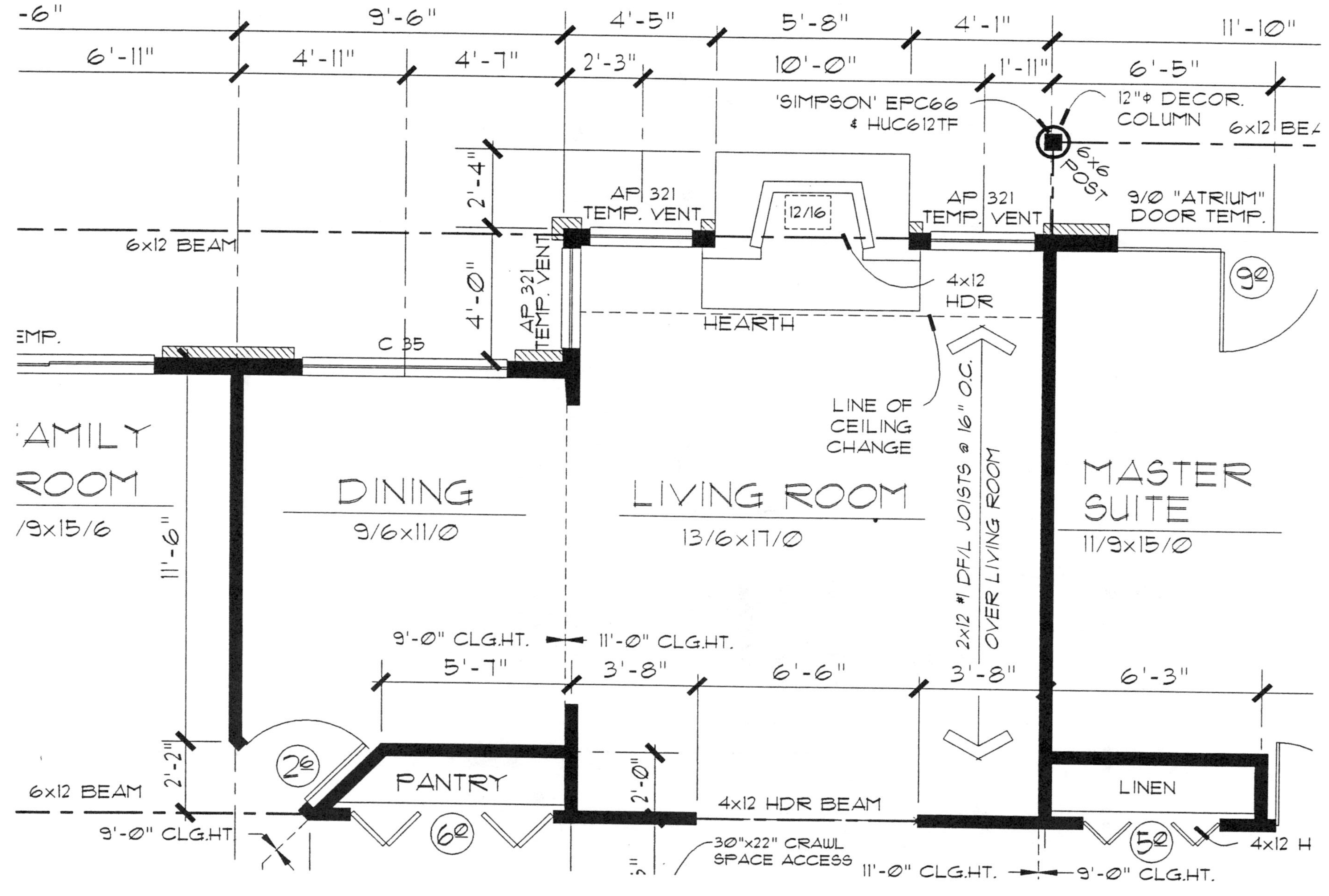

Problem 3–2. *Courtesy Piercy and Barclay Designers, Inc.*

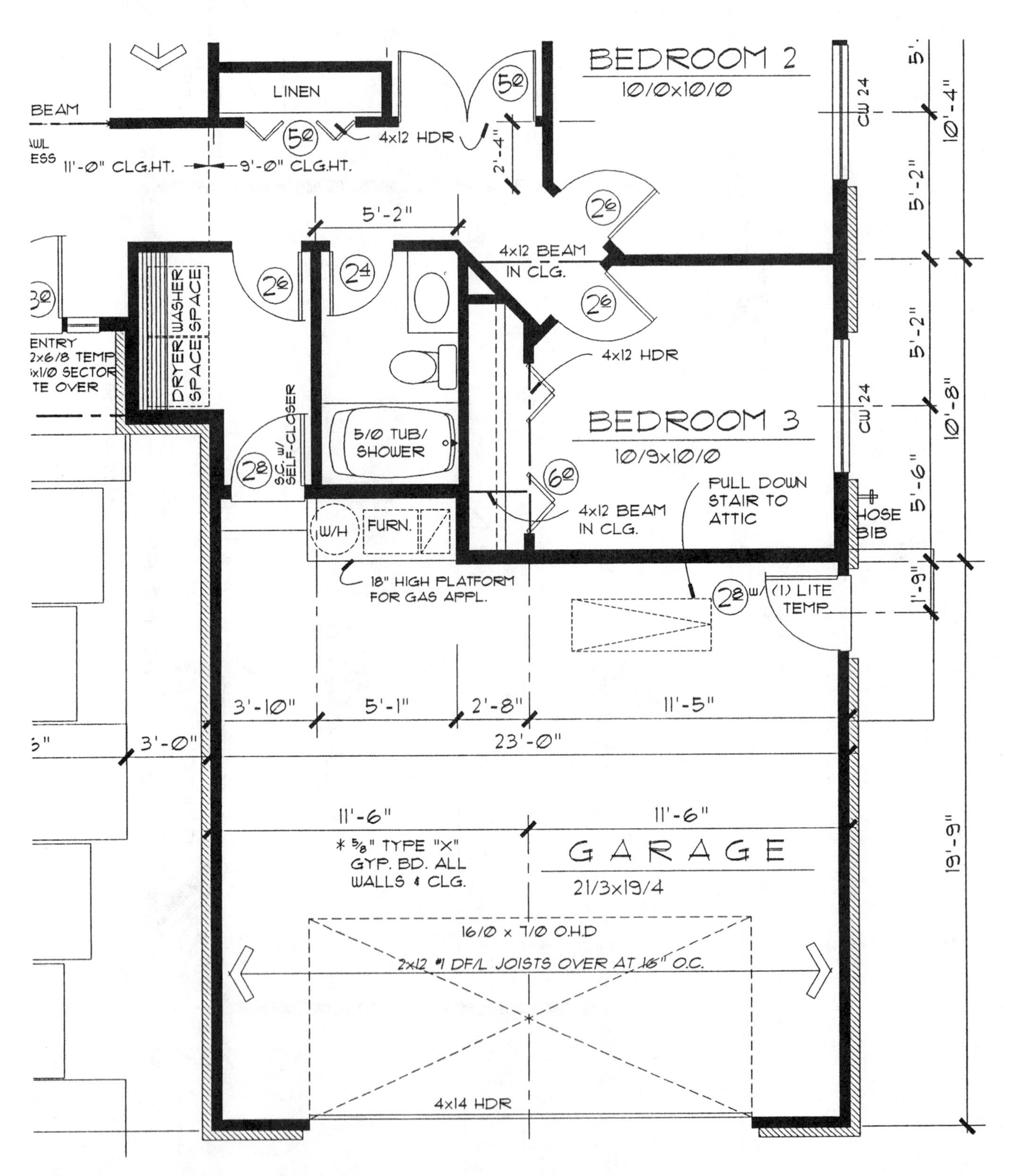

Problem 3–3. *Courtesy Piercy and Barclay Designers, Inc.*

Chapter 4

Reading Floor Plans, Part 3: Electrical, Plumbing, and HVAC

ELECTRICAL PLANS
PLUMBING PLANS
HEATING, VENTILATING, AND AIR CONDITIONING (HVAC) PLANS
CENTRAL VACUUM SYSTEMS
TEST
EXERCISES

ELECTRICAL PLANS

NATIONAL ELECTRICAL CODE requirements dictate the size of some circuits and the placement of certain outlets and switches within the home.

Electrical Symbols

Electrical symbols are used to show the lighting arrangement desired in the home. This includes all switches, fixtures, and outlets, as seen in Figure 4–1.

Switch symbols are generally placed perpendicular to the wall and read from the right side or bottom of the sheet. Look at Figure 4–2.

Figure 4–3 shows several typical electrical installations with switches to light outlets. The switch leg or electrical circuit line is usually dashed and shown in a curve.

When special characteristics are required, such as a specific size fixture, a location requirement, or any other specification, a local note that briefly describes the situation may be applied next to the outlet, as shown in Figure 4–4.

Figure 4–5 shows some examples of typical electrical layouts. Figure 4–6 shows a bath layout. Figure 4–7 shows a typical kitchen electrical layout.

PLUMBING PLANS

There are two classifications of piping: industrial and residential. Industrial piping is used to carry liquids and gases used in the manufacture of products. Steel pipe with welded or threaded connections and fittings is used in heavy construction.

Residential piping is called plumbing and carries fresh water, gas, or liquid and solid waste. The pipe used in plumbing may be made of copper, plastic, galvanized steel, or cast iron. Copper pipes have soldered joints and fittings and are used for carrying hot or cold water. Plastic pipes have glued joints and fittings and are used for vents and for carrying fresh water or solid waste. Very little galvanized steel pipe is used except where conditions require such installation. Steel pipe has threaded joints and fittings and is used to carry natural gas. Cast iron pipe is commonly used to carry solid and liquid waste as the sewer pipe that connects the structure with the local or regional sewer system.

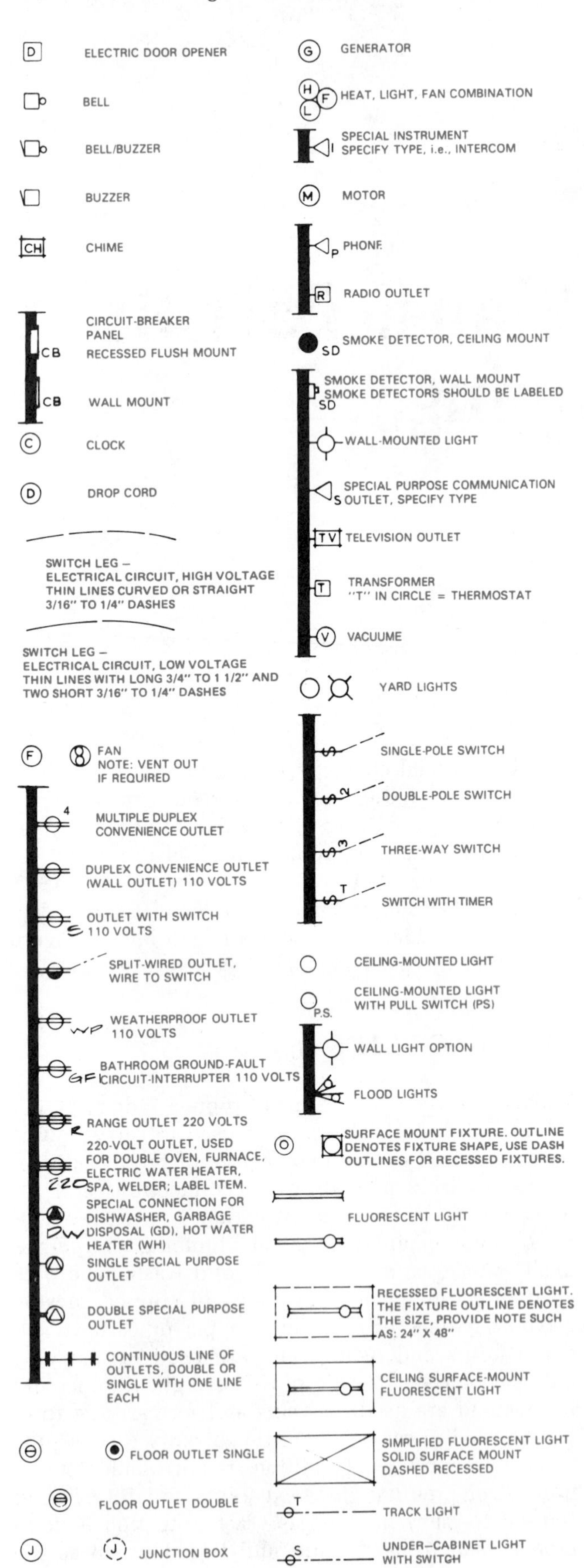

Figure 4–1 Common electrical symbols.

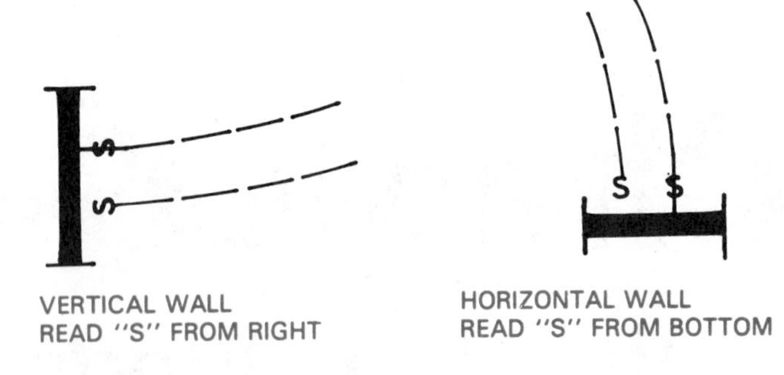

Figure 4–2 Switch symbols and their placement.

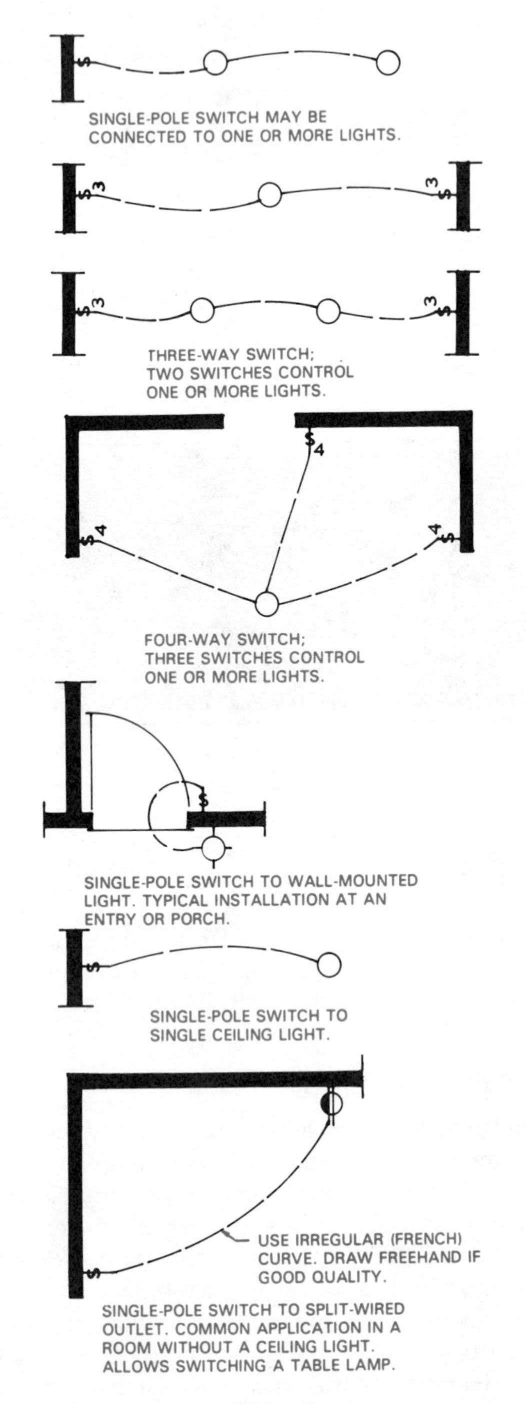

Figure 4–3 Typical electrical installations.

Figure 4–4 Special notes placed with electrical fixtures.

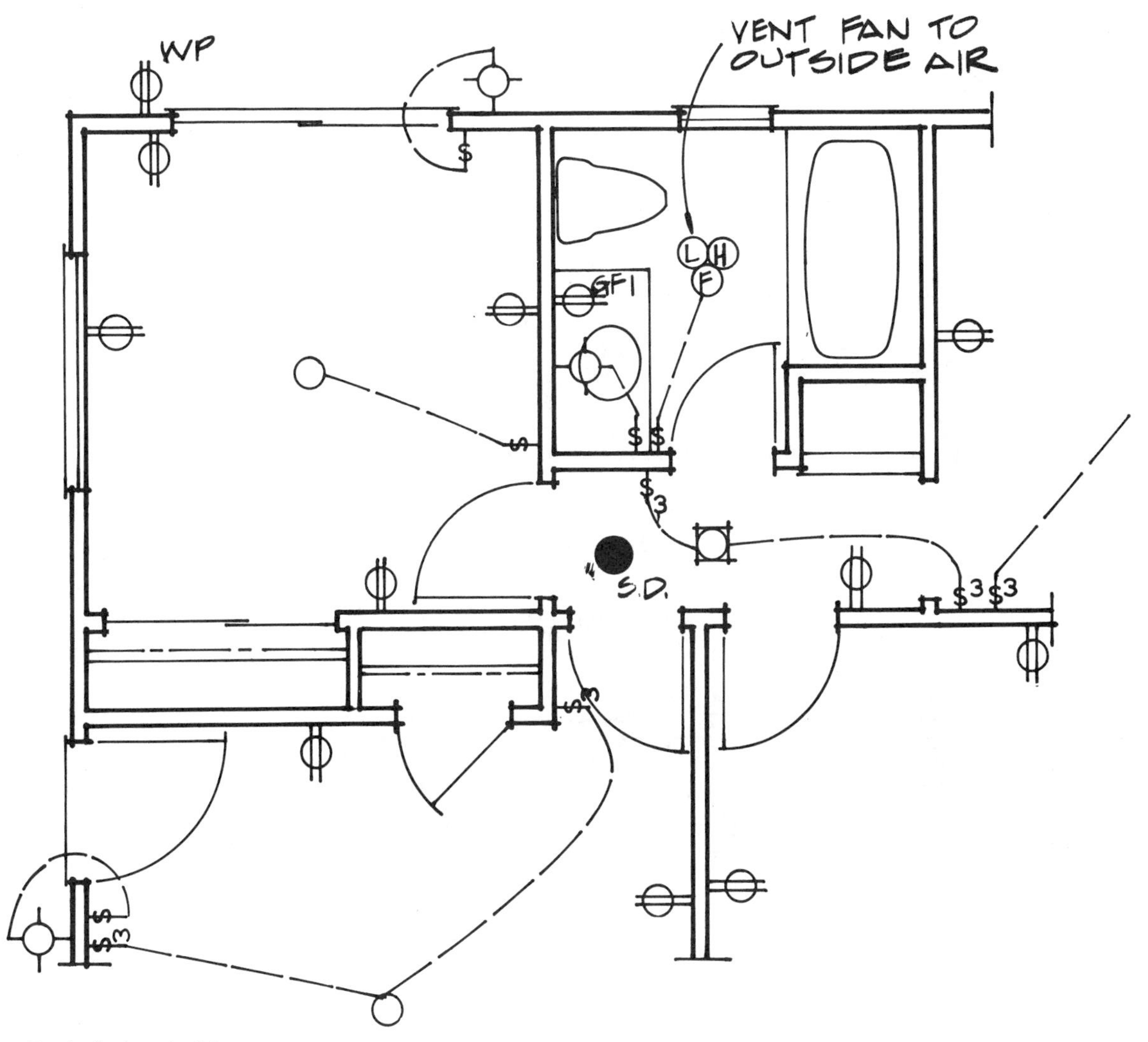

Figure 4–5 Typical electrical layouts.

Residential plans may not require a complete plumbing plan. In most cases, the plumbing requirements can be clearly provided on the floor plan in the form of symbols for fixtures and notes for specific applications or conditions.

Other plumbing items to be added to the floor plan include floor drains, vent pipes, and sewer or water connections. Floor drains are shown in their approximate location and are identified with a note identifying size, type, and slope to drain. Vent pipes are shown in the wall where they are to be located and labeled by size. Sewer and water service lines are located in relationship to the position in which these utilities enter the home. The service lines are commonly found on the plot plan. In the situation described here, where a very detailed plumbing layout is not provided, the plumbing contractor is required to install quality plumbing in a manner that meets local code requirements and is also economical. Figure 4–8 shows a complete floor plan including plumbing symbols. Figure 4–9 shows plumbing fixture symbols in plan, frontal, and profile views.

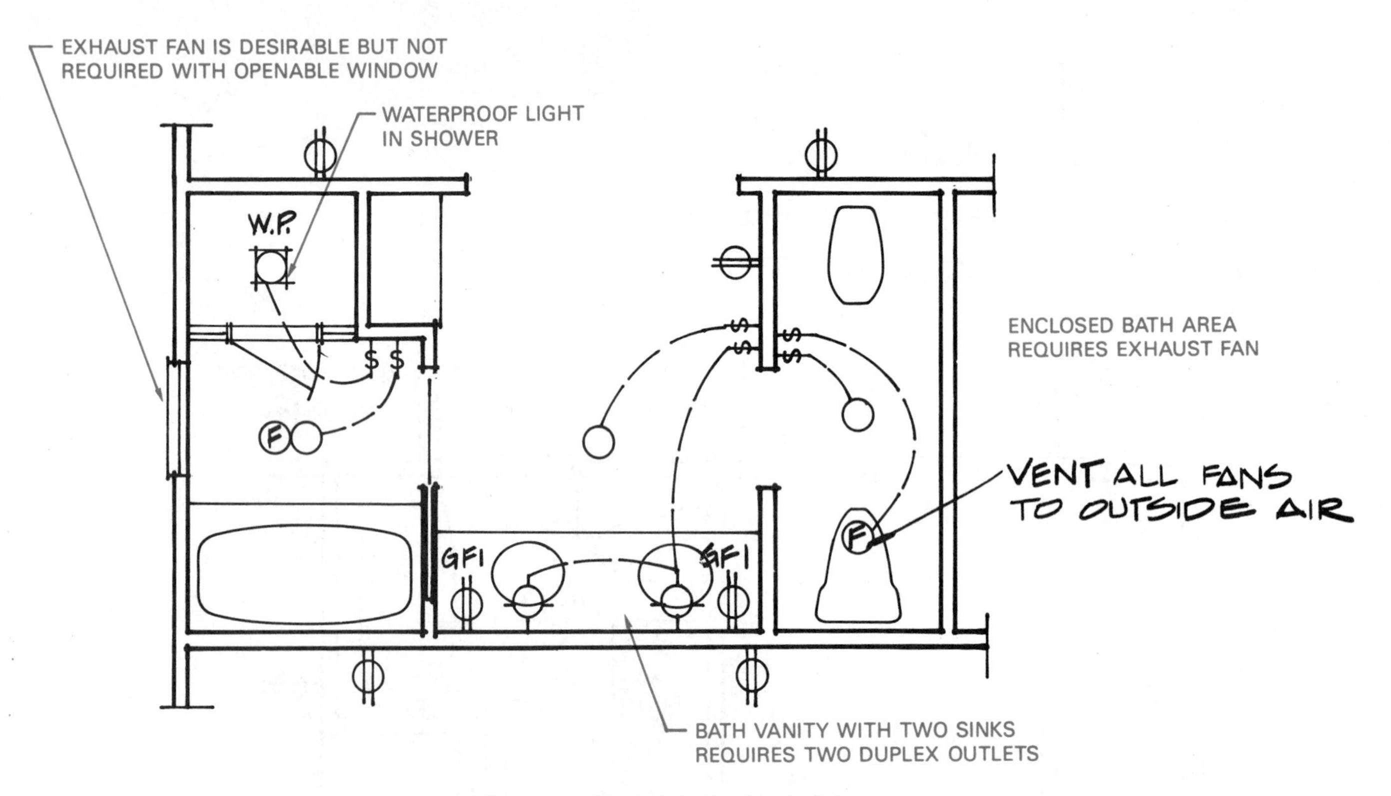

Figure 4–6 Typical bath electrical layout.

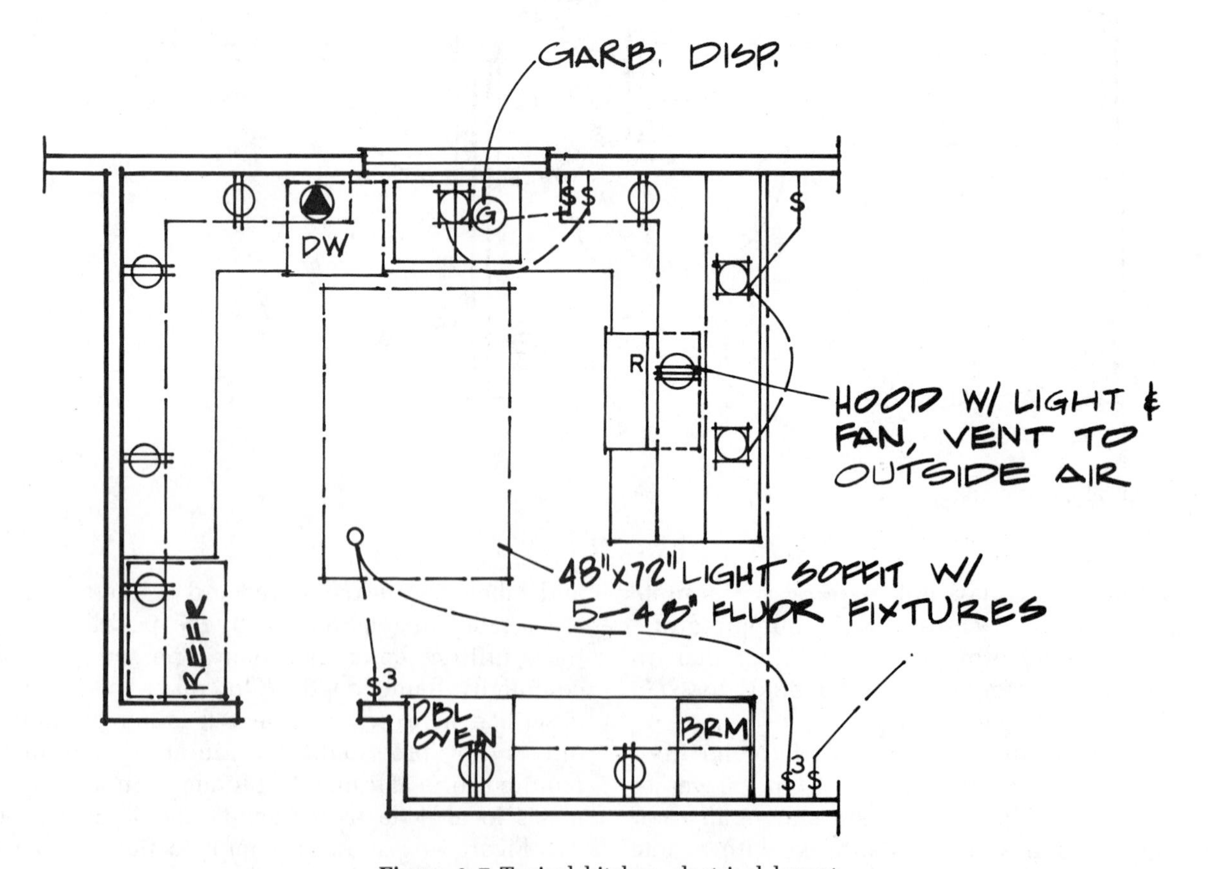

Figure 4–7 Typical kitchen electrical layout.

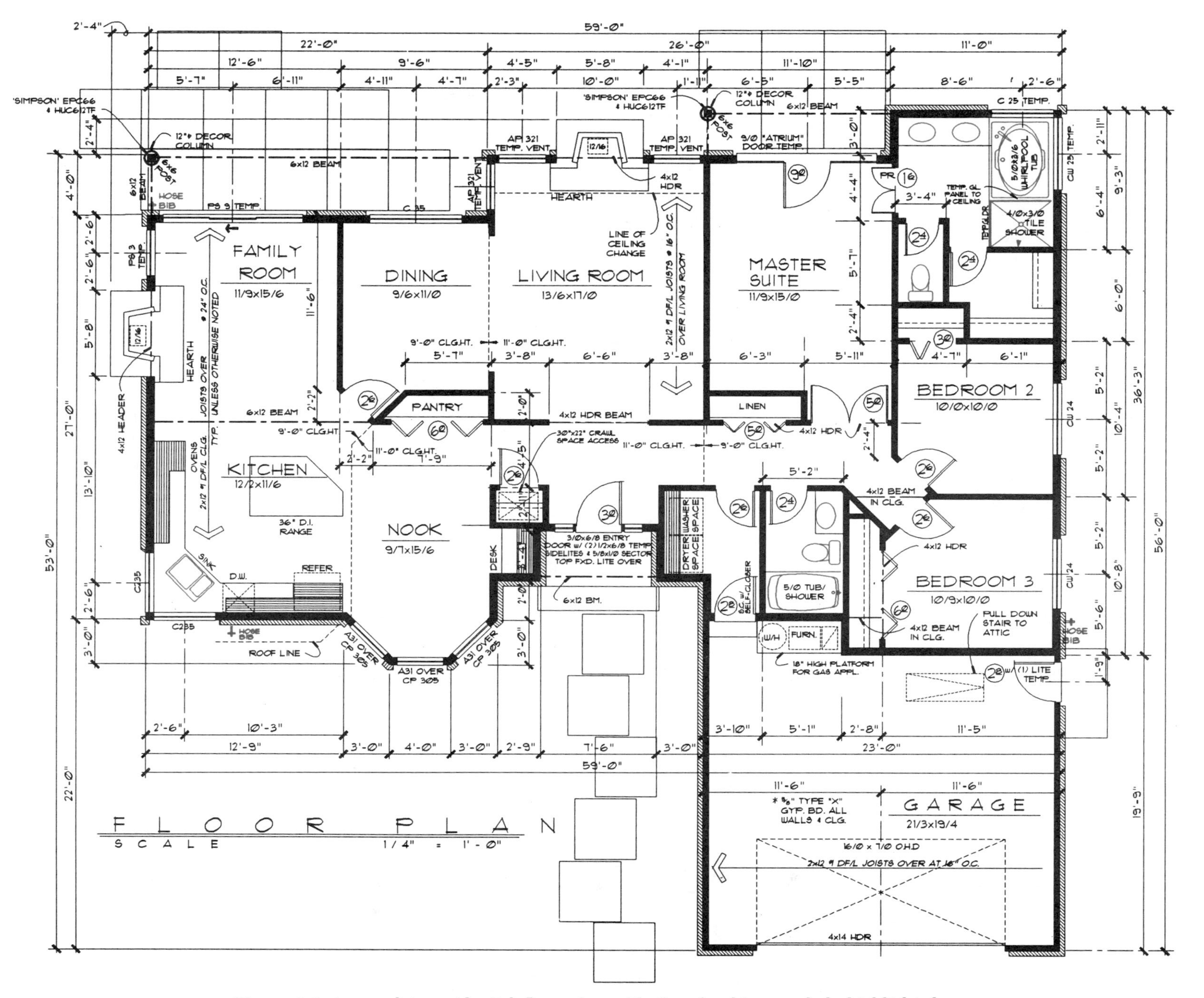

Figure 4–8 A complete residential floor plan with the plumbing symbols highlighted.
Courtesy Piercy and Barclay Designers, Inc.

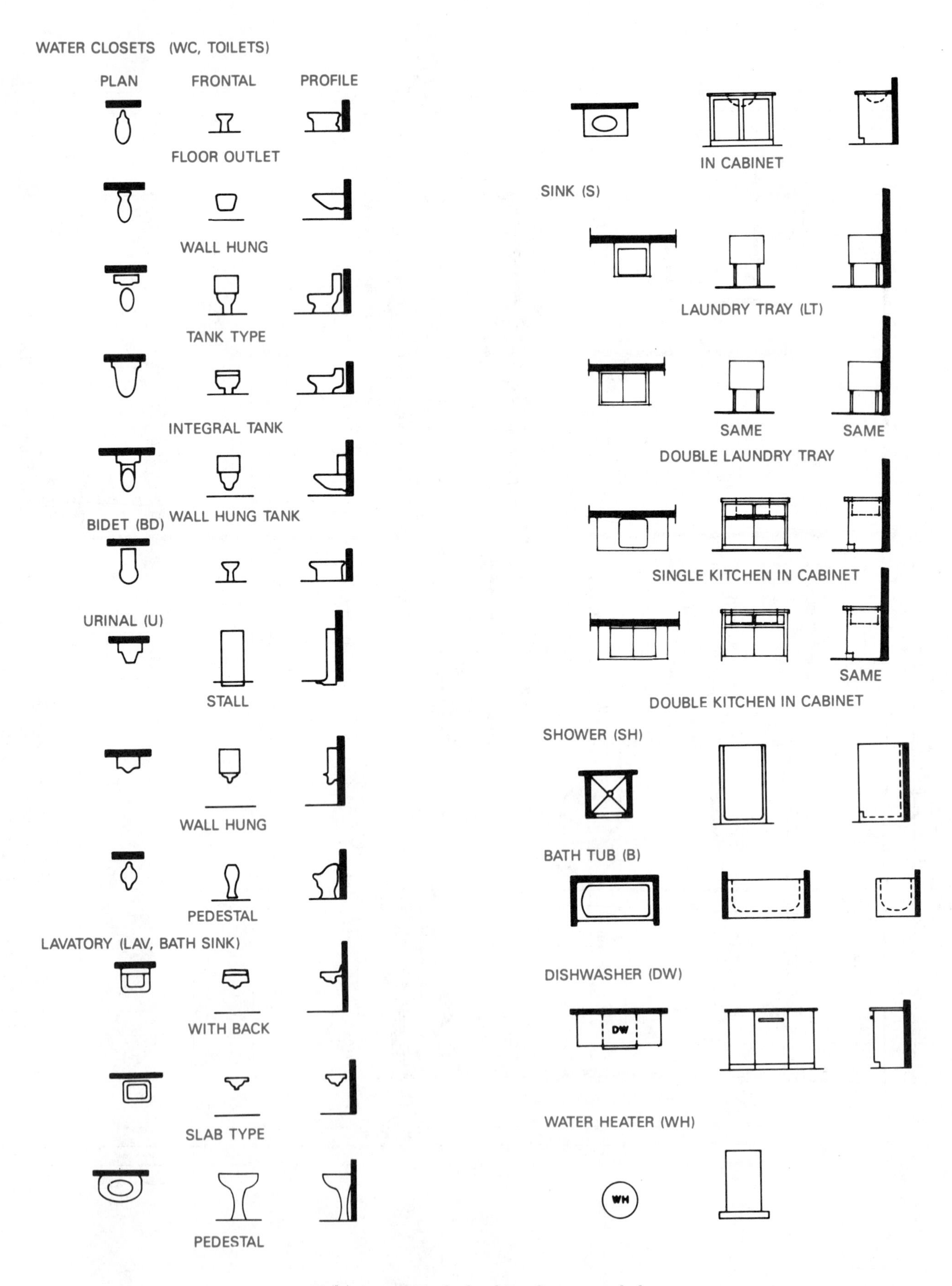

Figure 4–9 Typical plumbing fixture symbols.

Plumbing Schedules

Plumbing schedules are similar to door, window, and finish schedules. Schedules provide specific information regarding plumbing equipment, fixtures, and supplies. The information is condensed in a chart so the floor plan is not unnecessarily crowded with information. Figure 4–10 shows a typical plumbing fixture schedule.

Other schedules may include specific information regarding floor drains, water heaters, pumps, boilers, or radiators. These schedules generally key specific items to the floor plan with complete information describing size, manufacturer, type, and other specifications as appropriate.

PLUMBING FIXTURE SCHEDULE

LOCATION	ITEM	MANUFACTURER	REMARKS
MASTER BATH	36" F.G. SHR. COLOR BIDET C.I. SR. COLOR LAV. COLOR W. C.	HYTEC K-4868 K-2904 K-3402-PBR	M2620 BRASS K1940 BRASS M4625 BRASS PLAS. SEAT
BATH #2	KEG STYLE TUB C.I. COLOR PED. LAV. COLOR W.C.	KOHLER KOHLER K3402-PBR	M2850 BRASS M4625 BRASS PLAS. SEAT
BATH #3	URINAL F.G. SHOWER C.I. LAVS.	K-4980 HYTEC K2904	 M2620 BRASS M4625 BRASS
KITCHEN	C.I. 3 HOLE SINK	K5960	M7531 BRASS
WATER HTR.	82 GAL. ELEC.	MORFIO	P & T VALVE
UTILITY	F.G. LAUN. TRAY	24 X21	D2121 BRASS

Figure 4–10 A typical plumbing fixture schedule.

Mortgage lenders may require a complete description of materials for the structure. Part of the description often includes a plumbing section in which certain plumbing specifications are described as shown in Figure 4–11.

Plumbing Drawings

Plumbing drawings usually are not drawn on the same sheet as the complete floor plan. The only plumbing items shown on the floor plans are fixtures, as previously explained. The plumbing drawing is often placed over an outline of the floor plan showing all walls, partitions, doors, windows, plumbing fixtures, and utilities.

Plumbing drawings are prepared by the architectural office or in conjunction with a plumbing contractor. In some situations, when necessary, plumbing contractors work up their own rough sketches or field drawings. Plumbing drawings are made up of lines and symbols that show how liquids, gases, or solids are transported to various locations in the structure. Plumbing lines and features are drawn thicker than wall lines so they are clearly distinguishable. Symbols identify types of pipes, fittings, valves, and other components of the system. Sizes and specifications are provided in local or general notes. Figure 4–12 shows some typical plumbing symbols. Certain abbreviations are commonly used in plumbing drawings, as shown in Figure 4–13.

PLUMBING:

Fixture	Number	Location	Make	Mfr's Fixture Identification No.	Size	Color
Sink ______						
Lavatory ______						
Water closet ______						
Bathtub ______						
Shower over tub △ ______						
Stall shower △ ______						
Laundry trays ______						

△☐ Curtain rod △☐ Door ☐ Shower pan: material ______

Water supply: ☐ public; ☐ community system; ☐ individual (private) system.★

Sewage disposal: ☐ public; ☐ community system; ☐ individual (private) system.★

★*Show and describe individual system in complete detail in separate drawings and specifications according to requirements.*

House drain (inside): ☐ cast iron; ☐ tile; ☐ other ______ House sewer (outside): ☐ cast iron; ☐ tile; ☐ other ______

Water piping: ☐ galvanized steel; ☐ copper tubing; ☐ other ______ Sill cocks, number ______

Domestic water heater: type ______; make and model ______; heating capacity ______

______ gph 100° rise. Storage tank: material ______; capacity ______ gallons.

Gas service: ☐ utility company; ☐ liq. pet. gas; ☐ other ______ Gas piping: ☐ cooking; ☐ house heating.

Footing drains connected to: ☐ storm sewer; ☐ sanitary sewer; ☐ dry well. Sump pump; make and model ______

______; capacity ______; discharges into ______

Figure 4–11 A partial specification sheet showing the area for the plumbing description of materials.

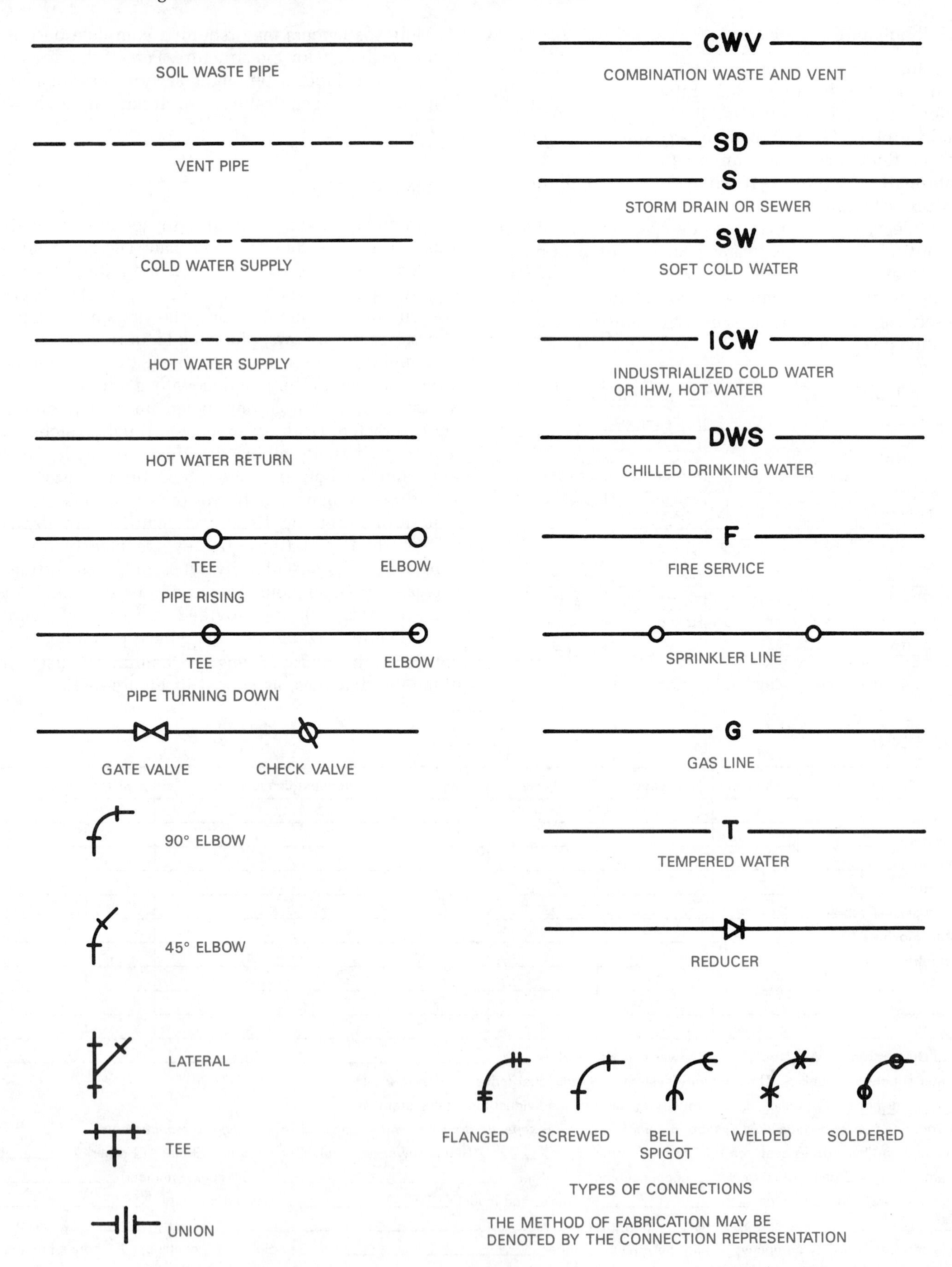

Figure 4–12 Typical plumbing piping symbols.

CURRENT		MCS	
CW	Cold-Water Supply	WC	Water Closet (Toilet)
HW	Hot-Water Supply	LAV	Lavatory (Bath Sink)
HWR	Hot-Water Return	B	Bathtub
HB	Hose Bibb	S	Sink
CO	Clean Out	U	Urinal
DS	Downspout	SH	Shower
RD	Rain Drain	DF	Drinking Fountain
FD	Floor Drain	WH	Water Heater
SD	Shower Drain	DW	Dishwasher
CB	Catch Basin	BD	Bidet
MH	Manhole	GD	Garbage Disposal
VTR	Vent Thru Roof		

Figure 4–13 Typical plumbing abbreviations.

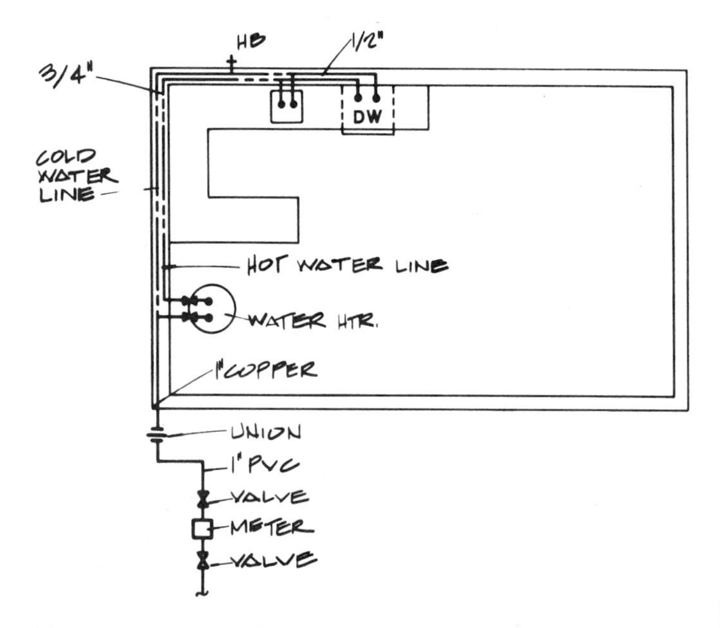

Figure 4–14 A typical water supply system.

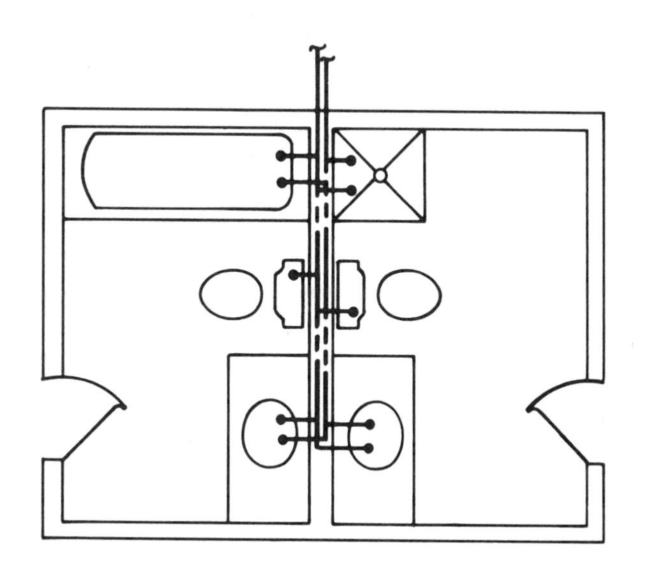

Figure 4–15 A common plumbing wall with back-to-back baths.

Water Systems

Water supply to a structure begins at a water meter for public systems or from a water storage tank for private well systems. The water supply to the home or business, known as the main line, is generally 1 in. plastic pipe. This size may vary in relationship to the service needed. The plastic main line joins a copper line within a few feet of the structure. The balance of the water system piping is usually copper pipe, although plastic pipe is increasing in popularity for cold-water applications. The 1 in. main supply often changes to ¾ in. pipe where a junction is made to distribute water to various specific locations. From the ¾ in. distribution lines, ½ in. pipe usually supplies water to specific fixtures, for example, the kitchen sink. Figure 4–14 shows a typical installation from the water meter of a house with distribution to a kitchen. The water meter location and main line representation are generally shown on the plot plan. Verify local codes regarding the use of plastic pipe.

There is some advantage to placing plumbing fixtures back to back when possible. This practice saves materials and labor costs. Plumbing fixtures one above the other aids in an economical installation for two-story buildings. Figure 4–15 shows a back-to-back bath situation. If the design allows, another common installation may be a bath and laundry room next to each other or provide a common plumbing wall. See Figure 4–16. If a specific water temperature is required, that specification can be applied to the hot-water line, as shown in Figure 4–17.

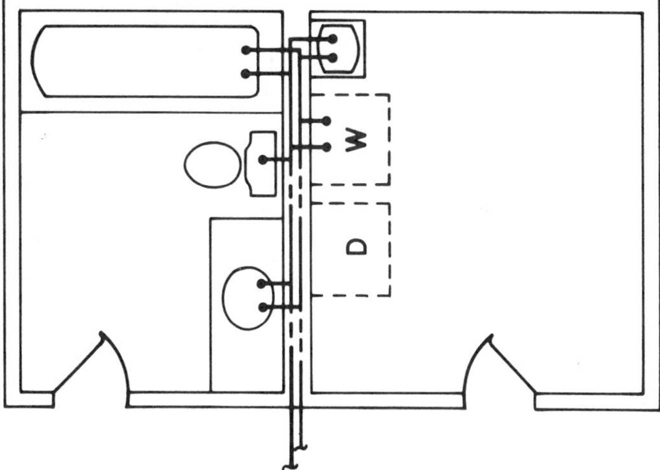

Figure 4–16 A common plumbing wall with a bath back to back with a laundry room.

120°

Figure 4–17 Hot-water temperature displayed in the plumbing pipe symbol.

Drainage and Vent Systems

The drainage system provides for the distribution of solid and liquid waste to the sewer line. The vent system allows a continuous flow of air through the system so gases and odors may dissipate and bacteria do not have an opportunity to grow. These pipes throughout the house are generally made of PVC, (polyvinyl chloride) plastic, although the pipe from the house to the concrete sewer pipe is commonly 3 or 4 in. cast iron. Drainage and vent systems, as with water systems, are drawn with thick lines using symbols, abbreviations, and notes. Figure 4–18 shows a sample drainage vent system. Figure 4–19 shows a house plumbing plan.

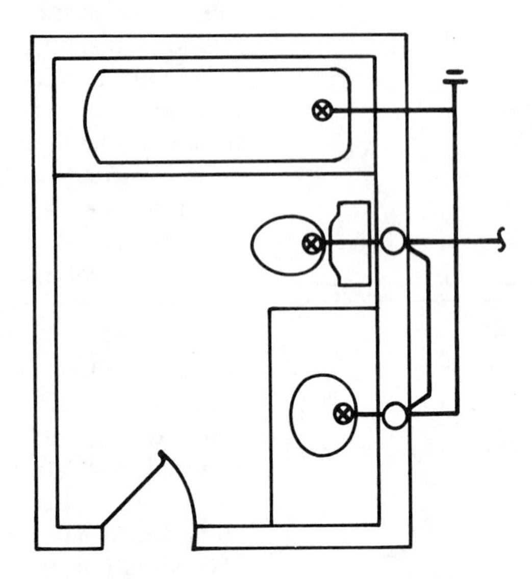

Figure 4–18 A typical drainage and vent drawing.

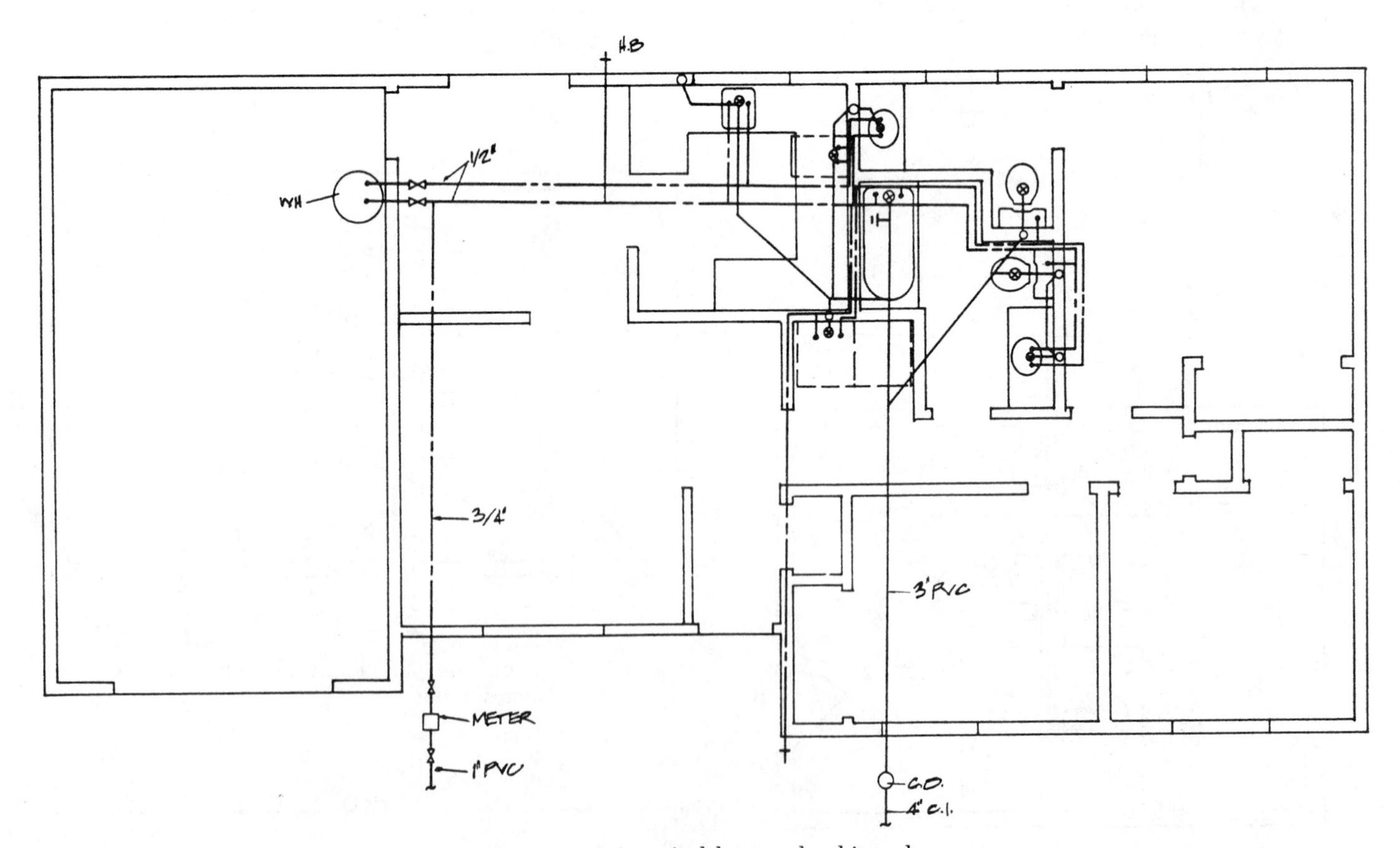

Figure 4–19 A typical house plumbing plan.

Isometric Plumbing Drawings

Isometric drawings may be used to provide a three-dimensional representation of a plumbing layout. Especially for a two-story structure, an isometric drawing provides an easy to understand pictorial drawing. Figure 4–20 shows an isometric drawing of the system shown in plan view in Figure 4–18. Figure 4–21 shows a detailed isometric drawing of a typical drain, waste, and vent system. Figure 4–22 shows a single-line isometric drawing of the detailed isometric drawing shown in Figure 4–21. Figure 4–23 shows a detailed isometric drawing of a hot- and cold-water supply system. Figure 4–24 shows a single-line isometric drawing of the same system.

Sewage Disposal

Public Sewers. Public sewers area available in and near most cities and towns. The public sewers are generally located under the street or an easement adjacent to the construction site. In some situations the sewer line may have to be extended to accommodate another home or business in a newly developed area. The cost of this extension may be the responsibility of the developer. The cost of this construction usually includes installation expenses, street repair, sewer tap, and permit fees. Figure 4–25 shows an illustration of a sewer connection and the plan view usually found on the plot plan.

Private Sewage Disposal: A Septic System. The septic system consists of a storage tank and an absorption field and operates as follows. Solid and liquid waste enters the septic tank where it is stored and begins decomposition into sludge. Liquid material, or effluent, flows from the tank outlet and is dispersed into a soil-absorption field, or drain field (also known as leach lines). When the solid waste has effectively decomposed, it also dissipates into the soil absorption field.

The characteristics of the soil must be verified for suitability for a septic system by a soil feasibility test, also known as a percolation test. This test will determine if the soil will accommodate septic system. The test should also identify certain specifications that should be followed for system installation. The U.S. Veterans Administration (VA) and the U.S. Federal Housing Administration (FHA) require a minimum of 240 ft of field line. Verify these dimensions with local building officials. When the soil characteristics do not allow a conventional system, there may be some alternatives, such as a sand filter system, which filters the effluent through a specially designed sand filter before it enters the soil absorption field. Check with your local code officials before installing such a system. Figure 4–26 shows a typical serial septic system. The serial system allows one drain field line to fill before the next line is used. The drain field lines must be level and must follow the contour of the land perpendicular to the slope. The drain field should be at least 100 ft from a water well, but verify the distance with local codes. There is usually no minimum distance to a public water supply.

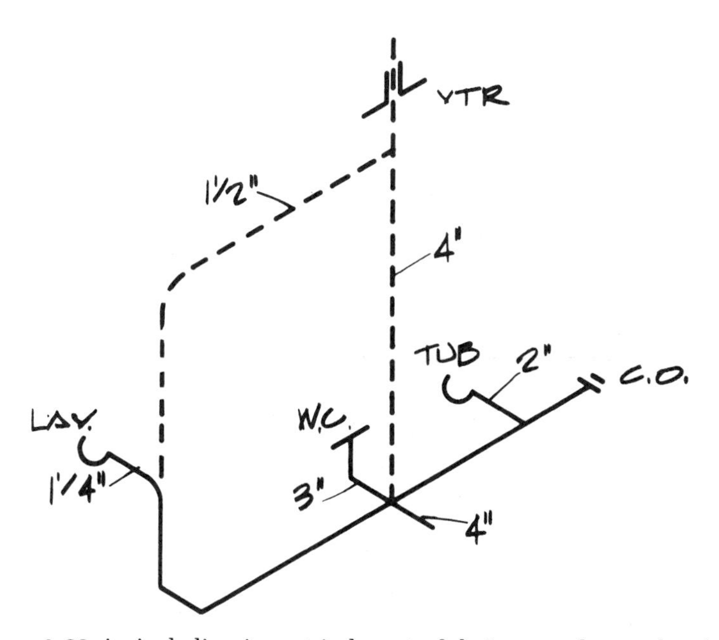

Figure 4–20 A single-line isometric layout of drainage and vent drawing.

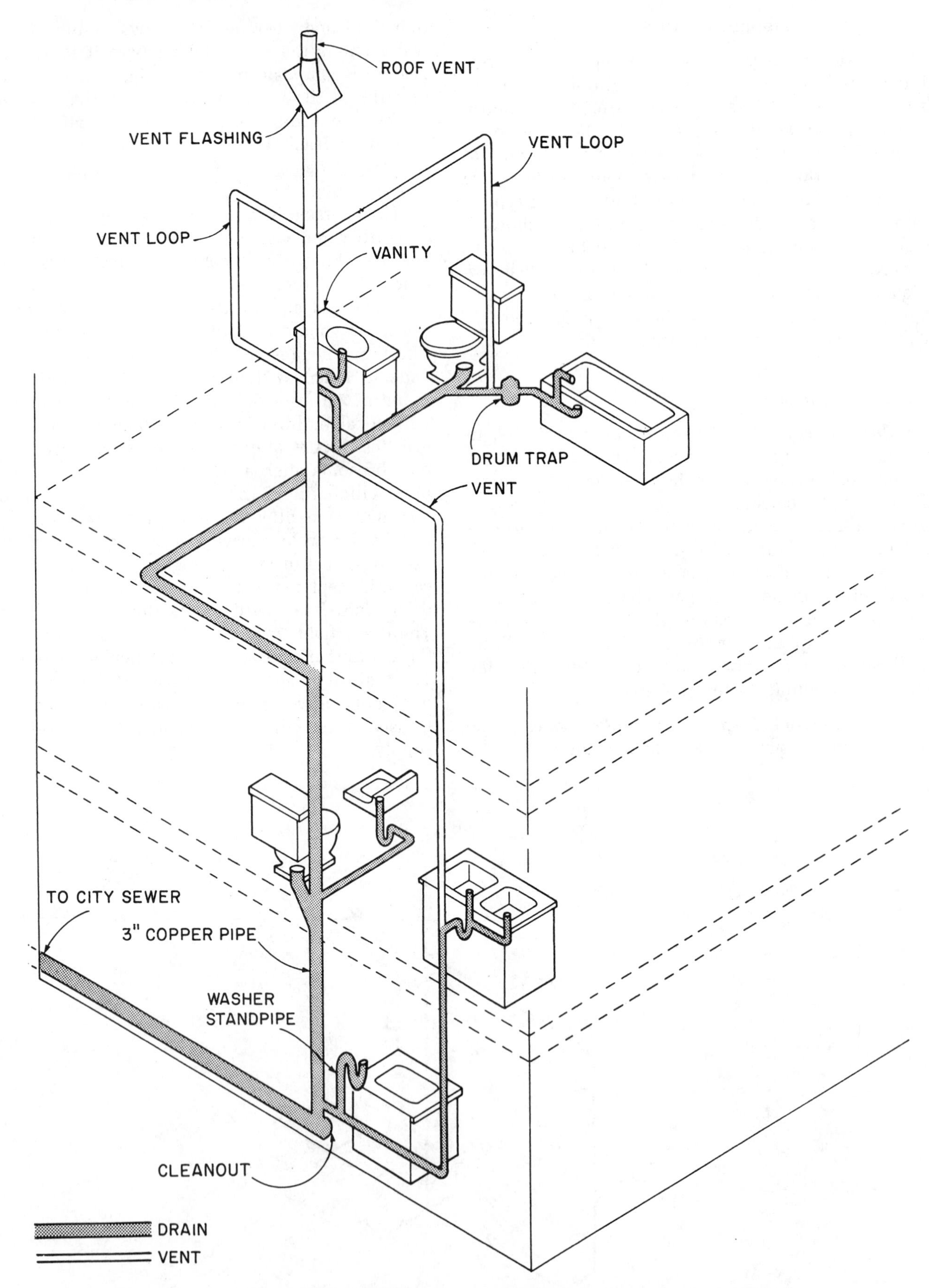

Figure 4–21 A detailed isometric drawing of drain, waste, and vent system. *From Huth,* Understanding Construction Drawings, *Delmar Publishers, Inc.*

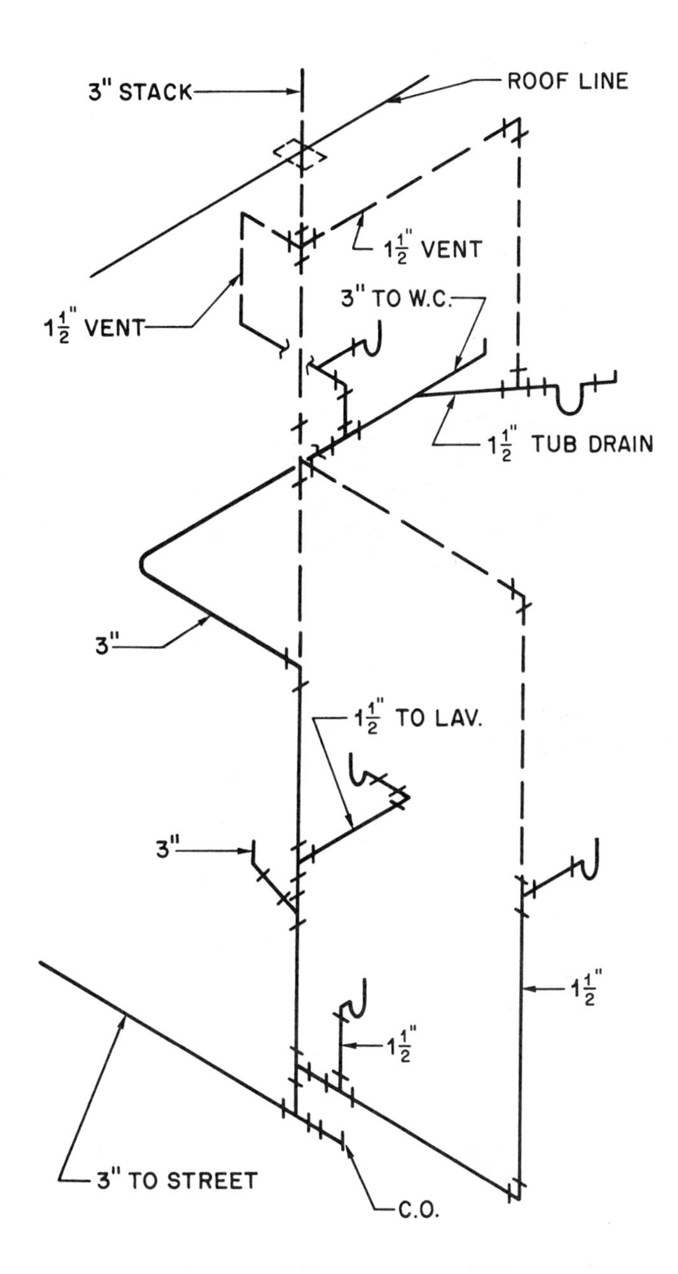

Figure 4–22 A single-line isometric drawing of drain, waste and vent system. *From Huth,* Understanding Construction Drawings, *Delmar Publishers, Inc.*

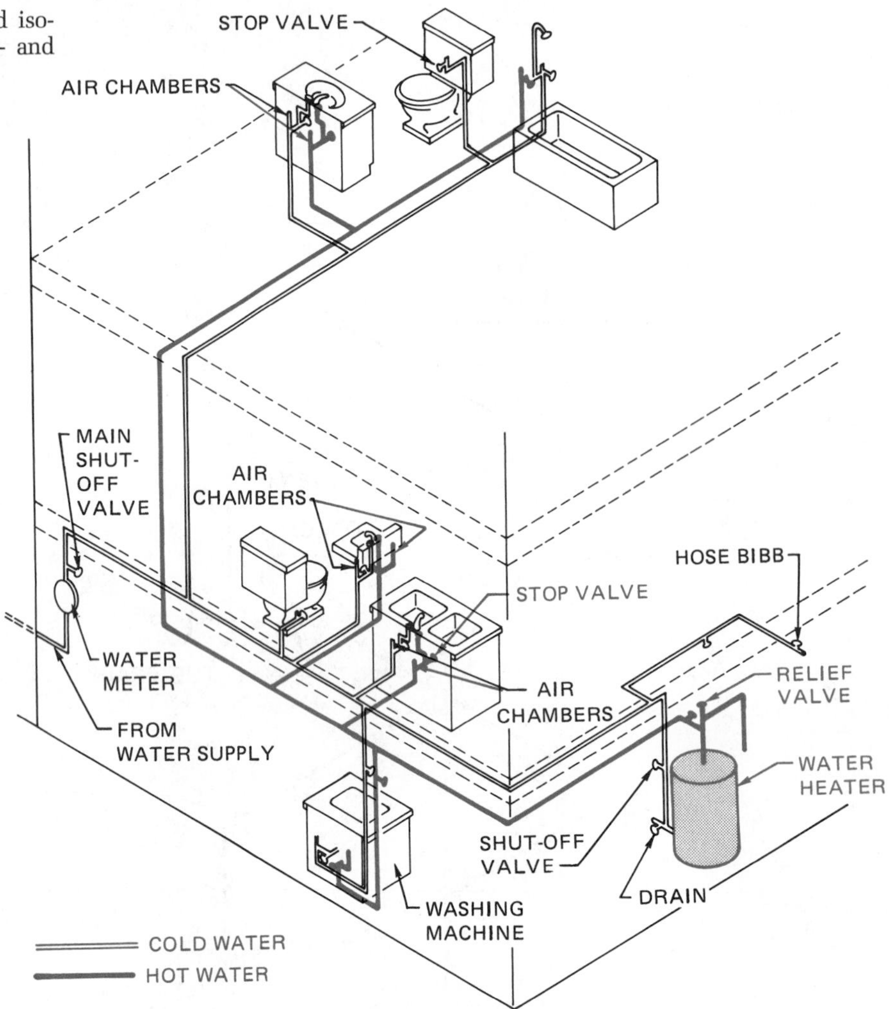

Figure 4–23 A detailed isometric drawing of hot- and coldwater system.

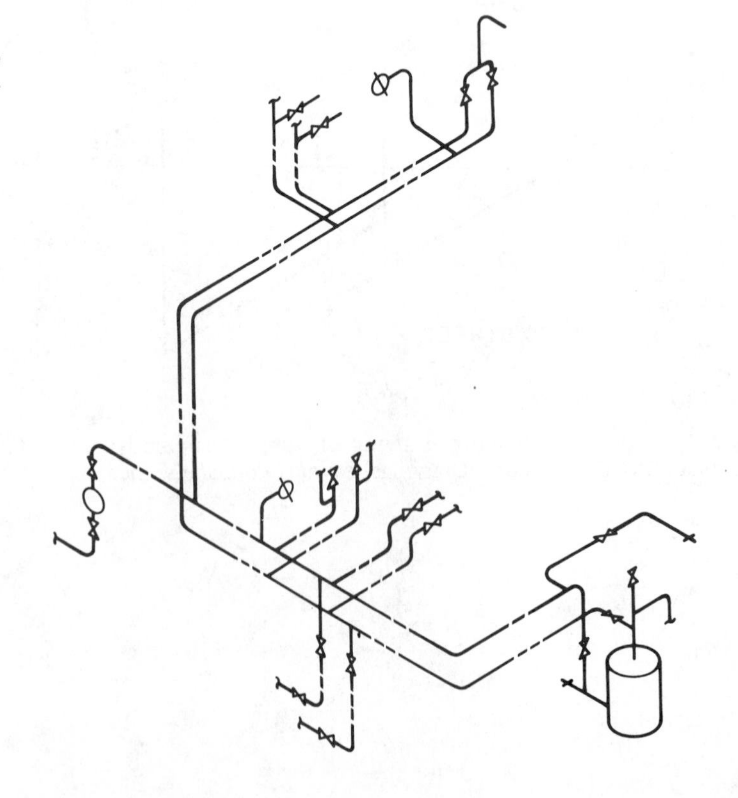

Figure 4–24 A single-line isometric drawing of hot- and cold-water system.

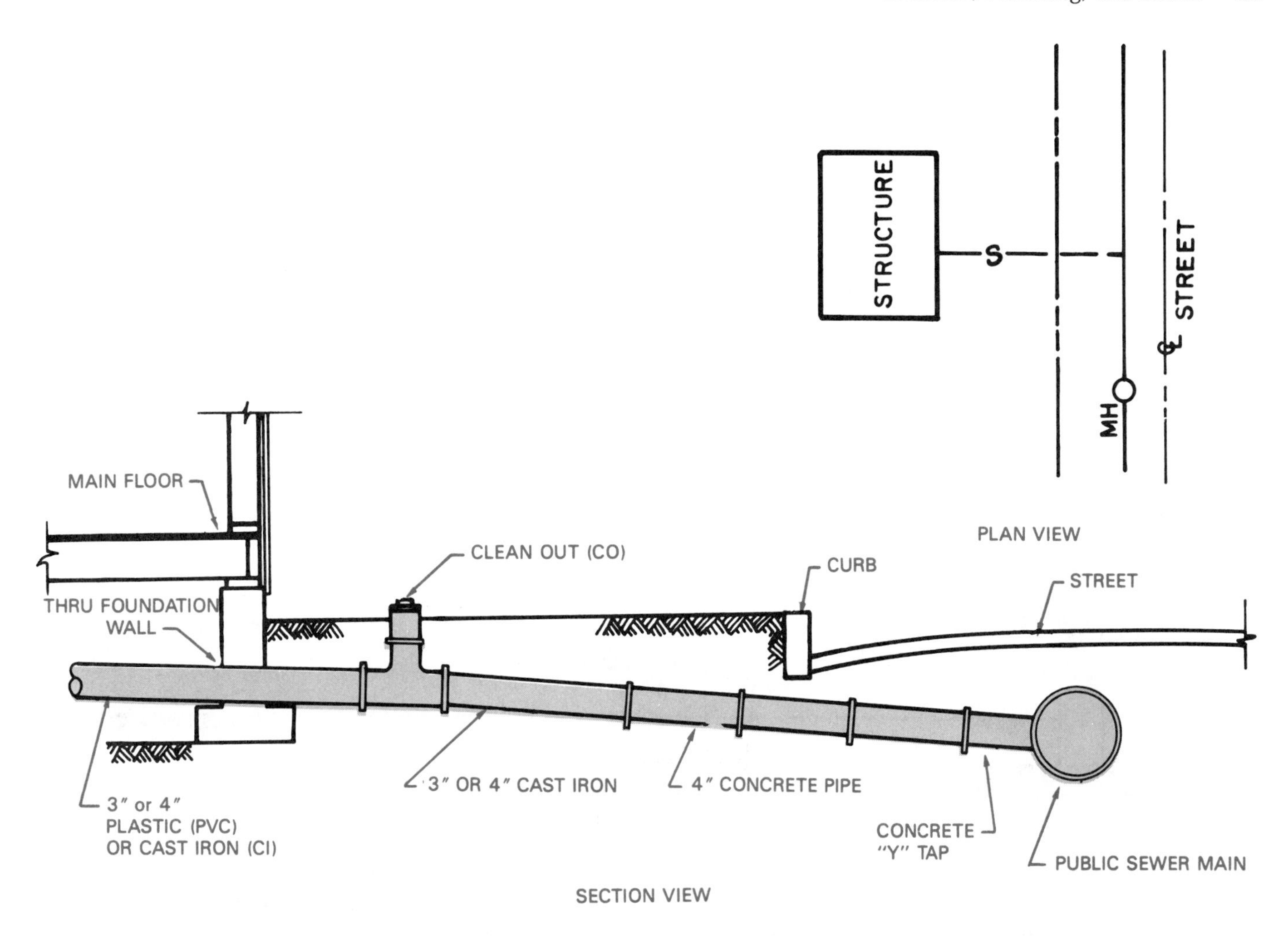

Figure 4–25 Public sewer system.

HEATING, VENTILATING, AND AIR CONDITIONING (HVAC) PLANS

Central Forced Air Systems

One of the most common systems for heating and air conditioning circulates the air from the living spaces through or around heating or cooling devices. A fan forces the air into sheet metal or plastic pipes called ducts, and the ducts connect to openings called diffusers, or air supply registers. Warm air (WA) or cold air (CA) passes through the ducts and registers to enter the rooms and either heats or cools them as needed.

Air then flows from the room through another opening into the return duct, or return air register (RA). The return duct directs the air from the rooms over the heating or cooling device. If warm air is required, the return air is passed over either the surface of a combustion chamber (the part of a furnace where fuel is burned) or a heating coil. If cool air is required, the return air passes over the surface of a cooling coil. Finally, the conditioned air is picked up again by the fan and the air cycle is repeated. Figure 4–27 shows the air cycle in a forced-air system.

Heating Cycle. If the air cycle just described is used for heating, the heat is generated in a furnace. Furnaces for residential heating produce heat by burning fuel oil or natural gas, or from electric heating coils. If the heat comes from burning fuel oil or natural gas, the combustion (burning) takes place inside a combustion chamber. The air to be heated does not enter the combustion chamber but absorbs heat from the outer surface of the chamber. The gases given off by combustion are vented through a chimney. In an

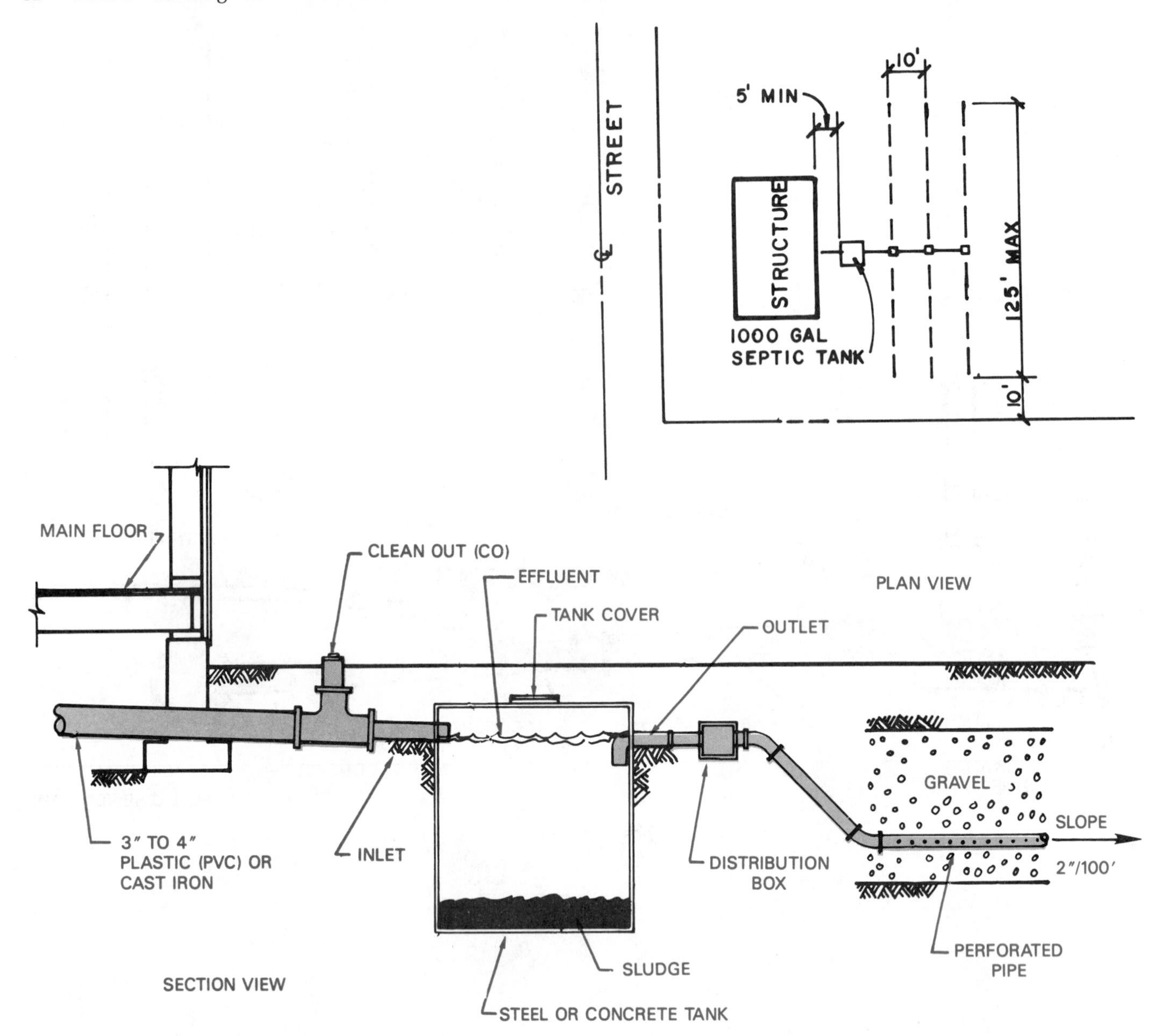

Figure 4–26 A septic sewer system.

electric furnace, the air to be heated is passed directly over the heating coils. This type of furnace does not require a chimney.

Cooling Cycle. If the air from the room is to be cooled, it is passed over a cooling coil. The most common type of residential cooling system is based on two principles:

1. As liquid changes to vapor, it absorbs large amounts of heat.
2. The boiling point of a liquid can be changed by changing the pressure applied to the liquid. This is the same as saying that the temperature of a liquid can be raised by increasing its pressure and lowered by reducing its pressure.

Common refrigerants can boil (change to a vapor) at very low temperatures, some as low as 21°F below zero.

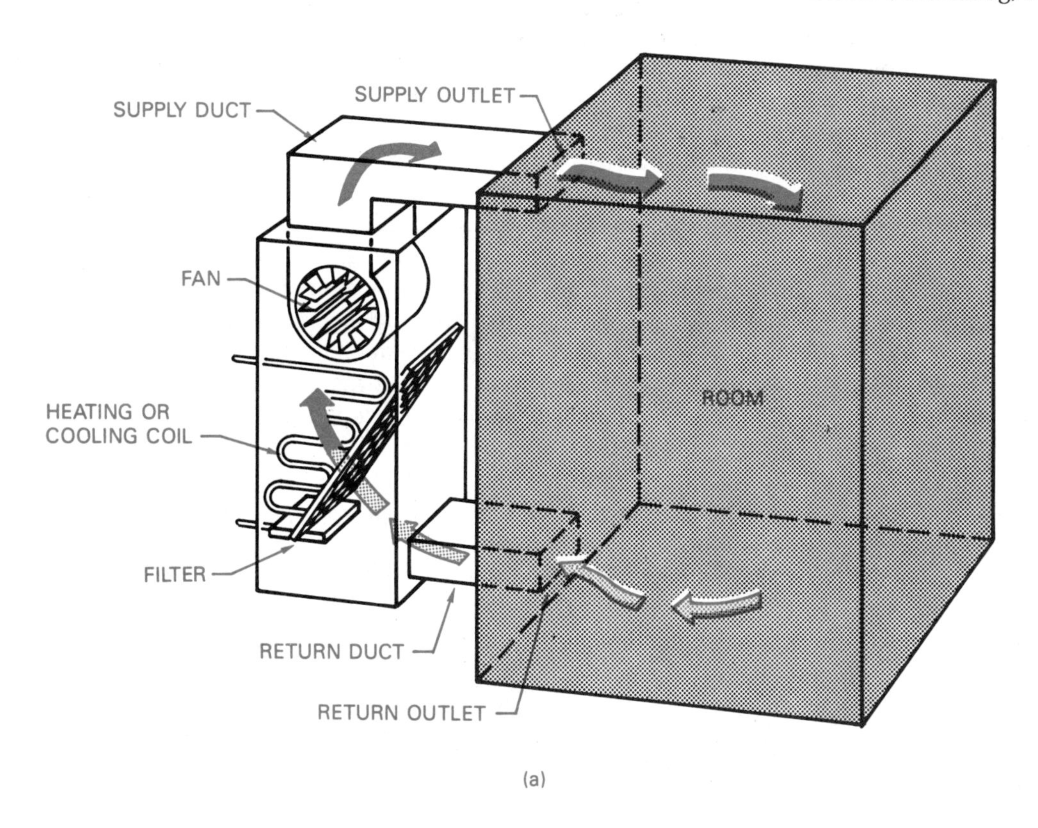

(a)

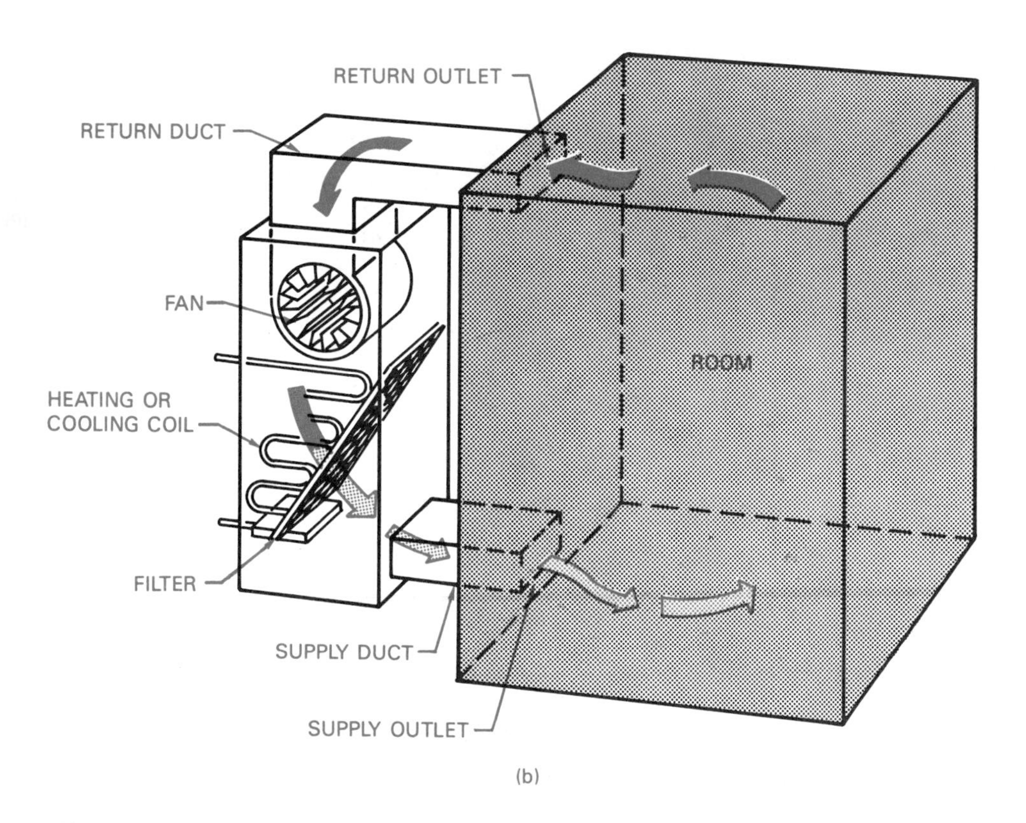

(b)

Figure 4–27 (a) A downdraft forced-air system heated air cycle. *From Huth,* Understanding Construction Drawings, *Delmar Publishers, Inc.* (b) updraft forced-air system heated cycle.

The principal parts of a refrigeration system are the cooling coil (evaporator), compressor, condenser, and expansion valve. Figure 4–28 shows a diagram of the cooling cycle. The cooling cycle operates as warm air from the ducts is passed over the evaporator. As the cold liquid refrigerant moves through the evaporator coil, it picks up heat from the warm air. As the liquid picks up heat, it changes to a vapor. The heated refrigerant vapor is then drawn into the compressor, where it is put under high pressure. This causes the temperature of the vapor to rise even more.

Next, the high-temperature, high-pressure vapor passes to the condenser, where the heat is removed. This is done by blowing air over the coils of the condenser. As the condenser removes heat the vapor changes to a liquid. It is still under high pressure, however. From the condenser, the refrigerant flows to the expansion valve. As the liquid refrigerant passes through the valve, the pressure is reduced which lowers the temperature of the liquid even further so that it is ready to pick up more heat.

The cold low-pressure liquid then moves to the evaporator. The pressure in the evaporator is low enough to allow the refrigerant to boil again and absorb more heat from the air passing over the coil of the evaporator.

Forced-Air Heating Plans. Complete plans for the heating system may be needed when applying for a building permit or a mortgage depending upon the requirements of the local building jurisdiction or the lending agency. If a complete heating layout is required, it is prepared by the architect, mechanical engineer, or heating contractor.

When forced-air electric, gas, or oil heating systems are used, the warm-air outlets and return air locations may be shown as in Figure 4–29. Notice the registers are normally placed in front of a window so that warm air is circulated next to the coldest part of the room. As the warm air rises, a circulation action is created as the air goes through the complete heating cycle. Cold-air returns are often placed in the ceiling or floor of a central location.

A complete forced-air heating plan shows the size, location, and number of BTU dispersed to the rooms from the warm-air supplies. BTU stands for British

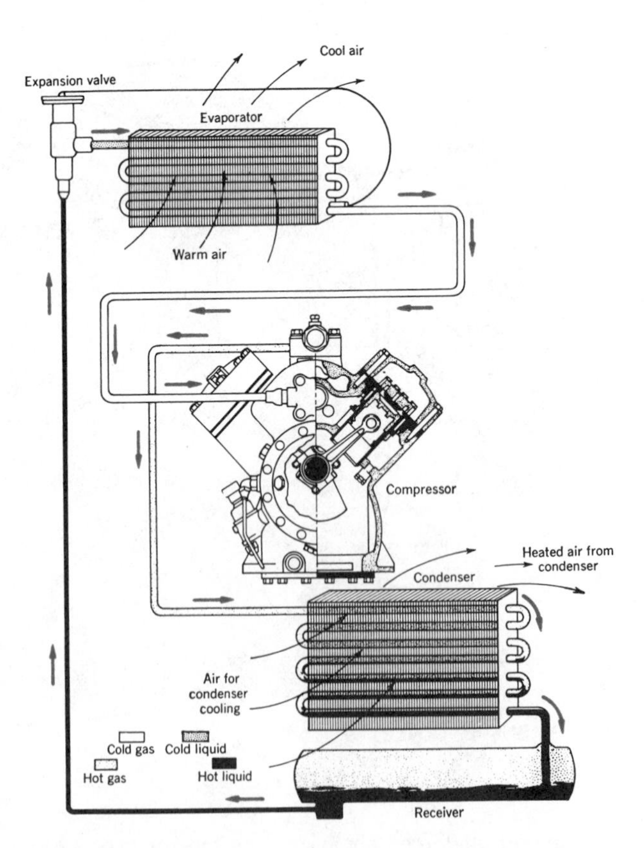

Figure 4–28 A schematic diagram of the cooling cycle. *From Lang,* Principles of Air Conditioning, *3rd ed., Delmar Publishers, Inc.*

Thermal Unit, which is a measure of heat. The location and size of the cold-air return and the location, type, and output of the furnace are also shown.

The warm-air registers are sized in inches, for example, 4 × 12. The size of the duct is also given, as shown in Figure 4–30. The note 20/24 identifies 20" × 24" register and a number 8 next to a duct labels an 8" diameter duct. This same system may be used as a central cooling system when cool air is forced from an air conditioner through the ducts and into the rooms. WA denotes warm air, and RA is return air. CFM is cubic feet per minute, the rate of airflow.

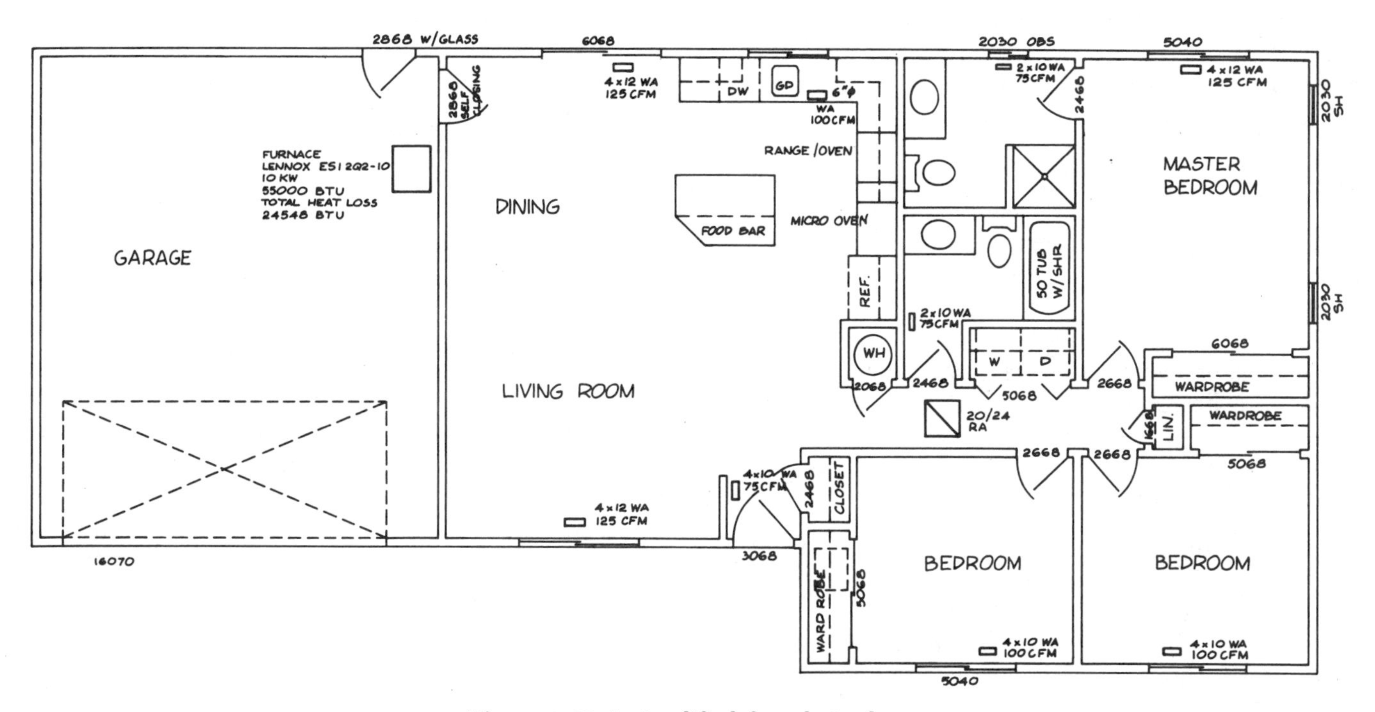

Figure 4–29 A simplified forced-air plan.

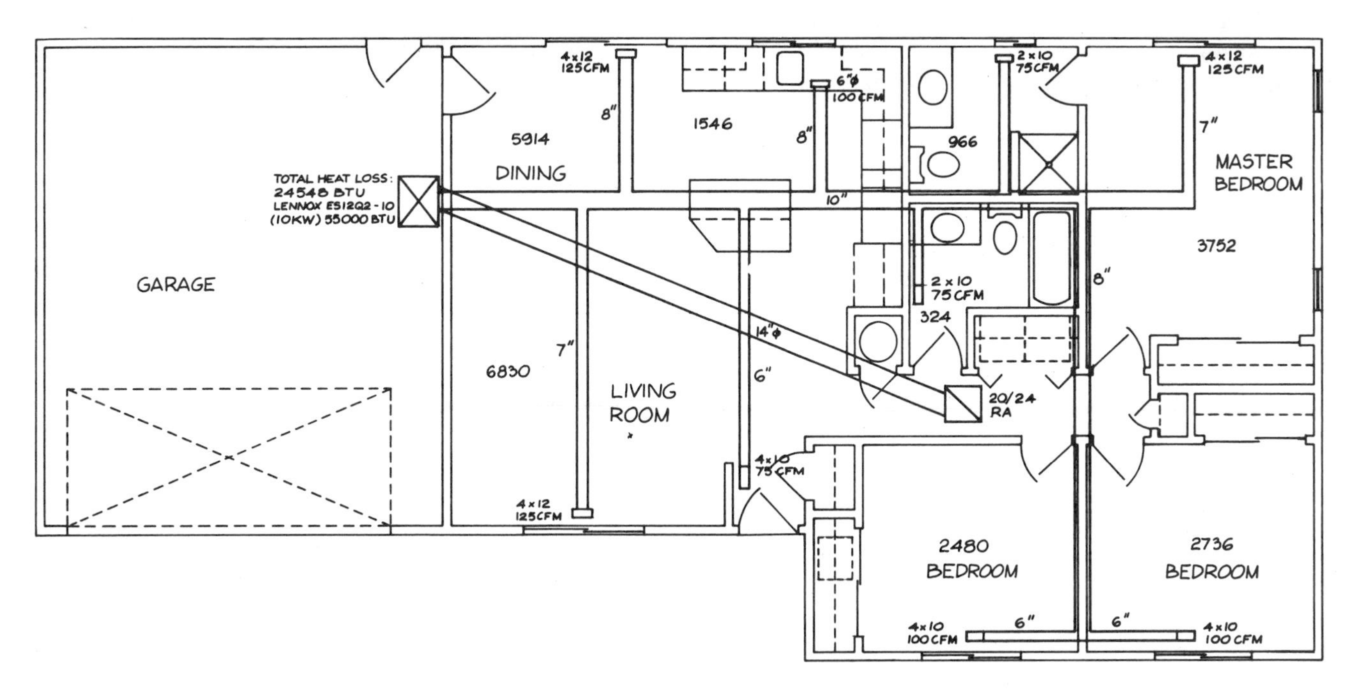

Figure 4–30 A detailed forced-air plan.

Hot-Water Systems. In a hot-water system, the water is heated in an oil- or gas-fired boiler and then circulated through pipes to radiators or convectors in the rooms. The boiler is supplied with water from the fresh water supply for the house. The water is circulated around the combustion chamber, where it absorbs heat.

In one-pipe system, one pipe leaves the boiler and runs through the rooms of the building and back to the boiler. The heated water leaves the supply, is circulated through the outlet, and is returned to the same pipe as shown in Figure 4–31. In a two-pipe system, two pipes run throughout the building. One pipe supplies heated water to all of the outlets. The other is a return pipe that carries the water back to the boiler for reheating, as seen in Figure 4–32.

Hot-water systems use a pump, called circulator, to move the water through the system. The water is kept at a temperature between 150 and 180°F in the boiler. When heat is needed, the thermostat starts the circulator, which supplies hot water to the convectors in the rooms.

HVAC Symbols

Over a hundred heating, ventilating, and air conditioning (HVAC) symbols may be used in residential and commercial heating plans. Only a few of the symbols are typically used in residential HVAC drawings. Figure 4–33 shows some common HVAC symbols.

Heat Pump Systems

The heat pump is a forced-air central heating and cooling system. It operates using a compressor and a circulating refrigerant system. Heat is extracted from the outside air and pumped inside the structure. The heat pump supplies up to three times as much heat per year for the same amount of electrical consumption as a standard electric forced-air heating system. In comparison, this can result in a 30 to 50 percent annual energy savings. in the summer the cycle is reversed and the unit operates as an air conditioner. In this mode, the heat is extracted from the inside air and pumped outside. On the cooling cycle the heat pump also acts as a dehumidifier. Figure 4–34 shows a graphic example of how heat pump works.

Zone Control Systems

A zoned heating system requires one heater and one thermostat per room. No ductwork is required, and only the heaters in occupied rooms need be turned on.

One of the major differences between a zonal and a central system is flexibility. A zonal heating system allows the home occupant to determine how many rooms are heated, how much energy is used, and how much money is spent on heat. A zonal system allows the home to be heated to the family's needs; a central system requires using all the heat produced. If the airflow is restricted, the efficiency of the central system is reduced. There is also a 10 to 15 percent heat loss through ductwork in central systems.

Regardless of the square footage in a house, its occupants normally use less than 40 percent of the entire area on a regular basis. A zoned system is very adaptable to heating the 40 percent of the home that is occupied using automatic controls that allow night setback, day setback, and nonheated areas. The homeowner can save as much as 60 percent on energy costs through controlled heating systems.

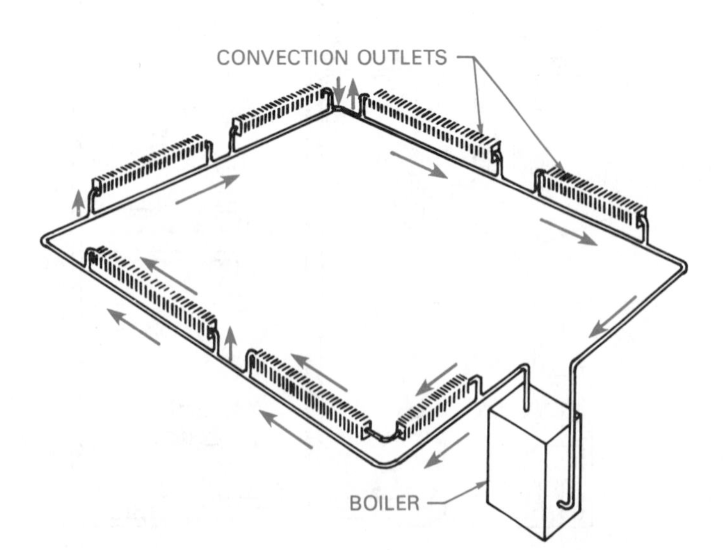

Figure 4–31 One-pipe hot-water system. *From Huth,* Understanding Construction Drawings, *Delmar Publishers, Inc.*

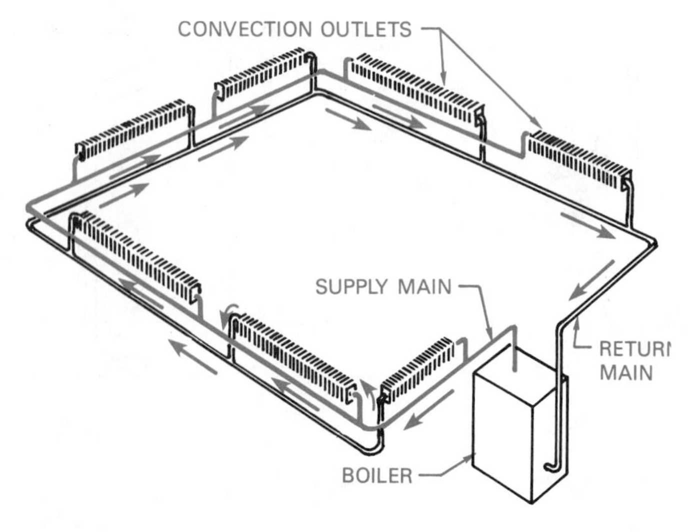

Figure 4–32 Two-pipe hot-water system. *From Huth,* Understanding Construction Drawings, *Delmar Publishers, Inc.*

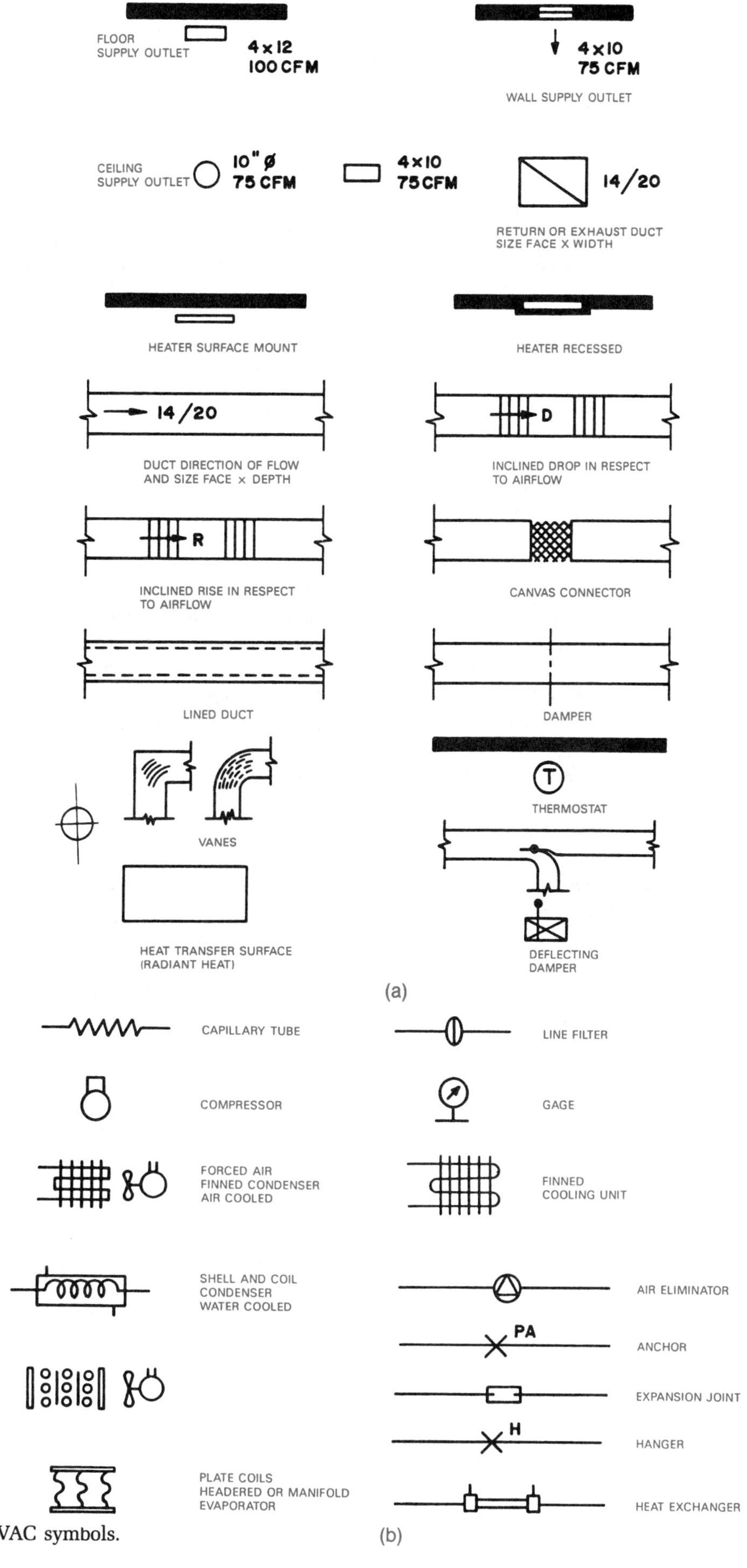

Figure 4–33 Typical HVAC symbols.

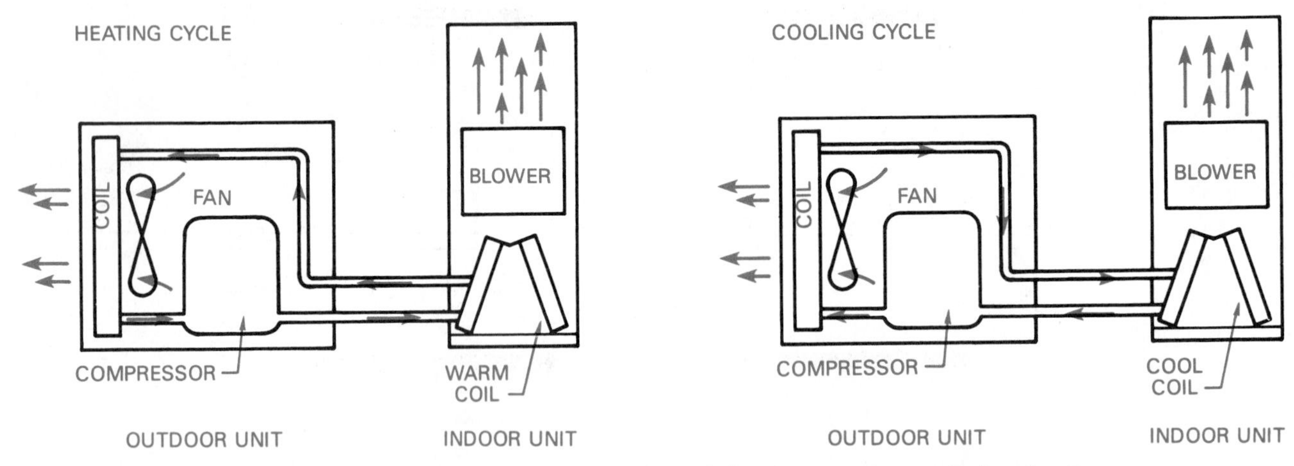

Figure 4–34 Heat pump heating and cooling cycle. *Courtesy Lennox Industries, Inc.*

There are typically two types of zone heaters: baseboard and fan. Baseboard heaters have been the most popular type for zoned heating systems for the past several decades. They are used in many different climates and under various operating conditions. No ducts, motors, or fans are required. Baseboard units have an electric heating element that creates a convection current as the air around the unit is heated. The heated air rises into the room and is replaced by cooler air that falls to the floor. Baseboard heaters should be placed on exterior walls under or next to windows.

Fan heaters are generally mounted in a wall recess. A resistance heater is used to generate the heat, and a fan circulates the heat into the room. These units should be placed to circulate the warmed air in each room adequately. Avoid placing the heaters on exterior walls because the recessed unit reduces or eliminates the insulation in that area.

Heat pumps may require a split system in very large homes. Split systems are also possible using zonal heat in part of the home and central heat in the balance of the structure. An alternative is to install a heat pump for the areas used most often and zoned heaters for the remainder of the house. This is also an option for additions to homes that have a central system. Zoned heat can be used effectively in an addition so that the central system is not overloaded or required to be replaced.

RADIANT HEAT

Radiant heating and cooling systems provide a comfortable environment by means of controlling surface temperatures and minimizing excessive air motion within the space. Warm ceiling panels are effective for winter heating because they warm the floor surfaces and glass surfaces by direct transfer of radiant energy. The surface temperature of well-constructed and properly insulated floors will be 2°F to 3°F above the ambient air temperature, and the inside surface temperature of glass will be increased significantly. As a result of these heated surfaces, down-drafts are minimized to the point where no discomfort is felt.

Radiant heat systems generate operating cost savings of 20 to 50 percent annually compared to conventional convective systems. This saving is accomplished through lower thermostat settings. Savings are also due to the superior, cost-effective design inherent in radiant heating products.

Radiant heat may be achieved with oil- or gas-heated hot water piping in the floor or ceiling, to electric coils, wiring, or elements either in or above the ceiling gypsum board, and transferred to metal radiator panels generally mounted by means of a bracket about an inch below the ceiling surface. A recent evolution of the radiant panel concept is a lightweight, quick response, totally zone-controlled panel system which may be mounted directly to the ceiling surface, on joists, or placed in a suspended ceiling grid. The radiant solid-state heating panels are available in a full range of sizes and voltages which are ideal for both remodeling and new construction applications as primary or auxiliary heating.

THERMOSTATS

The thermostat is an automatic mechanism for controlling the amount of heating or cooling given by a central or zonal heating or cooling system. The thermostat floor plan symbol is shown in Figure 4–35.

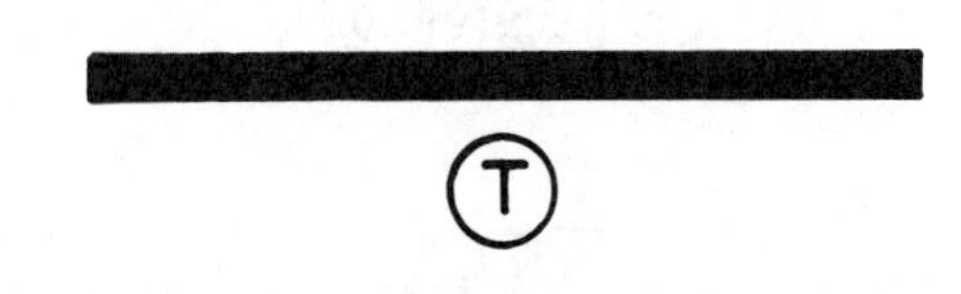

Figure 4–35 Thermostat floor plan symbol.

The location of the thermostat is an important consideration to the proper functioning of the system. For zoned heating or cooling units thermostats may be placed in each room or a central thermostat panel that controls each room may be placed in a convenient location. For central heating and cooling systems there may be one or more thermostats depending upon the layout of the system or the number of units required to service the structure. For example, a very large home or office building may have a split system that divides the structure into two or more zones. Each individual zone has its own thermostat.

Several factors contribute to the effective placement of the thermostat for a central system. A good location is near the center of the structure and close to a return air duct for a central forced-air system, where an average temperature reading can be achieved. There should be no drafts that would adversely affect temperature settings. The thermostat should not be placed in a location where sunlight or a heat register would cause an unreliable reading. A thermostat not be placed close to an exterior door where temperatures can change quickly. Thermostats should be placed on inside partitions rather than on outside walls where a false temperature reading could also be obtained. Avoid placing the thermostat near stairs or a similar traffic area where significant bouncing or shaking could cause the mechanism to alter the actual reading.

Programmable microcomputer thermostats are also available that effectively help reduce the cost of heating or cooling. Some units automatically switch from heat to cool while minimizing temperature deviation from the setting under varying load conditions. These computers can be used to alter heating and cooling temperature settings automatically for different days of the week or different months of the year.

Heat Recovery and Ventilation

Sources of Pollutants. Air pollution in a structure is the principal reason for installing a heat recovery and ventilation system. A number of sources that contribute to an unhealthy environment within a home or business:

- Moisture in the form of relative humidity can cause structural damage as well as health problems, such as respiratory problems. The source of relative humidity is the atmosphere, steam from cooking and showers, and individuals, who can produce up to 1 gal water vapor per day.
- Incomplete combustion from gas-fired appliances or wood-burning stoves and fireplaces can generate a variety of pollutants, including carbon monoxide, aldehydes, and soot.
- Humans and pets can transmit bacterial and viral diseases through the air.
- Tobacco smoke contributes chemical compounds to the air.
- Formaldehyde is found in carpets, furniture, and the glue used in construction materials, such as plywood and particle board, as well as some insulation products.
- Radon is a naturally occurring radioactive gas that breaks down into compounds that are *carcinogenic* (cancer causing) when large quantities are inhaled over a long period of time. Radon may be more apparent in a structure that contains a large amount of concrete or in certain areas of the country.
- Products, such as those available in aerosol spray cans, and craft materials, such as glues and paints, can contribute a number of toxic pollutants.

Air-to-air Heat Exchangers. Government energy agencies, architects, designers, and contractors around the country have been evaluating construction methods that are designed to reduce energy consumption. Some of the tests have produced superinsulated, vapor barrier-lined, airtight structures. The result has been a dramatic reduction in heating and cooling costs; however, the air quality in these houses has been significantly reduced and may even be harmful to health. In essence, the structure does not breathe and the stale air and pollutants have no place to go. A recent technology has emerged from this dilemma in the form of an air-to-air heat exchanger. In the past the air in a structure was exchanged by leakage through walls, floors, ceilings, and around openings. Although this random leakage was no insurance that the building was properly ventilated, it did ensure a certain amount of heat loss. Now, with the concern of energy conservation, it is clear the internal air quality of a home or business cannot be left to chance.

An air-to-air heat exchanger is a heat recovery and ventilation device that pulls polluted, stale warm air from the living space and transfers the heat in that air to fresh, cold air being pulled into the house. Heat exchangers do not produce heat: they only exchange heat from one airstream to another. The heat transfer takes place in the core of the heat exchanger, which is designed to avoid mixing the two airstreams to ensure that indoor pollutants are expelled. Moisture in the stale air condenses in the core and is drained from the unit. Figure 4–36 shows the function and basic components of an air-to-air heat exchanger.

The recommended minimum effective air change rate is 0.5 air changes per hour (ach). Codes in some areas of the country have established a rate of 0.7 ach. The American Society of Heating, Refrigeration, and

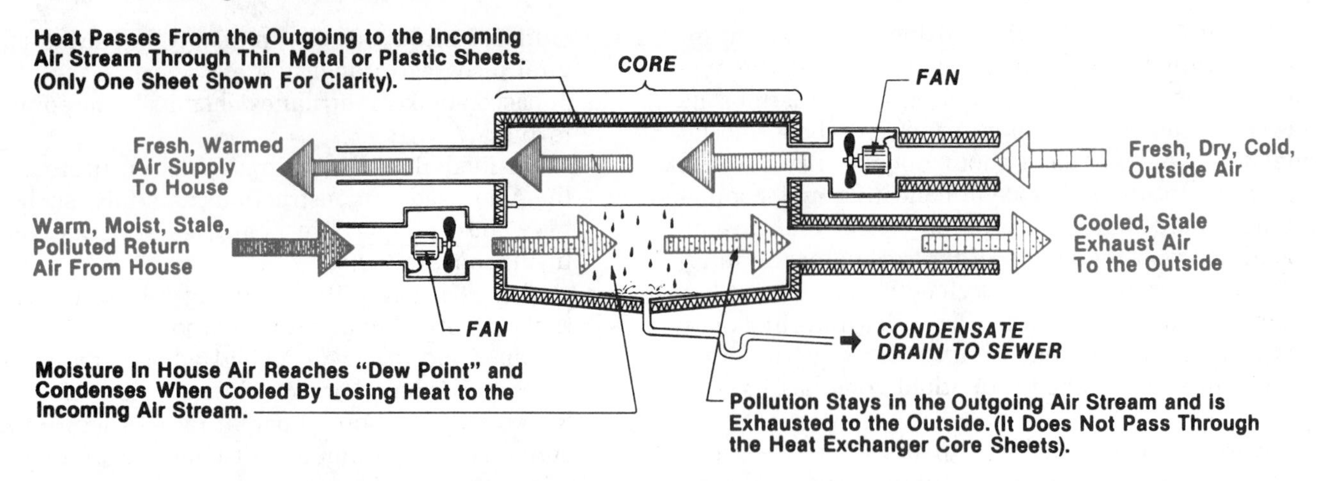

Figure 4–36 The components and function of an air-to-air heat exchanger. *Courtesy U.S. Department of Energy.*

Air Conditioning Engineers, Inc. (ASHRAE) recommends ventilation levels based on the amount of air entering a room. The recommended amount of air entering most rooms in 10 cubic ft per minute (cfm). The rate for kitchens is 100 cfm and bathrooms is 50 cfm. Mechanical exhaust devices vented to outside air should be added to kitchens and baths to maintain the recommended air exchange rate.

The minimum heat exchanger capacity needed for a structure can easily be determined. Assume a 0.5 ach rate in a 1,500 sq ft single level, energy efficient house and follow these steps:

1. Determine the total floor area in sq ft. Use the outside dimensions of the living area only.

   30' × 50' = 1,500 sq ft

2. Determine the total volume within the house in cubic ft by multiplying the total floor area by the ceiling height.

   1,500 sq ft × 8' = 12,000 cu ft

3. Determine the minimum exchanger capacity in cfm by first finding the capacity in cubic feet per hour (cfh). Multiply the house volume by the ventilation rate required from the exchanger.

   12,000 cu ft × 0.5 ach = 6,000 chf

4. Convert the cfh rate to cfm by dividing the cfh rate by 60 min.

   6,000 cfh ÷ 60 min = 100 cfm

There is a percentage of capacity loss due to mechanical resistance that should be considered by the system designer.

CENTRAL VACUUM SYSTEMS

A well-designed system requires only a few outlets to cover the entire home or business, including exterior use. The hose plugs into a wall outlet and the vacuum is ready for use. A central canister stores the dust and debris from the house or business and is generally located in the garage or in a storage area. The floor plan symbol for vacuum cleaner outlets is shown in Figure 4–37. The central unit may be found in the garage or storage area as a circle that is labeled Central Vacuum System.

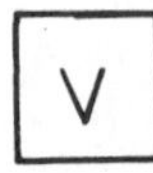

Figure 4–37 A vacuum outlet floor plan symbol.

CHAPTER 4 TEST

Fill in the blanks below with the proper word or short statement as needed to correctly complete the sentence or answer the question.

1. What does the abbreviation GFI mean? ____.
2. What is the purpose of a GFI? ____.
3. Each enclosed bath or laundry room should have an ______________________.
4. Describe how a special characteristic, such as a specific size of fixture, a location requirement, or any other specification, might be listed on the floor plan drawing with an element of the electrical layout. ____.
5. For capacity of service, the average single-family residence should be equipped with a __________ service entrance.
6. Name the two classifications of piping ____.
7. In most cases, the plumbing requirements can be clearly provided on the floor plan in the form of ______ for fixtures and ______________ for specific applications or conditions.
8. How are floor drains identified on the floor plan? ____.
9. Plumbing drawings are made up of lines and symbols that show how ______________ are transported to various locations in the structure.
10. Describe the function of the drainage system. ____.
11. Describe the function of the vent system. ____.
12. Name at least two types of sewage disposal systems. ____.
13. What does the abbreviation HVAC mean? ____.
14. Explain the basic principle behind the complete cycle of the central forced air system. ____.
15. Name two types of hot-water heating systems. ____.

16. In a forced-air system, why are registers normally placed in front of a window? ______________________________.

17. Describe a common location for the cold-air return in a central forced-air system. ______________________________.

18. Name the three items shown in a forced-air heating plan for dispersement to the rooms from the warm air supplies. ______________________________.

19. Describe the basic principle behind the heat pump system's heat and cooling cycles. ______________________________.

20. List at least four advantages of the zone control system compared to the central system. ______________________________.

21. How much heat loss commonly occurs through the ductwork in a central system? ______________________________

22. Radiant heating and cooling systems provide a comfortable environment by means of controlling surface temperatures and minimizing excessive ______________ within the space.

23. Define thermostat. ______________________________.

24. Several factors contribute to the effective placement of the thermostat for a central system; name at least four. ______________________________.

25. Identify at least five sources of pollutants in a structure. ______________________________.

26. Describe the function of an air-to-air heat exchanger. ______________________________.

27. List at least four advantages of a central vacuum system. ______________________________.

28. Mechanical exhaust devices vented to outside air should be added to ______________ and ______________ to maintain the recommended air exchange rate.

CHAPTER 4 EXERCISES

PROBLEM 4–1. Answer the following questions as you read the floor plan drawing on page 94.

1. Identify the locations of the ground fault interrupters. ____.
2. How many square surface-mounted fixtures are located in the kitchen? ______________________
3. How many three-way switch circuits are in this partial plan? ______________________
4. Describe the light fixture located in the utility room. ____.
5. Describe the location of the four-way switch, and explain the arrow at the end of the switch leg connected to that switch. ____.
6. Describe the outlet behind the dryer space. ____.
7. How many duplex convenience outlets are there in the kitchen? ______________________
8. What does the abbreviation D.W. mean? ______________________
9. What does the abbreviation P.C. mean? ______________________
10. Is the kitchen sink single or double? ______________________
11. What does the abbreviation WH mean? ______________________
12. Where is the WH located, and what are the specifications associated with it? ____.

PROBLEM 4–2. Answer the following questions as you read the floor plan drawing on page 95.

1. Identify the locations of the ground fault interrupters. ____.
2. Describe the difference between the exhaust fans in bath 1 and in bath 2. ____.
3. Describe the light fixture in bedroom 2. ____.
4. Describe how the switch in the master bedroom is designed to provide light in the room. ____.
5. Describe the location of the three switches in the four-way switch system. ____.

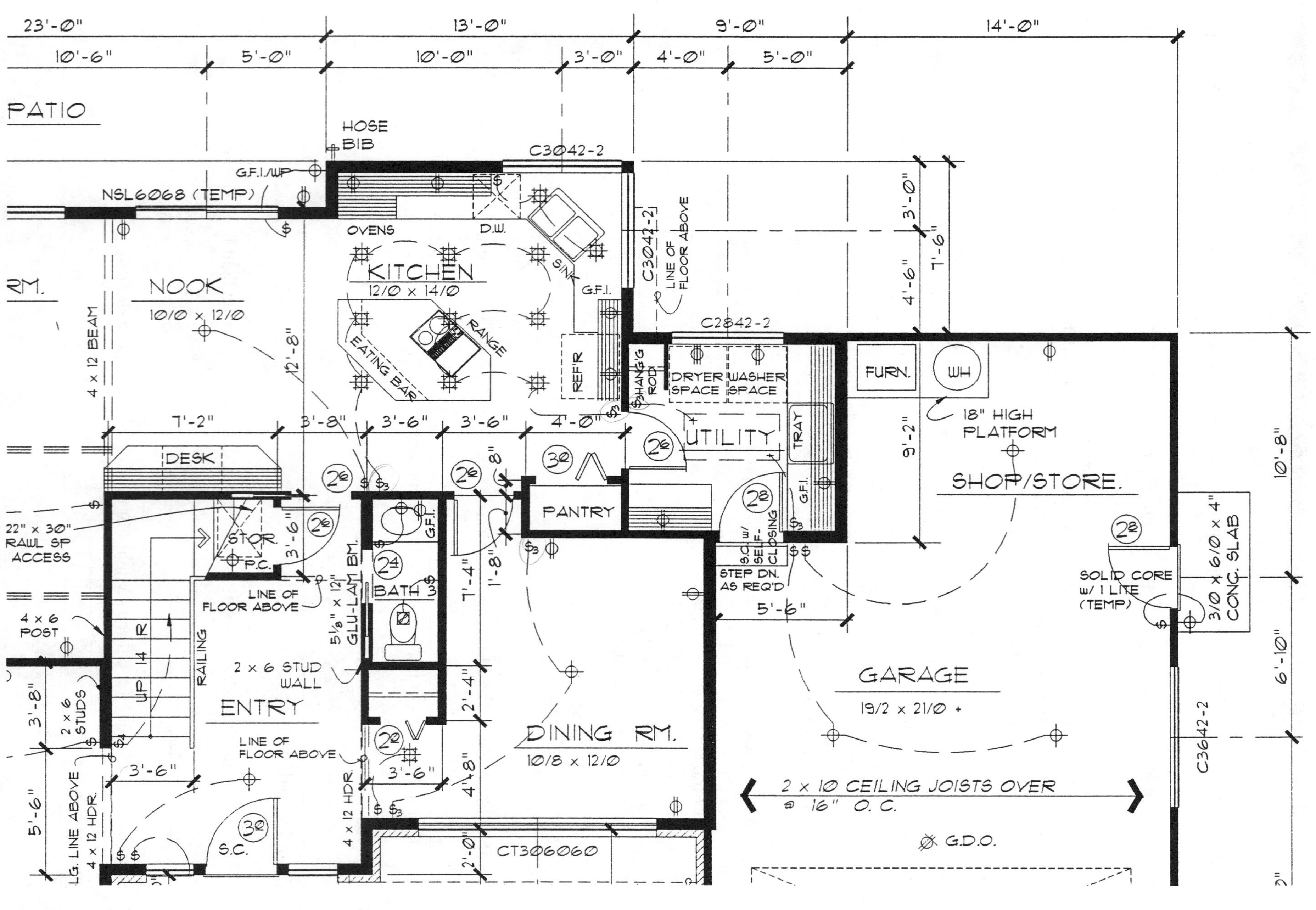

Problem 4–1. *Courtesy Piercy and Barclay Designers, Inc.*

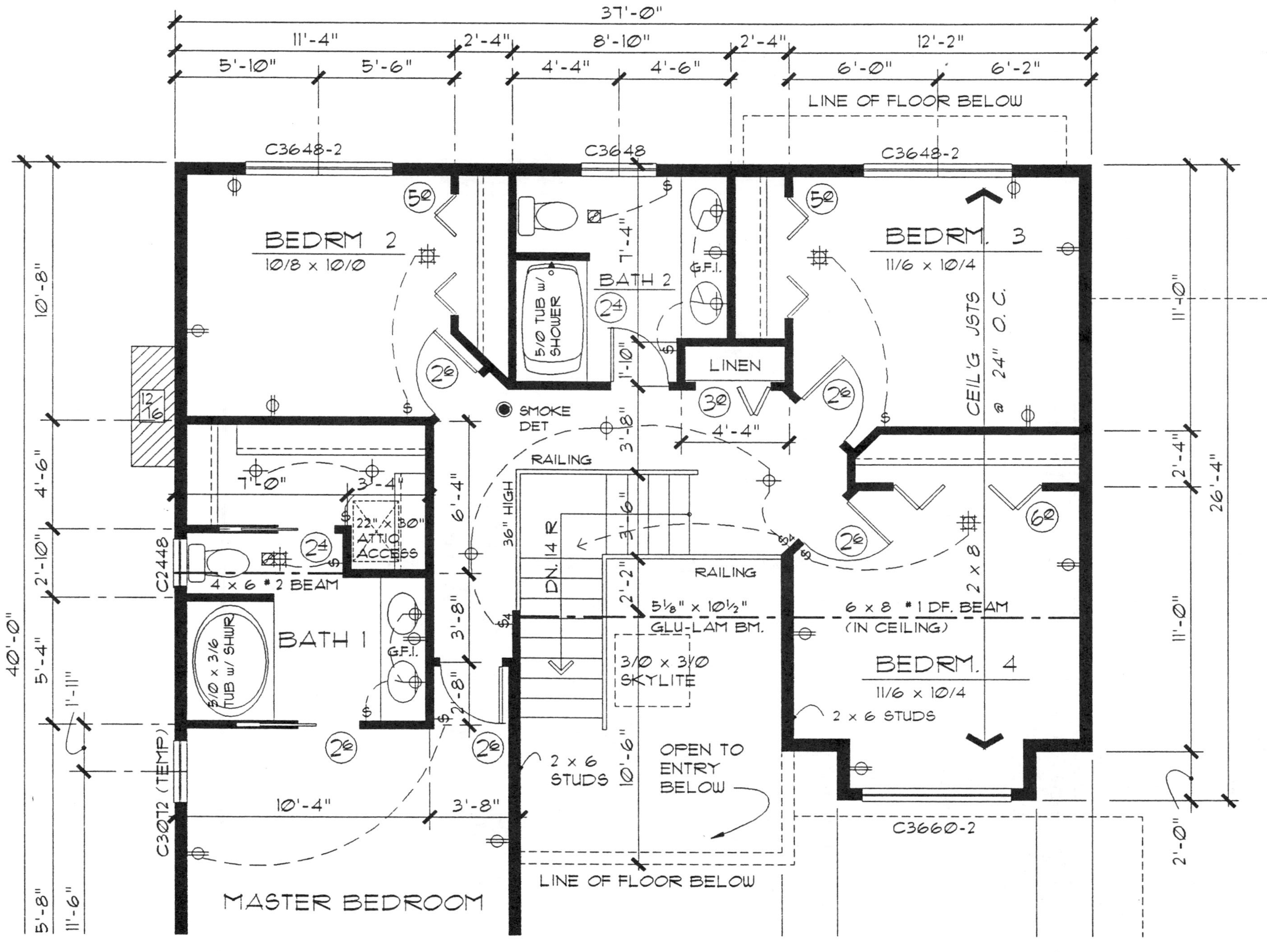

Problem 4–2. *Courtesy Piercy and Barclay Designers, Inc.*

6. Give the complete specifications associated with the tub in the master bath. (Change all dimensions and abbreviations to feet and inches, and whole words.) ______________________________

____

______________________________.

7. Give the complete specifications associated with the tub in bath 2. (Change all dimensions and abbreviations to feet and inches, and whole words.) ______________________________

____

______________________________.

8. How many water closets are there in this partial floor plan? ______________________________

9. Describe the location of the smoke detector. ______________________________

10. How many bathroom sinks are there, and what are their shapes? ______________________________

PROBLEM 4–3. Given the following plumbing pipe symbols on page 101, place the name of each symbol on the blank line provided below or to the side of the symbol.

PROBLEM 4–4. Answer the following questions as you read the partial plumbing plan below.

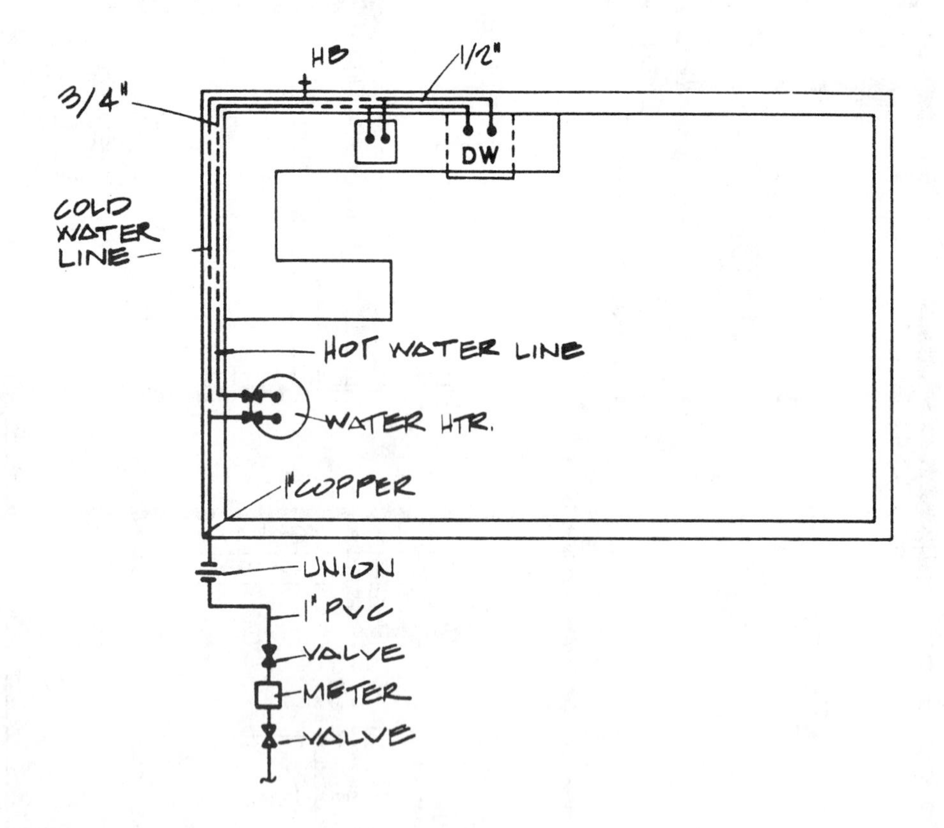

1. Describe the plumbing system shown in this partial floor plan. ______________________________

____

______________________________.

2. What is the size of the hot-water line? ______________________________

3. What is the size of the cold-water line? ______________________________

4. What is the size and type of pipe used from the meter to the union? ______________________________

5. What is the size and type of pipe from the union to the water heater junction? ______________________________

CWV

1) ____________

14) ____________

2) ____________

SD

S

15) ____________

SW

3) ____________

16) ____________

ICW

4) ____________

17) ____________

DWS

5) ____________

18) ____________

F

6a) ______ 6b) ______

6c) ____________

19) ____________

7a) ______ 7b) ______

7c) ____________

20) ____________

G

8a) ______ 8b) ______

21) ____________

9) ____________

T

22) ____________

10) ____________

23) ____________

11) ____________

12) ____________

24) ______ 25) ______ 26) ______ 27) ______ 28) ______

13) ____________

29) ____________

30) ____________

____

Problem 4–3.

PROBLEM 4–5. Describe the plumbing system shown in this partial floor plan. ______________________

____

____.

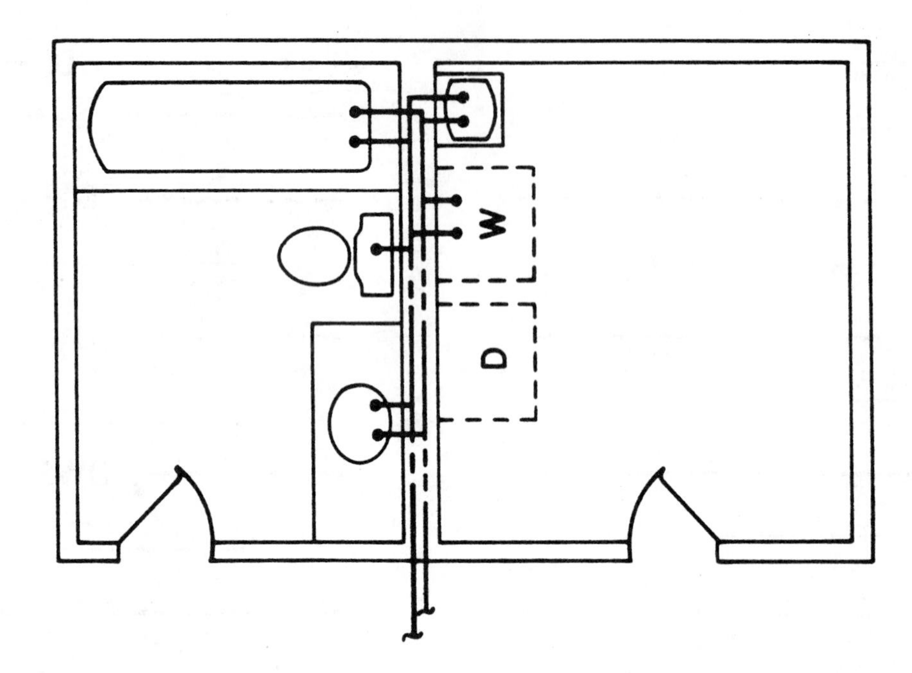

PROBLEM 4–6. Describe the plumbing system shown in this partial floor plan. ______________________

____

____.

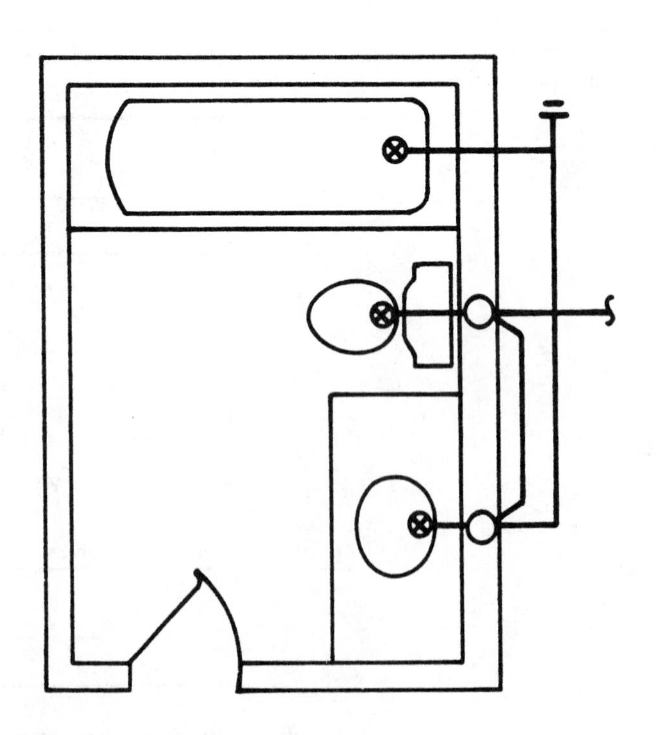

PROBLEM 4–7. Answer the following questions as you read the residential plumbing plan below.

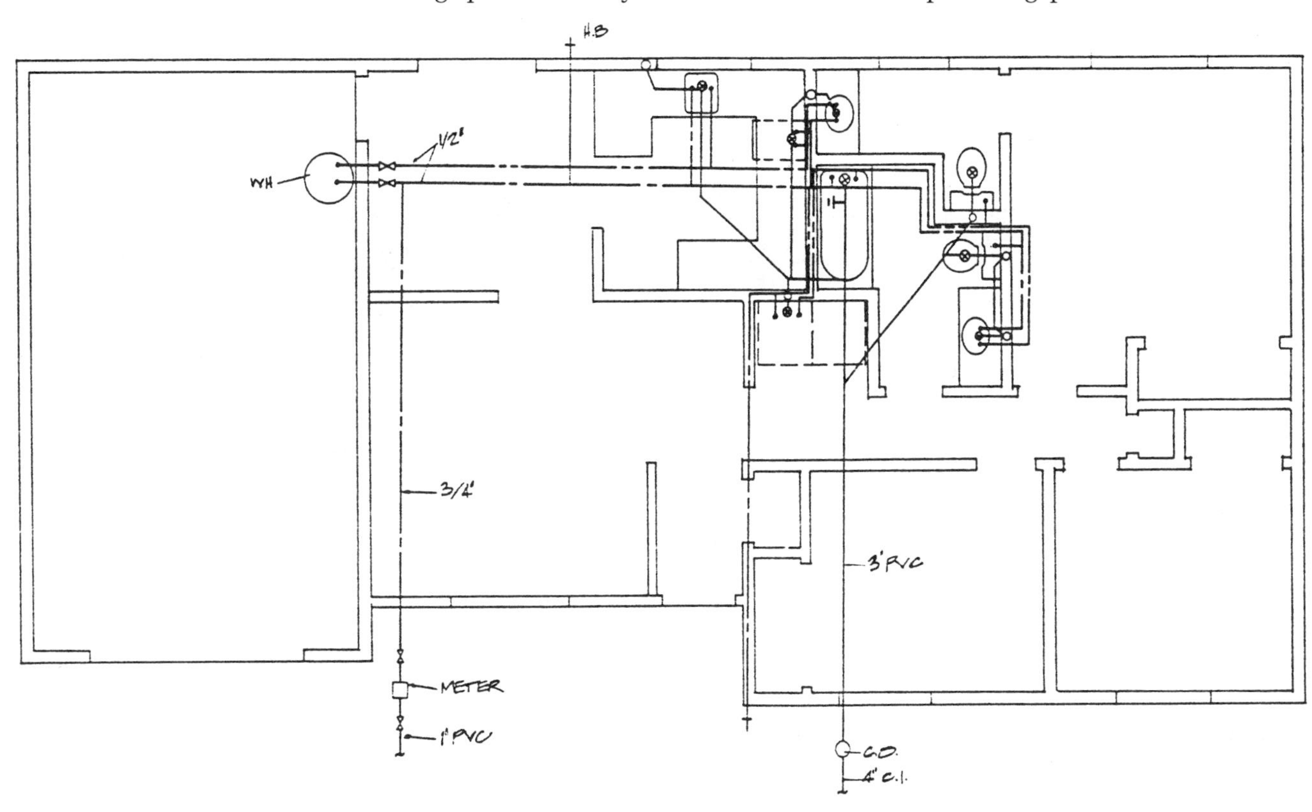

1. What is the size and type of pipe from the street to the meter? ____________________
2. What size pipe runs from the meter to the water heater junction? ____________________
3. What size pipe is used for the hot and cold water system? ____________________
4. What do the following abbreviations mean?

   WH ____________________

   HB ____________________

   HB ____________________

   CO ____________________

   CI ____________________
5. What is the size and type of pipe used in the drainage system to the CO? ____________________
6. What is the size and type of pipe that goes to the sewer? ____________________

PROBLEM 4–8. Answer the following questions as you read the residential HVAC plan below.

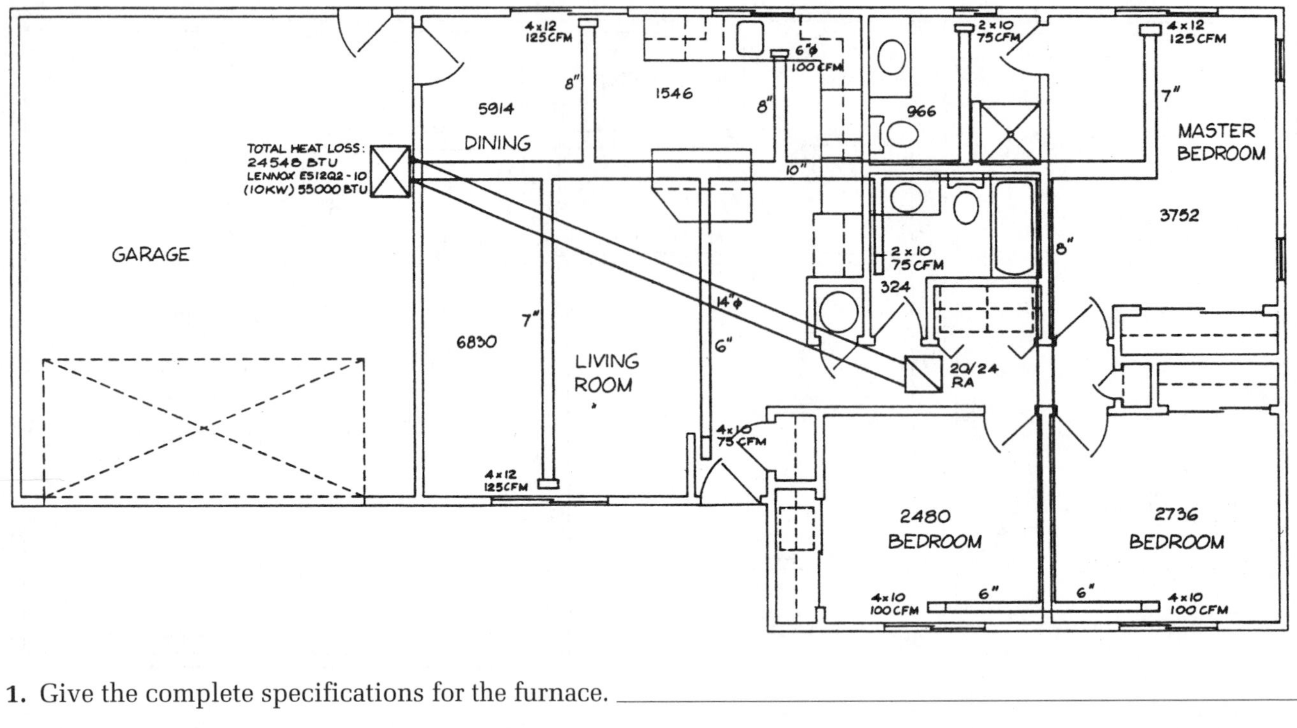

1. Give the complete specifications for the furnace. ______________________________
____
______________________________.
2. What is the total estimated heat loss in this home? ______________________________
3. What is the size of the main duct that runs from the furnace to the ducts that supply the individual areas? ______________________________
4. What are the sizes of the ducts that supply the following rooms:

Living room	____________	To both lower bedrooms	____________
Dining room	____________	To each bedroom	____________
Kitchen	____________	Master bedroom	____________
Entry	____________		

5. What is the size of the duct that runs from the return air register to the furnace? ____________
6. What is the size of the return air register? ______________________________
7. What is the size of the warm air registers in the following rooms?

Living room	____________	Master bedroom	____________
Dining room	____________	Main bath	____________
Kitchen	____________	Master bath	____________

8. What do the following abbreviations mean?

Btu	____________
cfm	____________
RA	____________

9. What is the estimated cfm at the warm air registers in the following rooms?

Living room	____________	Master bath	____________
Dining room	____________	Master bedroom	____________
Kitchen	____________		

10. What is the estimated heat loss in each of the following rooms?

Living room	____________	Master bedroom	____________
Dining room	____________	Master bath	____________
Kitchen	____________	Main bath	____________

PROBLEM 4–9. Given the following HVAC symbols, place the name or identification of the symbol on the blank line provided below or next to each symbol.

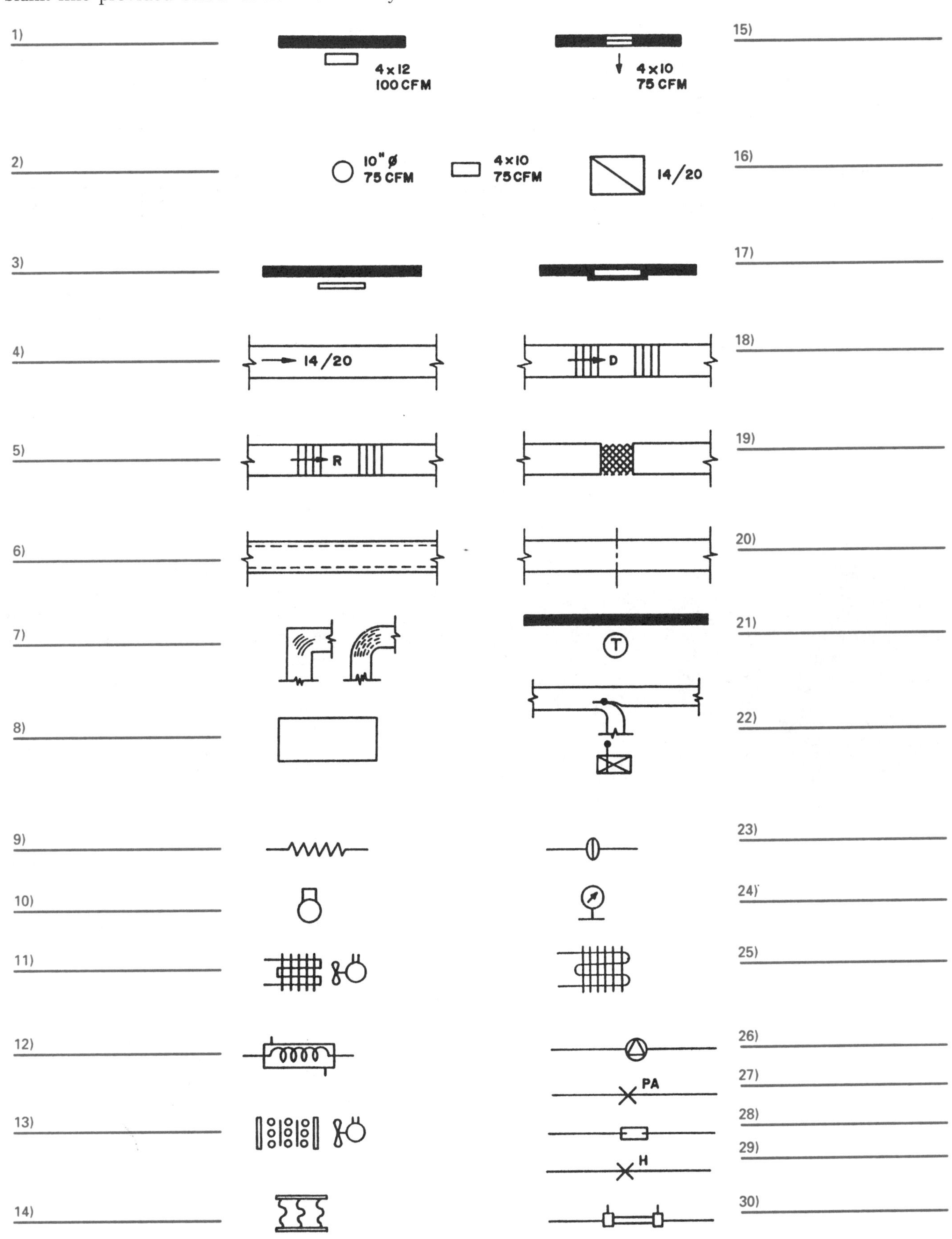

Chapter 5

Foundation and Floor Systems

FOUNDATION SYSTEMS
SOIL CONSIDERATIONS
TYPES OF FOUNDATIONS
DIMENSIONING FOUNDATION COMPONENTS
FLOOR SYSTEMS
CRAWL SPACE FLOOR SYSTEMS
READING A FOUNDATION PLAN
TEST
EXERCISES

FOUNDATION SYSTEMS

THE FOUNDATION provides a base to distribute the weight of the structure onto the soil. The weight, or loads, must be evenly distributed over enough soil to prevent them from compressing it.

In addition to resisting the loads from gravity, the foundation must resist floods, winds, and earthquakes. Where flooding is a problem, the foundation system must be constructed for the possibility that much of the supporting soil may be washed away. The foundation must also be able to resist any debris that may be carried by flood waters.

The forces of wind on a structure can cause severe problems for a foundation. If the structure is not properly anchored to the foundation, the walls can be ripped away by wind, which tries to push a structure not only sideways and upward. Because the structure is securely bonded together at each intersection, wind pressure builds under the roof overhangs and inside the structure and causes an upward tendency. Proper foundation construction resists this upward movement.

Depending on the risk of seismic damage, special design considerations may be required of a foundation. Although earthquakes cause both vertical and horizontal movement, it is the horizontal movement that causes the most damage to structures. The foundation system must be designed so that it can move with the ground yet keep its basic shape. Steel reinforcing and welded wire mesh are often required to help resist or minimize damage from the movement of the earth.

SOIL CONSIDERATIONS

The nature of the soil supporting the foundation must also be considered when reading a foundation plan. A concrete support may at first appear to be oversized until the nature of the soil is taken into account. The texture of the soil and the tendency of the soil to freeze influence the construction of the foundation system.

Soil Texture

The texture is generally classified from bedrock to silt, as shown in Figure 5–1. The soil material

TABLE NO. 29-B—ALLOWABLE FOUNDATION AND LATERAL PRESSURE

CLASS OF MATERIALS[2]	ALLOWABLE FOUNDATION PRESSURE LBS. SQ. FT.[3]	LATERAL BEARING LBS./SQ. FT./FT. OF DEPTH BELOW NATURAL GRADE[4]	LATERAL SLIDING[1] COEF-FICIENT[5]	RESISTANCE LBS./SQ. FT.[6]
1. Massive Crystalline Bedrock	4000	1200	.79	
2. Sedimentary and Foliated Rock	2000	400	.35	
3. Sandy Gravel and/or Gravel (GW and GP)	2000	200	.35	
4. Sand, Silty Sand, Clayey Sand, Silty Gravel and Clayey Gravel (SW, SP, SM, SC, GM and GC)	1500	150	.25	
5. Clay, Sandy Clay, Silty Clay and Clayey Silt (CL, ML, MH and CH)	1000[7]	100		130

[1]Lateral bearing and lateral sliding resistance may be combined.
[2]For soil classifications OL, OH and PT (i.e., organic clays and peat), a foundation investigation shall be required.
[3]All values of allowable foundation pressure are for footings having a minimum width of 12

Figure 5–1 Soil texture affects the ability of soil to resist the loads supported by a foundation. *Reproduced from the 1991 edition of the* Uniform Building Code,™ *copyright 1991, with the permission of the publisher, the International Conference of Building Officials.*

determines the bearing capacity. The soil bearing capacity is the amount of weight a square foot of soil can support. The type of soil can often be determined by the local building department although with commercial construction, a soils engineer is usually required to study the various types of soil at the job site and make recommendations for footing design. The soil bearing values are often listed on a foundation plan so that the construction crew can better understand the size of material selected.

Freezing

In addition to the texture, the tendency of freezing must also be considered. Even in the warmer southern states, ground freezing can be a problem. Soil expands as it freezes and then contracts as it thaws. The foundation must rest on stable soil so that the foundation does not crack.

TYPES OF FOUNDATIONS

The foundation is usually constructed of pilings, continuous footings, or grade beams.

Pilings

Piling foundations are used when conventional trenching equipment cannot be used safely or economically. A *piling* is a type of foundation system that uses a column to support the loads of the structure. Three common piling systems are shown in Figure 5–2. In each system beams are placed below

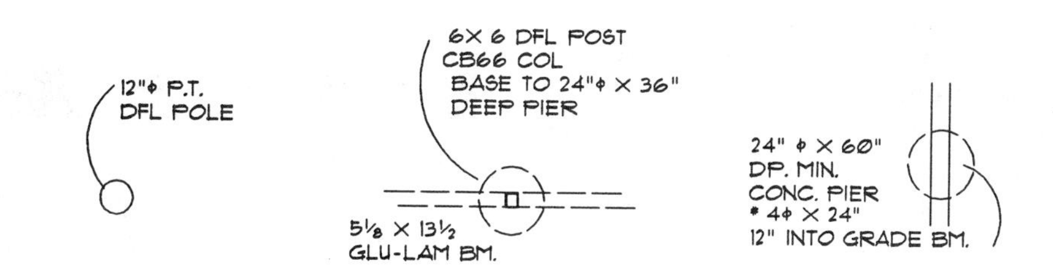

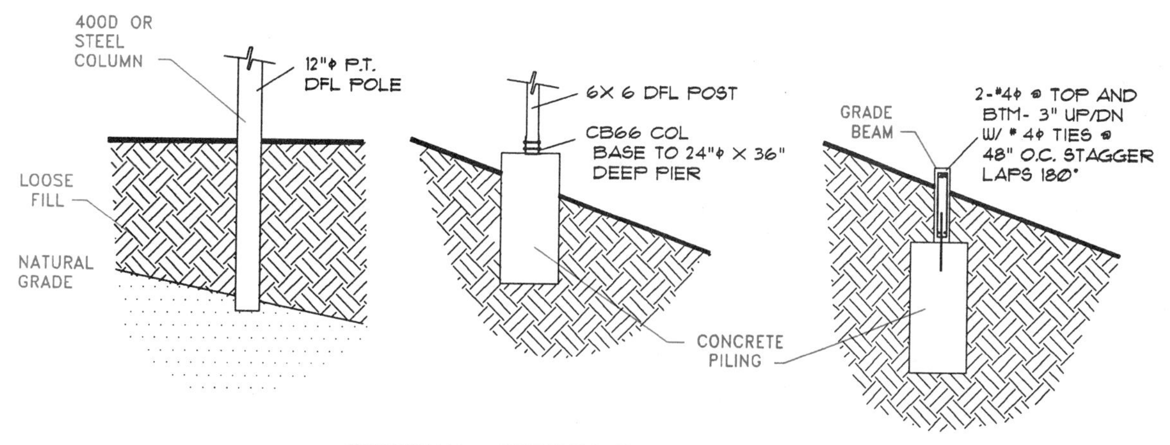

Figure 5–2 Common types of pilings used to transfer loads onto firm soil.

each bearing wall and are supported on a steel column or wood post that is in turn supported by a concrete piling. The piling is usually round and may extend several feet below the ground. In some areas, a pressure-treated wooden pole or steel column may be driven into the ground until it rests on solid soil, and this is also called a piling. This type of foundation system is typically used for hillside or beach residential construction where erosion might be a problem.

Continuous or Spread Foundations

The most typical type of foundation used in residential and light commercial construction is a continuous or spread foundation. This type of foundation consists of a footing and wall. A footing is the base of the foundation system and is used to displace the building loads over the soil. Figure 5–3 shows a typical footings and how they are usually presented on foundation plans. Footings are typically made of poured concrete and placed so that they extend below the freezing level. Figure 5–4 shows common footing sizes and depths as required by building codes. The strength of the concrete must also be specified based on the location of the concrete in the structure and the chance of freezing.

When footings are placed over areas of soft soil or fill material, reinforcement steel is often placed in the footing. Concrete is extremely durable when it supports a load and is compressed but is very weak when in tension. Steel is placed in the footing to help resist the forces of tension.

The material used to construct the foundation wall and the area in which the building is to be located affect how the wall and footing are tied together. If the wall and footing are made at different times, a *keyway* is placed in the footing. When the concrete for the wall is poured, it forms a key by filling in the keyway. If a stronger bond is desired, steel is often used to tie the footing to the foundation wall. Both methods of bonding the foundation wall to the footing can be seen in Figure 5–5. The reinforcing steel, or *rebar*, is usually not shown on a foundation plan but is specified in a note giving the size and

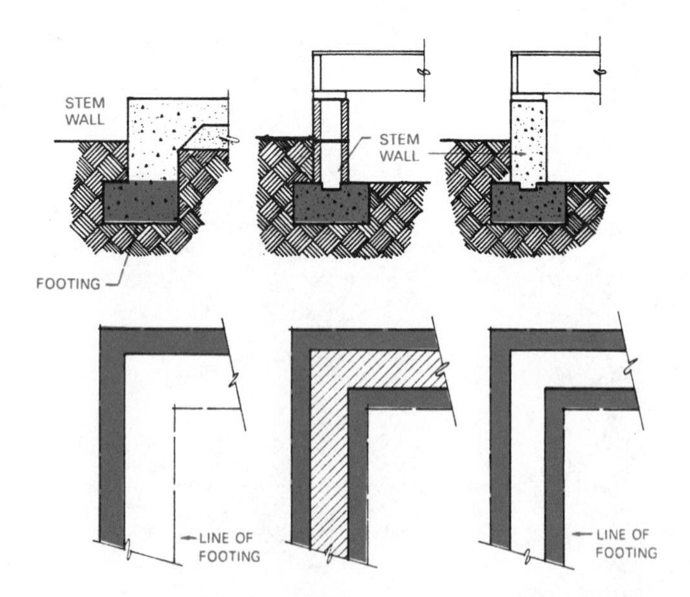

Figure 5–3 Typical representations for stem wall and footings.

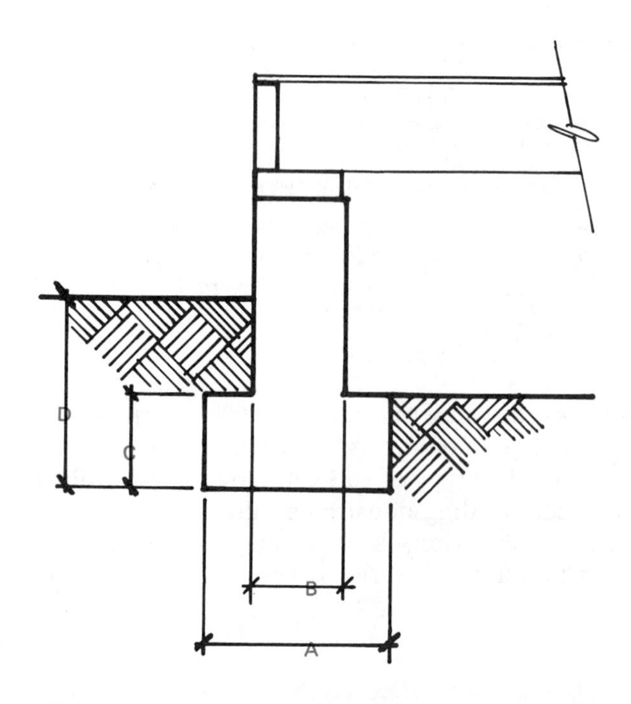

UNIFORM BUILDING CODE FOOTING REQUIREMENTS				
Building Height	Footing Width (A)	Wall Width (B)	Footing Height (C)	Depth into Undisturbed Grade (D)
1 Story	12″	6″	6″	12″
2 Story	15″	8″	7″	18″
3 Story	18″	10″	8″	24″

Figure 5–4 Minimum foundation sizes based on the Uniform Building Code and the Basic National Building Code.

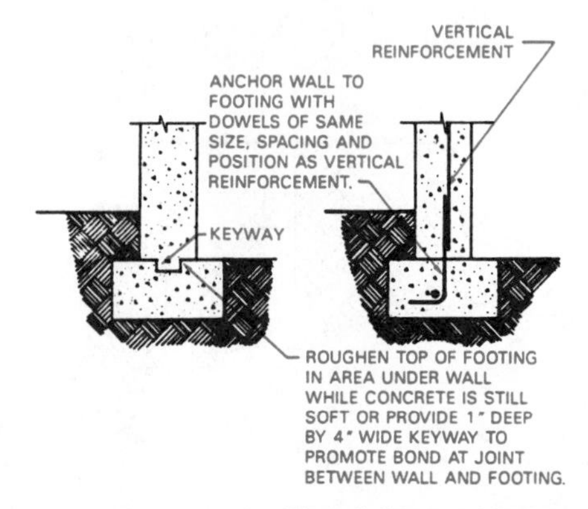

Figure 5–5 Common stem wall and footing intersections using key or steel reinforcing.

the quantity. The size is listed as a number that represents the diameter in eights of an inch from 1/4" through 2 1/2". A #4 bar is 4/8" in diameter, or 1/2".

Fireplace Footings

A masonry fireplace needs to be supported on a footing. Building codes usually require the footing to be a minimum of 12 in. deep. The footing is required to extend 6 in. past the face of the fireplace on each side. Figure 5–6 shows how fireplace footings are represented on the foundation plan.

Veneer Footings

If masonry veneer is used, the footing must be wide enough to provide adequate support for the veneer. Depending on the type of veneer material to be used, the footing is typically widened by 4 to 6 in. Figure 5–7 shows common methods of providing footing support for veneer.

Foundation Wall

The foundation wall is the vertical wall that extends from the top of the footing to the first floor

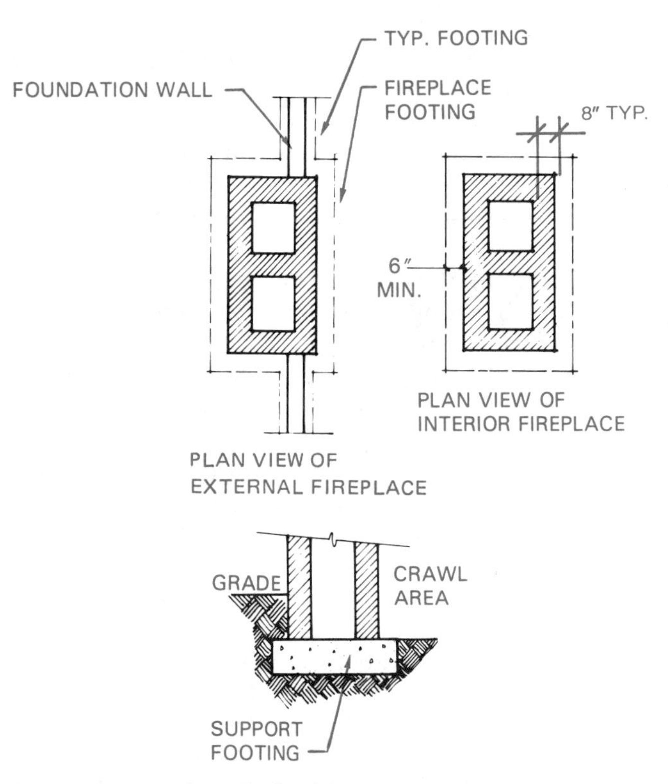

Figure 5–6 Typical methods of representing masonry chimney footings on a foundation plan.

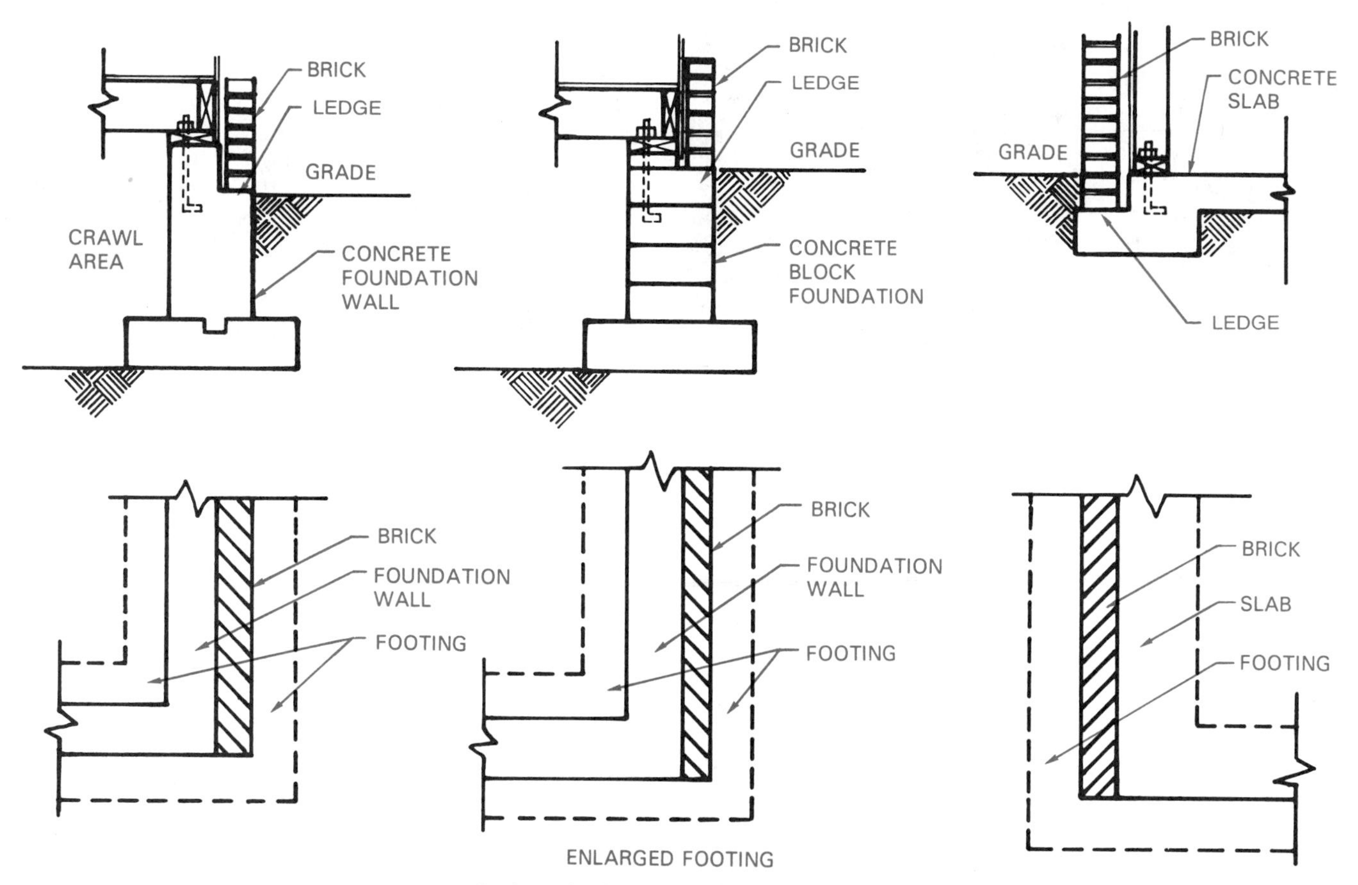

Figure 5–7 Standard methods of providing support for veneered walls.

level of the structure, as shown in Figure 5–3. The foundation wall is usually centered on the footing to help equally disperse the loads being supported. The height of the wall extends a minimum of 8 in. above the ground and reflects the minimum distance required between woodframing members and grade. The required width of the wall varies depending on the code used.

The top of the foundation wall must be level. When the building lot is not level, the foundation wall is often stepped. This helps reduce the material needed to build the foundation wall. As the ground slopes downward, the height of the wall is increased, as shown in Figure 5–8. Foundation walls may not step more than 24 in. in one step, and wood framing between the wall and any floor being supported may not be less than 14 in. in height. The wall height may also change for a room with a sunken floor. Figure 5–9 shows how wall changes are often shown on a foundation plan.

The foundation wall is usually made from either poured concrete or concrete blocks. Steel anchor bolts

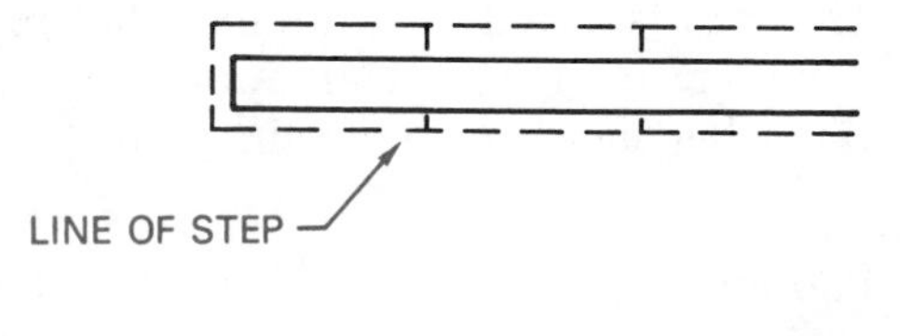

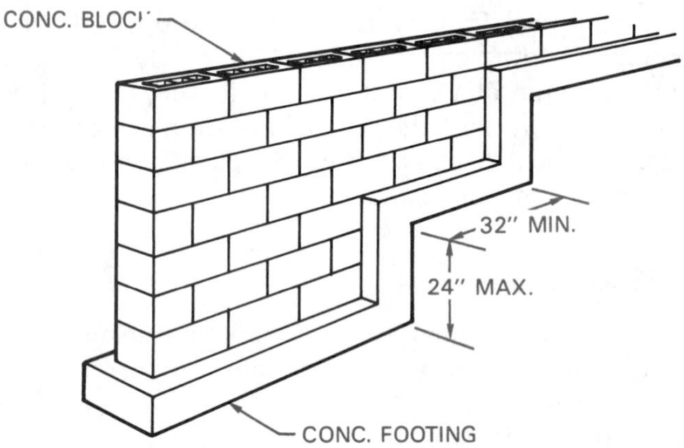

Figure 5–8 The footing and stem wall are often stepped on sloping lots to reduce framing and provide level support for the floor framing members.

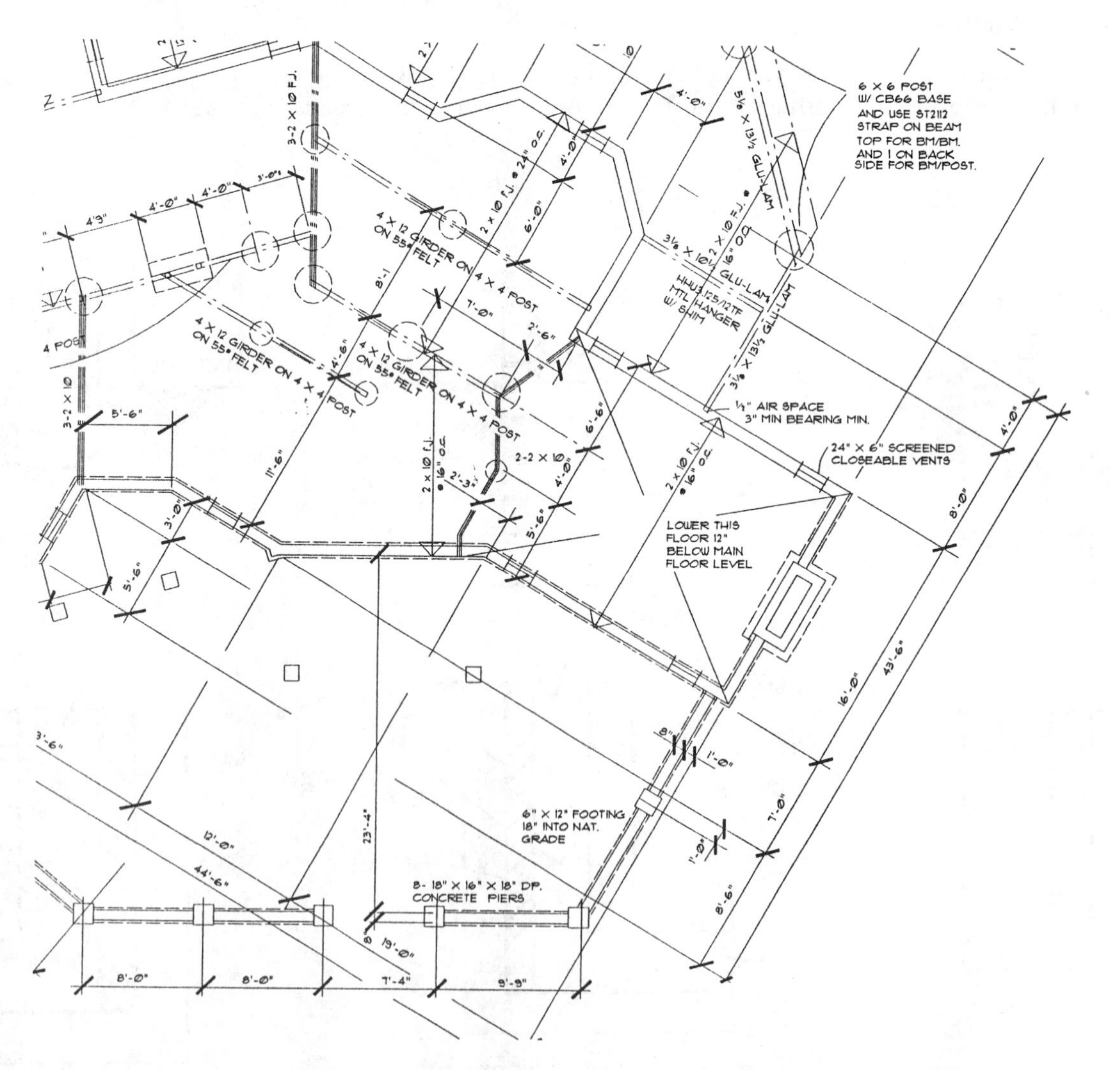

Figure 5–9 Floor elevation is often varied on the foundation. The floor on the right side is 12" lower than the floor on the left.

are placed in the top of the wall to secure the wood mudsill to the concrete. If concrete blocks are used, the cell in which the bolt is placed must be filled with grout. The mudsill is required to be pressure treated or made of some other water-resistant wood so that it does not absorb moisture from the concrete. A 2 in. round washer is placed over the bolt projecting through the mudsill before the nut is installed to increase the holding power of the bolt. The mudsill and anchor bolts are not drawn on a foundation plan but are specified in a local note near the stem wall. Examples of each can be seen in Figure 5–10.

The mudsill must also be protected from termites in many parts of the country. Among the most common methods of protection are the use of metal caps between the mudsill and the wall, chemical treatment of the soil around the foundation, and chemically treated wood near the ground. The metal shield is not drawn on a foundation plan but is specified in a local note near the stem wall.

If the house is to have a wood flooring system, some method of securing support beams to the foundation must be provided. Typically a cavity or beam pocket is built into the foundation wall. The cavity provides a method of supporting the beam and helps tie the floor and foundation system together. A 3 in. minimum amount of bearing surface must be provided for the beam. A ½ in. airspace is provided around the beam in the pocket for air circulation. Air must also be allowed to circulate under the floor system. To provide ventilation under the floor, vents must be set into the foundation wall. If an access opening is not provided in the floor, an opening must be provided in the foundation wall for access to the crawl space under the floor. Figure 5–10 shows how the crawl access, vents, and girder pockets are represented on a foundation plan.

In addition to supporting the loads of the structure, the foundation walls also must resist the lateral pressure of the soil pressing against the wall. When

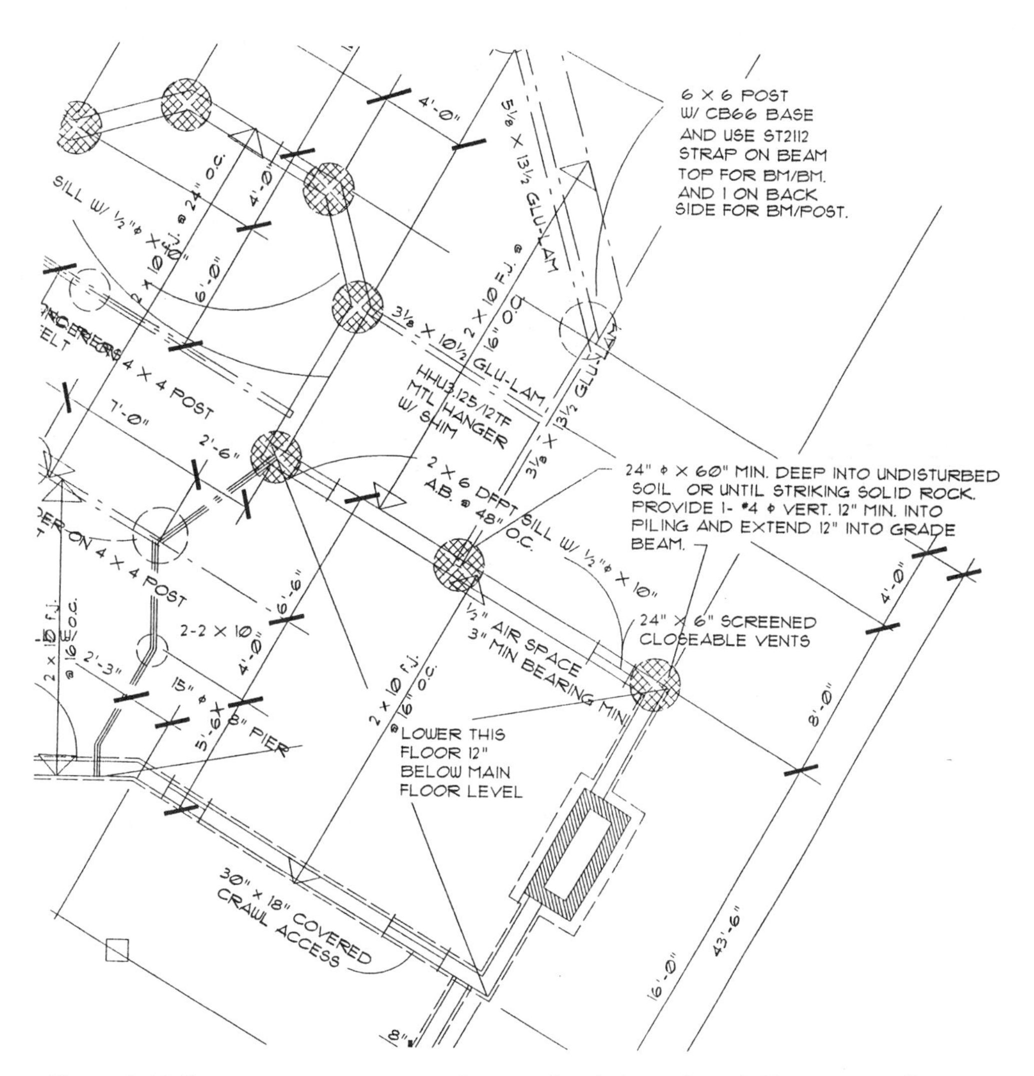

Figure 5–10 Common components, such as anchor bolts and mudsills, are typically not shown but are specified in a note. Framing members, vents, and access are usually drawn and noted.

the wall is over 24 in. in height, vertical steel is usually added to the wall to help reduce tension in the wall. Horizontal steel is also added to foundation walls in same seismic zones to help strengthen the wall. Wall steel is not shown on the foundation plan but is specified in a note similar to that used for footing steel.

Retaining Walls

Retaining or basement walls are primarily made of concrete blocks or poured concrete. The material used depends on labor trends in your area. Regardless of which material is used, basement walls serve the same functions as the shorter foundation walls. Because of the added height, the lateral forces acting on the side of these walls are magnified. This lateral soil pressure tends to bend the wall inward, thus placing the soil side of the wall in compression and the interior face of the wall in tension. To resist this tensile stress, steel reinforcing may be required by the building department. On a foundation plan this steel is specified in a note similar to the note used for wall steel.

The footing for a retaining wall is usually 16 in. wide and either 8 or 12 in. deep. The depth depends on the weight to be supported. Steel is typically extended from the footing into the wall. At the top of the wall, anchor bolts are placed in the wall in the same method as with a foundation wall. Anchor bolts for retaining walls are typically placed much more closely than for shorter walls, and metal angles are added to the anchor bolts to make the tie between the wall and the floor joist extremely rigid. Figure 5–11 shows how the connection might be represented on a foundation plan.

To reduce soil pressure next to the footing, a drain is installed. The drain is set at the base of the footing to collect and divert water from the face of the wall. The area above the drain is filled with gravel so that subsurface water percolates, easily down to the drain and away from the wall. By reducing the water content of the soil, the lateral pressure on the wall is reduced. The drain is usually not drawn on the foundation plan but must be specified in a note.

The basement wall should be waterproofed to help reduce moisture from passing through the wall into the living area. Adding windows to the basement can help cut down the moisture content of the basement. This sometimes requires adding a window well to pre-

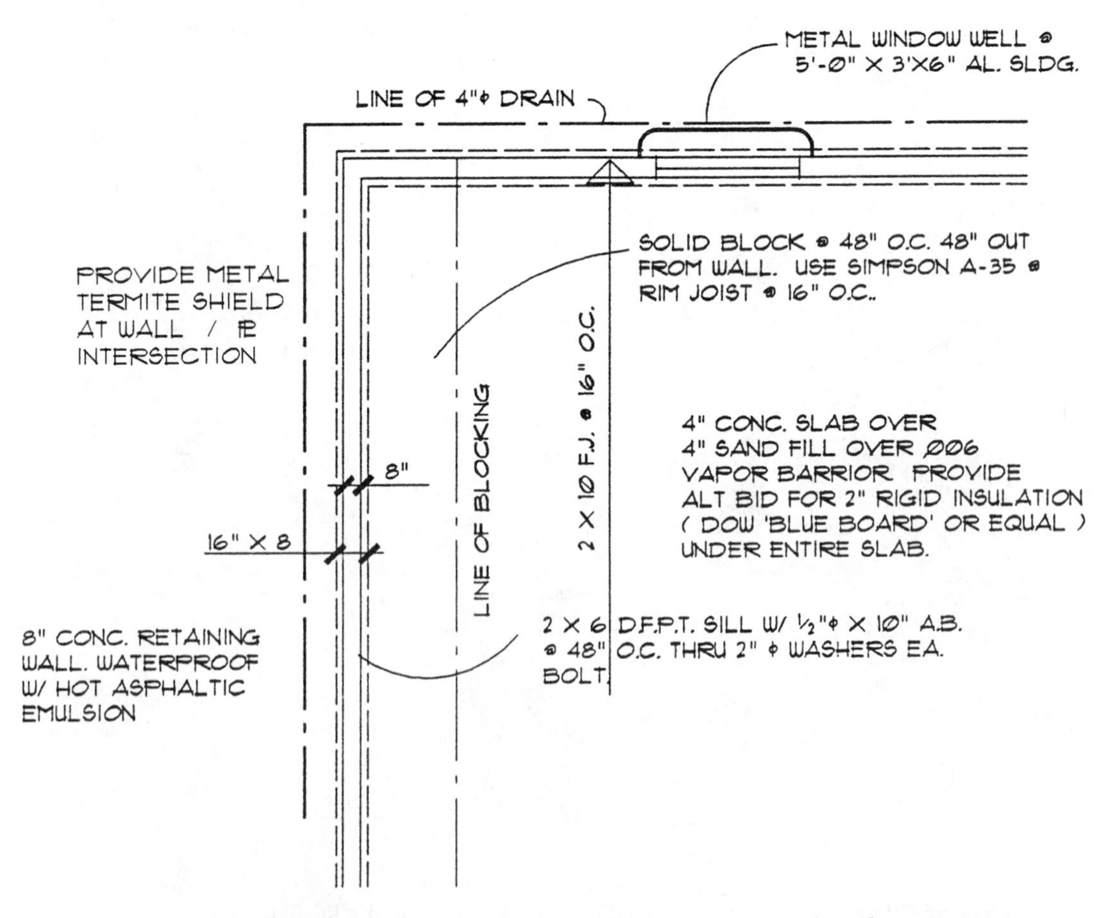

Figure 5–11 Window wells to allow light and emergency egress for a basement.

vent the ground from being pushed in front of the window. Figure 5–11 shows how a window well, or *areaway* as it is sometimes called, can be represented.

Restraining Walls

When a structure is built on a sloping site, the masonry wall may not need to be full height. Although less soil is retained than with a full-height wall, more problems are encountered. Figure 5–12 shows the tendencies for bending for this type of wall. Because the wall is not supported at the top by the floor, the soil pressure must be resisted through the footing. This requires a larger footing than for a full-height wall to resist the tendency to overturn. Depending on the slope of the ground being supported, a key may be required, as seen in Figure 5–13. The key is added to the bottom of the footing to help keep it from sliding as a result of soil pressure against the wall. Generally the key is not shown on the foundation plan but is shown in a detail of the wall.

Interior Supports

Foundation walls and footings are typically part of the foundation system that supports the exterior shape of the structure. Interior loads are generally supported on spot footings, or *piers*. Pier depth is generally required to match that of the footings. The placement of piers is determined by the type of floor system to be used. The size of the pier depends on the load being supported and the soil bearing pressure. Piers are usually drawn on the foundation plan with dotted lines, as shown in Figure 5–14.

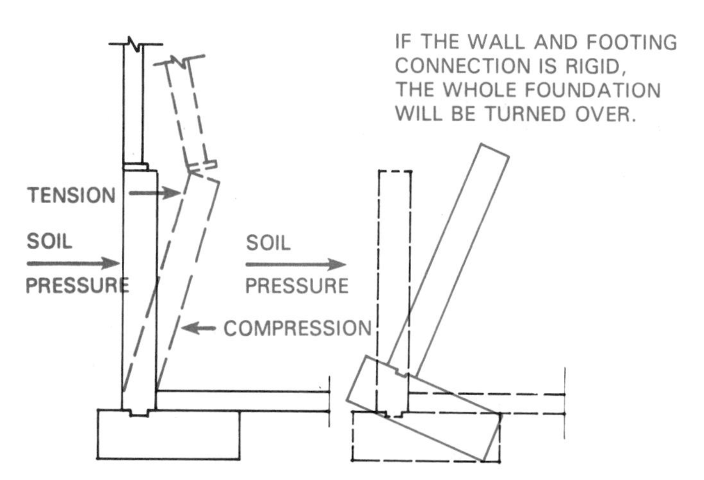

Figure 5–12 Stresses acting on a restraining wall. When a wall is not held in place at the top, the wall tends to tip inward.

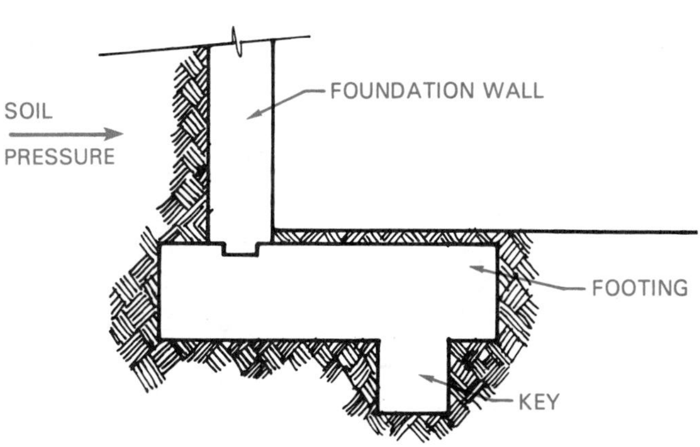

Figure 5–13 A restraining wall may require a key to keep the soil pressure from tipping the wall and footing.

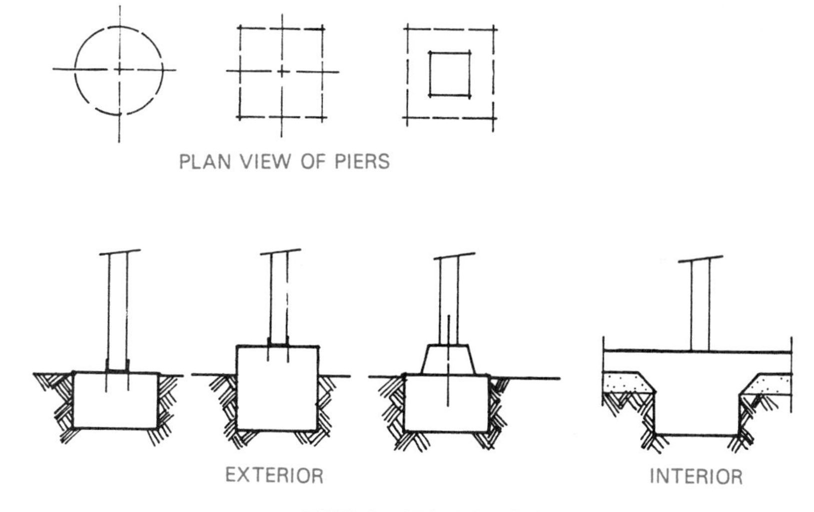

Figure 5–14 Concrete piers are used to support interior loads.

Metal Connectors

Metal connectors are often used at the foundation level. How the connector is used determines how it is specified. Figure 5–15 shows how these connectors might be specified on the foundation plan.

DIMENSIONING FOUNDATION COMPONENTS

The line quality for the dimension and leader lines is the same as was used on the floor plan. Jogs in the foundation wall are dimensioned using the same

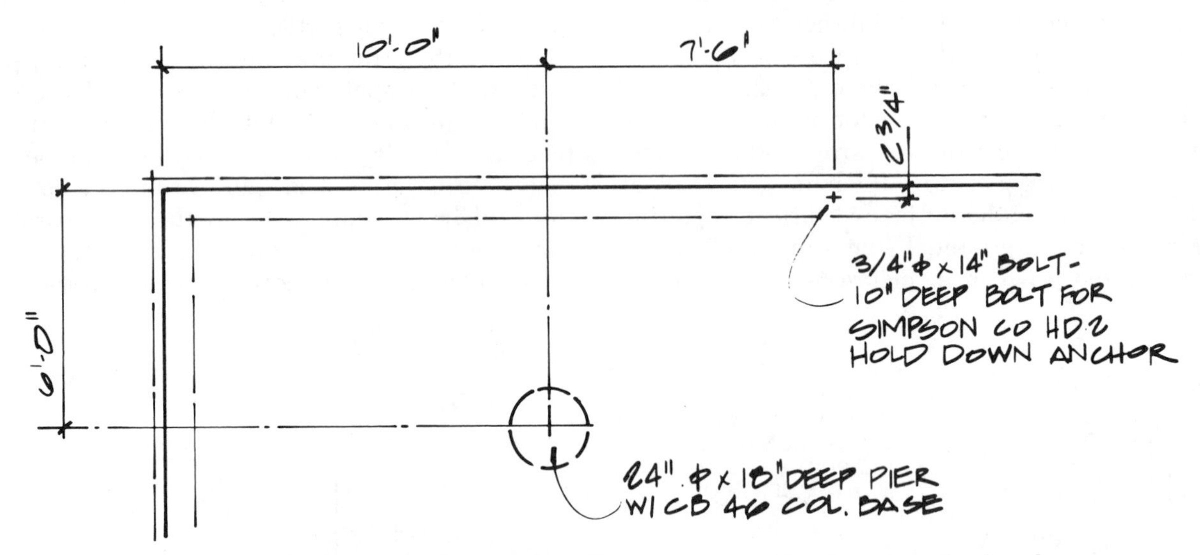

Figure 5–15 The location and type of metal connector must be shown on the foundation plan.

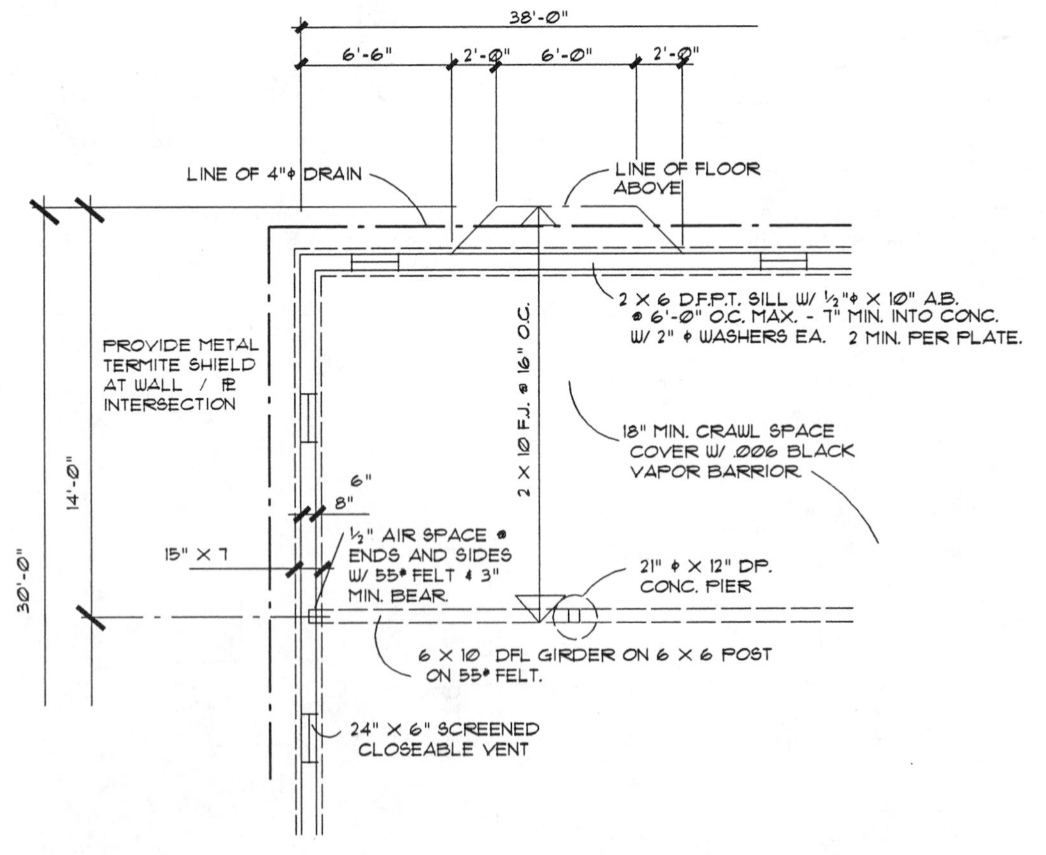

Figure 5–16 Typical dimensions required to locate a bay window or a cantilevered floor.

methods used on the floor plan. Most dimensions for major shapes are exactly the same as seen on the floor plan. The only variation is typically where a floor or bay window might cantilever past the foundation, as seen in Figure 5–16.

A different method is used to dimension the interior walls, however. Foundation walls are dimensioned from face to face rather than from face to center as on the floor plan. Footings widths are usually dimensioned center to center. Each type of dimension can be seen in Figure 5–17.

FLOOR SYSTEMS

The foundation plan not only shows the concrete footings and walls, but also the members that are used to form the floor. Two common types of floor systems are typically used in residential construction. These include floor systems with a crawl space or cellar below the floor system, and a floor system built at grade level. Each has its own components and information that must be put on a foundation plan.

ON GRADE FOUNDATIONS

A concrete slab is often used for the floor system of either residential or commercial structures. A concrete slab provides a firm floor system with little or no maintenance and generally requires less material and labor than conventional wood floor systems. The floor slab is usually poured as an extension of the foundation wall and footing as shown in Figure 5–18. Other common methods of pouring the foundation and floor system are shown in Figure 5–19.

A 3½ in. concrete slab is the minimum thickness allowed by major building codes for residential floor slabs. The slab is only used as a floor surface and is not used to support the weight of the walls or roof.

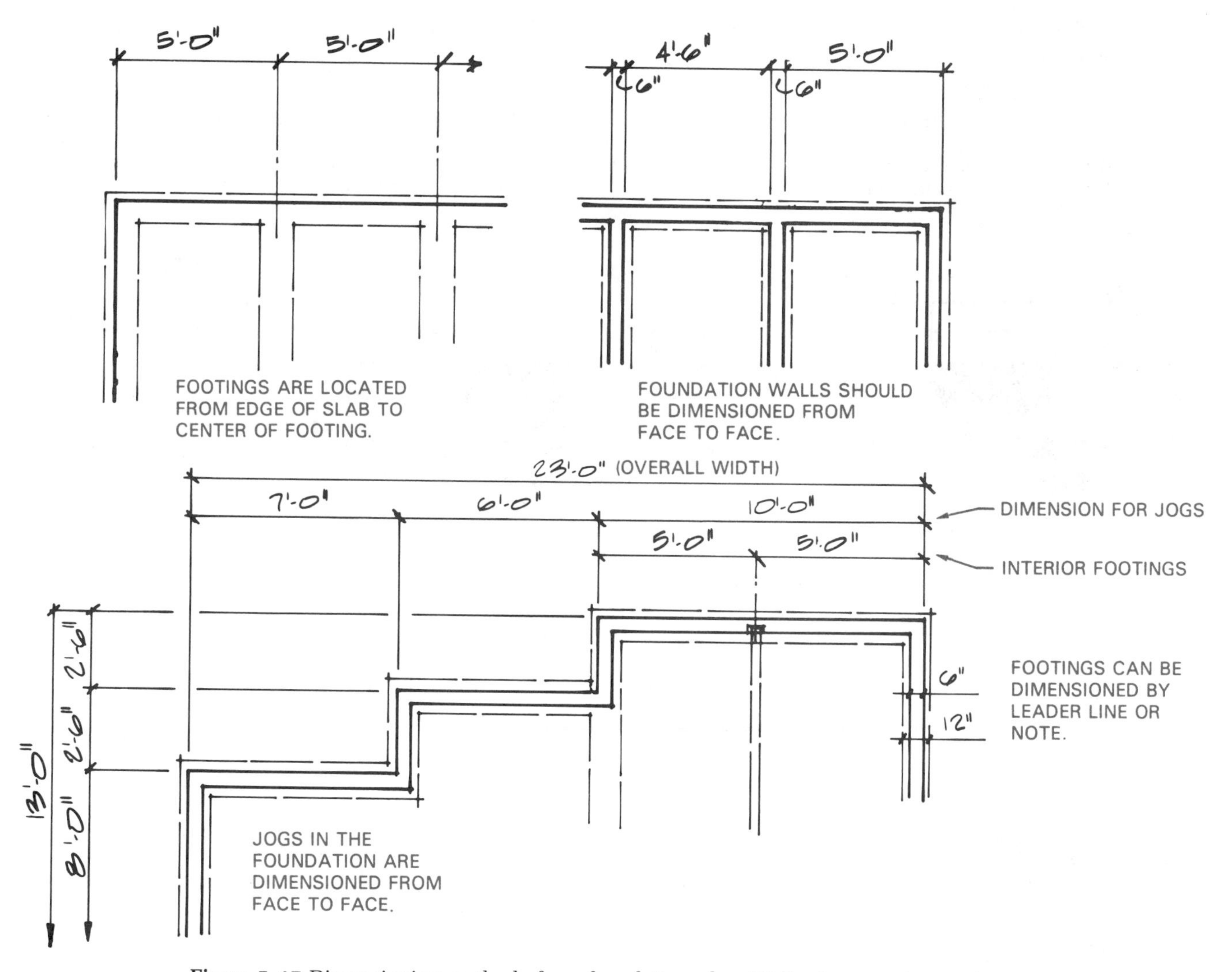

Figure 5–17 Dimensioning methods for a foundation plan. Walls above grade are always dimensioned to an edge. Footings located below grade are usually dimensioned to the center.

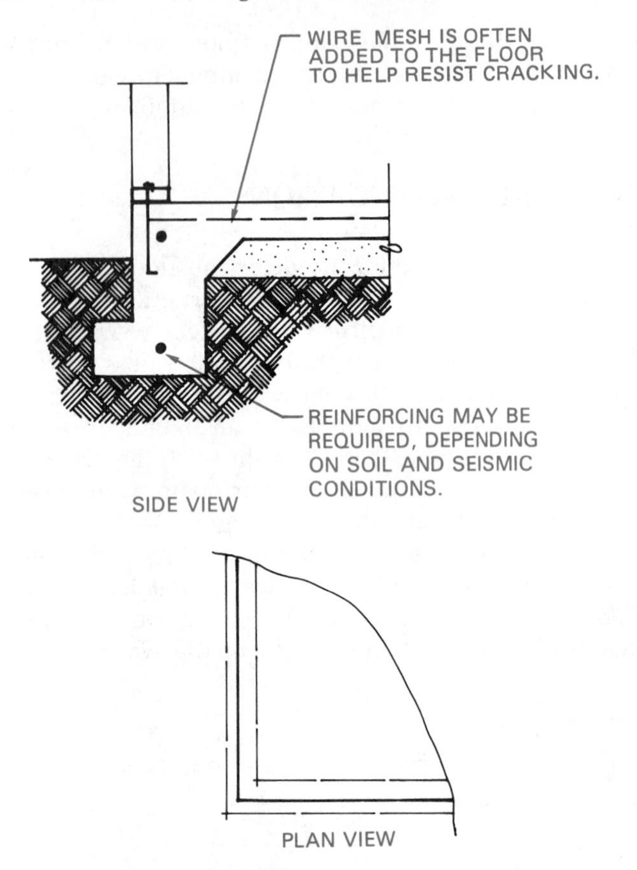

Figure 5–18 On-grade concrete slabs may be formed at the same time the footing and stem wall are poured.

If loadbearing walls must be supported, the floor slab must be thickened as shown in Figure 5–20. If a load is concentrated in a small area, a pier may be placed under the slab to help disperse the weight as seen in Figure 5–14. Depending on the size of the slab, joints may be placed in the slab. Joints are used so the construction crew can pour several small slabs instead of one large one and to help reduce the spread of cracks throughout the slab. By using control or expansion joints, any cracks that develop in one area do not spread to other areas of the slab.

The slab may be placed above, below, or at grade level. Slabs below grade are most commonly used in basements. When used at grade, the slab is usually placed 3 in. above finish grade to keep structural wood away from ground moisture.

Occasionally an exterior walkway or garage requires a concrete slab that cannot be built at grade. Hillside construction often requires a garage to be built over living space. In this situation, the garage has a floor made of light-weight concrete poured over a plywood or ribbed metal support base. The weight of the concrete depends on the amount of air pumped into the mixture during the manufacturing process. This material is usually noted on the foundation plan but not drawn. The foundation plan usually shows the supports required to support the increased weight of the floor. An example of an above-ground founda-

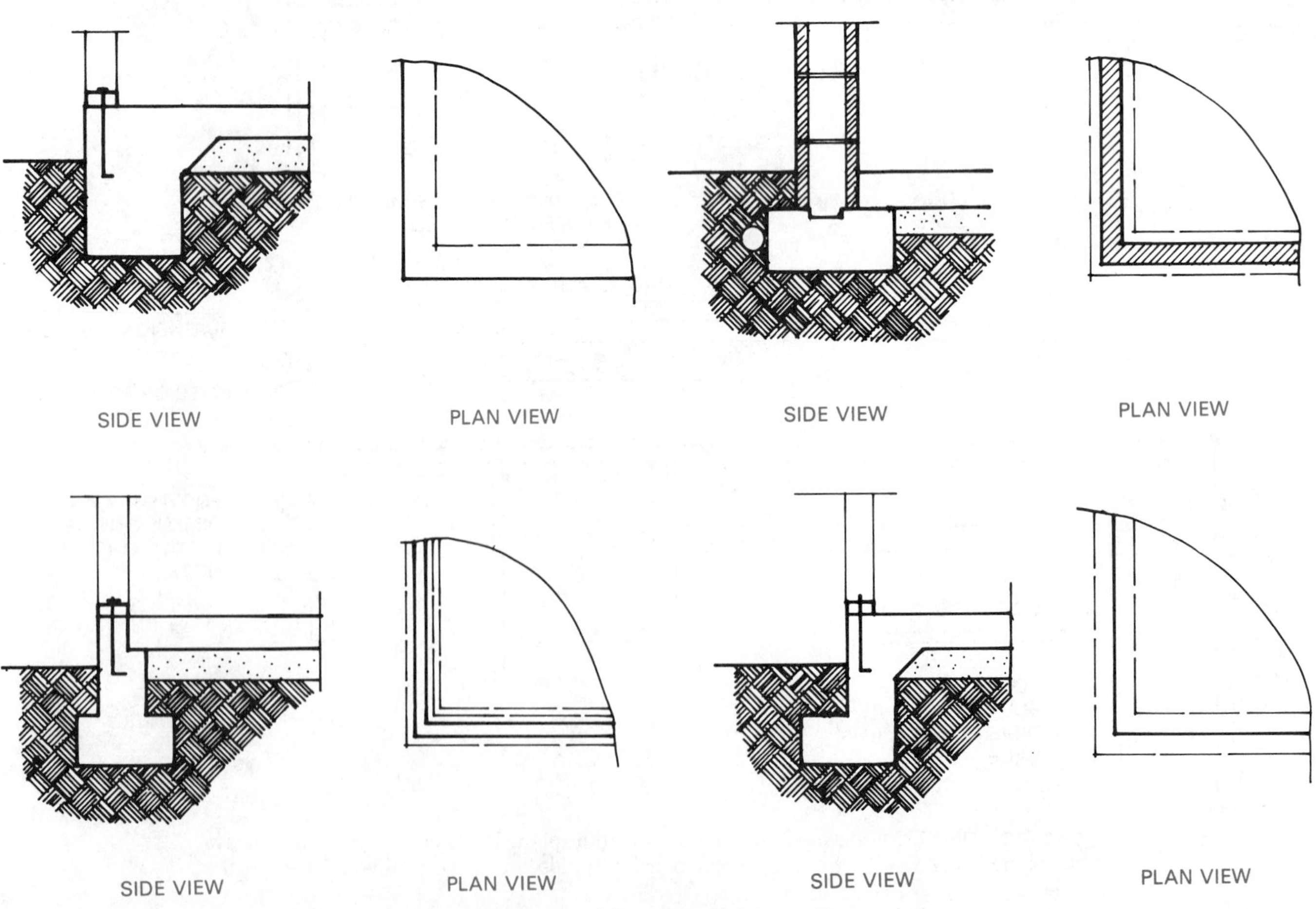

Figure 5–19 On-grade slab and wall intersection variations.

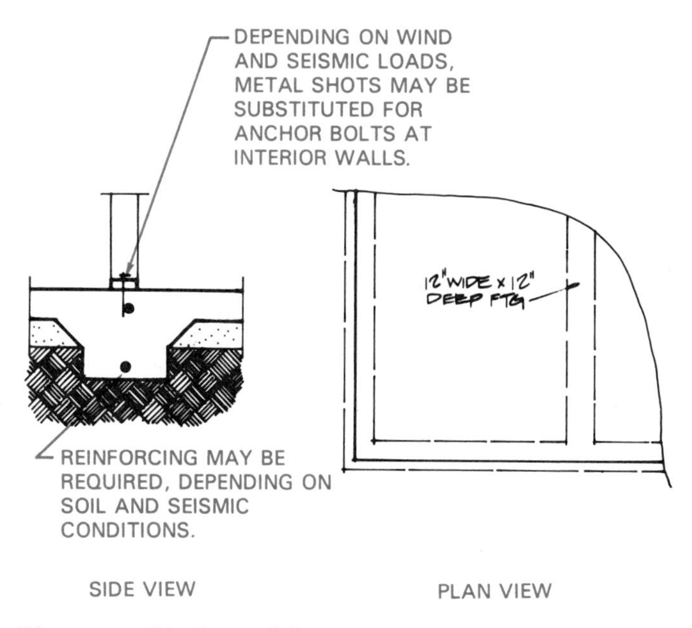

Figure 5–20 Footing and floor intersection at an interior load-bearing wall.

tion can be seen in Figure 5–21. The construction would also be shown in details or sections similar to those in Figure 5–22.

It must be placed on a 4 in. minimum base of compacted sand or gravel fill. The fill material provides a level base for the concrete slab and helps eliminate cracking in the slab caused by settling of the ground under the slab. The fill material is not shown on the foundation plan but is specified with a note.

When the slab is placed on more than 4 in. of uncompacted fill, mesh should be placed in the slab to help resist cracking. Typically a steel mesh of number 10 wire in 6 in. grids is used for residential floor slabs. Generally the mesh or steel is not shown on the foundation plan but is specified by note.

In most areas, the slab is required to be placed over 6 mil polyethylene sheet plastic to protect the floor from ground moisture. An alternative is to use 55 lb rolled roofing since visqueen cannot be used with gravel fill. This vapor barrier is not drawn but specified with a general note on the foundation plan.

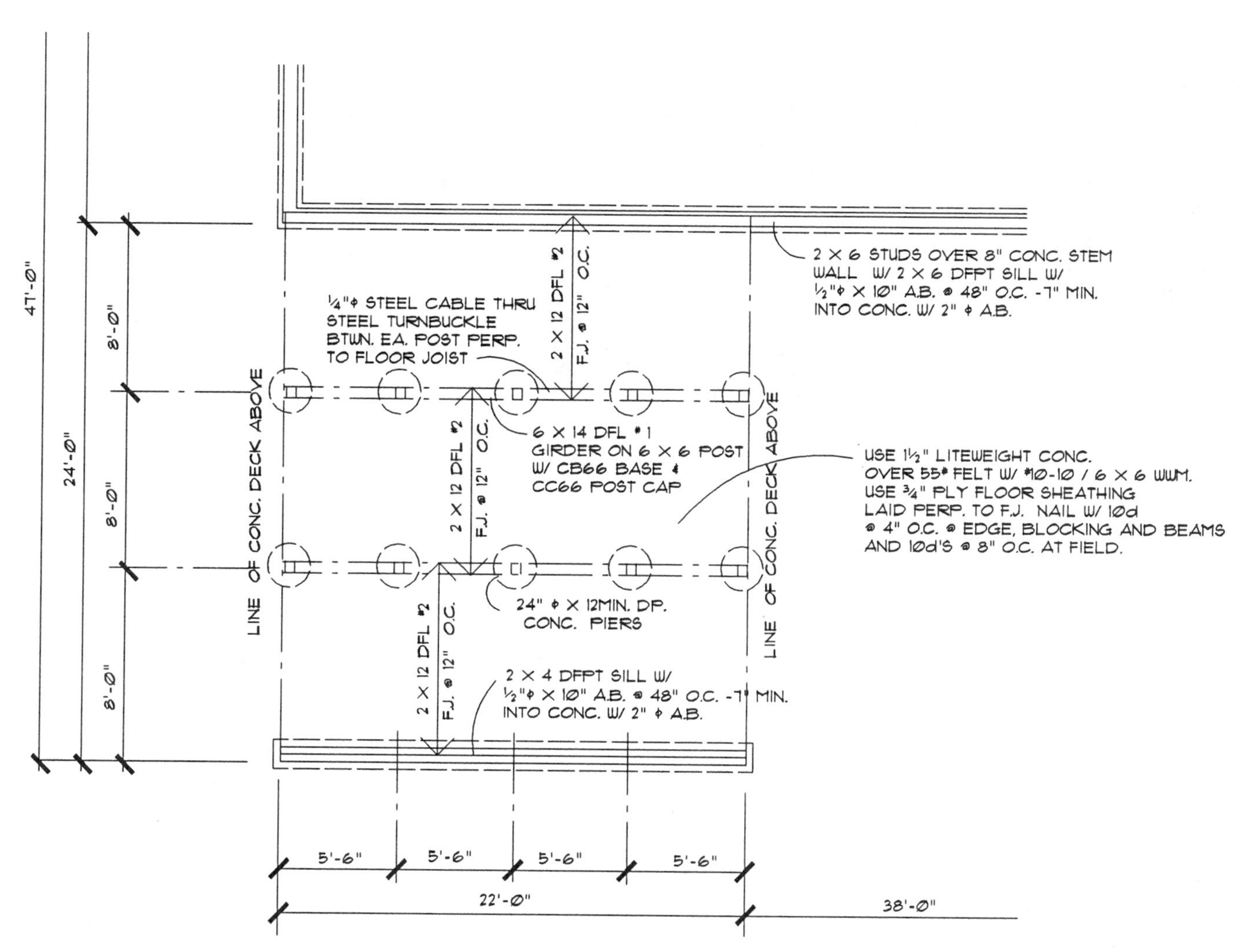

Figure 5–21 An above-grade concrete slab requires the framing members supporting the slab to be shown although the actual slab is not shown.

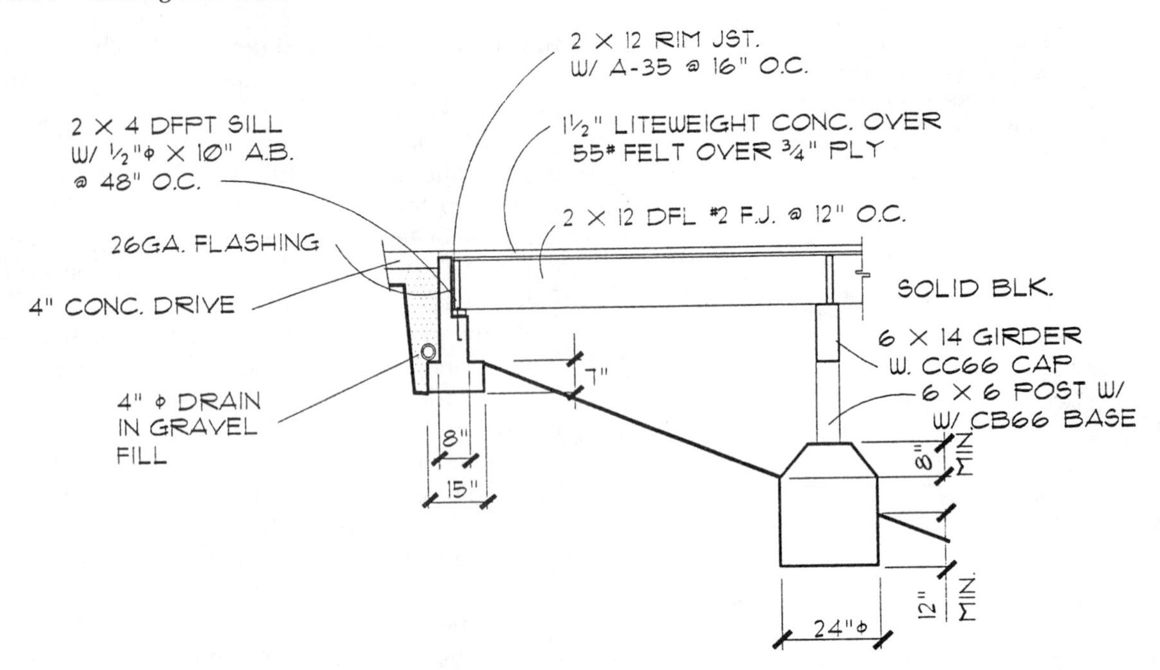

Figure 5–22 An above-ground slab is typically shown in a detail placed near the foundation plan.

Depending on the risk of freezing, a layer of rigid insulation may be placed under the slab at the perimeter to help reduce heat loss from the floor into the ground. The insulation is not shown on a foundation plan but is called out in a note, as seen in Figure 5–11. All items typically specified or drawn on a concrete foundation plan can be seen in Figure 5–23. An example of a concrete slab can be seen in Problem 5–1.

Plumbing and heating ducts must be placed under the slab before the concrete is poured. On residential plans plumbing is usually not shown on the foundation plan. Generally the skills of the plumbing contractor are relied on for the placement of required utilities. If heating ducts are to be run under the slab, they are usually drawn on the foundation plan. Drawings for commercial construction usually include both a plumbing and a mechanical plan.

CRAWL SPACE FLOOR SYSTEMS

The crawl space is the area formed between the floor system and the ground. Building codes require a minimum of 18 in. from the bottom of the floor to the ground and 12 in. from the bottom of beams to the ground. Two common methods of providing a crawl space below the floor are the conventional method using floor joists and the post-and-beam system. An introduction to each system is needed to complete the foundation. Both floor systems are discussed in detail in relationship to the entire structure in Chapter 7.

Joist Floor Framing

The most common method of framing a wood floor is with wood members called *floor joists.* Floor joists are used to span between the foundation walls. The floor joists are usually placed at 16 in. on center, but the spacing of joists may change depending on the span and the load to be supported.

To construct this type of floor, a pressure-treated sill is bolted to the top of the foundation wall with the anchor bolts that were placed when the foundation was formed. The floor joists can then be nailed to the sill. With the floor joists in place, plywood floor sheathing is installed to provide a base for the finish floor.

When the distance between the foundation walls is too great for the floor joists to span, a girder is used to support the joists. A *girder* is a horizontal load-bearing member that spans between two or more supports at the foundation level. Either a wood or steel member may be used for the girder and for the support post. The girder is usually supported in a beam pocket where it intersects the foundation wall. A concrete pier is placed under the post to resist settling. Figure 5–24 shows methods of representing the girders, posts, piers, and beam pocket on the foundation plan. An example of a joist foundation can be seen in Problem 5–2.

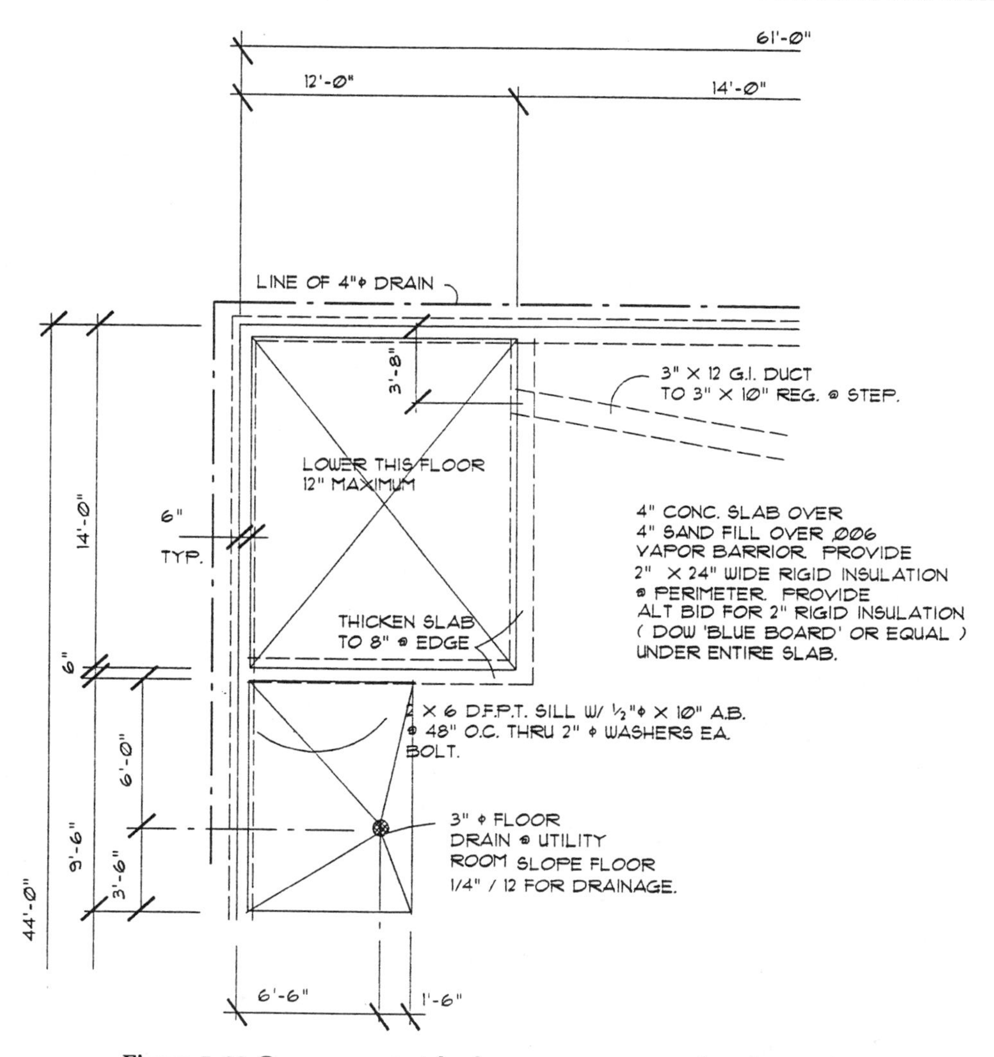

Figure 5–23 Common materials shown on a concrete foundation plan.

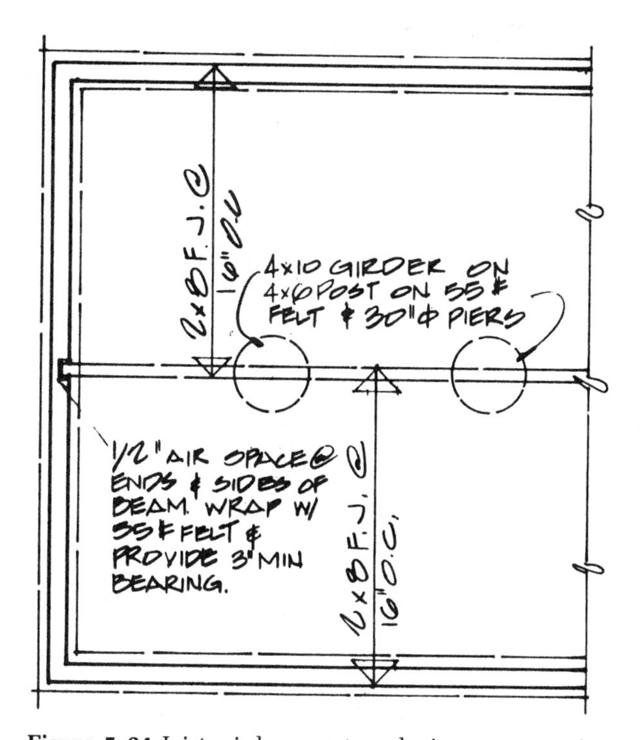

Figure 5–24 Joist, girders, post, and pier representation.

Post-and-Beam Floor Systems

A post-and-beam floor system is built using a standard foundation system. Rather than having floor joists span between the foundation walls, a series of beams is used to support the subfloor. The beams are usually placed at 48 in. on center, but the spacing can vary depending on the size of the floor decking to be used. Generally 2 in. thick material is used to span between the beams to form the subfloor.

The beams are supported by wooden posts where they span between the foundation walls. Posts are usually placed about 8 ft on center, but spacing may vary depending on the load to be supported. Each post is supported by a concrete pier. An example of a post-and-beam foundation can be seen in Problem 5–3.

Combined Floor Methods

Floor and foundation methods may be combined depending on the building site. This is typically done on partially sloping lots when part of a structure may be constructed with a slab and part of the structure with a joist floor system, as seen in Problem 5–4.

One component typically used when floor systems are combined is a ledger. A ledger is used to provide support for floor joist and subfloor when they intersect the concrete. Unless felt is placed between the concrete and the ledger, the ledger must be pressure-treated lumber. The ledger can be shown on the foundation plan as shown in Figure 5–25.

READING A FOUNDATION PLAN

Although you have been exposed to the basic concepts of foundation plans, interpreting a plan may still be a bit of a challenge. To understand a foundation plan fully, you must view other plans first. Check the floor plan or roof framing plan to see what type of framing system was used on the roof. If trusses where used, you can expect no bearing walls throughout the center of the house, and thus no girders are required to support roof loads.

Skim the plan to determine the type of floor framing system being used. Become familiar with above-grade concrete, below-grade concrete, and girders and their supports. A quick viewing of the sections may be helpful at this point.

Once the lower floor framing system is determined, you should begin to see members characteristics of the system. Look for general notes that may set standards for the entire job that are different from the norm. Once you understand the overall components of the plan, look at the local notes to see specific requirements for this specific job. Don't make the mistake of assuming that all post-and-beam or joist foundations are always the same.

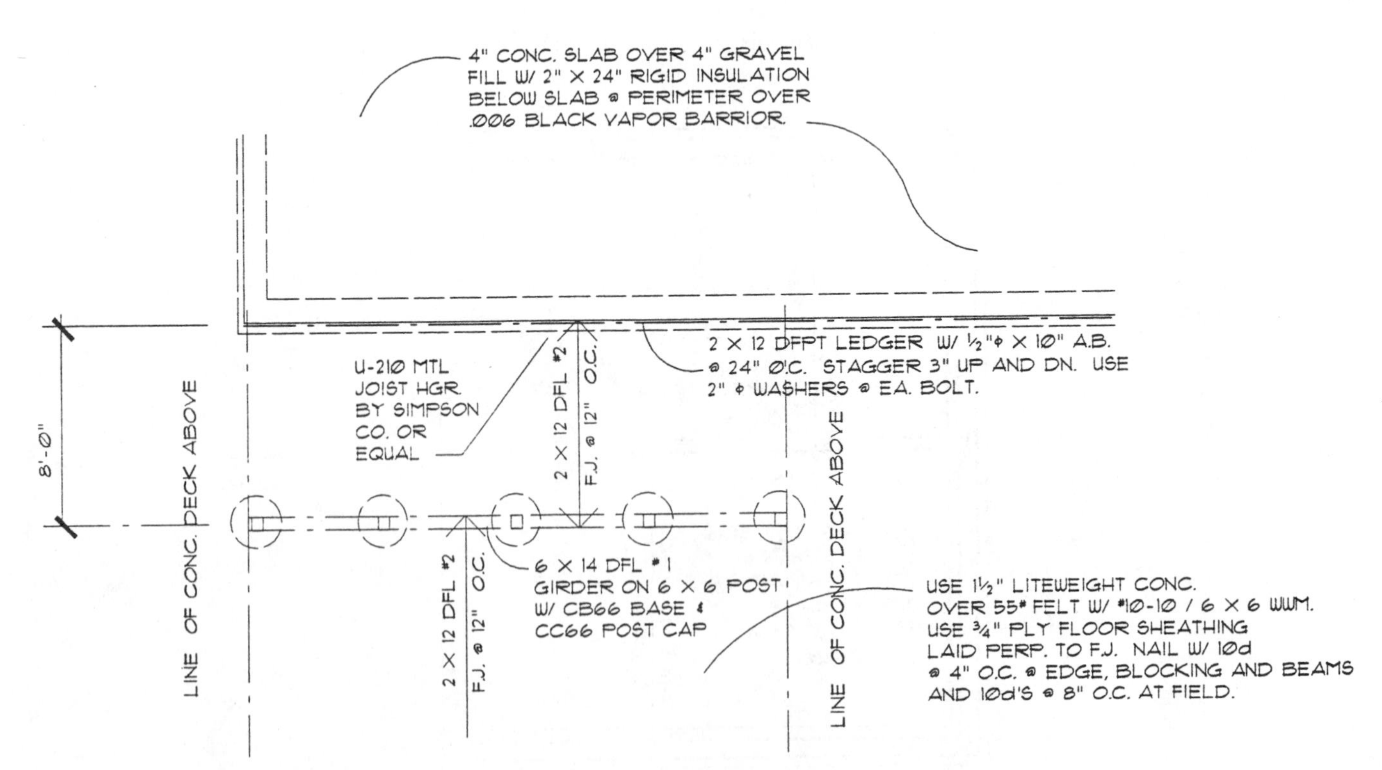

Figure 5–25 A wood ledger is used to connect a wood floor to a concrete slab. The line used to represent the ledger varies with each set of plans.

CHAPTER 5 TEST

1. What are the two major parts of the foundation system? ______________________________.
2. List five forces that a foundation must withstand. ______________________________.
3. How can the soil texture influence a foundation? ______________________________.
4. List the two major types of material used to build foundation walls. ______________________________
5. What type of stress requires steel to be placed in footings? ______________________________
6. What size footing should be used with a basement wall? ______________________________
7. What influences the size of piers? ______________________________.
8. Describe the difference between a retaining and restraining wall. ______________________________.
9. What is considered common soil bearing pressure? ______________________________.
10. Is steel placed near the soil or air side of a retaining wall? ______________________________.
11. What is the common spacing for girders in a post-and-beam floor system? ______________________________.
12. What can be done to reduce the size of a concrete slab and aid workers? ______________________________.
13. What keeps floor joists from lifting off the foundation? ______________________________.
14. What is the common spacing for floor joists. ______________________________.
15. What is the common spacing for piers supporting the girders in a post-and-beam foundation and a joist foundation? ______________________________.

CHAPTER 5 EXERCISES

PROBLEM 5–1. Use the drawing shown on page 120 to answer the following questions. Place the answer in the space provided.

1. What type of floor system is used? ______________________________
2. What size floor drain will be installed? ______________________________
3. What size will the garage slab be? ______________________________

4. What is the soil bearing pressure of the job site? ______
5. Give the size of the fireplace footing. ______
6. What is the depth of the living room recess? ______
7. How far are the footings to extend into the ground for the two-story and for the one-story potions of the house? ______
8. What is the overall size of the structure? ______
9. How many piers are required to support roof or wall loads within the structure? Give their sizes. ______
10. How many piers are required to support roof or wall loads on the exterior of the structure? Give their sizes. ______

PROBLEM 5–2. Use the drawing shown on page 121 to answer the following questions.

1. What type of floor system is used? ______
2. What is the soil bearing pressure of the job site? ______
3. What is the overall size of the garage? ______
4. How many piers are required to support roof or wall loads within the structure? Give their size. ______
______.
5. List the compressive strengths of concrete to be used on this structure. ______
______.
6. What is the spacing of the anchor bolts? ______
7. How many vents need to be supplied for this residence? ______
8. Determine the girder size and support required. ______
______.
9. What size members are used to span between the girder and the stem wall? ______
10. What type and grade of framing lumber will be used? ______
11. Show your work and determine the area of the crawl space if 6" walls were used. ______
______
______
______
______.
12. What will be the total amount of slope in the garage slab? ______
13. How much concrete is required to pour the slab? ______
14. If five extra joists are ordered for rim joist and blocking, how many 16' long floor joists need to be ordered? ______
______
______
15. How many linear feet of foundation will be poured for one level and two level? ______
______.

PROBLEM 5–3. Use the drawing shown on page 122 to answer the following questions.

1. What type of floor system is used? ______
2. What is the soil bearing pressure of the job site? ______
3. How many piers are required to support roof or wall loads within the structure? Give their size. ______
______.
4. Determine the girder size and support required to support the floor loads. ______
______
______.

5. What size door is on the left side of the garage? ______
6. What size of floor joist will be used with this floor system? ______
7. What size washers will be used with the anchor bolts? ______
8. What type of fill material will be used under the garage slab? ______
9. What is the minimum height of the crawl space? ______
10. At what scale was this print drawn? ______
11. How much concrete will be required to pour all of the interior piers? ______
12. If the lower left corner of the house had a finish grade elevation of 10'–2", what would be the elevation of the ground 10'–0" away from the structure? ______
13. If the lower left corner of the house had a finish grade elevation of 10'–2", what is the elevation of the floor if 1⅝" decking is used? ______
14. What is the area of the fireplace footing? ______
15. How many anchor bolts should be purchased? ______

PROBLEM 5–4. Use the drawing shown on page 123 to answer the following questions.

1. What type of floor system is used? ______
2. What is the soil bearing pressure of the job site? ______
3. Determine the girder size and support required to support the floor loads. ______
4. What size door is on the right side of the garage? ______
5. What size floor joist will be used with this floor system? ______
6. What spacing will be used with the anchor bolts around around the basement wall? ______
7. What type of fill material will be used under the garage slab? ______
8. What size steel could be used to reduce the stem wall thickness? ______
9. What are the window wells to be made of? ______
10. How thick is the wall on the right side of the basement? ______
11. What is the area of the basement slab? ______
12. How many blocks are required to be cut to support the concrete wall at the right side of the basement? ______
13. How many 16' long floor joists will need to be ordered to support the floor above the basement? ______
14. How many 14' long floor joist will need to be ordered to support the floor above the basement? ______
15. How many vents will need to be ordered? ______

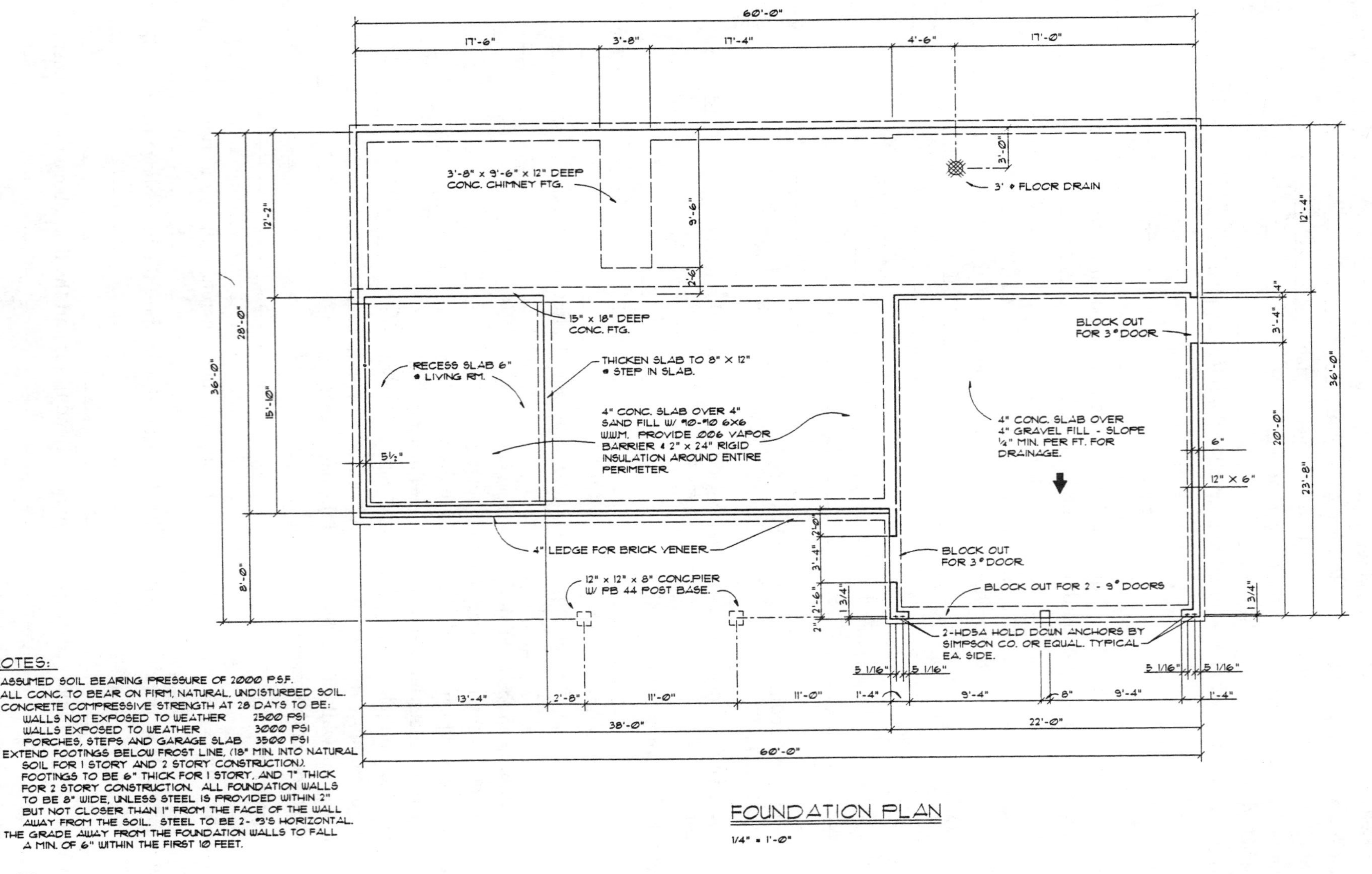

NOTES:

1. ASSUMED SOIL BEARING PRESSURE OF 2000 P.S.F.
2. ALL CONC. TO BEAR ON FIRM, NATURAL, UNDISTURBED SOIL.
3. CONCRETE COMPRESSIVE STRENGTH AT 28 DAYS TO BE:
   WALLS NOT EXPOSED TO WEATHER 2500 PSI
   WALLS EXPOSED TO WEATHER 3000 PSI
   PORCHES, STEPS AND GARAGE SLAB 3500 PSI
4. EXTEND FOOTINGS BELOW FROST LINE, (18" MIN. INTO NATURAL SOIL FOR 1 STORY AND 2 STORY CONSTRUCTION). FOOTINGS TO BE 6" THICK FOR 1 STORY, AND 7" THICK FOR 2 STORY CONSTRUCTION. ALL FOUNDATION WALLS TO BE 8" WIDE, UNLESS STEEL IS PROVIDED WITHIN 2" BUT NOT CLOSER THAN 1" FROM THE FACE OF THE WALL AWAY FROM THE SOIL. STEEL TO BE 2- #3'S HORIZONTAL.
5. THE GRADE AWAY FROM THE FOUNDATION WALLS TO FALL A MIN. OF 6" WITHIN THE FIRST 10 FEET.

Problem 5–1.

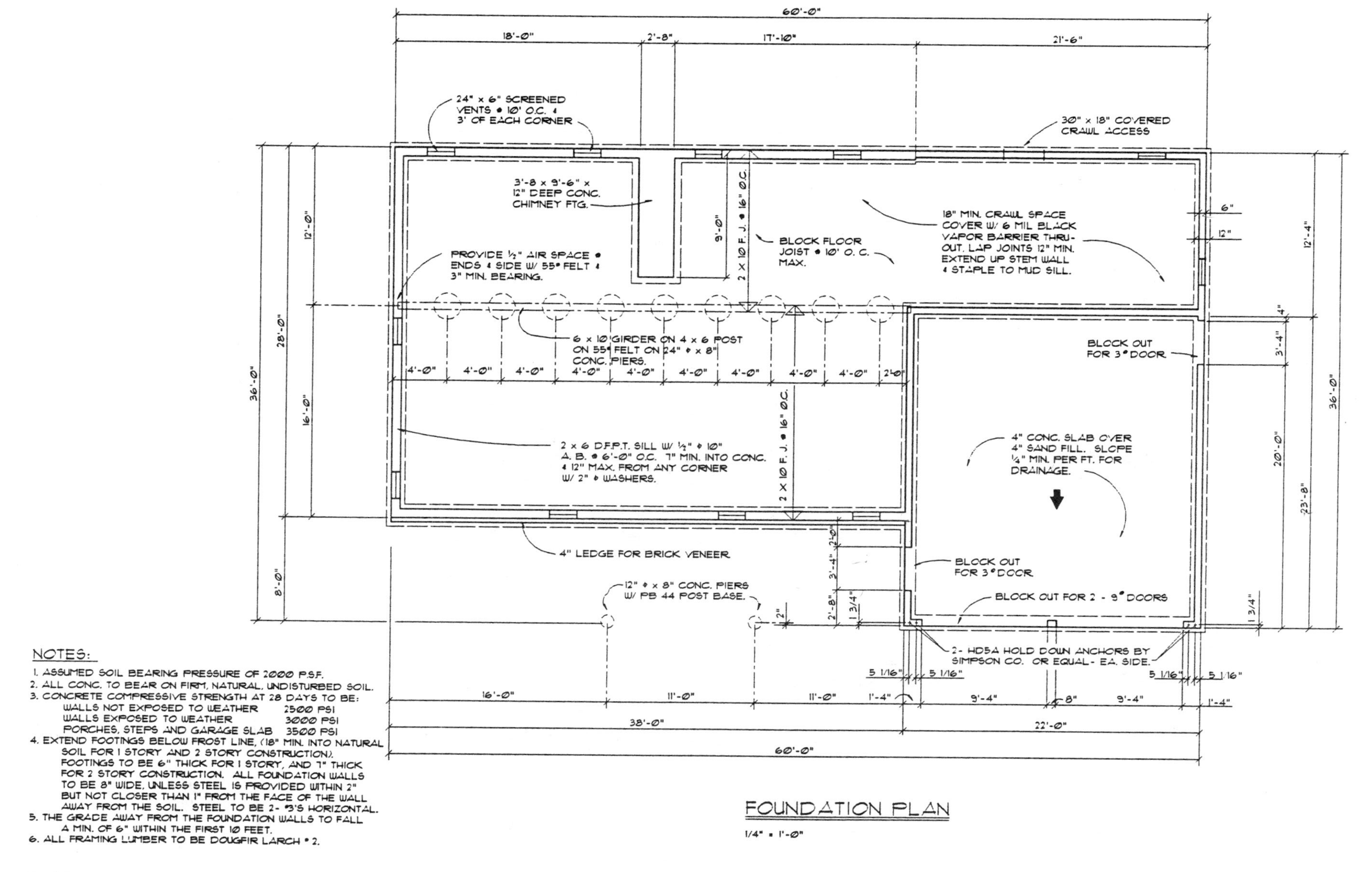

NOTES:

1. ASSUMED SOIL BEARING PRESSURE OF 2000 P.S.F.
2. ALL CONC. TO BEAR ON FIRM, NATURAL, UNDISTURBED SOIL.
3. CONCRETE COMPRESSIVE STRENGTH AT 28 DAYS TO BE:
   WALLS NOT EXPOSED TO WEATHER 2500 PSI
   WALLS EXPOSED TO WEATHER 3000 PSI
   PORCHES, STEPS AND GARAGE SLAB 3500 PSI
4. EXTEND FOOTINGS BELOW FROST LINE, (18" MIN. INTO NATURAL SOIL FOR 1 STORY AND 2 STORY CONSTRUCTION). FOOTINGS TO BE 6" THICK FOR 1 STORY, AND 7" THICK FOR 2 STORY CONSTRUCTION. ALL FOUNDATION WALLS TO BE 8" WIDE, UNLESS STEEL IS PROVIDED WITHIN 2" BUT NOT CLOSER THAN 1" FROM THE FACE OF THE WALL AWAY FROM THE SOIL. STEEL TO BE 2- #3'S HORIZONTAL.
5. THE GRADE AWAY FROM THE FOUNDATION WALLS TO FALL A MIN. OF 6" WITHIN THE FIRST 10 FEET.
6. ALL FRAMING LUMBER TO BE DOUGFIR LARCH # 2.

Problem 5–2.

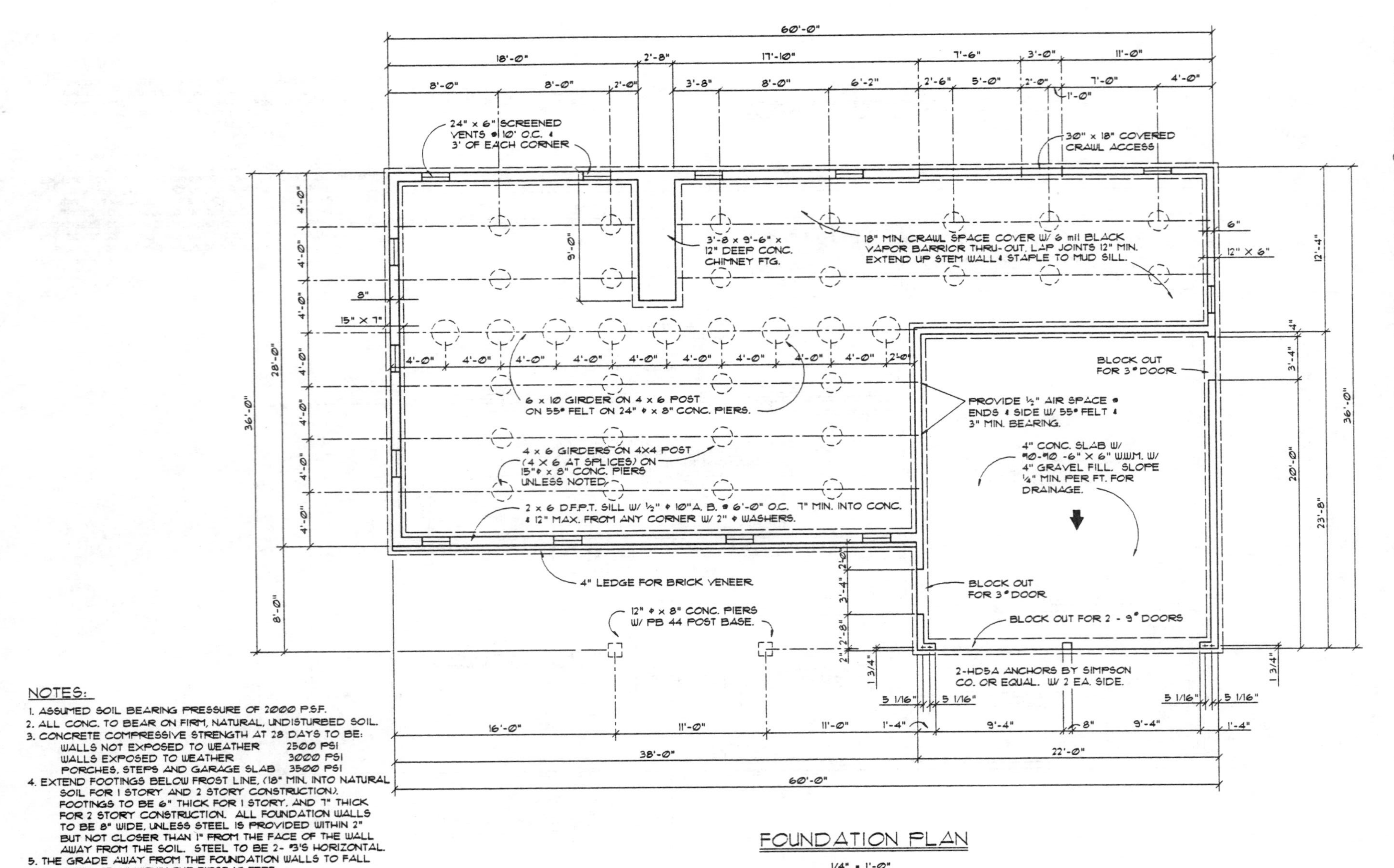
60'-0"
18'-0"
2'-8"
17'-10"
7'-6"
3'-0"
11'-0"
8'-0"
8'-0"
2'-0"
3'-8"
8'-0"
6'-2"
2'-6"
5'-0"
2'-0"
7'-0"
4'-0"
1'-0"
24" x 6" SCREENED VENTS @ 10' O.C. & 3' OF EACH CORNER
30" x 18" COVERED CRAWL ACCESS
3'-8 x 9'-6" x 12" DEEP CONC. CHIMNEY FTG.
18" MIN. CRAWL SPACE COVER W/ 6 mil BLACK VAPOR BARRIOR THRU-OUT. LAP JOINTS 12" MIN. EXTEND UP STEM WALL & STAPLE TO MUD SILL.
9'-0"
6"
12" X 6"
12'-4"
8"
15" X 7"
4'-0"
2'-0"
6 x 10 GIRDER ON 4 x 6 POST ON 55# FELT ON 24" ø x 8" CONC. PIERS.
BLOCK OUT FOR 3' DOOR
PROVIDE ½" AIR SPACE @ ENDS & SIDE W/ 55# FELT & 3" MIN. BEARING.
4" CONC. SLAB W/ #10-#10 -6" X 6" W.W.M. W/ 4" GRAVEL FILL. SLOPE ¼" MIN. PER FT. FOR DRAINAGE.
4 x 6 GIRDERS ON 4X4 POST (4 X 6 AT SPLICES) ON 15"ø x 8" CONC. PIERS UNLESS NOTED.
2 x 6 D.F.P.T. SILL W/ ½" ø 10"A. B. @ 6'-0" O.C. 7" MIN. INTO CONC. & 12" MAX. FROM ANY CORNER W/ 2" ø WASHERS.
36'-0"
28'-0"
3'-4"
20'-0"
23'-8"
36'-0"
4"
4" LEDGE FOR BRICK VENEER
BLOCK OUT FOR 3' DOOR
12" ø x 8" CONC. PIERS W/ PB 44 POST BASE.
BLOCK OUT FOR 2 - 9' DOORS
8'-0"
2'-8"
2"
1 3/4"
2-HD5A ANCHORS BY SIMPSON CO. OR EQUAL. W/ 2 EA. SIDE.
5 1/16"
16'-0"
11'-0"
11'-0"
1'-4"
9'-4"
8"
9'-4"
1'-4"
38'-0"
22'-0"
60'-0"
FOUNDATION PLAN
1/4" = 1'-0"
NOTES:
1. ASSUMED SOIL BEARING PRESSURE OF 2000 P.S.F.
2. ALL CONC. TO BEAR ON FIRM, NATURAL, UNDISTURBED SOIL.
3. CONCRETE COMPRESSIVE STRENGTH AT 28 DAYS TO BE:
WALLS NOT EXPOSED TO WEATHER 2500 PSI
WALLS EXPOSED TO WEATHER 3000 PSI
PORCHES, STEPS AND GARAGE SLAB 3500 PSI
4. EXTEND FOOTINGS BELOW FROST LINE, (18" MIN. INTO NATURAL SOIL FOR 1 STORY AND 2 STORY CONSTRUCTION). FOOTINGS TO BE 6" THICK FOR 1 STORY, AND 7" THICK FOR 2 STORY CONSTRUCTION. ALL FOUNDATION WALLS TO BE 8" WIDE, UNLESS STEEL IS PROVIDED WITHIN 2" BUT NOT CLOSER THAN 1" FROM THE FACE OF THE WALL AWAY FROM THE SOIL. STEEL TO BE 2- #3'S HORIZONTAL.
5. THE GRADE AWAY FROM THE FOUNDATION WALLS TO FALL A MIN. OF 6" WITHIN THE FIRST 10 FEET.
6. ALL FRAMING LUMBER TO BE DOUG. FIR LARCH # 2.

Problem 5–3.

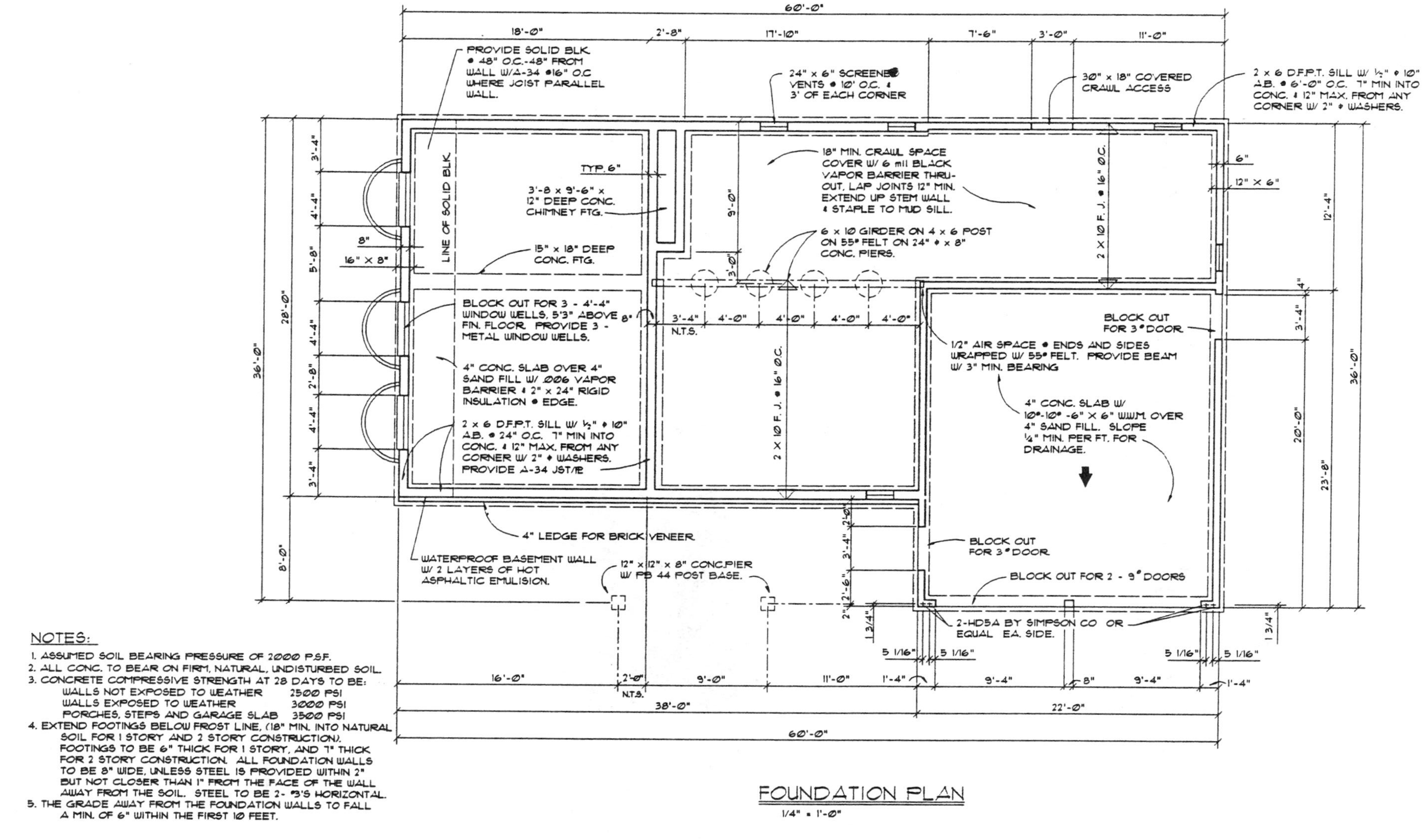

NOTES:

1. ASSUMED SOIL BEARING PRESSURE OF 2000 P.S.F.
2. ALL CONC. TO BEAR ON FIRM, NATURAL, UNDISTURBED SOIL.
3. CONCRETE COMPRESSIVE STRENGTH AT 28 DAYS TO BE:
   WALLS NOT EXPOSED TO WEATHER 2500 PSI
   WALLS EXPOSED TO WEATHER 3000 PSI
   PORCHES, STEPS AND GARAGE SLAB 3500 PSI
4. EXTEND FOOTINGS BELOW FROST LINE, (18" MIN. INTO NATURAL SOIL FOR 1 STORY AND 2 STORY CONSTRUCTION). FOOTINGS TO BE 6" THICK FOR 1 STORY, AND 7" THICK FOR 2 STORY CONSTRUCTION. ALL FOUNDATION WALLS TO BE 8" WIDE, UNLESS STEEL IS PROVIDED WITHIN 2" BUT NOT CLOSER THAN 1" FROM THE FACE OF THE WALL AWAY FROM THE SOIL. STEEL TO BE 2- #3'S HORIZONTAL.
5. THE GRADE AWAY FROM THE FOUNDATION WALLS TO FALL A MIN. OF 6" WITHIN THE FIRST 10 FEET.

Problem 5–4.

Chapter 6

Introduction to Elevations

REQUIRED ELEVATIONS
TYPES OF ELEVATIONS
ELEVATION SCALES
SURFACE MATERIALS IN ELEVATION
TEST
EXERCISES

ELEVATIONS ARE an essential part of the design and drafting process. The *elevations* are a group of drawings that show the exterior of a building. An elevation is an orthographic drawing that shows one side of a building. Elevations are projected as shown in Figure 6–1. Elevations are drawn to show exterior shapes and finishes as well as the vertical relationships of the building levels. By using the elevations, sections, and floor plans, the exterior shape of a building can be determined.

REQUIRED ELEVATIONS

Typically, four elevations are required to show the features of a building. On a simple building only the front, rear, and one side elevation may be provided. On a building of irregular shape, several additional elevations may be provided to show features that may have been hidden in one of four basic elevations. If a building has walls that are not at 90° to each other, a true orthographic drawing could be very confusing. In the orthographic projection, part of the elevation is distorted. Elevations of irregular shape structures are usually expanded so that a separate elevation of each different face is drawn. Figure 6–2 shows the overall elevation with the left half of the structure distorted. Figure 6–3 shows the true shape of the left portions of the structure. Because of the shape of structure, the roof may not appear in true shape in either of these views. The print reader must use both the elevations and the roof framing plan to determine the shape of the roof.

TYPES OF ELEVATIONS

Elevations can be drawn as either presentation drawings or working drawings. Presentation drawings are drawings that are a part of the initial design process and may range from sketches to very detailed drawings intended to help the owner and lending institution understand the basic design concepts. See Figure 6–4.

Working elevations are part of the working drawings and are drawn to provide information for the building team. This includes information on roofing, siding, openings, chimneys, land shape, and sometimes even the depth of footings, as shown in Figure 6–5. When used with the floor plans, the elevations

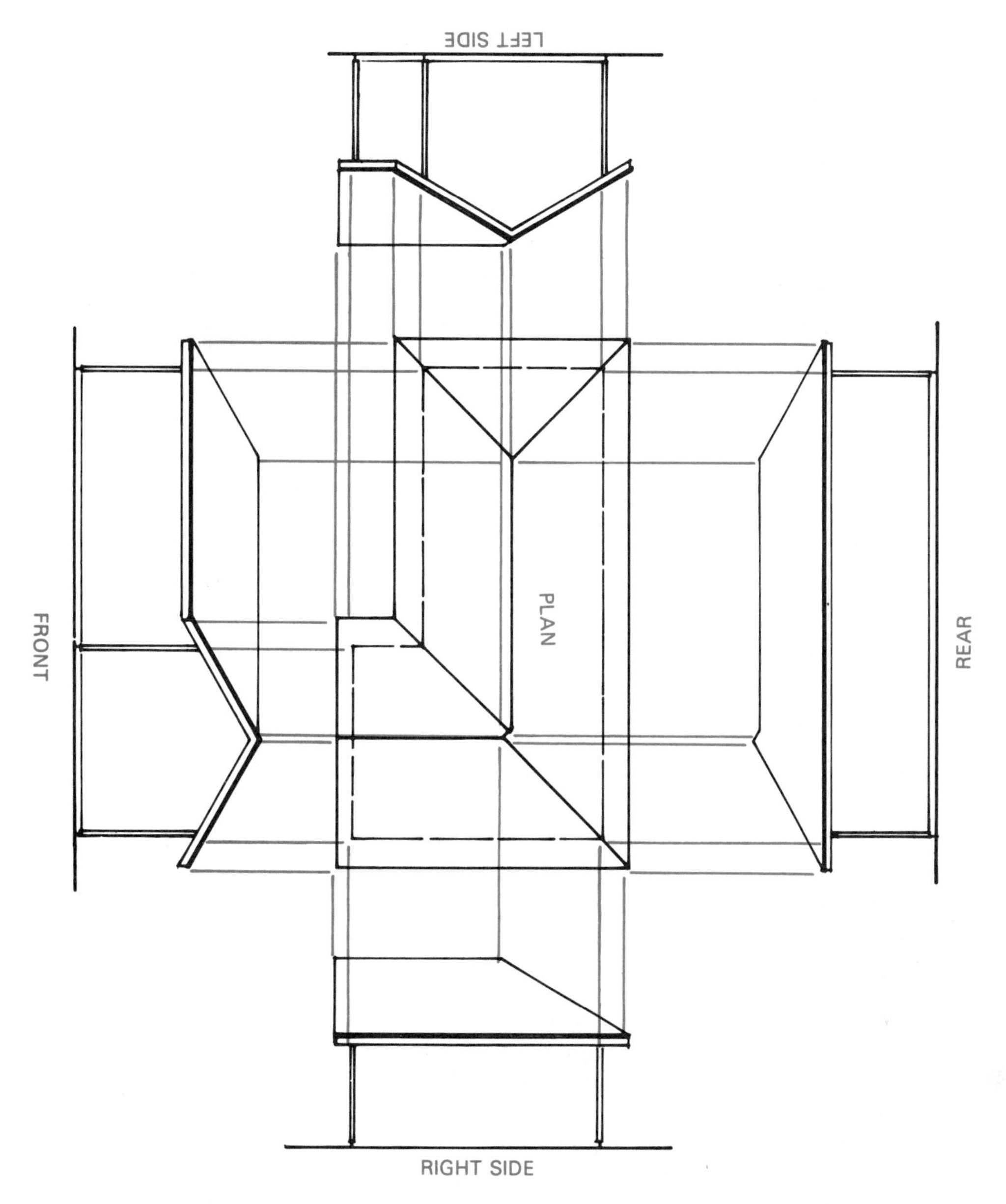

Figure 6–1 Elevations are orthographic projections showing each side of a structure.

provide information for the contractor to determine surface areas. Once surface areas are known, exact quantities of material can be determined. The elevations are also necessary when making heat loss calculations. The elevations are used to determine the surface area of walls and wall openings for the required heat loss formulas.

ELEVATION SCALES

Elevations are typically drawn at the same scale as the floor plan. For most plans, this means a scale of 1/4" = 1'–0" is used. Some floor plans for multifamily and commercial projects may be laid out at a scale of 1/16" = 1'–0" or even as small as 1/32" = 1'–0". When a scale of 1/8 in. or less is used, generally very little detail is placed on the drawing, as seen in Fig. 6–6. Depending on the complexity of the project or the amount of space on a page, the front elevation may be drawn at 1/4 in. scale and the balance of the elevations drawn at a smaller scale.

SURFACE MATERIALS IN ELEVATION

The materials that are used to protect the building from the weather need to be shown on the elevations. This information is considered under the categories of roofing, wall coverings, doors, and windows.

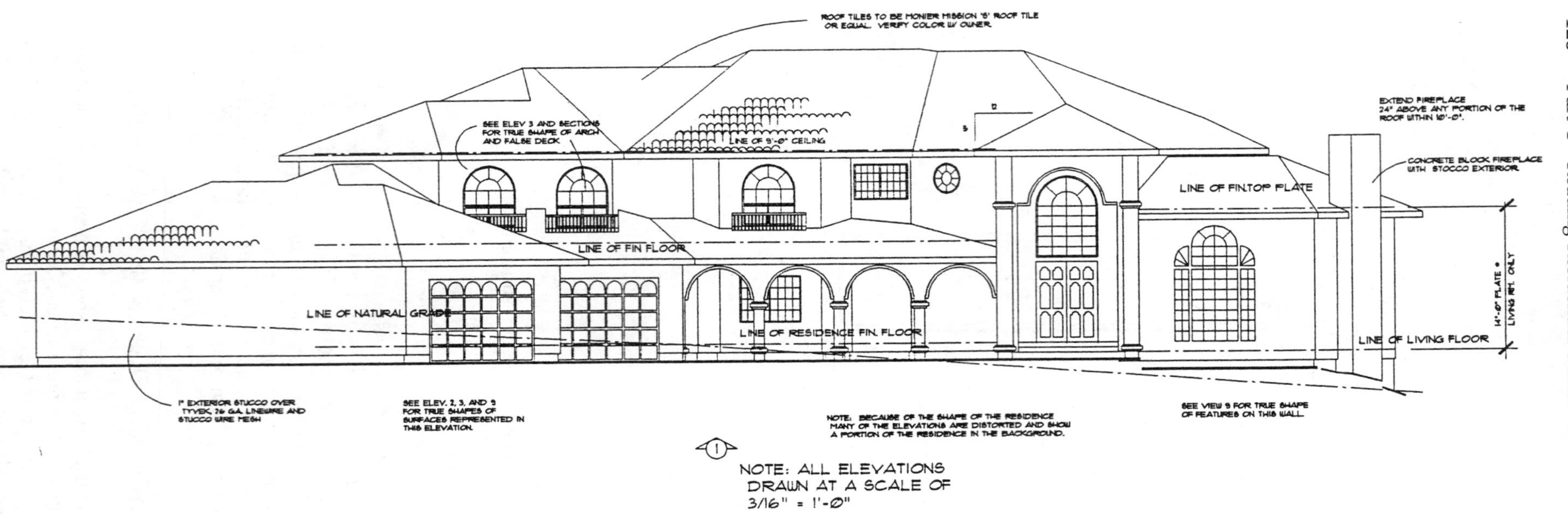

Figure 6–2 Using true orthographic projection methods for a structure with an irregular shape may result in distortion of part of the elevation.

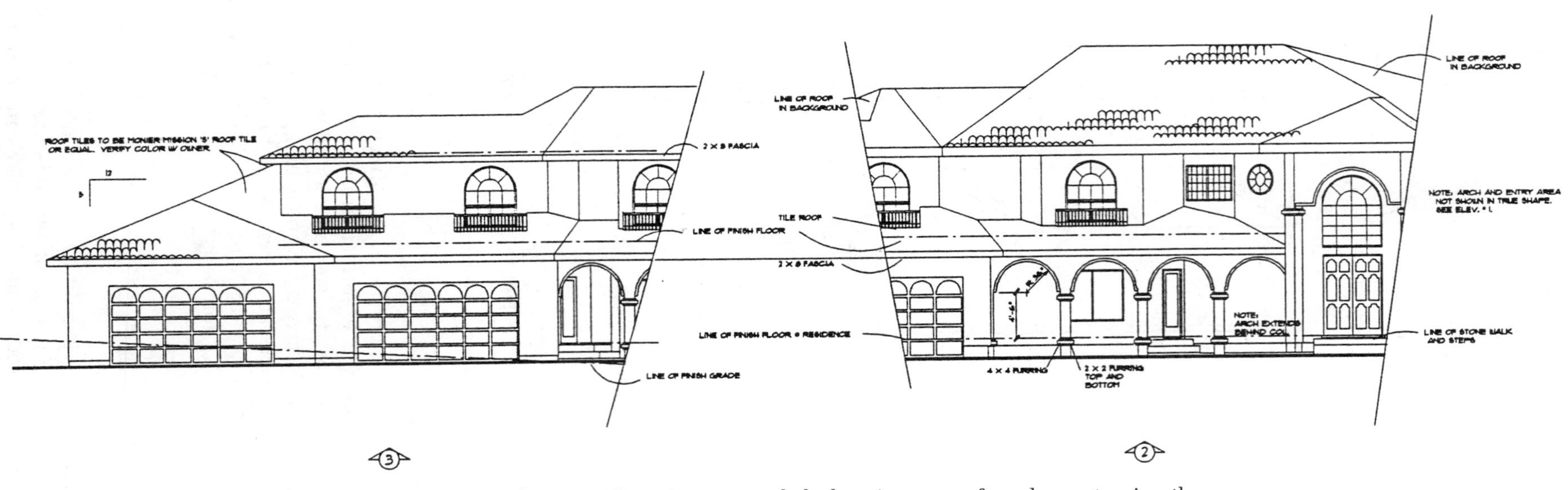

Figure 6–3 To eliminate distortion, expanded elevations are often drawn to give the print reader a true view of a specific portion of a structure.

Figure 6–4 Presentation elevations are highly detailed drawings with little material specified. Exterior building materials, landscaping, and shading are typically shown. *Courtesy of Piercy and Barclay Designers, Inc.*

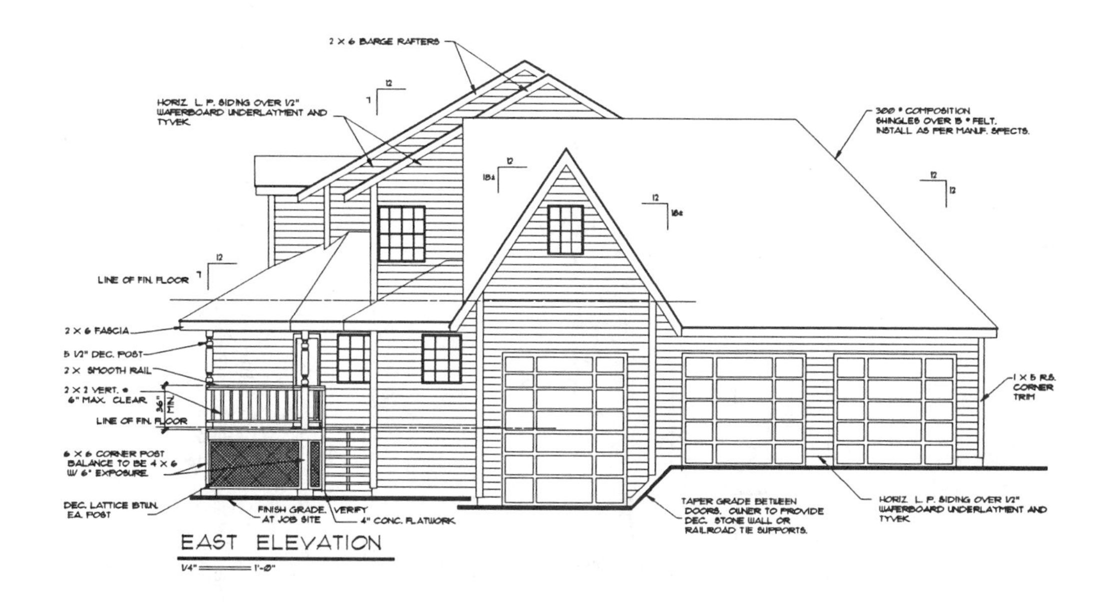

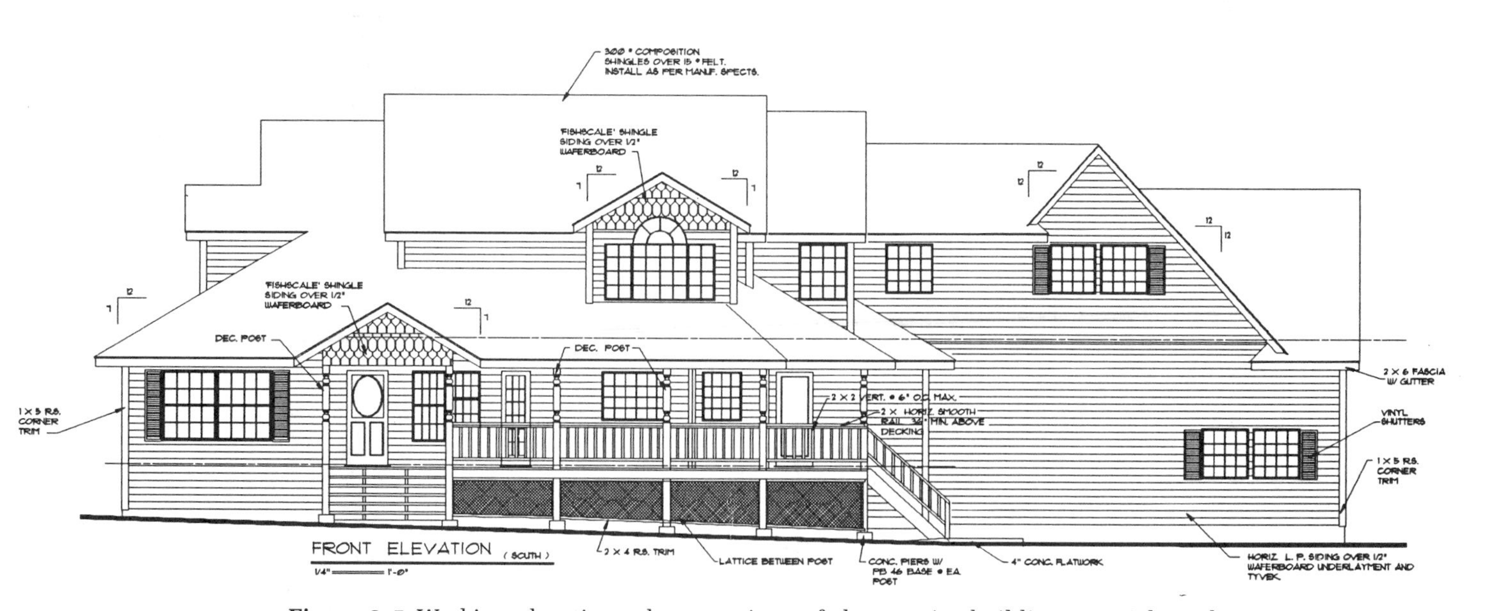

Figure 6–5 Working elevations show portions of the exterior building materials and other information needed by the construction crews.

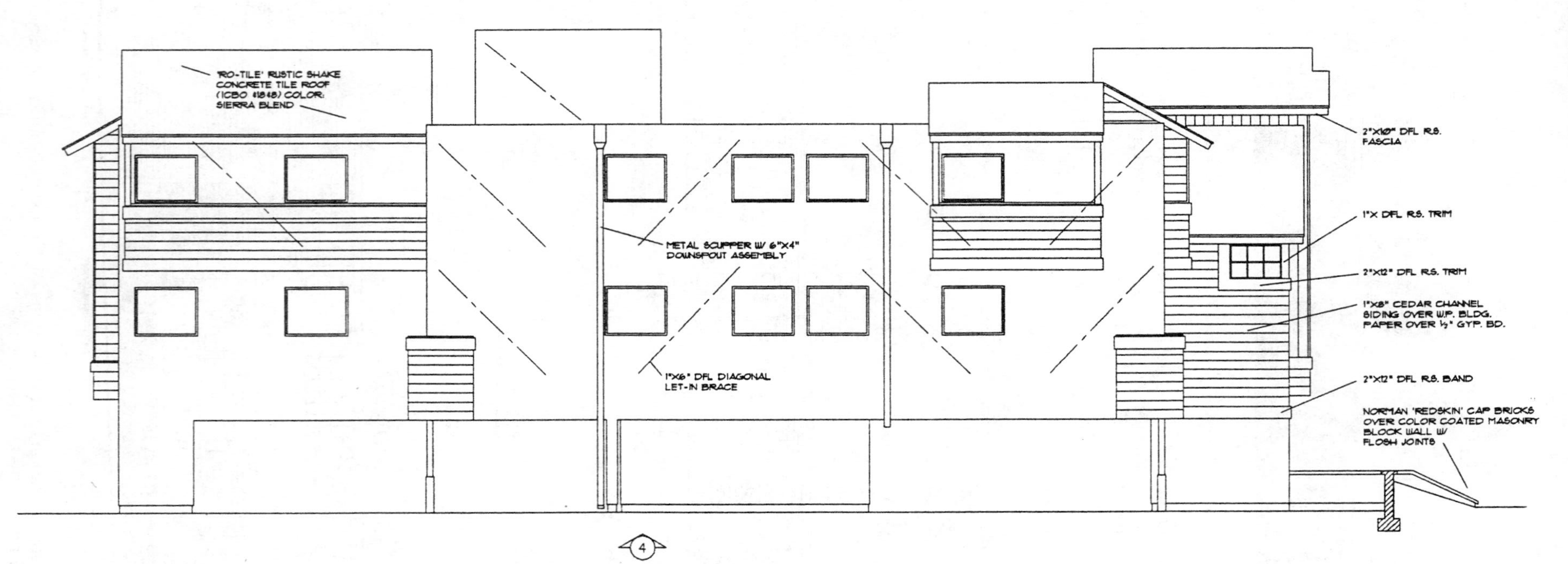

Figure 6–6 Elevations of commercial structures are often drawn at a scale of ⅛" = 1'–0" and show very little detail because of the small scale. Details are often provided to supplement the elevations. *Courtesy of Structureform.*

Roofing Materials

Several common materials are used to protect the roof. Among the most frequently used are asphalt shingles, wood shakes and shingles, clay and concrete tiles, metal sheets, and built-up roofing materials.

Asphalt Shingles. Asphalt shingles come in a variety of colors and patterns. Also known as composition shingles, they are typically made of fiberglass backing and covered with asphalt and a filler with a coating of finely crushed particles of stone. The asphalt waterproofs the shingle and the filler provides fire protection. The standard shingle is a three tab rectangular strip. The upper portion of the strip is coated with self-sealing adhesive and is covered by the next row of shingles. The lower portion of a three tab shingle is divided into three flaps that are exposed to the weather.

Composition shingles are also available in random width and thickness to give the appearance of cedar shakes. Both types of shingles can be used in a variety of conditions on roofs having a minimum slope of 2/12. The lifetime of shingles range from 20-, 25-, 30-, and 40-year guarantees.

Shingles are typically specified on drawings in note form listing the material, the weight, and the underlayment. The color and manufacture may also be specified. This information is often omitted in residential construction to allow the contractor to purchase a suitable brand at the best cost. A typical callout might be:

- 235 # COMPOSITION SHINGLES OVER 15 # FELT.
- 300 # COMPOSITION SHINGLES OVER 15 # FELT.
- ARCHITECT 80 'DRIFTWOOD' CLASS 'A' FIBERGLASS SHINGLES BY GENSTAR WITH 5⅝" EXPOSURE OVER 15# FELT UNDERLAYMENT WITH 30-YEAR WARRANTY.

Asphalt and fiberglass shingles are usually shown on drawings as seen in Figure 6–7.

Wood Shakes and Shingles. Wood shingles are typically made of cedar and may cost as much as one-third more than composition shingles. Wood shingles typically have a lifetime of 20 years, but this varies if the shingles are not made of old-growth timber. Other materials, such as Masonite, are also used to simulate shakes. These materials are typically specified on plans in note form listing the material, underlayment, and the amount of shingle exposed to the weather. A typical specification for wood shingles:

MEDIUM CEDAR SHAKES OVER 15# FELT W / 15# × 18" WIDE FELT BETWEEN EACH COURSE. LAY WITH 10½" EXPOSURE.

Wood shakes are often represented on drawings as shown in Figure 6–8.

Tile. Tile is the material most often used for homes on the high end of the price scale or where the risk of fire is extreme. Although tile may cost twice as much as the better grades of asphalt shingle, it offers a lifetime guarantee. Tile is available in a variety of colors, materials, and patterns. Clay, concrete, and metal are the most common materials. Clay and concrete tiles can weigh as much as four times more than asphalt tiles and require larger framing members to support their weight. In elevation, the tile is typically called out in a note listing the manufacturer, tile pattern and shape, material, color, and underlayment. A typical note on the elevations might resemble:

MONIER BURNT TERRA COTTA MISSION S ROOF TILE OVER 15# FELT AND 1 × 3 SKIP SHEATHING. USE A 3" MIN. HEAD LAP AND INSTALL AS PER MANUF. SPECIFICATIONS.

Tile is often represented on elevations as seen in Figures 6–9 and 6–10.

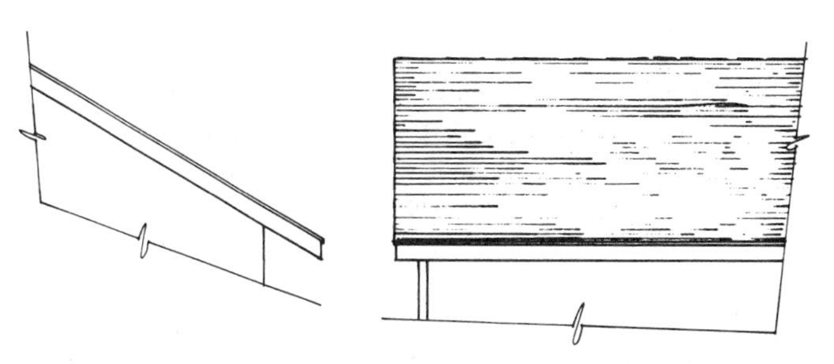

Figure 6–7 Representing asphalt and composition shingles.

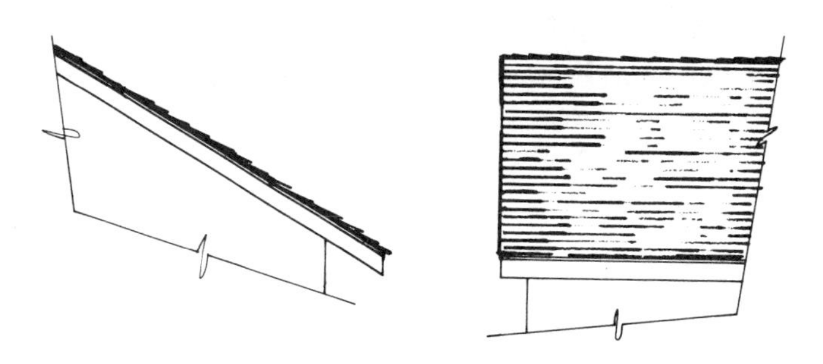

Figure 6–8 Wood shakes and shingles.

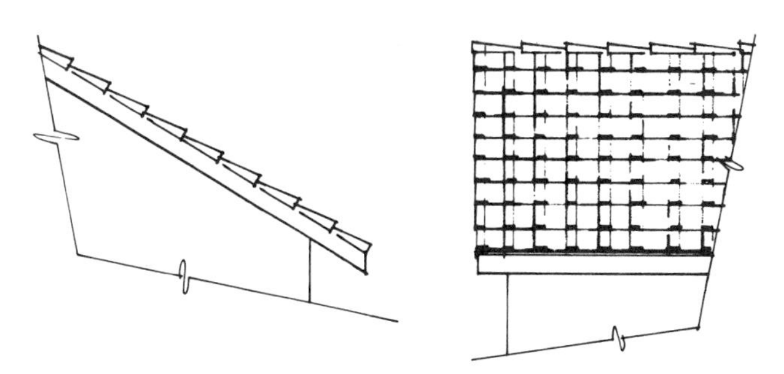

Figure 6–9 Flat tile representation on elevations.

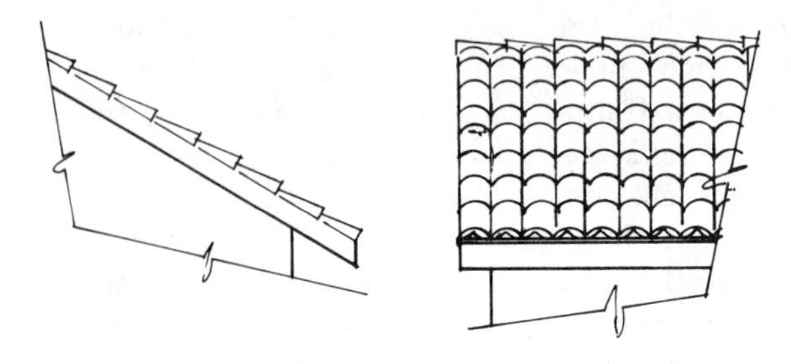

Figure 6–10 Spanish-style tile roofing.

Metal. Metal shingles and panels are common on many types of roofs. Metal shingles would usually be represented in a manner similar to asphalt shingles. Metal panels come in many styles.

Built-up Roofs. Built-up roofing, or hot tar roofs, are used on very low pitched roofs. Because of the low pitch and the lack of surface texture, built-up roofs are usually outlined and left blank. Occasionally a built-up roof will be covered with 2- or 3-in.-diameter rock. The drawing technique for this roof can be seen in Figure 6–11.

Skylights. Skylights may be made of either flat glass or domed plastic. Although they come in a variety of shapes and styles, skylights usually resemble those shown in Figure 6–12. Depending on the pitch of the roof, skylights may or may not be drawn. Unless the roof is very steep, a rectangular skylight will appear almost square. The flatter the roof, the more distortion there will be in the size of the skylight.

Wall Coverings

Exterior wall coverings are usually made of wood, wood substitutes, masonry, metal, plaster or stucco. Each has its own distinctive look in elevation.

Wood. Wood siding can either be installed in large sheets or in individual pieces. Plywood sheets are a popular wood siding because of the low cost and ease of installation. Individual pieces of wood provide an attractive finish but usually cost more than plywood. This higher cost results from differences in material and the labor to install each individual piece.

Plywood siding can have many textures, finishes, and patterns. Textures and finishes are not shown on the elevations but may be specified in a general note. Patterns in the plywood are usually shown. The most common patterns in plywood are T-1-11 reverse board and batten and plain or rough-cut plywood. Plywood siding is typically specified with a note to specify the thickness, pattern, and underlayment. A typical note on the elevations:

5/8" T-1-11 PLY SIDING WITH GROOVES AT 4" O.C. OVER 15 # FELT.

See Figure 6–13 for common methods of showing plywood siding. Lumber siding comes in several types and is laid in many patterns. Among the most common lumber for sidings are cedar, redwood, pine, fir, spruce, and hemlock. Common styles of lumber siding are tongue and groove, bevel, and channel siding. Each of these materials can be installed in a vertical, horizontal, or diagonal position. The material and type of siding must be specified in a general note on the elevations. The pattern in which the siding is to be installed must be shown on the elevations in a manner similar to that in Figure 6–14. The type of siding and the position in which it is laid affect how the siding appears at a corner. Figure 6–15 shows two common methods of corner treatment.

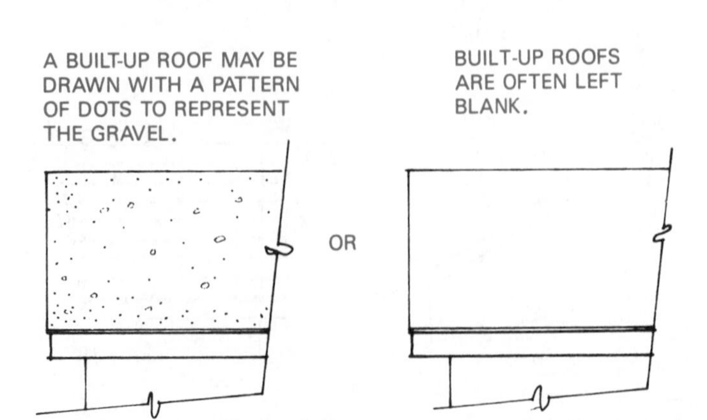

Figure 6–11 Built-up roofing.

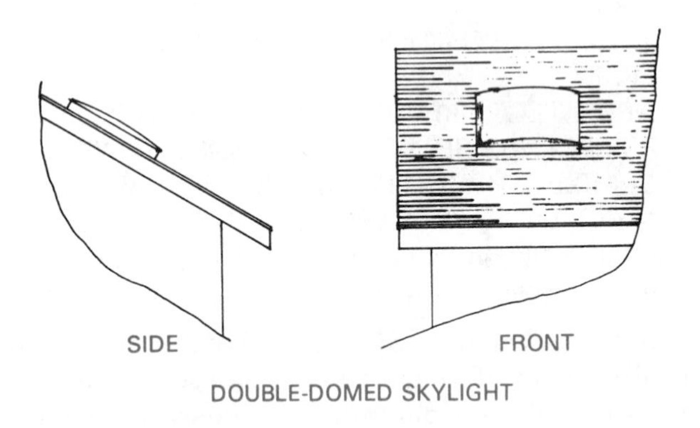

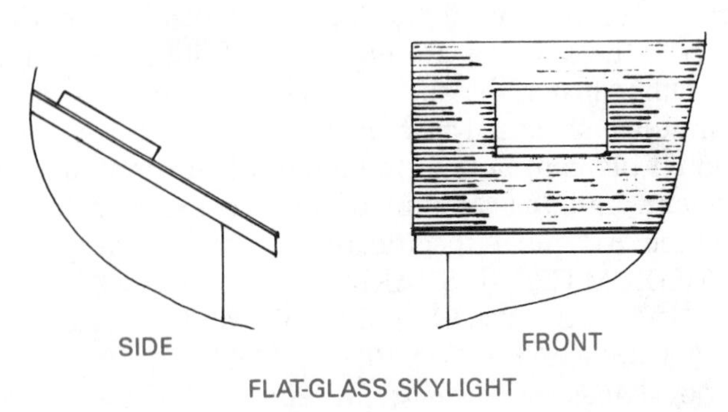

Figure 6–12 Common methods of representing glass and plastic skylights.

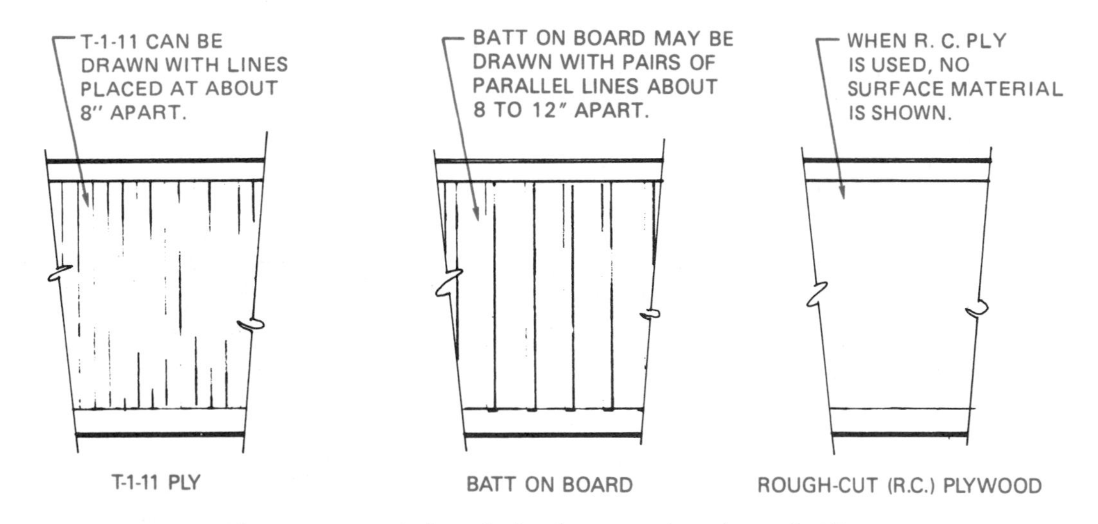

Figure 6–13 Typical methods of representing plywood siding.

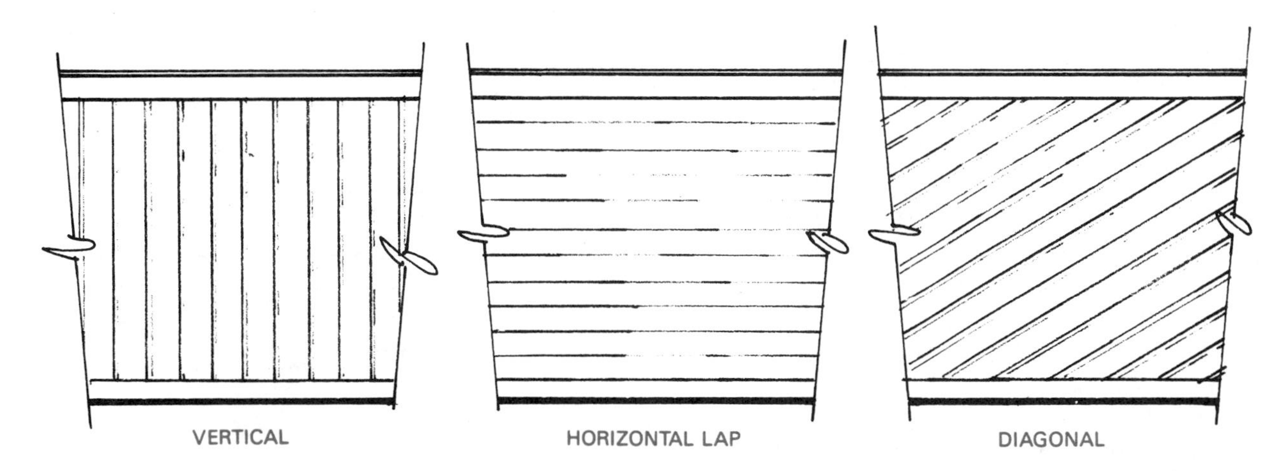

Figure 6–14 Typical methods of representing wood siding.

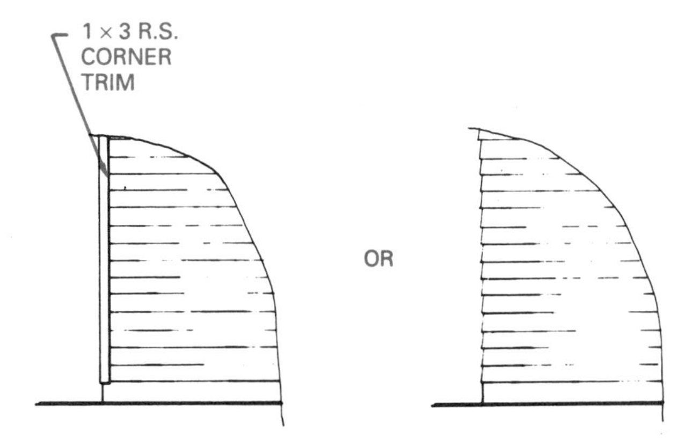

Figure 6–15 Common methods of corner treatment.

Wood shingles can be installed either individually or in panels. Shingles are often shown on plans, as seen in Figure 6–16. Hardboard, aluminum, and vinyl siding can be produced to resemble lumber siding. Hardboard siding is generally installed in large sheets similar to plywood but often has more detail than plywood or lumber siding. It is typically represented using methods similar to those used for showing lumber sidings. Aluminum and vinyl sidings also resemble lumber siding and are drawn similar to their lumber counterpart.

Masonry. Masonry finishes may be made of brick, concrete block, or stone.

Bricks and concrete blocks come in a variety of sizes, patterns, and textures. The elevations represent the pattern on the drawing and the material and texture in the written specifications. Methods for drawing bricks are shown in Figure 6–17. Figure 6–18 shows methods of representing some of the various concrete blocks.

Stone or rock finishes also come in a wide variety of sizes and shapes and are laid in a variety of patterns, as shown in Figure 6–19. Stone or rock may be either natural or artificial. Both appear the same in elevation. Stone, brick, and concrete are each specified by a note to describe the material, the airspace required, and the backing material. Typical notes that might describe each material are as follows:

- STONE VENEER OVER 1" AIRSPACE, WITH 15# FELT AND 26 GA. METAL TIES @ 24" O.C. EACH STUD.
- BRICK VENEER OVER 1" AIRSPACE, WITH TYVEK AND 26 GA. METAL TIES @ 24" O.C. EACH STUD.
- 8 × 8 × 16 GRADE 'A' CONC. SCORED BLOCK. REINFORCE W / #4 DIA. @ 18" O.C. EACH WAY.

Although primarily a roofing material, metal can be used as an attractive wall covering. Metal in elevation will look similar to lumber siding and will be specified in a note.

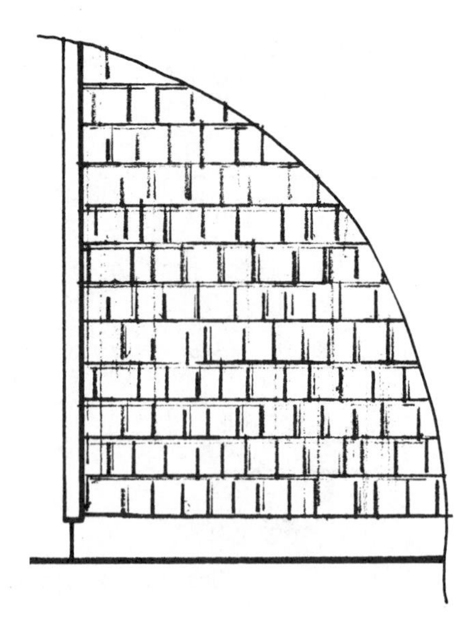

Figure 6–16 Shingle siding.

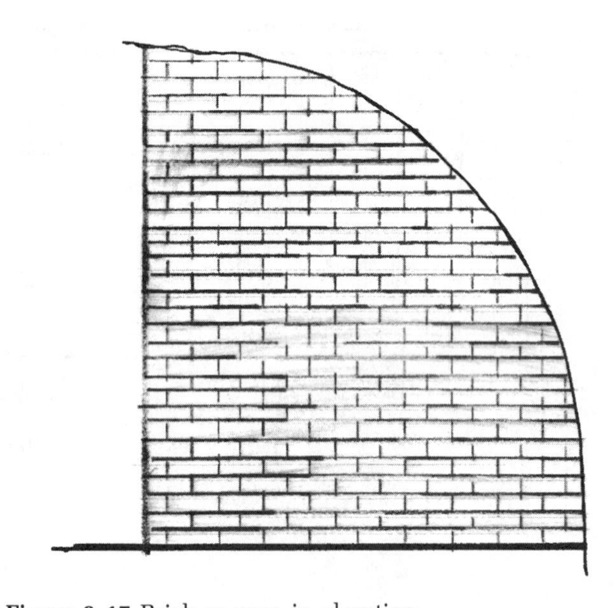

Figure 6–17 Brick as seen in elevation.

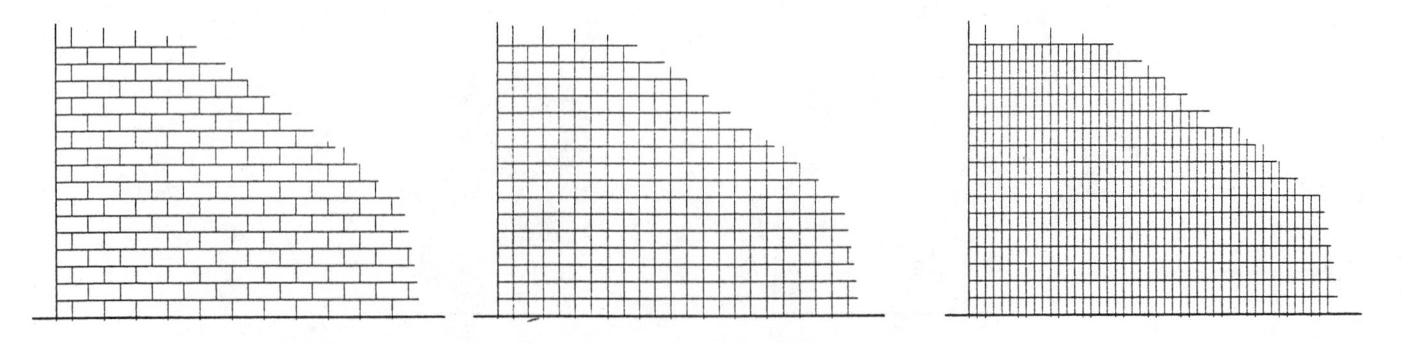

Figure 6–18 Concrete block represented in elevation.

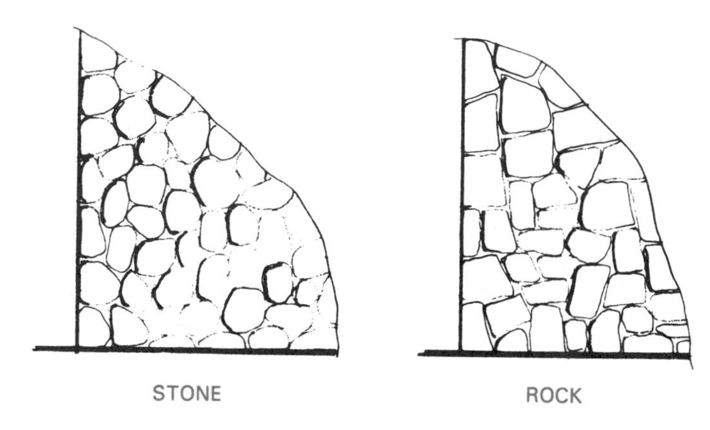

Figure 6–19 Stone and rock in elevation.

Plaster or Stucco. Although primarily used in areas with little rainfall, plaster or stucco can be found throughout the country. No matter what the pattern, stucco is typically drawn as shown in Figure 6–20.

Stucco often has a metal corner bead at each corner to reinforce the stucco. This bead may be omitted to provide a rustic look. A common note often found on elevations is as follows:

1" MISSION TEXTURE STUCCO WITH NO CORNER BEAD OVER 26 GA. LINEWIRE @ 16" O.C. WITH STUCCO WIRE MESH OVER 15# FELT.

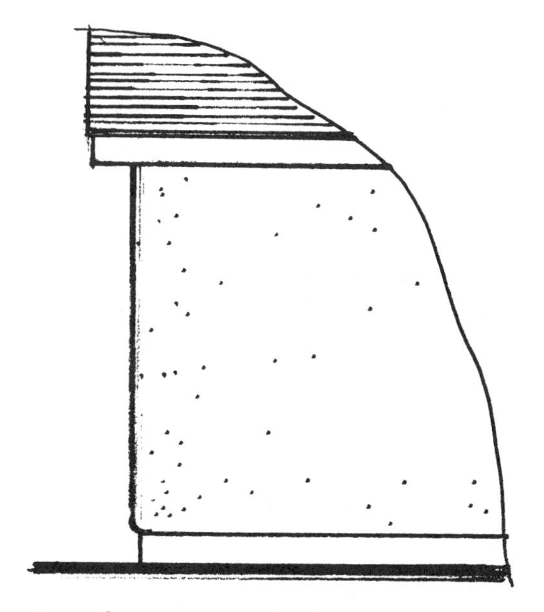

Figure 6–20 Plaster or stucco in elevations.

Doors

On the elevations, doors are drawn to resemble the type of door specified in the door schedule. Figure 6–21 shows how common types of doors are often seen in elevation.

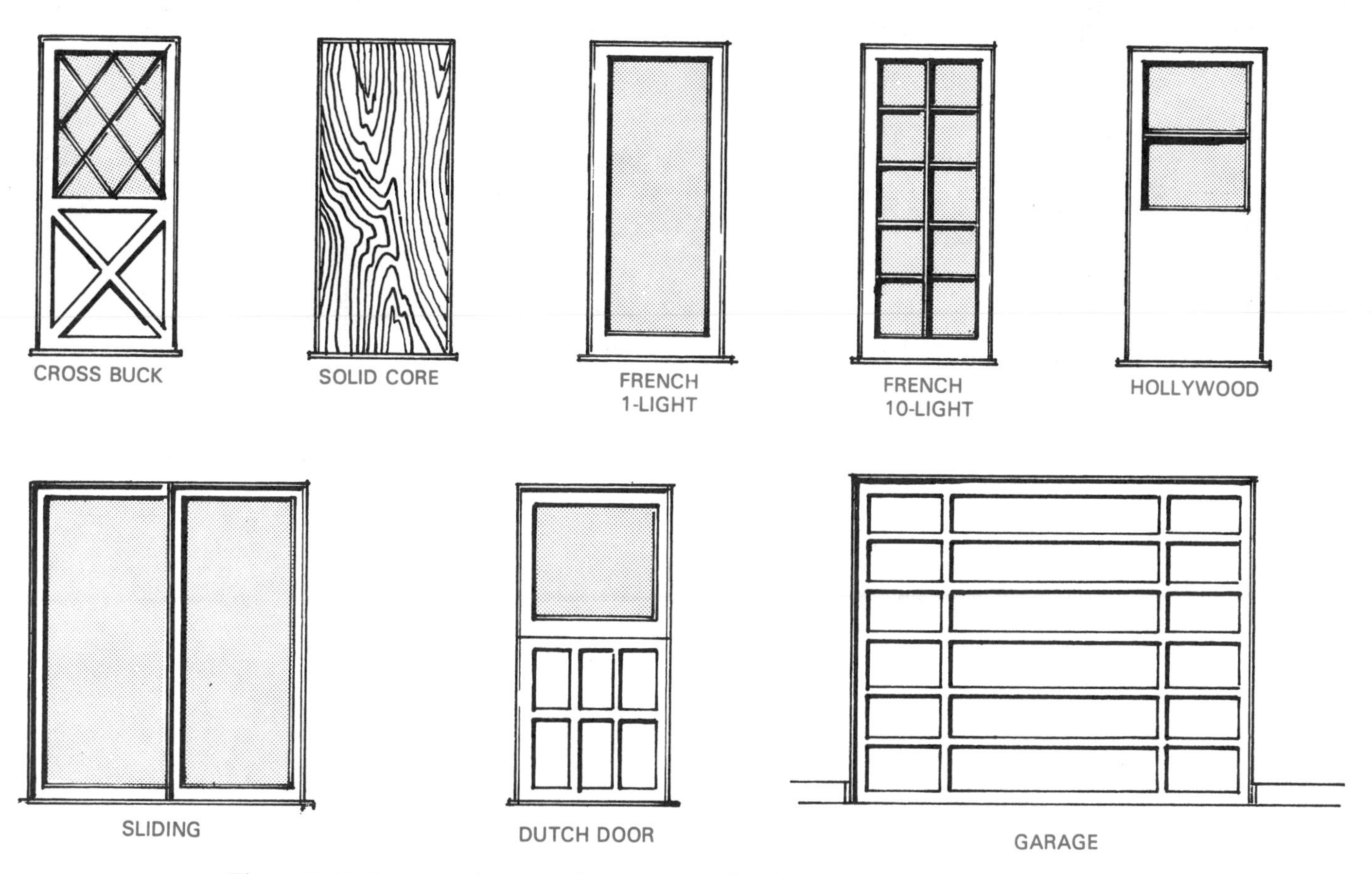

Figure 6–21 Common doors as they are typically shown in elevation. The floor plan and schedules contain all the technical information.

Windows

Windows are typically represented in elevation by the type of material of which the frame is made. Aluminum, vinyl, and wood are the most common frame materials. Figure 6–22 shows how these materials are often represented. In addition to the frame material, the style of window can also be determined from the elevation. Figure 6–23 shows common types of windows that might be found on elevations. Window information is typically not specified on elevations unless the window is located in a vaulted area that could not be represented on a floor plan.

Rails

Rails can be either solid to match the wall material or open. Open rails are typically made of wood or wrought iron. Although spacing varies, vertical rails are usually required to be no more than 6 in. clear. Rails are often drawn as shown in Figure 6–24.

Chimney

Several different methods can be used to represent a chimney. Figure 6–25 shows examples of wood and masonry chimneys. The chimney material is specified in a similar manner to the exterior siding. A metal chimney cap or spark arrester is often specified on the elevations.

Footings

If the structure is built on a level lot, the line of the footing is usually omitted. On a sloping job site, the line of footing can help the foundation crew visualize the grading required to place the foundation. Although the line type may vary, the footing is usually represented as shown in Figure 6–26.

Dimensions

Dimensions are usually kept to a minimum on elevations. Most offices use the elevations to show sizes of major components, with the majority of dimensions placed on the sections where construction materials can be better seen because of the larger scale at which the sections are drawn.

Dimensions most often seen on elevations show the vertical relationship of the floor levels to each other. Other major changes in shape, such as cantilevers and overhangs, may also be shown. The relationship of the chimney to the roof or other portions of the structure within 10 ft of the chimney are usually specified on the elevations. Examples of dimensions to show major shapes in elevations can be seen in Figure 6–26.

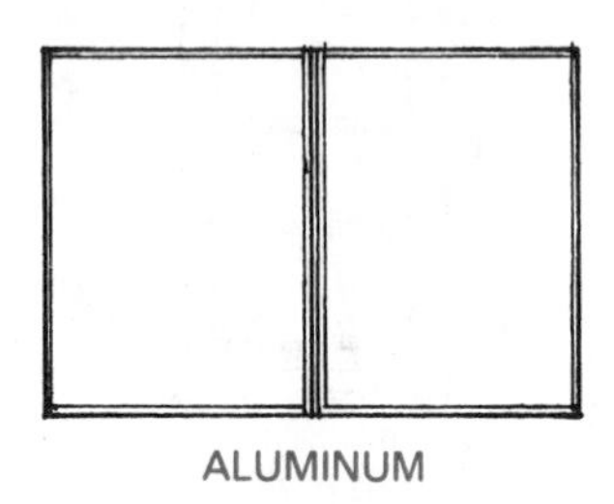

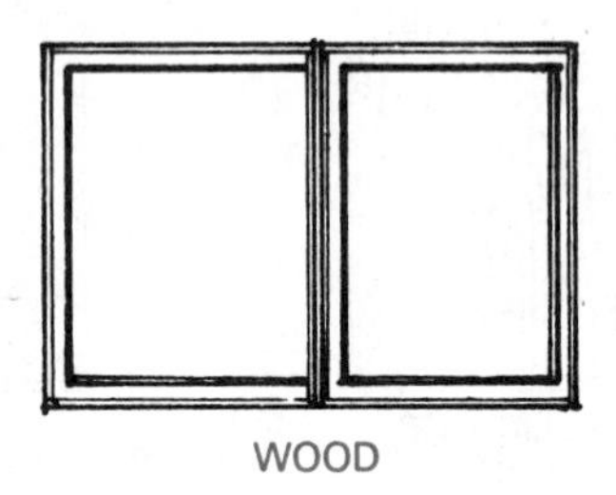

Figure 6–22 Representing wood and aluminum frame windows.

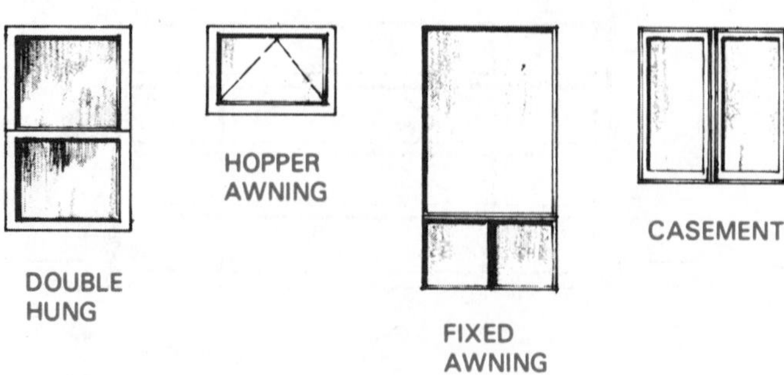

Figure 6–23 Common types of windows shown on elevations.

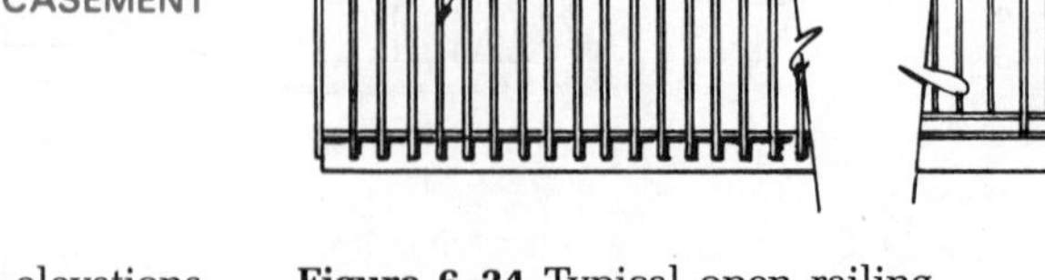

Figure 6–24 Typical open railing.

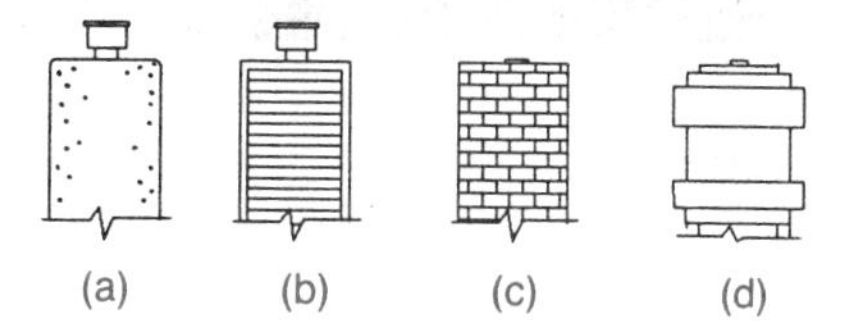

Figure 6–25 Common methods of representing chimneys in elevation: (a) stucco, (b) wood siding, (c) brick, and (d) decorative masonry.

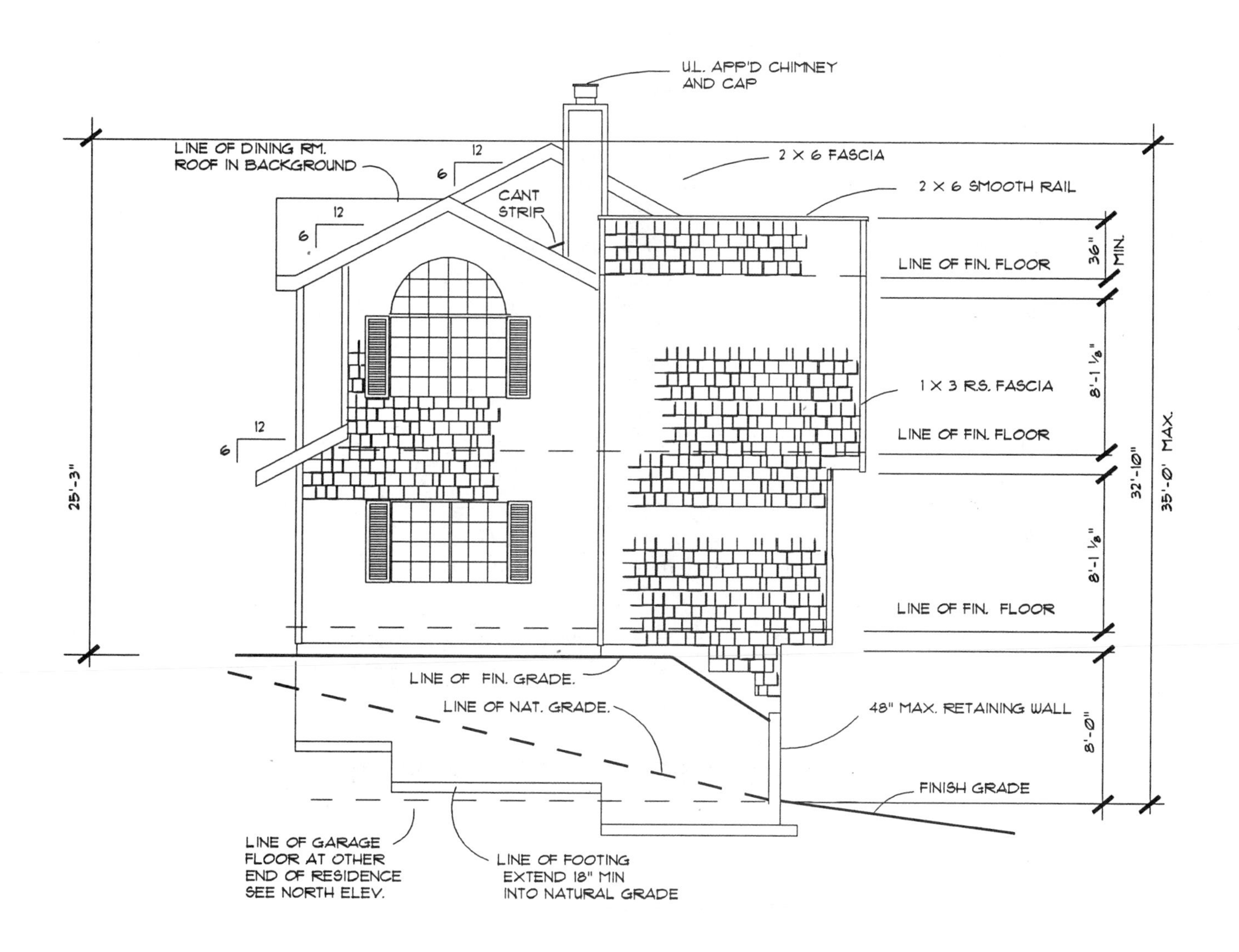

Figure 6–26 Dimensions and footings are sometimes shown to aid the construction crew.

CHAPTER 6 TEST

1. What is the purpose of exterior elevations? ____
____
____
____.

2. Of the two major types of elevations, which will you be most likely to encounter as a construction worker.
____.

3. What would be the most common scale at which you would expect the front elevation to be drawn?
   a. 1/8" = 1'–0"
   b. 1" = 10'
   c. 1/4' = 1'–0"
   d. same scale as the floor plan

4. What are the two major types of wood siding? ____
____.

5. List four major types of roofing. ____
____
____
____.

6. What material will need to be specified to order a tile roof? ____
____.

7. List four dimensions that may be found on an elevation. ____
____.

8. What are the two major weights of composition shingles? ____
____.

9. List four major types of siding that may be found on elevations. ____
____
____
____.

10. List situations in which fewer than four and more than four elevations would be drawn. ____
____
____
____.

CHAPTER 6 EXERCISES

PROBLEM 6–1. Use the floor plan and elevation shown on page 139 to answer the following questions.

1. What size corner trim is used? ____
2. What size of fascia will be used? ____
3. What type of roofing will be used? ____
4. What underlayment will be needed to support the siding? ____
____
____.

5. What underlayment will be used for the roof? ____
____.

6. What is the ceiling height in the basement? ______.
7. What thickness of roof plywood sheathing will be used? ______.
8. What do the dashed lines represent below grade line? ______.
9. What is the height of the main floor? ______.
10. Describe two different treatments used to dress up the windows in this house from the house in Problem 6–1. ______.
11. Determine the surface area of the heated wall area shown in the front elevation. ______.
12. What is the area of door and window openings located in the heated walls? ______.
13. A 1 gal can of sealer will cover 50 sq ft. How many cans will be required to seal the basement wall shown in the front elevation and to seal the total basement wall? ______.
14. What is the percentage of openings to wall area for heat loss of the front elevation? ______.
15. How many feet of fascia will be required for the front of the house if 12" overhangs are used? ______.

PROBLEM 6–2. Use the floor plan and elevation shown on page 140 to answer the following questions.

1. What is the thickness of the exterior siding? ______.
2. Who is the manufacturer of the finish roofing? ______
3. How far will the stucco walls extend past the windows? ______
4. List the various sizes of trim to be used. ______.
5. What type of siding will be used to accent the upper windows? ______.
6. What gage of line wire will be used to support the stucco? ______.
7. Why is the height of the living room ceiling? ______.

8. How will the decorative trim be framed? ____
____.

9. Why are the stucco corners rounded? ____
____.

10. This residence is being built in a subdivision with a height limitation of 25'–0" max. If the roof is built at a 6/12 pitch, what is the height of the ridge? ____
____
____
____.

11. Determine how many feet of each size of trim will need to be purchased for the front elevation. ________
____
____
____
____
____.

12. The roof tile is 13" wide. How many will be required to form the first course of the upper and lower roofs?
____
____.

13. How many sheets of T-1-11 plywood will be required? ____
____
____.

14. If the horizontal siding is laid with a 5" exposure, how many individual pieces will be required between each window? ____
____
____
____.

15. Would there be less scrap with 8', 10', or 12' lumber when installing the siding between the bath and bedroom window? ____
____
____
____.

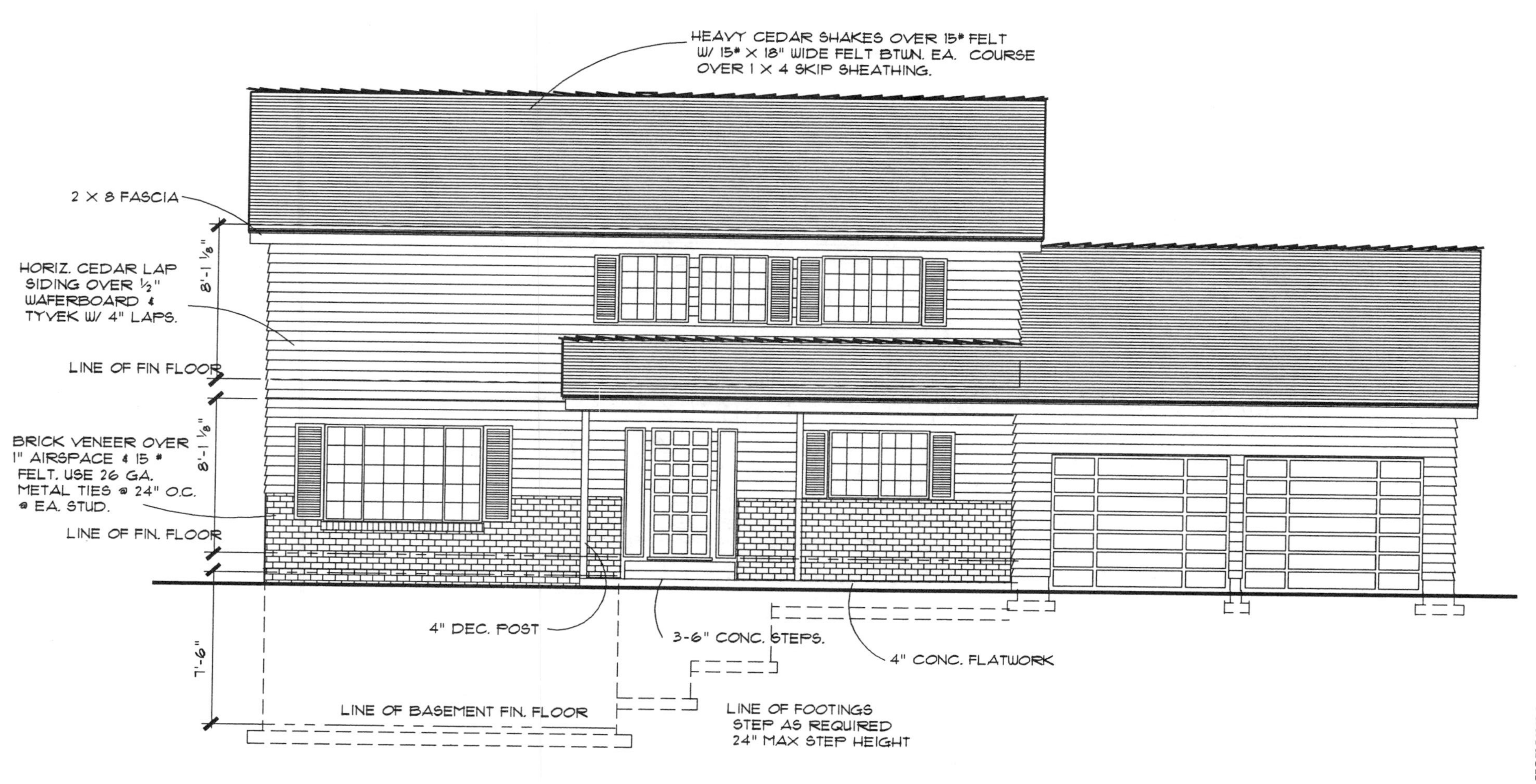

HEAVY CEDAR SHAKES OVER 15# FELT W/ 15# X 18" WIDE FELT BTWN. EA. COURSE OVER 1 X 4 SKIP SHEATHING.
2 X 8 FASCIA
HORIZ. CEDAR LAP SIDING OVER 1/2" WAFERBOARD & TYVEK W/ 4" LAPS.
8'-1 1/8"
LINE OF FIN FLOOR
BRICK VENEER OVER 1" AIRSPACE & 15 # FELT. USE 26 GA. METAL TIES @ 24" O.C. @ EA. STUD.
8'-1 1/8"
LINE OF FIN. FLOOR
4" DEC. POST
3-6" CONC. STEPS.
4" CONC. FLATWORK
7'-6"
LINE OF BASEMENT FIN. FLOOR
LINE OF FOOTINGS STEP AS REQUIRED 24" MAX STEP HEIGHT

Problem 6–1.

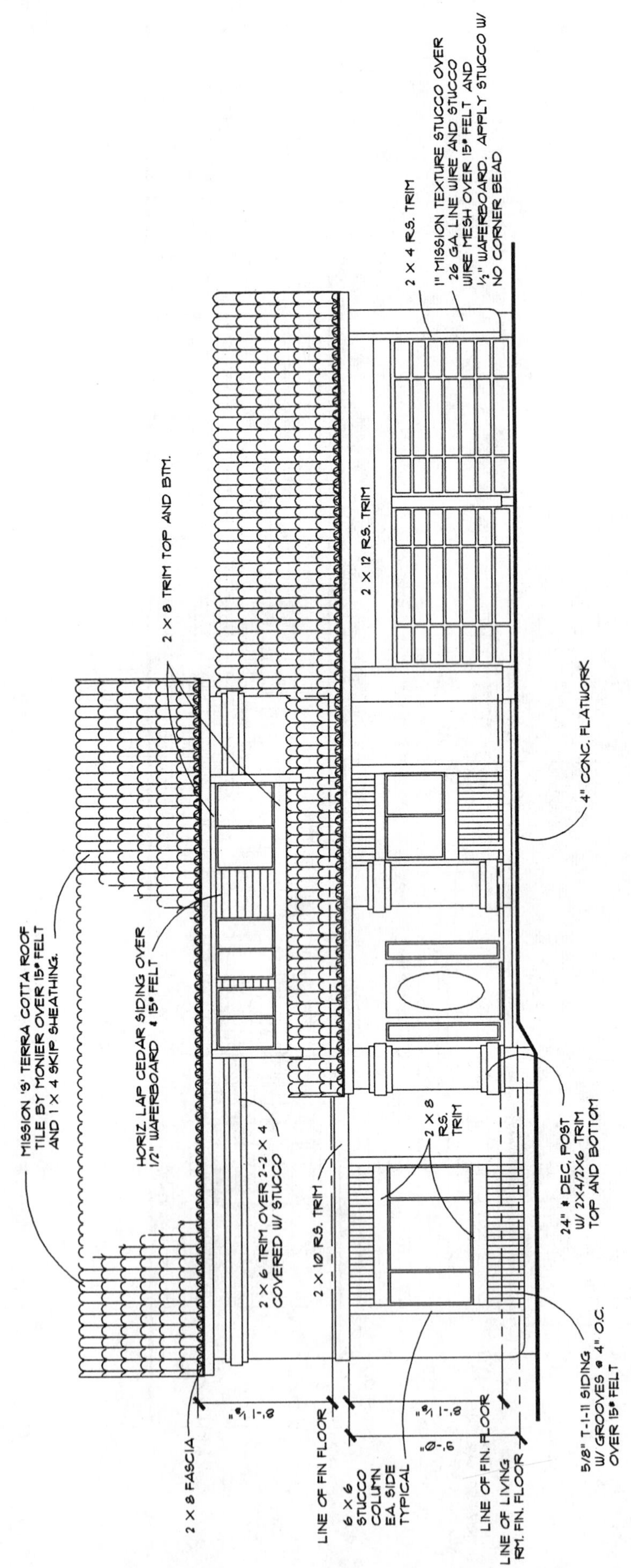
MISSION 'S' TERRA COTTA ROOF TILE BY MONIER OVER 15# FELT AND 1 X 4 SKIP SHEATHING.
HORIZ. LAP CEDAR SIDING OVER 1/2" WAFERBOARD & 15# FELT
2 X 8 TRIM TOP AND BTM.
2 X 12 R.S. TRIM
2 X 4 R.S. TRIM
1" MISSION TEXTURE STUCCO OVER 26 GA. LINE WIRE AND STUCCO WIRE MESH OVER 15# FELT AND 1/2" WAFERBOARD. APPLY STUCCO W/ NO CORNER BEAD
4" CONC. FLATWORK
2 X 6 TRIM OVER 2-2 X 4 COVERED W/ STUCCO
2 X 10 R.S. TRIM
2 X 8 R.S. TRIM
24" Ø DEC. POST W/ 2X4/2X6 TRIM TOP AND BOTTOM
5/8" T-1-11 SIDING W/ GROOVES @ 4" O.C. OVER 15# FELT
2 X 8 FASCIA
8'-1 1/8"
LINE OF FIN FLOOR
6 X 6 STUCCO COLUMN EA. SIDE TYPICAL
9'-0"
8'-1 1/8"
LINE OF FIN. FLOOR
LINE OF LIVING RM. FIN. FLOOR

Problem 6–2.

Chapter 7

Framing Methods and Structural Components

FRAMING METHODS
STRUCTURAL COMPONENTS
TEST

FRAMING METHODS

MORE SO than any other drawing, reading sections requires a thorough understanding of the materials and the process of construction. To read the sections and see how the building is to be constructed requires an understanding of basic construction principles.

In Chapter 5 you were introduced to different types of foundation systems. There are also different types of framing systems. Wood, masonry, and concrete are the most common materials used in the construction of residential and small office buildings. With each material several different framing methods can be used to assemble the components. Wood is the most widely used material for the framing of houses and apartments. The most common framing systems used with wood include balloon, platform, and post and beam. There have also been several variations of platform framing in recent years to provide better energy efficiency.

Balloon Framing

Although balloon framing is not widely used, a knowledge of this system will prove helpful if an old building is being remodeled. With the balloon or Eastern framing method, the exterior studs run from the top of the foundation to the top of the highest level. Because the wall members are continuous from the foundation to the roof, fewer horizontal members are used. Because of the long pieces of lumber needed, a two-story structure was the maximum that could be built easily using balloon framing. Floor framing at the midlevel was supported by a ledger set into the studs. Structural members were usually spaced at 12, 16, or 24 in. on center. Figure 7–1 shows typical construction methods used in the balloon framing system.

Although the length of the wood gave the building stability, it also caused the demise of the system. The major flaw with balloon framing is fire danger. A fire starting in the lower level could race quickly through the cavities formed in the wall or floor systems of the building.

Platform Framing

Platform or Western platform framing is the most common framing system now in use. The system gets

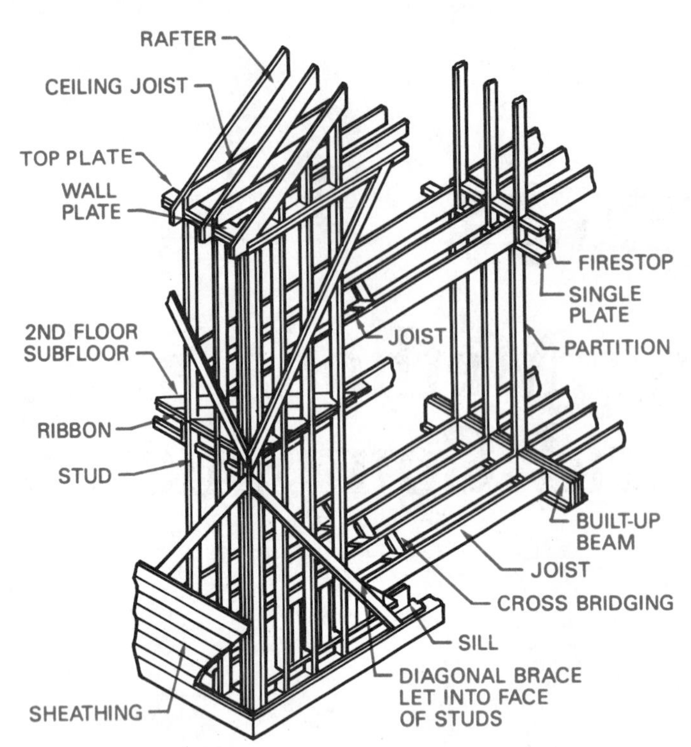

Figure 7–1 Structural members of the balloon framing system are typically spaced at 12, 16, or 24" in. O.C.

its name from the platform created by each floor as the building is being framed. The framing crew is able to use the floor as a platform to assemble the walls for that level. Platform framing grew out of the need for the fireblocks in the balloon framing system. The fireblocks that were placed individually between the studs in balloon framing became continuous members placed over the studs to form a solid bearing surface for the floor or roof system.

Building with the Platform System. With the foundation is in place, the framing crew sets the girders and floor members.

Once the floor platform is in place, the framing materials for the walls can be laid out on the floor. The wall members are aligned, the windows are framed, and even the siding can be installed while the wall is lying on the floor. Once assembled, the wall can be slid into position and tilted into its vertical position. With all the walls in position that will be used to support the upper level, the next floor level can be started. Figure 7–2 shows materials typically used with Western platform construction.

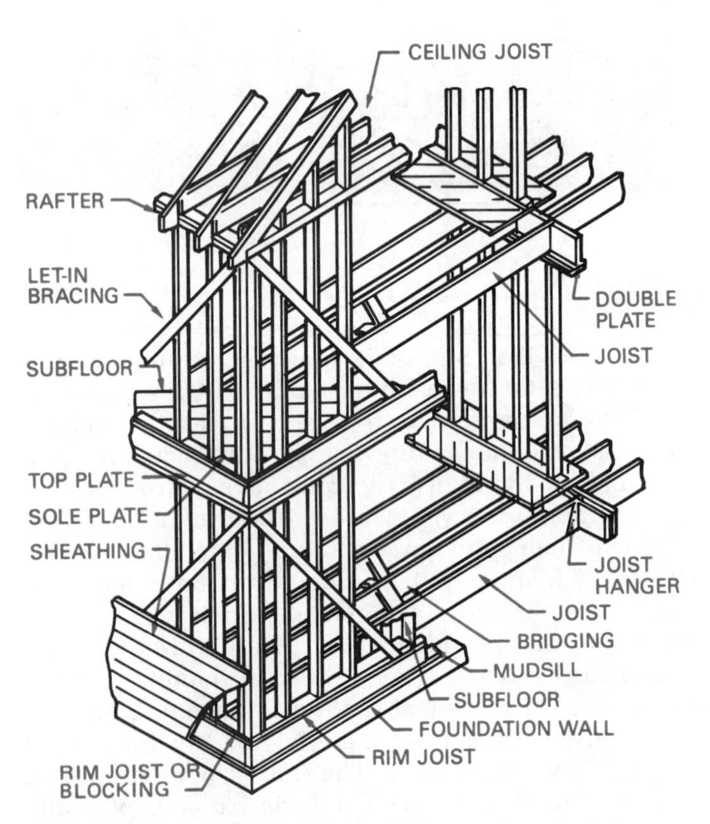

Figure 7–2 Structural members of the western platform system.

Post-and-Beam Framing

Post-and-beam construction places framing members at greater distances apart than platform methods. In residential construction, posts and beams are usually spaced at a minimum of 48 in. on center. Although the system uses less lumber than other methods, the members required are larger. Sizes vary depending on the span, but beams 4 or 6 in. and wider are typically used. The subflooring and roofing over the beams is commonly T&G plywood.

Post-and-beam construction can offer great savings in both lumber and nonstructural materials. This savings results from careful planning of the locations of the posts and the doors and windows located between them. Savings also result by having the building conform to the modular dimensions of the material being used. Figure 7–3 shows common structural member of post-and-beam framing.

Although an entire home can be framed with this method, many contractors use the post-and-beam system for supporting (1) the lower floor when no basement is required and (2) roof systems and then use conventional framing methods for the walls and upper levels.

Brick and Stone Construction

Brick and stone have been used for thousands of years as building materials for structures, providing a durable, maintenance-free exterior. Depending on availability, stone can be used to provide a very rustic wall material. Brick can be purchased almost anywhere and provides a classic upscale finish to a structure. The use of both is restricted by the price of the material and the labor needed to install each material. Both products also pose structural problems where earthquakes occur.

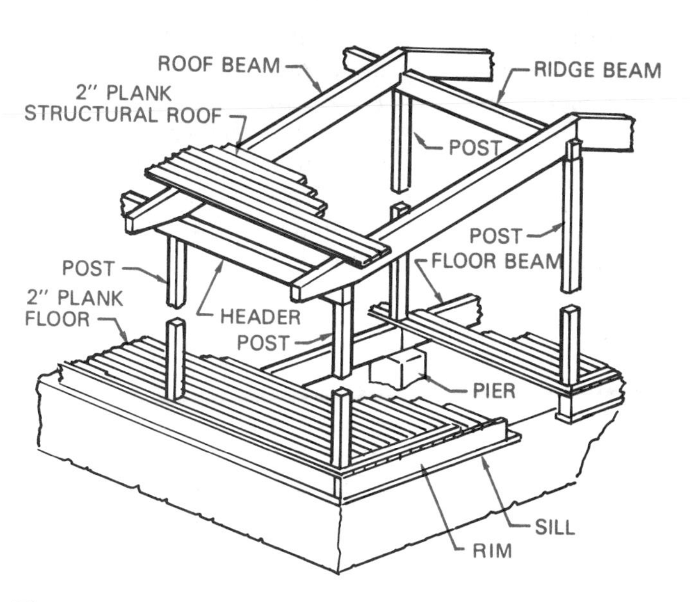

Figure 7–3 Post-and-beam structural members are typically spaced at 48" O.C. In commercial uses spacings may be 20 ft or greater.

A common method of using brick and stone in residential construction is as a veneer. Veneer is a non-load-bearing material. Using brick as a veneer over wood frame construction offers the charm and warmth of brick with a lower construction cost than structural brick. Brick veneer can be used with any of the common methods of frame construction. Figure 7–4 shows a typical method of attaching brick veneer to wood-framed walls.

Energy-Efficient Construction

No matter what system of construction is used, energy efficiency can be a part of the construction process. Some of the following construction techniques may seem like excessive protection, and depending on the area of the country, some of the methods would be inappropriate. The examples given are simply examples of various methods that have been used and found effective for special conditions. The goal of energy-efficient construction is to decrease the dependence on the heating system. This is best done by the use of caulking, vapor barriers, and insulation.

Caulking. Caulking normally consists of filling small seams in the siding or the trim to reduce air

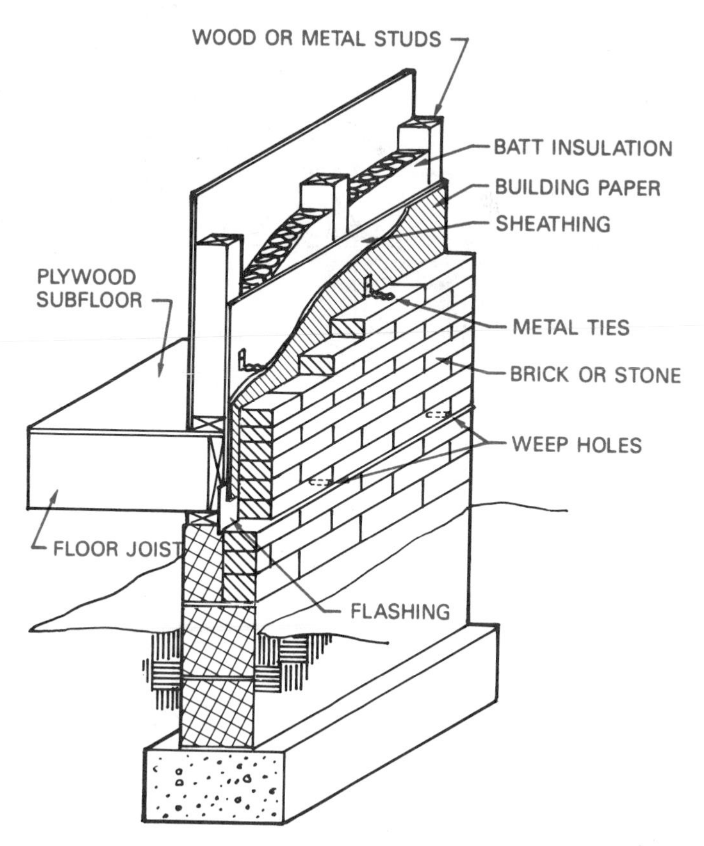

Figure 7–4 Brick is typically used as a non-load-bearing veneer attached to a wood wall in residential applications.

drafts. In energy-efficient construction, caulking is added during construction at all seams or intersections of floors and walls to reduce further the chance of air infiltration. Figure 7–5 shows typical areas where caulking can be added. These beads of caulk keep air from leaking between joints in construction materials. It may seem like extra work for a small effect, but caulking is well worth the effort. Caulking involves minimal expense for material and labor.

Caulking is typically specified in note form rather than being shown on drawings. Most plans include a set of written specifications that dictate the type of caulking and its location.

Vapor Barriers. For an energy-efficient system, airtightness is critical. The ability to eliminate air infiltration through small cracks is imperative if heat loss is to be minimized. Vapor barriers are a very effective method of decreasing heat loss. Building codes require 6 mil thick plastic to be placed over the earth in the crawl space and paper to be placed on the interior surfaces of the wall and ceiling insulation. Many energy-efficient construction methods use these precautions and add a continuous vapor barrier to the walls. This added vapor barrier is designed to keep exterior moisture from the walls and insulation. To be effective, the vapor barrier must be lapped and sealed to keep air from penetrating through the seams in the plastic. Care should be taken to seal the laps in the vapor barrier.

The vapor barrier can be installed in the ceiling, walls, and floor system for effective air control to prevent small amounts of heated air from escaping. All the effort required to keep the vapor barrier intact at the seams must be continued wherever an opening in the wall or ceiling is required.

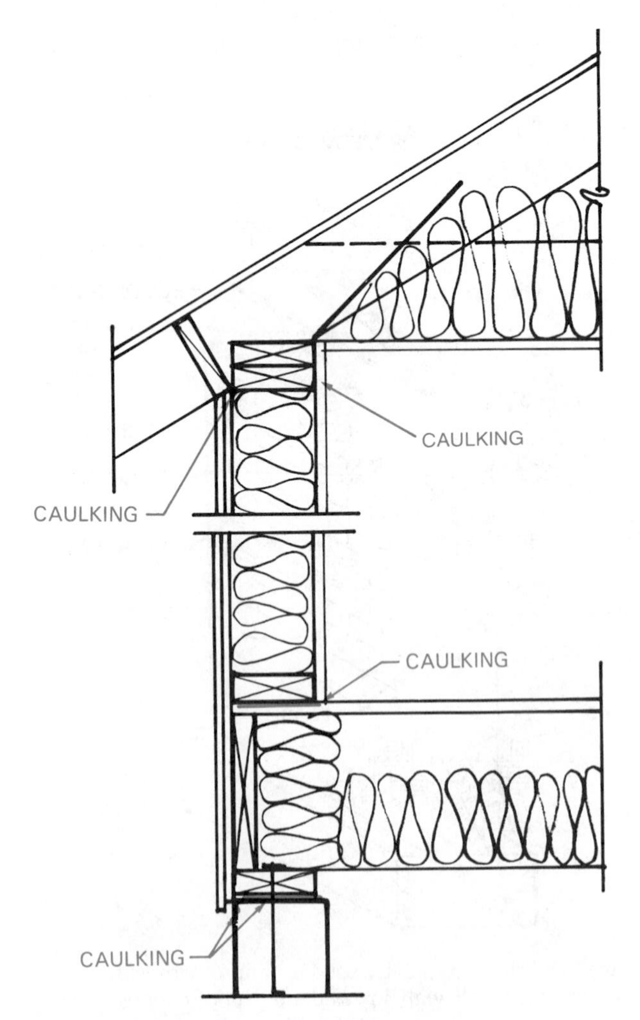

Figure 7–5 Caulking is an important key to eliminating drafts and air infiltration.

Insulation. Many energy-efficient construction methods depend on added insulation to help reduce air infiltration and heat loss. Some of these systems require not only adding more insulation to the structure, but also adding more framing material to contain the insulation. Insulation requirements vary for each area of the country and depending on which national code is enforced. Insulation requirements are increasingly being tied to the efficiency of the heating or cooling unit. Most codes now set up insulation standards for furnaces with 80 percent efficiency or greater. A sample of insulation requirements based on CABO is as follows:

	Less than 80 percent	Greater than 80 percent
Flat ceiling	38	30
Vaulted ceilings	30	19
Walls	19	11
Floors	19	19

Wall openings are also limited by some codes, placing a maximum of between 17 and 22 percent of the heated wall area in openings. These openings include doors, windows, and skylights. Typically a set of plans indicates the amount of openings either on the floor plan or on the list of specifications. It is important for the print reader to find the existing opening to wall percentage before making substitutions in window sizes.

STRUCTURAL COMPONENTS

As with every phase of print reading, sections include their own terminology. The terms referred to in this chapter are basic for structural components of sections. The terms cover floors, walls, and roofs.

Floor Construction

Two common methods of framing the floor system are conventional joist and post and beam.

Conventional Floor Framing. Conventional, or stick, framing involves the use of 2 in. wide members placed one at a time in a repetitive manner. Basic terms of the system include mudsill, floor joist, girder, and rim joist. Each can be seen in Figure 7–6 and throughout this chapter.

The mudsill, sill, or base plate is the first of the structural lumber used in framing the home was introduced in Chapter 5. The *mudsill* is the plate that rests on the foundation and provides a base for all framing. Mudsills are set along the entire perimeter of the foundation and attached to the foundation with anchor bolts. The size of the mudsill is typically 2 × 6.

With the mudsills in place the girders can be set to support the floor joists. Floor joists are usually set on top of the girder, as shown in Figure 7–7, but they also may be hung from the girder with joist hangers.

Posts are used to support the girders. Steel columns may be used in place of a wooden post depending on the load to be transferred to the foundation.

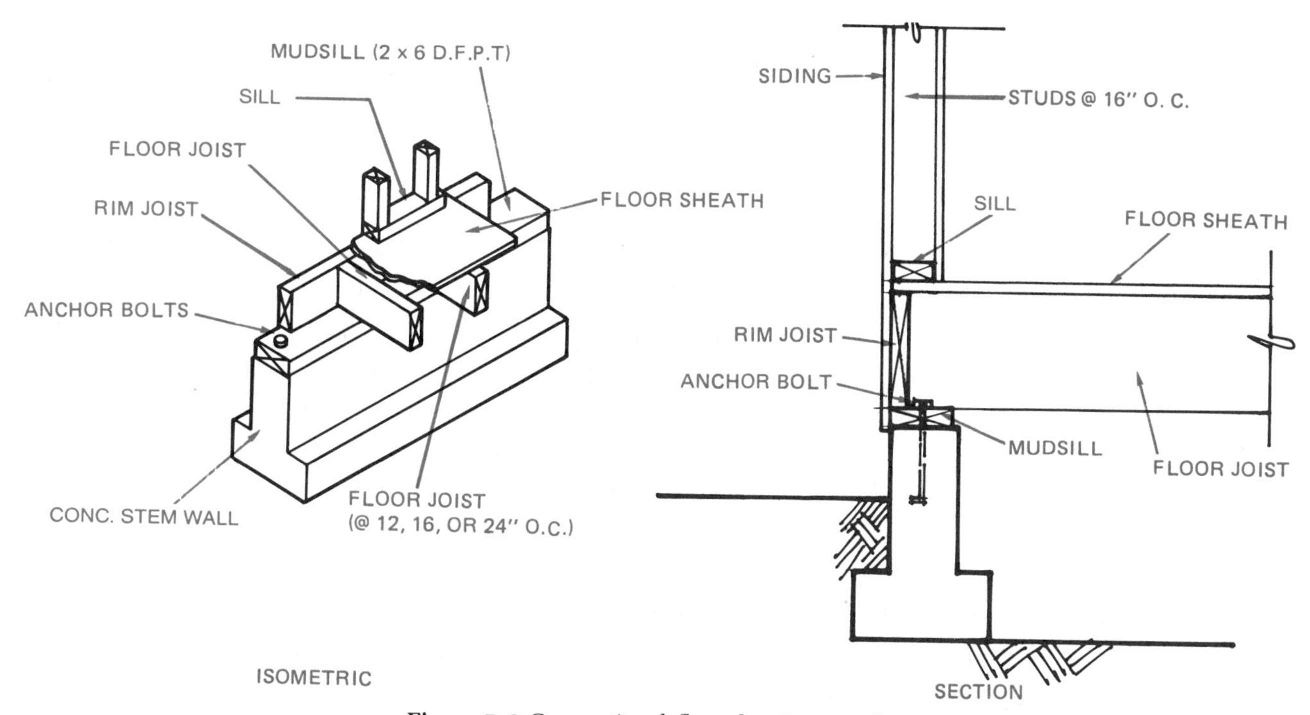

Figure 7–6 Conventional floor framing members.

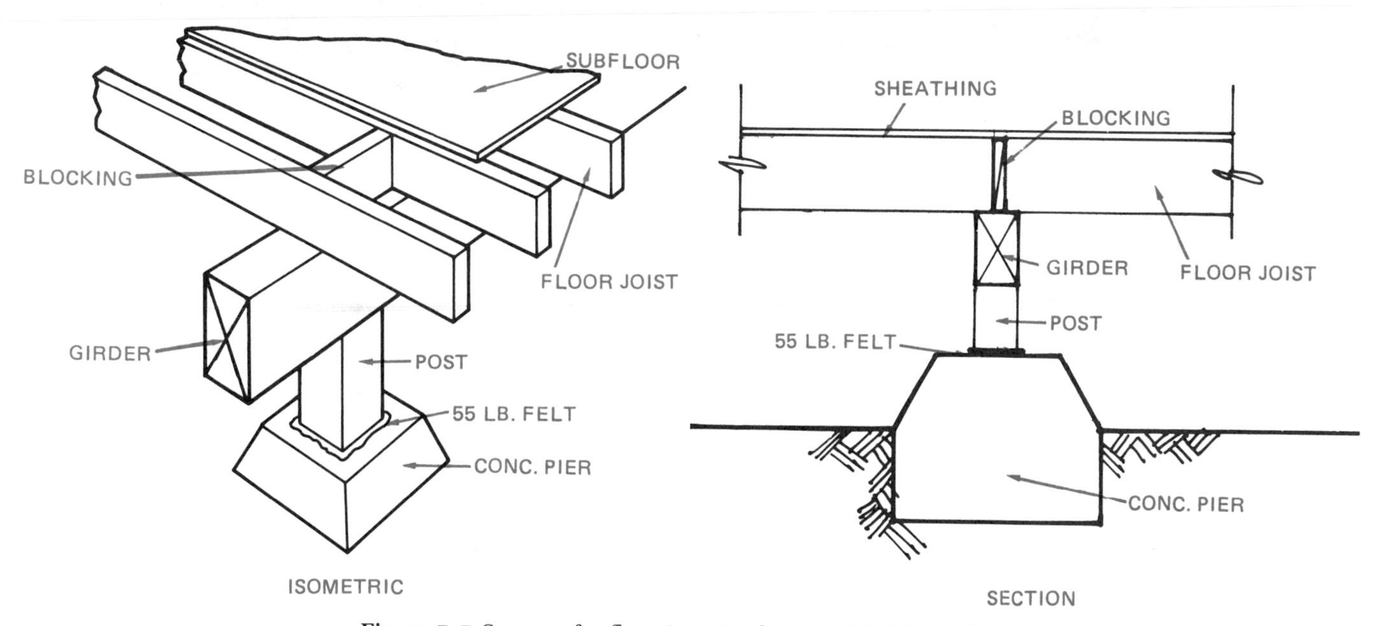

Figure 7–7 Support for floor joist is often provided by girders.

Because a wooden post draws moisture out of the concrete foundation it must be underlaid by 55 lb felt. Sometimes an asphalt roofing shingle is used between the post and the girder.

Once the framing crew has set the support system, the floor joists can be set in place. *Floor joists* are the structural members used to support the subfloor, or rough floor. Floor joists usually span between the foundation and a girder, but as shown in Figure 7–8, a joist may extend past its support. This extension is known as a *cantilever*. Floor joists range in size from 2 × 6 through 2 × 14 and may be spaced at 12, 16, or 24 in. on center. The 16 in. on center is the most common spacing. The spacing varies depending on the span of the joist and the load being supported. Residential floors are typically designed to support a total of 50 lb per sq ft (psf). Commercial structures are typically expected to support 100 lb or more. Floor joists can also be made from plywood. These joists, called truss joists, allow the members to be lighter and longer than conventional floor joists. See Figure 7–9. Trusses can also be used for floor support when a large area must be spanned.

A rim or band joist (or header, as it is sometimes called) is usually aligned with the outer face of the foundation and mudsill. Some framing crews set a

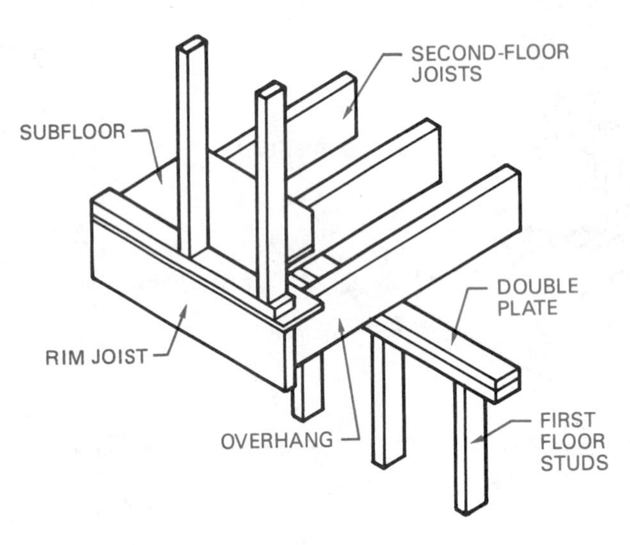

Figure 7–8 A structural member that extends past its support is cantilevered.

Figure 7–9 Plywood truss joist provide a lightweight, economical alternative to using wood floor joist. *Courtesy of Truss Joist MacMillan.*

rim joist around the entire perimeter and then end-nail the floor joists. An alternative to the rim joist is solid blocking at the sill between each floor joist. See Figure 7–10.

Blocking in the floor system is also used to keep floor joists from rolling over on their sides. One use of blocking is at the center of the joist span, as shown in Figure 7–7. Spans longer than 10 ft must be blocked at the center to help transfer lateral forces from one joist to another and then to the foundation system. These blocks help keep the entire floor system rigid. Blocking is also used to reduce the spread of fire and smoke through the floor system.

Floor sheathing is installed over the floor joists to form the subfloor. The sheathing provides a surface on which the base plate of the wall can rest. Plywood or waferboard are the most common floor sheathings. Depending on the spacing of the joist, 1/2, 5/8, or 3/4 in. material is used. Plywood is usually laid so that the face grain on each surface is perpendicular to the floor joist. This provides a rigid floor system without having to block the edges of the plywood.

The plywood is typically covered with hardboard to provide a smooth base for carpet or sheet vinyl. The underlayment is not installed until the walls, windows, and roof are in place, making the house weathertight. The underlayment provides a smooth surface on which to install the finished flooring. Underlayment is usually 3/8 or 1/2 in. hardboard, and may be omitted if the holes in the plywood are filled.

Post-and-Beam Construction. Terms to be familiar with when working with post-and-beam floor systems include girder, post, decking, and finished floor. Notice there are no floor joists with this system.

With post-and-beam construction the girder supports the floor decking instead of floor joists. Girders are usually 4 × 6 beams spaced at 48 in. O.C., but the size and distance may vary depending on the loads to be supported. Similar to conventional methods, posts are used to support the girders. Typically a 4 × 4 is the minimum size used for a post, with a 4 × 6 post used to support joints in the girders. With the support system in place, the floor system can be installed.

Decking is the material laid over the girders to form the subfloor. Typically decking is 1 1/8 in. T&G plywood. The decking is usually finished in fashion similar to conventional decking with a hardboard overlay.

Framed Wall Construction

You will be concerned with two types of walls, bearing and nonbearing. A bearing wall supports not only itself but also the weight of the roof or other floors constructed above it. A bearing wall requires some type of support under it at the foundation or lower floor level in the form of a girder or another bearing wall. A nonbearing wall serves no structural purpose. It is used to divide rooms and can be removed without causing damage to the building.

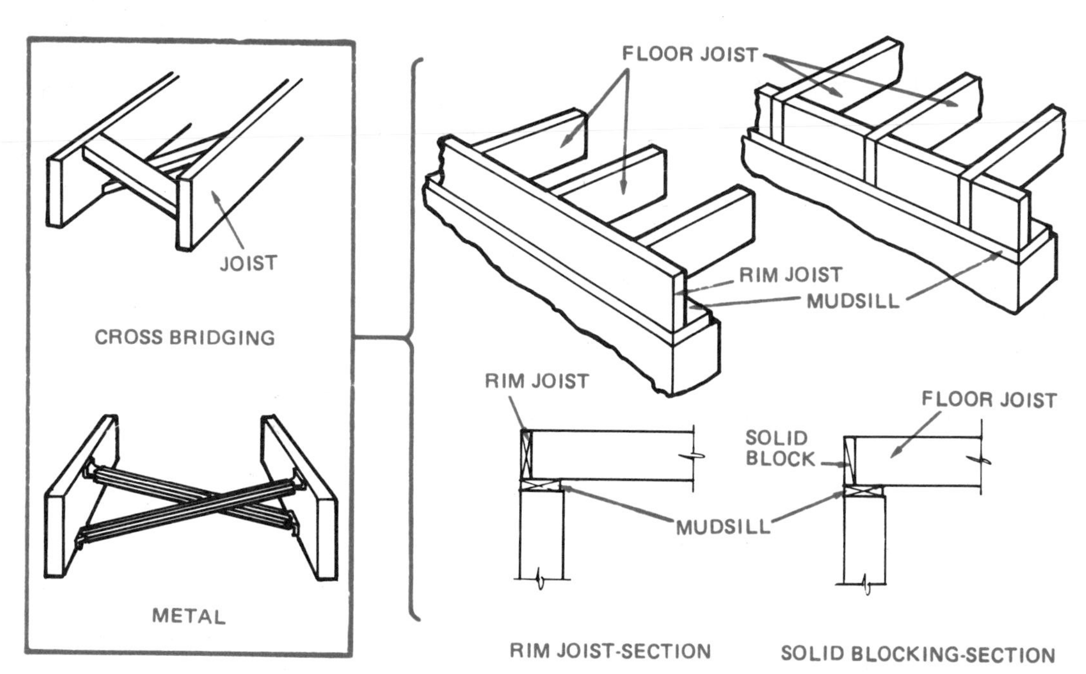

Figure 7–10 Solid or cross blocking is used to prevent floor joist from rolling over. Blocking at the exterior joist ends may be replaced by a continuous rim joist.

Bearing and nonbearing walls are both constructed with similar materials using a sole plate, studs, and a top plate. Each can be seen in Figure 7–11. The sole, or bottom, plate is used to help disperse the loads from the wall studs to the floor system. The sole plate also holds the studs in position as the wall is being tilted into place. Studs are the vertical framing members used to transfer loads from the top of the wall to the floor system. Typically studs are spaced at 16 in. O.C. and provide a nailing surface for the wall sheathing on the exterior side and the wallboard on the interior side. The top plate is located on top of the studs and is used to hold the wall together. Two top plates are required on bearing walls, and each must lap the other a minimum of 48 in. This lap distance provides a continuous member on top of the wall to keep the studs from separating. An alternative to the double top plate is to use one plate with a steel strap at each joint in the plate. The top plate also provides a bearing surface for the floor joists from an upper level or for the roof members.

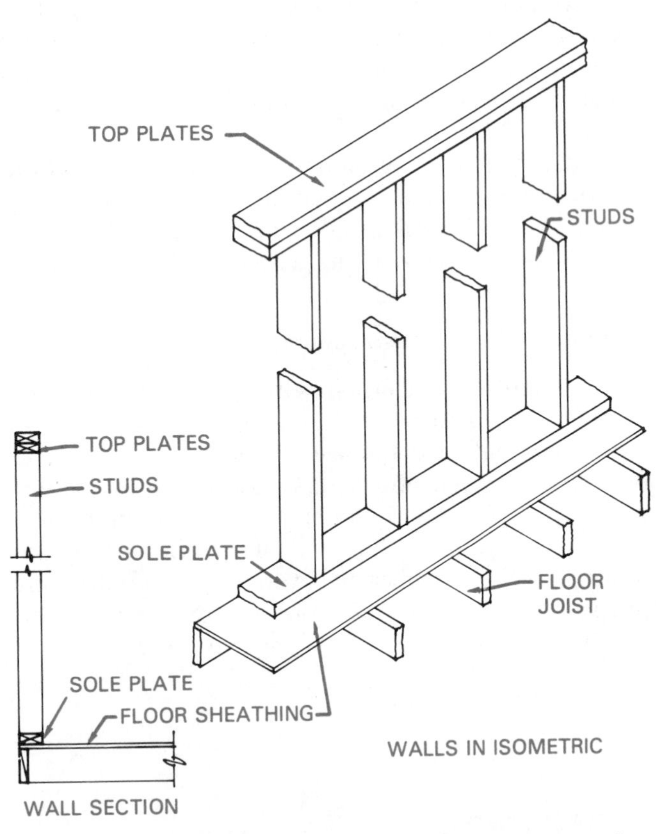

Figure 7–11 Walls are typically made with a double top plate, vertical studs, and a sole plate.

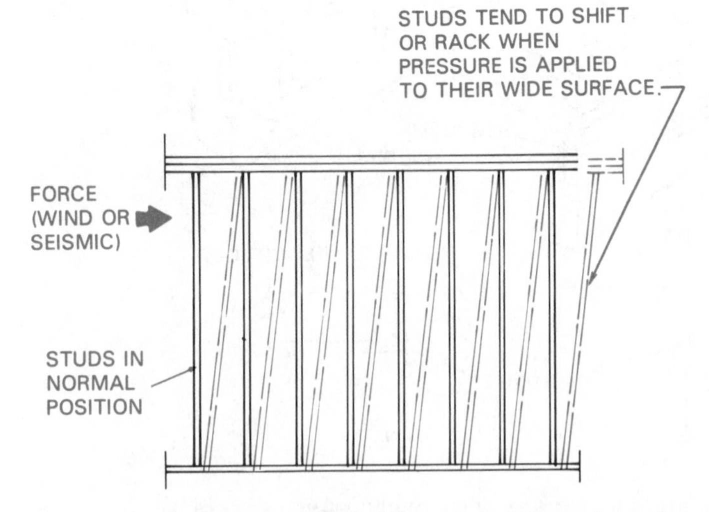

Figure 7-12 Wall racking occur when natural forces force the studs out of their vertical position. Plywood sheathing or 1 × 4 diagonal let-in braces keep the studs square with the floor.

Plywood sheathing or ½" waferboard is primarily used as an insulator against the weather and also as a backing for the exterior siding. Sheathing may be considered optional depending on the area of the country. When sheathing is used on exterior walls, it provides *double-wall construction*. With single-wall construction, the siding is attached over building paper or vapor barrier to the studs. The cost of the home and its location have a great influence on whether wall sheathing is used. In areas where wood is plentiful, a minimum of ⅜ in. plywood is used on exterior walls as an underlayment.

Plywood is also used for its ability to resist the tendency of a wall to twist or rack. Racking can be caused from wind or seismic forces. Plywood used to resist these forces is called a shear panel. See Figure 7–12 for an example of racking and how plywood can be used to resist this motion.

In areas where plywood is not required for shear panels or for climatic reasons, the studs can be kept from racking by using let-in braces. A notch is cut into the studs and a 1 × 4 is laid flat in this notch at a 45° angle to the studs.

Before the mid-1960s, blocking was common in walls to help provide stiffness. Blocking for structural or fire reasons is now no longer required unless a wall exceeds 10 ft in height. Blocking is often installed for a nailing surface for mounting cabinets and plumbing fixtures. Blocking is sometimes used to provide extra strength in some seismic zones.

Wall Openings. A header in a wall is used over an opening such as a door or window. When an opening is made in a wall, one or more studs must be omitted. A header is used to support the weight the missing studs would have carried. The header is supported on each side by a trimmer. Depending on the weight the header supports, double trimmers may be required. The trimmers also provide a nailing surface for the window and the interior and exterior finishing materials. A king stud is placed beside each trimmer and extends from the sill to the top plates. It provides support for the trimmers so that the weight imposed from the header can only go downward, not sideways. Each member can be seen in Figure 7–13.

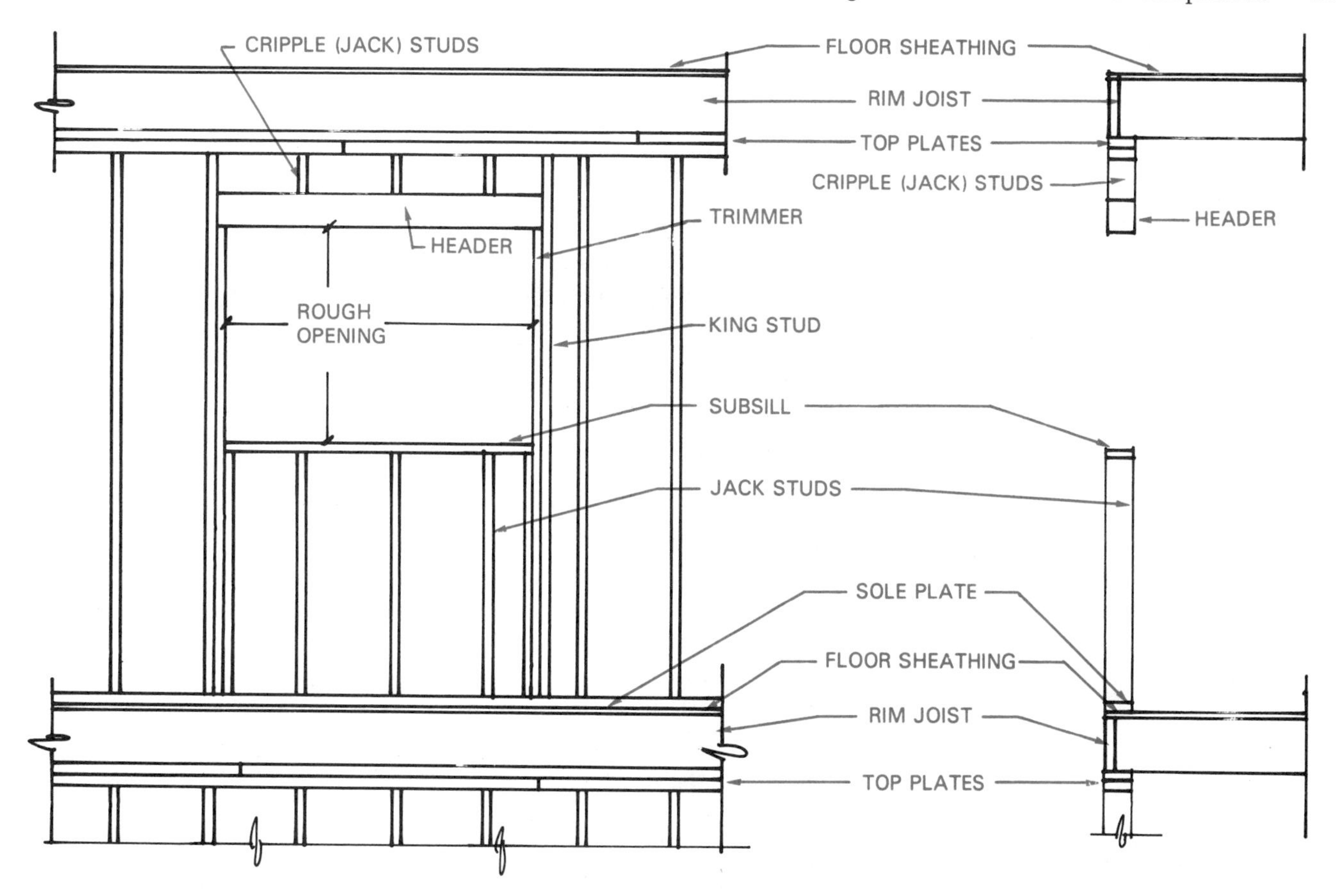

Figure 7–13 Construction components of a wall opening.

Between the trimmers is a subsill located on the bottom side of a window opening. It provides a nailing surface for the interior and exterior finishing materials. *Jack studs*, or cripples, are studs that are not full height. They are placed between the sill and the sole plate and between a header and the top plates.

Masonry Wall Construction

Concrete block and brick construction offers many advantages. A structure made of masonry is practically maintenance free and can last for centuries. Although masonry is only used in some areas of the country for foundation and basement walls, concrete masonry units (CMU) are an excellent building material in warmer, drier climates. Units come in a variety of textures and sizes. The most common size has a nominal size of 8" × 8" × 16". Walls formed with concrete blocks are often represented in sections and details as seen in Figure 7–14.

Blueprints showing concrete block walls typically specify the texture and block pattern on the elevation, not on the sections or details. Sections of CMU are more concerned with the actual construction and attachment methods and will typically specify the

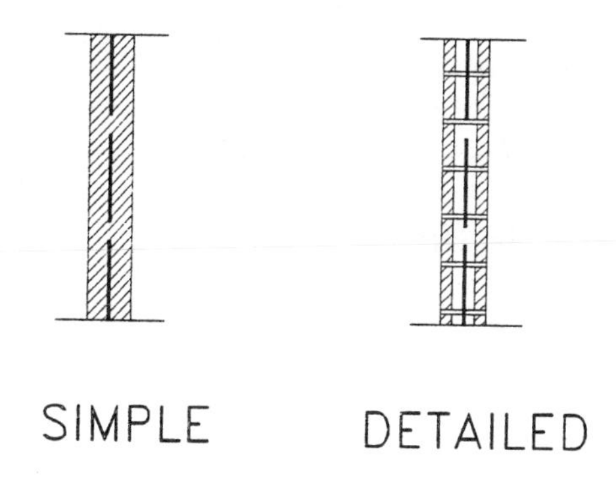

Figure 7–14 Common methods of representing concrete block on sections.

size of units, steel reinforcing pattern, and attachment of the roof or floor to the wall. Figure 7–15 shows a portion of a section for a concrete wall.

Brick. More common than CMU throughout the country is the use of brick. One of the most popular features of brick is the wide variety of positions and patterns in which it can be placed. These patterns are achieved by placing the brick in various positions in relation to each other. The position in which the brick

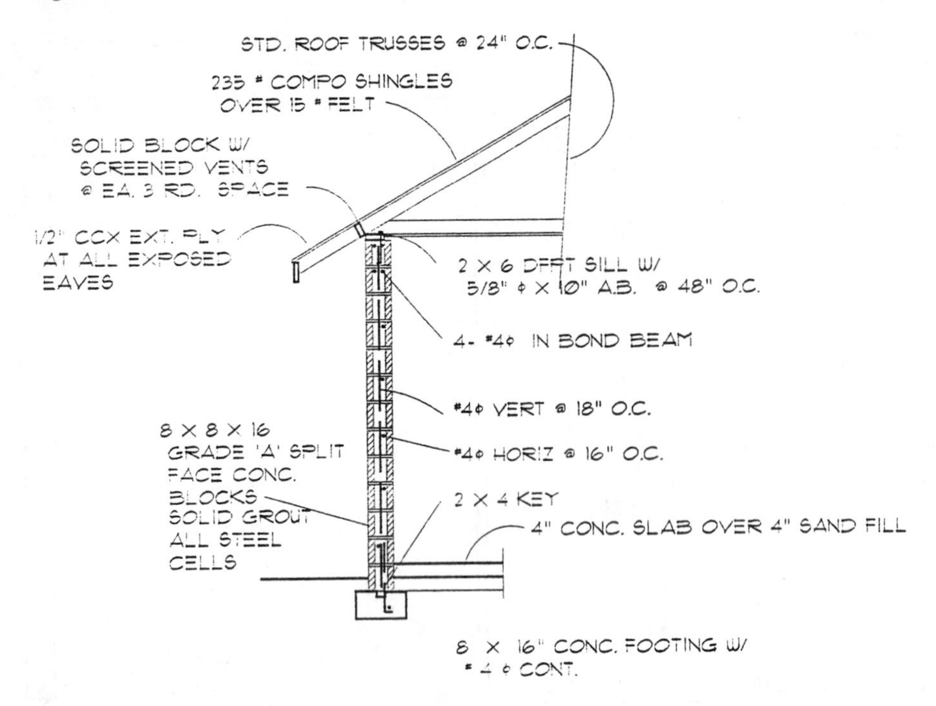

Figure 7–15 Typical use of concrete block in residential construction.

is placed may alter what the brick is called. Figure 7–16 shows the names of common brick positions. Many patterns can be used when laying brick, and Figure 7–17 shows some of the most common. The pattern depends on the desired effect and the economy associated with constructing the wall.

The most common method of using brick is as a veneer on a wood-framed structure. Using brick as veneer allows the amount of brick to be cut in half, resulting in a large financial saving while still providing an attractive exterior appearance. Great care must be taken with brick veneer to protect the wood frame from moisture and to keep the brick from pulling away from the wood. To protect the wood from moisture a vapor barrier of 15# felt or 6 mil

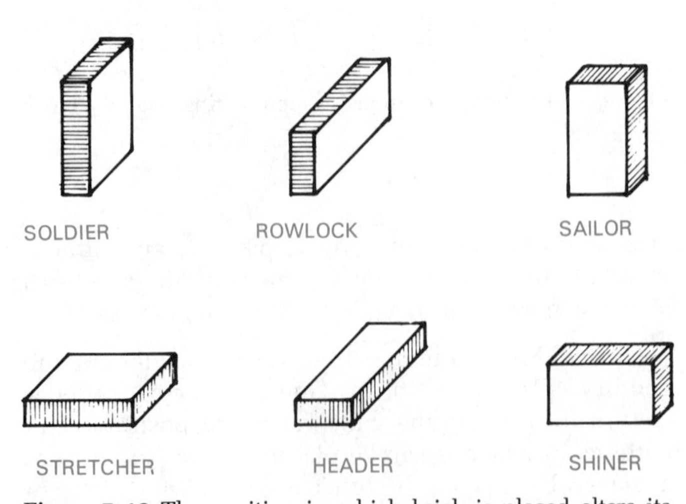

Figure 7–16 The position in which brick is placed alters its name.

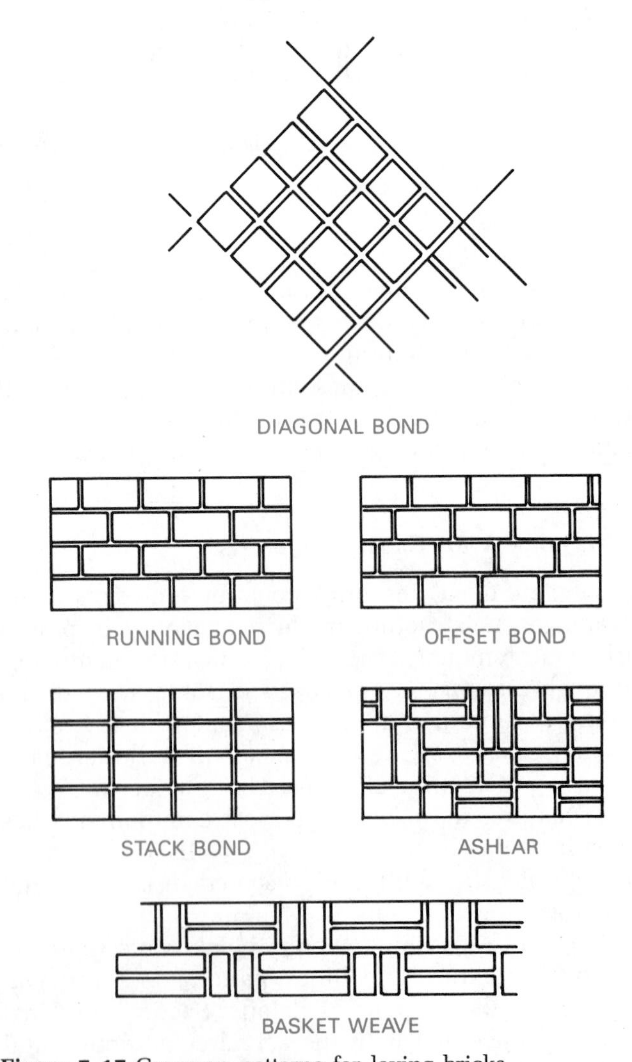

Figure 7–17 Common patterns for laying bricks.

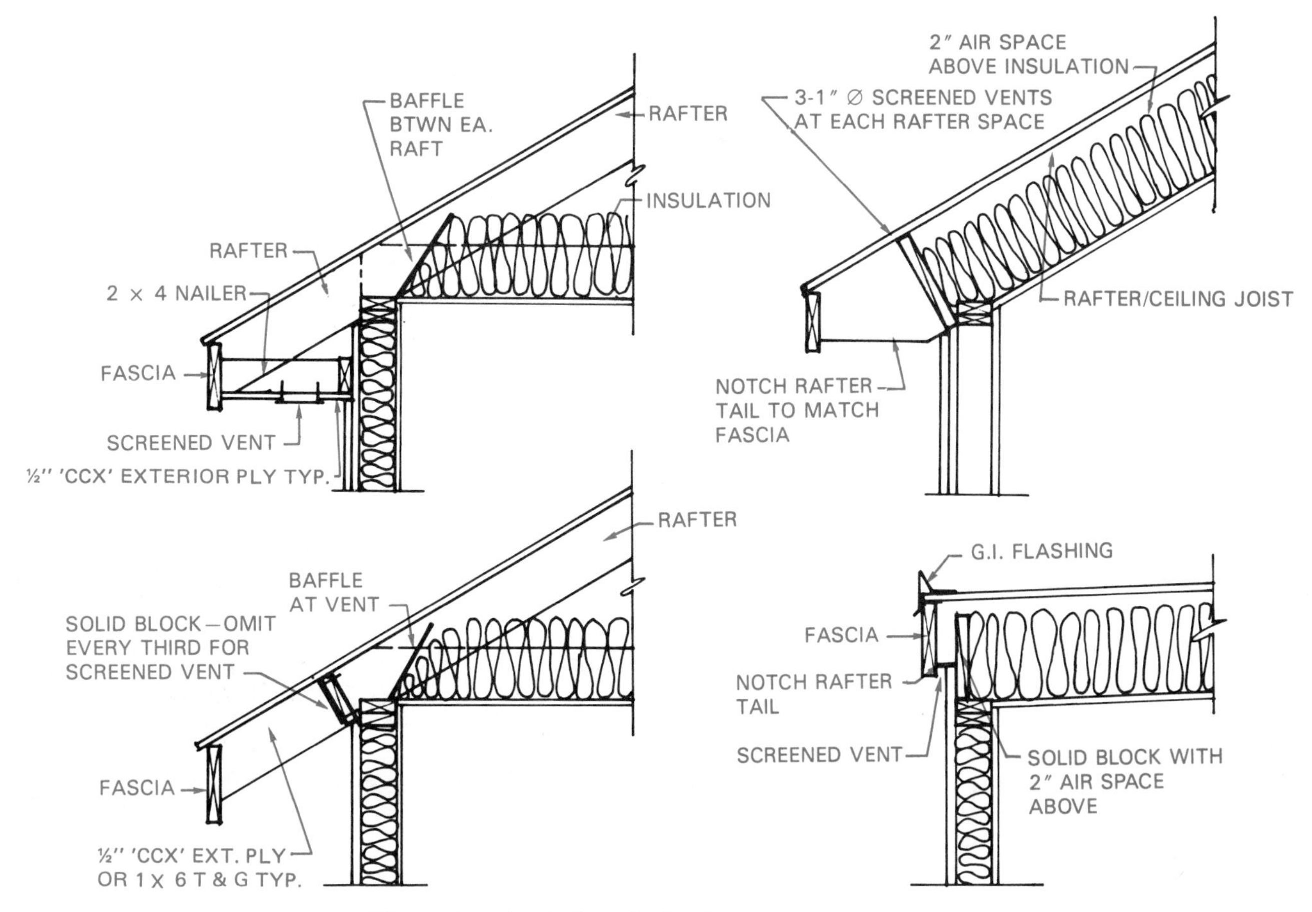

Figure 7–18 Typical methods of constructing a cornice.

plastic is installed in a 1 in. minimum airspace between the brick and the wood frame. The brick is normally attached to the wall with galvanized metal straps at 24 in. on center at each stud.

Roof Construction

Roof framing includes both conventional and truss framing methods. Each has its own special terminology, but many terms apply to both systems. These common terms are described first, followed by the terms for conventional and truss framing methods.

Basic Roof Terms. Roof terms common to conventional and trussed roofs include eave, cornice, eave blocking, fascia, ridge, sheathing, finishing roofing, flashing, and roof pitch dimensions.

The *eave* is the portion of the roof that extends beyond the walls. The *cornice* is the covering that is applied to the eaves. Common methods for constructing the cornice can be seen in Figure 7–18. *Eave* or *bird blocking* is a spacer block placed between the rafters or truss tails at the eave. This block keeps the spacing of the rafters or trusses uniform and keeps small animals from entering the attic. It also provides a cap to the exterior siding.

A *fascia* is a trim board placed at the end of the rafters or truss tails and usually perpendicular to the building wall. It hides the truss or rafter tails from sight and also provides a surface where the gutters may be mounted. The fascia is typically 2 in. deeper than the rafters or truss tails. The fascia may be omitted for economic reasons and replaced with a deeper gutter to hide the rafter tails. At the opposite end of the rafter from the fascia is the ridge. The ridge is the highest point of a roof formed by the intersection of the rafters or the top chords of a truss. See Figure 7–19.

Roof sheathing is similar to wall and floor sheathing but may be either solid or skip. The area of the country and the finished roofing to be used determine which type of sheathing is used. For solid sheathing, ½ in. thick CDX plywood or ½" waferboard is generally used. CDX is the specification given by the American Plywood Association to designate standard-grade plywood. It provides an economical, strong covering for the framing, as well as an even base for installing the finished roofing.

Skip sheathing is generally used with either tile or shakes. Typically 1 × 4's are laid perpendicular to the rafters with a 4 in. space between each piece of sheathing. A water-resistant sheathing must be used

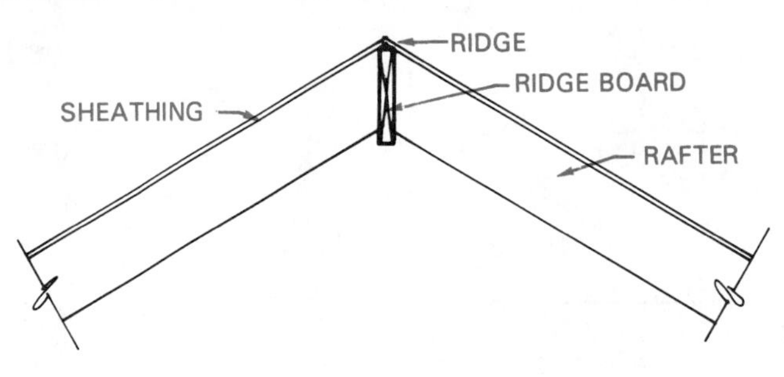

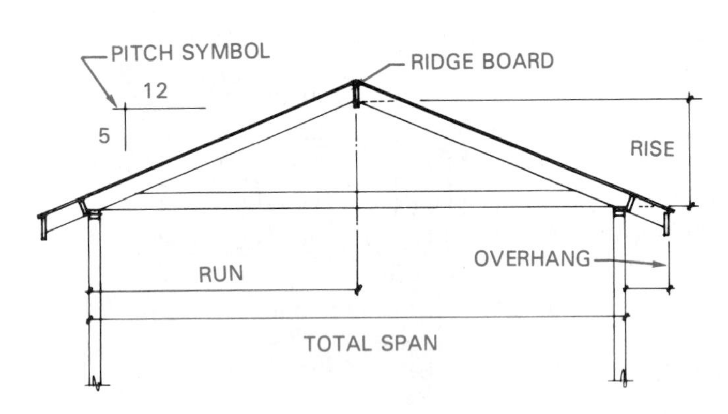

Figure 7–20 Roof dimensions needed for construction.

Figure 7–21 Roof members in conventional construction.

when the eaves are exposed to weather. This usually consists of plywood rated CCX or 1 in. T&G decking. CCX is the specification for exterior use plywood. The finished roofing is the weather protection system. Typically roofing might include built-up roofing, asphalt shingles, fiberglass shingles, cedar, tile, or metal panels. For a complete discussion of roofing materials see Chapter 6. Flashing is generally 20 to 26 gage metal used at wall and roof intersections to keep out water.

Pitch, span, and overhang are dimensions needed to define the *angle*, or steepness, of the roof. Each can be seen in Figure 7–20. Pitch is used to describe the slope of the roof. *Pitch* is the ratio between the horizontal run and the vertical rise of the roof. The *run* is the horizontal measurement from the outside edge of the wall to the centerline of the ridge. The *rise* is the vertical distance from the top of the wall to the highest point of the rafter being measured. The *span* is the measurement between the outside edges of the supporting walls. The *overhang* is the horizontal measurement between the exterior face of the wall and the end of the rafter tail.

Conventionally Framed Roof Terms. Conventional, or stick, framing methods involve the use of wood members placed in repetitive fashion. Stick framing involves the use of such members as ridge board, rafter, and ceiling joists. See Figure 7–21 for their location and relationship to each other.

The ridge board is the horizontal member at the ridge that runs perpendicular to the rafters. The ridge board is centered between the exterior walls when the pitch on each side is equal. The ridge board resists the downward thrust resulting from gravity trying to force the rafters into a V shape between the walls. The ridge board does not support the rafters but is used to align the rafters so that their forces are pushing against each other.

Rafters are the sloping members used to support the roof sheathing and finished roofing. Rafters are typically spaced at 12, 16, or 24 in. O.C. with 24" spacing most common. There are various kinds of rafters, including common, hip, valley, and jack rafters. Each can be seen in Figure 7–21.

A *common rafter* is used to span and support the roof loads from the ridge to the top plate. Common rafters run perpendicular to both the ridge and the wall supporting them. The upper end rests squarely against the ridge board, and the lower end receives a notch and rests on the top plate of the wall. The notch, or bird's mouth, is cut in the rafter at the point where the rafter intersects a beam or bearing wall. This notch increases the contact area of the rafter by placing more rafter surface against the top of the wall.

Hip rafters are used when adjacent slopes of the roof meet to form an inclined ridge. The hip rafter extends diagonally across the common rafters and forms an exterior corner. The hip is inclined at the same pitch as the rafters. A valley rafter is similar to a hip rafter. It is inclined at the same pitch as the common rafters that it supports. *Valley rafters* get their name because they are located where adjacent roof slopes meet to form a valley. *Jack rafters* span from a wall to a hip or valley rafter. They are similar to a common rafter but span a shorter distance. Typically a section specifies only common, hip, and valley rafters.

Rafters tend to settle because of the weight of the roof and because of gravity. As the rafters settle, they push supporting walls outward. These two actions, downward and outward, require special members to resist these forces. These members include ceiling joists, ridge bracing, collar ties, purlins, and purlin blocks and braces. Each can be seen in Figure 7–22.

Ceiling joists span between the top plates of bearing walls to resist the outward force placed on the walls from the rafters. Ceiling joists also support the finished ceiling. *Collar ties* are also used to help resist the outward thrust of the rafters. They are usually the same size as the rafters and placed in the upper third of the roof.

Ridge braces are used to support the downward action of the ridge board. The brace is usually a 2 × 4 spaced at 48 in. O.C. maximum set at 45° maximum to the ceiling joist. A *purlin* is a brace used to provide support for the rafters as they span between the ridge and the wall. The purlin is usually the same size as the rafter and is placed below the rafter to reduce the span. As the rafter span is reduced, the size of the rafter can be reduced. A *purlin brace* is similar to a ridge brace and is used to transfer weight from the purlin to a supporting wall. A scrap block of wood is used to keep the purlin from sliding down the brace. When there is no wall to support the ridge brace, a strongback is added. A *strongback* is a beam placed over the ceiling joist to support the ceiling and roof loads.

If you are working with a set of plans that has a vaulted ceiling, you must be familiar with two other terms. These are rafter/ceiling joist and ridge beam. Both can be seen in Figure 7–23. A *rafter/ceiling joist*, or *rafter joist*, is a combination of rafter and ceiling joist. The rafter/ceiling joist is used to support both the roof loads and the finished ceiling.

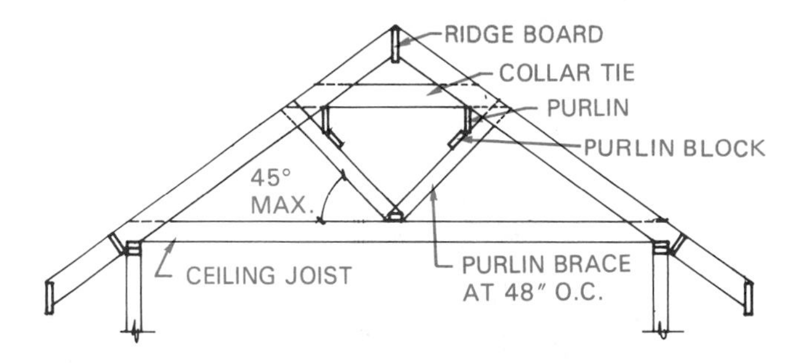

Figure 7–22 Common roof supports.

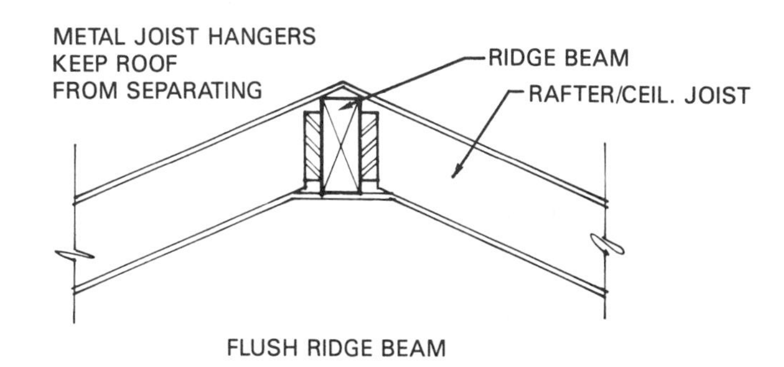

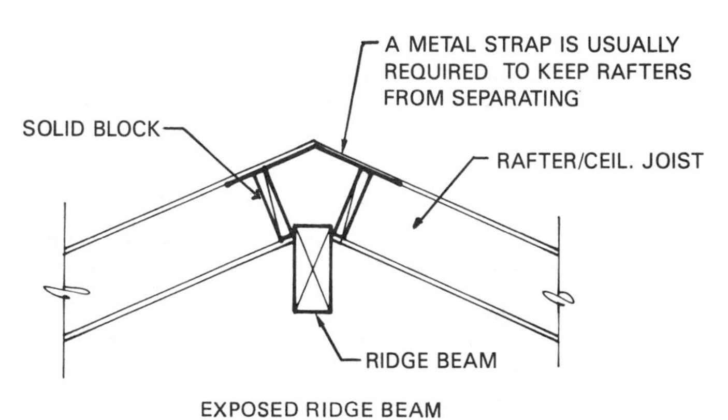

Figure 7–23 Common connections between the ridge beam and rafters. The ridge may be either exposed or hidden.

Typically a 2 × 10 or 12 rafter/ceiling joist is used. This allows room for 8" or 10" of insulation and 2 in. of airspace above the insulation.

A ridge beam is used to support the upper end of the rafter/ceiling joist. Since there are no horizontal ceiling joists, metal joist hangers must be used to keep the rafters from separating from the ridge beam.

The final terms that you need to be familiar with for a stick roof are header and trimmer. Both are terms that are used in wall construction, and they have a similar function when used as roof members. See Figure 7–24. A *header* at the roof level consists of two members nailed together to support rafters around an opening, such as a skylight or chimney. *Trimmers* are two rafters nailed together to support the roofing on the inclined edge of an opening.

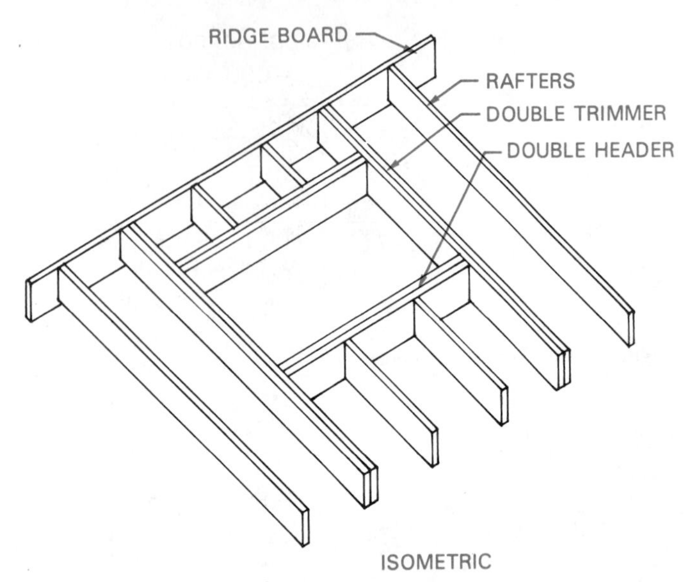

Figure 7–24 Framing around an opening in the roof requires the use of a header and trimmers.

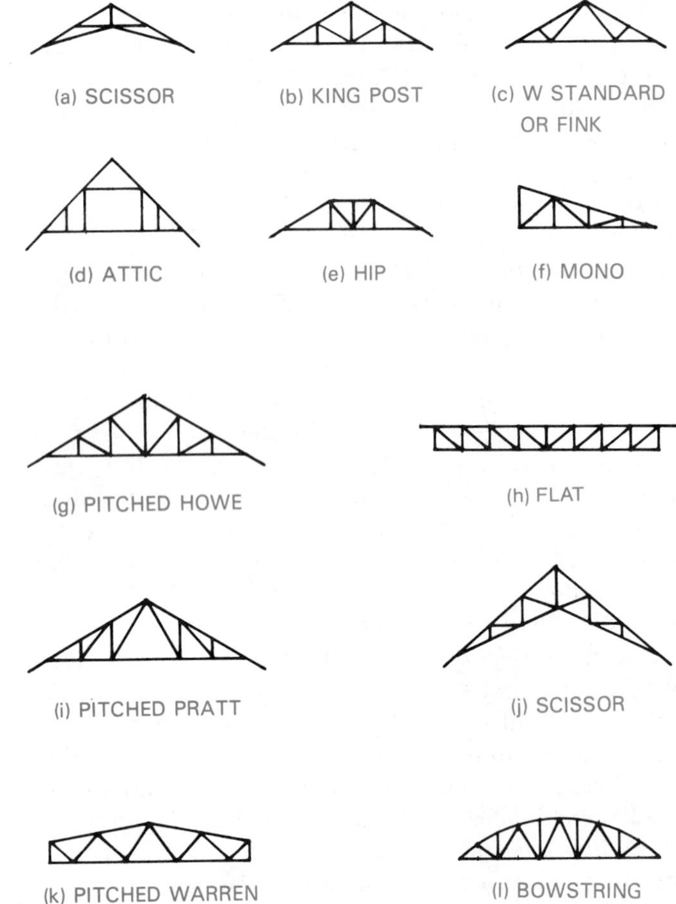

Figure 7–25 Common types of trusses used in construction.

Truss Roof Construction Terms. Truss construction is generally considered non-conventional construction. A *truss* is a component used to span large distances without intermediate supports. Trusses can be either prefabricated or job built. Prefabricated trusses are commonly used in residential construction. Assembled at the truss company and shipped to the job site, the truss roof can be set in place quickly. The size of the material used in trusses is smaller than with conventional frames. Typically, residential truss members need only be 2 × 4's set at 24 in. on center.

Job-built trusses are similar to prefabricated trusses except the framing crew builds the trusses on-site. Job-built trusses are laid out on the floor and then lifted into place. Although this method is only slightly faster than conventional framing, it has the advantage of using smaller members to build the trusses.

Several shapes and types of trusses are available for residential and commercial use. Each can be seen in Figure 7–25. The most commonly used truss in residential construction is the W or Fink truss. On many plans it is commonly called a *standard truss.* The *scissor truss* provides the option of a vaulted ceiling with the speed of the truss installation. A standard and scissor truss may be combined to provide a flat ceiling in one area and a vaulted ceiling in another portion of the structure. When these two trusses are combined this is typically specified on the framing plan as a STD./SCISSOR TRUSS. Usually the plans also contain a view showing the truss manufacturer where the shape changes occur. An example can be seen in Figure 7–26.

An alternative to using a combination standard and scissor truss is to use a scissor truss and have the framing crew add the required members to frame the flat ceiling area. This practice is often referred to as *scabbing on.* An example of a scabbed-on ceiling can be seen in Figure 7–27.

The *mono truss* is very useful in blending a one-level structure into a two-story structure. An example can be seen in Figure 7–28.

It is important to remember, when a truss roof is used, that there are no ceiling joists, rafters, or any of the other members seen in Figure 7–22. Instead of these common members, a truss is formed from top and bottom chords and webs. Other members associated with a truss roof are ridge blocks, hurricane ties, and truss connectors. Each can be seen in Figure 7–29.

The *top chord* is the inclined member used to support the sheathing and finished roof material. The *bottom chord* is the horizontal member that resists the outward forces on the top chord and supports the finished ceiling.

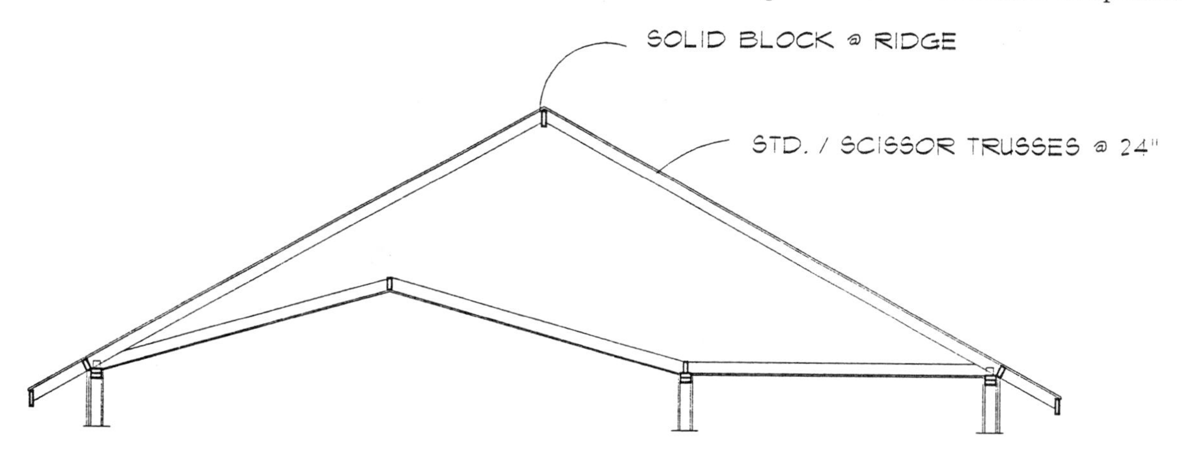

Figure 7–26 A drawing of a standard/scissor truss is usually provided to show the manufacture where to vault the roof.

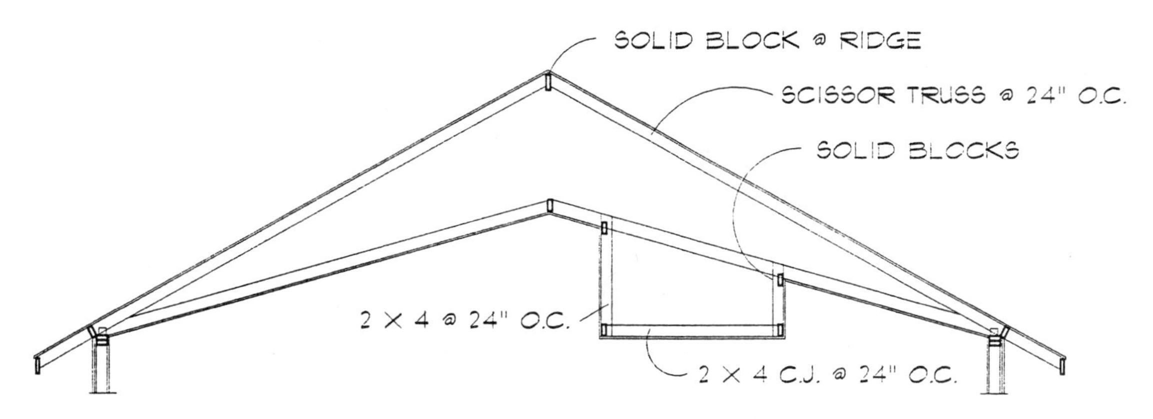

Figure 7–27 An alternative to a std./scissor truss is to "scab on" a false ceiling at the job site.

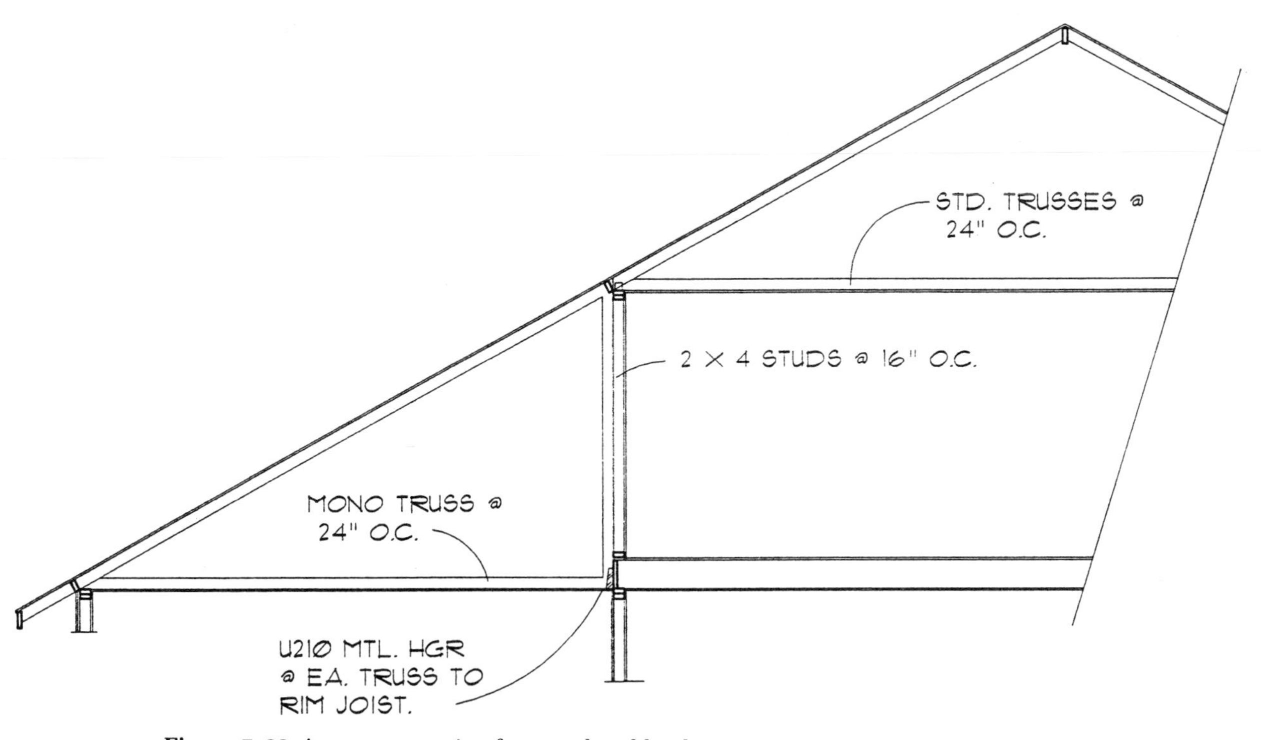

Figure 7–28 A mono truss is often used to blend a one-story area into a two-story area.

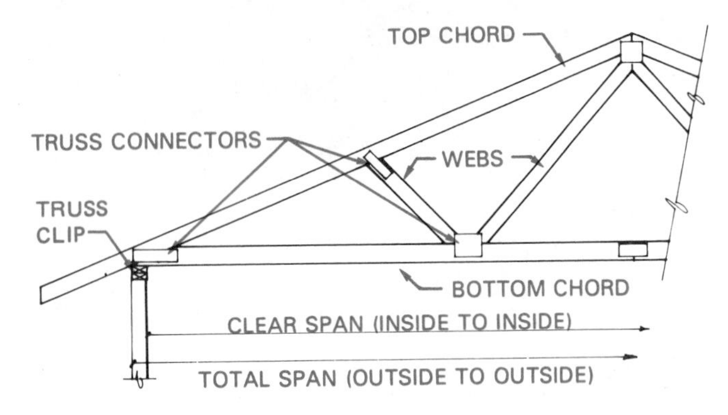

Figure 7–29 Common members used in truss construction.

Since a truss is a self-contained unit, no ridge board is required. Instead, a block is placed between each truss to help maintain uniform spacing and provide a nailing surface for the plywood at the ridge. Figure 7–30 shows truss connections typically shown on sections.

Truss clips or *hurricane tie*, resists the tendency of lift the truss and roof off the wall from wind pushing upward on the eave. This clip transfers the force of any wind uplift acting on the trusses down through the walls and into the foundation.

The truss connector is typically not shown or specified on a set of drawings. These are shown only on the drawings produced by the truss manufacturer. The plans drawn by the architect or designer show what type of truss is desired. The drawings provided by the truss manufacture show how the truss is constructed, with lumber sizes and the types of connectors to be used.

Metal Hangers. Metal hangers are used on floor, ceiling, and roof members. These hangers are typically used to keep structural members from separating. Several common types of connectors are used in light construction. These connectors keep beams from lifting off posts, keep posts from lifting off foundations, or hold one beam or joist to another. The methods used to represent these items on blueprints can be seen in Figure 7–30.

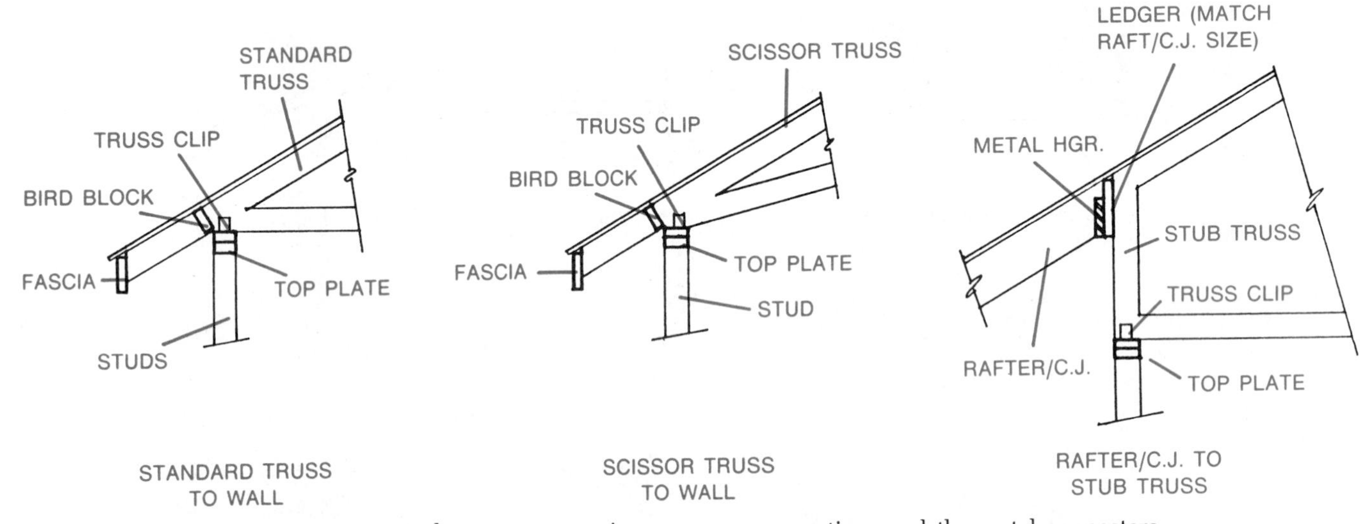

Figure 7–30 Typical trusses connections are seen on sections and the metal connectors used to hold the trusses.

CHAPTER 7 TEST

1. What framing system has studs that extend from the foundation to the ceiling at the tallest level? ____.

2. What are the disadvantages in the framing system of question 1? ____.

3. What framing system uses timbers placed at approximately 48 in. on center? ________________.
4. What is the most common method of using brick in residential construction? ________________
____
________________.
5. What is the nominal size of common concrete blocks? ________________
6. What factors influence how much reinforcing will be used in a wall made of CMU? ________________
____
________________.
7. List three common uses for concrete blocks. ________________
____
________________.
8. Identify the framing system that provides a level base for workers to construct walls. ________________.
9. What is the main goal of energy-efficient construction? ________________
____
________________.
10. What are the three main components of energy-efficient construction? ________________
________________.
11. What is used to keep the structure from lifting off the foundation? ________________
________________.
12. What is the purpose of the mudsill? ________________
____
________________.
13. What angle and spacing are ridge and purlin braces set at? ________________
____
________________.
14. List the three main structural members of a stick-framed roof. ________________
____
________________.
15. What is the typical method used to increase the bearing surface of a rafter where it rests on a wall? ____
____
________________.
16. The horizontal member above a window is called a ________________ and the horizontal member below a window is a ________________.
17. List and describe two methods of keeping a wall from wracking. ________________
____
________________.
18. List the three main members of a roof truss. ________________
____
________________.
19. A horizontal member used to decrease the span of a rafter is called a ________________.
20. List two components of a truss used at a connection. Describe what the member connects and where it is typically specified. ________________
____
____
________________.
21. List two types of walls. ________________
22. A horizontal member placed above the ceiling joist to support roof loads is called a ________________.
23. A ________________ is used to support the horizontal support member below a window.

24. What is the purpose of a purlin? ______

25. A ______ is used to support rafters at openings in a roof, such as a skylight.
26. List two types of studs that are used to support the structural member over a window. ______.
27. List and describe the three main members of a wall. ______

28. Describe the difference between a wythe and a course of a brick or masonry wall. ______

29. What structural member is used to support the floor joists? ______

30. What member is used to support floor sheathing in a post-and-beam floor? ______.
31. List eight patterns of laying brick in a wall. ______

32. The member used to keep furry vermin out of the attic where the rafters and wall intersect is a ______.

33. What is used to keep a floor joist from rolling over? ______

34. What are the two actions of rafters, and what members are used to provide a reaction? ______

35. Describe the difference between ridge blocking and a ridge board. ______

36. A floor joist that extends past its support is a ______ joist.
37. What three members are used in the support of a girder to transfer the loads into the soil? ______

38. A framing member that combines the role of a ceiling joist and a rafter is a ______.
39. What is the minimum lap of top plates in a wall? ______.
40. When the exterior siding is applied directly to the studs the process is known as ______.
41. What is used to divert water from intersections of roofs to walls? ______.
42. The highest part of a roof is a ______.
43. What keeps the purlin from sliding down the brace? ______

44. List four common types of trusses in residential construction. ______

45. List the required R value for the following items assuming an 80 percent efficient furnace:

Walls ______

Floor ______

Flat ceiling ______

Vaulted ceiling ______

Chapter 8
Reading Sections and Details

TYPES OF SECTIONS
SCALE
SECTION ALIGNMENT
READING SECTIONS
TEST
EXERCISES

SECTIONS ARE drawn to show the vertical relationships of the structural materials called for on the floor, framing, and foundation plans. The sections show the methods of construction for the framing crew. Before drawing sections it is important to understand the different types of sections, their common scales, and the relationship of the cutting plane to the section.

TYPES OF SECTIONS

Three types of sections may be drawn for a set of plans. These include full sections, partial sections, and details.

For a simple structure, only one section might be required to explain fully the types of construction to be used. A full section is a section that cuts through the entire building. A full section can be seen in Figure 8–1. A structure may include several full sections depending on the different types of construction materials and techniques used. An alternative to using several full sections is to use a full section showing typical framing techniques and a partial section. The partial section can be used to show atypical construction methods, such as variations of the roof or foundation. Figure 8–2 shows a partial section. Depending on the complexity of the project, details or enlargements of a section may also be required. An area of the section where several items intersect or where many small items are required are examples of when a detail might be required. An example of a foundation detail can be seen in Figure 8–3. Often when details are drawn, a section is drawn with very little information placed on it. This section serves as a reference map to indicate how the details relate to each other. An example of this type of section can be seen in Figure 8–4 with a portion of the related details seen in Figure 8–5. Each detail is specified by a letter over a number. The bottom number indicates on which page the detail can be found. The upper letters specify the detail location on the page. Some companies use numbers instead of letters, but the principle remains the same.

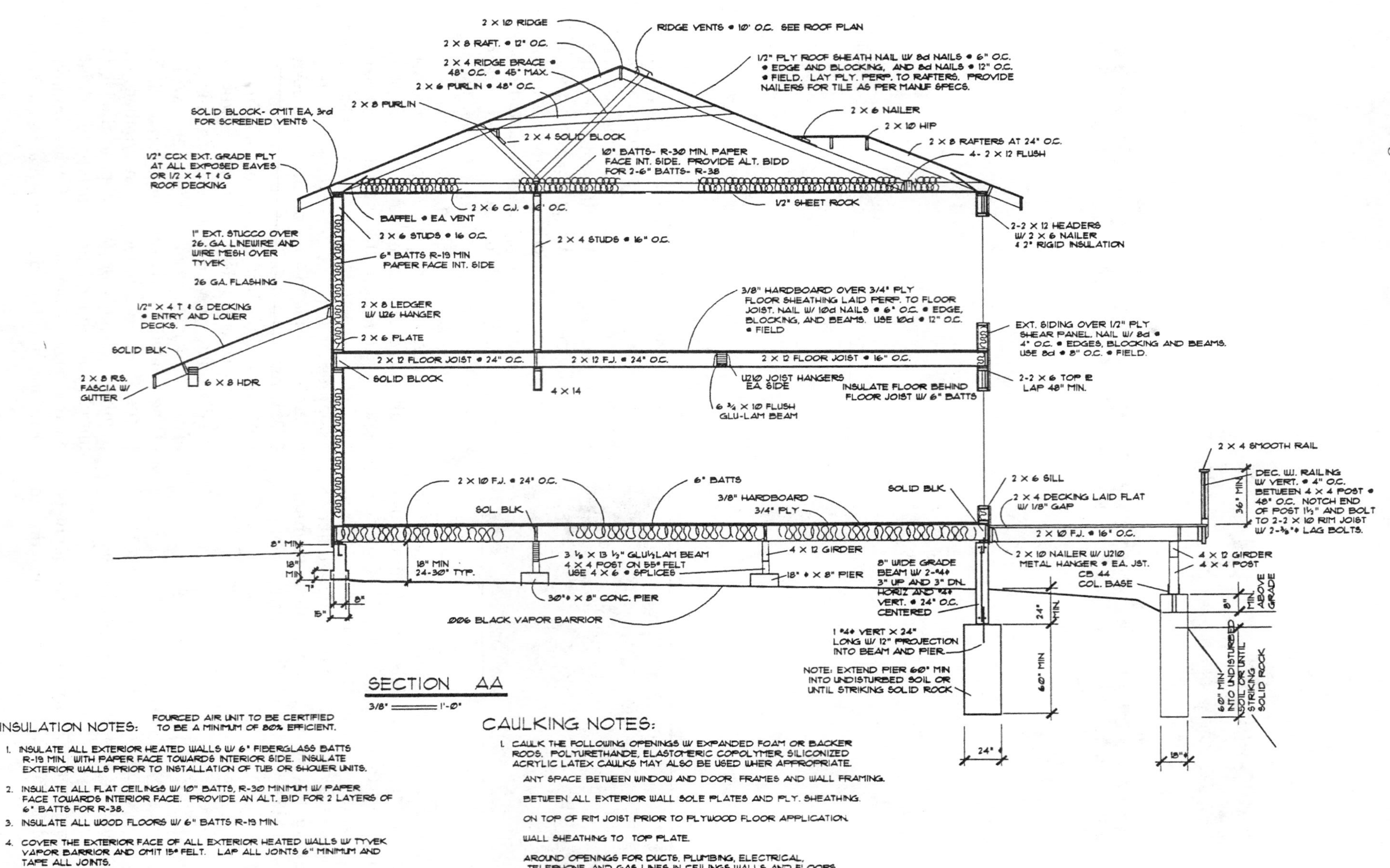

Figure 8–1 A full section is used to show the vertical relationships within a structure.

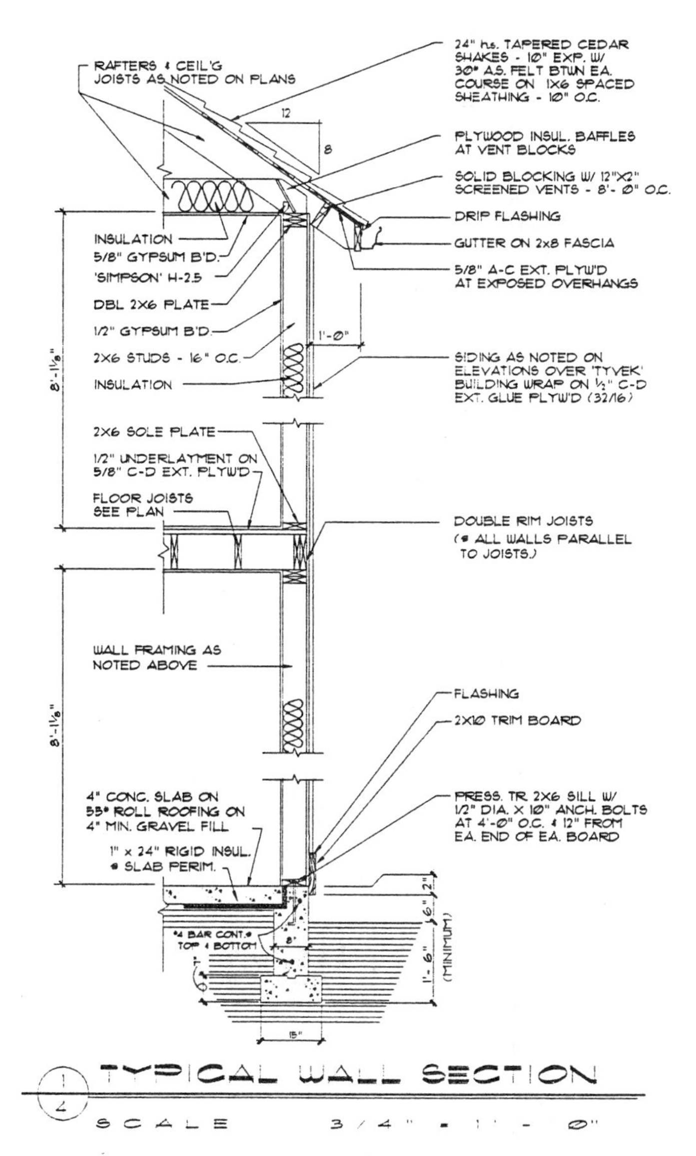

Figure 8–2 A partial section will show construction methods for a specific area of a structure to supplement a full section. *Courtesy Piercy & Barclay Designers, Inc.*

By combining the information on the framing plans and the sections, the contractor should be able to make accurate estimates of the amount of material required and the cost of completing the project. To help make the sections easier to read, sections have become somewhat standardized in several areas. These include the areas of scales and alignment.

SCALE

Sections are typically drawn at a scale of ⅜" = 1'–0". Scales of ⅛ or ¼ in. may be used for supplemental sections requiring little detail. A scale of ¾" = 1'–0" or larger may be required to draw some construction details.

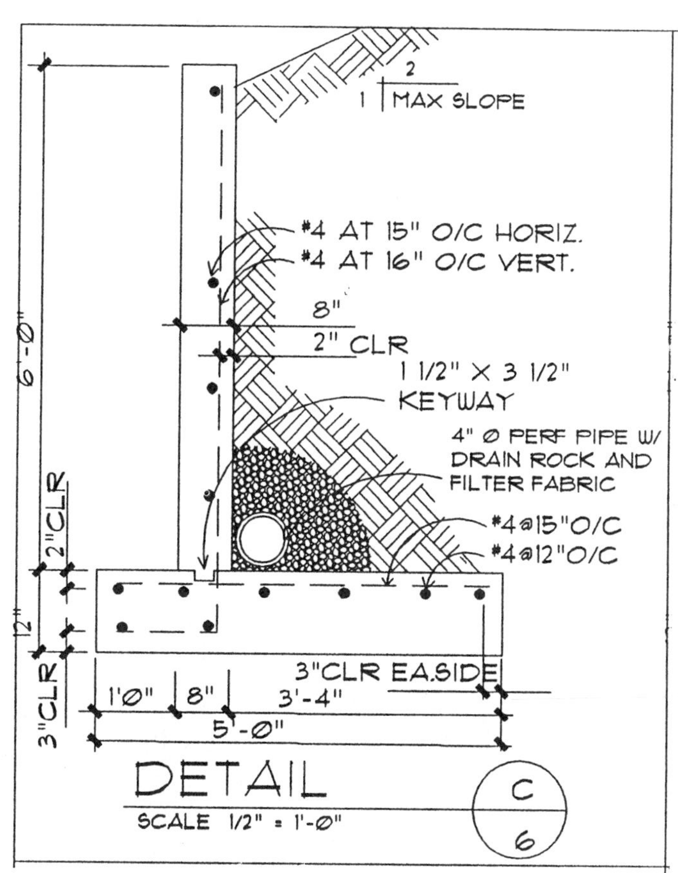

Figure 8–3 Many parts of a structure will need more clarity than can be gained in a full or partial section. Details are typically part of a job to provide clarity. *Courtesy Sunridge Design, Portland, OR.*

The primary section is typically drawn at a scale of ⅜" = 1'–0". The main advantage of using this scale is the ease of distinguishing each structural member. At a smaller scale separate members, such as the finished flooring and the rough flooring, are difficult to read. Without good clarity, problems can arise at the job site. Often, if more than one section must be drawn, the primary section is drawn at ⅜" = 1'–0" and the other sections are drawn at ¼" = 1'–0". By combining drawings at these two scales, typical information can be placed on the ⅜ in. section, and the ¼ in. sections are used to show variations with little detail.

SECTION ALIGNMENT

Reading sections, as with other parts of the plans, the drawing is read from the bottom or right side of the page. The cutting plane on the floor or framing plan shows which way the section is being viewed. The arrows of the cutting plane normally point to the top or left side of the paper, depending on the area of the building being sectioned. See Figure 8–6.

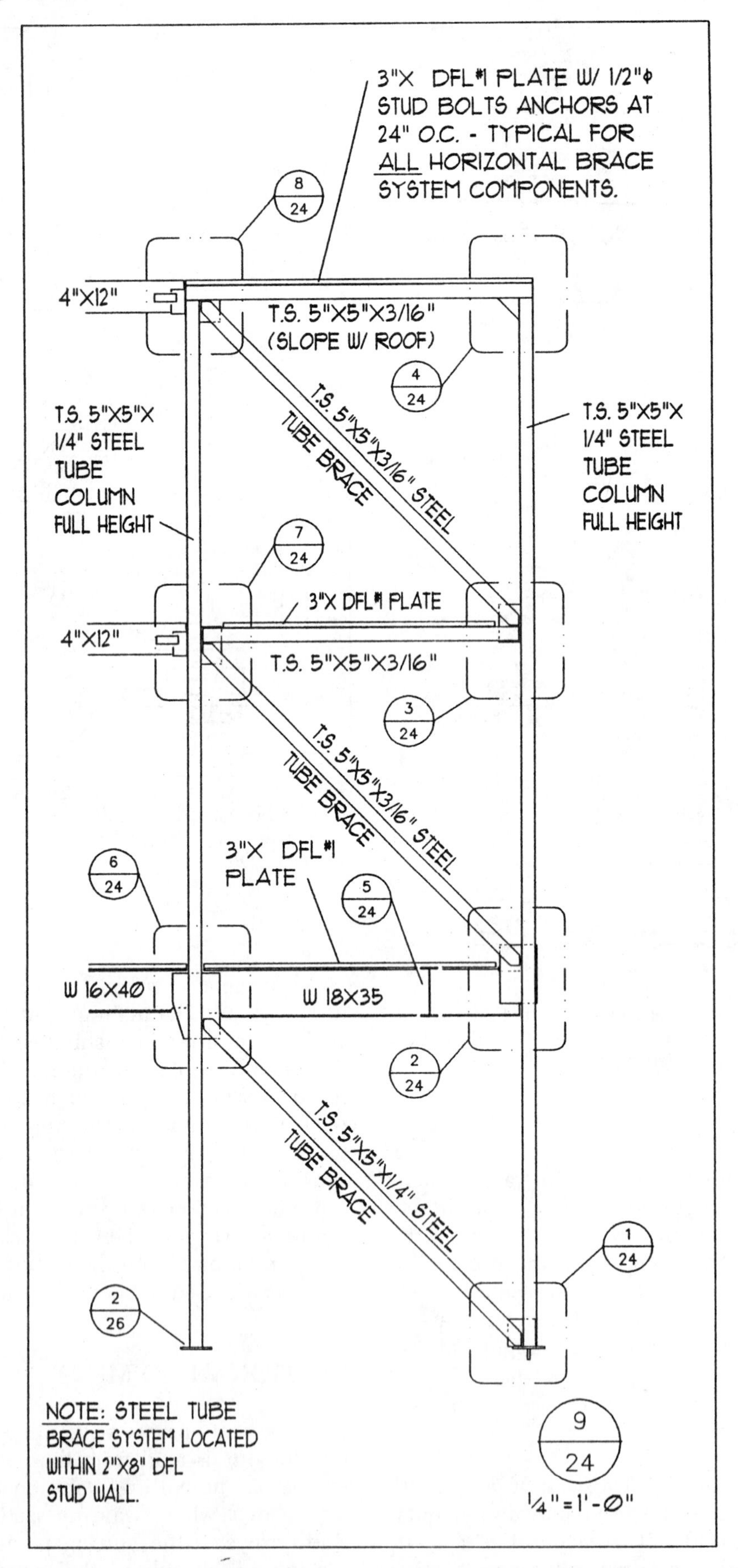

Figure 8–4 On a complicated project, a section may be used to provide a reference map to show how construction details relate to the overall project. *Courtesy Structureform.*

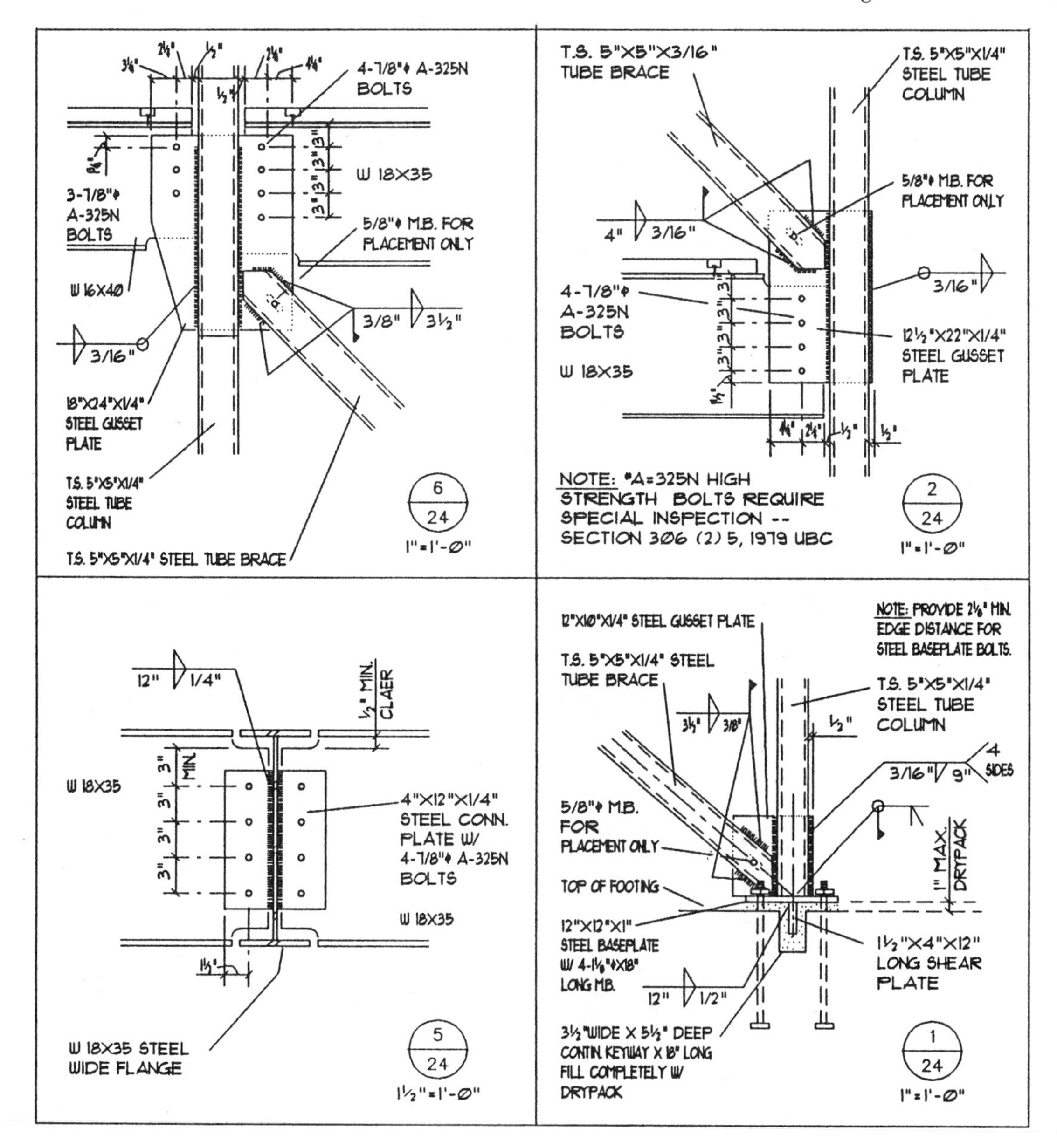

Figure 8–5 Details can provide an enlarged view of construction methods. *Courtesy Structureform.*

READING SECTIONS

Before trying to gain information from the sections, study the floor, foundation, and framing plans and elevations to gain a general understanding of the structure. Look for general information such as how many floors will be constructed. For each level in the sections, a floor or framing plan should be available to provide sizes and layout. Determine the main material used to construct walls. The use of wood or metal studs, brick or concrete block will affect what you should be expecting to find as you enter the sections.

As you study the sections, determine how many have been drawn, and make sure you can find a cutting plane on the floor or framing plan.

Once you feel comfortable with the basic layout of the structure, examine the main section and look for basic materials used to construct the floor, walls, and roofs. Now compare the main section to the other sections and try to determine why the other sections were drawn. The sections may only be trying to show shape variation from one area of the building to another. There may be a variety of building materials that could only be seen by drawing these additional sections.

Having determined general knowledge of the structure relating to its shape and basic materials, you can now search the sections for specific information. The sections will show information about connections, intersections of structural members, and other information necessary to construct the structure.

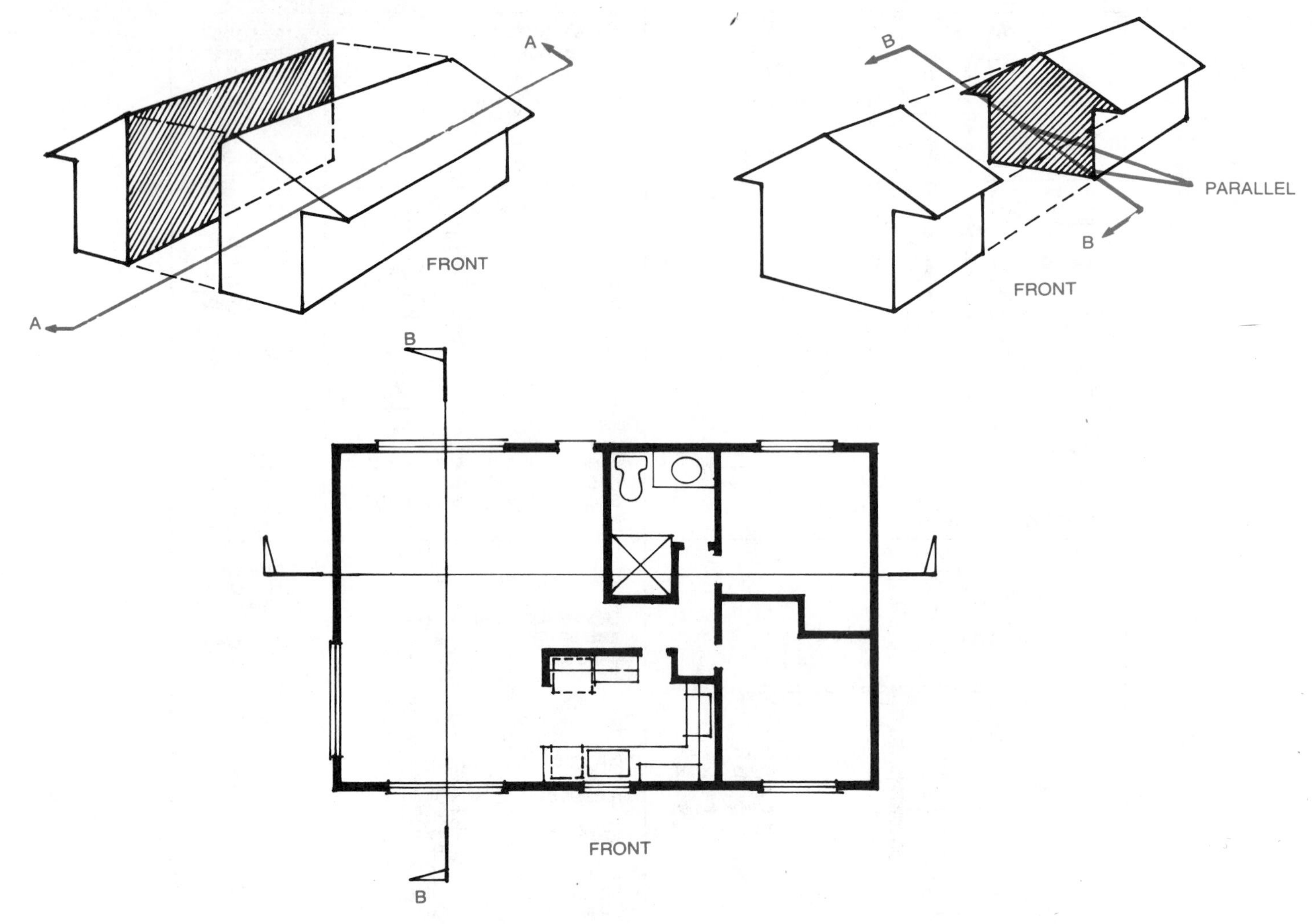

Figure 8–6 The section view is represented on a floor or framing plan by a cutting plane. Normally the plane points to the left or the top of the drawing.

CHAPTER 8 TEST

Place the name of the framing member in the space provided on the drawings on this page and page 166. Do not give specific sizes, such as ½" plywood, but answer in general terms, such as "roof sheathing."

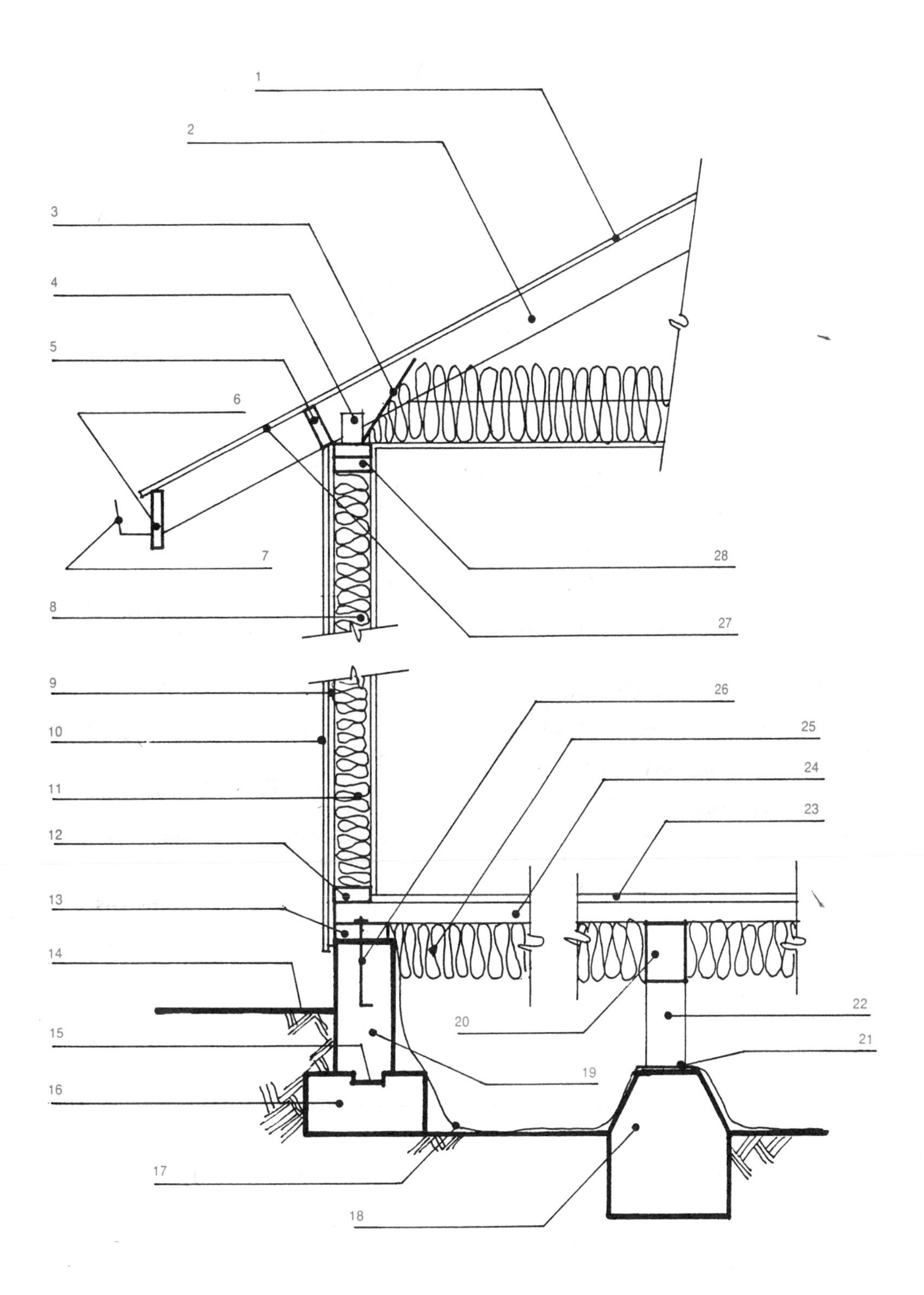

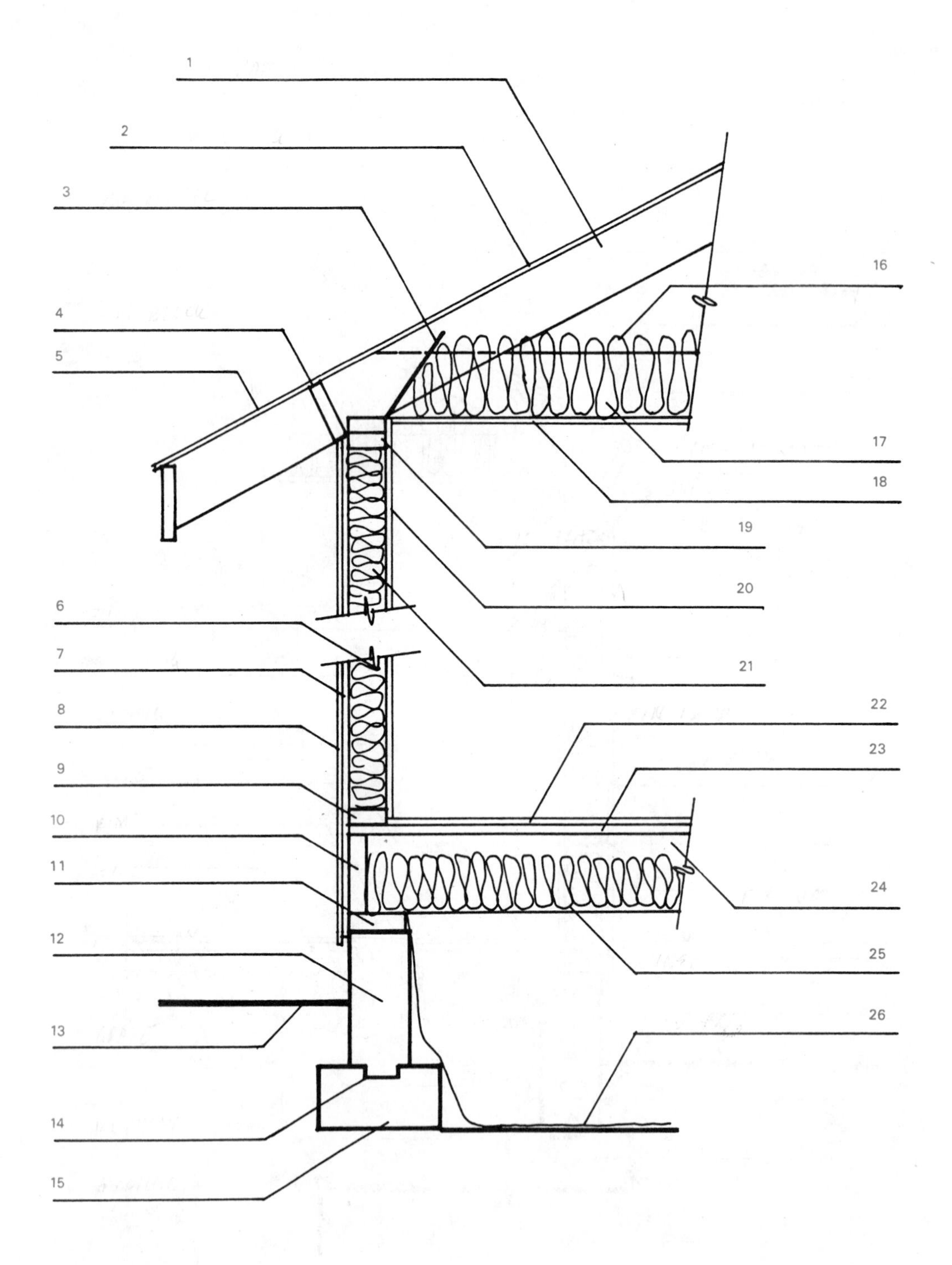
1
2
3
4
5
6
7
8
9
10
11
12
13
14
15
16
17
18
19
20
21
22
23
24
25
26

CHAPTER 8 EXERCISES

PROBLEM 8–1. Use Drawing 'AA' shown on page 172 to provide the answer to your questions.

1. How many types of floors will be constructed for this structure? ______
2. What size are the rafters for the main roof? ______
3. What size rafters are used for the porch roof? ______
4. What size and type of roof insulation will be used? ______
5. What is the plate height for each level? ______
6. What size footing will be used at the interior load-bearing wall? ______
7. How deep are the footings to be set? ______
8. What will the concrete slab be built over? ______
9. Give the nail size and spacing for the floor sheathing. ______
10. Give the nail size and spacing for the roof sheathing. ______
11. How will the foundation be insulated? ______
12. What will be the lap of the top plates? ______
13. What sizes of fascias will be used? ______
14. What is the roof pitch? ______
15. If the roof is 30 feet wide, how high above the top plate will the roof be at the ridge? ______
16. How far below the top plate will the bottom of the trusses extend? ______

17. What will prevent uplift of the front porch? ______________________________

18. What will keep the brick from falling away from the stud wall? ______________________________

19. Why would the 26 ga. at the front porch flashing be specified? ______________________________

20. The kitchen will have a lowered ceiling. How will this ceiling be supported? ______________________________

PROBLEM 8–2. Use Drawing 'BB' shown on page 173 to answer the following questions.

1. What type of roofing will be used? ______________________________

2. What will the siding be installed over? ______________________________

3. What will be the underlayment for the brick? ______________________________

4. What size floor joist will be used on the upper floor? ______________________________

5. What size floor joist will be used on the lower floor? ______________________________

6. What will the crawl space be covered with? ______________________________

7. How wide will the stem walls be? ______________________________

8. What size post will be used to support the front porch? ______________________________

9. If the window headers are set at 6'–8", will they be o.k.? ______________________________

10. What thickness of floor sheathing will be used on the upper floor? ______________________________

11. What thickness of floor sheathing will be used on the lower floor? ______________________________

12. List the R values for the following members of this house:

Roof ______________

Upper floor ______________

Lower floor ______________

Walls ______________

13. What size girders are to be used? ______________________________

14. Will 15# felt be suitable underlayment for the walls? Explain your answer. ________________.

15. What size and spacing of anchor bolts will be used? ________________.

PROBLEM 8–3. Use Drawing 'CC' shown on page 174 to answer the following questions.

1. What type of floor system will be used on the lower floor? ________________.

2. What type of roof system will be used? ________________

3. What spacing is used on the roof trusses? ________________

4. What size of studs will be used? ________________

5. List the size of the subfloor. ________________

6. What will support the main girder? ________________

7. How will the wallboard be supported by the 4" headers? ________________.

8. The sill is DFPT. What do these letters stand for? ________________.

9. What size and spacing are the rafters over the porch? ________________.

10. What will be used to control airflow through the 4" headers? ________________.

11. What will the exterior siding be placed over? ________________.

12. How will the upper ceiling be finished? ________________.

13. What will provide ventilation to the upper rafters? ________________.

14. What is the height of the kitchen ceiling? ______
15. What is the exposure of the finished roofing? ______
16. How will the cantilever be protected? ______
17. What is the spacing of the upper floor joists under the area with the flat ceiling? ______
18. Why would the upper floor joist on the left side be at a different spacing than those on the right? ______
______
______
______.
19. Will solid blocking or a rim joist be used? ______
20. What will hold the upper end of the porch rafters? ______
______
______
______.

PROBLEM 8–4. Use Section 'DD' shown on page 175 to answer the following questions.

1. A circle is drawn in the lower left-hand corner. What do the numbers in the circle mean? ______
______
______
______.
2. How will the retaining walls be reinforced? ______
______
______
______.
3. What grade of concrete blocks will be used to build the basement wall? ______
4. Can the 2 × 4 stud wall in the basement be built right next to the concrete block wall? ______
______
______
______.
5. How will the CMU containing rebar be treated? ______
______
______
______.
6. This house is being built in an area of the country that gets 36" of rain a year. Is the basement prepared? Explain your answer. ______
______
______
______.
7. What size studs will be used in the basement bearing walls? ______
______.
8. The lower floor is to be blocked. What are the specifications for the blocking? ______
______
______
______.
9. Two braces are shown in the attic without any mention of what angle to build them. What is the maximum angle for such braces? ______
______.

10. How will exposed eaves be protected from the weather? ______________________________.

11. What size purlin will be used? ______________________________

12. How will the rim joist be connected to the mudsill? ______________________________.

13. What will support the ceiling joist on the right side of the roof? ______________________________.

14. Determine the size and spacing of the collar ties. ______________________________

15. If the residence is 38' long, how many collar ties will be needed? ______________________________.

16. If the residence is 38' long, how many sheets of plywood will be needed to cover the floor cantilever? ______________________________.

17. How will the load-bearing walls in the basement be attached to the floor? ______________________________.

18. What size pier will support the porch roof? ______________________________

19. What type of lumber will be used to frame this structure? ______________________________.

20. What will be used to keep the lower floor from cracking? ______________________________.

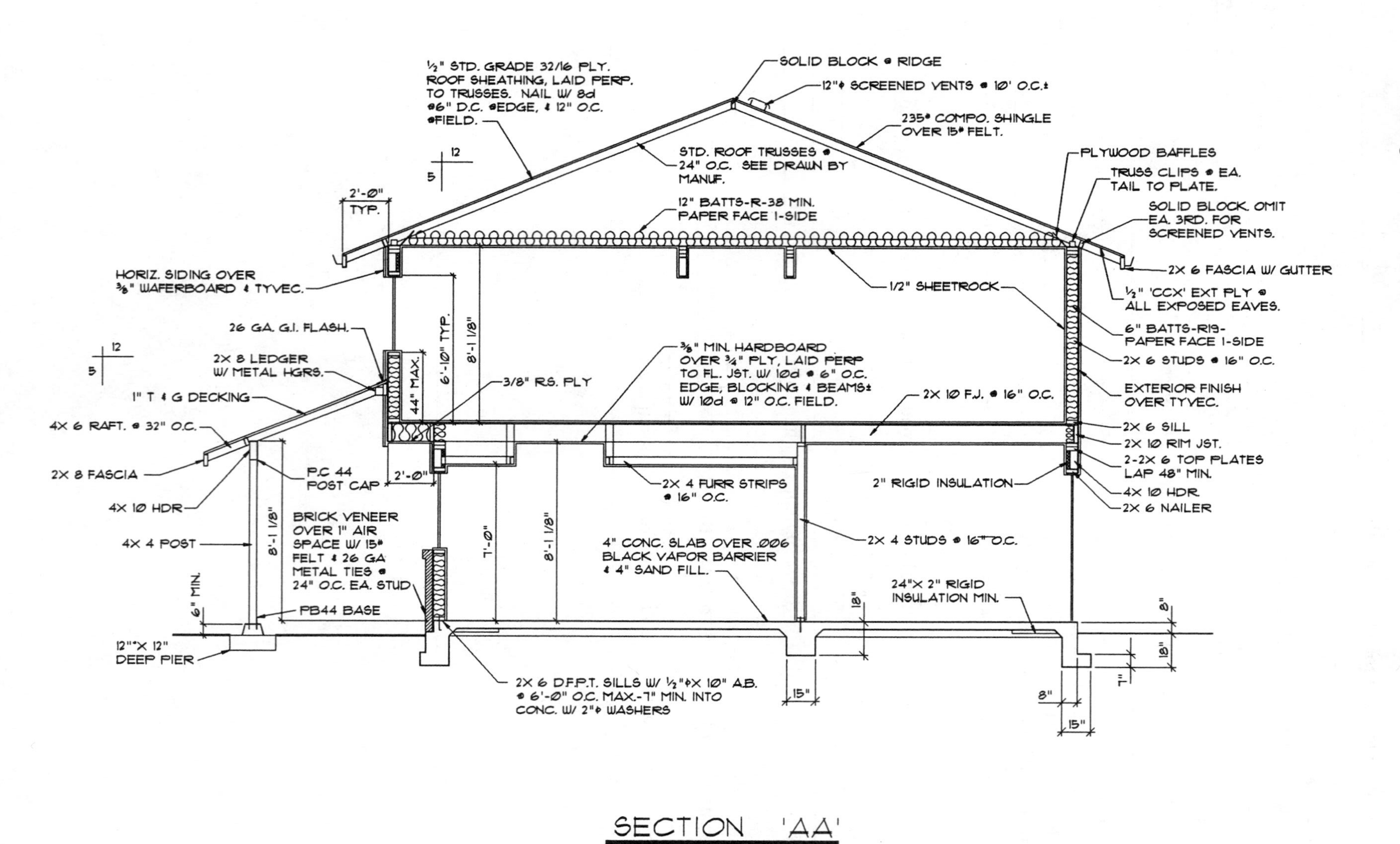

½" STD. GRADE 32/16 PLY. ROOF SHEATHING, LAID PERP. TO TRUSSES. NAIL W/ 8d @6" O.C. @EDGE, & 12" O.C. @FIELD.
SOLID BLOCK @ RIDGE
12"ϕ SCREENED VENTS @ 10' O.C.±
235# COMPO. SHINGLE OVER 15# FELT.
STD. ROOF TRUSSES @ 24" O.C. SEE DRAWN BY MANUF.
12" BATTS-R-38 MIN. PAPER FACE 1-SIDE
PLYWOOD BAFFLES
TRUSS CLIPS @ EA. TAIL TO PLATE.
SOLID BLOCK. OMIT EA. 3RD. FOR SCREENED VENTS.
2X 6 FASCIA W/ GUTTER
½" 'CCX' EXT PLY @ ALL EXPOSED EAVES.
6" BATTS-R19-PAPER FACE 1-SIDE
2X 6 STUDS @ 16" O.C.
EXTERIOR FINISH OVER TYVEC.
2X 6 SILL
2X 10 RIM JST.
2-2X 6 TOP PLATES LAP 48" MIN.
4X 10 HDR.
2X 6 NAILER
2'-0" TYP.
12
5
HORIZ. SIDING OVER ⅜" WAFERBOARD & TYVEC.
26 GA. G.I. FLASH.
2X 8 LEDGER W/ METAL HGRS.
1" T & G DECKING
4X 6 RAFT. @ 32" O.C.
2X 8 FASCIA
4X 10 HDR
4X 4 POST
12"ϕX 12" DEEP PIER
P.C 44 POST CAP
2'-0"
44" MAX.
6'-10" TYP.
8'-1 1/8"
3/8" RS. PLY
1/2" SHEETROCK
⅜" MIN. HARDBOARD OVER ¾" PLY, LAID PERP TO FL. JST. W/ 10d @ 6" O.C. EDGE, BLOCKING & BEAMS± W/ 10d @ 12" O.C. FIELD.
2X 10 F.J. @ 16" O.C.
2X 4 FURR STRIPS @ 16" O.C.
2" RIGID INSULATION
BRICK VENEER OVER 1" AIR SPACE W/ 15# FELT & 26 GA METAL TIES @ 24" O.C. EA. STUD
PB44 BASE
6" MIN.
7'-0"
4" CONC. SLAB OVER .006 BLACK VAPOR BARRIER & 4" SAND FILL.
2X 4 STUDS @ 16" O.C.
24"X 2" RIGID INSULATION MIN.
2X 6 D.F.P.T. SILLS W/ ½"ϕX 10" A.B. @ 6'-0" O.C. MAX.-7" MIN. INTO CONC. W/ 2"ϕ WASHERS
18"
15"
8"
7"
SECTION 'AA'
3/8" = 1'-0"
ALL FRAMING LUMBER TO BE DFL #2 OR, BETTER

Problem 8–1.

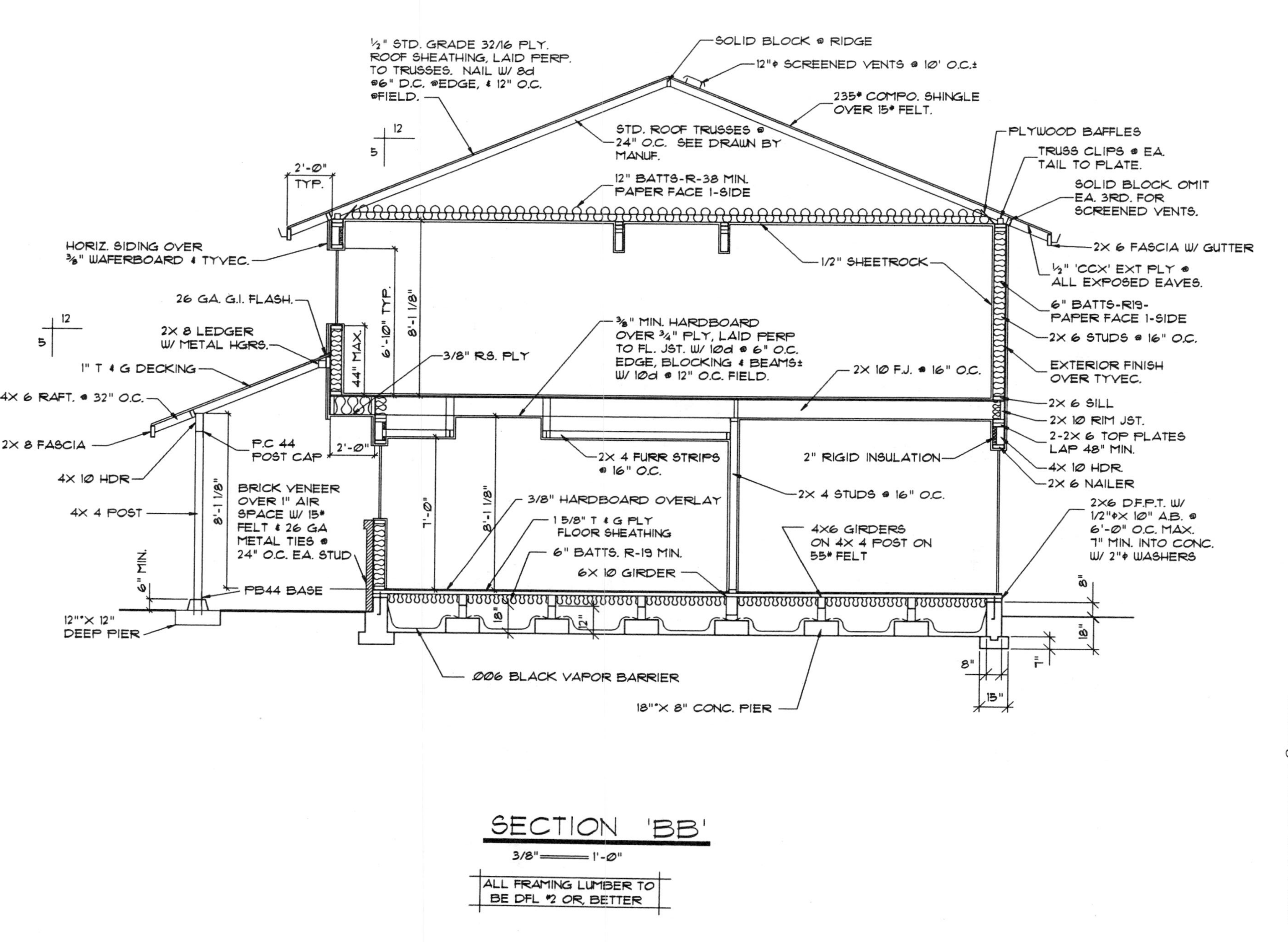

Problem 8–2. *Courtesy of Piercy and Barclay Designers, Inc.*

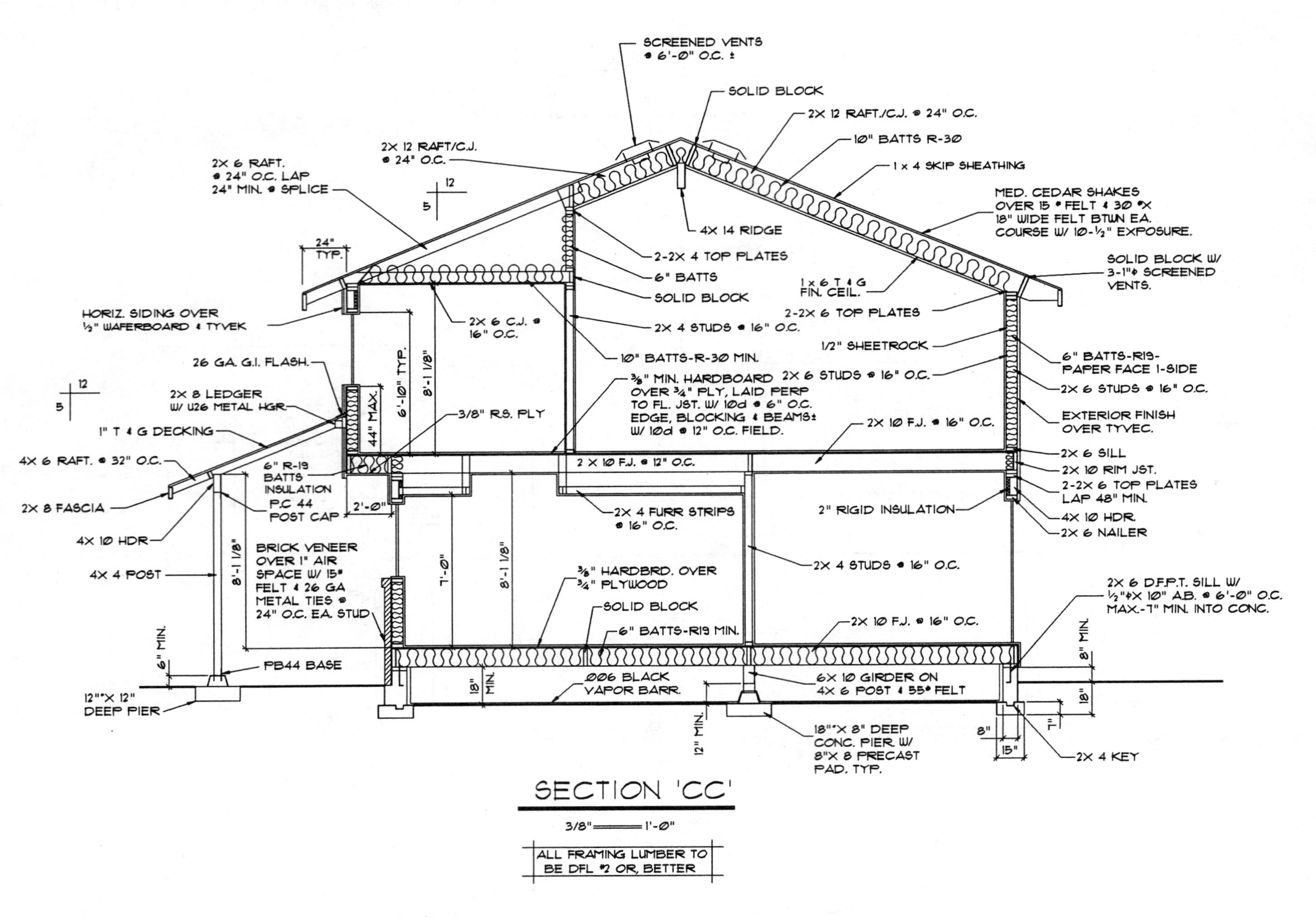

Problem 8–3. *Courtesy of Sunridge Design, Portland, OR.*

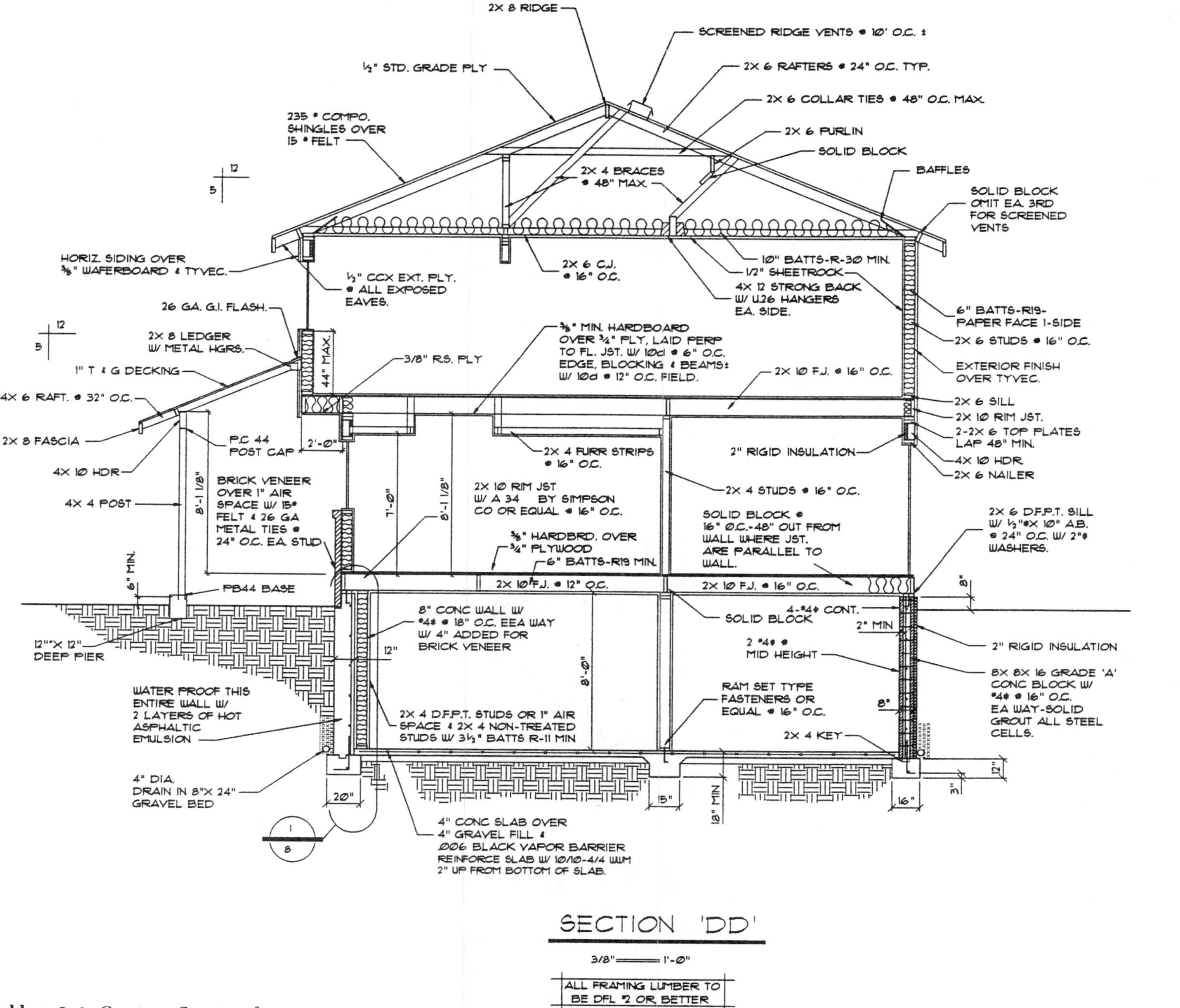

Problem 8–4. *Courtesy Structureform.*

Chapter 9
Reading Stair Drawings

STAIRS ARE typically represented on two types of drawings within the blueprints: plan views, such as floor or framing plans, and sections. The basic shape, number of treads, and rail locations can usually be determined from the floor plan, as seen in Figure 9–1. The stair section is typically used to show vertical relationships and the structural material to be used.

STAIR TERMINOLOGY

You must be familiar with several basic terms when reading plans relating to stairs. Each can be seen in Figure 9–2.

The *run* is the horizontal distance from end to end of the stairs.

The *rise* is the vertical distance from top to bottom of the stairs.

The *tread* is the horizontal step of the stairs. Tread width is the measurement from the face of the riser to the nosing. The *nosing* is the portion of the tread that extends past the riser.

The *riser* is the vertical backing between the treads. It is usually made from 1 in. material for enclosed stairs and is not used on open stairs.

The *stringer*, or *stair jack*, is the support for the treads. A 2 × 12 notched stringer is typically used for enclosed stairs. For an open stair 4 × 14 is common, but sizes vary greatly. Figure 9–2 shows the stringers, risers, and treads.

The *kick block*, or *kicker*, is used to keep the bottom of the stringer from sliding on the floor when downward pressure is applied to the stringer.

Headroom is the vertical distance measured from the tread nosing to a wall or floor above the stairs. Building codes specify a minimum size.

The *handrail* is the railing that you slide your hand along as you walk down the stairs.

The *guardrail* is the railing placed around an opening for the stairs.

Type X gypsum (GYP.) board 5/8 in. thick is required by the UBC for enclosing all usable storage space under the stairs. This is a gypsum board that has a 1 h fire rating. Figure 9–3 shows common stair dimensions from four major codes.

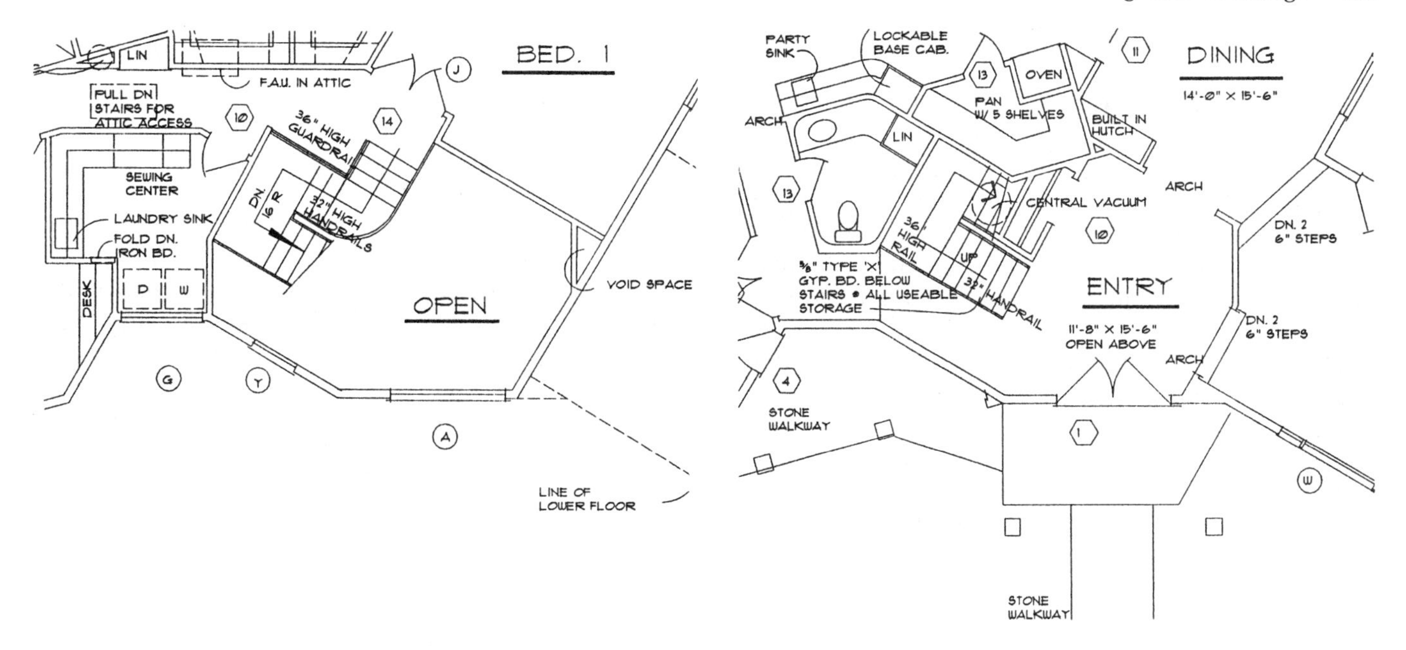

Figure 9–1 The basic shape and the number of steps can usually be determined from the floor plan.

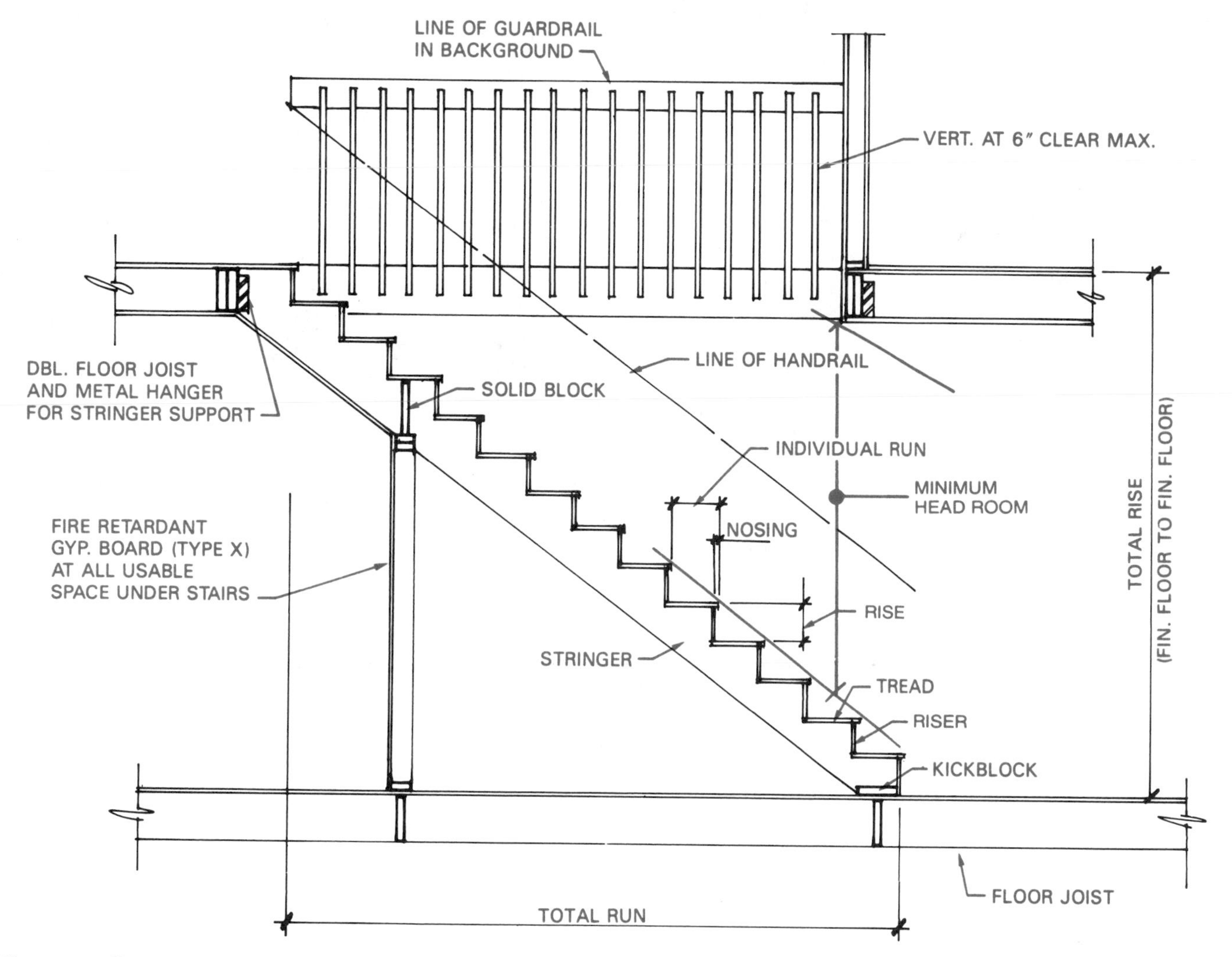

Figure 9–2 Common stair terms.

DETERMINING RISE AND RUN

Building codes dictate the maximum rise of the stairs. To determine the actual rise, the total height from floor to floor must be known. By measuring the floor to ceiling height, the depth of the floor joist, and the depth of the floor covering, the total rise can be found. The total rise can then be divided by the maximum allowable rise to determine the number of steps required.

Once the rise is known, the required number of treads can be found easily since there is always one less tread than the number of risers. Thus, a typical stair for a house with 8'–0" ceilings has 14 risers and 13 treads. If each tread is 10½ in. wide, the total run can be found by multiplying 10½ in. (the width) by 13 (the number of treads required).

STRAIGHT STAIR CONSTRUCTION

The straight-run stair is a common type of stair seen in blueprints. It is a stair that goes from one floor to another in one straight run. An example of a straight-run stair can be seen in Figure 9–3.

This basic stair layout supports the weight of the stair materials and those using it by the use of stringers. The stringers are usually supported at the upper end by metal hangers that transfer the weight to the upper floor framing. These hangers also resist the natural tendency of gravity to pull down the stringer. Since the stairs cannot go downward, force on the stringer tends to make the stringer slide away from the upper support. A kicker or metal angle is used to keep the stringer from sliding along the floor. Each of these members can be seen in Figure 9–2.

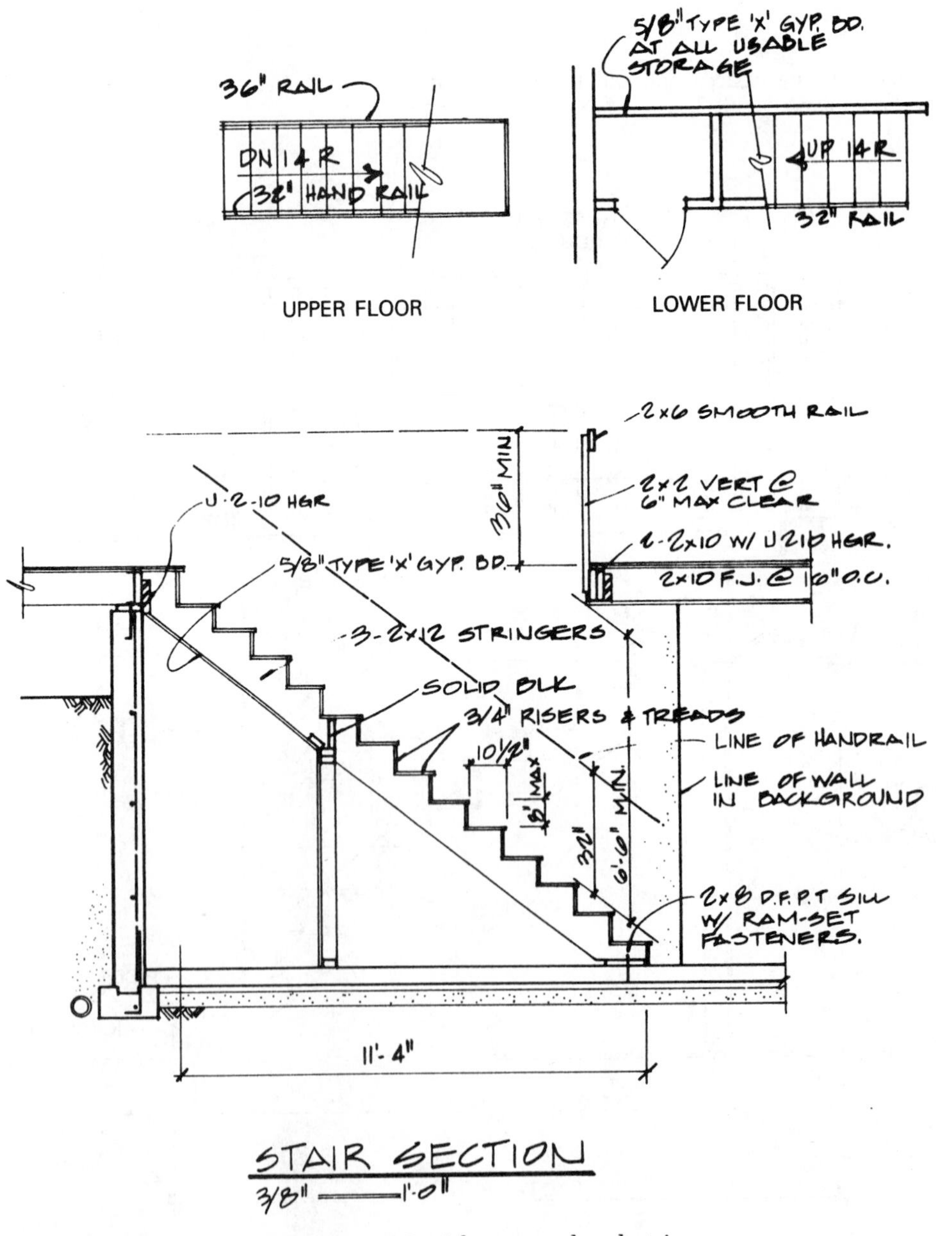

Figure 9–3 Straight-run enclosed stairs.

OPEN STAIR CONSTRUCTION

An open stairway is similar to the straight enclosed stairway. It goes from one level to the next in a straight run. The major difference is that with the open stair, there are no risers between the treads. This allows viewing from one floor to the next, creating an open feeling. Figure 9–4 shows an open stairway. The open stairway functions in a similar manner to an enclosed stairway. Rather than having three- 3 × 12 stringers support the tread, the tread of an open stair is supported between the stringers. The tread can be supported by metal L brackets underneath the tread, or a notch may be cut into the stringer to provide support.

U-SHAPED STAIRS

The U-shaped stair is often used. Rather than going up a whole flight of steps in a straight run, this stair layout introduces a landing. The landing is usually located at the midpoint of the run, but it can be offset depending on the amount of room allowed for stairs on the floor plan. Figure 9–5 shows what a U-shaped stair looks like on the floor plan and in section.

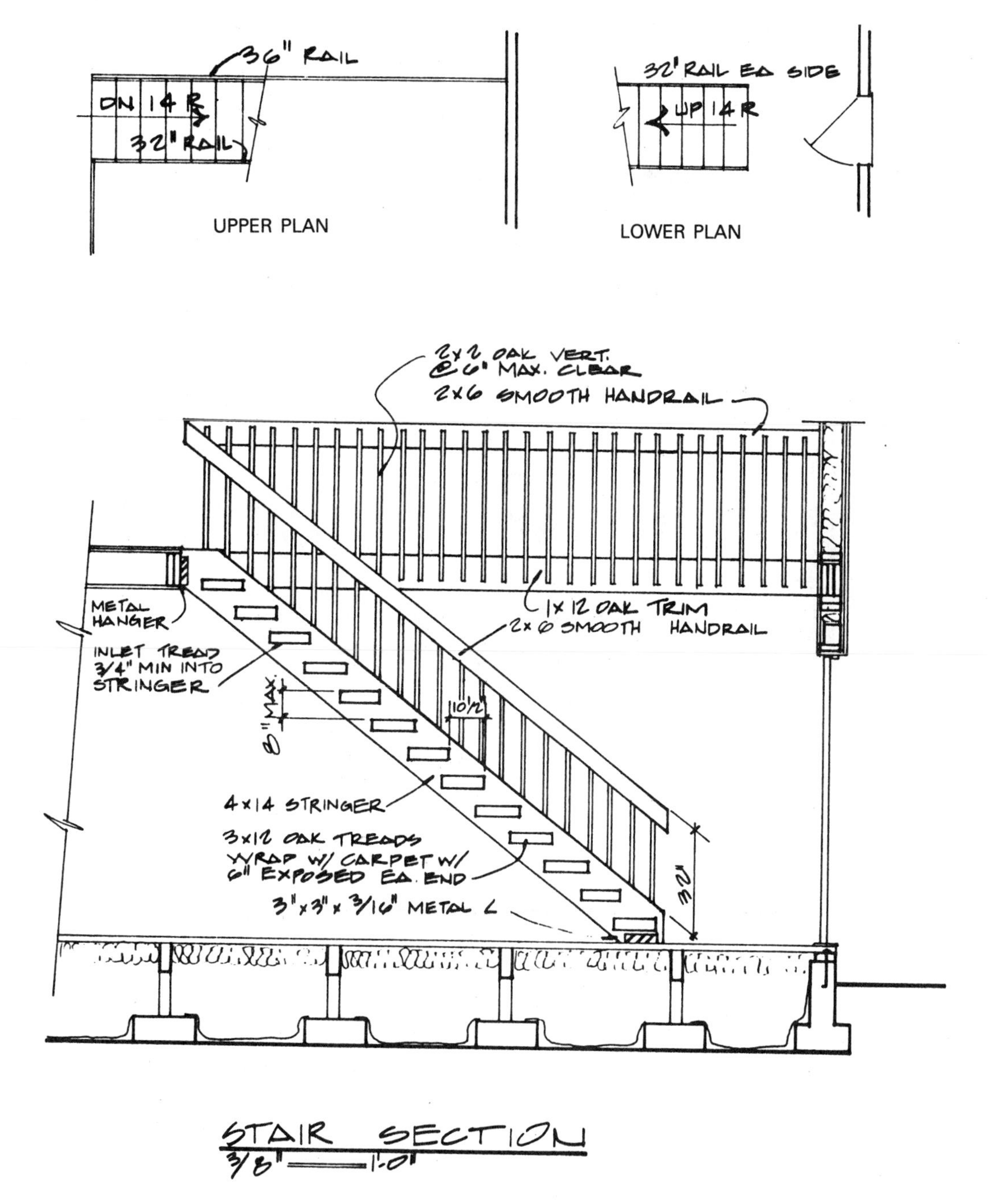

Figure 9–4 Open-tread stairs.

The stairs may be either open or enclosed, depending on the location. The construction of the stair is similar to the layout of the straight-run stair.

MULTILEVEL STAIRS

Another type of stair that may be encountered is a multilevel stair. An example of a section view of a multilevel stair can be seen in Figure 9–6. This type of stair places the steps connecting one level over a set of steps for another level. This stacking of stairs flights is typical of multilevel structures to make efficient use of floor space.

This stacking effect creates a vertical tunnel normally referred to as a *stairwell.* The height limits of the stairwell are typically seen in section view. While reading the drawings it is important to determine the means of protecting the stairwell from the spread of fire. If a fire were to start at a lower level, it could

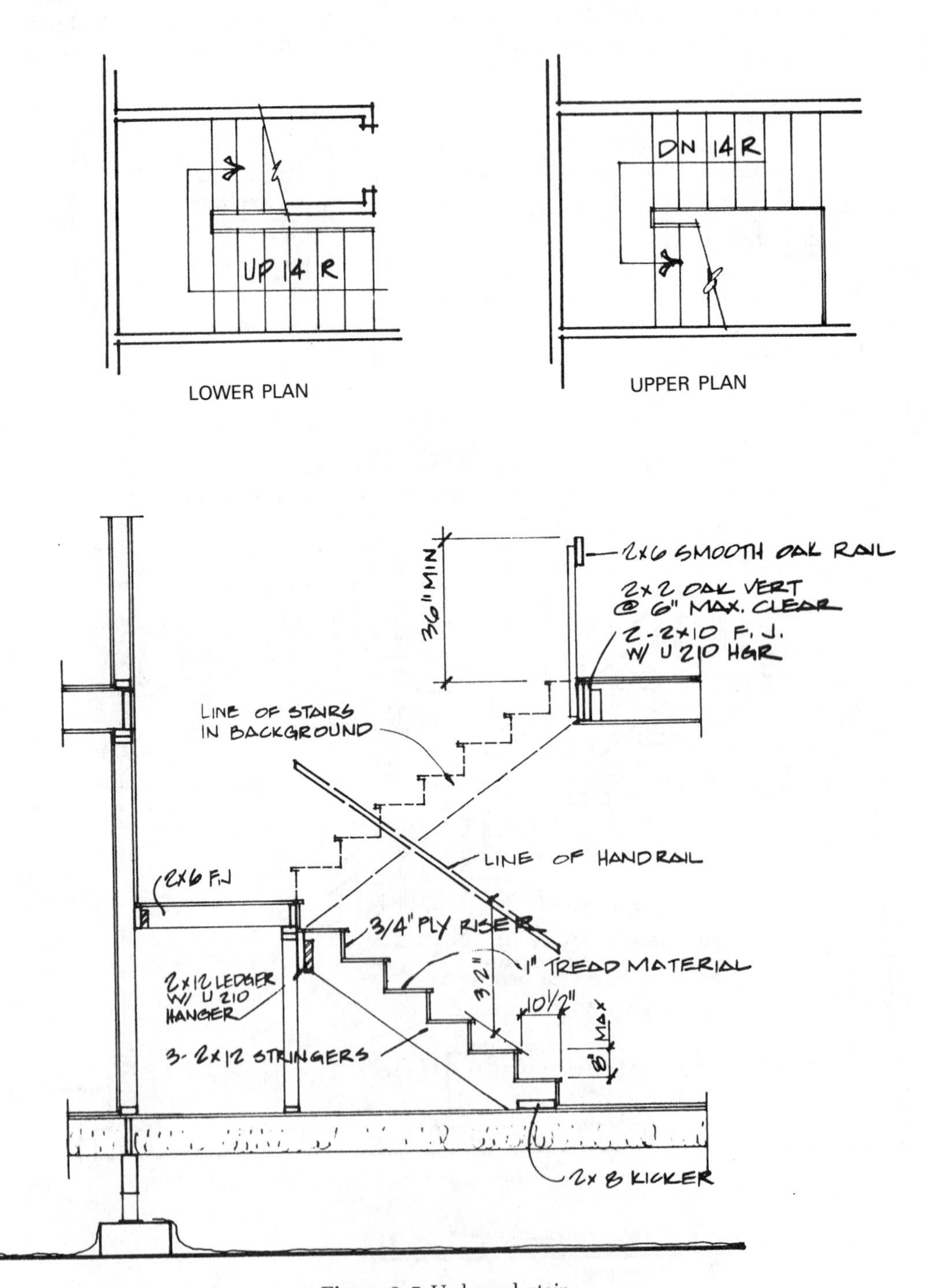

Figure 9–5 U-shaped stair.

race through the structure through the stairwell. Access to the stairs at each level is typically controlled by a fire-rated door to protect the stairs from fire within the structure and to protect the structure from fire in the stairwell. The stairwell is also typically required to be protected by a sprinkler system in multilevel structures other than residential use.

Multilevel stairs may be constructed of either straight or U-shaped stair flights, but the two configurations should not be mixed. The materials of construction can be mixed. It is common to enclose stairs to a basement but to build open stairs to the upper floors. This allows the view into one area to be restricted or for the control of heat flow.

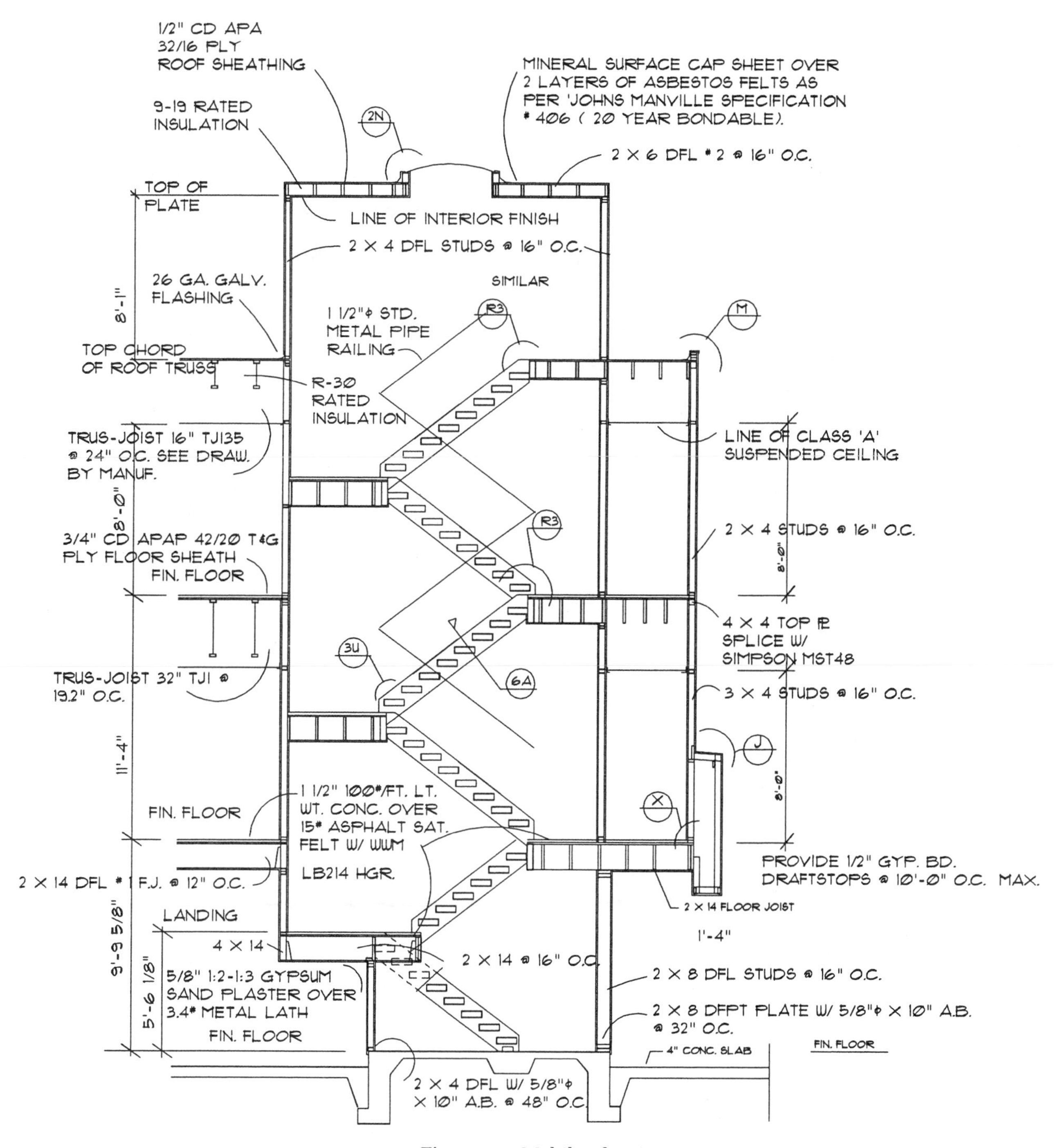

Figure 9–6 Multilevel stair.

EXTERIOR STAIRS

It is quite common to need sections of exterior stairs for multilevel homes. Figure 9–7 shows two different types of exterior stairs. Although there are many variations, these two options are common. Both can be constructed by following the procedure for straight-run stairs. There are some major differences in the finishing materials. Notice there is no riser on the wood stairs, and the tread is thicker than the tread of an interior step. Usually the same material that is used on the deck is used for tread material. In many parts of the country a nonskid material is also called for to cover the treads.

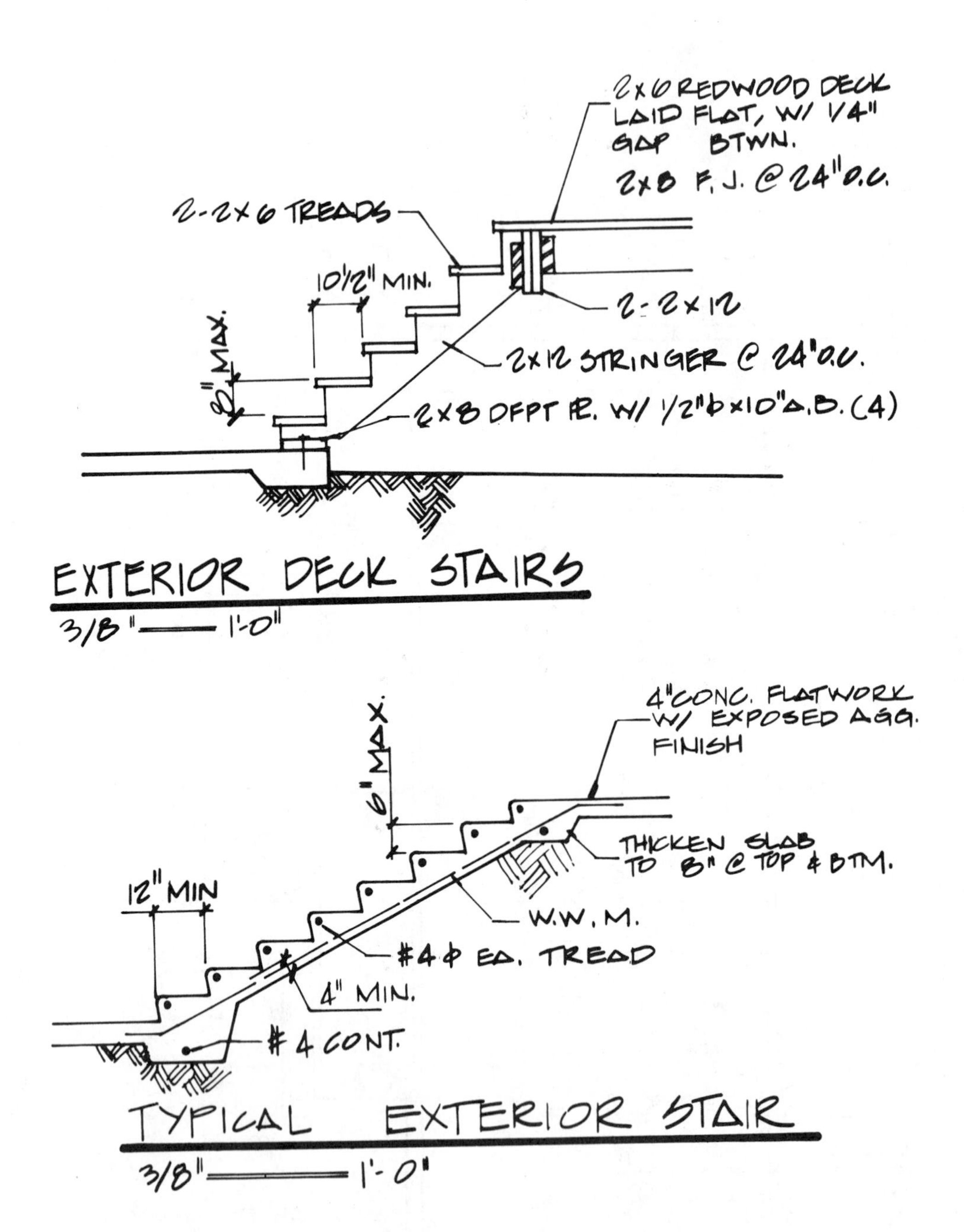

Figure 9–7 Common exterior stairs.

CHAPTER 9 TEST

1. What is the legal minimum height for a handrail? ______.
2. List the minimum run for a stair according to the four major national codes. ______.
3. What member is placed between the stringers to limit vibrations? ______.
4. What is the minimum required headroom over stairs? ______.
5. List the minimum spacing between vertical rails of a handrail or guardrail. ______
6. How can the lateral movement of the stairs along the floor be resisted? ______
______
______
______.
7. Describe how a stringer is kept in place at the upper end. ______
______
______
______.
8. What term describes the horizontal member of a stair? ______.
9. List the maximum rise according to UBC. ______.
10. What is the minimum radius allowed for a spiral stair? ______
______
______
______.

CHAPTER 9 EXERCISES

PROBLEM 9–1. Use Figure 9–1 shown on page 177 to answer the following questions.

1. Describe the guardrail required for the main stair. ______.
2. List the number of required risers and treads. ______.
3. How many risers will be framed in the upper portion of this stair? ______.
4. If this stair is governed by the UBC, approximately how high is the main landing above the lower floor?
______.

PROBLEM 9–2. Use Figure 9–3 shown on page 178 to answer the following questions.

1. What type of stair is this? ______
2. Describe how the bottom of the stair will be protected from fire. ______
______
______
______.
3. What will hold the stringer to the upper floor? ______
______
______
______.

4. What will support the floor joist where the stairwell is formed? ____.

5. What type of handrail will be used? ________________________.
6. How many treads will be required? ________________________.
7. If the owner changes the tread to 10¾", what will be the total run? ________________________.
8. What material will be used for the risers? ________________________.
9. If the owner wants 7¾" risers, how many risers will be required if the lower floor has 8' ceilings? ________________________.

10. What will support the lower end of the stringer? ____.

PROBLEM 9–3. Use Figure 9–4 shown on page 179 to answer the following questions.

1. What type of stair is to be used? ________________________.
2. List the size of the stringer. ________________________.
3. Describe the treads. ____.

4. How will the treads be supported? ____.

5. How will the stringer be supported at the bottom? ____.

PROBLEM 9–4. Use Figure 9–5 shown on page 180 to answer the following questions.

1. What type of stair will be framed? ________________________.
2. How will the guardrail be supported? ________________________.
3. Determine the thickness of the tread material. ________________________.
4. What will support the lower end of the lower stringer? ________________________.
5. Describe the stringers. ________________________.
6. What do the dotted lines in the section view indicate? ____.

7. How will the stringers be supported at the upper end? ____.

8. What would be the required run if a 7" rise were used? Assume an 8'–0" ceiling. ____.

9. How will the landing be supported? ____.

10. What will be used to frame the landing? ________________________

Chapter 10
Reading Fireplace Drawings

FIREPLACE TERMS
FLOOR PLAN REPRESENTATIONS
EXTERIOR ELEVATIONS
FOUNDATION PLAN
SECTIONS
FIREPLACE ELEVATION
TEST
EXERCISES

THE FIREPLACE is typically first encountered on floor plans and elevations. Information may also be contained on the foundation plan. These three drawings contain information about the shape style and support of the fireplace. The fireplace section contains information about the materials to be used for construction. Before examining each type of drawing, an understanding of common terms is needed.

FIREPLACE TERMS

Figure 10–1 shows the parts of a fireplace and chimney. Each of these parts should be understood to help you read fireplace drawings.

Fireplace Parts

Hearth. The *hearth* is the floor of the fireplace and consists of an inner and outer hearth. The inner hearth is the floor of the firebox. The hearth is made of fire-resistant brick and holds the burning fuel. The outer hearth may be made of any incombustible material. The outer hearth protects the combustible floor around the fireplace. Figure 10–2 shows the minimum sizes for the outer hearth.

Ash Dump. The *ash dump* is an opening in the hearth into which the ashes can be dumped. The ash dump normally is covered with a small metal plate that can be removed to provide access to the ash pit.

Ash Pit. The *ash pit* is the space below the fireplace where the ashes can be stored.

Fireplace Opening. The *fireplace opening* is the area between the side and top faces of the fireplace. Although the size of the opening is very important, it may not be shown on the drawings. The size of the opening is important for the appearance and operation of the fireplace. If the opening is too small, the fireplace does not produce a sufficient amount of heat. If the opening is too large, the fire can make a room too hot. Figure 10–3 shows suggested fireplace opening sizes compared to room size. The ideal dimensions for a single-face fireplace have been determined to be 36 in. wide and 26 in. high.

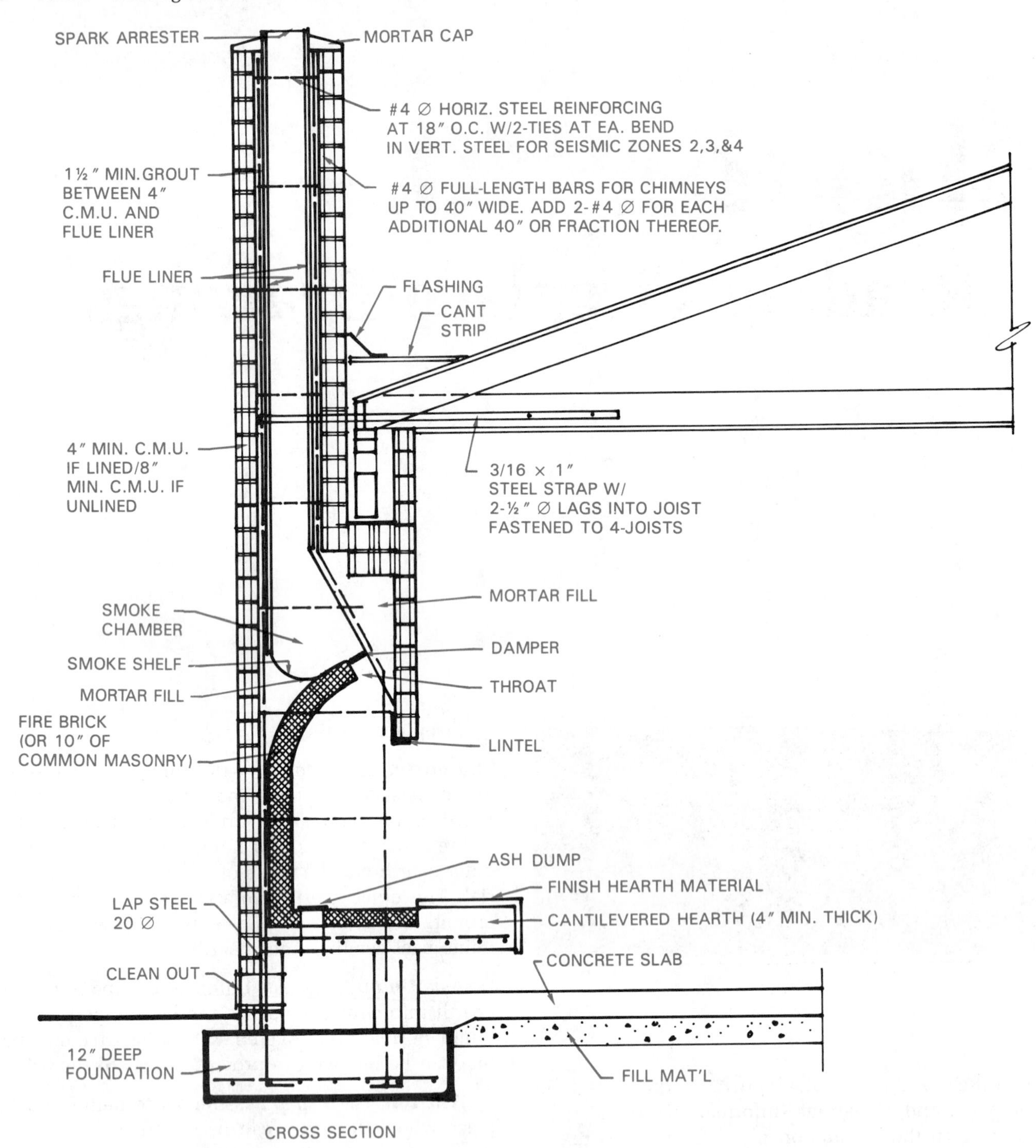

Figure 10–1 Common components of a masonry fireplace and chimney.

Firebox. The *firebox* is where the combustion of the fuel occurs. The side should be slanted slightly to radiate heat into the room. The rear wall should be curved to provide an upward draft into the upper part of the fireplace and chimney. The firebox is usually constructed of fire-resistant brick set in fire-resistant mortar. Figure 10–2 shows minimum wall thickness for the firebox.

The firebox depth should be proportional to the size of the fireplace opening. With an opening of 36" × 26", a depth of 20" should be provided for a single-face fireplace. Figure 10–4 lists recommended fireplace opening-depth proportions.

Lintel. The *lintel* is a steel angle above the fireplace opening that supports the fireplace face.

GENERAL CODE REQUIREMENTS

ITEM	Letter	Uniform Building Code * 1979 & 1982 Edition	FHA & VA	Local or Special Requirements
Hearth Slab Thickness	A	4"	4"	
Hearth Slab Width (Each side of opening)	B	8" Fireplace opg. <6 sq.ft. 12" Fireplace opg. ≥6 sq. ft.	8"	
Hearth Slab Length (Front of opening)	C	16" Firepl. opg. <6 sq. ft. 20" Firepl. opg. ≥6 sq. ft.	16"	
Hearth Slab Reinforcing	D	Reinforced to carry its own weight and all imposed loads.	Required if cantilevered in connection with raised wood floor construction.	
Thickness of Wall of Firebox	E	10" common brick or 8" where a fireback lining is used. Jts. in fireback ¼" max.	8" including minimum 2" fireback lining—12" when no lining is provided.	
Distance from Top of Opening to Throat	F	6"	6" min.; 8" recommended.	
Smoke Chamber Edge of Shelf Rear Wall—Thickness Front & Side wall—Thickness	G	 6" 8"	½" offset. 6" plus paraging may be omitted if wall thickness is 8" or more of solid masonry. Form damper is required.	
Chimney Vertical Reinforcing	**H	Four #4 full length bars for chimney up to 40" wide. Add two #4 bars for each additional 40" or fraction of width or each additional flue.	Four #4 bars full length, no splice unless welded.	
Horizontal Reinforcing	J	¼" ties at 18" and two ties at each bend in vertical steel.	¼" bars at 24"	
Bond Beams	K	No specified requirements. L.A. City requirements are good practice	Two ¼" bars at top bond beam 4" high. Two ¼" bars at anchorage bond beam 5" high.	
Fireplace Lintel	L	Incombustible material	2½" x 3" x 3/16" angle with 3" end bearing	
Walls with Flue Lining	M	Brick with grout around lining. 4" min. from flue lining to outside face of chimney.	Brick with grout around lining. 4" min. from outside flue lining to outside face of chimney.	
Walls with Unlined Flue	N	8" Solid masonry	8" Solid masonry	
Distance Between Adjacent Flues	O	4" including flue liner	4" wythe for brick	
Effective Flue Area (Based on Area of Fireplace Opening)	P	Round lining—1/12 or 50 sq.in.min. Rectangular lining 1/10 or 64 sq. in min. Unlined or lined with firebrick—1/8 or 100 sq. in. min.	1/10 for chimneys over 15' high and over 1/8 for chimneys less than 15' high	
Clearances Wood Frame Combustible Material Above Roof	R	1" when outside of wall or ½" gypsum board 1" when entirely within structure 6" min. to fireplace opening. 12" from opening when material projecting more than 1/8 for ea. 1" 2' at 10'	¾" from subfloor or floor or roof sheathing. 2" from framing members 3½" to edge of fireplace 12" from opening when projecting more than 1½" 2' at 10'	
Anchorage Strap Number Embedment into chimney Fasten to Bolts	S	3/16" x 1" 2 12" hooked around outer bar w/6"ext. 4 Joists Two ½" Dia.	1/4" x 1" 2 18" hooked around outer bar 3 joists Two ½" Dia.	
Footing Thickness Width	T	12" min. 6" each side of fireplace wall	8" min. for 1 story chimney 12" min. for 2 story chimney 6" each side of fireplace wall.	
Outside Air Intake	U	Optional	Required	6 sq. in. minimum area (California Energy Commission Requirement)
Glass Screen Door		Optional	Required but shall not interfere with energy conservation improving devices	Required but shall not interfere with energy conservation improving devices

*Applies to Los Angeles County and Los Angeles City requirements.

**H EXCEPTION. Chimneys constructed of hollow masonry units may have vertical reinforcing bars spliced to footing dowels, provided that the splice is inspected prior to grouting of the wall.

Figure 10–2 General code requirements for fireplace and chimney construction. The letters in the second column can be used to find a specific item in Figure 10–9. *Reprinted from* Residential Fireplace and Chimney Construction Details and Specifications, *with permission from Masonry Institute of America, Los Angeles, CA.*

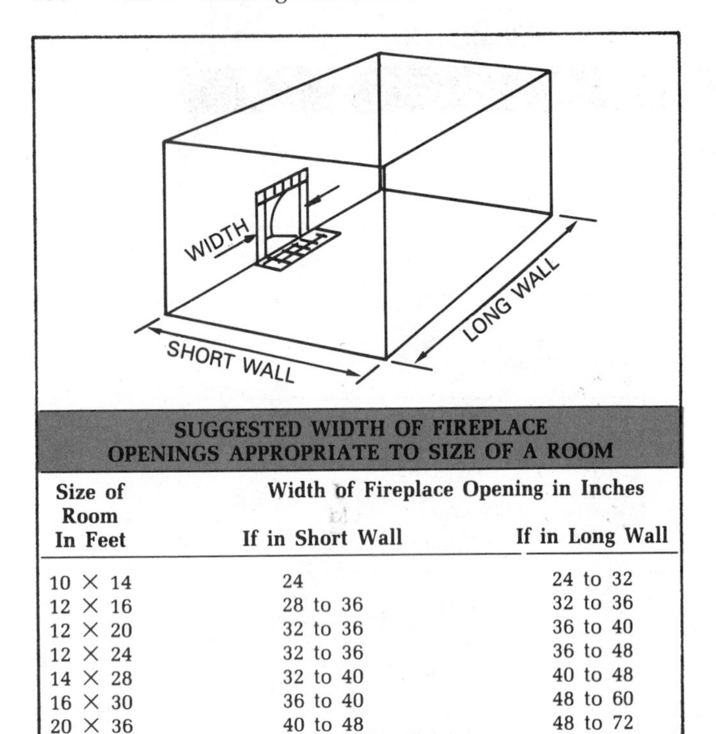

SUGGESTED WIDTH OF FIREPLACE OPENINGS APPROPRIATE TO SIZE OF A ROOM		
Size of Room In Feet	Width of Fireplace Opening in Inches	
	If in Short Wall	If in Long Wall
10 × 14	24	24 to 32
12 × 16	28 to 36	32 to 36
12 × 20	32 to 36	36 to 40
12 × 24	32 to 36	36 to 48
14 × 28	32 to 40	40 to 48
16 × 30	36 to 40	48 to 60
20 × 36	40 to 48	48 to 72

Figure 10–3 The size of the firebox opening should be proportioned to the size of the room. *Reprinted from* Residential Fireplace and Chimney Construction Details and Specifications, *with permission from Masonry Institute of America, Los Angeles, CA.*

Fireplace Type	Width of Opening w, inches	Height of Opening h, inches	Depth of Opening d, inches
Single Face			
	28	24	20
	30	24	20
	30	26	20
	36	26	20
	36	28	22
	40	28	22
	48	32	25

Figure 10–4 Firebox opening-depth proportion guide. *Reprinted from* Residential Fireplace and Chimney Construction Details and Specifications, *with permission from Masonry Institute of America, Los Angeles, CA.*

Throat. The *throat* of a fireplace is the opening at the top of the firebox that opens into the chimney. The throat should be able to be closed when the fireplace is not in use. This is done by installing a damper.

Damper. The *damper* extends the full width of the throat and is used to prevent heat from escaping up the chimney when the fireplace is not in use.

Smoke Chamber. The *smoke chamber* acts as a funnel between the firebox and the chimney. The shape of the smoke chamber should be symmetrical so that the chimney draft pulls evenly and creates an even fire in the firebox.

Smoke Shelf. The *smoke shelf* is located at the bottom of the smoke chamber behind the damper. The smoke shelf prevents down drafts from the chimney from entering the firebox.

Chimney. The *chimney* is the upper extension of the fireplace and is built to carry off the smoke from the fire. Figure 10–2 shows the minimum wall thickness for chimneys.

Chimney Parts

Flue. The *flue* is the opening inside the chimney that allows the smoke and combustion gases to pass from the firebox away from the structure. A flue may be constructed of masonry products or may be covered with a flue liner. Flue sizes are generally required to equal either 1/8 or 1/10 of the fireplace opening. Figure 10–5 shows recommended areas for residential fireplaces of various shapes.

Chimney Liners. A *chimney liner* is usually built of fire clay or terra-cotta. The smooth surface of the liner helps to reduce the buildup of soot, which can cause a chimney fire.

Chimney Anchors. The *anchors* are steel straps that connect the masonry chimney to the framing members at each floor and ceiling level. Steel straps are embedded into the grout of the chimney or

Type of Fireplace	Width of Opening w In.	Height of Opening h In.	Depth of Opening d In.	Area of Fireplace opening for flue determination sq. in.	Flue Size Required at 1/10 Area of fireplace opening	Flue Size Required at 1/8 Area of fireplace opening (FHA Requirement)*
	28	24	20	672	8½ x 13	8½ x 17
	30	24	20	720	8½ x 17	13" round
	30	26	20	780	8½ x 17	10 x 18
	36	26	20	936	10 x 18	13 x 17
	36	28	22	1008	10 x 18	10 x 21
	40	28	22	1120	10 x 18	10 x 21
	48	32	25	1536	13 x 21	17 x 21
	60	32	25	1920	17 x 21	21 x 21

Figure 10–5 Recommended flue areas for residential masonry fireplaces. *Reprinted from* Residential Fireplace and Chimney Construction Details and Specifications, *with permission from Masonry Institute of America, Los Angeles, CA.*

wrapped around the reinforcing steel and then bolted to the framing members of the structure.

Chimney Reinforcement. A minimum of four ½ in. diameter (#4) vertical reinforcing bars should be used in the chimney, extending from the foundation to the top of the chimney. Typically these vertical bars are supported at 18 in. intervals with ¼ in. horizontal rebars known as *ties.*

Chimney Cap. The *chimney cap* is the sloping surface on the top of the chimney. The slope prevents rain from collecting on the top of the chimney.

Chimney Hood. The *chimney hood* is a covering that may be placed over the flue for protection from the elements. The hood can be made of either masonry or metal. The masonry cap is built so that openings allow the prevailing wind to blow through the hood and create a draft in the flue. The metal hood can usually be rotated by wind pressure to keep the opening of the hood downwind and thus prevent rain or snow from entering the flue.

Spark Arrester. The *spark arrester* is a screen placed at the top of the flue to prevent combustibles from leaving the flue.

FLOOR PLAN REPRESENTATIONS

Three common types of wood-burning appliances are likely to be encountered when reading plans: the masonry fireplace, 0-clearance metal fireplace, and wood stove. The representations for each can be seen in Figure 10–6.

The floor plan is typically used to show the type, location, venting, and floor protection around the fireplace. On a masonry fireplace, dimensions are typically not shown on the floor plan. Because they are surrounded by wood walls, the 0-clearance fireplace and wood stove are located with dimensions. For all three types some type of venting is normally specified to meet the demands of the major codes. The floor protection around the fireplace is provided by the hearth. The outline of the hearth is shown on the floor plan, but typically no specific sizes for the hearth are given.

EXTERIOR ELEVATIONS

The exterior elevations will be helpful for determining the height of the chimney and its relationship to the roof. The chimney is required by code to extend a minimum of 24" above any portion of the structure within 10'–0". The elevations are helpful in determining the height of materials that affect the chimney height. The elevations are also useful in determining the exterior finish and shape of the chimney. An example of the fireplace in exterior elevations can be seen in Figure 10–7.

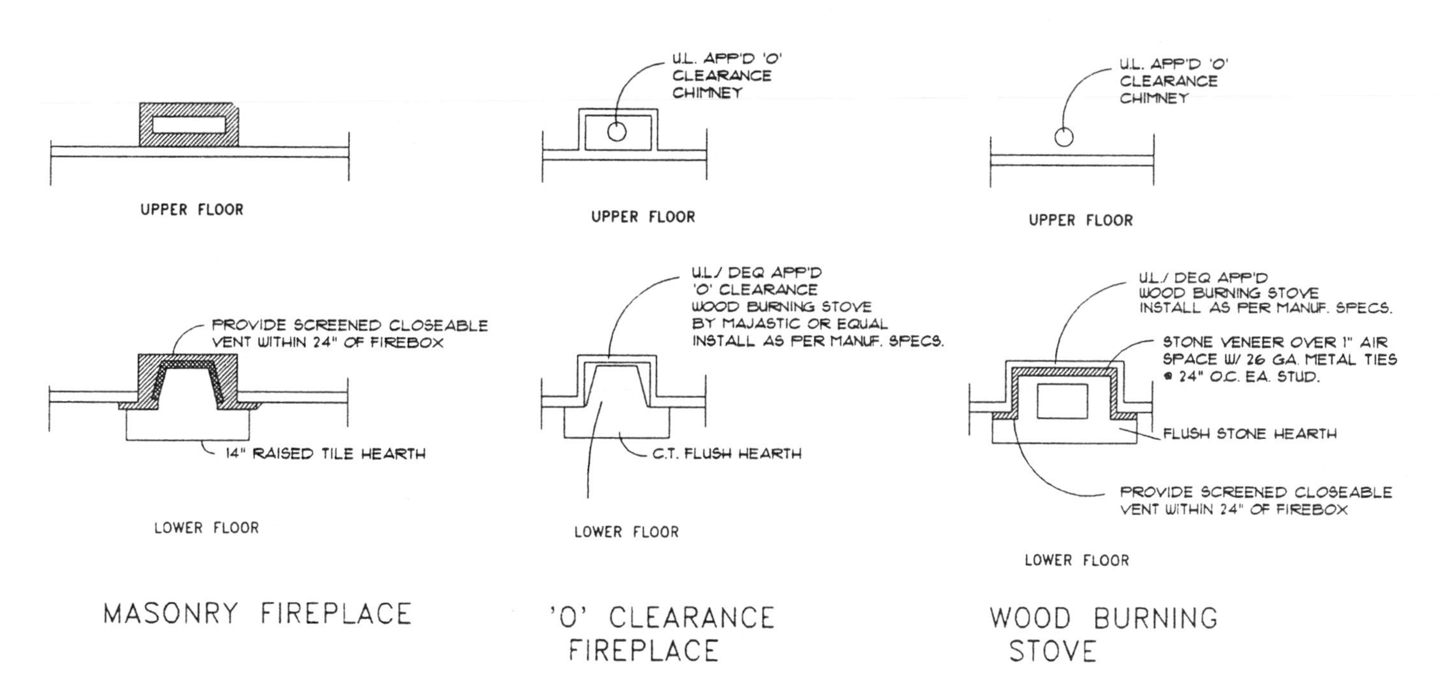

Figure 10–6 Common methods of representing fireplaces and wood stoves on floor plans.

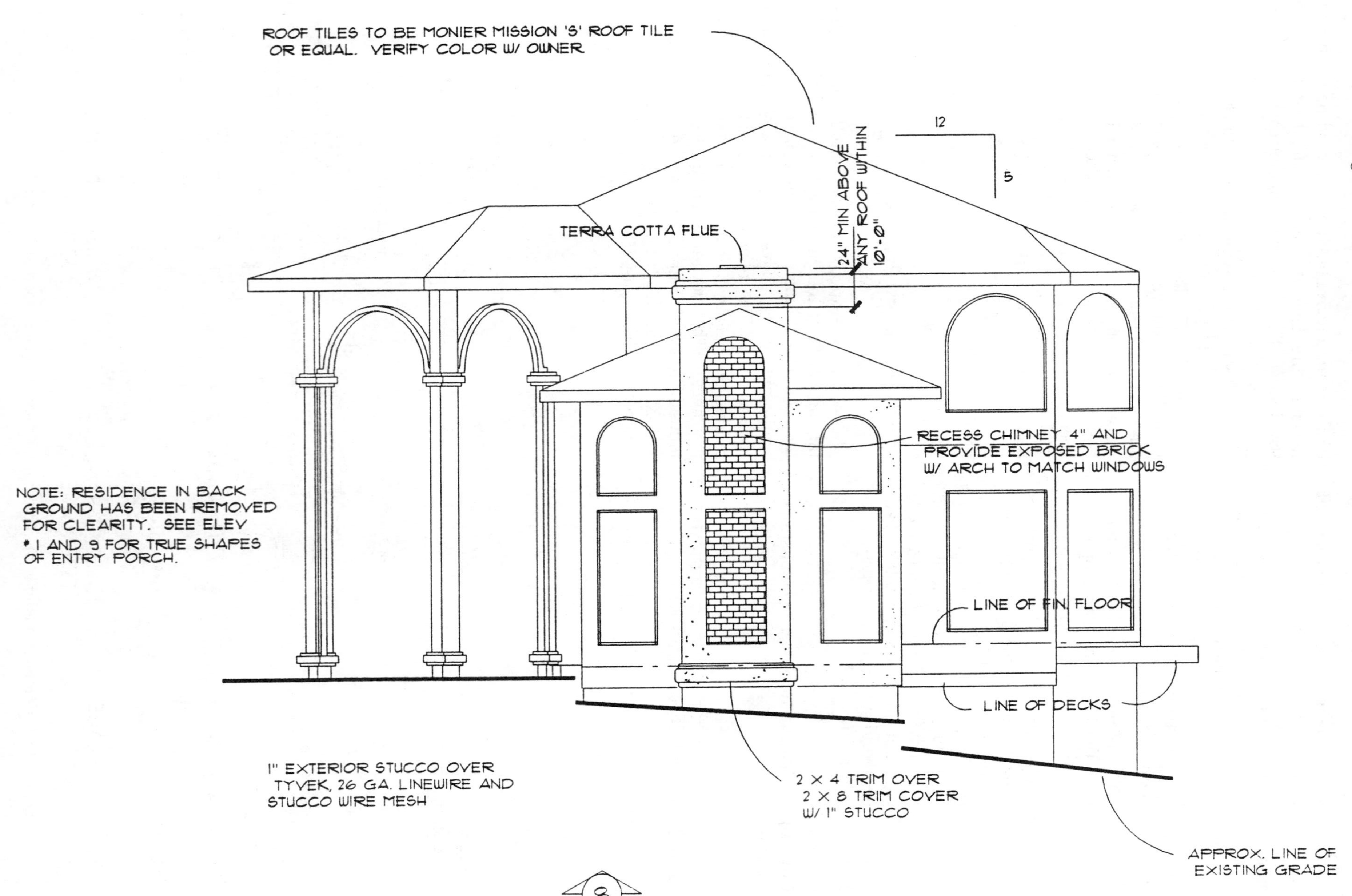

Figure 10–7 A chimney in elevation. *Courtesy of Residential Designs.*

FOUNDATION PLAN

A masonry fireplace is always represented on the foundation plan. Figure 10–8 shows how a foundation for a masonry fireplace is typically represented. The foundation plan shows the size of the footing to support the fireplace and the location of the chimney.

A metal fireplace or wood stove typically requires no extra support and is not shown on the foundation plan. If either is built in a chase that projects outside the foundation wall, the projection is shown on the foundation.

SECTIONS

The sections are typically the most useful drawings for determining construction materials for the fireplace. Figure 10–9 shows how a masonry fireplace can be represented in a section drawing. Typically a metal fireplace or woodstove is not represented in a section.

The section is used to provide a location for dimensions and common materials. The section is the primary drawing for determining all clearances between the chimney and the structure, reinforcement requirements, and general construction guidelines. Always check the scale of the section since the fireplace section is often drawn at a larger scale than the other sections. The section may be a stock detail, as seen in Figure 10–9. A *stock detail* is a drawing used to provide general guidelines for construction, but the designer depends on the knowledge of the construction crew to build the fireplace.

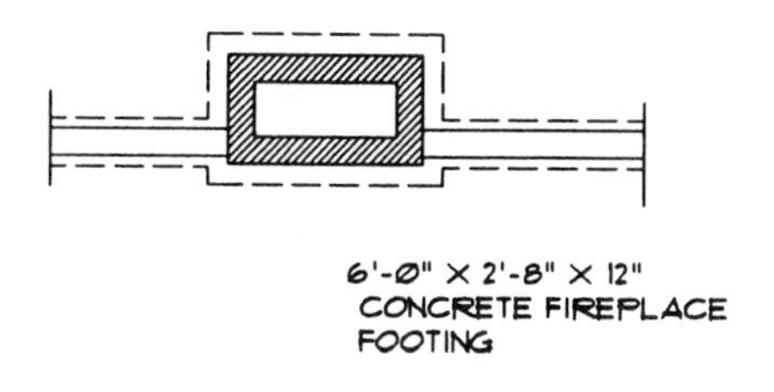

FIREPLACE ON
EXTERIOR WALL

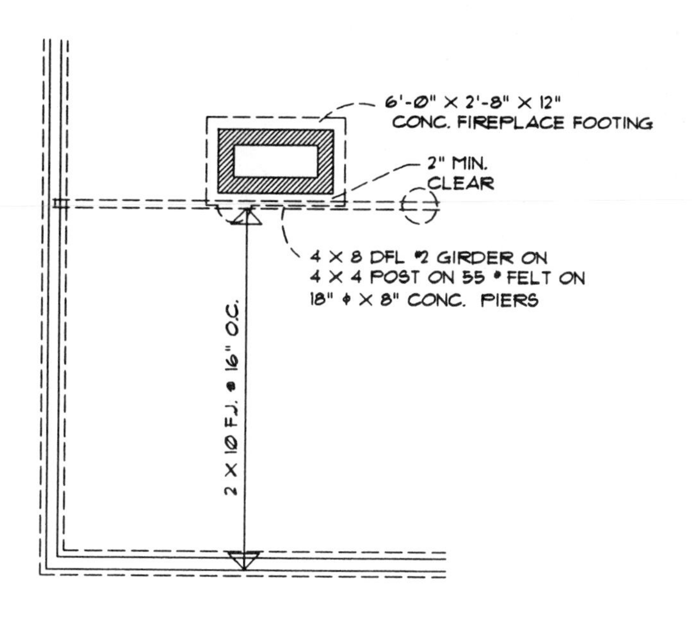

FIREPLACE IN
FOUNDATION INTERIOR

Figure 10–8 Masonry chimney representations on the foundation plan.

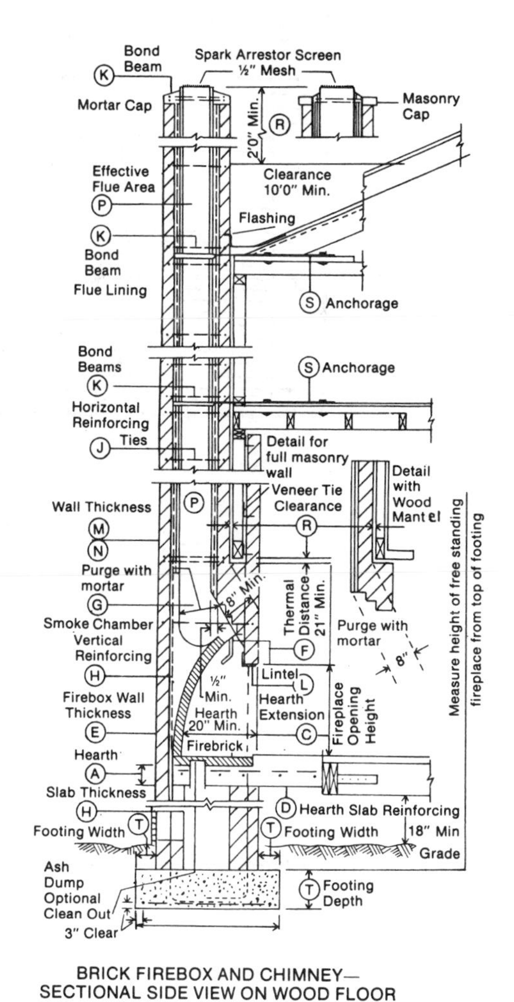

Figure 10–9 A stock fireplace detail is often attached to a set of plans to show general construction guidelines. The circled letters refer to code minimums shown in Figure 10–2. *Reprinted from* Residential Fireplace and Chimney Construction Details and Specifications, *with permission from Masonry Institute of America, Los Angeles, CA.*

FIREPLACE ELEVATION

An interior elevation of the fireplace and the surrounding area is often part of a well-drawn set of plans. This drawing shows the finished relationship of the hearth to the floor, the fireplace to the hearth, and the fireplace opening to the mantel. This drawing typically shows all sizes for the trim used to decorate the fireplace. Figure 10–10 shows a fireplace elevation.

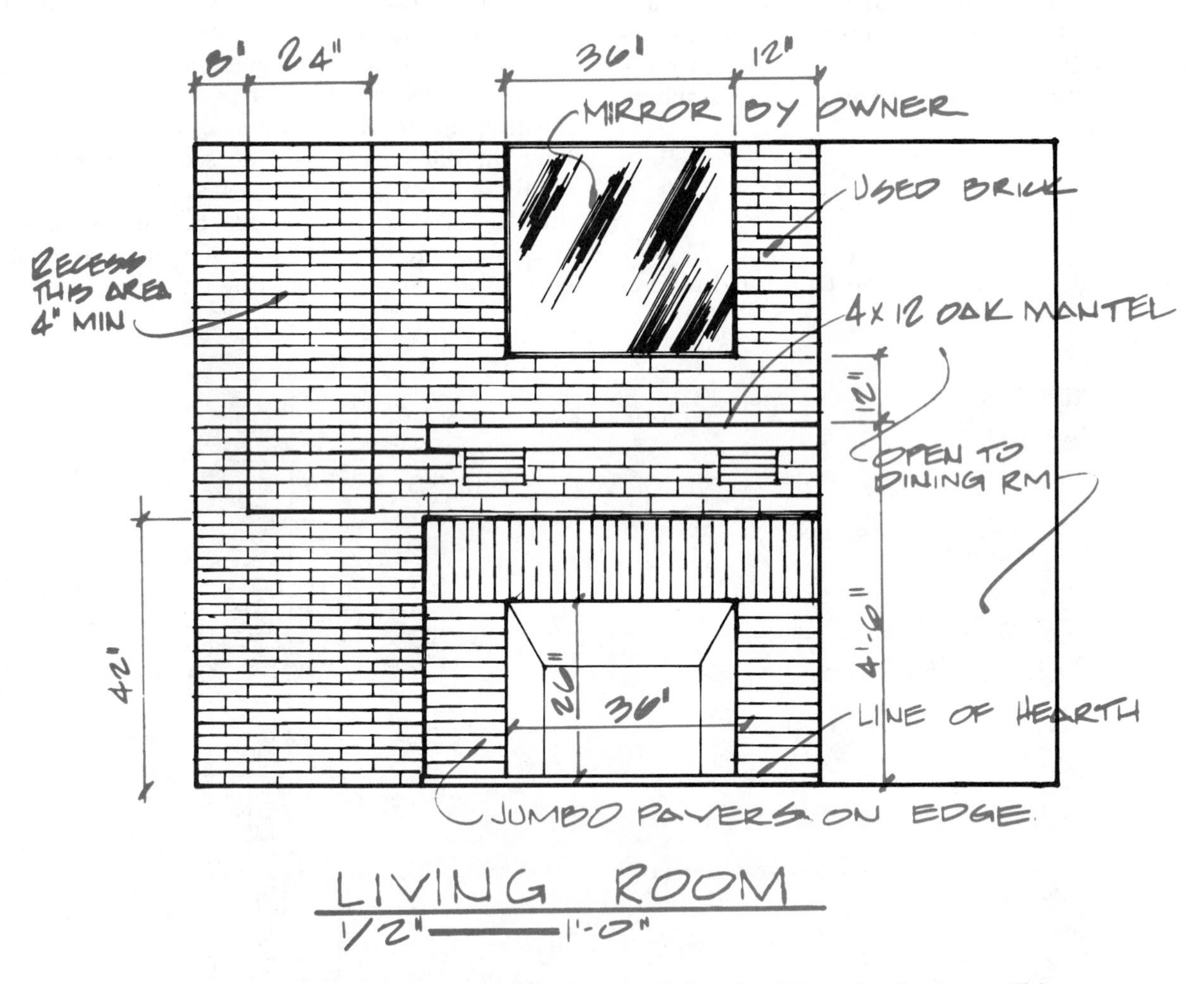

Figure 10–10 An interior elevation may be provided to detail how the fireplace will be completed.

CHAPTER 10 TEST

1. On which drawing would you expect to see the hearth represented without the size specified? ________________.
2. On which drawing would you expect to see the hearth information specified? ________________.
3. What term is used to describe the area where the combustion occurs? ________________.
4. How does the size of the fireplace opening affect the fire? ________________.
5. What are the dimensions of a typical firebox? ________________.
6. The opening from the firebox to the chimney is called a ________________.
7. How is a chimney anchored to a structure at each floor? ________________.
8. If a fireplate is to be located on the short wall of a 13'–6" × 18'–9" family room, how wide would you make the firebox if no size is indicated on the plan? ________________.
9. A roof is built at a 5/12 pitch. The ridge is 12 ft from the chimney. How tall should the chimney be if the floor is 24" above the ground and 8' high walls are used? Show your work. ________________
10. If the chimney in the last question is to be constructed of concrete block, how many courses in height will it be? ________________.

CHAPTER 10 EXERCISES

PROBLEM 10–1. Use the following drawing shown on page 195 to answer the following questions.

1. What is the clearance between the chimney and the upper floor? ________________.
2. What will keep rain from running down the chimney? ________________.
3. What size header will support the upper floor? ________________.
4. Give the size of the lintel. ________________.
5. How will the hearth be reinforced? ________________.
6. Will the chimney interior be unlined? ________________.
7. What is the spacing of the horizontal reinforcing? ________________.
8. To how many joists will the chimney strap be connected? ________________.
9. What material will be used to finish the hearth? ________________.
10. How will the the footing be reinforced? ________________.

PROBLEM 10–2. Use the following drawing shown on page 196 to answer these questions.

1. What is the depth of the firebox? ______________________________.
2. How will the flue be finished? ______________________________.
3. What code governs this fireplace? ______________________________.
4. What will control the flow of smoke from the firebox? ______________________________
____
______________________________.
5. What are the options for finishing the hearth? ______________________________
____
______________________________.
6. How can the rafter size be determined? ______________________________
____
______________________________.
7. What is the height of the firebox? ______________________________.
8. How will the hearth be reinforced? ______________________________
____
______________________________.
9. How will the support strap be tied to the chimney? ______________________________
____
______________________________.
10. If the chimney is wider than 84", how many vertical rebars will be required? ______________________________
____
______________________________.

PROBLEM 10–3. Use Figure 10–10 shown on page 192 to answer the following questions.

1. What type of brick will be used on each side of the firebox? ______________________________
____
______________________________.
2. What is the size of the firebox? ______________________________.
3. What is the height of the mantel above the finish floor? ______________________________.
4. What is the mantel made of? ______________________________.
5. What will surround the fireplace? ______________________________

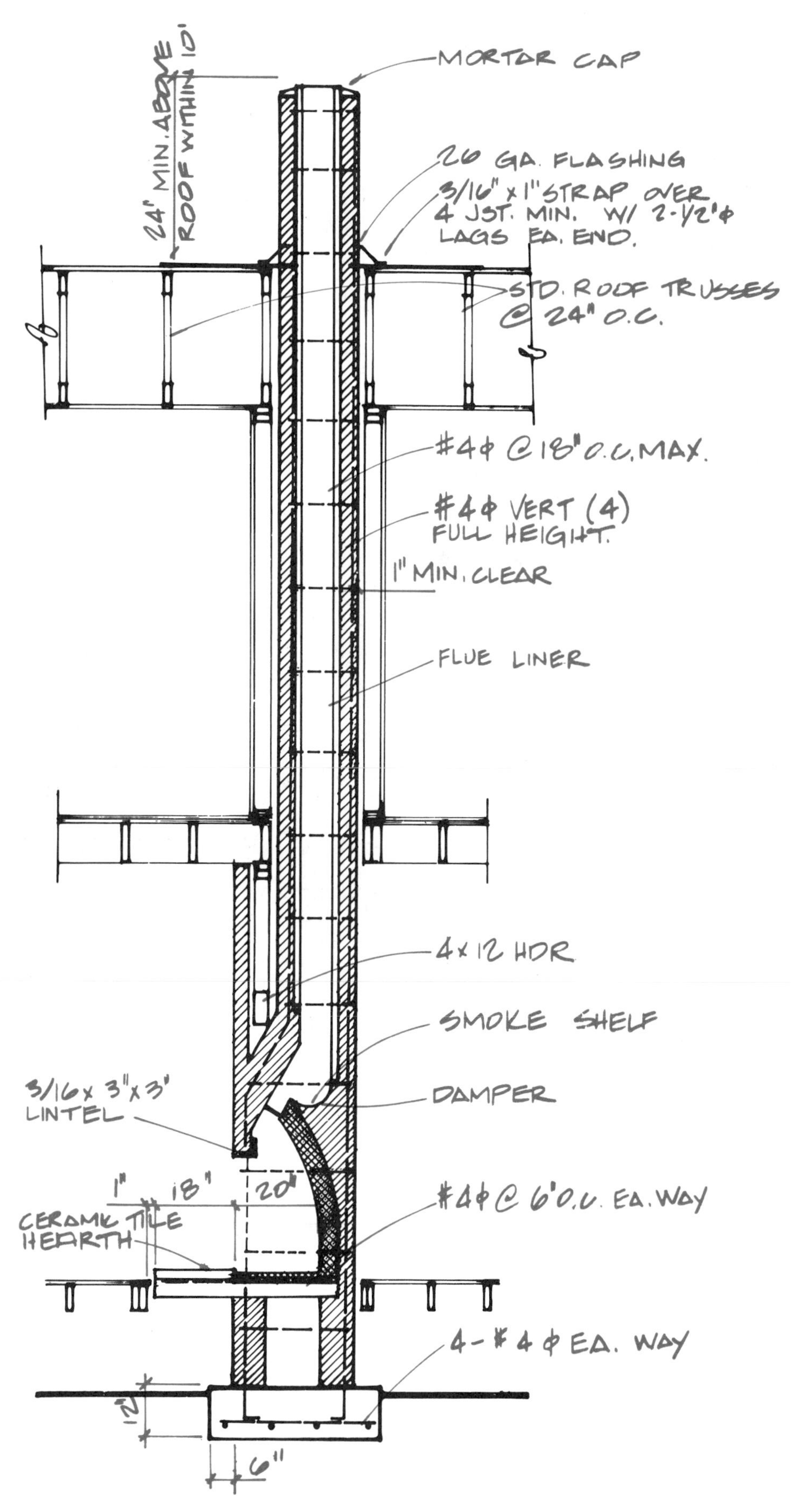
24" MIN. ABOVE ROOF WITHIN 10'
MORTAR CAP
26 GA. FLASHING
3/16" x 1" STRAP OVER 4 JST. MIN. W/ 2-1/2"φ LAGS EA. END.
STD. ROOF TRUSSES @ 24" O.C.
#4φ @ 18" O.C. MAX.
#4φ VERT (4) FULL HEIGHT.
1" MIN. CLEAR
FLUE LINER
4x12 HDR
SMOKE SHELF
3/16 x 3" x 3" LINTEL
DAMPER
1"
18"
20"
CERAMIC TILE HEARTH
#4φ @ 6' O.C. EA. WAY
4-#4φ EA. WAY
12"
6"

Problem 10–1.

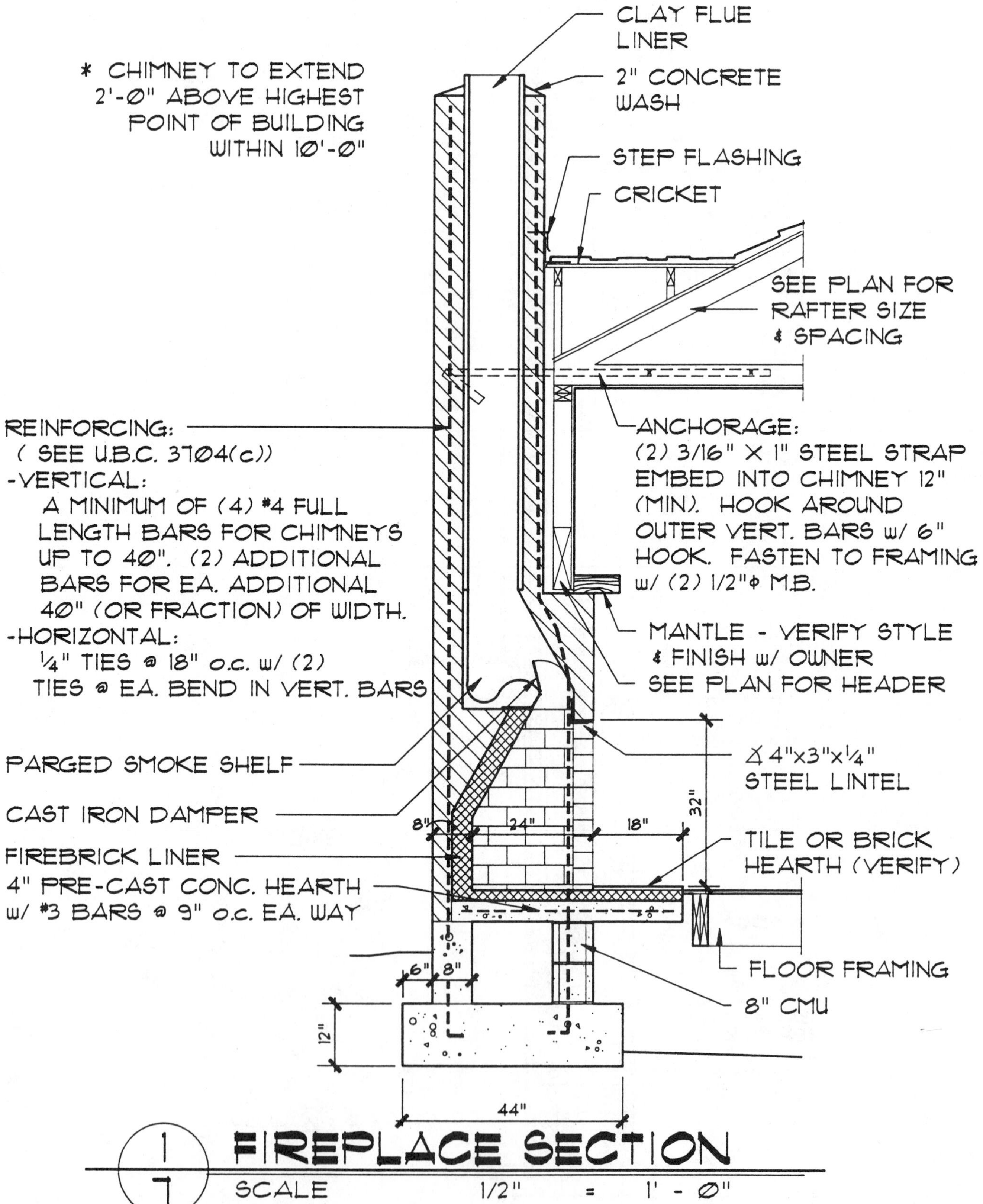

CLAY FLUE LINER

2" CONCRETE WASH

* CHIMNEY TO EXTEND 2'-Ø" ABOVE HIGHEST POINT OF BUILDING WITHIN 1Ø'-Ø"

STEP FLASHING

CRICKET

SEE PLAN FOR RAFTER SIZE & SPACING

REINFORCING:
(SEE U.B.C. 37Ø4(c))
-VERTICAL:
A MINIMUM OF (4) #4 FULL LENGTH BARS FOR CHIMNEYS UP TO 4Ø". (2) ADDITIONAL BARS FOR EA. ADDITIONAL 4Ø" (OR FRACTION) OF WIDTH.
-HORIZONTAL:
¼" TIES @ 18" o.c. w/ (2) TIES @ EA. BEND IN VERT. BARS

ANCHORAGE:
(2) 3/16" X 1" STEEL STRAP EMBED INTO CHIMNEY 12" (MIN). HOOK AROUND OUTER VERT. BARS w/ 6" HOOK. FASTEN TO FRAMING w/ (2) 1/2"ϕ M.B.

MANTLE - VERIFY STYLE & FINISH w/ OWNER

SEE PLAN FOR HEADER

PARGED SMOKE SHELF

∡ 4"x3"x¼" STEEL LINTEL

32"

CAST IRON DAMPER

8"

24"

18"

FIREBRICK LINER

TILE OR BRICK HEARTH (VERIFY)

4" PRE-CAST CONC. HEARTH w/ #3 BARS @ 9" o.c. EA. WAY

6"

8"

FLOOR FRAMING

8" CMU

12"

44"

1
7

FIREPLACE SECTION

SCALE 1/2" = 1' - Ø"

Problem 10–2.

Chapter 11
Reading Roof Plans

ROOF-FRAMING METHODS AND DRAWINGS
ROOF PITCHES
ROOF SHAPES
ROOFING MATERIALS
LINES AND SYMBOLS
TEST
EXERCISES

INFORMATION CONCERNING the framing and finishing of the roof is contained on several drawings. The floor plan, elevations, sections, and roof plans all contain some information about the roof structure. The type of roof-framing method must be considered before looking for specific roofing information.

ROOF-FRAMING METHODS AND DRAWINGS

In Chapters 7 and 8 you were introduced to framing methods. Floor plans, elevations, sections and roof plans each reflect the different framing method used, the finishing materials, and the shape of the roof.

Floor Plans

On the floor plan, either the ceiling joist or the trusses are typically specified as shown in Figure 11–1. If ceiling joists are specified on the floor plan, it indicates a stick roof is to be used. This size of members and the spacing and span for ceiling members can be determined from the floor plan. If a ceiling is to be vaulted, the rafters/ceiling joist are typically specified on the floor plan. Information is then determined from the sections and roof plan about the rafters, purlins, and other support members of the upper roof area.

When trusses are used to frame a roof, their location is often shown only on a floor plan. Because of the simplicity of construction, the direction, spacing, and type of truss can be specified on the floor plan and the truss manufacturer then provides the information needed to build the trusses. The intersections for a girder truss, hip, or Dutch hip roof may also be shown on a floor plan, as seen in Figure 11–2.

Elevations

Although no structural information is usually shown, the elevations can be used to gain valuable information to help understand the shape of the roof.

The elevations show the shape, angle, and finishing materials of the roof. Hips, valleys, and other roof configurations are often easily visualized by looking at the elevations rather than by looking at the roof plans. Figure 11–3 shows how roof intersections can be seen on elevations.

Sections

Roof construction for a specific area can best be seen in the sections. These drawings show framing member size and type for each major area of the roof. Sections also show the vertical relationships of each roof area and how the surfaces intersect with each

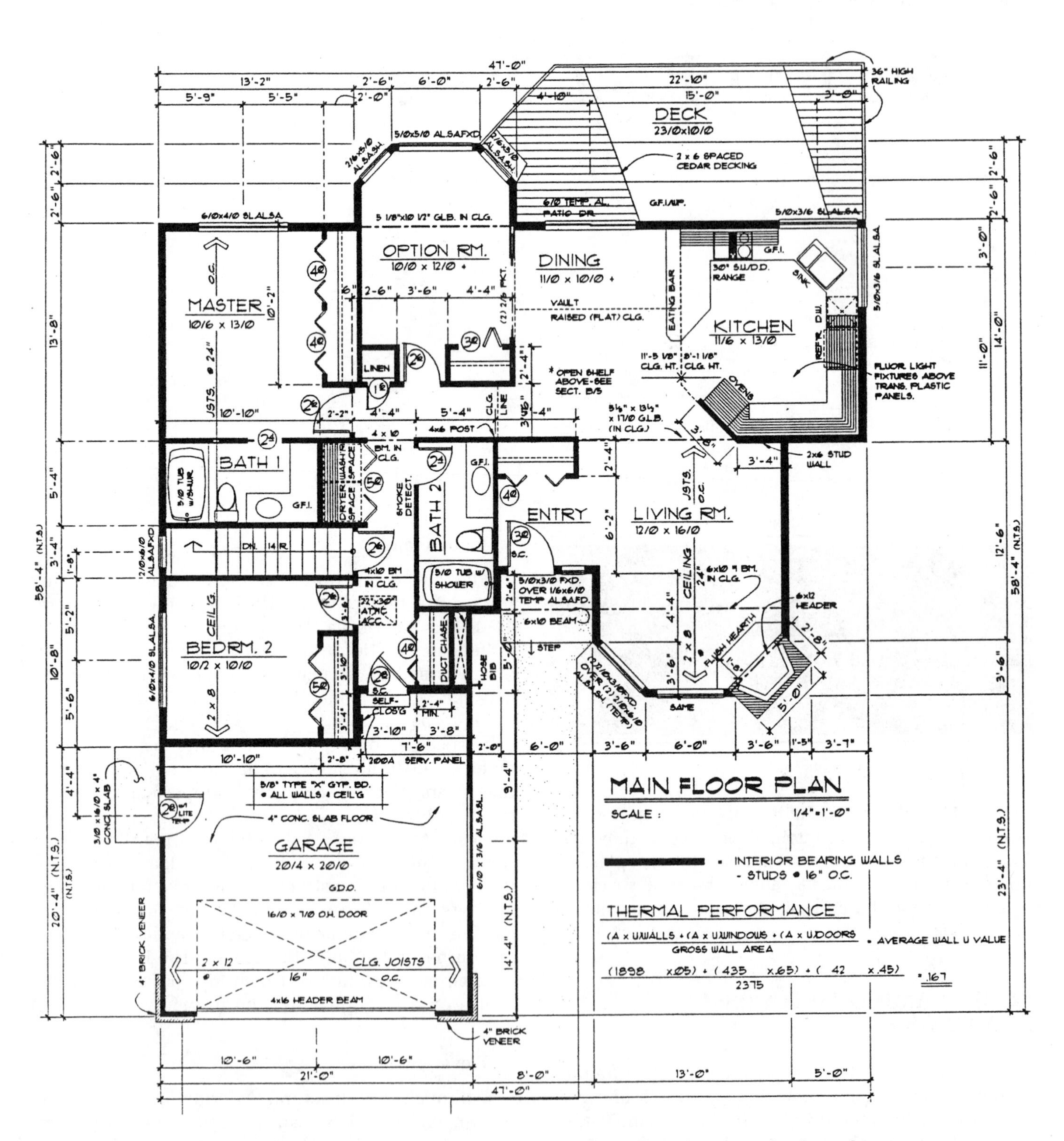

Figure 11–1 On a simple structure the roof framing is often shown on the floor plan. *Courtesy of Piercy and Barclay, Inc.*

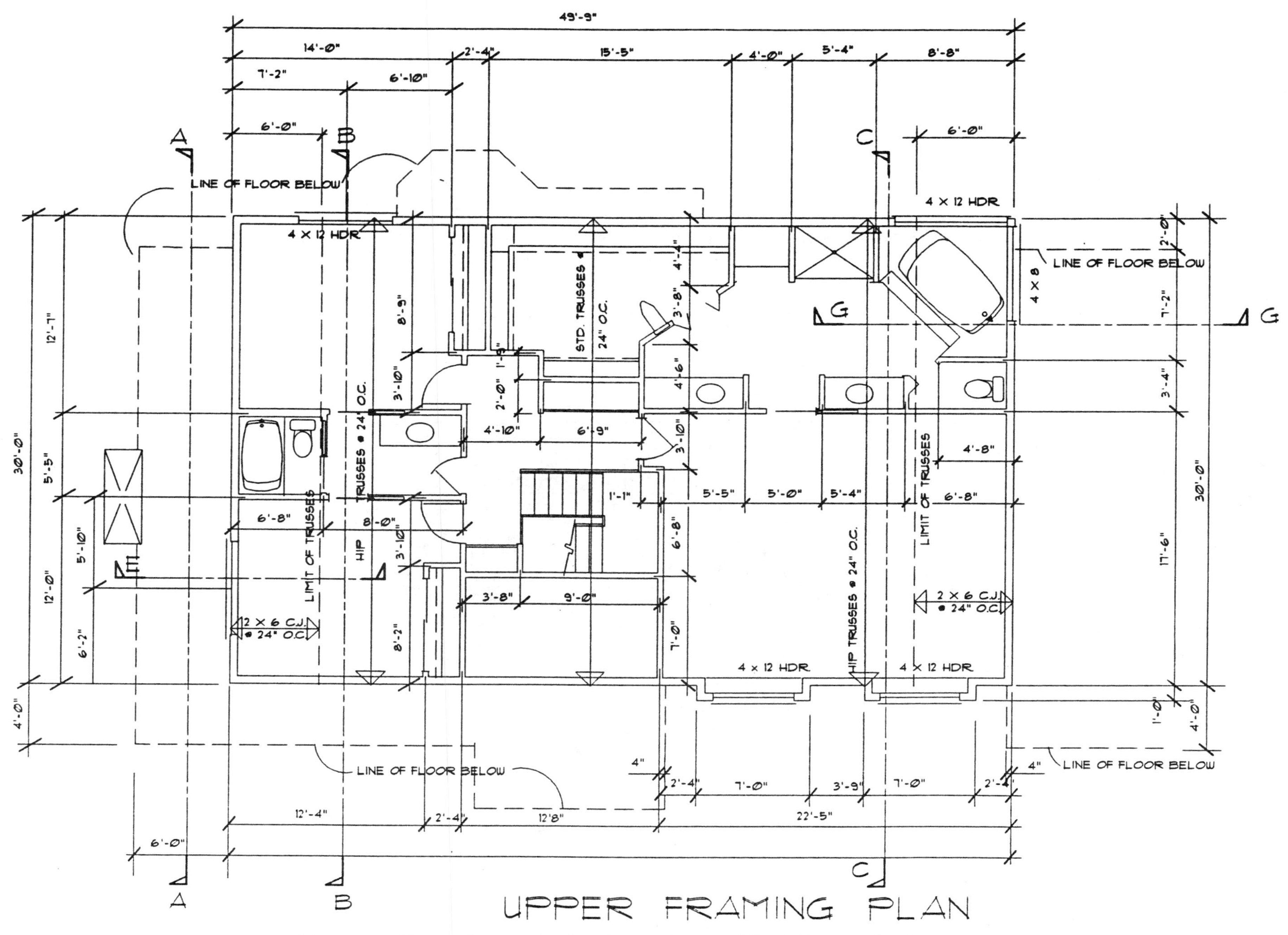

Figure 11–2 The members used to form hips, valleys, and Dutch hips may be shown on a floor plan.

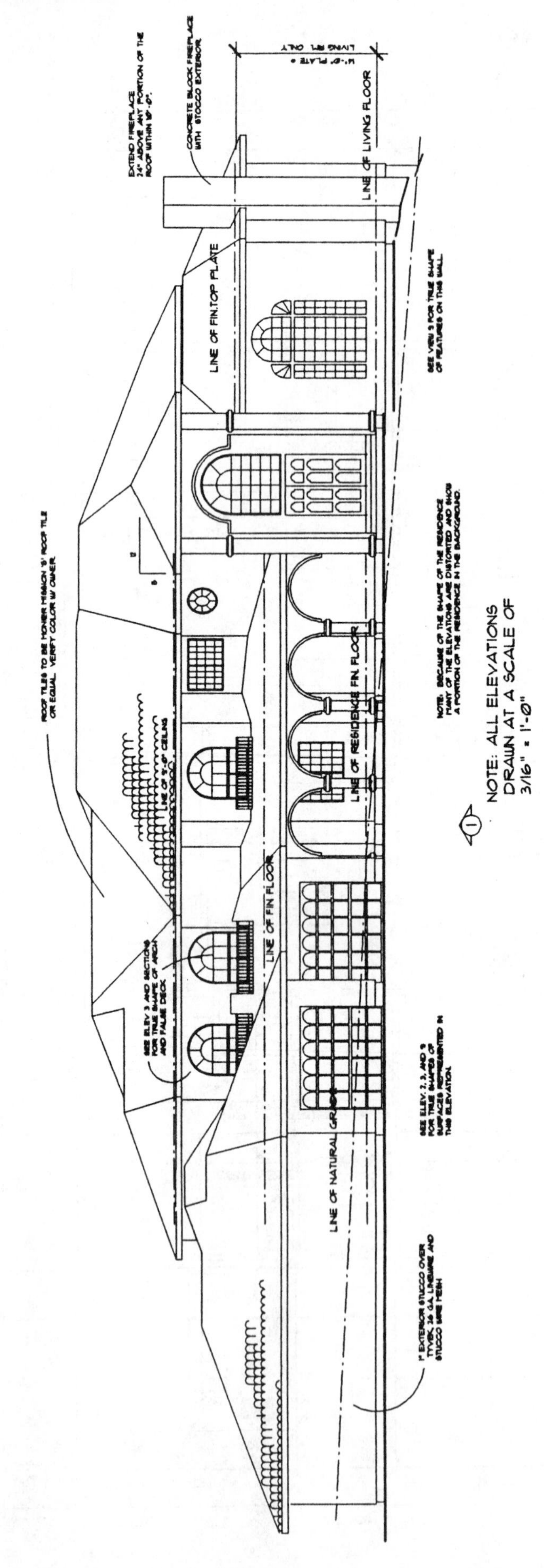

Figure 11–3 The elevations can be helpful in viewing the shapes shown on a roof plan. See Problem 11–4 for roof plan of this elevation. *Courtesy of Residential Designs.*

other. The sections are often used for a simple roof system in place of a roof plan. By showing the size and direction of members on the floor plan and the intersections and vertical relationships on the section, a roof plan may be unnecessary.

Roof Plans

The roof plans provide a view looking down on the entire structure. The plan of the roof area may be either a roof plan or a roof-framing plan. For some types of roofs a roof drainage plan may also be drawn.

A roof plan is used to show the shape of the roof and the size and direction of its major construction materials. Other materials, such as the roofing material, vents and their location, and the type of underlayment, are also typically specified on the roof plan and can be seen in Figure 11–4. Roof plans are typically drawn at a scale smaller than the scale used for the floor plan.

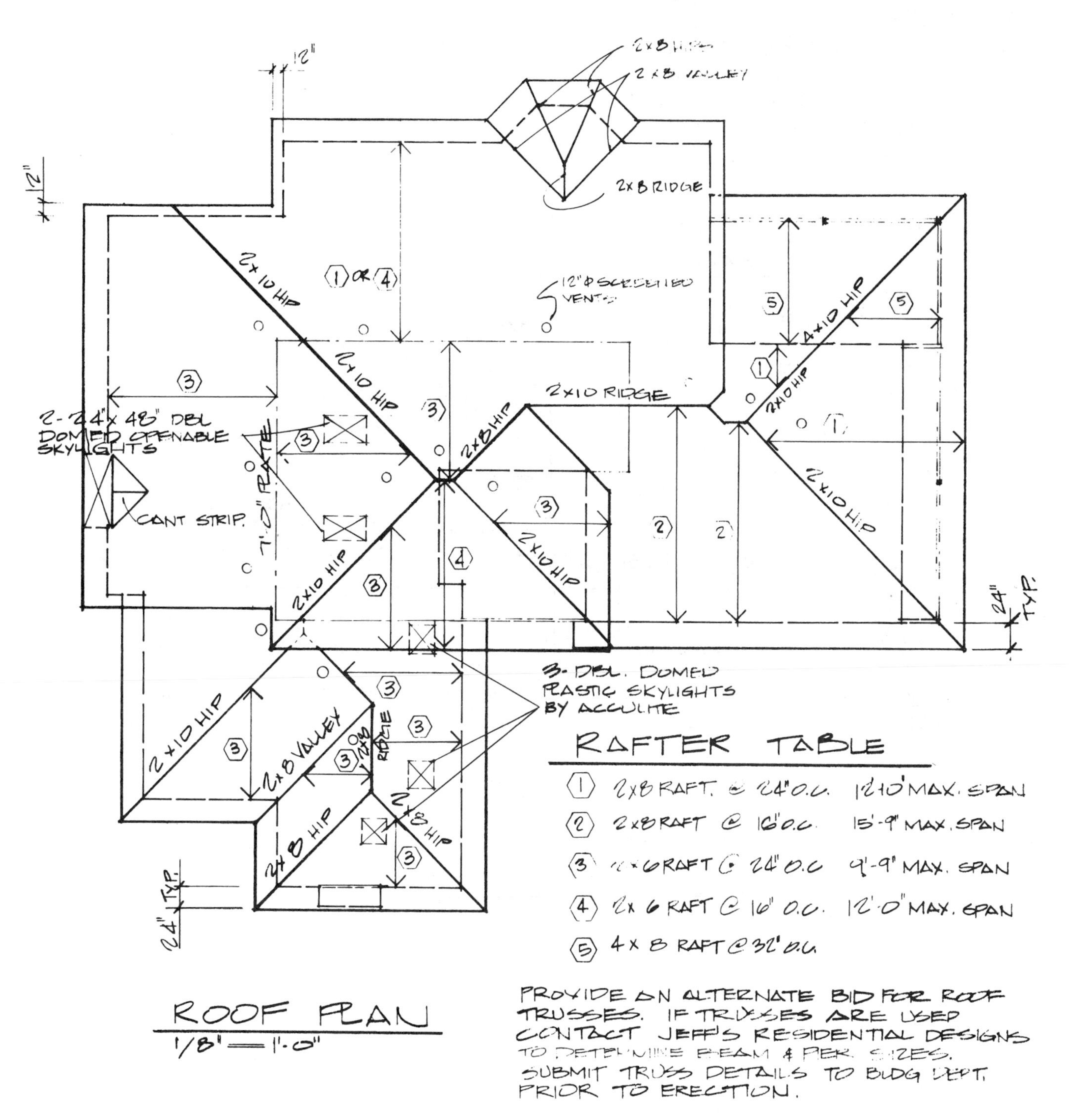

Figure 11–4 A roof plan is used to describe the roof shape and major framing members. *Courtesy of Residential Designs.*

Roof-framing Plans

Roof-framing plans are usually required on complicated residential roof shapes and with most commercial projects. A roof-framing plan shows the size and direction of every construction member required to frame the roof. Figure 11–5 shows an example of a roof-framing plan. A residential roof-framing plan typically shows each member required to frame the roof.

Roof Drainage Plans

A roof drainage plan is a plan showing how water will be diverted over and away from the roof system. The roof drainage plan typically shows ridges or valleys in the roof, roof drains, and downspouts. On a residence this information may be placed on the roof plan. Because of its simplicity, the plan is usually drawn at a scale much smaller than the floor plan.

ROOF PITCHES

Roof *pitch*, or *slope*, is a description of the angle of the roof that compares the horizontal run to the vertical rise. The slope is shown on the elevations and sections. The intersections that result from various roof pitches can be seen on the roof plan. On a roof plan for a structure with equal roof pitches, the intersection, or *ridge*, is formed halfway between the two walls supporting the rafters. When the roof pitches

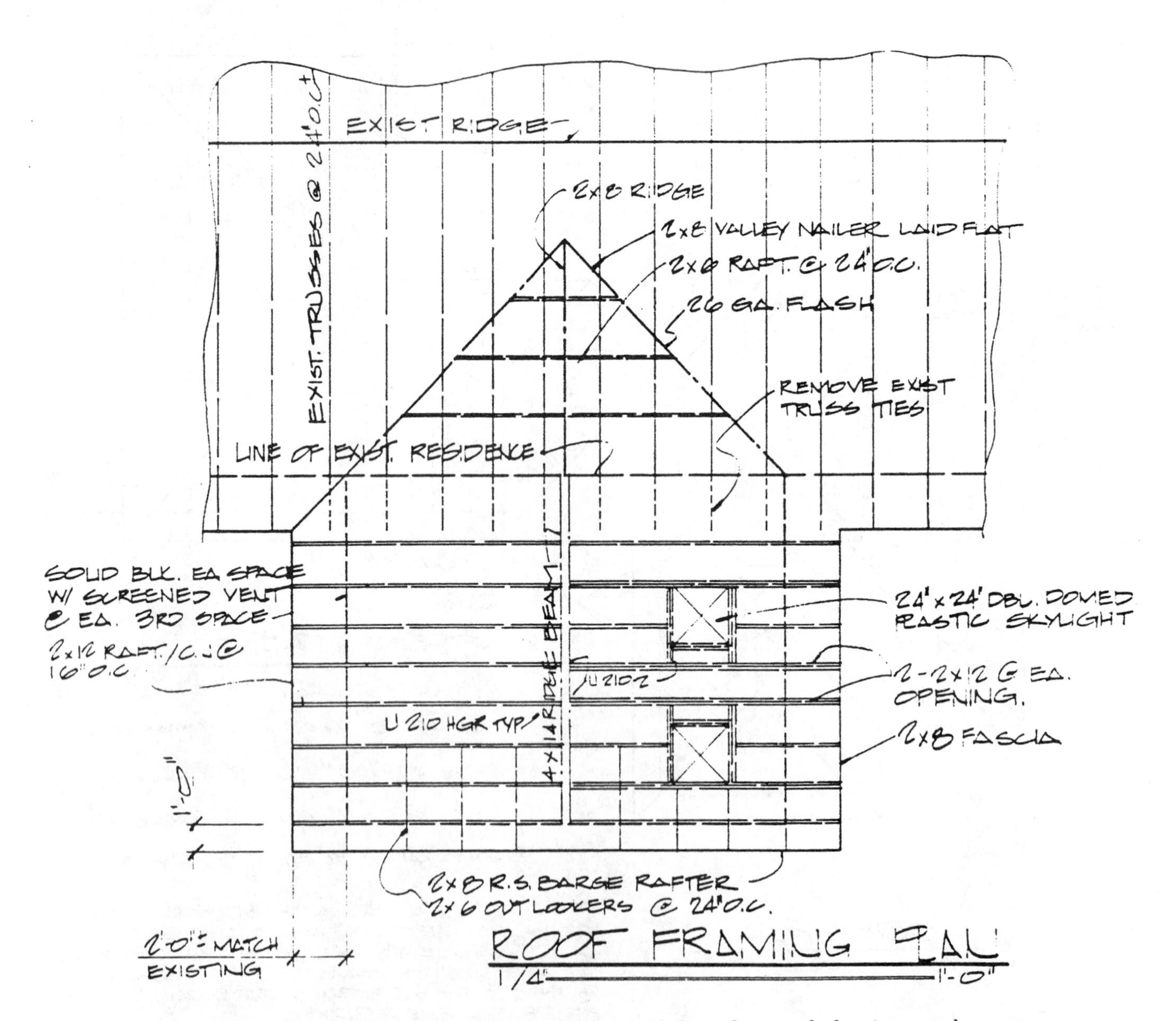

Figure 11–5 A roof-framing plan is often provided to show each framing member required. *Courtesy of Residential Designs.*

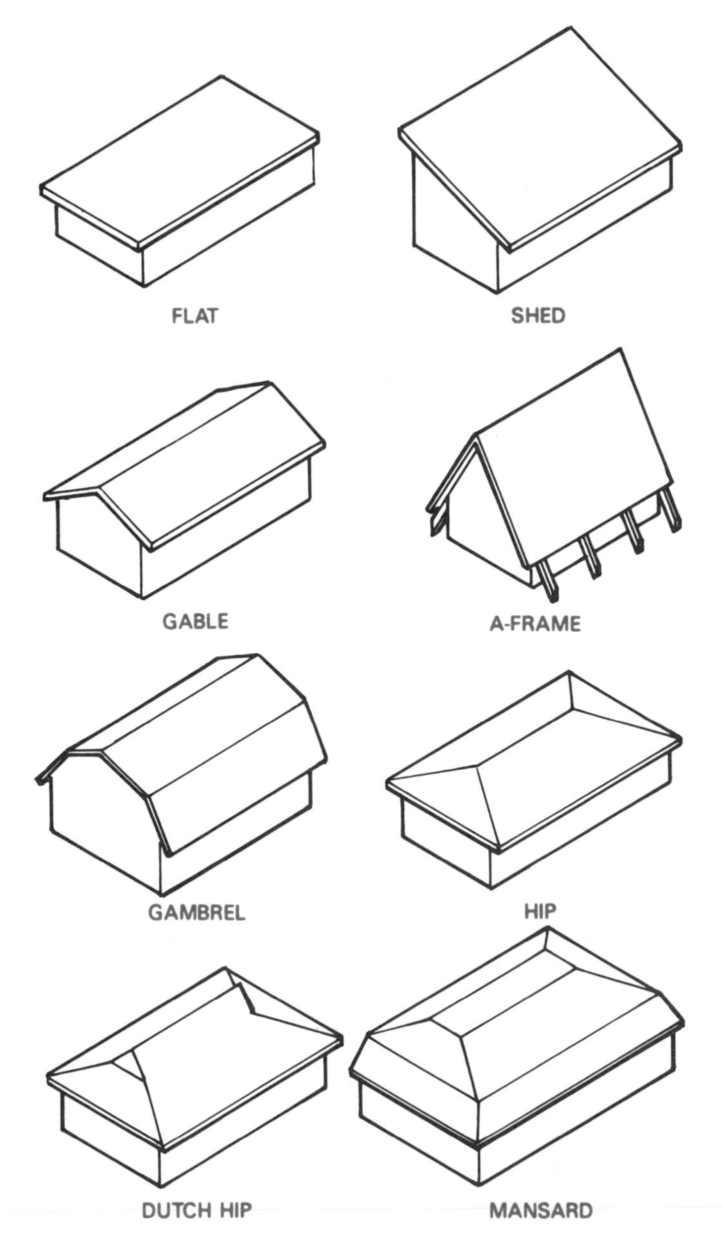

Figure 11–6 Common roof shapes for residential structures.

are unequal or the supporting walls are at different heights, the ridge is not in the center of the support walls. The sections and elevations are necessary to help visualize how the roof planes will be formed. The pitch also can determine the size of the rafter to be used. When a rafter is at an angle of 30° or less from vertical, the roof member can be considered a wall member rather than a rafter.

ROOF SHAPES

By changing the roof pitch the shape of the roof may be changed. Common roof shapes include flat, shed, gable, A frame, gambrel, hip, Dutch hip, and mansard. Each can be seen in Figure 11–6.

Flat Roofs

The flat roof is a very common style of roof in areas with little rain or snow. The flat roof is economical to construct because ceiling joists are eliminated and rafters are used to support both the roof and ceiling loads. Figure 11–7 shows the common materials used to frame a flat roof. Figure 11–8 shows how a flat roof can be represented on the roof plan.

Often the flat roof has a slight pitch in the rafters. A pitch of 1/8 in. per ft is often used to help prevent water from ponding on the roof. As water flows to the edge, a metal diverter is usually placed at the eave to prevent dripping on walkways. A flat roof often has a *parapet*, or false wall, surrounding the perimeter of the roof. This wall can be used for decoration or for protection of mechanical equipment.

Shed Roofs

The shed roof offers the same simplicity and economical construction methods of a flat roof but does

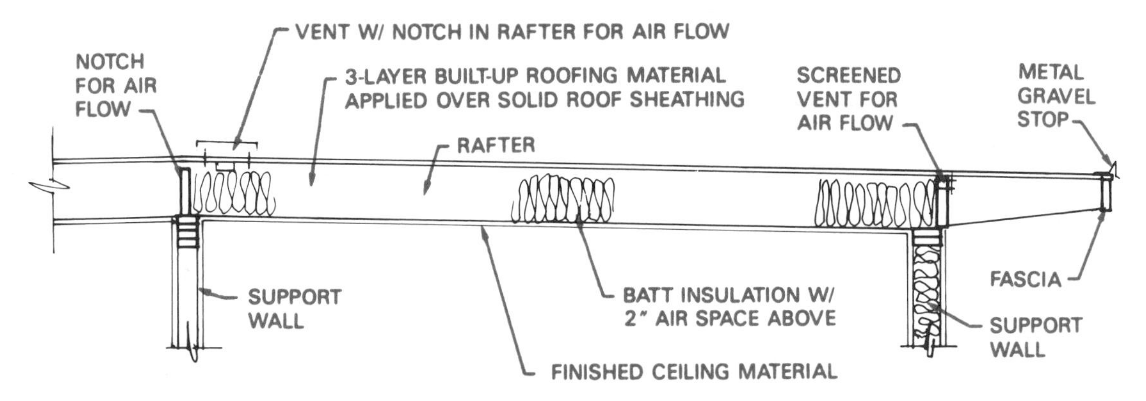

Figure 11–7 Common construction materials for a flat roof.

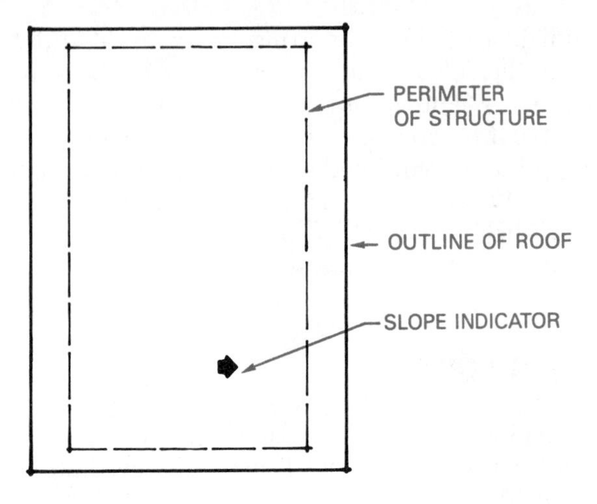

Figure 11–8 A flat roof in plan view.

not have the drainage problems associated with a flat roof. Figure 11–9 shows the construction methods for shed roofs. The shed roof may be constructed at any pitch. When seen in plan view, the shed roof resembles the flat roof.

Gable Roofs

A gable roof is one of the most common types of roof used in residential construction. Figure 11–10 shows the construction of a gable roof system. The gable can be constructed at any pitch, with the choice of pitch limited only by the roofing material. Figure 11–11 shows how a gable roof is typically represented in plan view. Many plans use two or more gables at 90° angles to each other. The intersections of gable surfaces are called either *hips* or *valleys*. Typically the valley and hip are specified on the roof plan.

A-frame Roofs

An A frame is a method of framing walls as well as a system of framing roofs. An A-frame structure uses rafters to form its supporting walls, as shown in Figure 11–12. The roof plan for an A frame is very similar to the plan for a gable roof. However, the materials and rafter sizes are usually quite different.

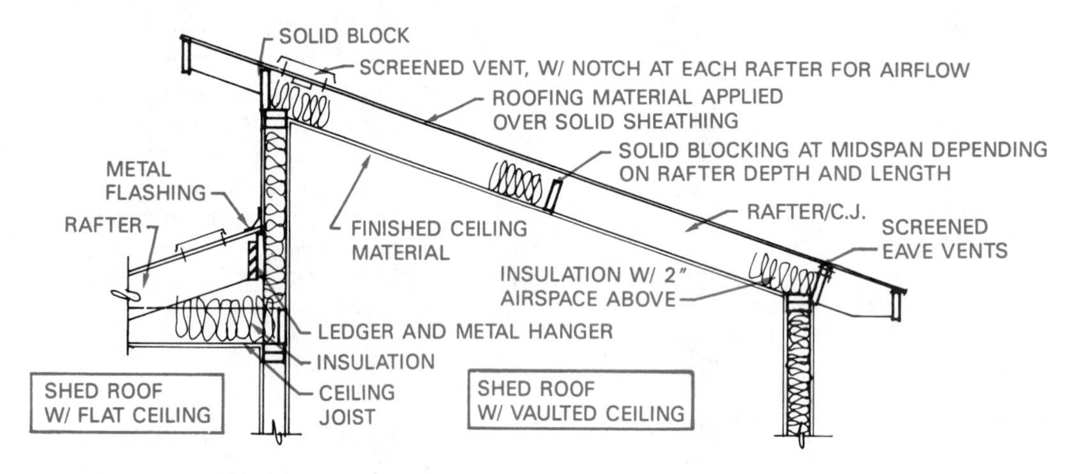

Figure 11–9 Common construction materials for a shed roof.

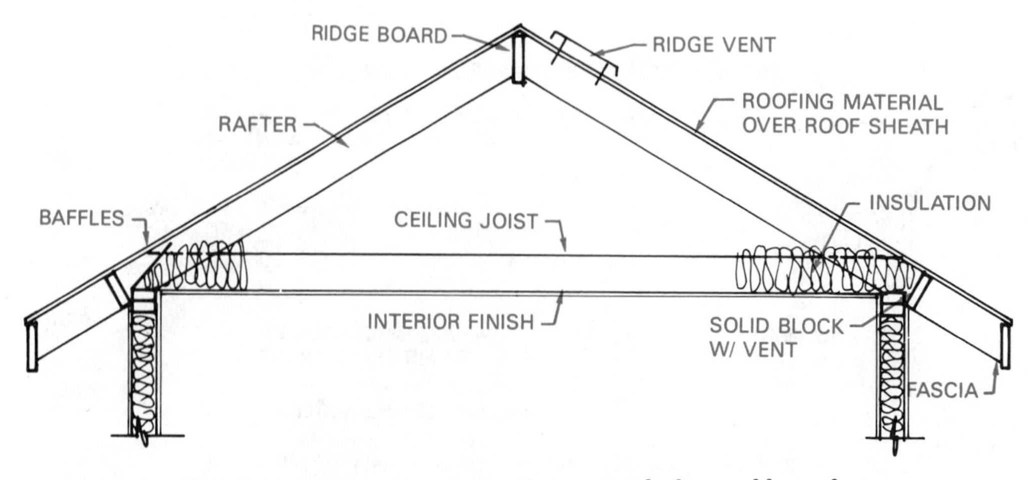

Figure 11–10 Common construction materials for a gable roof.

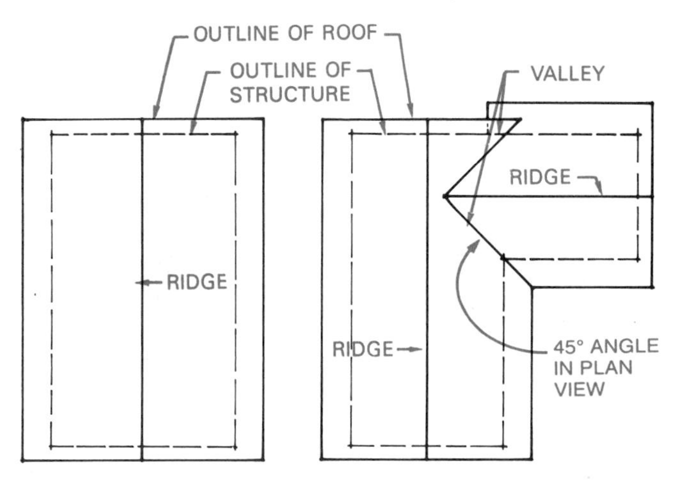

Figure 11–11 A gable roof in plan view.

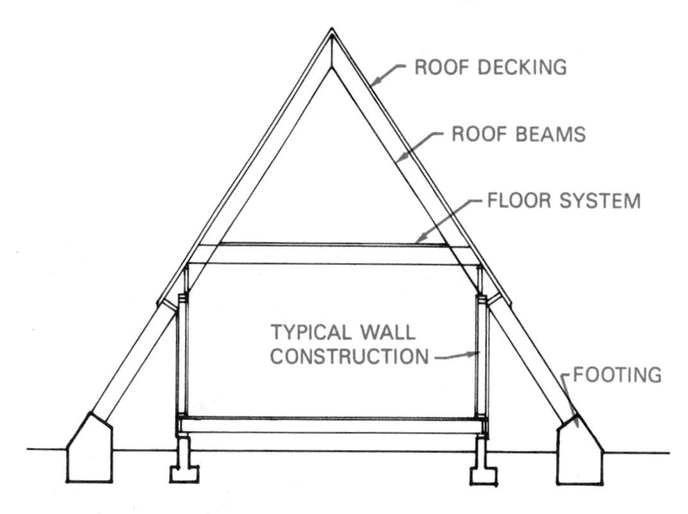

Figure 11–12 Common components of A-frame construction.

Gambrel Roofs

A gambrel roof can be seen in Figure 11–13. The gambrel roof is a very traditional roof shape that dates back to the colonial period. The upper level is covered with a steep roof surface that connects to a roof system with a slighter pitch. By covering the upper level with roofing material rather than siding, the height of the structure appears shorter than it is. This roof system can also be used to reduce the cost of siding materials because it uses less expensive roofing materials. Figure 11–14 shows a plan view of a gambrel roof.

Hip Roofs

The hip roof of Figure 11–15 is a traditional roof shape. A hip roof has many similarities to a gable roof but instead of having two surfaces the hip roof has four. The intersection between each surface is called a *hip*. If built on a rectangular structure, the hips form two points with a ridge spanning between them. When hips are placed over an L- or T-shaped structure, an interior intersection is formed that is called a *valley*. The valley of a hip roof is the same as the valley for a gable roof. Hip roofs can be seen in plan view in Figure 11–16.

Figure 11–13 A gambrel roof is often used to enhance a traditional residence. *Courtesy of Cabot's Stains, design by Architect Russell Swinton Oatman, R. A. Princeton, MA.*

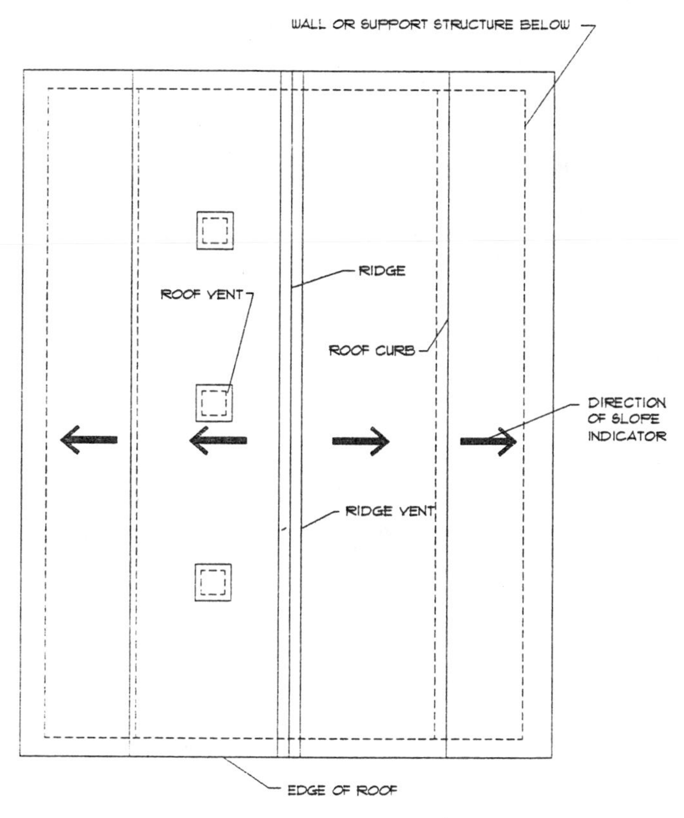

Figure 11–14 Gambrel roof in plan view.

Figure 11–15 A hip roof is a traditional roof shape used to provide shade on all sides of a structure. *Courtesy Masonite Corporation.*

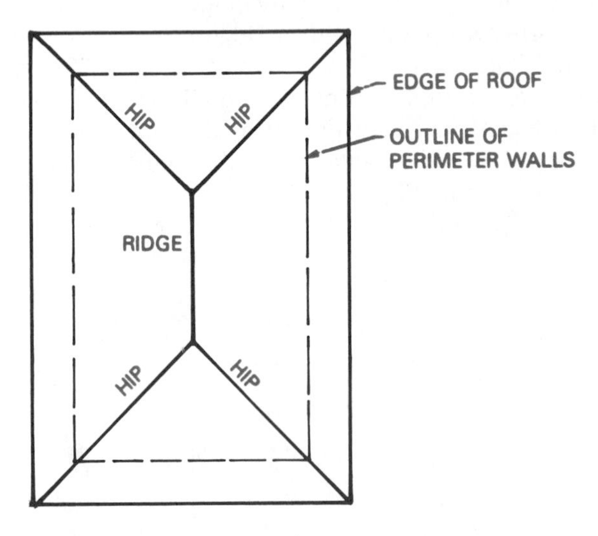

Figure 11–16 A hip roof in plan view.

Dutch Hip Roofs

The Dutch hip roof is a combination of a hip and a gable roof. The center section of the roof is framed in a method similar to that for a gable roof. The ends of the roof are framed with a partial hip that blends into the gable. A small wall is formed between the hip and the gable roofs, as seen in Figure 11–17. On the roof plan, the shape, distance, and wall location must be shown, similar to the plan in Figure 11–18.

Mansard Roofs

The mansard roof is similar to a gambrel roof but has the angled lower roof on all four sides rather than just two. The mansard roof is often used as a parapet wall to hide mechanical equipment on the

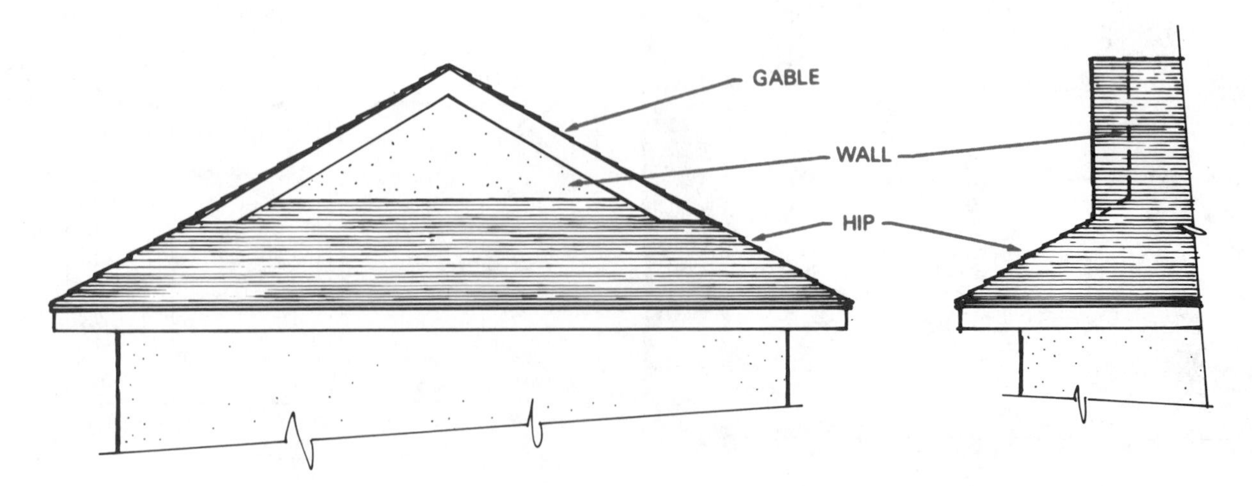

Figure 11–17 A Dutch hip roof combines a gable and a hip roof.

roof, or it can be used to help hide the height of the upper level of a structure. Mansard roofs can be constructed in many different ways. Figure 11–19 shows two common methods of constructing a mansard roof. The roof plan for a mansard roof resembles the plans shown in Figure 11–20.

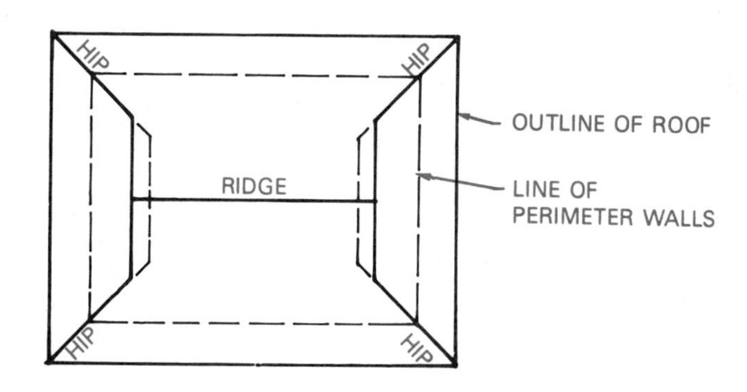

Figure 11–18 Common members of a Dutch hip roof in plan view.

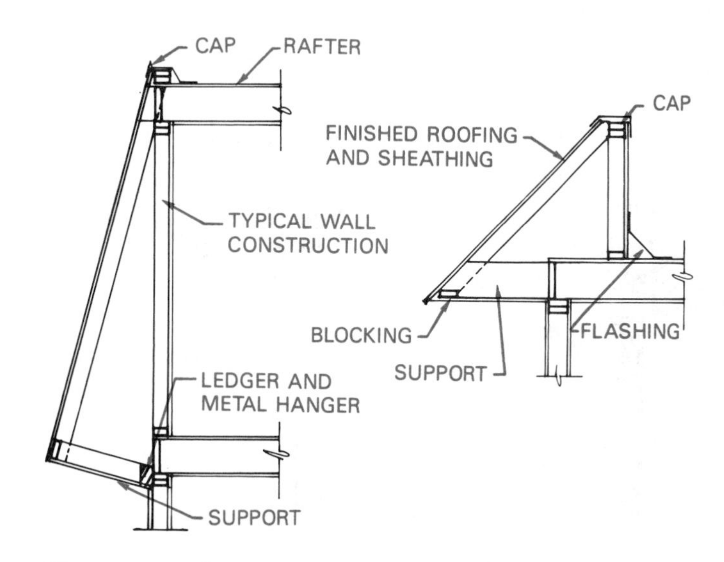

Figure 11–19 Common construction members for a mansard.

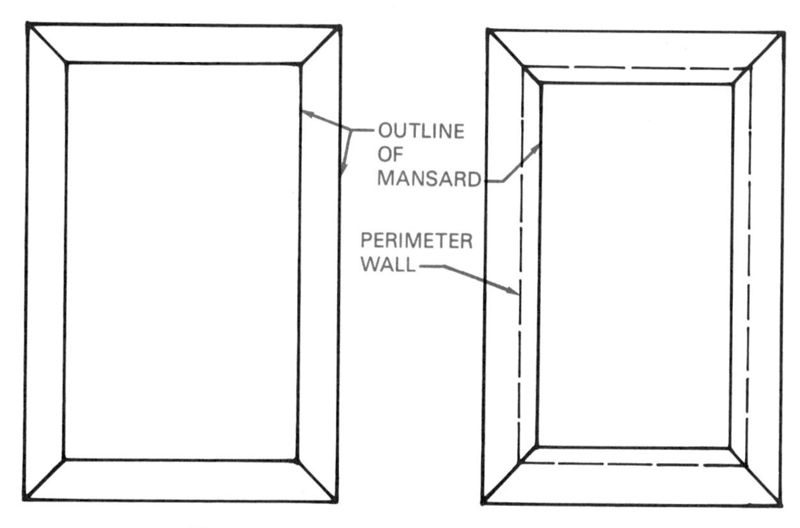

Figure 11–20 The mansard in plan view.

Dormers

A *dormer* is an opening framed in the roof to allow window placement. Dormers are most frequently used on traditional style roofs, such as the gable or hip. Figure 11–21 shows one of the many ways that dormers can be constructed. Dormers are usually shown on the roof plan, as seen in Figure 11–22.

ROOFING MATERIALS

The material to be used on the roof is dependent on the pitch, exterior style, weather, and cost of the structure. Common roofing materials include built-up roofing, composition and wood shingles, and clay and cement tiles. When ordering or specifying these materials, the term "square" is often used. A *square* is used to describe an area of roofing that covers 100 sq ft. See Chapter 7 for full review of roofing materials.

LINES AND SYMBOLS

Several different types of lines and symbols must be understood when reading roof plans. These include the lines used to represent roof supports, roof shapes, structural materials, nonstructural materials, and written specifications.

Supports

Exterior walls, interior bearing walls, purlins, beams, and outlookers are the supports for the roof structure. Each may be shown on the roof plan. Exterior walls are generally shown with bold dashed lines. Interior bearing walls are often represented using thin dashed lines similar to those used to represent the exterior walls. Purlins are often represented with thin dashed lines in a long-short-long pattern. Beams and outlookers are usually represented with two thin parallel lines. Figure 11–23 shows these methods for drawing each type of roof support.

Roof Shape

Changes in the shape of the roof, such as ridges, hips, Dutch hips, and valleys, are also represented with solid lines. Each can be seen in Figure 11–24.

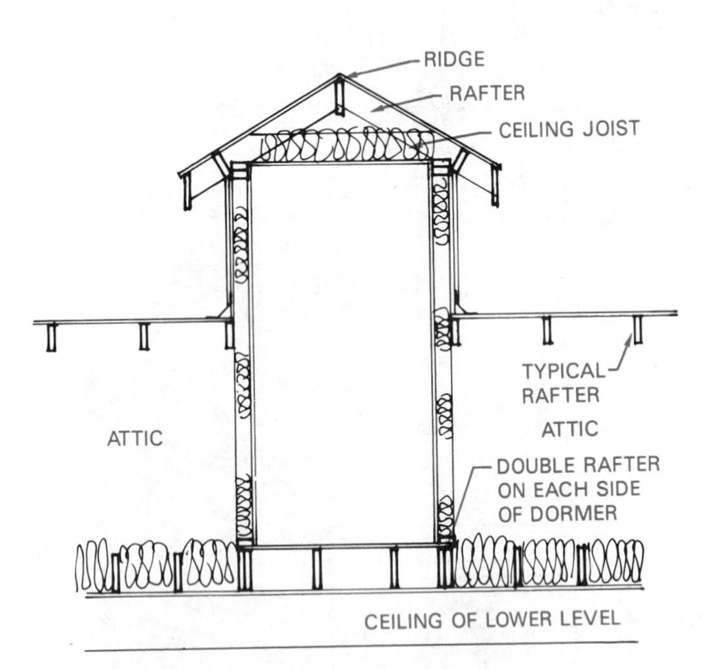

Figure 11–21 Typical structural members for a dormer.

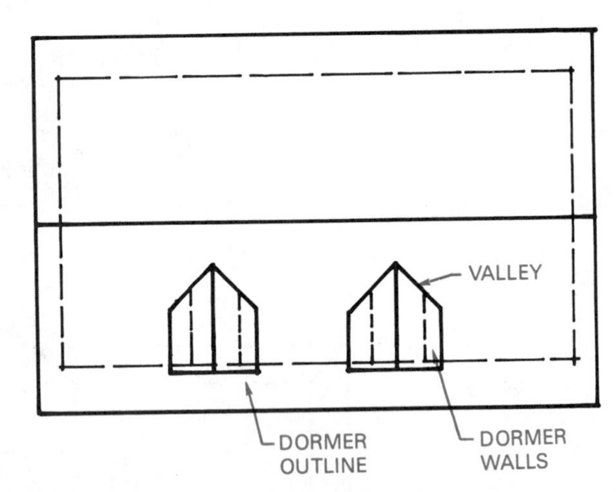

Figure 11–22 Dormers shown in plan view.

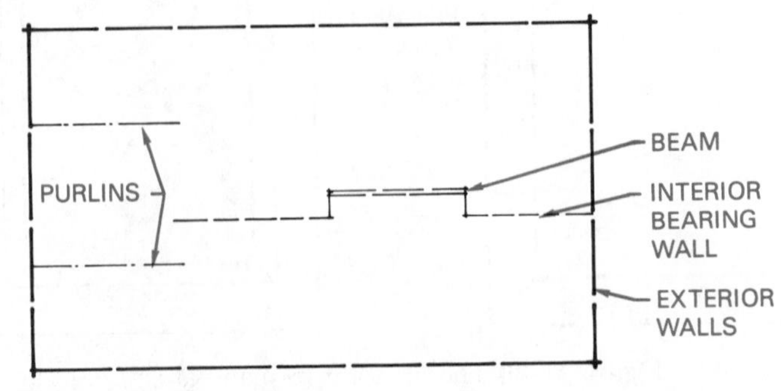

Figure 11–23 Common methods of representing roof support in plan view.

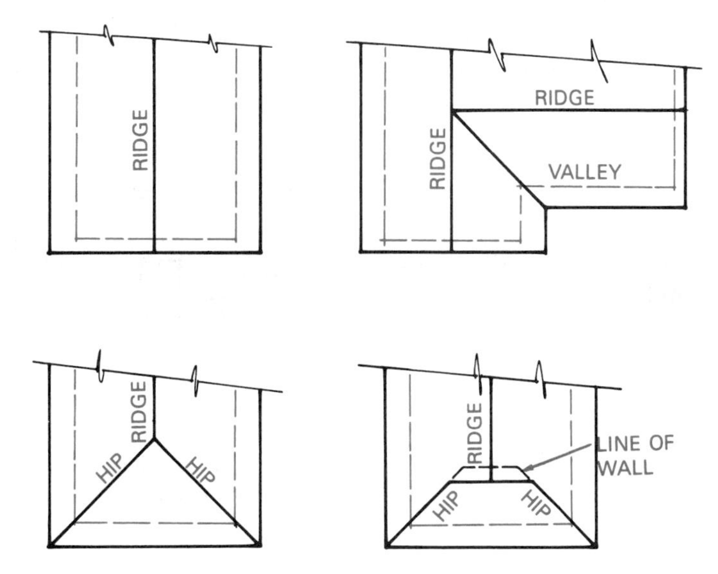

Figure 11–24 Changes in the roof shape, such as eaves, hips, valleys, and ridges, are typically represented on roof plans with solid lines.

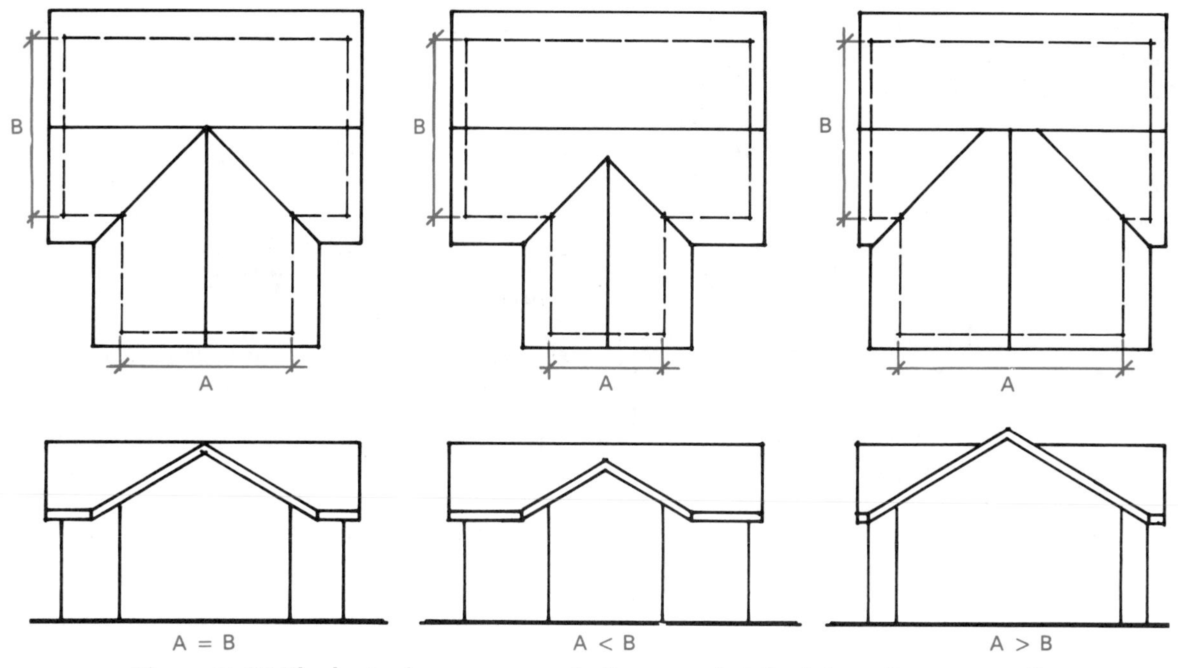

Figure 11–25 The basic shape may remain the same, but the intersections vary as the distance between the walls is varied.

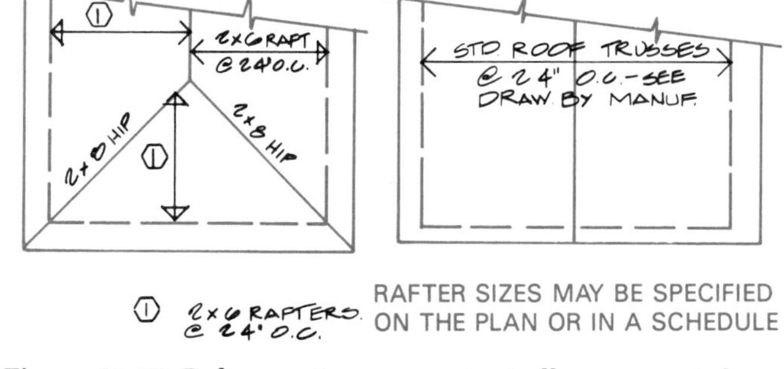

Figure 11–26 Rafters or trusses are typically represented on the roof plan by a thin line showing proper direction and span.

Notice in Figure 11–25 that the shape of the roof changes dramatically as the width between the walls is changed.

Structural Materials

The type of plan you are working with determines the method used to represent structural material. If you are working with a roof plan the rafters or trusses are typically represented as seen in Figure 11–26. A thin line is typically used to represent the direction of the roof members, and a note specifies the size and spacing. If a roof-framing plan is drawn, thin lines in a long-short-long pattern are often used to represent the rafters. Beams and other structural members are usually shown with thin dashed lines, as seen in Figure 11–27.

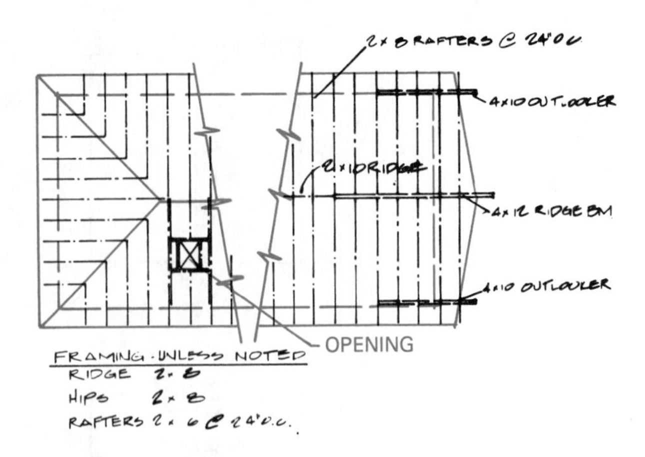

Figure 11–27 A typical method of representing members on a roof-framing plan.

Nonstructural Materials

Vents, chimneys, skylights, solar panels, diverters, cant strips, slope indicators, and downspouts are the most common materials shown on the roof plan. The *cant strip*, or *saddle*, is a small gable built behind the chimney to divert water away from the chimney. Common methods of representing each of the nonstructural materials can be seen in Figure 11–28. These materials are usually not shown on a roof-framing plan.

Roof Ventilation and Access

Attic ventilation is typically specified on the roof plan, but the exact number of vents is often determined at the job site by the roofing crew. The *attic* is the space formed between the ceiling joists and the rafters. The UBC, SBC, and BOCA each requires that the attic space be provided with vents that are covered with 1/4 in. screen mesh. These vents must have an area equal to 1/150 of the attic area. This area can be reduced to 1/300 of the attic area if a vapor barrier is provided on the heated side of the attic floor and half of the required vents are placed in the upper half of the roof area.

The method used to provide the required vents varies throughout the country. Vents may be placed in the gabled end walls near the ridge. This allows the roof surface to remain vent free. In other areas, a continuous vent is placed in the eaves or a vent may be placed in each third rafter space.

The plan must also specify how to get into the attic space. The actual opening into the attic is usually shown on the floor plan. The size of the access opening varies depending on the code that you are using. The UBC requires an opening that is 22" × 30" with 30 in. minimum headroom. The SBC requires an access opening of 22" × 36" with 24 in. minimum headroom.

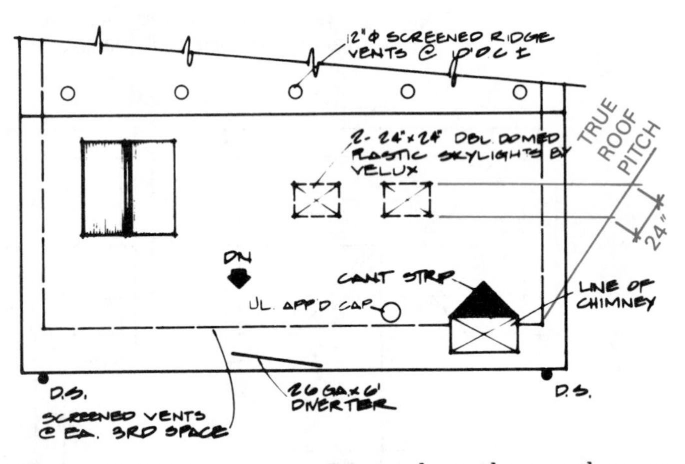

Figure 11–28 Common nonstructural members as they are often represented on a roof plan.

Dimensions

The roof plan and framing plan typically show very few dimensions. Only the overhangs and openings are dimensioned. These may even be specified in note form rather than with dimensions.

Notes

As with the other drawings, notes on the roof or framing plans can be divided into general and local notes. General notes might include the following:

- Vent notes
- Sheathing information
- Roof covering
- Eave sheathing
- Pitch
- Rafter size and spacing if uniform

Material often specified in local notes includes the following:

- Skylight type, size, and material
- Metal anchors
- Chimney caps
- Solar panel type and size
- Framing materials
- Cant strips
- Metal diverters

CHAPTER 11 TEST

1. What are two uses for a mansard roof? ____ .
2. What roof shape has four equal planes? ______________________ .
3. What roof shape combines hip and gable roof features? ______________________ .
4. What is an interior corner formed by two intersecting roofs called? ______________________ .
5. The area constructed to frame a window into an attic area is called a ______________________ .
6. What is an exterior corner formed by two intersecting roofs called? ______________________ .
7. List five areas where information about a roof can be found. ____ .
8. The horizontal member of the roof system used to resist the outward thrust of the rafters is a ________ .
9. What is the minimum pitch that should be used with a flat roof? ______________________ .
10. On what drawing would you expect to see all rafters shown? ______________________ .
11. What information would you expect to find on a roof plan? ____ .
12. What roof shape uses the roof to form the exterior walls? ______________________ .
13. What member is used with flat roofs to connect the rafters to a wall that extends above the wall? ______________________ .
14. When shed roofs are used, what keeps water from running down the wall at the wall/roof intersection? ______________________ .
15. How would you expect the roof shape to be represented on a roof plan? ______________________ .

CHAPTER 11 EXERCISES

PROBLEM 11–1. Use the attached drawings shown on page 213 to answer the following questions.

1. Identify the type of drawings that accompany this quiz.
   A. ______________________ .
   B. ______________________ .
2. Two different framing systems are provided with this plan. What are they?
   A. ______________________ .
   B. ______________________ .
3. What size roof vents will be used? ______________________ .
4. What will prevent water from building up behind the chimney? ______________________ .
5. What size ridges will be used? ______________________ .
6. Other than the ridges, what members are listed to provide support for the rafters? ______________________ .

7. What is the size of the overhang at the gable end walls? ______.
8. What size and type of skylight will be used? ______.
9. What type and spacing of trusses will be used? ______.
10. Who will supply the solar panels? ______.

PROBLEM 11–2. Use the drawings shown on page 214 to answer the following questions.

1. What roof shape will be used for this residence? ______.
2. How many roof downspouts will be required? ______.
3. What will the finish roofing material be? ______.
4. List the required overhang sizes. Which is the most typical? ______.
5. What size rafters will be used? ______.
6. How many skylights must be purchased? ______.
7. At what scale were these drawings originally drawn? ______.
8. What size members will be used to frame the skylights? ______.
9. What type and grade rafters will be used? ______.
10. What size will the ridge boards be? ______.

PROBLEM 11–3. Use the drawings shown on page 215 and 216 to answer the following questions.

1. What size rafters will be used over the living room? ______.
2. What will support the ceiling joist between the family room and the kitchen? ______.
3. What size fascias will be used? ______.
4. What spacing of ceiling joist will be used in the kitchen? ______.
5. All hips will be ______.
6. How many rafters will be required to frame the garage? ______.
7. How many ridge vents will need to be ordered? ______.
8. List two sources to find the roofing material. ______.
9. How many downspouts will be required? ______.
10. What length of ceiling joist will be used over the garage? ______.

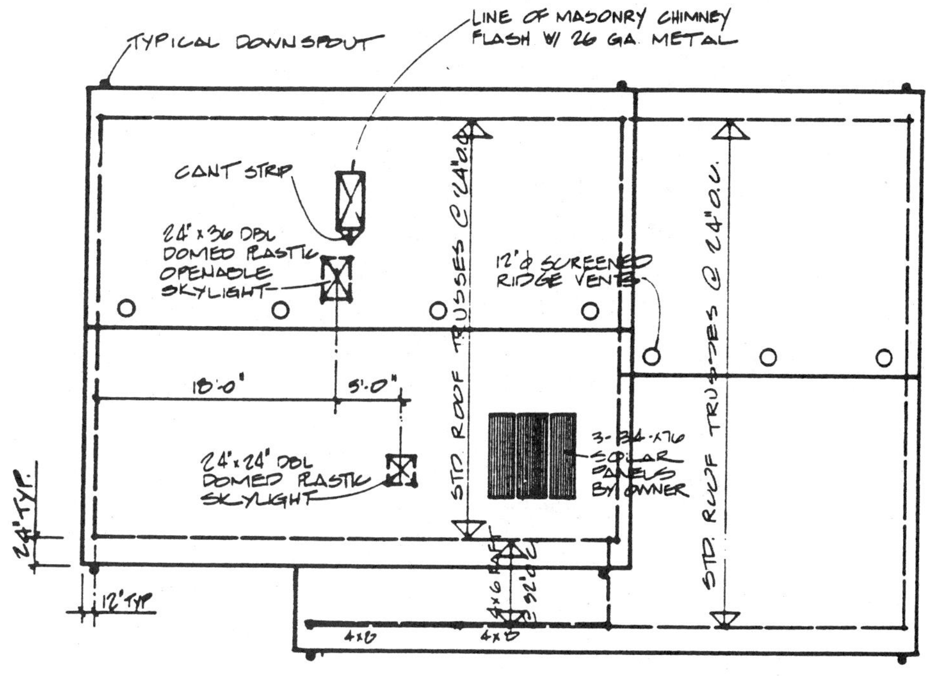

GENERAL NOTES

- PROVIDE SCREENED VENTS @ EA 3RD JST SPACE @ ALL ATTIC EAVES.
- PROVIDE SCREENED ROOF VENT @ 12'-0" O.C.
- USE 1/2" 'CCX' PLY @ ALL EXPOSED EAVES.
- USE 235# COMPO. SHINGLES OVER 15# FELT.
- SUBMIT TRUSS CALCS. TO BLDG DEPT PRIOR TO TRUSS ERECTION.

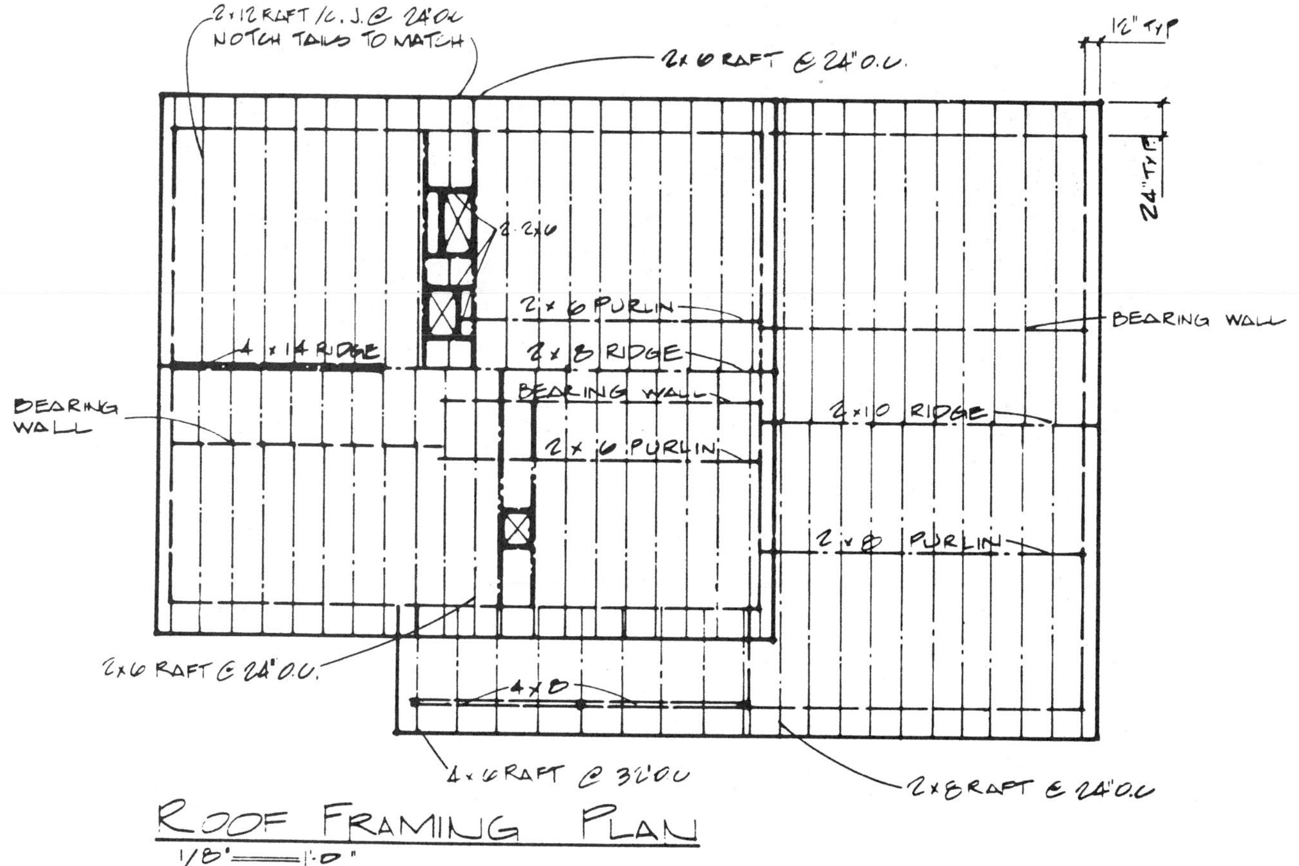

ALL FRAMING LUMBER TO BE DFL #2 UNLESS NOTED.

Problem 11–1.

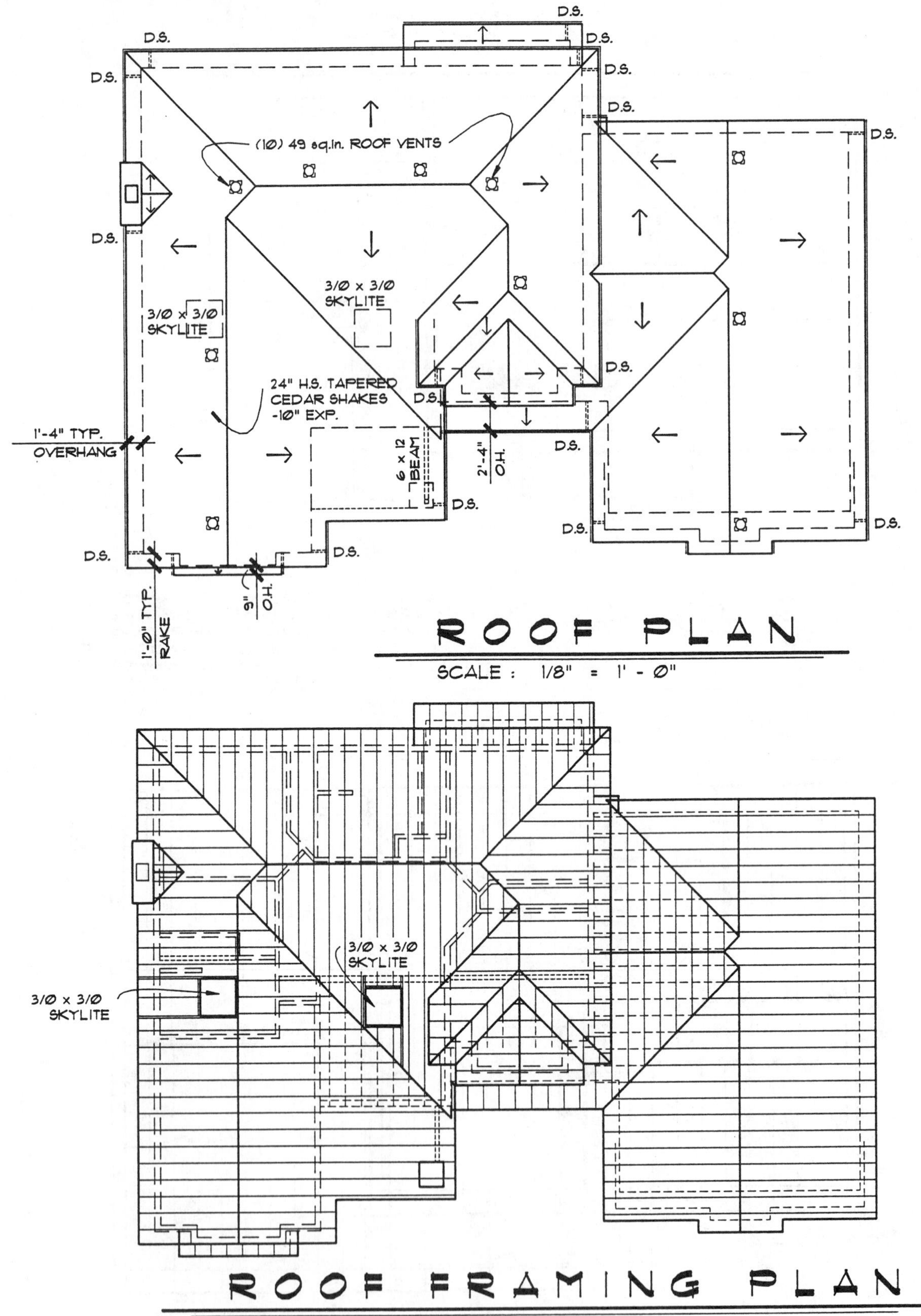

(b)

Problem 11–2.

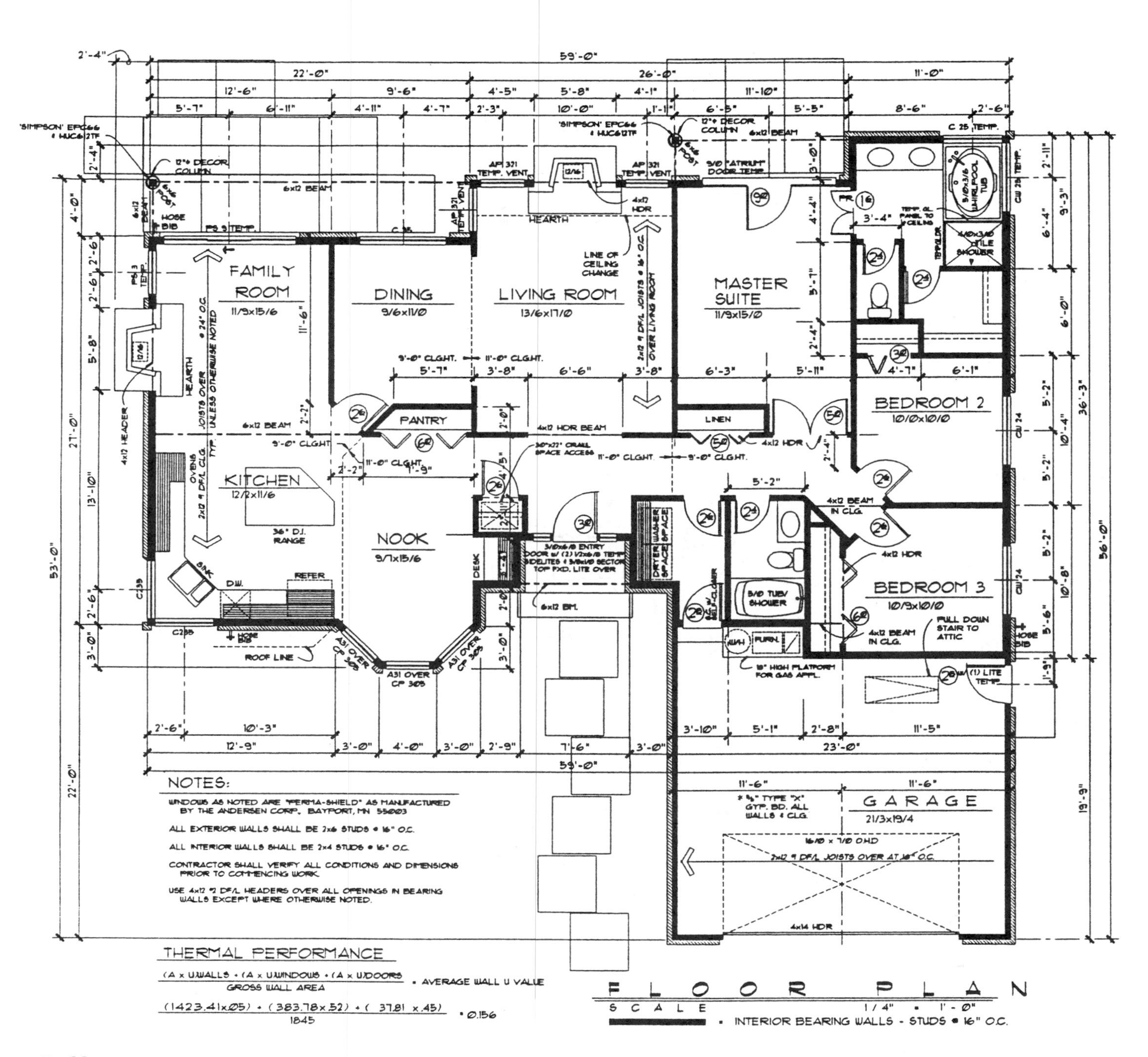

Problem 11–3. *Courtesy Piercy & Barclay Designers, Inc.*

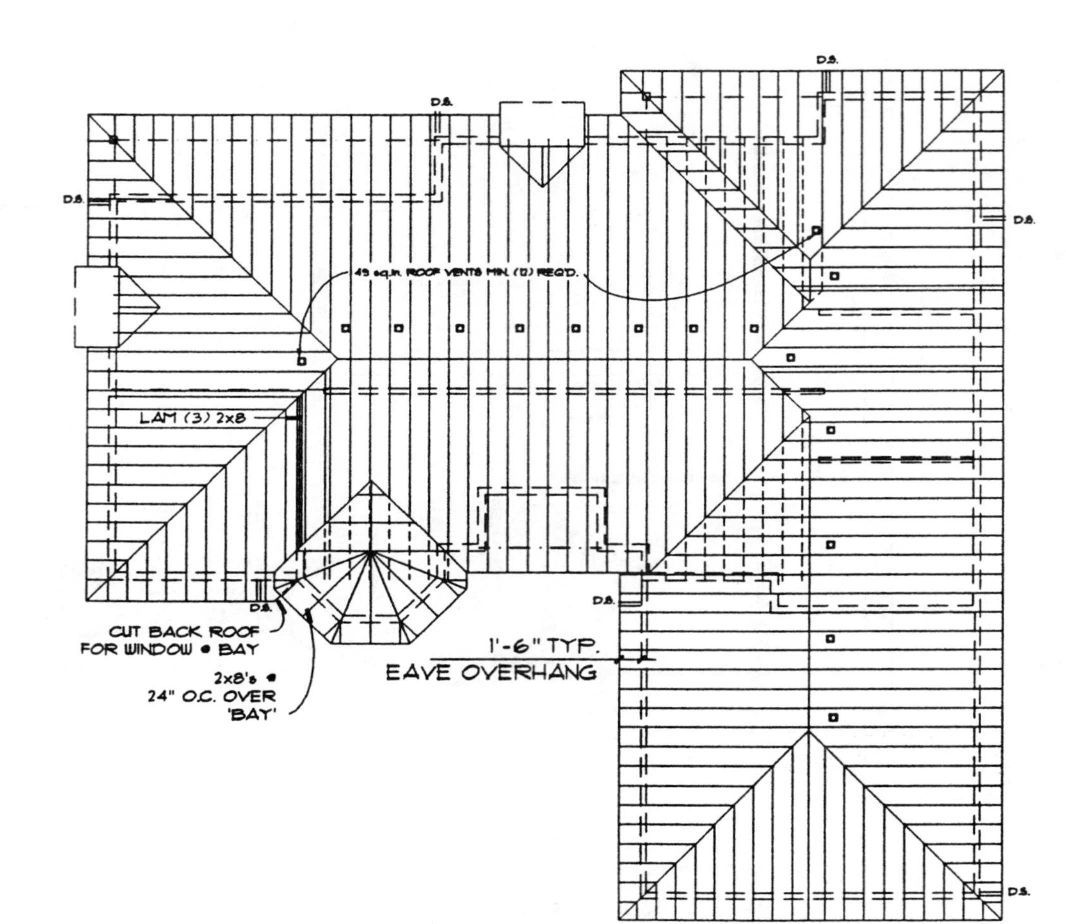

NOTES:

1. ALL RAFTERS TO BE 2 x 8 #2 DF/L @16" o.c. EXCEPT WHERE OTHERWISE NOTED.
2. ALL RIDGE BOARDS AND HIP RAFTERS SHALL BE 2 x 10 #2 DF/L.
3. ALL CEILING JOISTS SHALL BE 2 x 8 #2 DF/L @ SAME SPACINGS AS THE RAFTERS ABOVE.
4. REFER TO FLOOR PLAN FOR BEAM SIZES AND SPECIFICATIONS.
5. ALL FASCIA BOARDS TO BE 2 x 8 SELECT WESTERN RED CEDAR.

ROOF FRAMING PLAN

SCALE 1/4" = 1'-0"

Problem 11–3 (continued). *Courtesy Piercy & Barclay Designers, Inc.*

Chapter 12

Reading Cabinet Drawings

TYPES OF MILLWORK
READING CABINET DRAWINGS
CABINET ELEVATIONS AND LAYOUT
KEYING CABINET ELEVATIONS TO FLOOR PLANS
TEST
EXERCISES

CABINET CONSTRUCTION is an element of the final details of a structure known as millwork. *Millwork* is any item that is considered finish trim or finish woodwork. For example, the custom plans for a residential or commercial structure may show very detailed and specific drawings of the finish woodwork in the form of plan views, elevations, construction details, and written specifications. Plans of a house that is one of many to be built with a group of houses may show only floor plan views of cabinetry, giving the building contractor the flexibility of being able to use available millwork without strict requirements placed on selection. Other factors that determine the extent of millwork drawings on a set of plans are the requirements of specific lenders and local code jurisdictions.

TYPES OF MILLWORK

Millwork may be designed for appearance or for function. When designed for appearance, ornate and decorative millwork may be created with a group of shaped wooden forms placed together to capture a style of architecture. There are also vendors that manufacture a wide variety of prefabricated millwork moldings that are available at less cost than custom designs. See Figure 12–1. Millwork that is designed for function may be very plain in appearance and is also less expensive than standard sculpted forms. In some situations wood millwork may be replaced with plastic or ceramic products. For example, in a public restroom or in a home laundry room, a plastic strip may be used around the wall at the floor to protect the wall. This material stands up to abuse better than wood. In some cases drawings are made for specific millwork applications. There are as many possible details as there are design ideas. The following discussion provides a general example of the items to help define the terms.

Baseboards

Baseboards are placed at the intersection of walls and floors and are generally used to protect the wall

Figure 12–1 Prefabricated millwork installed in a restaurant. *Courtesy Cumberland Woodcraft Co., Inc., Carlisle, PA.*

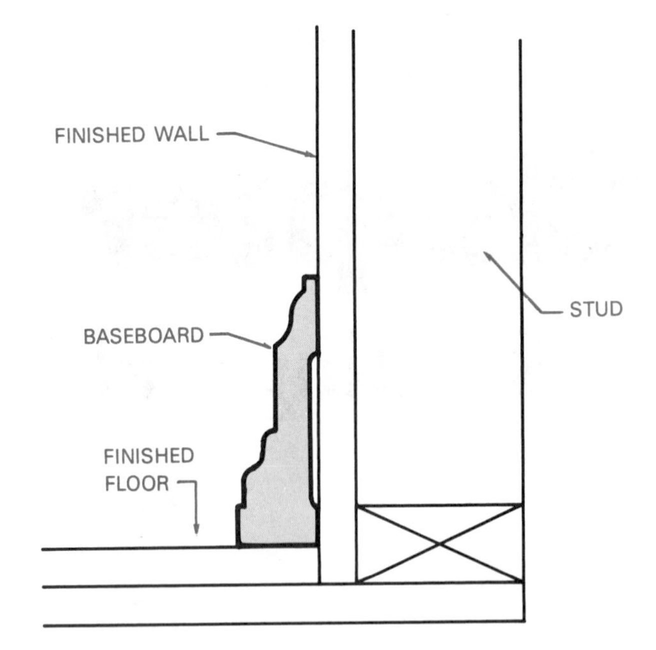

Figure 12–2 Typical baseboard installation.

from damage as shown in Figure 12–2. Figure 12–3 shows some standard molded baseboards.

Wainscots

For a wall finish in which the material on the bottom portion of the wall is different from that on the upper portion, the lower portion is the *wainscot* and the material used is *wainscoting.* Wainscots may be used on the interior or exterior of the structure. Exterior wainscoting is often brick veneer. Interior wainscoting may be any material that is used to divide walls into two visual sections. For example, wood paneling, plaster texture, ceramic tile, wallpaper, or masonry may be used as wainscoting. Figure 12–4 shows the detail of wood wainscot with plywood panels. The plywood panel may have an oak or other hardwood outer veneer to match the surrounding hardwood material. This is a less expensive method of constructing an attractive wood wainscot than with the hardwood panels shown in Figure 12–5.

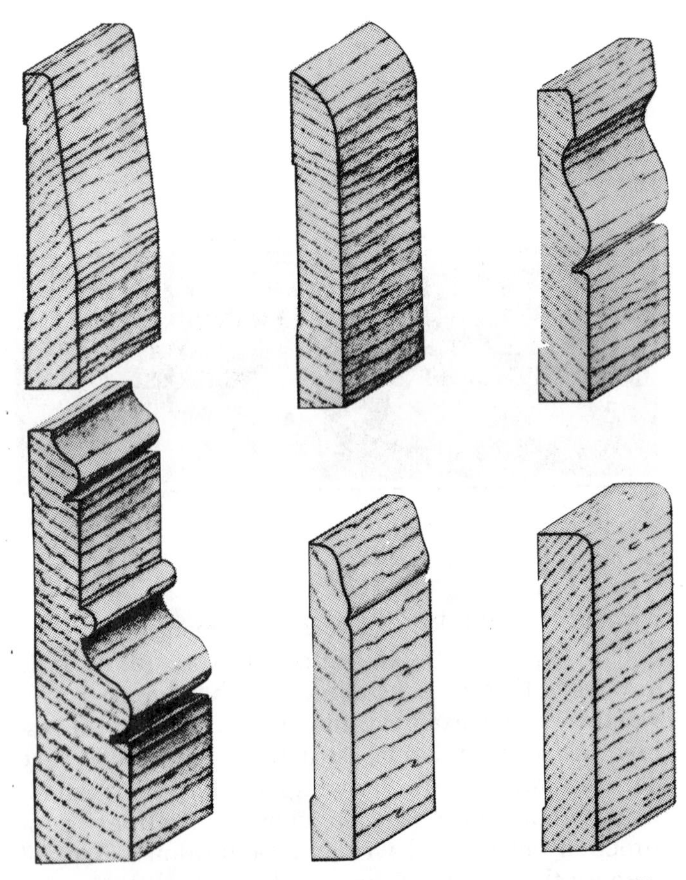

Figure 12–3 Examples of standard baseboards. *Courtesy Hillsdale Sash & Door Co., Inc.*

Chair Rail

The *chair rail* has traditionally been placed horizontally on the wall at a height where chair backs would otherwise damage the wall. See Figure 12–6. Chair rails are found in the dining room, den, office, or in other areas where chairs are frequently moved against a wall. Chair rails may be used in conjunction with wainscoting. In some applications the chair rail is an excellent division between two different materials or wall textures. Figures 12–7 shows sample chair rail moldings.

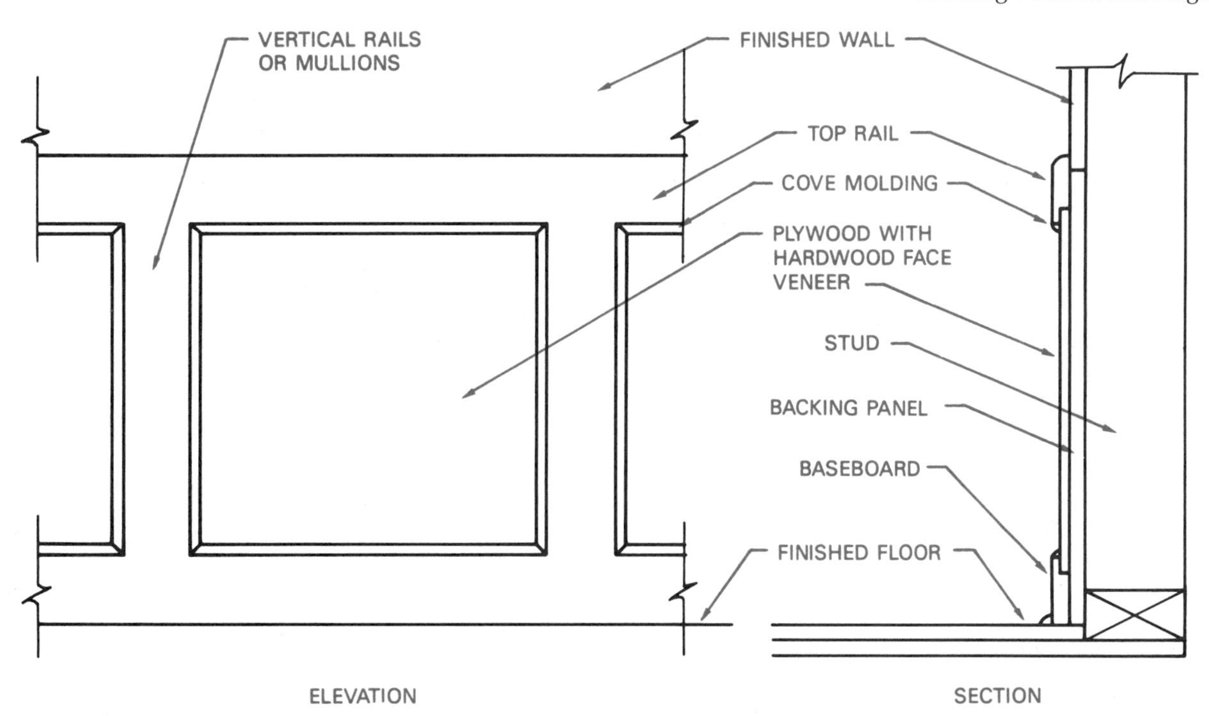

Figure 12–4 Plywood panel wood wainscot installation.

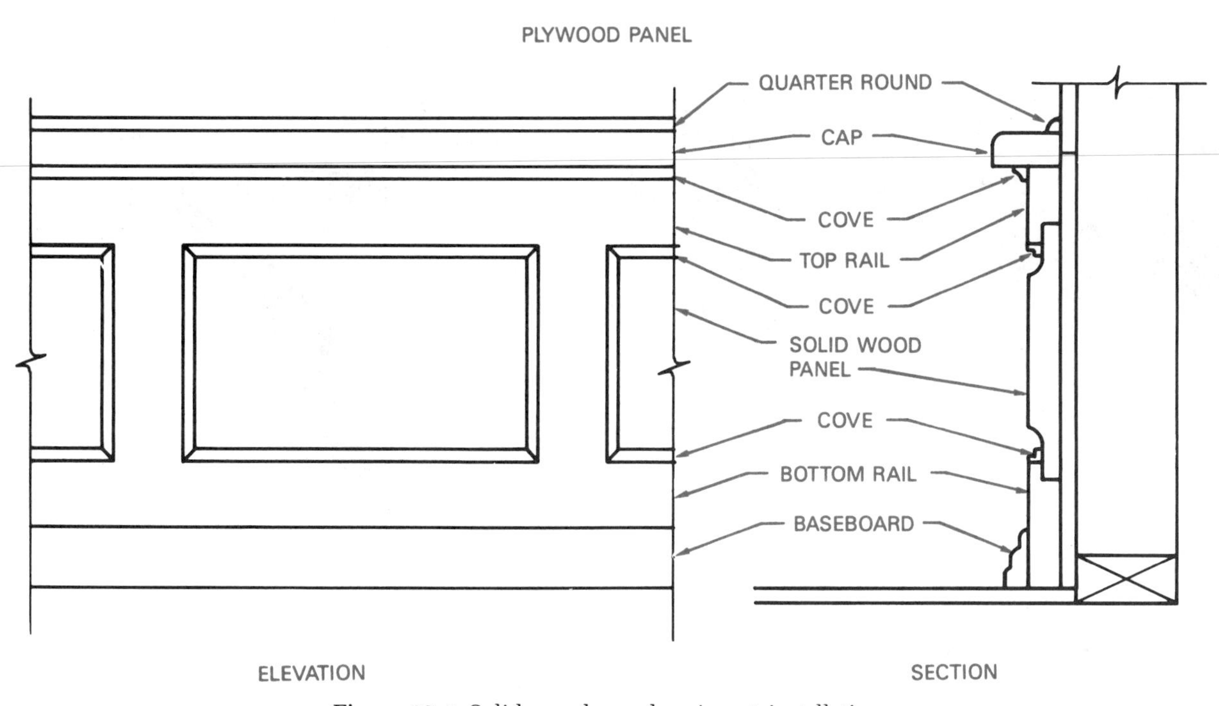

Figure 12–5 Solid wood panel wainscot installation.

Cornice

The *cornice* is decorative trim placed in the corner where the wall meets the ceiling. A cornice may be a single shaped wood member, called *cove* or *crown* molding, as seen in Figure 12–8, or the cornice may be a more elaborate structure made up of several individual wood members, as shown in Figure 12–9. Cornice boards traditionally fit into specific types of architectural styles, such as English Tudor, Victorian, or colonial. Figure 12–10 shows some stan

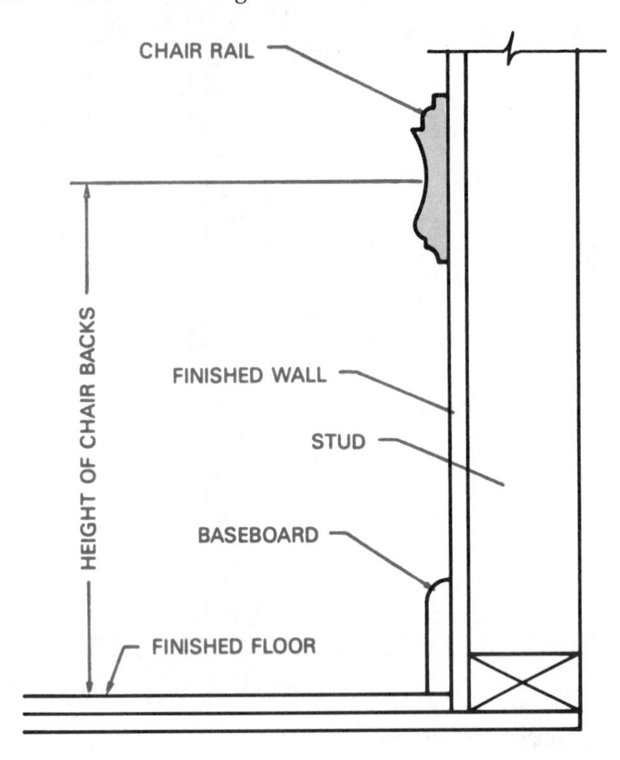

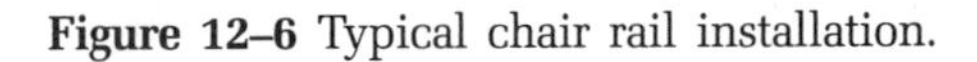
Figure 12–6 Typical chair rail installation.

JOIST
STUD
FINISHED CEILING
FINISHED WALL

Figure 12–8 A typical individual-piece cornice installation.

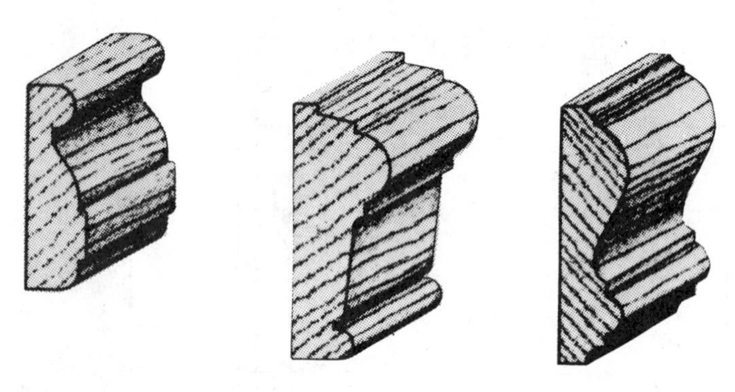

Figure 12–7 Standard chair rails. *Courtesy Hillsdale Sash & Door Co., Inc.*

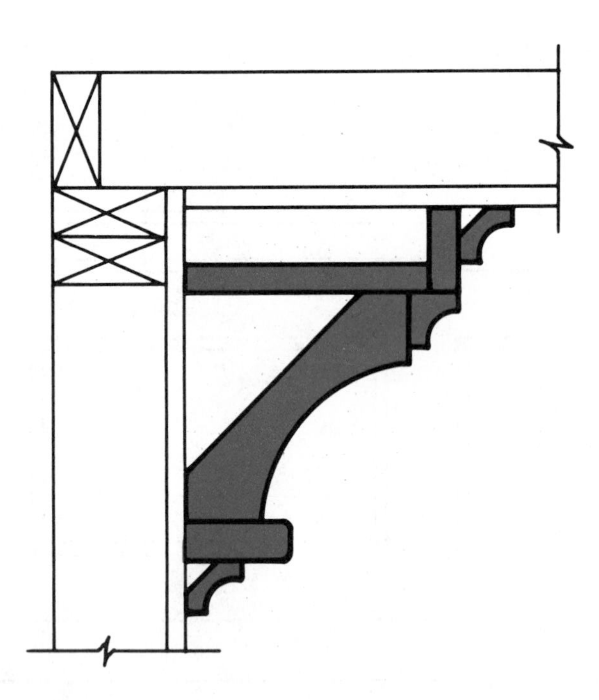

Figure 12–9 A multiple-piece cornice installation.

dard cornice moldings. In most construction, where contemporary architecture or cost saving is important, wall-to-ceiling corners are left square.

Casings

Casings are the members that are used to trim around windows and doors. Casings are attached to the window or doorjamb (frame) and to the adjacent wall, as shown in Figure 12–11. Casings may be decorative to match other moldings or plain to serve the functional purpose of covering the space between the door or window jamb and the wall. Figure 12–12 shows a variety of standard casings.

Mantels

The *mantel* is an ornamental shelf or structure that is built above a fireplace opening, as seen in Figure 12–13. Mantel designs vary with individual preference. Mantels may be made of masonry as part of the fireplace structure, of ornate decorative wood moldings, or even of a rough-sawn length of lumber bolted to the fireplace face. Figure 12–14 shows a traditional mantel application.

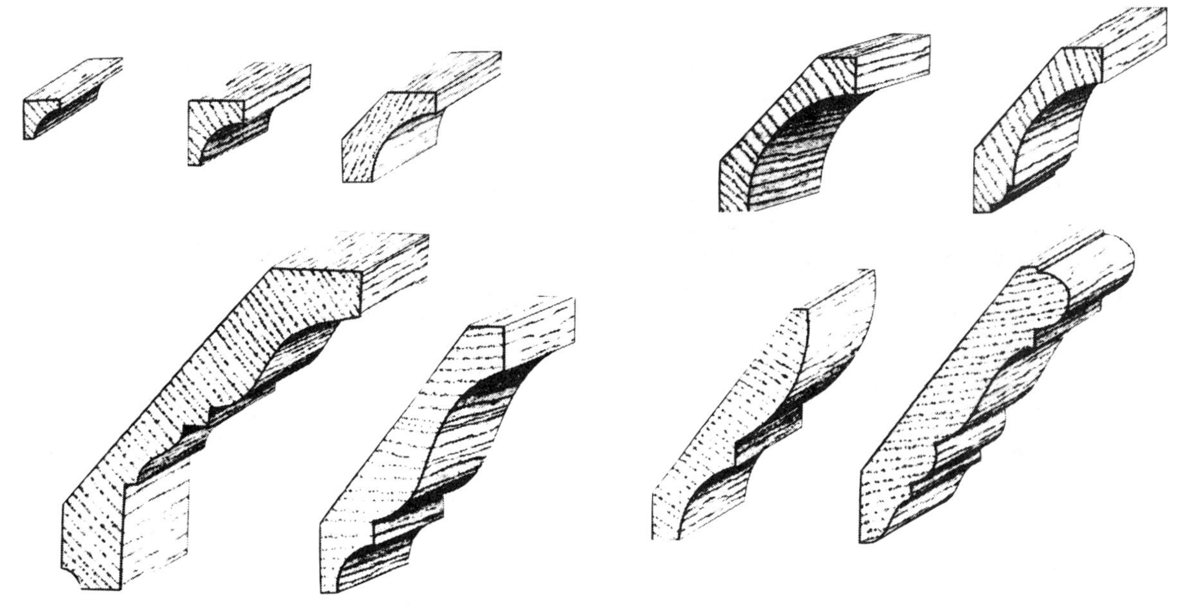

Figure 12–10 Standard cornice (cove and ceiling) moldings. *Courtesy Hillsdale Pozzi Co.*

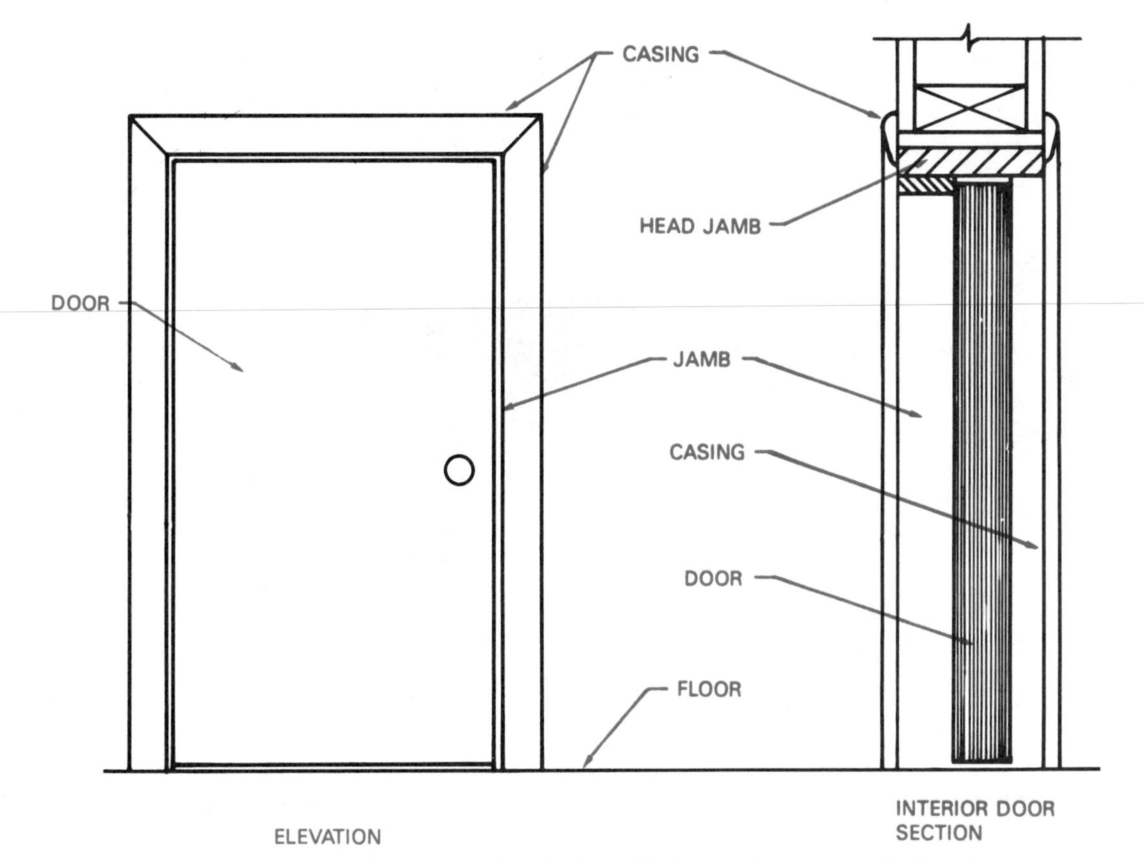

Figure 12–11 Typical door casing installation. Window casings are applied in the same manner.

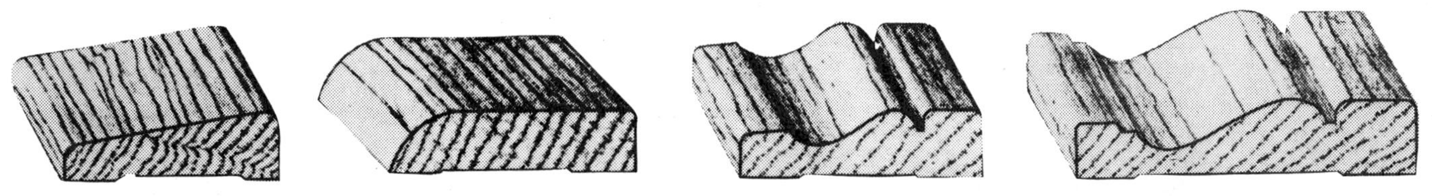

Figure 12–12 Standard casings. *Courtesy Hillsdale Sash & Door Co., Inc.*

Figure 12–13 A fireplace mantel and bookshelf cabinetry. *Photo by Bruce Davies, courtesy Park Place Wood Products, Inc.*

Bookshelves

Book and display shelves may have simple construction with metal brackets and metal or wood shelving, or they may be built detailed as fine furniture. Bookshelves are commonly found in such rooms as the den, library, office, or living room. Shelves that are designed to display items other than books are also found in almost any room of the house. A common application is placing shelves on each side of a fireplace, as seen in Figure 12–13. Shelves are also used for functional purposes in storage rooms, linen closets, laundry rooms, and any other location where additional storage is needed.

Railings

Railings are used for safety at stairs, landings, decks, and open balconies from which people might fall. Railings may also be used as decorative room dividers or for special accents. Railings may be built enclosed or open and are constructed of wood or metal. Enclosed rails are often the least expensive to build because they require less detailed labor than open rails. A decorative wood cap is typically used to trim the top of enclosed railings. See Figure 12–15. Open railings can be one of the most attractive elements of interior design. Open railings may be as detailed as the designer's or craftsperson's imagination. Detailed open railings built of exotic hardwoods can be one of the most expensive items in the structure. Figure 12–16 shows some standard railing components.

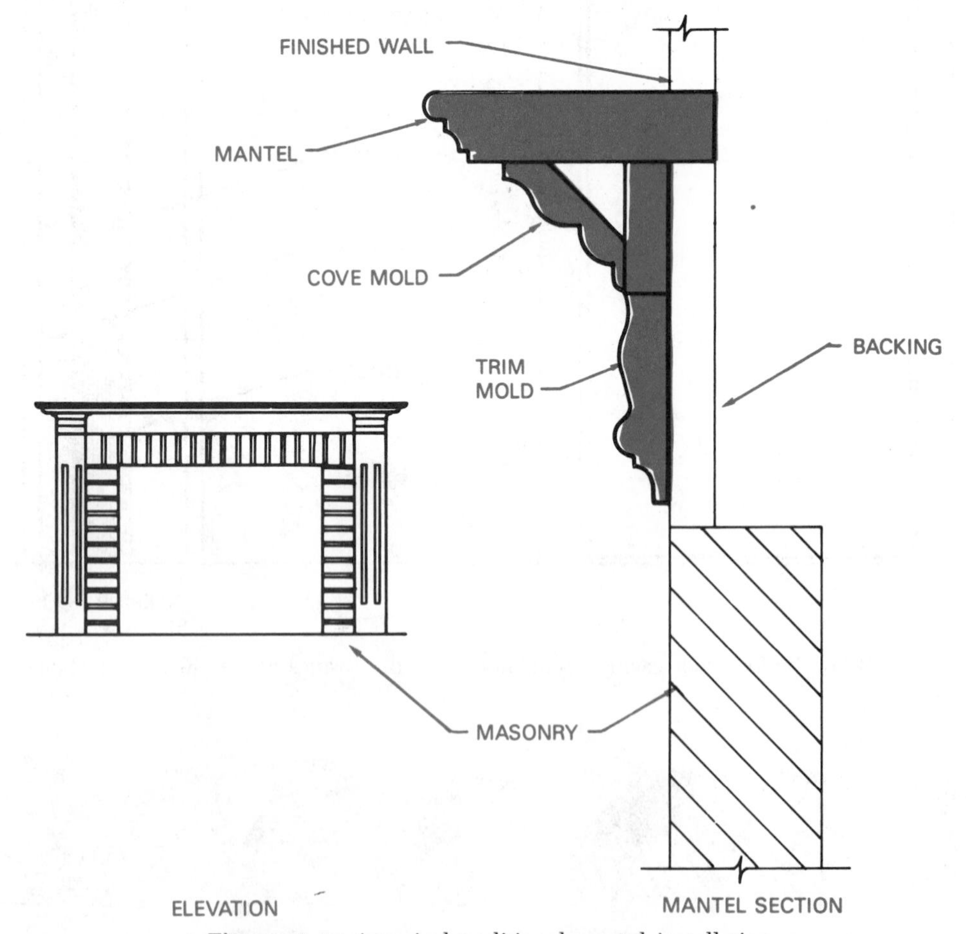

Figure 12–14 A typical traditional mantel installation.

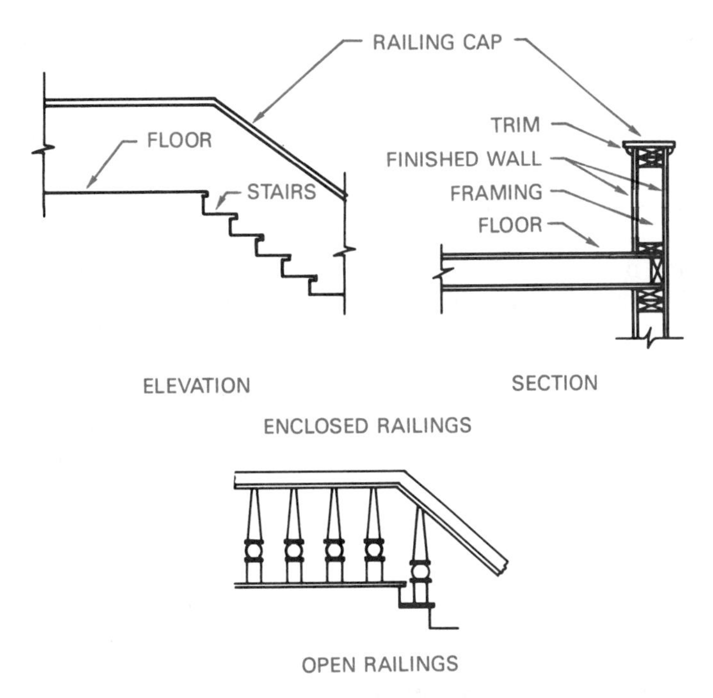

Figure 12–15 Typical railing installation and terminology.

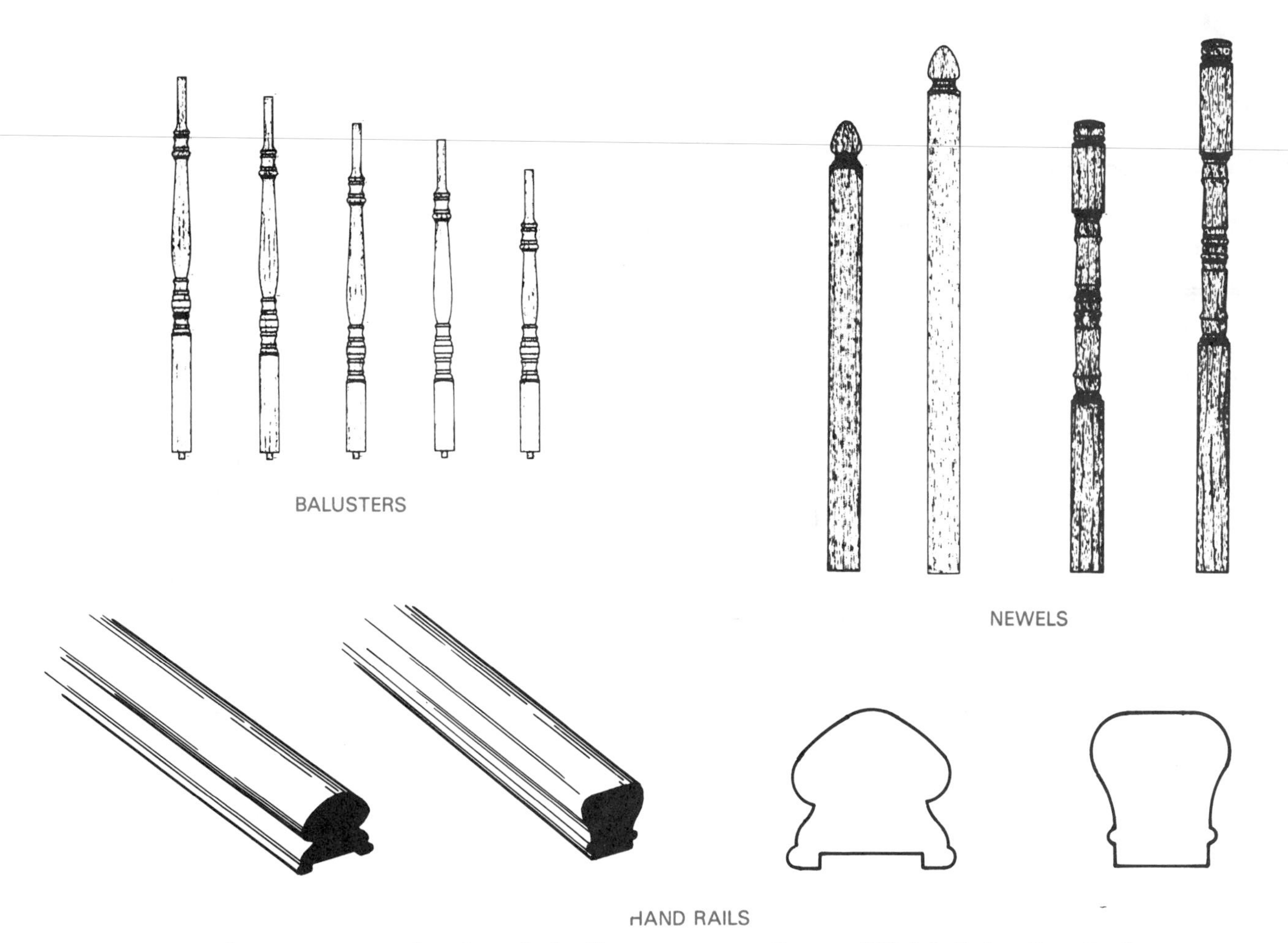

Figure 12–16 Examples of standard railing components. *Courtesy Hillsdale Sash & Door Co., Inc.*

READING CABINET DRAWINGS

One of the most important items for buyers of a new home is the cabinets. The quality of cabinetry can vary greatly. Cabinets are used for storage and as furniture. Kitchen cabinets have two basic elements: the base cabinet and the upper cabinet. See Figures 12–17 and 12–18. Drawers, shelves, cutting boards, pantries, and appliance locations are found in the kitchen cabinets. Bathroom cabinets are called vanities, linen cabinets, and medicine cabinets. See Figure 12–19. Other cabinetry may be found throughout a house, such as in utility or laundry rooms for storage. For example, a storage cabinet above a washer and dryer is common.

Cabinet Types

There are as many cabinet styles and designs as the individual can imagine. However, there are only two general types of cabinets based upon their method of construction. These are modular, or prefabricated cabinets, and custom cabinets.

Modular Cabinets

The term "modular" refers to prefabricated cabinets because they are constructed in specific sizes called modules. The best use of modular cabinets is when a group of modules can be placed side by side in a given space. Most modular cabinet vendors offer different door styles, wood species, or finish colors on their cabinets. Modular cabinets are sized in relationship to standard or typical applications, although many modular cabinet manufacturers can make cabinet components that fit nearly every design situation.

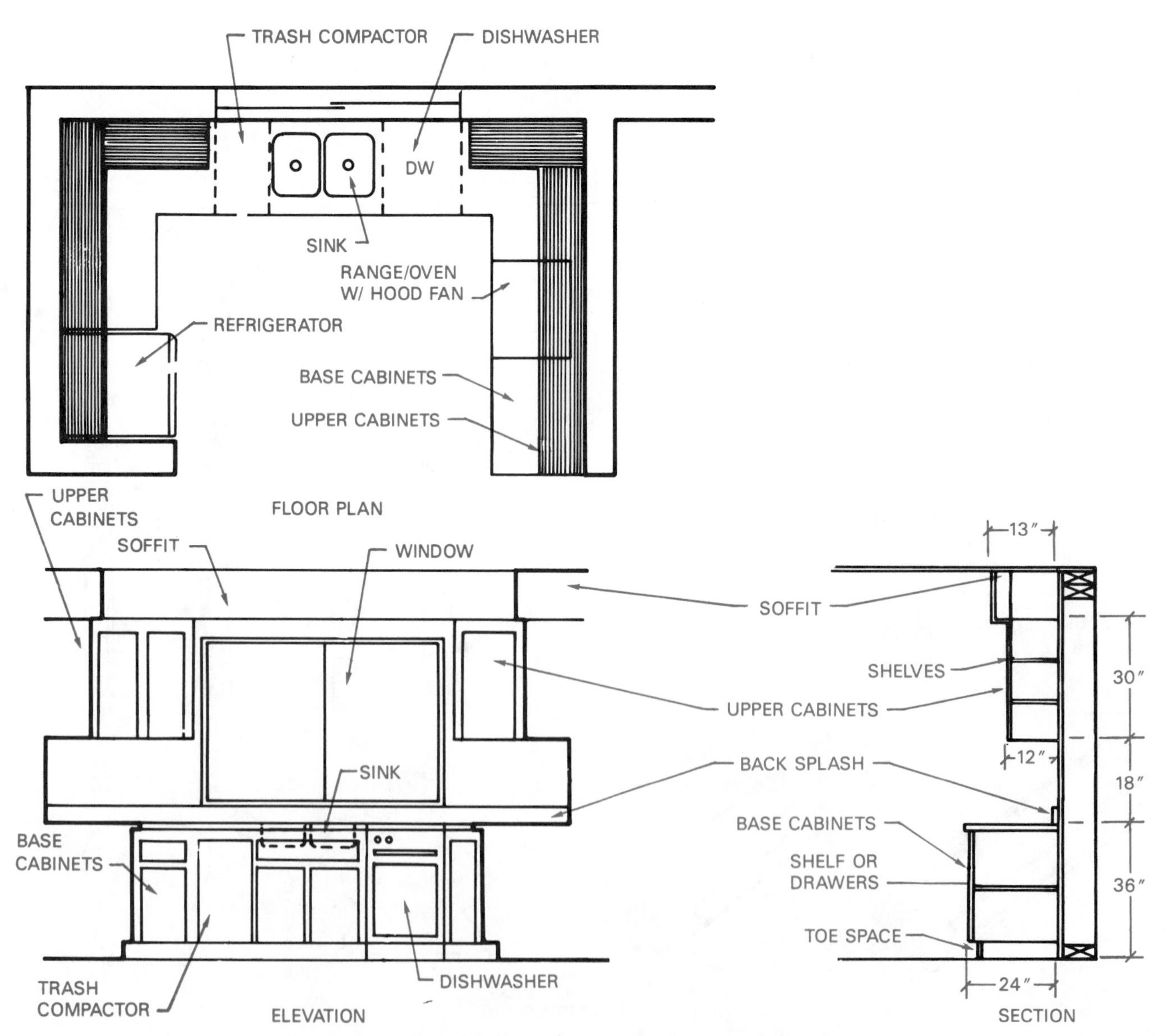

Figure 12–17 Basic kitchen cabinet components and dimensions.

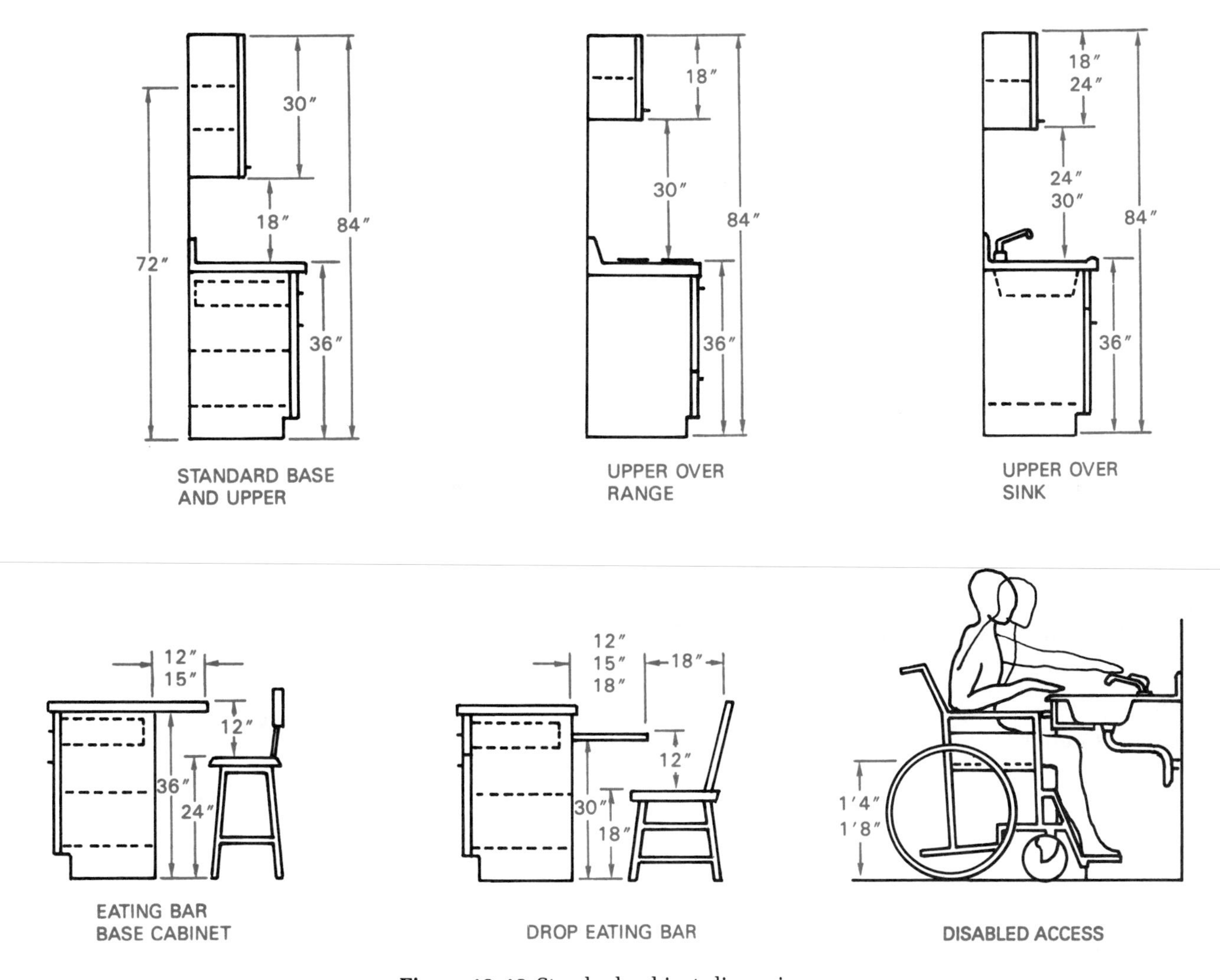

Figure 12–18 Standard cabinet dimensions.

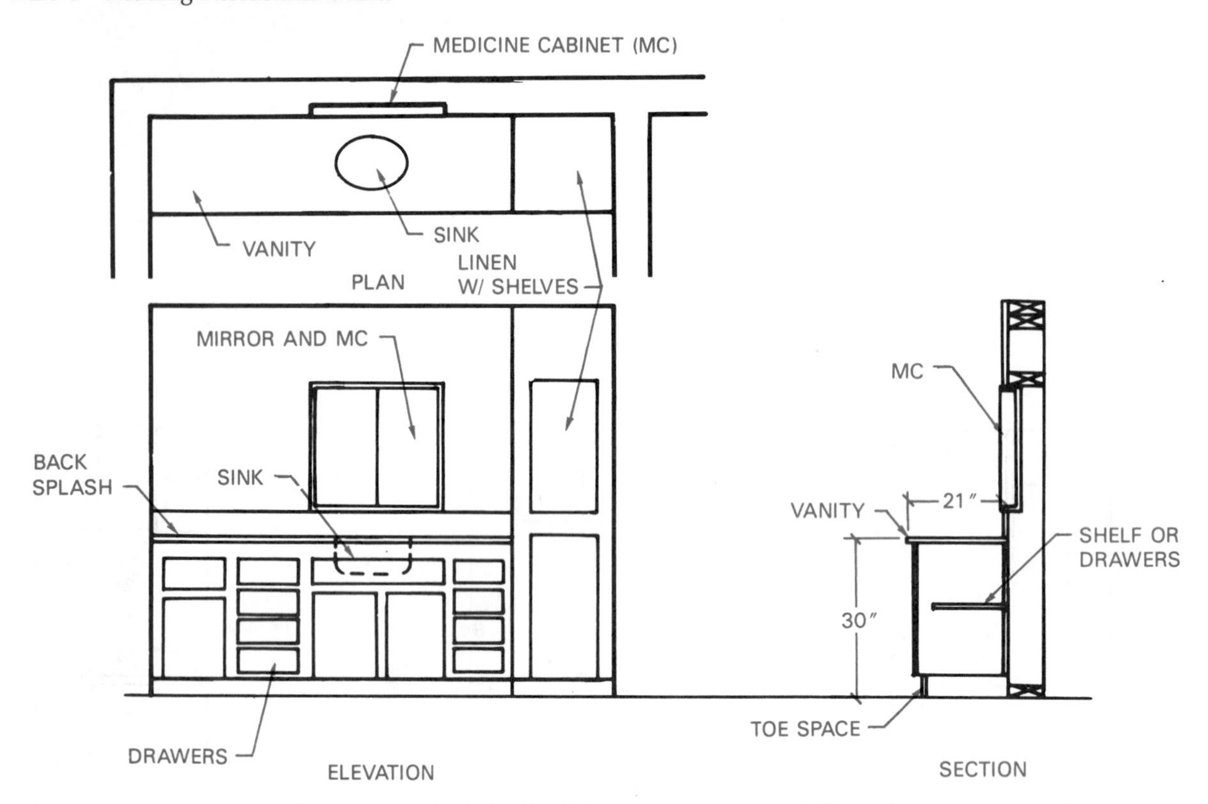

Figure 12–19 Basic bath cabinet components and dimensions.

The modular cabinet layout begins with standard manufactured cabinet components, which include wall, or upper, cabinets, base cabinets, vanity cabinets, tall cabinets, and accessories. Modular cabinets incorporate the individual units into the floor plan based upon the manufacturer's sizes and specifications. The actual floor plans look the same as any of the cabinet representations presented in Chapter 4.

Custom Cabinets

The word *custom* means made to order. The big difference between custom cabinets and modular cabinets is that custom cabinets are generally fabricated locally in a shop to the specifications of an architect or designer, and after construction they are delivered to the job site and installed in large sections. Modular cabinets are often manufactured nationally and delivered in modules.

One of the advantages of custom cabinets is that their design is limited only by the imagination of the architect and the cabinet shop. Custom cabinets may be built for any situation, such as for any height, space, type of exotic hardwood, type of hardware, geometric shape, or other design criteria.

Cabinet Options

Some of the cabinet design alternatives that are available from either custom or modular cabinet manufacturers include the following:

- Door styles, materials, and finishes
- Self-closing hinges
- A variety of drawer slides, rollers, and hardware
- Glass cabinet fronts for a more traditional appearance
- Wooden or metal range hoods
- Specially designed pantries, appliance hutches, and a lazy Susan or other corner cabinet design for efficient storage
- Bath, linen storage, and kitchen specialties

Cabinet elevations or exterior views are developed directly from the floor plan drawings. The purpose of the elevations is to show how the exterior of the cabinets will look when completed and to give general dimensions, notes, and specifications. See Figure 12–20.

KEYING CABINET ELEVATIONS TO FLOOR PLANS

Several methods may be used to key the cabinet elevations to the floor plan. In Figure 12–20, the designer keyed the cabinet elevations to the floor plans with room titles such as KITCHEN CABINET ELEVATIONS and BATH AND UTILITY ELEVATIONS. In Figure 12–21 an arrow with a letter inside can also be used to correlate the elevation to the floor plan. For example, in Figure 12–22 the E and F arrows pointing to the vanities are keyed to letters E and F that appear below the vanity elevations.

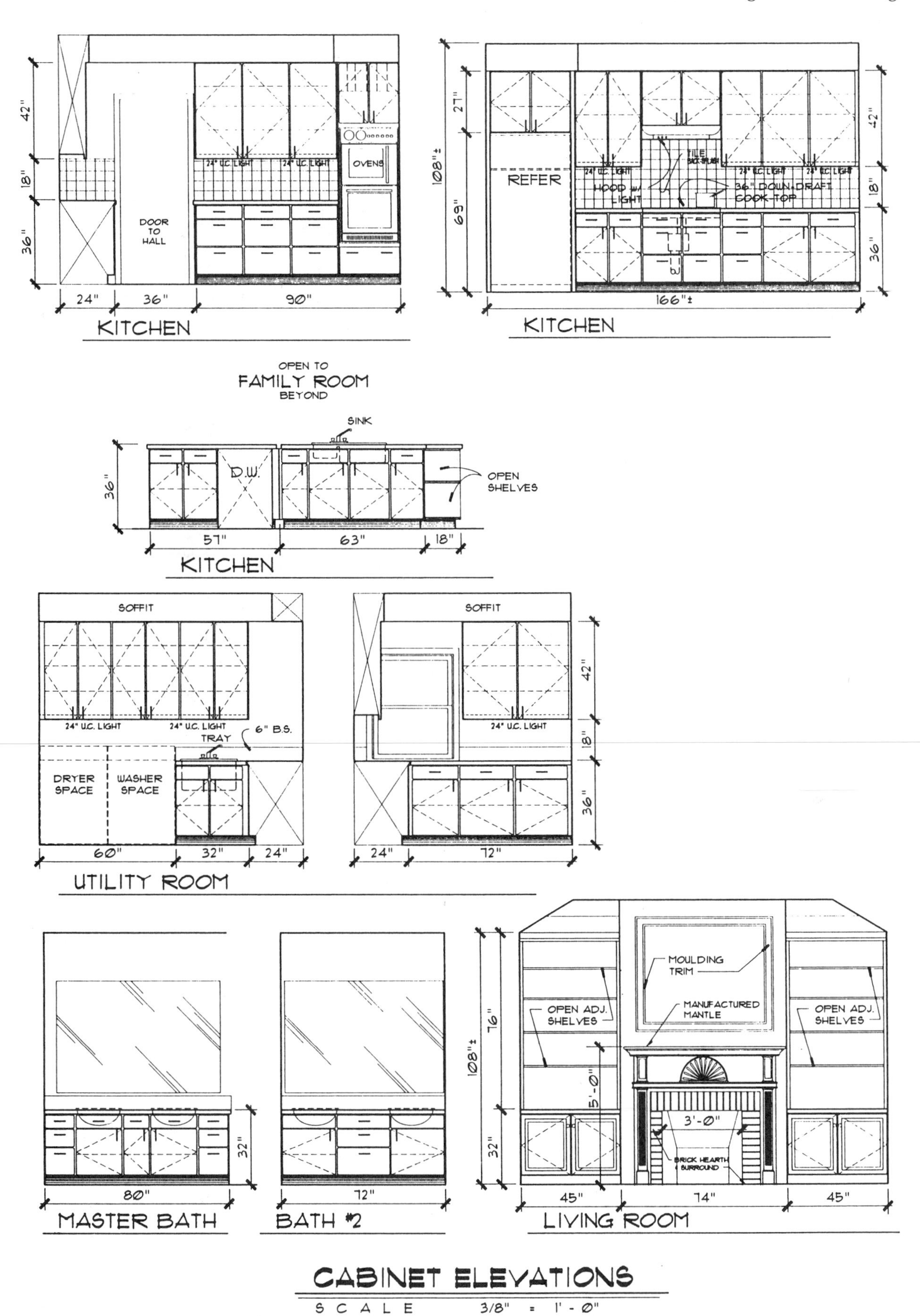

Figure 12–20 Cabinet elevations keyed to the floor plans with room titles. *Courtesy Piercy & Barclay Designers, Inc.*

Figure 12–21 Cabinet elevations with floor plan and cabinet elevation location symbol.

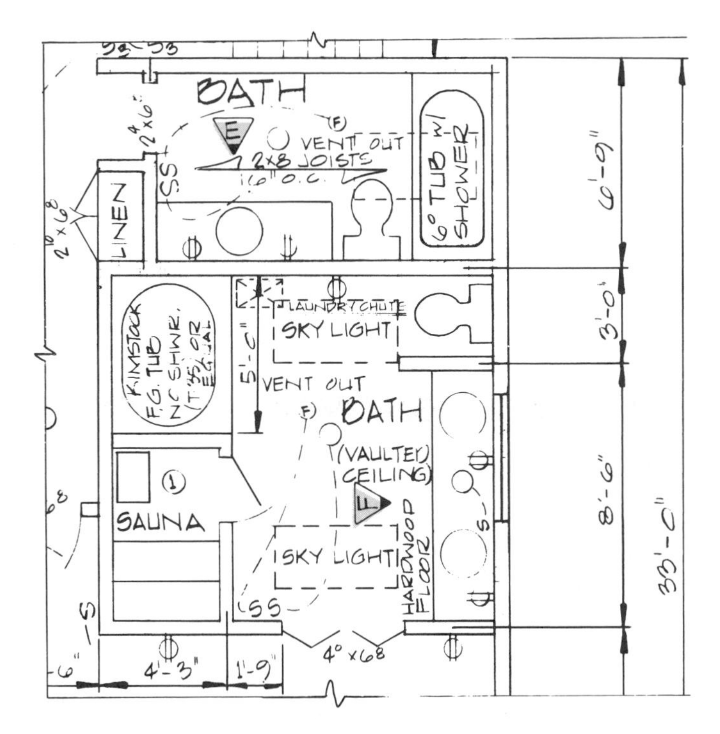

Figure 12–22 The floor plan and cabinet elevation location symbol placed on the floor plan. Look at Figure 12–21 to see the cabinet location symbol on the related cabinet elevations. *Courtesy Madsen Designs.*

CHAPTER 12 TEST

1. Cabinet construction is an element of the final details of a structure known as ______________.
2. Define millwork. ____
____
____.
3. The millwork placed at the intersection of walls and floors and used to protect the wall from damage is called the ______________________________.
4. A ______________ the material on the bottom portion of the wall when the wall finish of the bottom portion is different from the material on the upper portion.
5. Name the millwork item that has traditionally been placed horizontally on the wall at a height where chair backs would otherwise damage the wall. ______________.
6. The millwork that is a decorative trim placed in the corner where the wall meets the ceiling is called the ______________________________.
7. The members that are used to trim around doors and windows are called ______________.
8. An ornamental shelf or structure that is built above a fireplace opening is called a ______________.
9. Rails are recommended at any rise that measures ______________ or is ______________ or more stair risers.
10. Name the two basic elements of kitchen cabinets. ______________.
11. Bathroom cabinets where the sink is located are called ______________.
12. Describe modular cabinets. ______________________________
______________________________.
13. Describe custom cabinets. ______________________________
______________________________.

14. List at least three purposes of cabinet elevations. ____
____
____.

15. Explain at least two ways that cabinet elevations may be keyed to the floor plan. ________________
____
____
____.

CHAPTER 12 EXERCISES

PROBLEM 12–1. Given the following millwork component drawings, place the name of the type of millwork in the blank provided below each drawing.

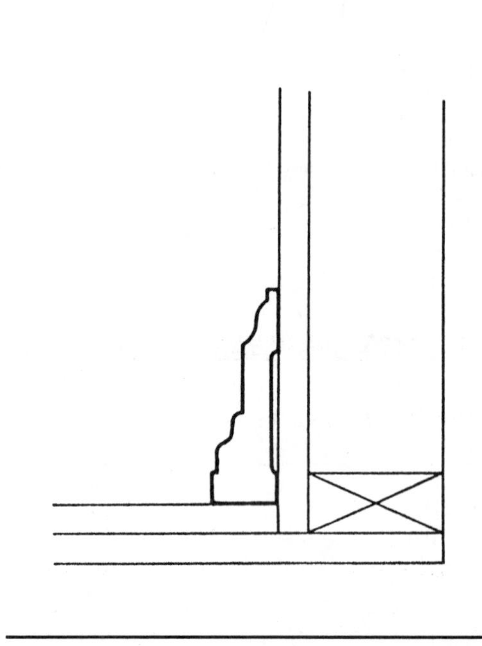

____

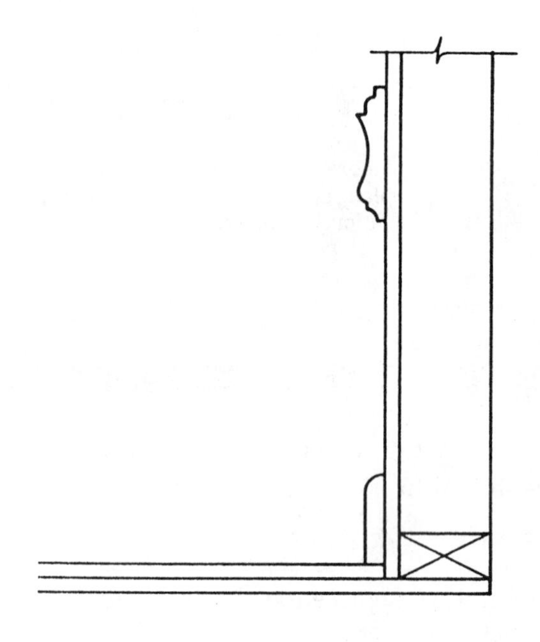

____

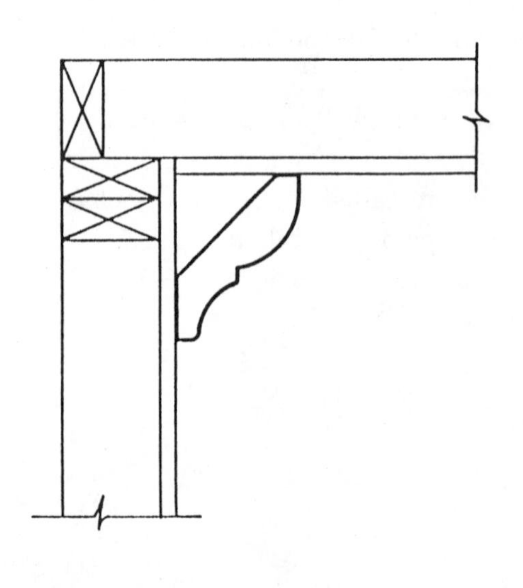

____

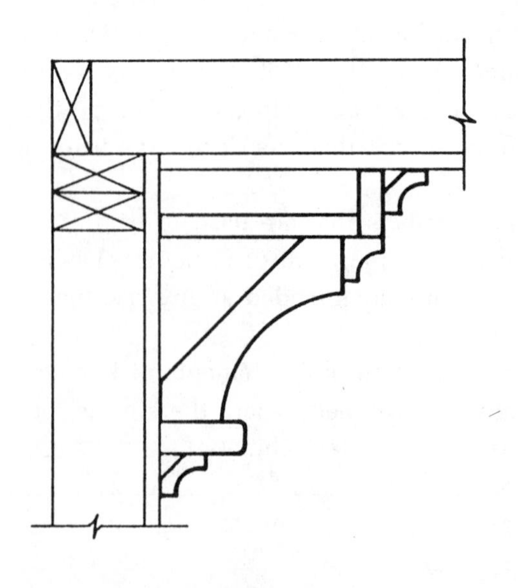

____

PROBLEM 12–2. Given the kitchen floor plan, cabinet elevation, and section through the cabinets shown, place the name of each component and the correct standard cabinet dimensions on the blanks provided.

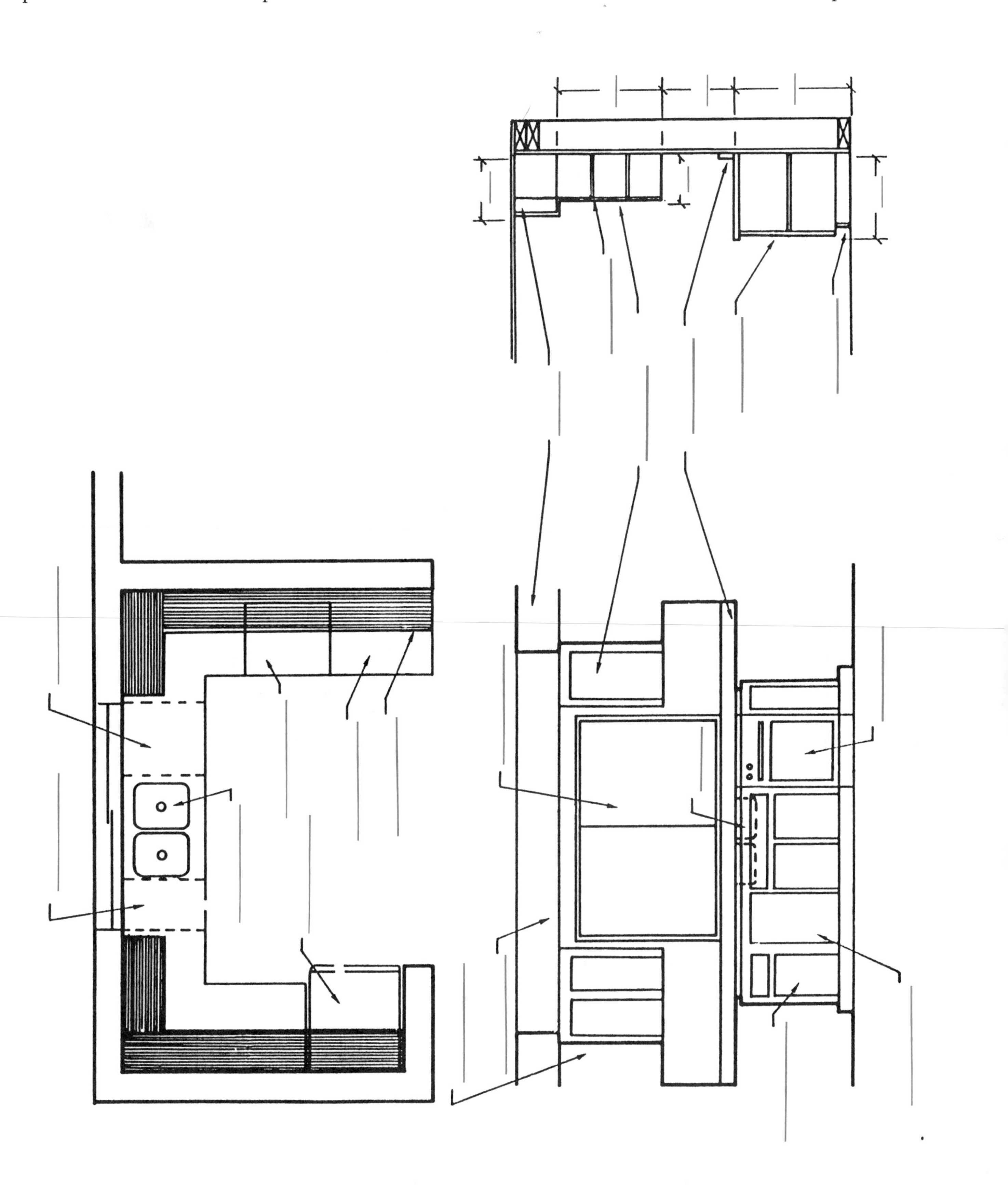

PROBLEM 12–3. Given the bathroom floor plan, cabinet elevation, and section through the cabinets shown, place the name of each component and the correct standard cabinet dimensions on the blanks provided.

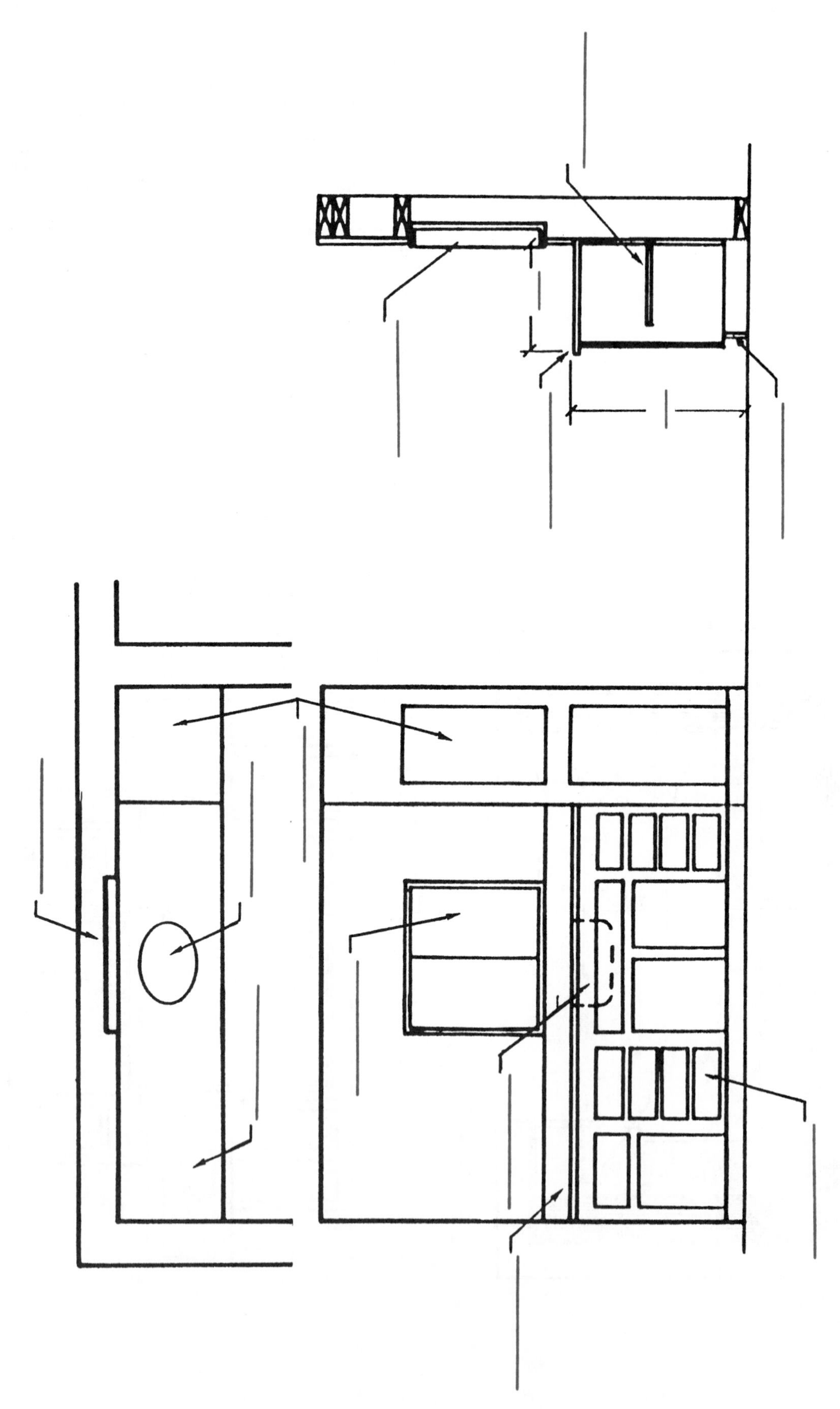

PROBLEM 12–4. Given the standard base and upper cabinet detail shown, provide the standard dimensions in the blanks provided.

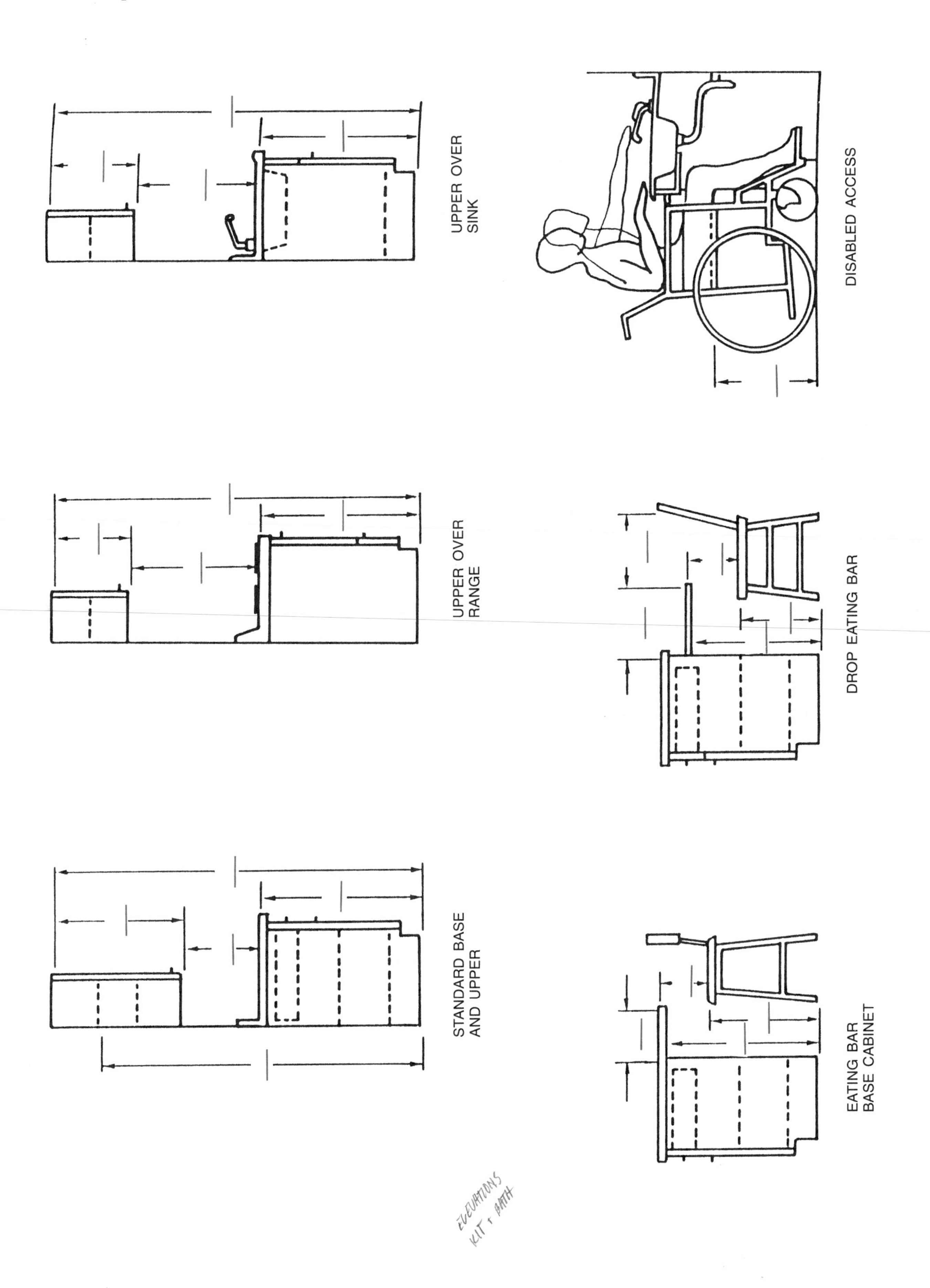

PROBLEM 12–5. Given the standard upper cabinet detail over the range shown, provide the standard dimensions in the blanks provided.

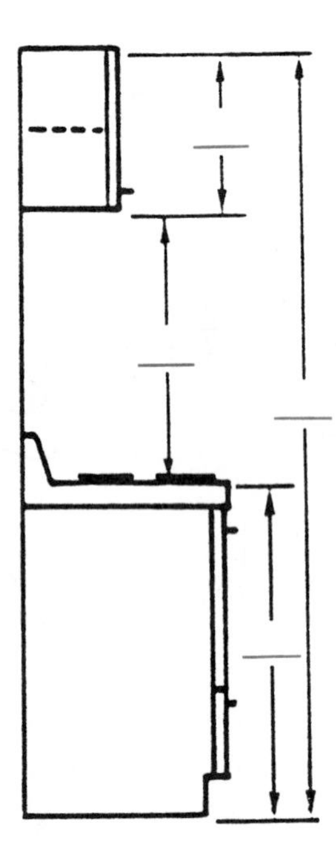

PROBLEM 12–6. Given the standard upper cabinet detail over the sink shown, provide the standard dimensions in the blanks provided.

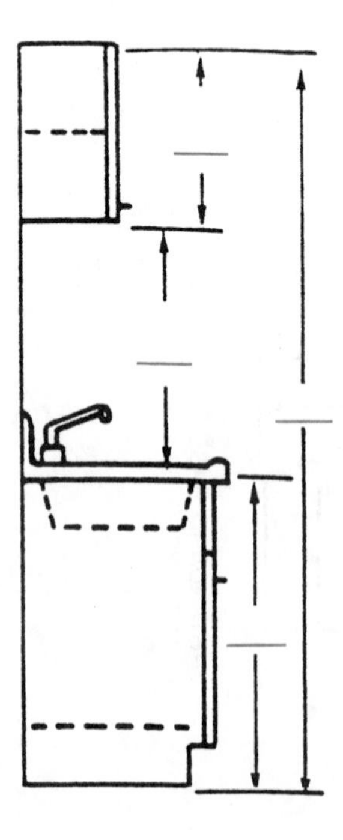

PROBLEM 12–7. Given the partial floor plans and correlated cabinet elevations shown on pages 236–238, answer the following questions:

1. How are the cabinet elevations keyed to the floor plans? ____
2. List the appliances found in the kitchen. ____
3. How far does the eating bar project out from the base cabinet? ______________________.
4. Describe the kitchen sink. ____
5. How many drawers are in the kitchen cabinets? (Note: The units that look like drawers at the range and sink are blank fronts: drawers cannot be placed in an area occupied by such appliances or fixtures as a sink or range.) ______________________.
6. What is specified about the sides of the ovens? ____.
7. What is the dimension of the soffit above the kitchen upper cabinets? ______________________.
8. What is the total dimension from the floor to the ceiling in the kitchen? (show your calculations) ______________________.
9. How do you know whether the refrigerator is included in the construction or is to be purchased later by the buyer? ____.
10. What scale was used to draw the cabinet elevations? ______________________.
11. What is the backsplash specification given in the kitchen? ____.
12. Which bathroom has a built-in medicine cabinet with mirror? ______________________.
13. How many bathroom sinks (lavatories) are there in the three bathrooms? ______________________.
14. How many drawers are there in bathroom 1? ______________________.
15. How many cabinet doors are there in bathroom 1? ______________________.
16. How many drawers are there in bathroom 2? ______________________.
17. How many cabinet doors are there in bathroom 2? ______________________.
18. How many drawers are there in bathroom 3? ______________________.
19. How many cabinet doors are there in bathroom 3? ______________________.
20. Describe the mirrors in bathroom 3? ____.
21. What is the purpose of the KNEE SPACE in bathroom 3? ____.
22. Give the specification provided for bathroom vanity and backsplash material. ____.

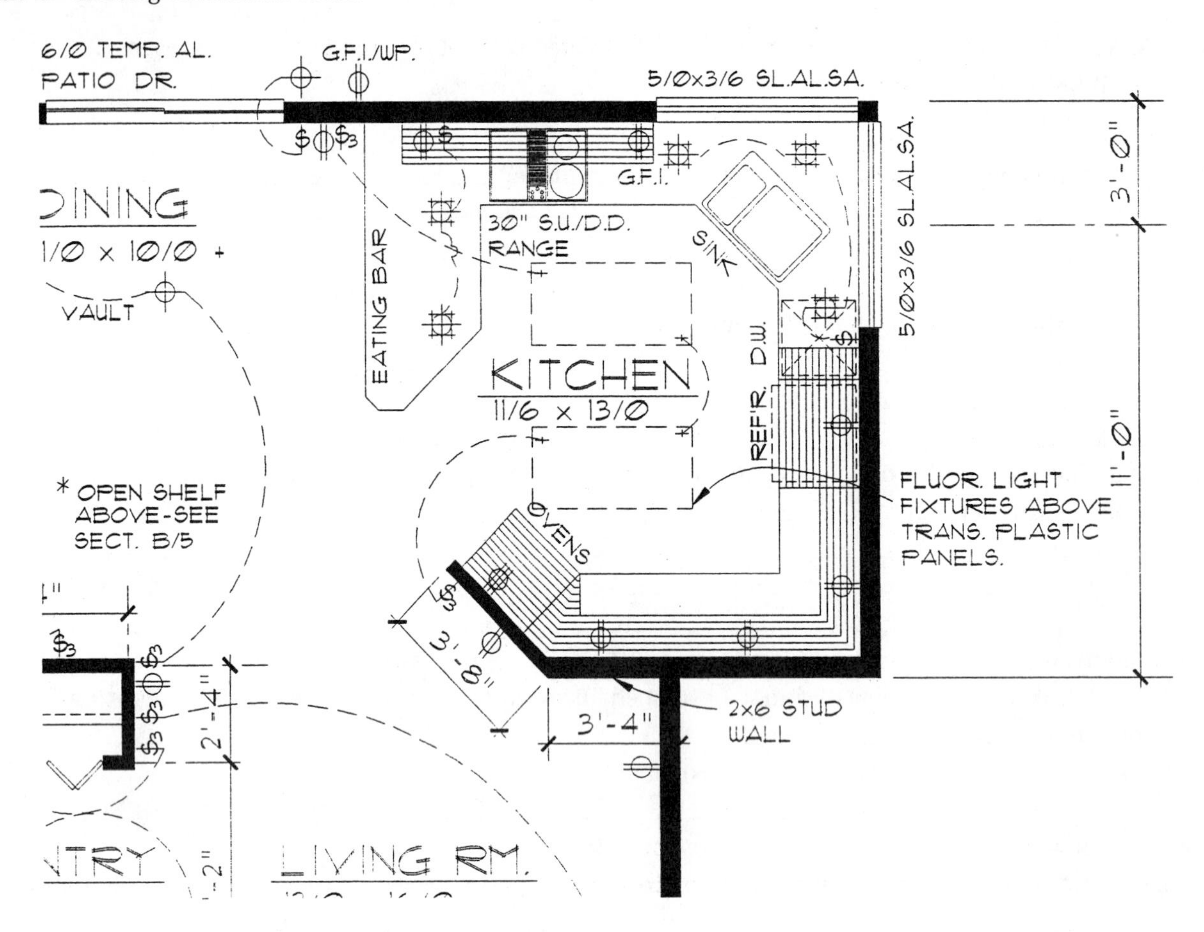

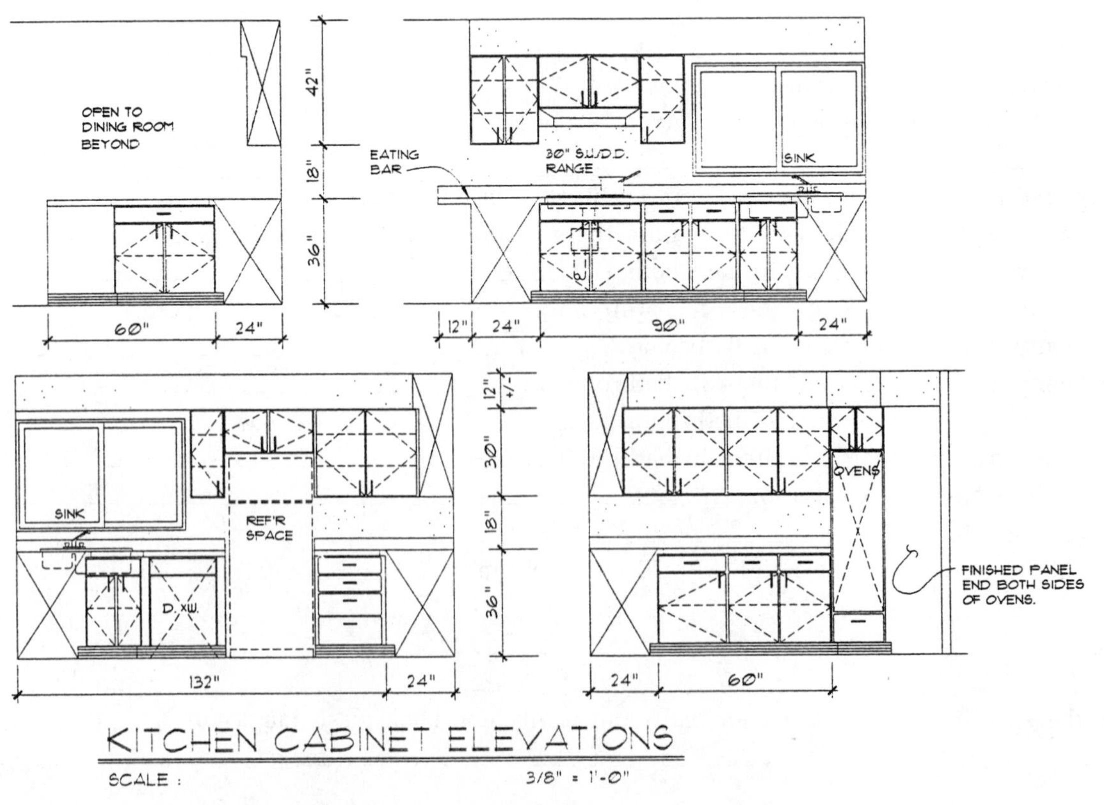

Problem 12–7. *Courtesy Piercy & Barclay Designers, Inc.*

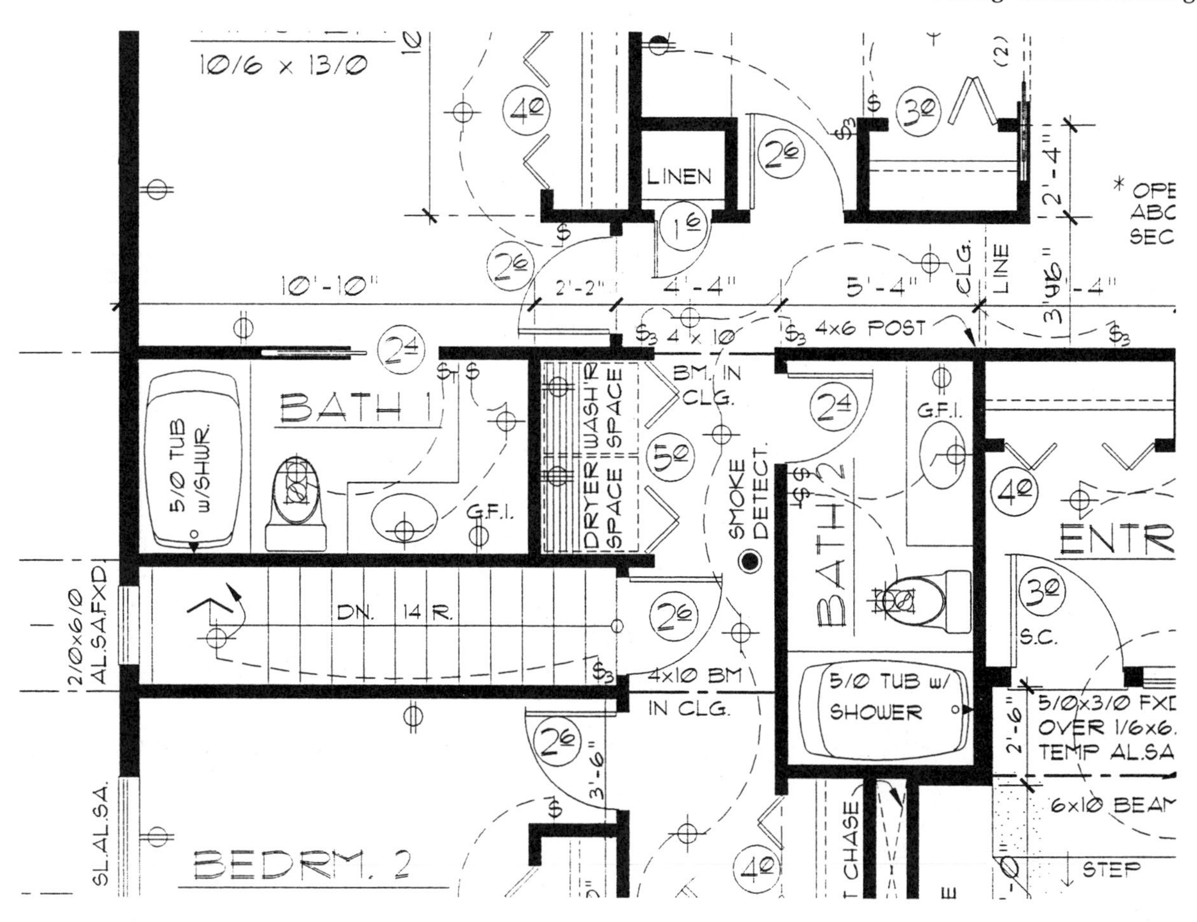

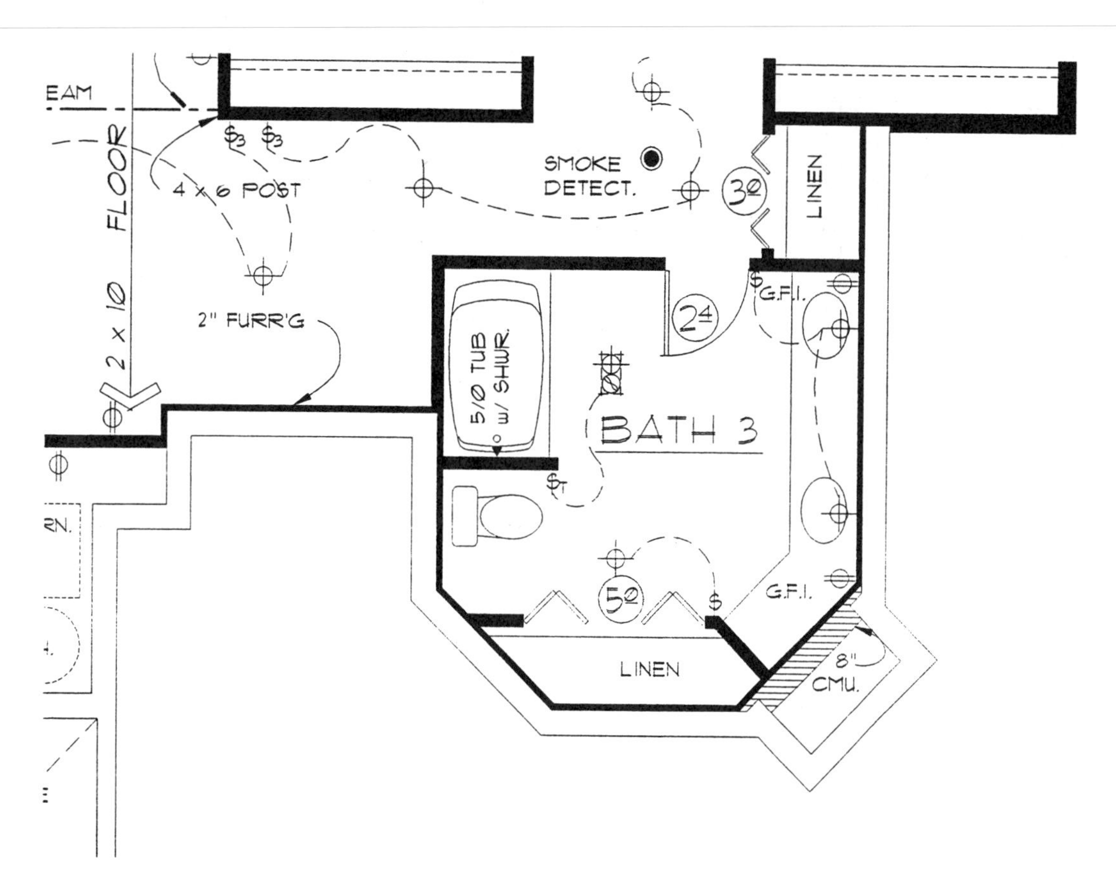

Problem 12–7 (continued). *Courtesy Piercy & Barclay Designers, Inc.*

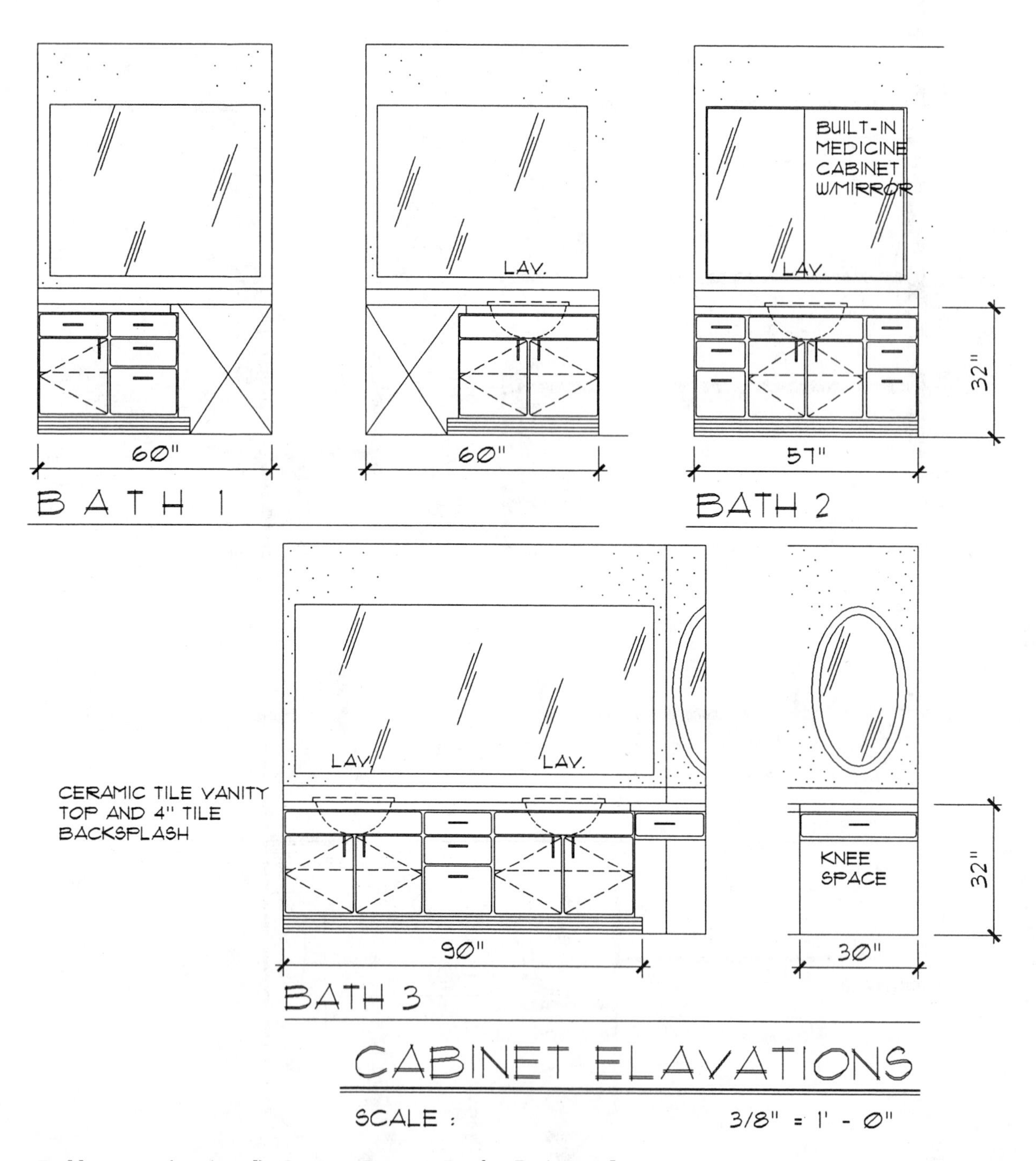

Problem 12–7 (continued). *Courtesy Piercy & Barclay Designers, Inc.*

Chapter 13
Reading Plot Plans

UNDERSTANDING LEGAL DESCRIPTIONS

VIRTUALLY EVERY piece of property in the United States is described for legal purposes. Legal descriptions of properties are filed in local jurisdictions, generally the county or parish courthouse. Legal descriptions are public record and may be reviewed at any time. This section deals with plot plan characteristics and requirements. A plot is an area of land generally one lot or construction site in size. The term plot is synonymous with lot. A plat is a map of part of a city or township showing some specific area such as a subdivision made up of several individual plots or lots. There are usually many plots in a plat. Some dictionary definitions, however, do not differentiate between plot and plat. There are three basic types of legal descriptions: metes and bounds, rectangular system, and lot and block.

METES AND BOUNDS SYSTEM

Metes, or measurements, and bounds, or boundaries may be used to identify the perimeters of any property. The metes are measured in feet, yards, rods (rd), or surveyor's chains (ch). There are 3 feet in 1 yard, 5.5 yards or 16.5 feet in one rod, and 66 feet in one

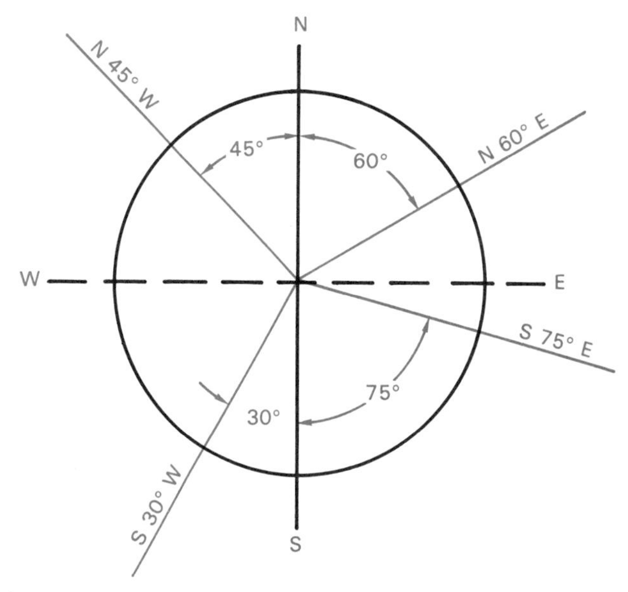

Figure 13–1 Determining bearings.

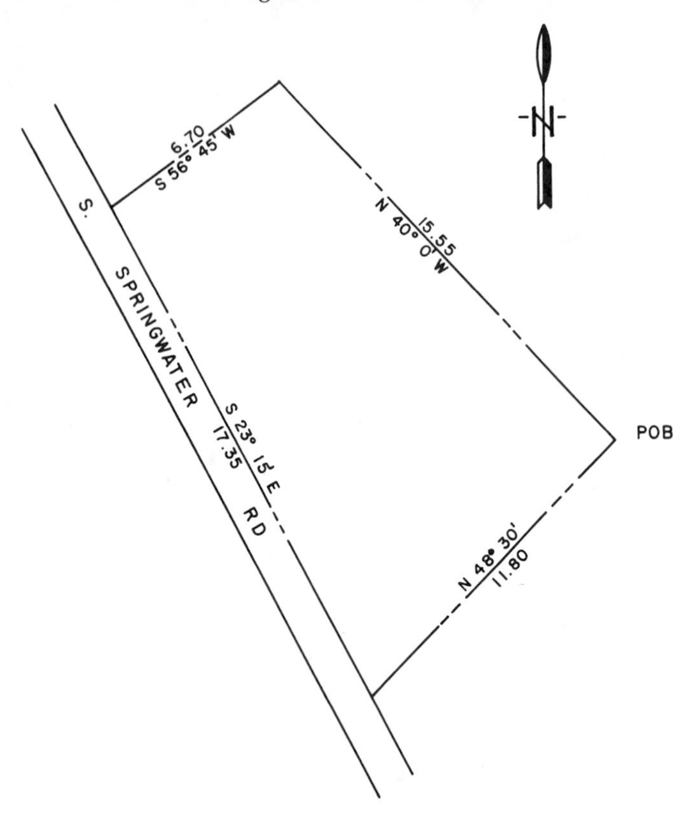

Beginning at a point 20 chains north 40° 0′ west from the southeast corner of the Asa Stone Donation Land Claim No. 49, thence north 40° 0′ west 15.55 chains to a pipe, thence south 56° 45′ west 6.70 chains to center of road, thence south 23° 15′ east 17.35 chains, thence north 48° 30′ east 11.80 chains to place of beginning.

Figure 13–2 A metes and bounds plot plan and legal description.

surveyor's chain. The boundaries may be a street, fence, creek, or river. Boundaries are also established as bearings. Bearings are directions with reference to one of the quadrants of the compass. There are 360° in a circle or compass, and each quadrant has 90°. Degrees are divided into minutes and seconds. There are 60 minutes (60') in 1 degree and there are 60 seconds (60") in 1 minute. Bearings are measured clockwise or counterclockwise from either north or south. For example, a reading 45° from north toward west is labeled N 45° W. See Figure 13–1. If a bearing reading requires great accuracy, fractions of a degree are used. For example, S 30° 20' 10" E, reads from south 30 degrees 20 minutes 10 seconds toward east.

The metes and bounds land survey begins with a monument known as the point-of-beginning. This point is a fixed location and in times past has been a pile of rocks, a large tree or an iron rod driven into the ground. Figure 13–2 shows an example of a plot plan that is laid out using metes and bounds and the legal description for the plot.

RECTANGULAR SYSTEM

The states in an area of the United States starting with the western boundary of Ohio, and including some southeastern states, to the Pacific Ocean, were described as public land sites. Within this area, the United States Bureau of Land Management devised a system for describing land known as the rectangular system.

Parallels of latitude and meridians of longitude were used to establish areas known as great land surveys. The point of beginning of each great land survey is where two basic reference lines cross. The lines of latitude, or parallels, are called the base lines and the lines of longitude, or meridians, are called principal meridians. There are 31 sets of these lines in the continental United States with three in Alaska. At the beginning the principal meridians were numbered, and the numbering system ended with the sixth principal meridian passing through Nebraska, Kansas and Oklahoma. The remaining principal meridians were given local names. The meridian through one of the last great land surveys near the west coast is named the Willamette Meridian because of its location in the Willamette Valley of Oregon. The principal meridians and base lines of the great land surveys are shown in Figure 13–3.

Townships

The great land surveys were, in turn, broken down into smaller surveys known as townships. The base lines and meridians were divided into blocks called townships. Each township measures 6 miles square. The townships are numbered by tiers running east-west. The tier numbering system is established either north or south of a principal base line. For example, the fourth tier south of the base line is labeled, Township Number 4 South, or abbreviated T. 4 S. Townships are also numbered according to vertical meridians, known as ranges. Ranges are established either east or west of a principal meridian. The third range east of the principal meridian is called Range Number 3 East, or abbreviated R. 3 E. Now, if we combine T. 4 S. and R. 3 E. we have located a township or a piece of land 6 mi by 6 mi or a total of 36 sq mil. Figure 13–4 shows the township just described.

Sections

To further define the land within a 6 mi square township, the area was divided into units 1 mi square. These 1-mi by 1-mi areas are called sections. Sections in a township are numbered from 1 to 36. Section 1 always begins in the upper right corner and consecutive numbers are arranged as shown in Figure 13–5.

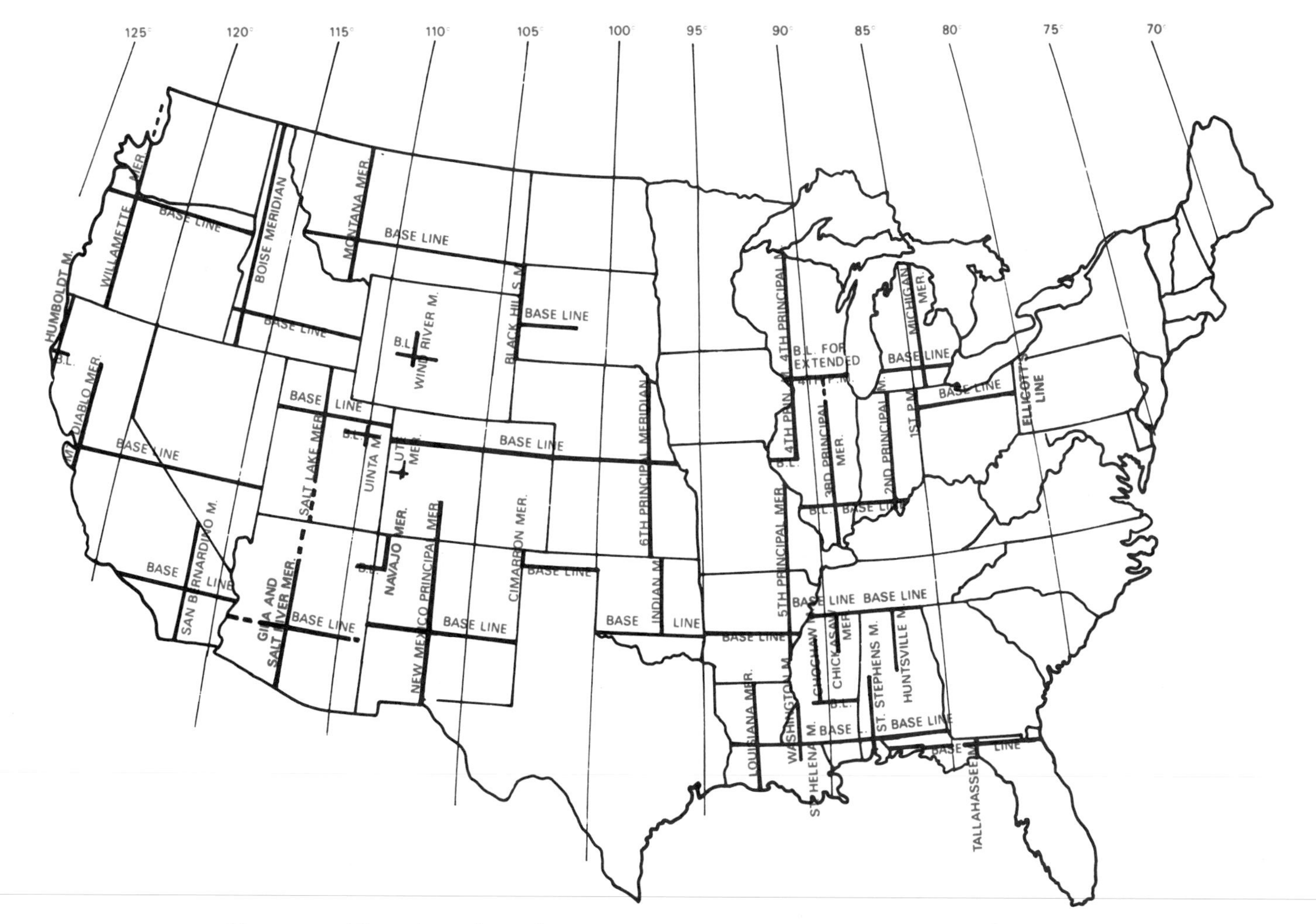

Figure 13–3 The principal meridians and base lines of the great land survey (not including Alaska).

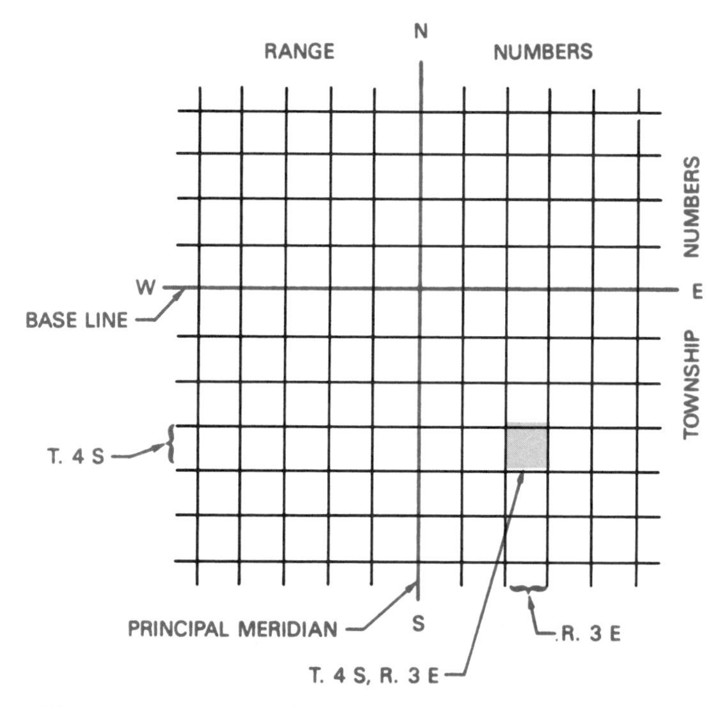

Figure 13–4 How townships are established.

6	5	4	3	2	1
7	8	9	10	11	12
18	17	16	15	14	13
19	20	21	22	23	24
30	29	28	27	26	25
31	32	33	34	35	36

Figure 13–5 How a township is divided into sections.

The legal descriptions of land can be carried one stage further. For example, Section 10 in the township given would be described as Sec. 10, T. 4 S., R. 3 E. This is an area of land 1-mi square. Sections are divided into acres. One acre equals 43,560 sq ft, and 1 section of land contains 640 acres.

In addition to dividing sections into acres, sections are also divided into quarters as shown in Figure

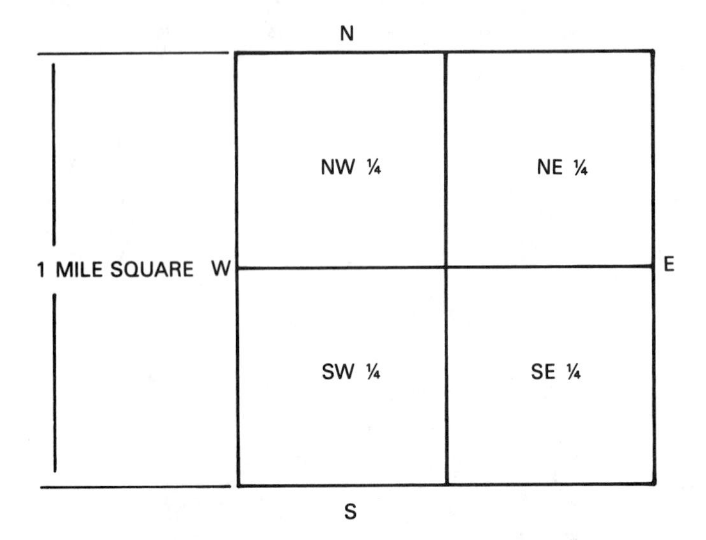

Figure 13–6 Splitting a section into quarters.

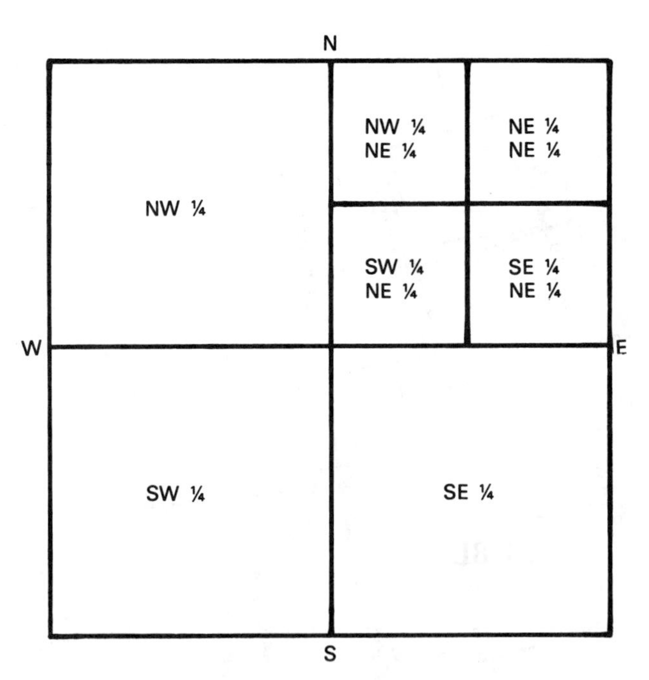

Figure 13–7 Dividing a quarter section.

13–6. The northeast one-quarter of Section 10 is 160-acre piece of land described as NE ¼, Sec. 10, T. 4 S., R. 3 E. When this section is keyed to a specific meridian, then it can only be one specific 160-acre area. The section can be broken further by dividing each quarter into quarters as shown in Figure 13–7. If the SW ¼ of the NE ¼ of Section 10 were the desired property, you have 40 acres known as

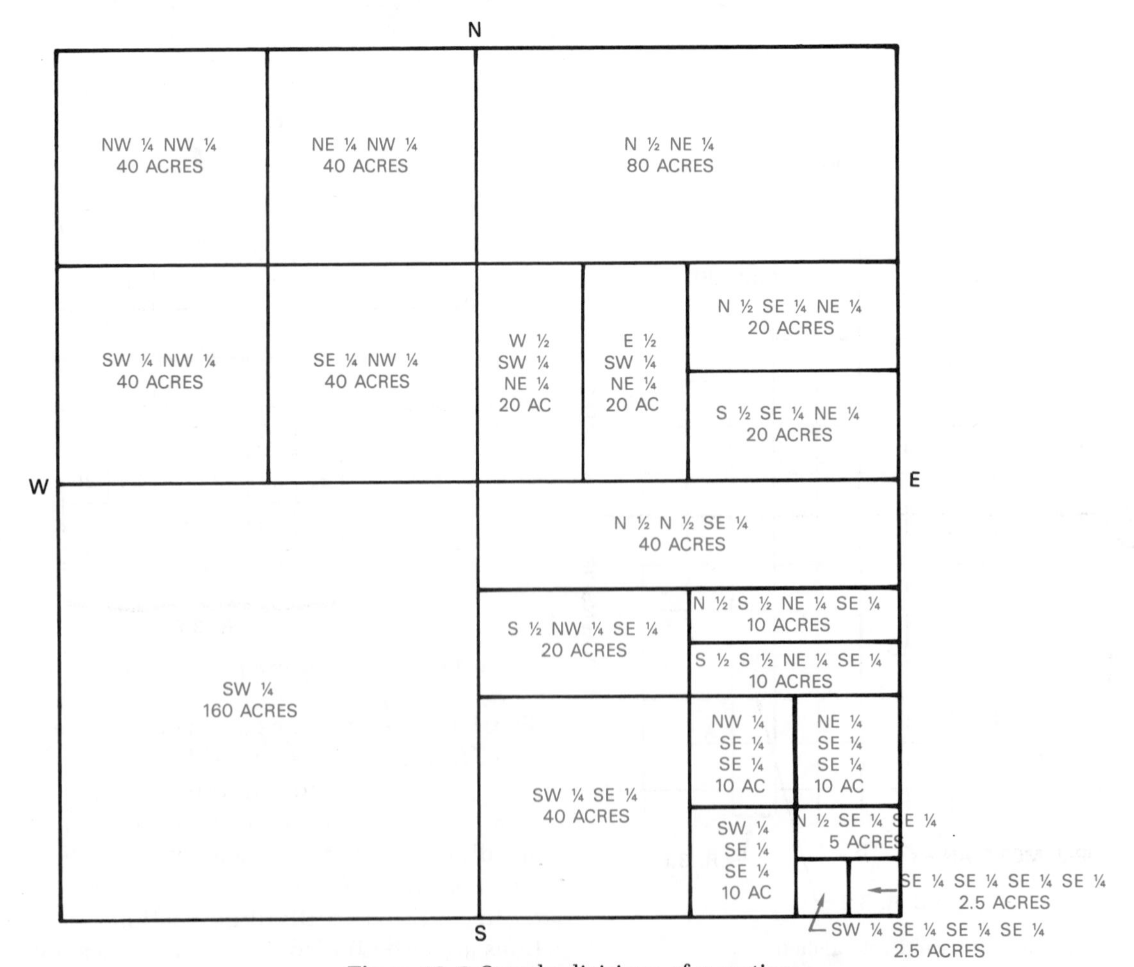

Figure 13–8 Sample divisions of a section.

SW ¼, NE ¼, Sec. 10, T. 4 S., R. 3 E. The complete rectangular system legal description of a 2.5 acre piece of land in Section 10 reads: SW ¼, SE ¼, SE ¼, SE ¼, Sec. 10, T. 4 N., R. 8 W. of the San Bernardino Meridian, in the County of Los Angeles, State of California. See Figure 13–8.

The rectangular system of land survey may be used to describe very small properties by continuing to divide the section of a township. Oftentimes, the township section legal description may be used to describe the location of the point of beginning of a metes and bounds legal description especially when the surveyed land is an irregular plot within the rectangular system.

LOT AND BLOCK SYSTEM

The lot and block legal description system can be derived from either the metes and bounds or the rectangular systems. Generally when a portion of land is subdivided into individual building sites the subdivision must be established as a legal plat and recorded as such in the local county records. The subdivision is given a name and broken into blocks of lots. A subdivision may have several blocks each divided into a series of lots. Each lot may be 50' × 100', for example, depending on the zoning requirements of the specific area. Figure 13–9 shows an example of a typical lot and block system.

READING PLATS AND PLOT PLANS

A plot plan, also known as a lot plan, is a map of a piece of land that may be used for any number of purposes. Plot plans may show a proposed construction site for a specific property. Plots may show topography with contour lines or the numerical value of land elevations may be given at certain locations.

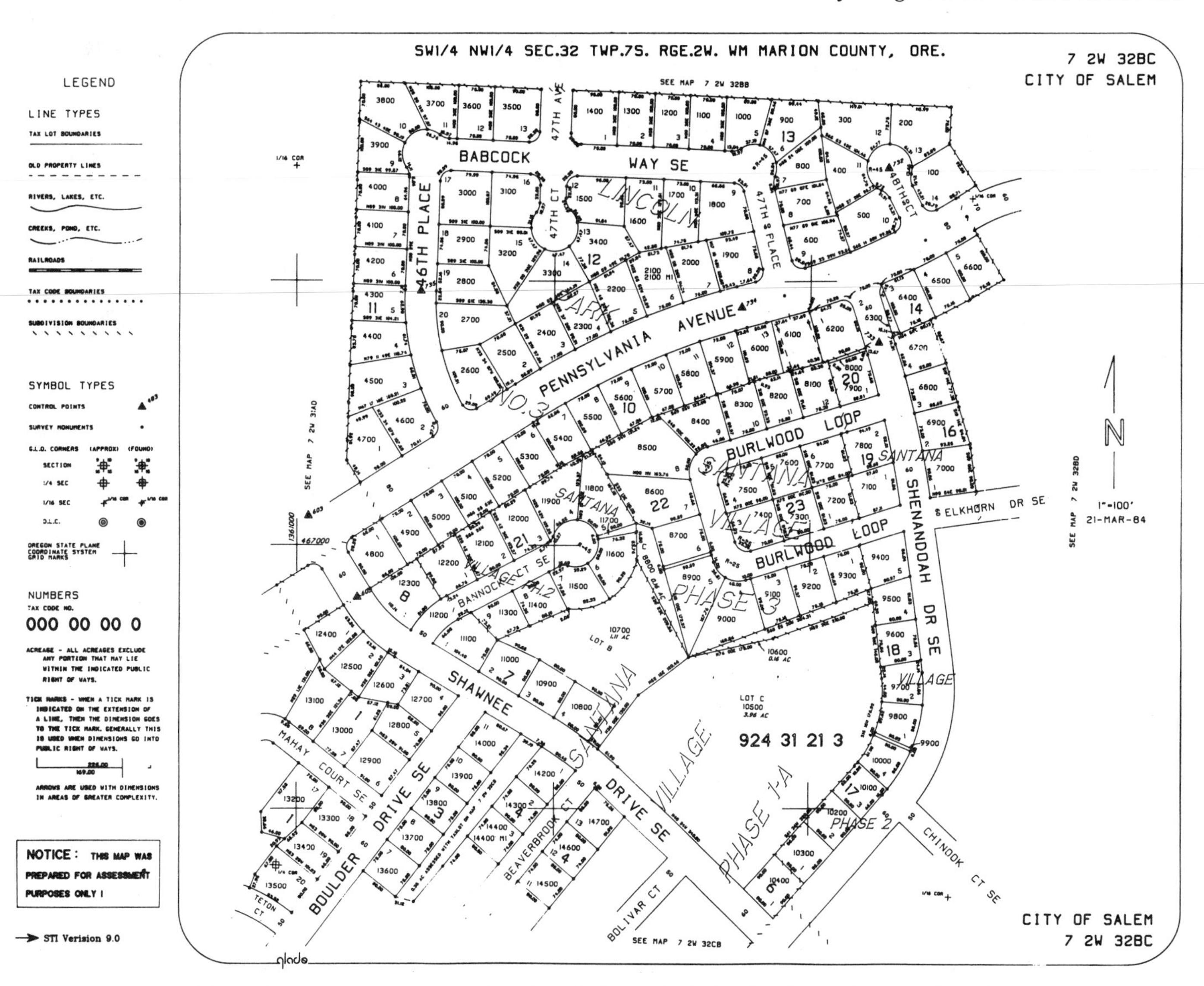

Figure 13–9 A computer generated plat of a lot and block subdivision. *Courtesy GLADS, a joint Marion County/City of Salem remapping project.*

Plot plans are also used to show how a construction site will be excavated and are then known as a grading plan.

Plats are maps that are used to show an area of a town or township. They show several or many lots and may be used by a builder to show a proposed subdivision of land.

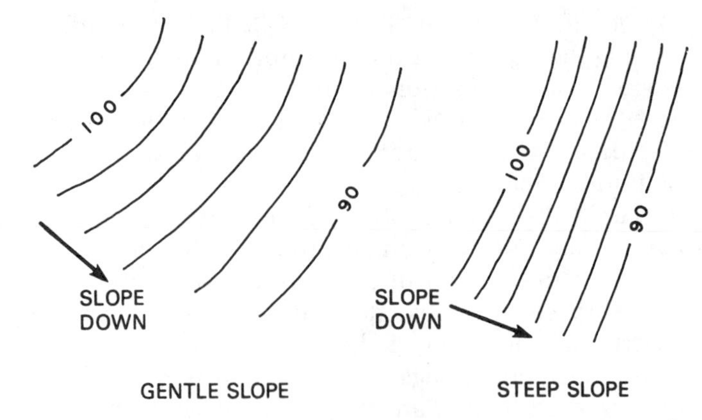

Figure 13–10 Examples of contour lines showing both gentle and steep slope.

TOPOGRAPHY

Topography is a physical description of land surface showing its variation in elevation, known as relief, and locating other features. Surface relief can be shown with graphic symbols that use shading methods to show the character of land or the differences in elevations can be shown with contour lines. Plot plans that require surface relief identification generally use contour lines. These lines connect points of equal elevation and help show the general lay of the land.

The vertical distance between contour lines is known as contour interval. When the contour lines are far apart, the contour interval shows relatively flat or gently sloping land. When the contour lines are close together, the contour interval shows a land that is much steeper. Drawn contour lines are broken periodically and the numerical value of the contour elevation above sea level is inserted. Figure 13–10 shows sample contour lines. Figure 13–11 shows a graphic example of land relief in pictorial form and contour lines of the same area.

Plot plans do not always require contour lines showing topography. In most stances, the only contour-related information required is property corner elevations, street elevation at a driveway, and the elevation of the finished floor levels of the structure. Additionally, slope may be defined and labeled with an arrow.

Figure 13–11 Land relief pictorial above and the contour lines of the same land area below. *Courtesy U.S. Department of the Interior, Geological Survey.*

READING PLOT PLANS

Plot plan requirements vary from one local jurisdiction to the next although there are elements of plot plans that are similar around the country. Typical plot plan items include the following:

- Plot plan scale
- Legal description of the property
- Property line bearings and dimensions
- North direction
- Existing and proposed roads
- Driveways, patios, walks, and parking areas
- Existing and proposed structures
- Public or private water supply
- Public or private sewage disposal
- Location of utilities
- Rain and footing drains, and storm sewers or drainage
- Topography including contour lines or land elevations at lot corners, street centerline, driveways, and floor elevations
- Setbacks, front, rear, and sides
- Specific items on adjacent properties may be required
- Existing and proposed trees may be required

Figure 13–12 shows a plot plan layout that is used as an example at a local building department. Figure 13–13 shows a basic plot plan for a proposed residential addition.

The method of sewage disposal is generally an important item shown on a plot plan drawing. The plot plan representation of a public sewer connection is shown in Figure 13–14. A private septic sewage disposal system is shown in a plot plan example in Figure 13–15.

READING THE GRADING PLAN

Grading plans are plots of construction sites that generally show existing and proposed topography. The outline of the structure may be shown with elevations at each building corner and the elevation given for each floor level. Figure 13–16 shows a detailed grading plan for a residential construction site. Notice that the legend identifies symbols for existing and finished contour lines. This particular grading plan provides retaining walls and graded slopes to accommodate a fairly level construction site from the front

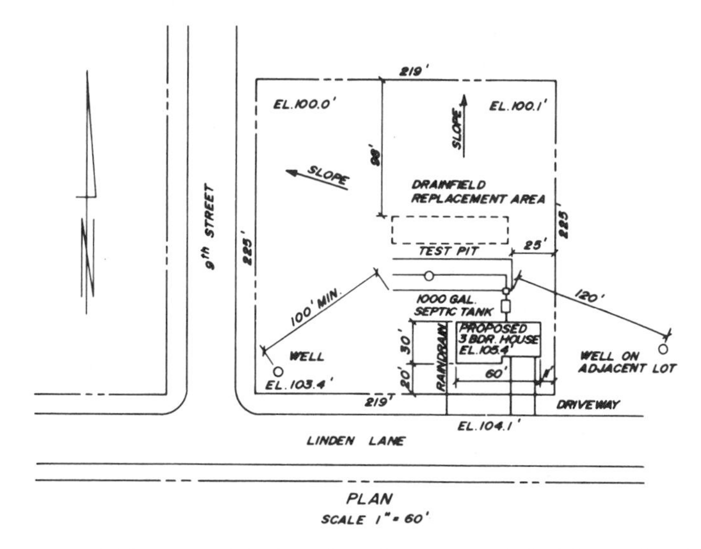

Figure 13–12 The typical information that you will find in a plot plan.

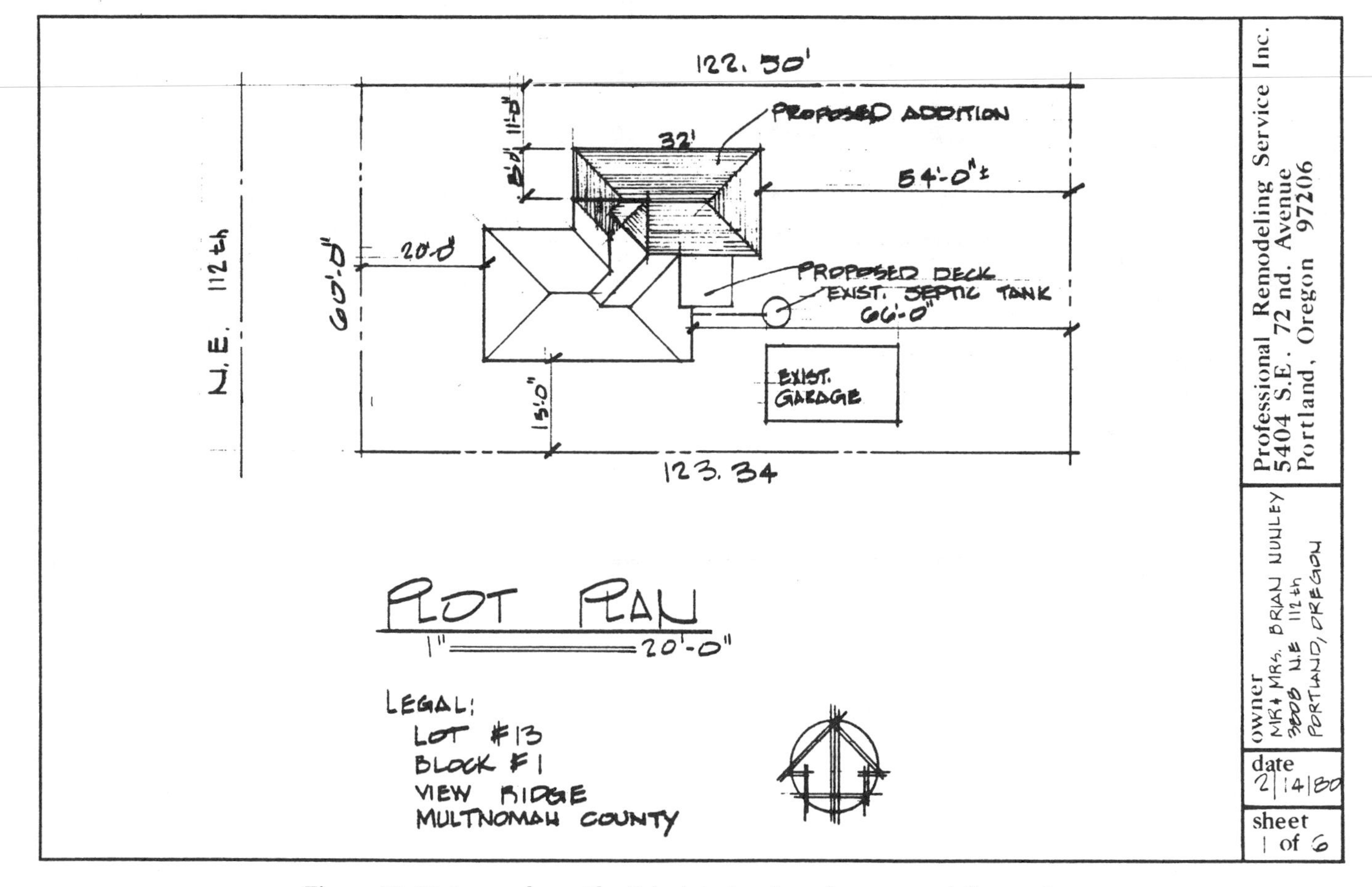

Figure 13–13 A sample residential plot plan for a home remodeling project.

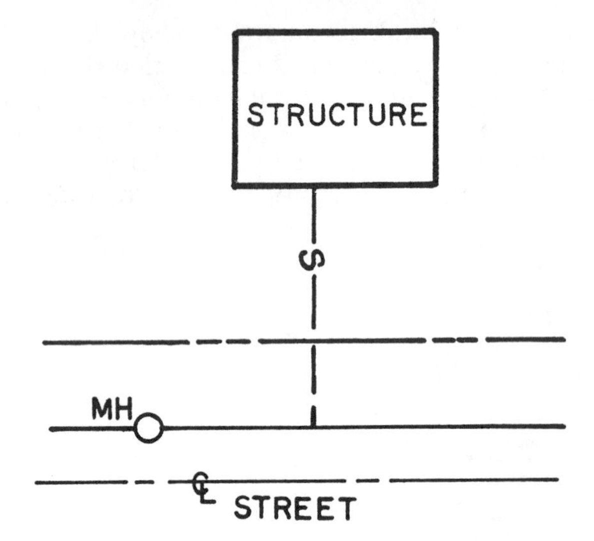

Figure 13–14 A plot plan showing a public sewer connection.

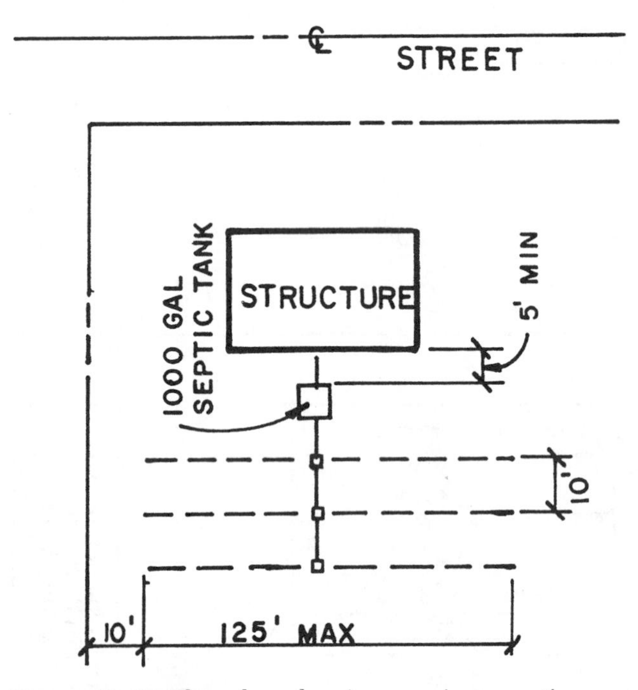

Figure 13–15 Plot plan showing a private septic sewage system.

of the structure to the extent of the rear yard. The finished slope represents an embankment that establishes the relationship of the proposed contour to the existing contour. This particular grading plan also shows a proposed irrigation and landscaping layout. Grading plan requirements may differ from one location to the next. Some grading plans may also show a cross section through the site at specified intervals or locations to evaluate the contour more fully.

READING SUBDIVISION PLATS

There may be some occasions when you need to read information on a subdivision plat that is made up of several individual plots. Keep in mind that each plot has the information already discussed; although, the subdivision shows only the lot information without regards to location of a structure. The information related to the location and construction of the building is normally only shown on the plot plan. Some of the items that may be included on the plat or in separate over are the following:

1. The name, address and phone number of the property owner, applicant, and engineer or surveyor
2. Source of water
3. Method of sewage disposal
4. Existing zoning
5. Proposed utilities
6. Calculations justifying the proposed density
7. Name of the major partitions or subdivision
8. Date the drawing was made
9. Legal description
10. North arrow
11. Vicinity sketch showing location of the subdivision
12. Identification of each lot or parcel and block by number
13. Gross acreage of property being subdivided or partitioned
14. Dimensions and acreage of each lot or parcel
15. Streets abutting the plat including name, direction of drainage, and approximate grade
16. Streets proposed including names, approximate grades, and radius of curves.
17. Legal access to subdivision or partition other than public road
18. Contour lines at 2-foot interval for slopes of 10 percent or less, 5-foot interval if exceeding a 10 percent slope
19. Drainage channels, including width, depth, and direction of flow
20. Existing and proposed easements locations
21. Location of all existing structures, driveways, and pedestrian walkways
22. All areas to be offered for public dedication
23. Connected property under the same ownership, if any
24. Boundaries of restricted areas, if any
25. Significant vegetative areas such as major wooded areas or specimen trees

Figure 13–17 shows an example of a major subdivision plat.

Look at Figure 13–17 and notice that only property boundaries, streets, and utilities are shown. Now, as you look at Figure 13–17 find Lot 2. The plot plan with building site information for Lot 2 is shown in Figure 13–18. By comparing the plat and plot plan, you can see the kind of information that is found on both types of prints.

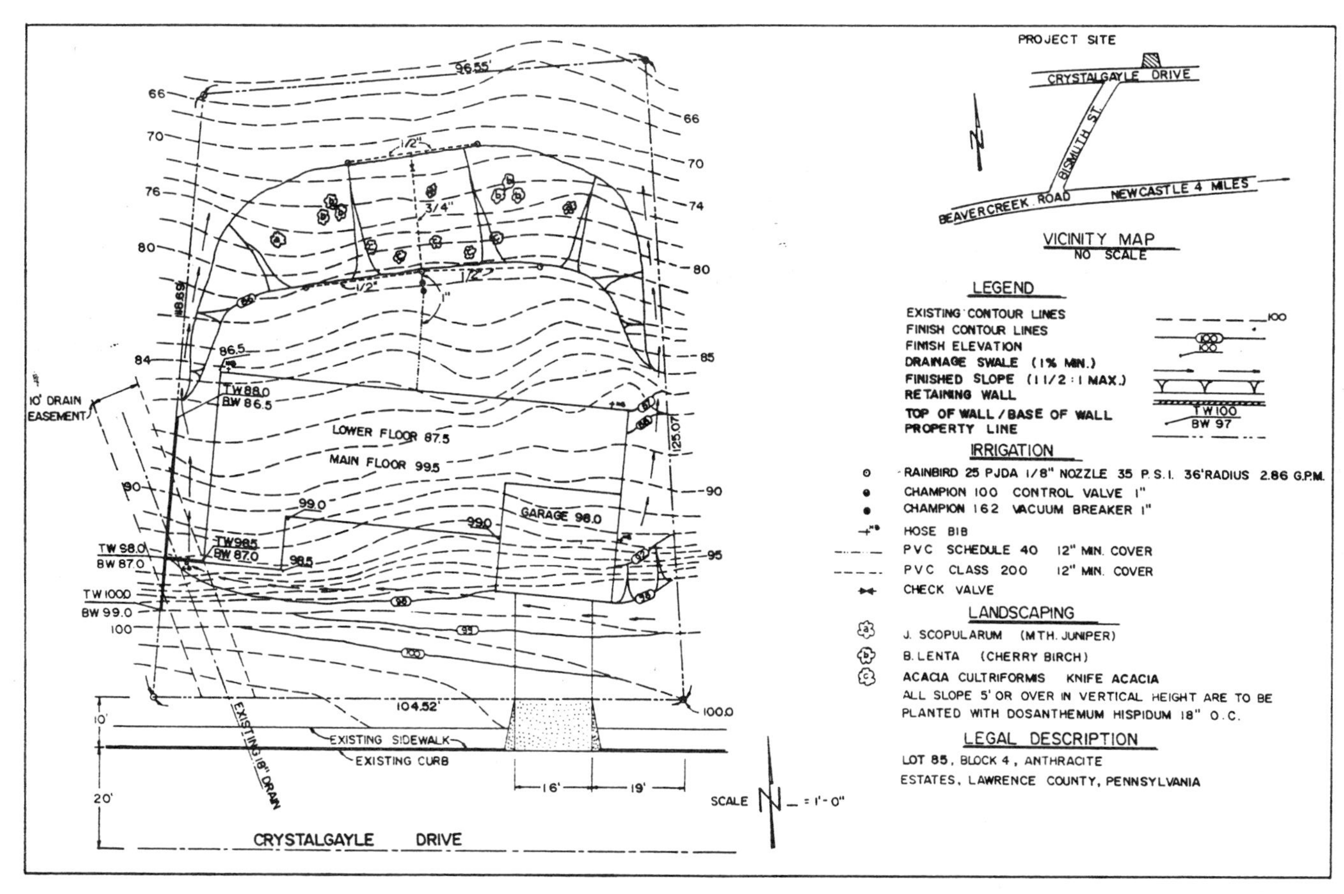

Figure 13–16 A grading plan. Notice the Vicinity Map which shows the project location in relation to the nearby streets.

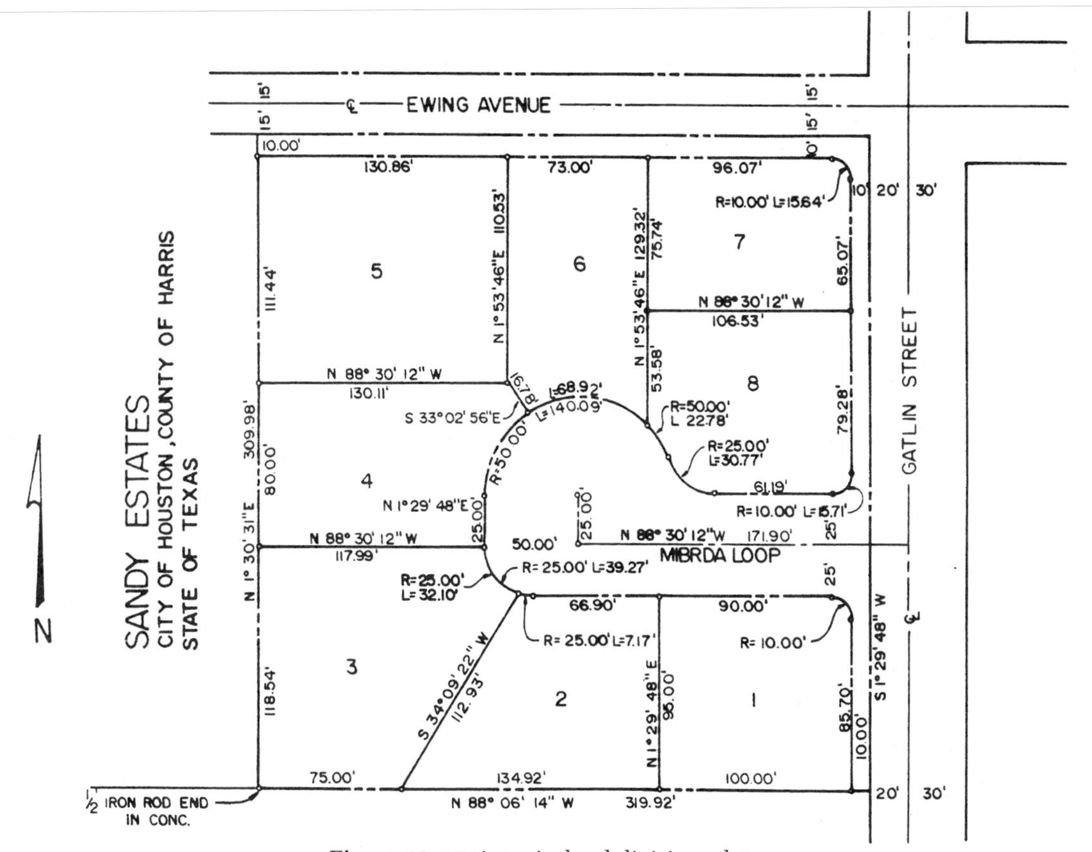

Figure 13–17 A typical subdivision plat.

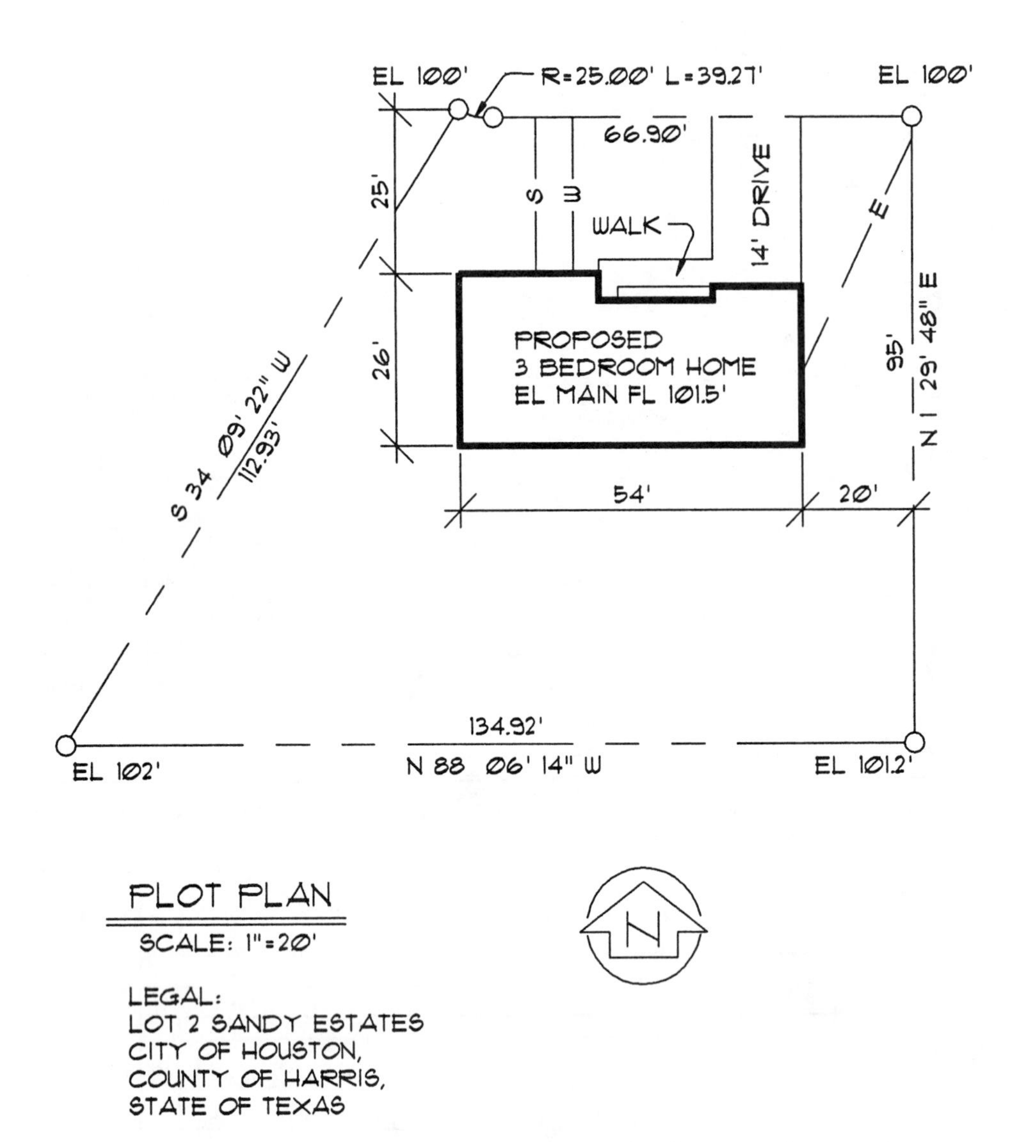

Figure 13–18 A plot plan from the subdivision plat shown in Figure 13–17.

CHAPTER 13 TEST

Fill in the blanks with short complete statements or words as needed:

1. Define plot plan as discussed in this chapter. ______________________________
____
______________________________.

2. Name the three types of legal descriptions: ______________________________
______________________________.

3. Metes, or ____________, and bounds, or ____________, may be used to identify the perimeters of any property.

4. Define bearings as discussed in this chapter. ____
____
____.
5. How would you label a bearing that is 30° from north toward the east? ______________________
____.
6. How would you label a bearing that is 47° 30' 12" from south toward the west? ______________.
7. The metes and bounds land survey begins with a ______________ known as the point-of-beginning.
8. A township, in the rectangular system of describing land, is a piece of land ______________ by ______________, or a total of ______________ square miles.
9. How many sections are there in a township? ______________________
10. How many acres are there in one section? ______________________
11. Give the proper label and the number of acres contained in the division of a section highlighted in the drawing below: ______________________
12. Give the proper label and the number of acres contained in the division of a section highlighted in the drawing below: ______________________
____.
13. Name the type of legal description that is associated with a lot within a named subdivision of land which may be divided into several blocks, and each block divided into lots. ______________
14. Define topography as discussed in this chapter. ______________________
____
____.
15. Explain the difference in slope of the land when contour lines are spaced far apart as compared to when they are spaced close together. ______________________
____
____.
16. Define grading plan. ______________________
____.

CHAPTER 13 EXERCISES

PROBLEM 13–1. Refer to the Plot Plan shown on page 251 and answer the following questions:

1. What is the legal description? ______________________
____.
2. What are the following setbacks to the proposed structure?
   Front ______________
   West side ______________
   East side ______________
   Rear ______________
3. What are the following minimum required setbacks?
   Front ______________
   West side ______________
   East side ______________
   Rear ______________
4. What is the scale of this plot plan? ______________________
5. Describe the structure to be built on this property. ______________________
____.

6. Give the finish floor elevation. ______________________________
7. Give the elevations at the following property corners:
   Northwest ______________
   Northeast ______________
   Southwest ______________
   Southeast ______________
8. What is the name of the street that runs in front of this proposed home? ______________.
9. What is the length of the South property line? ______________________________
10. What is the length of the West property line? ______________________________

PROBLEM 13–2. Refer to the Grading Plan shown on page 252 and answer the following questions:

1. What is the legal description for this property? ______________________________
____
____
______________________________.
2. What is the minimum front set back? ______________________________
3. What are the minimum side yard set backs? ______________________________
4. What is the width of the public sanitary sewer easement? ______________________________
5. Give the west property line length and bearing. ______________________________
______________________________.
6. Give the east property line length and bearing. ______________________________
7. Give the south property line length and bearing. ______________________________
______________________________.
8. What is the contour interval? ______________________________
9. How often are contour lines labeled? ______________________________
10. What is the difference between the contour lines that are labeled and the ones that are not labeled? ______
____
____
______________________________.
11. What is the entry floor elevation? ______________________________.
12. What is the lower finish floor elevation? ______________________________.
13. What is the upper finish floor elevation? ______________________________.
14. How do the site work contour lines differ from the natural contour lines? ______________
____
____
____
______________________________.
15. What is the height of the highest retaining wall? ______________________________
16. What is the difference in elevation between the garage finish floor and the entry finish floor? ______
17. What is the difference in elevation between the upper finish floor and the lower finish floor? ______.
18. What are the driveway specifications? ______________________________
______________________________.
19. What is the elevation of the highest contour line shown on this plan? ______________.
20. What is the elevation of the lowest contour line shown on this Plan? ______________.

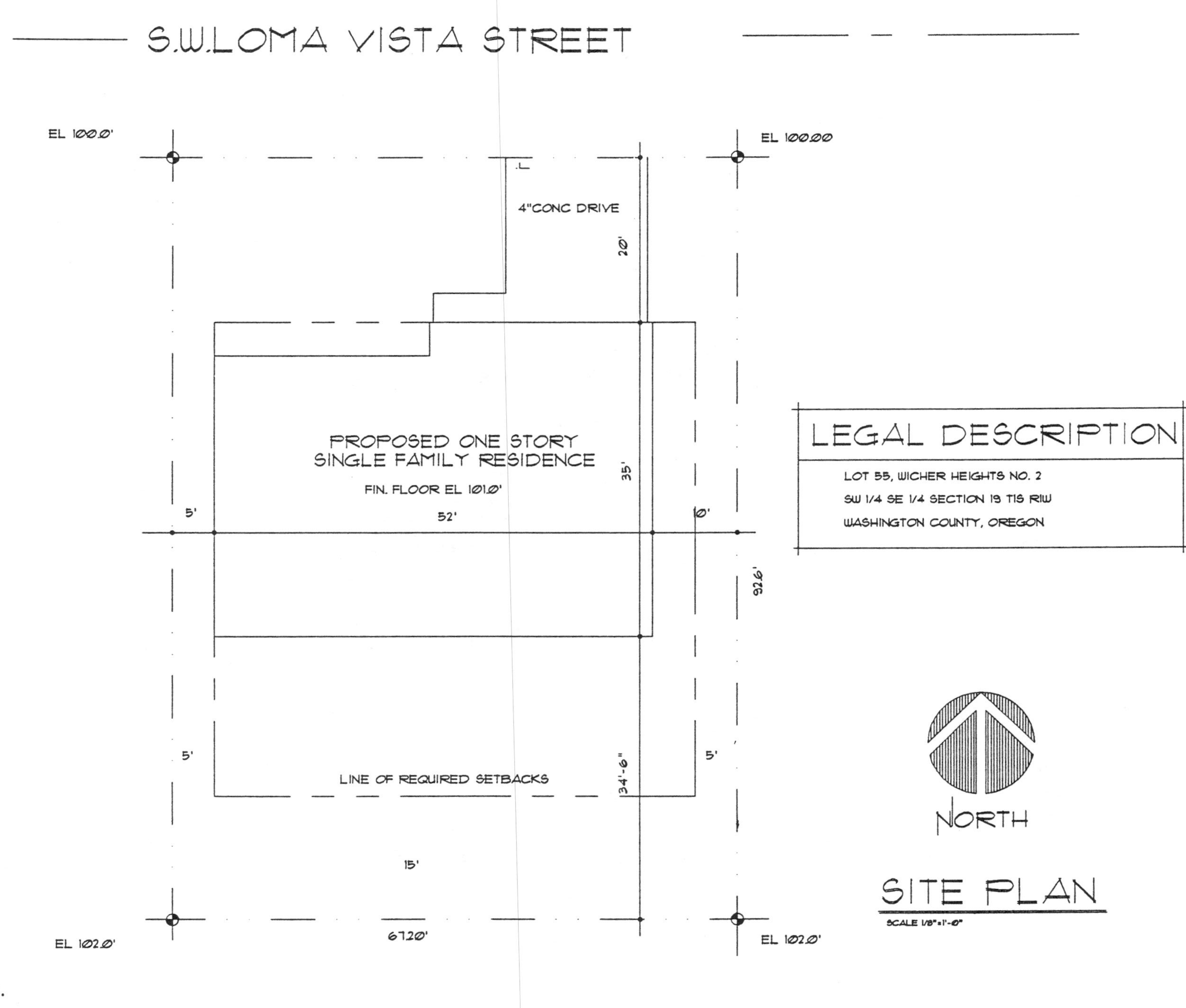
S.W. LOMA VISTA STREET
EL 100.0'
EL 100.00
4"CONC DRIVE
20'
PROPOSED ONE STORY
SINGLE FAMILY RESIDENCE
FIN. FLOOR EL 101.0'
35'
5'
52'
10'
LEGAL DESCRIPTION
LOT 55, WICHER HEIGHTS NO. 2
SW 1/4 SE 1/4 SECTION 13 T1S R1W
WASHINGTON COUNTY, OREGON
92.6'
5'
LINE OF REQUIRED SETBACKS
34'-6"
5'
NORTH
15'
SITE PLAN
SCALE 1/8"=1'-0"
EL 102.0'
67.20'
EL 102.0'

Problem 13–1.

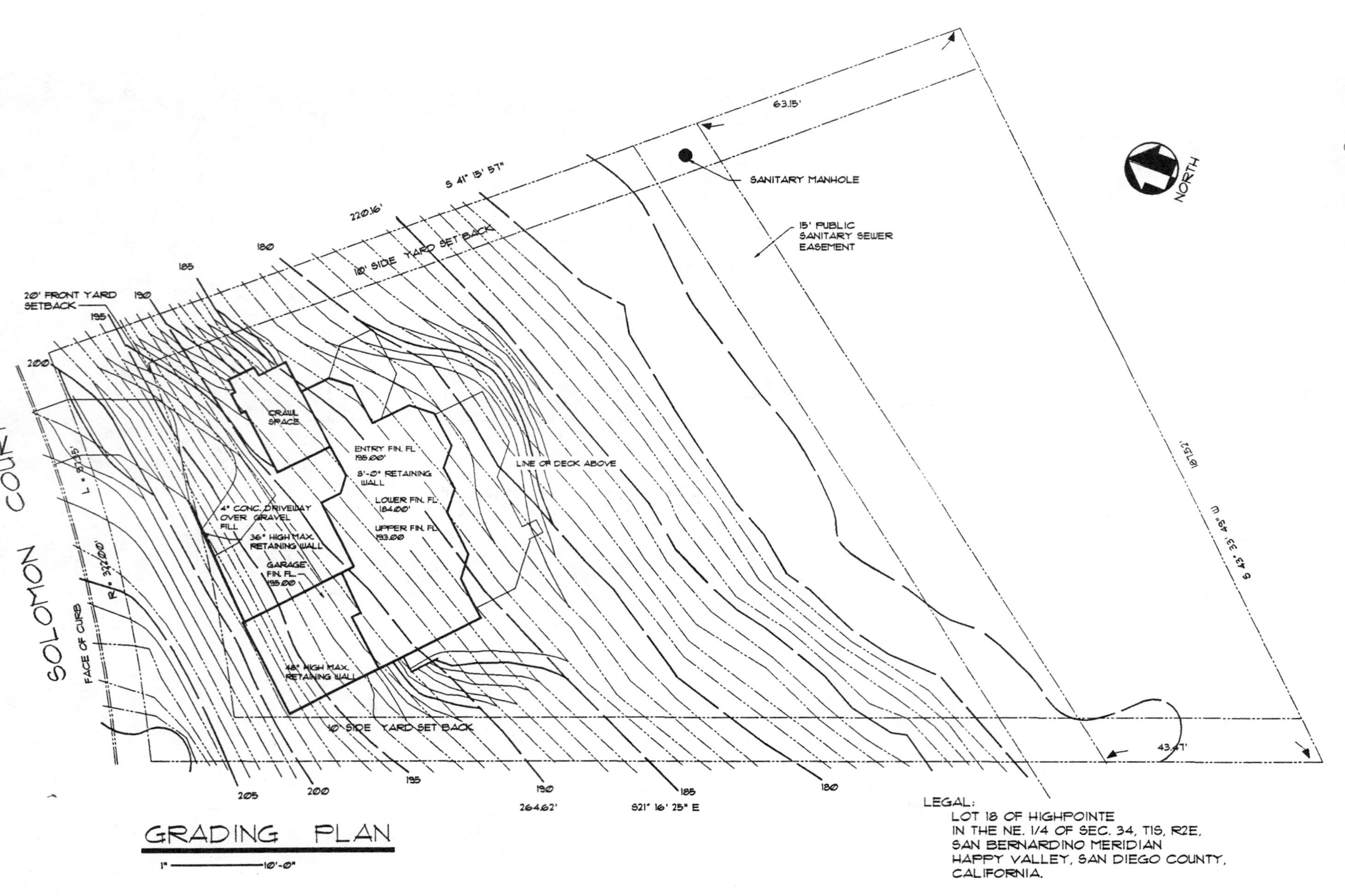
63.15'
SANITARY MANHOLE
NORTH
15' PUBLIC
SANITARY SEWER
EASEMENT
S 41° 15' 57"
220.16'
180
185
10' SIDE YARD SET BACK
20' FRONT YARD
SETBACK
190
195
200
CRAWL
SPACE
ENTRY FIN. FL
195.00'
8'-0" RETAINING
WALL
LOWER FIN. FL
184.00'
UPPER FIN. FL
193.00
LINE OF DECK ABOVE
4" CONC. DRIVEWAY
OVER GRAVEL
FILL
36" HIGH MAX.
RETAINING WALL
GARAGE
FIN. FL
195.00
48" HIGH MAX.
RETAINING WALL
187.52'
S 43° 33' 49" W
SOLOMON COURT
FACE OF CURB
R = 332.00'
10' SIDE YARD SET BACK
43.47'
205
200
195
190
264.62'
185
S21° 16' 25" E
180
LEGAL:
LOT 18 OF HIGHPOINTE
IN THE NE. 1/4 OF SEC. 34, T1S, R2E,
SAN BERNARDINO MERIDIAN
HAPPY VALLEY, SAN DIEGO COUNTY,
CALIFORNIA.
GRADING PLAN
1" = 10'-0"

Problem 13–2.

Chapter 14

General Construction Specifications

CONSTRUCTION SPECIFICATIONS
TYPICAL MINIMUM CONSTRUCTION SPECIFICATIONS
TEST
EXERCISES

CONSTRUCTION SPECIFICATIONS

BUILDING PLANS, including all of the elements that make up a complete set of residential or commercial drawings, contain most general and some specific information about the construction of the structure. Information that cannot be clearly or completely provided on the drawing or in schedules are provided in construction specifications. Specifications are an integral part of any set of plans. Specifications may not be required when applying for a building permit for a residential structure, but they are generally needed for loan approval when financing is required. Most lenders have their own format for residential construction specifications, although these forms are generally similar. The Federal Housing Administration (FHA) or the Federal Home Loan Mortgage Corporation (FHLMC) has a specification format entitled, Description of Materials. This specifications form is used widely in a revised or identical manner by most residential construction lenders. The same form is used by Farm Home Administration (FmHA) and by the Veterans Administration (VA). Figure 14–1 shows a FHA Description of Materials form for a typical structure. The set of plans, construction specifications, and building contract together become the legal documents for the construction project. These documents should be prepared very carefully in cooperation with the architect, client, and contractor. Any deviation from these documents should be approved by all three parties. When brand names are used, a clause specifying, "or equivalent," may be added. This means that another brand of equivalent value to the one specified may be substituted with the construction supervisor's approval.

TYPICAL MINIMUM CONSTRUCTION SPECIFICATIONS

Minimum construction specifications as established by local building officials vary from one location to the next and their contents are dependent on specific local requirements, climate, codes used, and the extent of coverage. Verify the local requirements for a construction project as they may differ from those given here. The following are some general classifications of construction specifications.

VETERANS ADMINISTRATION, U.S.D.A. FARMERS HOME ADMINISTRATION, AND
U.S. DEPARTMENT OF HOUSING AND URBAN DEVELOPMENT
HOUSING - FEDERAL HOUSING COMMISSIONER OMB Approval No. 2502-0192 (Exp. 6-30-87)

For accurate register of carbon copies, form may be separated along above fold. Staple completed sheets together in original order.

☐ Proposed Construction

DESCRIPTION OF MATERIALS No. ________ *(To be inserted by HUD, VA or FmHA)*

☐ Under Construction

Property address ________ City ________ State ________

Mortgagor or Sponsor ________ (Name) ________ (Address)

Contractor or Builder ________ (Name) ________ (Address)

INSTRUCTIONS

1. For additional information on how this form is to be submitted, number of copies, etc., see the instructions applicable to the HUD Application for Mortgage Insurance, VA Request for Determination of Reasonable Value, or FmHA Property Information and Appraisal Report, as the case may be.
2. Describe all materials and equipment to be used, whether or not shown on the drawings, by marking an X in each appropriate check-box and entering the information called for each space. If space is inadequate, enter "See misc." and describe under item 27 or on an attached sheet. THE USE OF PAINT CONTAINING MORE THAN THE PERCENTAGE OF LEAD BY WEIGHT PERMITTED BY LAW IS PROHIBITED.
3. Work not specifically described or shown will not be considered unless required, then the minimum acceptable will be assumed. Work exceeding minimum requirements cannot be considered unless specifically described.
4. Include no alternates, "or equal" phrases, or contradictory items. (Consideration of a request for acceptance of substitute materials or equipment is not thereby precluded.)
5. Include signatures required at the end of this form.
6. The construction shall be completed in compliance with the related drawings and specifications, as amended during processing. The specifications include this Description of Materials and the applicable Minimum Property Standards.

1. EXCAVATION:

Bearing soil, type ________

2. FOUNDATIONS:

Footings: concrete mix ________; strength psi ________ Reinforcing ________
Foundation wall: material ________ Reinforcing ________
Interior foundation wall: material ________ Party foundation wall ________
Columns: material and sizes ________ Piers: material and reinforcing ________
Girders: material and sizes ________ Sills: material ________
Basement entrance areaway ________ Window areaways ________
Waterproofing ________ Footing drains ________
Termite protection ________
Basementless space: ground cover ________; insulation ________; foundation vents ________
Special foundations ________
Additional information: ________

3. CHIMNEYS:

Material ________ Prefabricated *(make and size)* ________
Flue lining: material ________ Heater flue size ________ Fireplace flue size ________
Vents *(material and size)*: gas or oil heater ________; water heater ________
Additional information: ________

4. FIREPLACES:

Type: ☐ solid fuel; ☐ gas-burning; ☐ circulator *(make and size)* ________ Ash dump and clean-out ________
Fireplace: facing ________; lining ________; hearth ________; mantel ________
Additional information: ________

5. EXTERIOR WALLS:

Wood frame: wood grade, and species ________ ☐ Corner bracing. Building paper or felt ________
Sheathing ________; thickness ________; width ________; ☐ solid; ☐ spaced ________" o. c.; ☐ diagonal; ________
Siding ________; grade ________; type ________; size ________; exposure ________"; fastening ________
Shingles ________; grade ________; type ________; size ________; exposure ________"; fastening ________
Stucco ________; thickness ________"; Lath ________; weight ________ lb.
Masonry veneer ________ Sills ________ Lintels ________ Base flashing ________
Masonry: ☐ solid ☐ faced ☐ stuccoed; total wall thickness ________"; facing thickness ________"; facing material ________
Backup material ________; thickness ________"; bonding ________
Door sills ________ Window sills ________ Lintels ________ Base flashing ________
Interior surfaces: dampproofing, ________ coats of ________; furring ________
Additional information: ________
Exterior painting: material ________; number of coats ________
Gable wall construction: ☐ same as main walls; ☐ other construction ________

6. FLOOR FRAMING:

Joists: wood, grade, and species ________; other ________; bridging ________; anchors ________
Concrete slab: ☐ basement floor; ☐ first floor; ☐ ground supported; ☐ self-supporting; mix ________; thickness ________";
reinforcing ________; insulation ________; membrane ________
Fill under slab: material ________; thickness ________". Additional information: ________

7. SUBFLOORING: *(Describe underflooring for special floors under item 21.)*

Material: grade and species ________; size ________; type ________
Laid: ☐ first floor; ☐ second floor; ☐ attic ________ sq. ft.; ☐ diagonal; ☐ right angles. Additional information: ________

8. FINISH FLOORING: *(Wood only. Describe other finish flooring under item 21.)*

Location	Rooms	Grade	Species	Thickness	Width	Bldg. Paper	Finish
First floor							
Second floor							
Attic floor ________ sq. ft.							

Additional information: ________

Previous Edition is Obsolete

1

DESCRIPTION OF MATERIALS
HUD-92005(10-84) HUD HB 4145.1
VA Form 26-1852 Form FmHA 424-2

Figure 14–1 FHA Description of Materials form. *From Huth, Understanding Construction Drawings, Delmar Publishers, Inc.*

DESCRIPTION OF MATERIALS

9. PARTITION FRAMING:
Studs: wood, grade, and species ______ size and spacing ______ Other ______
Additional information: ______

10. CEILING FRAMING:
Joists: wood, grade, and species ______ Other ______ Bridging ______
Additional information: ______

11. ROOF FRAMING:
Rafters: wood, grade, and species ______ Roof trusses (see detail): grade and species ______
Additional information: ______

12. ROOFING:
Sheathing: wood, grade, and species ______; ☐ solid; ☐ spaced ______" o.c.
Roofing ______; grade ______; size ______; type ______
Underlay ______; weight or thickness ______; size ______; fastening ______
Built-up roofing ______; number of plies ______; surfacing material ______
Flashing: material ______; gage or weight ______; ☐ gravel stops; ☐ snow guards
Additional information: ______

13. GUTTERS AND DOWNSPOUTS:
Gutters: material ______; gage or weight ______; size ______; shape ______
Downspouts: material ______; gage or weight ______; size ______; shape ______; number ______
Downspouts connected to: ☐ Storm sewer; ☐ sanitary sewer; ☐ dry-well. ☐ Splash blocks: material and size ______
Additional information: ______

14. LATH AND PLASTER
Lath ☐ walls, ☐ ceilings: material ______; weight or thickness ______ Plaster: coats ______; finish ______
Dry-wall ☐ walls, ☐ ceilings: material ______; thickness ______; finish ______;
Joint treatment ______

15. DECORATING: *(Paint, wallpaper, etc.)*

Rooms	Wall Finish Material and Application	Ceiling Finish Material and Application
Kitchen		
Bath		
Other		

Additional information: ______

16. INTERIOR DOORS AND TRIM:
Doors: type ______; material ______; thickness ______
Door trim: type ______; material ______ Base: type ______; material ______; size ______
Finish: doors ______; trim ______
Other trim *(item, type and location)* ______
Additional information: ______

17. WINDOWS:
Windows: type ______; make ______; material ______; sash thickness ______
Glass: grade ______; ☐ sash weights; ☐ balances, type ______; head flashing ______
Trim: type ______; material ______ Paint ______; number coats ______
Weatherstripping: type ______; material ______ Storm sash, number ______
Screens: ☐ full; ☐ half; type ______; number ______; screen cloth material ______
Basement windows: type ______; material ______; screens, number ______; Storm sash, number ______
Special windows ______
Additional information: ______

18. ENTRANCES AND EXTERIOR DETAIL:
Main entrance door: material ______; width ______; thickness ______". Frame: material ______, thickness ______"
Other entrance doors: material ______; width ______; thickness ______". Frame: material ______; thickness ______"
Head flashing ______ Weatherstripping: type ______; saddles ______
Screen doors: thickness ______"; number ______; screen cloth material ______ Storm doors: thickness ______"; number ______
Combination storm and screen doors: thickness ______"; number ______; screen cloth material ______
Shutters: ☐ hinged; ☐ fixed. Railings ______, Attic louvers ______
Exterior millwork: grade and species ______ Paint ______; number coats ______
Additional information: ______

19. CABINETS AND INTERIOR DETAIL:
Kitchen cabinets, wall units: material ______; lineal feet of shelves ______; shelf width ______
Base units: material ______; counter top ______; edging ______
Back and end splash ______ Finish of cabinets ______; number coats ______
Medicine cabinets: make ______; model ______
Other cabinets and built-in furniture ______
Additional information: ______

20. STAIRS:

Stair	Treads		Risers		Strings		Handrail		Balusters	
	Material	Thickness	Material	Thickness	Material	Size	Material	Size	Material	Size
Basement										
Main										
Attic										

Disappearing: make and model number ______
Additional information: ______

Figure 14–1 Continued

21. SPECIAL FLOORS AND WAINSCOT: *(Describe Carpet as listed in Certified Products Directory)*

Floors	Location	Material, Color, Border, Sizes, Gage, Etc.	Threshold Material	Wall Base Material	Underfloor Material
	Kitchen				
	Bath				

Wainscot	Location	Material, Color, Border, Cap. Sizes, Gage, Etc.	Height	Height Over Tub	Height in Showers (From Floor)
	Bath				

Bathroom accessories: ☐ Recessed; material ____; number ____; ☐ Attached; material ____; number ____
Additional information: ____

22. PLUMBING:

Fixture	Number	Location	Make	Mfr's Fixture Identification No.	Size	Color
Sink						
Lavatory						
Water closet						
Bathtub						
Shower over tub△						
Stall shower△						
Laundry trays						

△☐ Curtain rod △☐ Door ☐ Shower pan: material ____
Water supply: ☐ public; ☐ community system; ☐ individual (private) system.★
Sewage disposal: ☐ public; ☐ community system; ☐ individual (private) system.★
★*Show and describe individual system in complete detail in separate drawings and specifications according to requirements.*
House drain (inside): ☐ cast iron; ☐ tile; ☐ other ____ House sewer (outside): ☐ cast iron; ☐ tile; ☐ other ____
Water piping: ☐ galvanized steel; ☐ copper tubing; ☐ other ____ Sill cocks, number ____
Domestic water heater: type ____; make and model ____; heating capacity ____
____ gph. 100° rise. Storage tank: material ____; capacity ____ gallons.
Gas service: ☐ utility company; ☐ liq. pet. gas; ☐ other ____ Gas piping: ☐ cooking; ☐ house heating.
Footing drains connected to: ☐ storm sewer; ☐ sanitary sewer; ☐ dry well. Sump pump; make and model ____
____; capacity ____; discharges into ____

23. HEATING:

☐ Hot water. ☐ Steam. ☐ Vapor. ☐ One-pipe system. ☐ Two-pipe system.
☐ Radiators. ☐ Convectors. ☐ Baseboard radiation. Make and model ____
Radiant panel: ☐ floor; ☐ wall; ☐ ceiling. Panel coil: material ____
☐ Circulator. ☐ Return pump. Make and model ____; capacity ____ gpm.
Boiler: make and model ____ Output ____ Btuh.; net rating ____ Btuh.
Additional information: ____
Warm air: ☐ Gravity. ☐ Forced. Type of system ____
Duct material: supply ____; return ____ Insulation ____, thickness ____ ☐ Outside air intake.
Furnace: make and model ____ Input ____ Btuh.; output ____ Btuh.
Additional information: ____
☐ Space heater; ☐ floor furnace; ☐ wall heater. Input ____ Btuh.; output ____ Btuh.; number units ____
Make, model ____ Additional information: ____
Controls: make and types ____
Additional information: ____
Fuel: ☐ Coal; ☐ oil; ☐ gas; ☐ liq. pet. gas; ☐ electric; ☐ other ____; storage capacity ____
Additional information: ____
Firing equipment furnished separately: ☐ Gas burner, conversion type. ☐ Stoker: hopper feed ☐; bin feed ☐
Oil burner: ☐ pressure atomizing; ☐ vaporizing ____
Make and model ____ Control ____
Additional information: ____
Electric heating system: type ____ Input ____ watts; @ ____ volts; output ____ Btuh.
Additional information: ____
Ventilating equipment: attic fan, make and model ____; capacity ____ cfm.
kitchen exhaust fan, make and model ____
Other heating, ventilating, or cooling equipment ____

24. ELECTRIC WIRING:

Service: ☐ overhead; ☐ underground. Panel: ☐ fuse box; ☐ circuit-breaker; make ____ AMP's ____ No. circuits ____
Wiring: ☐ conduit; ☐ armored cable; ☐ nonmetallic cable; ☐ knob and tube; ☐ other ____
Special outlets: ☐ range; ☐ water heater; ☐ other ____
☐ Doorbell. ☐ Chimes. Push-button locations ____ Additional information: ____

25. LIGHTING FIXTURES:

Total number of fixtures ____ Total allowance for fixtures, typical installation, $ ____
Nontypical installation ____
Additional information: ____

3 DESCRIPTION OF MATERIALS

Figure 14–1 Continued

DESCRIPTION OF MATERIALS

26. INSULATION:

Location	Thickness	Material, Type, and Method of Installation	Vapor Barrier
Roof			
Ceiling			
Wall			
Floor			

27. MISCELLANEOUS: *(Describe any main dwelling materials, equipment, or construction items not shown elsewhere; or use to provide additional information where the space provided was inadequate. Always reference by item number to correspond to numbering used on this form.)*

HARDWARE: *(make, material, and finish.)*

SPECIAL EQUIPMENT: *(State material or make, model and quantity. Include only equipment and appliances which are acceptable by local law, custom and applicable FHA standards. Do not include items which, by established custom, are supplied by occupant and removed when he vacates premises or chattles prohibited by law from becoming realty.)*

PORCHES:

TERRACES:

GARAGES:

WALKS AND DRIVEWAYS:

Driveway: width ______ ; base material ______ ; thickness ______"; surfacing material ______ ; thickness ______"

Front walk: width ______ ; material ______ ; thickness ______". Service walk: width ______ ; material ______ ; thickness ______"

Steps: material ______ ; treads ______"; risers ______". Cheek walls ______

OTHER ONSITE IMPROVEMENTS:

(Specify all exterior onsite improvements not described elsewhere, including items such as unusual grading, drainage structures, retaining walls, fence, railings, and accessory structures.)

LANDSCAPING, PLANTING, AND FINISH GRADING:

Topsoil ______" thick: ☐ front yard; ☐ side yards; ☐ rear yard to ______ feet behind main building.

Lawns *(seeded, sodded, or sprigged)*: ☐ front yard ______ ; ☐ side yards ______ ; ☐ rear yard ______

Planting: ☐ as specified and shown on drawings; ☐ as follows:

______ Shade trees, deciduous. ______" caliper.

______ Low flowering trees, deciduous, ______' to ______'

______ High-growing shrubs, deciduous, ______' to ______'

______ Medium-growing shrubs, deciduous, ______' to ______'

______ Low-growing shrubs, deciduous, ______' to ______'

______ Evergreen trees. ______' to ______', B & B.

______ Evergreen shrubs. ______' to ______', B & B.

______ Vines, 2-year ______

Identification.—This exhibit shall be identified by the signature of the builder, or sponsor, and/or the proposed mortgagor if the latter is known at the time of application.

Date ______ Signature ______

Signature ______

Figure 14–1 Continued

Room Dimensions

- Minimum room size is to be 70 sq ft.
- Ceiling height minimum is to be 7'–6" in 50 percent of area except 7'–0" may be used for bathrooms and hallways.

Light and Ventilation

- Minimum window area is to be 1/10 floor area with not less than 10 sq ft for habitable rooms and 3 sq ft for bathrooms and laundry rooms. Not less than one-half of this required window area is to be openable. Every sleeping room is required to have a window or door for emergency exit. Windows with an openable area of not less than 5 sq ft with no dimension less than 22 in. meet this requirement, and the sill height is to be not more than 44 in. above the floor.
- Glass subject to human impact is to be tempered glass.
- Glass doors in shower and tub enclosures are to be tempered glass or fracture-resistant plastic.
- Attic ventilation is to be a minimum of 1/300 of the attic area, one-half in the soffit and one-half in the upper area.
- Bathroom and kitchen fans and dryer are to vent directly outside.

Foundation

- Concrete mix is to have a minimum ultimate compressive strength of 2000 psi at 28 days and shall be composed of 1 part cement, 3 parts sand, 4 parts of 1 in. maximum size rock, and not more than 7½ gallons of water per sack of cement.
- Foundation mud sills, plates, and sleepers are to be pressure treated or of foundation-grade redwood. All footing sills shall have full bearing on the footing wall or slab and shall be bolted to the foundation with ½" × 10" bolts embedded at least 7 in. into the concrete or reinforced masonry, or 15 in. into unreinforced grouted masonry. Bolts shall be spaced not to exceed 6 ft on center with bolts not over 12 in. from cut end of sills.
- Crawl space shall be ventilated by an approved mechanical means or by openings with a net area not less than 1½ sq ft for each 25 linear ft of exterior wall. Openings shall be covered with not less than ¼ in. or more than ½ in. of corrosion-resistant wire mesh. If the crawl space is to be heated, closeable covers for vent openings shall be provided. Water drainage and 6-mil black ground cover shall be provided in the crawl space.
- Access to crawl space is to be a minimum of 18" × 24".
- Basement foundation walls with a height of 8 ft or less supporting a well-drained porous fill of 7 ft or less, with soil pressure not more than 30 lb per sq ft equivalent fluid pressure, and with the bottom of the wall supported from inward movement by structural floor systems may be of plain concrete with an 8-in. minimum thickness and minimum ultimate compressive strength of 2500 psi at 28 days. Basement walls supporting backfill and not meeting these criteria shall be designed in accordance with accepted engineering practices.
- Concrete forms for footings shall conform to the shape, lines, and dimensions of the members as called for on the plans and shall be substantial and sufficiently tight to prevent leakage of mortar and slumping out of concrete in the ground contact area.

Framing

- Lumber. All joists, rafters, beams, and posts 2–4 in. thick shall be No. 2 Grade Douglas fir-larch or better. All posts and beams 5 in. and thicker shall be No. 1 Grade Douglas fir-larch or better.
- Beams (untreated) bearing in concrete or masonry wall pockets shall have air space on sides and ends. Beams are to have not less than 4 in. of bearing on masonry or concrete.
- Wall Bracing. Every exterior wood stud wall and main cross partition shall be braced at each end and at least every 25 ft of length with 1 × 4 diagonal let-in braces or equivalent.
- Joists are to have not less than 1½ in. of bearing on wood or metal nor less than 3 in. on masonry.
- Joists under bearing partitions are to be doubled.
- Floor joists are to have solid blocking at each support and at the ends except when the end is nailed to a rim joist or adjoining studs. Joists 2 × 14 or larger are to have bridging at maximum intervals of 8 ft.
- Two in. clearance is required between combustible material and the walls of an interior fireplace or chimney. One in. clearance is required when the chimney is on an outside wall. (½-in. moisture-resistant gypsum board may be used in lieu of the 1-in. clearance requirement.)
- Rafter purlin braces are to be not less than 45° to the horizontal.
- Rafters, when not parallel to ceiling joists, are to have ties that are 1 × 4 minimum spaced not more than 4 ft on center.
- Provide a double top plate with a minimum 48 in. lap splice.
- Metal truss tie-downs are to be required for manufactured trusses.

- Plant manufactured trusses (if used) shall be of an approved design with an engineered drawing.
- Fire blocking shall be provided for walls over 10'–0" in height, also for horizontal shafts 10'–0" on center, and for any concealed draft opening.
- Garage walls and ceiling adjacent to or under dwelling require one-hour fire-resistant construction on the garage side. A self-closing door between the garage and dwelling is to be a minimum 1⅜ in. solid core construction.
- Ceramic tile, or other approved material, is to be used in a water-splash area.
- Building paper, or other approved material, is to be used under siding.
- Framing in the water-splash area is to be protected by waterproof paper, waterproof gypsum, or other approved substitute.
- Post-and-beam connections. A positive connection shall be provided between beam, post, and footing to ensure against uplift and lateral displacement. Untreated posts shall be separated from concrete or masonry by a rust-resistant metal plate or impervious membrane and be at least 6 in. from any earth.

Stairways

- Maximum rise is to be 8 in., minimum run 9 in., minimum head room to be 6'–6", and minimum width to be 30 in.
- Winding and curved stairways are to have a minimum inside tread width of 6 in.
- Enclosed usable space under stairway is to be protected by one-hour fire-resistant construction (⅝-in. type X gypsum board)
- Handrails are to be from 30 to 34 in. above tread nosing and intermediate rails are to be such that no object of 5 in. diameter can pass through.
- Generally, for commercial or public structures, all unenclosed floor and roof openings, balconies, decks, and porches more than 30 in. above grade shall be protected by a guardrail not less than 42 in. high with intermediate rails or dividers such that no object of 9 in. diameter can pass through. Generally, guardrails for residential occupancies may be not less than 36 in. high. Specific applications are subject to local or national building codes.

Roof

- Composition shingles on roof slopes between 4–12 and 7–12 shall have an underlayment of not less than 15 lb felt. For slopes from 2–12 to less than 4–12, Building Department approval of roofing manufacturers' low-slope instructions is required.
- Shake roofs require solid roof sheathing (in lieu of solid sheathing, spaced sheathing may be used but shall not be less than 1 × 4 with not more than 3 in. clearance between) with an underlayment of not less than 15 lb felt with an interlace of not less than 30 lb felt. For slopes less than 4–12, special approval is required.
- Attic scuttle is to have a minimum of 22 × 30 in. of headroom above.

Chimney and Fireplace

- Reinforcing. Masonry constructed chimneys extending more than 7 ft above the last anchorage point (example: roof line) must have not less than four number 4 steel reinforcing bars placed vertically for the full height of the chimney with horizontal ties not less than ¼ in. diameter spaced at not over 18 in. intervals. If the width of the chimney exceeds 40 in., two additional number 4 vertical bars shall be provided for each additional flue or for each additional 40 in. in width or fraction thereof.
- Anchorage. All masonry chimneys over 18 ft high shall be anchored at each floor and/or ceiling line more than 6 ft above grade, except when constructed completely within the exterior walls of the building.

Thermal Insulation and Heating

- Thermal designs employing the R factor must meet minimum R factors as follows:
  - **a.** Ceiling or roof: R-38, vaulted: R-30.
  - **b.** Walls: R-21, vapor barrier required. Minimum one permeability rating.
  - **c.** Floors over unheated crawl space or basements: R-25 including reflective foil.
  - **d.** Foundation walls: R-11 to ½ ft below exterior finished grade line.
  - **e.** Slab-on-Grade: R-11 around perimeter a minimum of 18 in. horizontally or vertically.
- Thermal glazing. Heated portions of buildings located in the 5,000 or less degree-day zone do not require thermal glazing on that portion of the glazing that is less than 20 percent of the total area of exterior walls including doors and windows. Heated portions of buildings located in zones over 5,000 degree days shall be provided with special thermal glazing in all exterior wall areas.
- Duct insulation. Supply and return air ducts used for heating and/or cooling located in unheated attics, garages, crawl spaces, or other unheated spaces other than between floors or interior walls shall be insulated with an R-3.5 minimum.

- Heating. Every dwelling unit and guest room shall be provided with heating facilities capable of maintaining a room temperature of 70°F at a point 3 ft above the floor.

Fire Warning System

- Every dwelling shall be provided with approved detectors of products of combustion mounted on the ceiling or a wall within 12 in. of the ceiling at a point centrally located in the corridor or area giving access to and not over 12 ft from rooms used for sleeping. Where sleeping rooms are on an upper level, the detector shall be placed at the high area of the ceiling near the top of the stairway.

CHAPTER 14 TEST

Fill in the blanks below with the proper word or short statement as needed to correctly complete the sentence or answer the question.

1. Information that cannot be clearly or completely provided on the drawings or in schedules are provided in construction ______
2. What do the following acronyms mean?
   FHA ______
   FHLMC ______
   FmHA ______
   VA ______
3. List the three items that become the legal documents for the construction project. ______
   ______.
4. When a specific product or equivalent is listed in the construction specifications, what does the "or equivalent" mean? ______
   ______
   ______.
5. List at least three items that may affect the contents of the minimum construction specifications as established by local building officials from one location to another. ______
   ______
6. Give the recommended minimum requirements for the following general classifications of construction specifications:
   - Minimum room size ______
   - Minimum ceiling height in 50% of the area ______
   - Bathrooms and hallways may have a ceiling height of ______
   - Minimum window area is to be ______ the floor area with not less than ______ for habitable rooms and ______ for bathrooms and laundry rooms.
   - Every sleeping area is required to have a window or door for ______.
   - Windows with an openable area of not less than ______ with no dimension less than ______ meet this requirement, and the sill height is to be not more than ______ above the floor.
   - Glass subject to human impact is to be ______ glass.
   - Foundation mud sills, and sleepers are to be ______ or of ______ redwood.
   - Crawl space shall be ventilated by approved mechanical means or by openings with a net of not less than ______ for each ______ of exterior wall.

- Joists under bearing partitions are to be ______________________________.
- The clearance required between combustible material and the walls of an interior fireplace or chimney is ______________________________.
- Provide a double top plate with a minimum ________________ lap splice.
- Fire blocking shall be provided for walls over ________________ in height.
- ________________, or other approved material, is to be used under siding.
- Enclosed usable space under stairway is to be protected by ____________ fire resistant construction.
- Handrails for residential construction are to be from ____________ above thread nosing and intermediate rails are to be such that no object __________ in diameter can pass through.
- Thermal designs employing the R factor must meet minimum R factors as follows: Ceiling or roof ____________; Walls ____________, vapor barrier required with one minimum permeability rating; floors over unheated crawl space or basements ____________ including reflective foil; Foundation walls ____________ to 6 in. below exterior finished grade; slab-on-grade ____________ around perimeter a minimum of ____________ horizontally or vertically.
- Every dwelling unit and guest room shall be provided with heating facilities capable of maintaining a room temperature of ____________ at a point 3 ft. above the floor.
- Every dwelling shall be provided with approved detectors of products of combustion mounted on the ceiling or a wall within ____________ of the ceiling at a point centrally located in the corridor or area giving access to and not over ____________ from rooms used for sleeping. Where sleeping rooms are on an upper level, the detector shall be placed at the high area of the ceiling near the top of the ______________________________.

CHAPTER 14 EXERCISES

PROBLEM 14–1. Given the set of plans shown on pages 262–267, complete the FHA Description of Materials on pages 254–257. Begin by taking information from the prints to start completing the forms. When you have taken as much information as possible from the prints, complete the rest of the information required by using specific materials found in your home, in manufacturer's catalogues available with your instructor (if any) or by finding manufacturer's catalogues from local suppliers. You might consider making copies of the Description of Materials forms to use as a rough draft. The final Description of Materials should be completed by typing or neatly hand lettered.

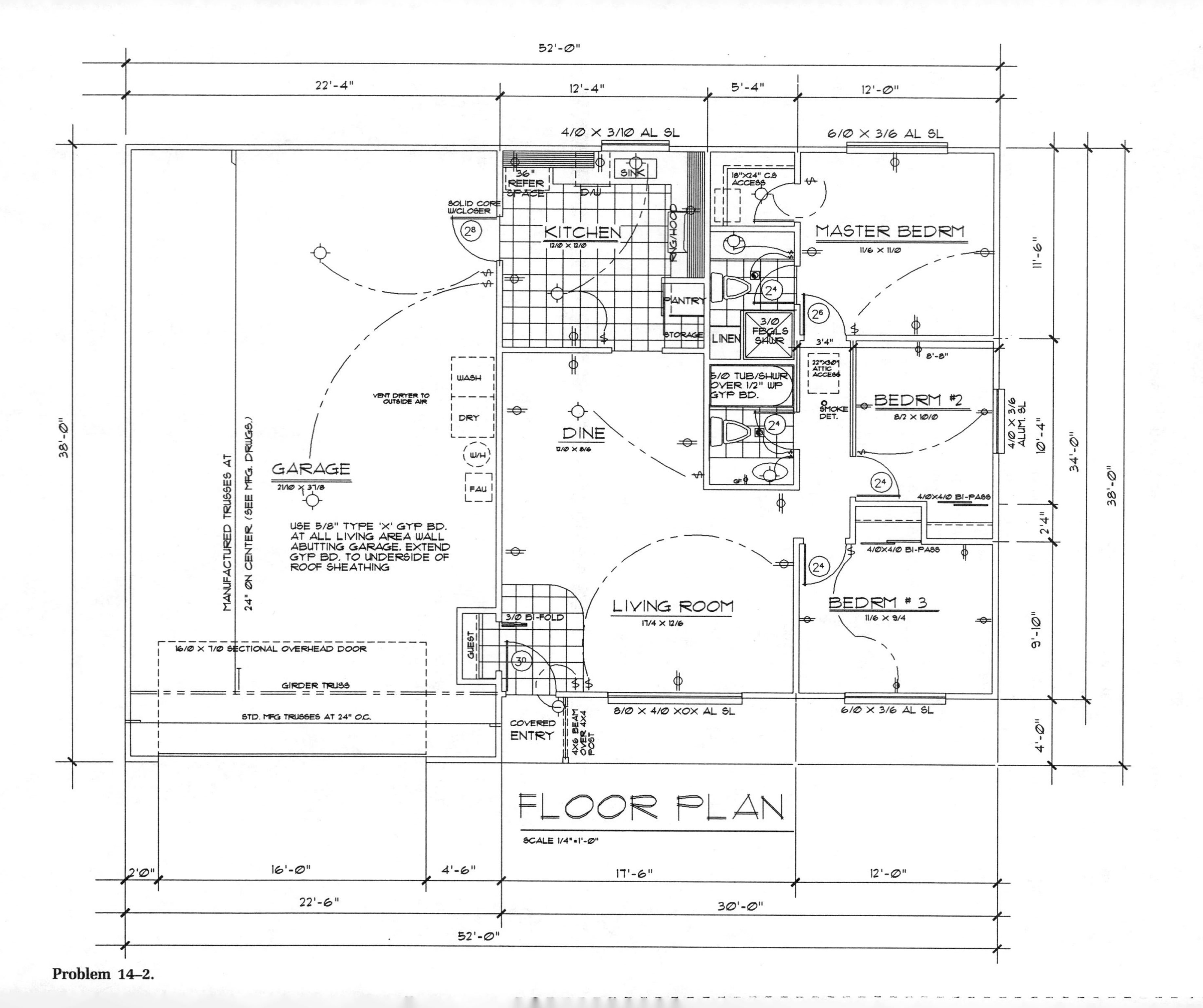
52'-0"
22'-4"
12'-4"
5'-4"
12'-0"
4/0 X 3/10 AL SL
6/0 X 3/6 AL SL
36" REFER SPACE
SINK
D/W
RNG/HOOD
18"X24" C.S. ACCESS
SOLID CORE W/CLOSER
KITCHEN
12/0 X 12/0
MASTER BEDRM
11/6 X 11/0
11'-6"
PANTRY
STORAGE
LINEN
3/0 FBGLS SHWR
3'4"
22"X30" ATTIC ACCESS
8'-8"
5/0 TUB/SHWR OVER 1/2" WP GYP BD.
SMOKE DET.
BEDRM #2
8/2 X 10/0
4/0 X 3/6 ALUM. SL
10'-4"
34'-0"
38'-0"
WASH
DRY
VENT DRYER TO OUTSIDE AIR
DINE
12/0 X 8/6
W/H
FAU
GARAGE
21/10 X 37/8
4/0X4/0 BI-PASS
2'4"
USE 5/8" TYPE 'X' GYP BD. AT ALL LIVING AREA WALL ABUTTING GARAGE. EXTEND GYP BD. TO UNDERSIDE OF ROOF SHEATHING
MANUFACTURED TRUSSES AT 24" ON CENTER (SEE MFG. DRWGS.)
LIVING ROOM
17/4 X 12/6
BEDRM # 3
11/6 X 9/4
9'-10"
3/0 BI-FOLD
GUEST
16/0 X 7/0 SECTIONAL OVERHEAD DOOR
GIRDER TRUSS
STD. MFG TRUSSES AT 24" O.C.
8/0 X 4/0 XOX AL SL
6/0 X 3/6 AL SL
COVERED ENTRY
4X6 BEAM OVER 4X4 POST
4'-0"
FLOOR PLAN
SCALE 1/4"=1'-0"
2'0"
16'-0"
4'-6"
17'-6"
12'-0"
22'-6"
30'-0"
52'-0"

Problem 14-2.

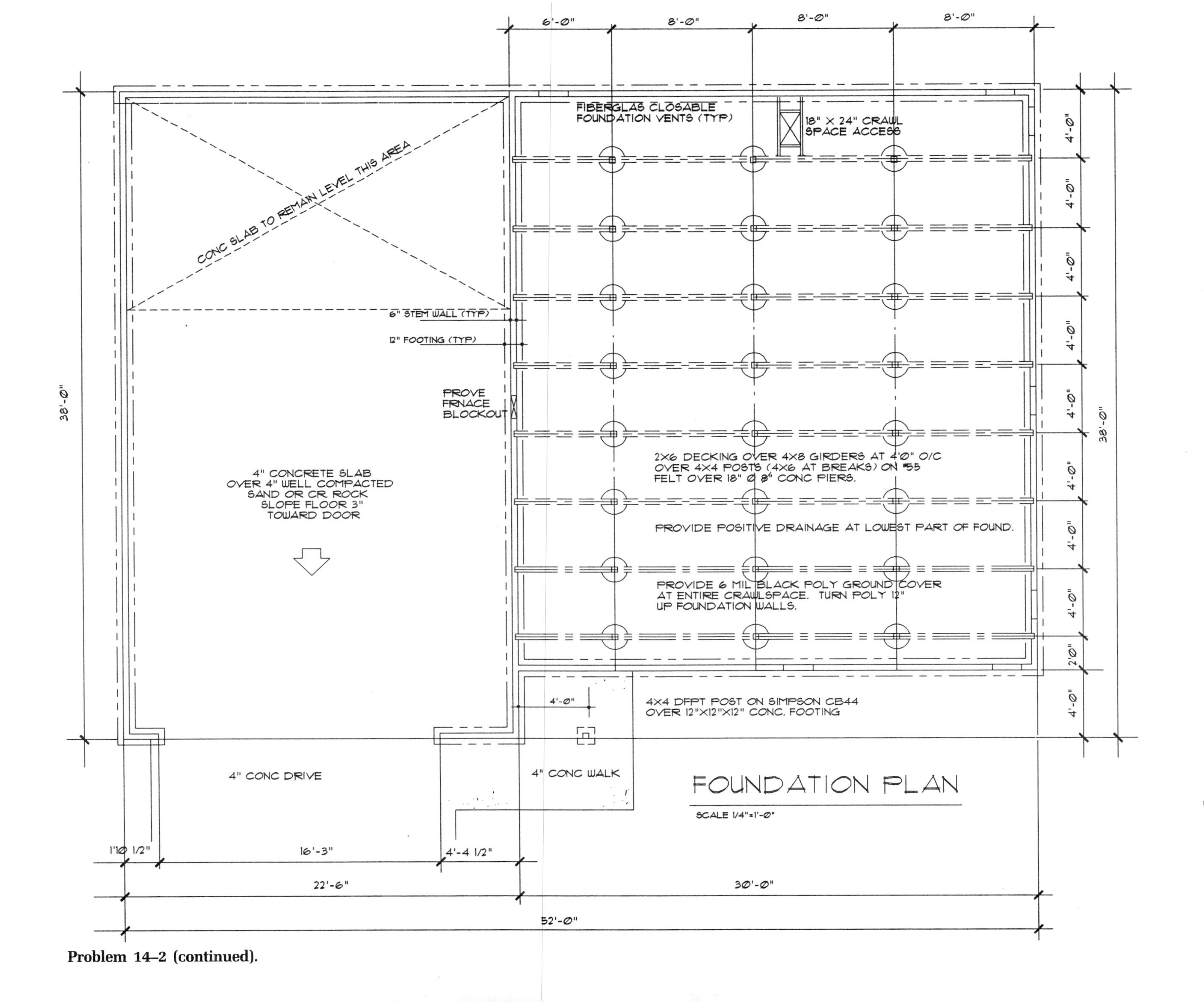

Problem 14–2 (continued).

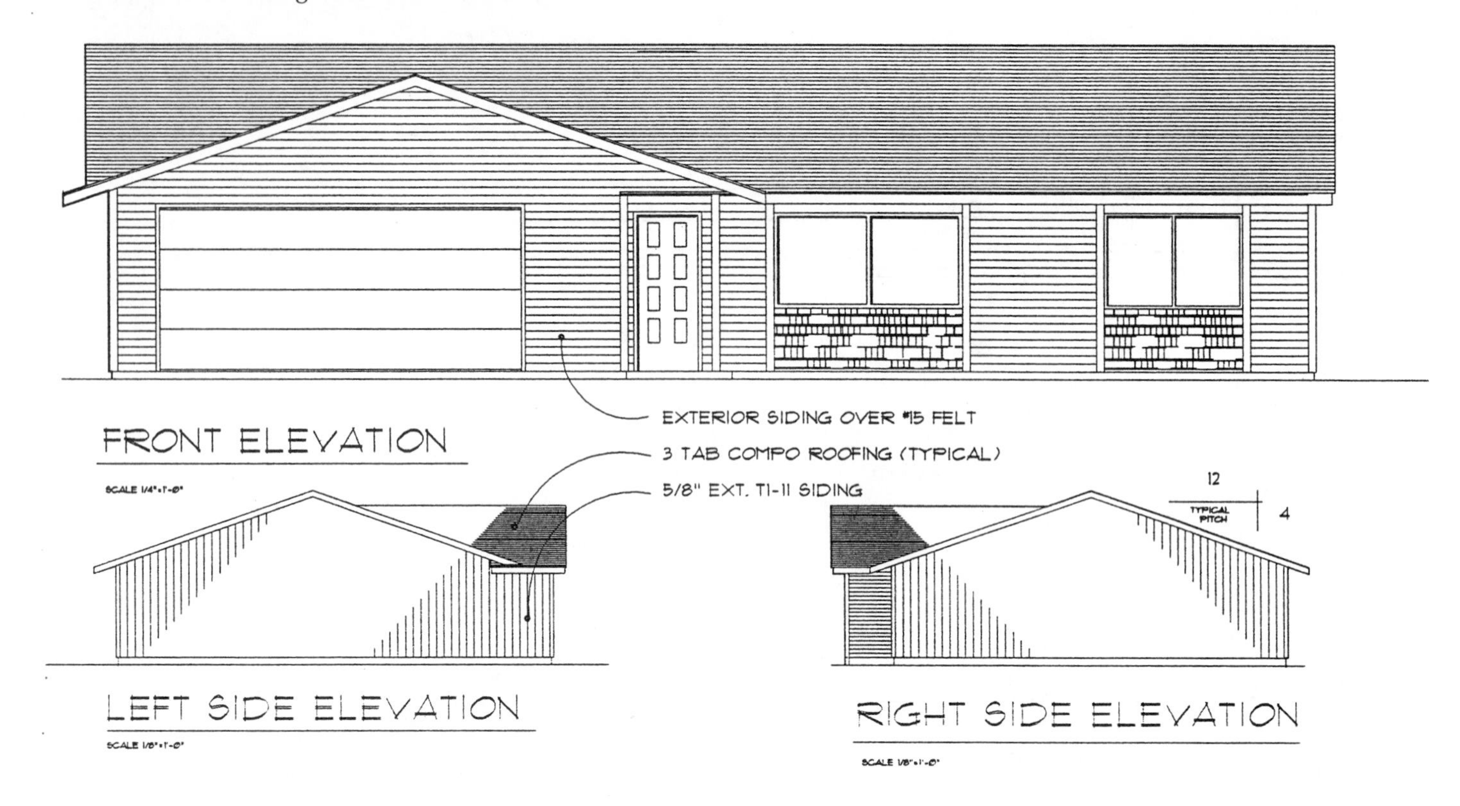

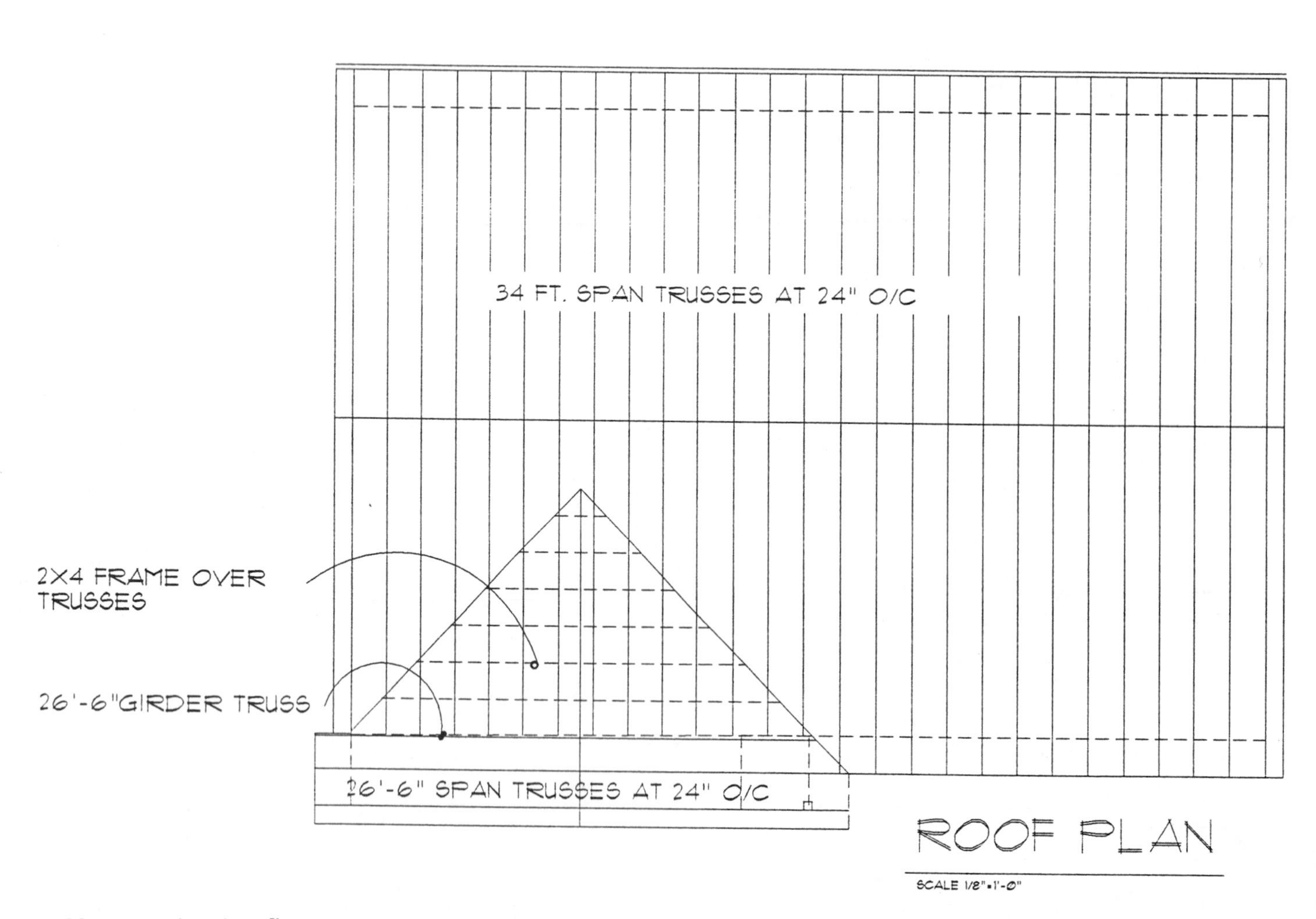

Problem 14–2 (continued).

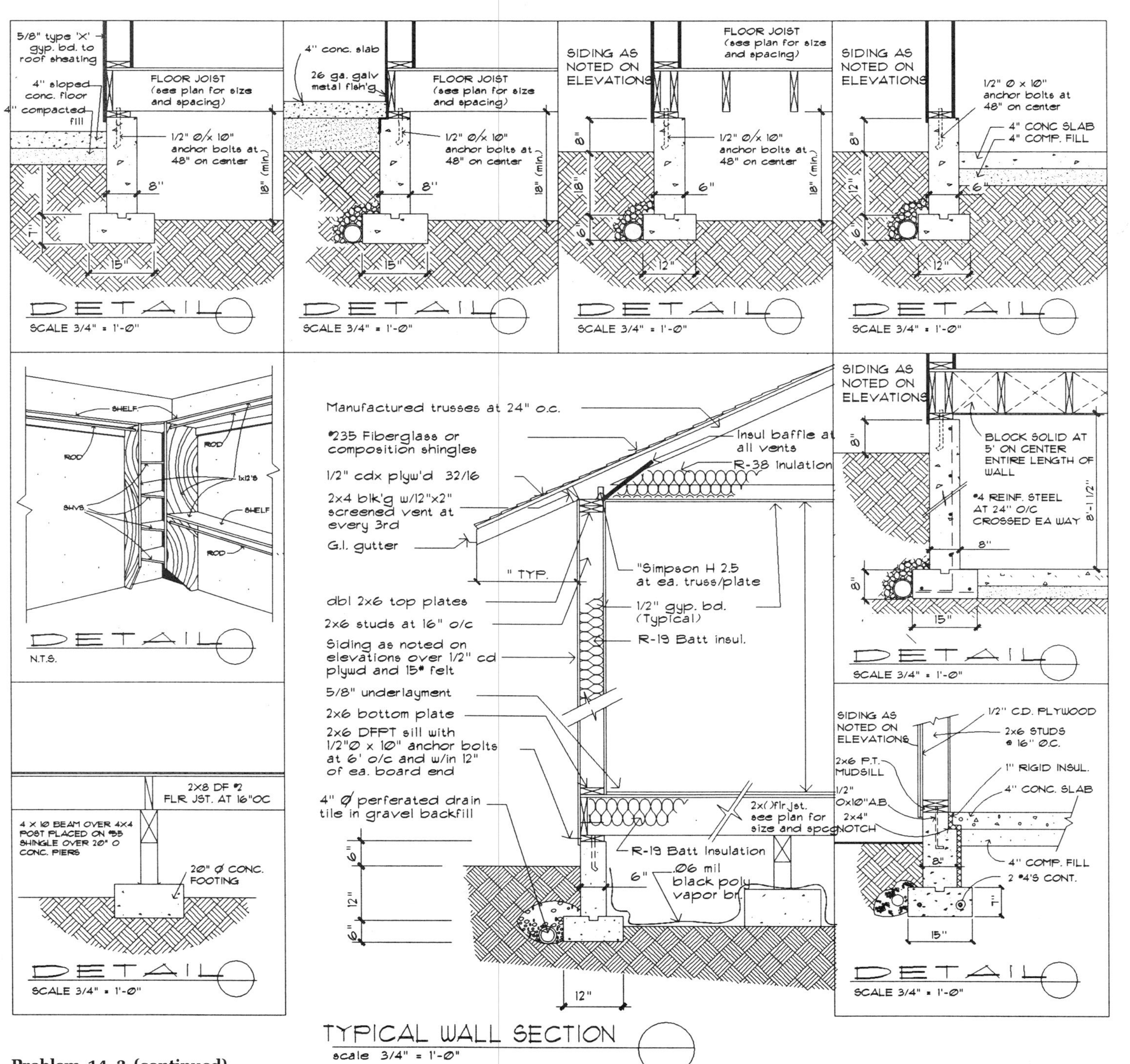

Problem 14–2 (continued).

GENERAL NOTES ..

FOUNDATION

1. FOOTINGS SHALL BEAR ON FIRM UNDISTURBED SOIL WITH MIN DEPTH BELOW FINAL GRADE OF 1'-6" FOR ONE AND TWO STORY AND 2'-0" FOR THREE STORY, UNLESS SHOWN OTHERWISE.
2. FOUNDATION SIZES ARE BASED ON A MINIMUM TOTAL BEARING CAPACITY OF 1500 PSF (ASSUMED)
3. DO NOT EXCAVATE CLOSER THAN 1 1/2 TO 2 SLOPE BELOW FOOTINGS.

CONCRETE

1. ALL CONCRETE USED AT EXTERIOR STEPS, PORCHES, OR CARPORTS SHALL BE 3500 PSI WITH 5-7% AIR ENTRAINMENT. ALL VERTICAL CONCRETE EXPOSED TO WEATHER SHALL BE 3500 PSI WITH 5-7% OF VOLUME AIR ENTRAINED. ALL OTHER CONCRETE EXPOSED TO FREEZE/THAW WEATHER CONDITIONS SHALL BE 2500 PSI WITH 5-7% AIR ENTRAINMENT.
2. CONCRETE METHODS, FORMS, AND SHORES SHALL BE IN ACCORDANCE WITH LATEST A.C.I. STANDARDS.

REINFORCING STEEL

1. REINFORCING BARS (WHEN CALLED OUT) SHALL BE DEFORMED BARS CONFORMING TO A.S.T.M. SPECS.
2. ALL WELDED WIRE FABRIC SHALL CONFORM TO CURRENT A.S.T.M. SPECS.

WOOD

1. ALL SAWN LUMBER SHALL BE DOUGLAS FIR LARCH INSTALLED AS REQUIRED ON NAILING SCHEDULE, ON PLANS, DETAILS AND ON SPECIFICATIONS.
2. GRADING SHALL BE IN ACCORDANCE WITH CURRENT WWPA STANDARD GRADING RULES.

POSTS AND BEAMS GRADE NO. 1

JOISTS AND RAFTERS GRADE NO 2

SILLS, PLATES & BLKG GRADE NO 3

STUDS DOUG FIR STUD GRADE

FLOORS OVER JOIST 5/8" CD DF PLYWD

2X6 T&G SUB-FLOOR GRADE NO. 3

ROOF AND WALL SHTG 1/2" CD PLYWD 32/16

FLOOR UNDERLAYMENT 1/2" PART. BD.

GLUE-LAM BEAMS (fb2200) PER A.I.T.C. INDUSTRIAL GRADE WITH DRY-USE ADHESIVE USE ARCH GRADE FOR EXPOSED BEAMS.

NOTE: SOLID INTERIOR BEAMS VISUALLY EXPOSED SHALL BE "CLEAR" GRADE, FREE OF HEART CENTER. ALL INTERIOR AND EXTERIOR BEARING WALL OPENINGS SHALL HAVE 4 X 12 DF# 1 HEADERS UNLESS NOTED OTHERWISE.

3. PLYWOOD- ALL STRUCTURAL SHEATHING SHALL BE DF PLYWD 32/16 AS INDICATED ON PLANS.

INSULATION

1. ROOFS/ATTICS R-19 WITH VAPOR BARRIER ON WINTER WARM SIDE AT VAULTS
R-38 WITH VAPOR BARRIER ON WINTER WARM SIDE AT ATTIC SPACES
2. WALLS R-19 (AT EXTERIOR & GARAGE COMMON WALL) WITH VAPOR BARRIER
3. FLOOR OVER UNHEATED SPACE R-19 W/VAPOR BARRIER ON WINTER WARM SIDE
4. BASEMENT WALLS (IF ANY) R-11 TO 12" BELOW EXTERIOR GRADE
5. CONCRETE SLAB ON GRADE R-11 TO 12" BELOW EXTERIOR GRADE
6. FURNACE DUCTS IN UNHEATED CRAWL SPACES R-35
7. CRAWLSPACE INSULATION TO HAVE A FLAME SPREAD RATING OF .25 (MAX.) AND A SMOKE DENSITY RATING OF 450 (MAX.) PER UBC 1713(C)

Problem 14–2 (continued).

NAILING SCHEDULE

CONNECTION — NAILING

1. JOIST TO SILL OR GIRDER, TOE NAIL 3-8d
2. BRIDGING TO JOIST, TOE NAIL EACH END 2-8d
3. 2" SUBFLOOR TO JOIST OR GIRDER, BLIND & FACE WALL . 2-16d
4. TOP PLATE TO STUD, END NAIL 2-16d
5. SOLE PLATE TO JOIST OR BLOCKING, FACE NAIL . . . . . . .16d AT 16"OC
6. STUD TO SOLE PLATE . . . . 4-8d TOE NAIL OR 2-16d END NAIL
7. DOUBLE STUDS, FACE WALL . . . . 16d AT 16"OC
8. DOUBLE TOP PLATES, FACE NAIL . . . . 16d AT 16"OC
9. TOP PLATES, LAP & INTERSECTIONS, FACE NAIL 2-16d
10. CONTINUOUS HEADER, 2 PIECES . . . . 16d AT 16"OC ALONG EACH END
11. CEILING JOIST TO PLATE, TOE NAIL 3-8d
12. CEILING JOIST LAPS OVER PARTITIONS, FACE NAIL 3-16d
13. CEILING JOIST TO PARALLEL RAFTERS 3-16d
14. RAFTER TO PLATE, TOE NAIL 3-8d
15. BUILT-UP CORNER STUDS . . . . 16d AT 24"OC
16. PLYWOOD SUBFLOOR . . . . 8d COMMON AT 6"OC AT EDGE & 10"OC AT INT.
17. PLYWOOD WALL SHEATHING . . . . 8d COMMON AT 6"OC AT EDGE & 12"OC AT INT.
18. PLYWOOD ROOF SHEATHING . . . . 8d COMMON AT 6" OC AT EDGE & 12"OC AT INT.

MISCELLANEOUS

1. EACH SLEEPING ROOM TO HAVE AT LEAST ONE EGRESSABLE WINDOW WITH A MIN NET OPENING OF 20" X 24", 5.7 SQUARE FT. AND A MAX NET SILL HEIGHT OF 44", MEASURED FROM FINISHED FLOOR.
2. PROVIDE OUTSIDE COMBUSTION AIR TO ALL FIREPLACES AND STOVES.
3. ALL TUB AND SHOWERS TO HAVE 1/2" WATERPROOF GYP. BD. AT WALLS AND A HARD MOISURE RESISTANT SURFACE TO A HEIGHT OF 6'0"
4. EXHAUST ALL FANS, RANGES, AND CLOTHES DRYERS TO OUTSIDE AIR
5. ALL WINDOWS AND EXTERIOR GLAZED DOORS SHALL BE GLAZED WITH DUAL PANED INSULATED GLASS AND TEMPERED WHERE SUBJECT TO HUMAN IMPACT.
6. GENERAL CONTRACTOR SHALL BE RESPONSIBLE FOR VERIFYING ALL DIMENSIONS AND CONDITIONS WITH PLANS AND REPORTING ANY DISCREPANCIES TO SUNRIDGE DESIGN BEFORE STARTING WORK
7. THIS STRUCTURE SHALL BE ADEQUATELY BRACED FOR WIND LOADS UNTIL THE ROOF AND WALLS HAVE BEEN PERMANENTLY ATTATCHED TOGETHER AND SHEATHED
8. ALL LUMBER IN PERMANENT CONTACT WITH CONCRETE SHALL BE PRESSURE TREATED WITH A WATER-BORNE PRESERVATIVE.
9. ALL SMOKE DETECTORS CALLED OUT SHALL BE CONNECTED TO HOUSE POWER.(HARD-WIRED)
10. ALL FEDERAL, STATE, AND LOCAL CODES, ORDINANCES AND REGULATIONS SHALL BE CONSIDERED AS PART OF SPECIFICATIONS FOR THIS BUILDING AND SHALL TAKE PREFERENCE OVER ANYTHING SHOWN, DESCRIBED OR IMPLIED WHERE VARIANCES OCCUR.
11. ANY MODIFICATIONS TO THE DESIGN OR COMPONENTS OF THIS BUILDING MAY AFFECT IT'S STRUCTURAL INTEGRITY AND SHOULD BE REVIEWED BY AN ENGINEER PRIOR TO CONSTRUCTION.

DESIGN LOADS

ROOF . . . . 30 PSF LIVE LOAD
FLOORS . . . . 40 PSF LIVE LOAD
STAIRS . . . . 100 PSF LIVE LOAD
GARAGE FLOORS . . . . 50 PSF LIVE LOAD
DECKS . . . . 60 PSF LIVE LOAD
WIND LOADS . . . . 90 PSF EXPOSURE "C"
SEISMIC . . . . ZONE II

Problem 14–2 (continued).

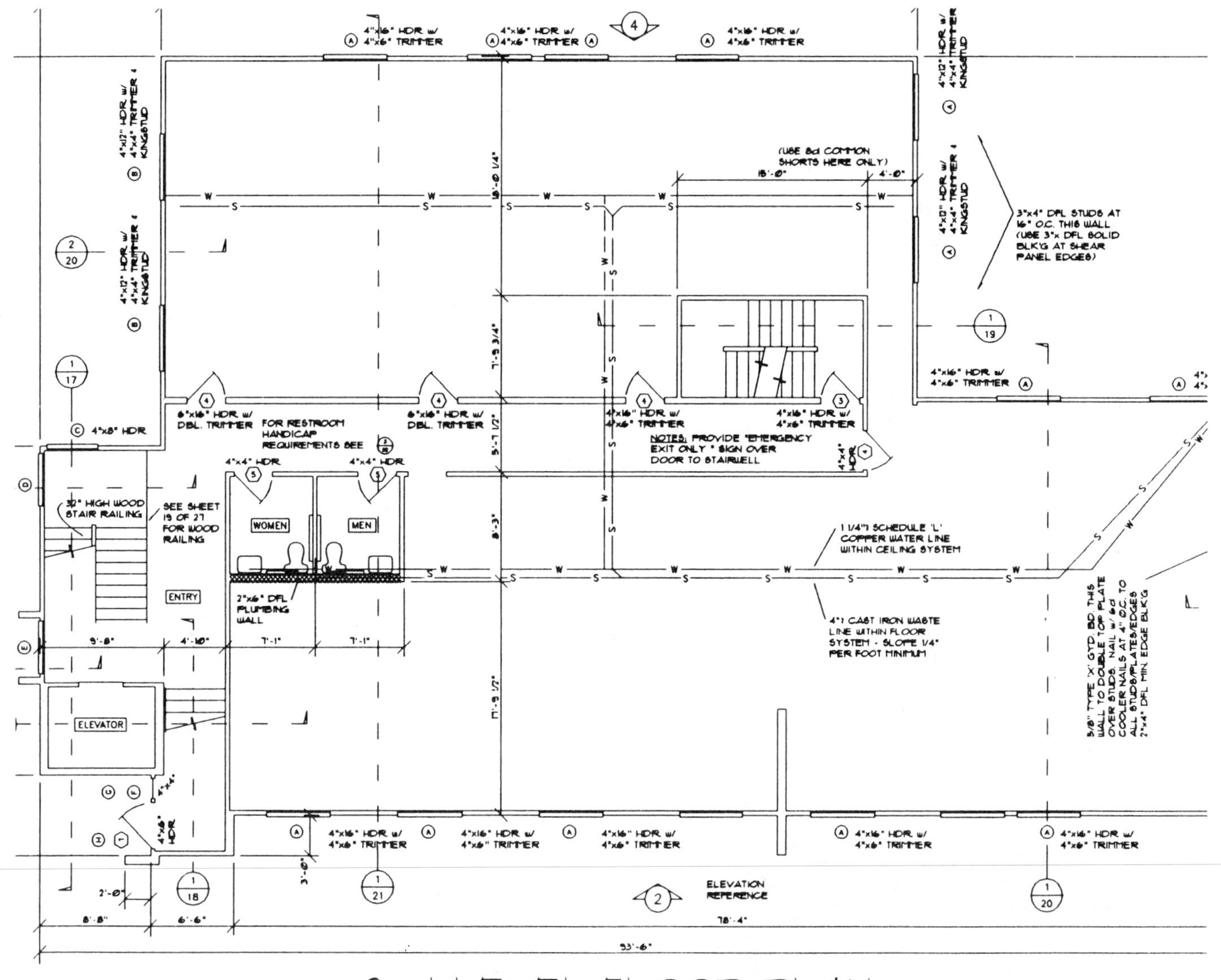

Courtesy of Ken Smith, Structureform Masters.

Part 2

Reading Commercial Prints

Chapter 15

Introduction to Commercial Construction Projects

TYPES OF DRAWINGS

SO FAR in this text, you have only been exposed to residential construction techniques and plans. Residential construction actually makes up a small percentage of the total building industry. Most jobs in the industry are found in the field of commercial construction. This would include projects ranging from multifamily units, office complexes, concrete tilt-up projects, and retail sales facilities. Each of these areas have a wide range of size and materials that could be encountered.

Commercial drawing projects have many similarities to the residential drawings to which you have already been exposed. The five basic residential drawings are also a part of a commercial set of drawings. The plot, floor, and foundation plans, elevations, and sections form the basis of commercial drawings. In addition to these drawings, there are several other drawings that are related that must be understood to compete successfully in the construction industry.

Floor Plans

The floor plan of a commercial project is used to show the locations of materials in the same way that a residential floor plan does. The difference in the floor plan comes in the type of material being specified. Figure 15–1 shows an example of a floor plan for an apartment project. This plan is very similar to a residential plan. Information common to all units of this type is placed on the typical floor plan. To show how units fit together within the project, a building floor plan is also drawn as seen in Figure 15–2. This plan is usually drawn at a scale of 1/8" = 1'–0" or smaller.

Another common type of floor plan would be the plan for an office complex. Figure 15–3 shows the floor plan for this type of development. Using methods similar to those of an apartment project, the overall floor plan is drawn at a scale of 1/8" = 1'–0" or smaller. Areas of the floor plan which are complex are then drawn at 1/4" = 1'–0" as seen in Figure 15–4. Figure 15–4 shows the floor plan for a 7-ELEVEN convenience store. Many of the drawings of this chapter are based on this plan.

Several drawings in a commercial set of plans use the floor plan as a base to show related information. Residential drawings will usually include the information contained on these drawings on the floor plan.

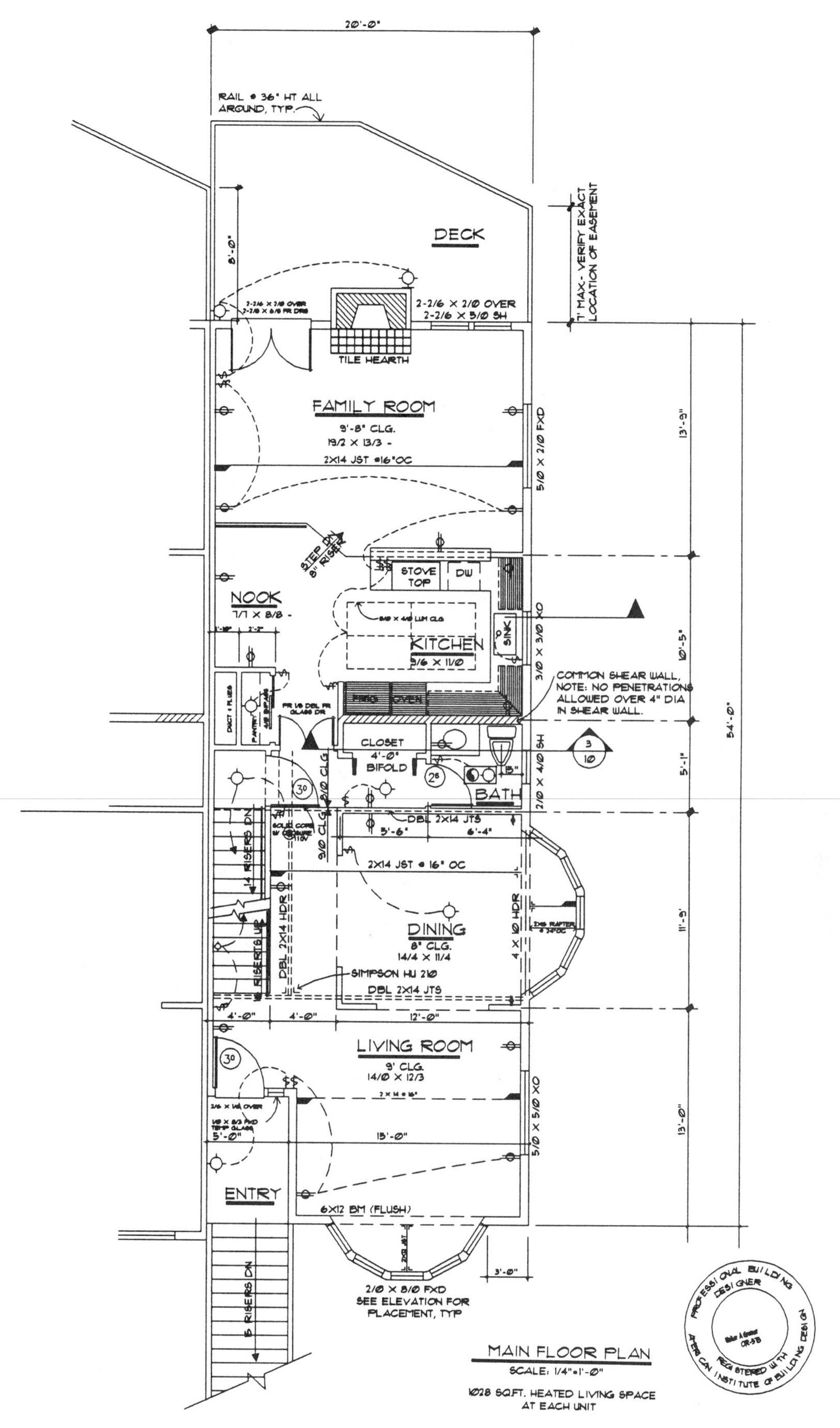

Figure 15–1 Because of their size, commercial projects such as apartments will typically have an enlarged floor plan to show how each unit is to be constructed. *Courtesy of Sunridge Designs, Portland, OR.*

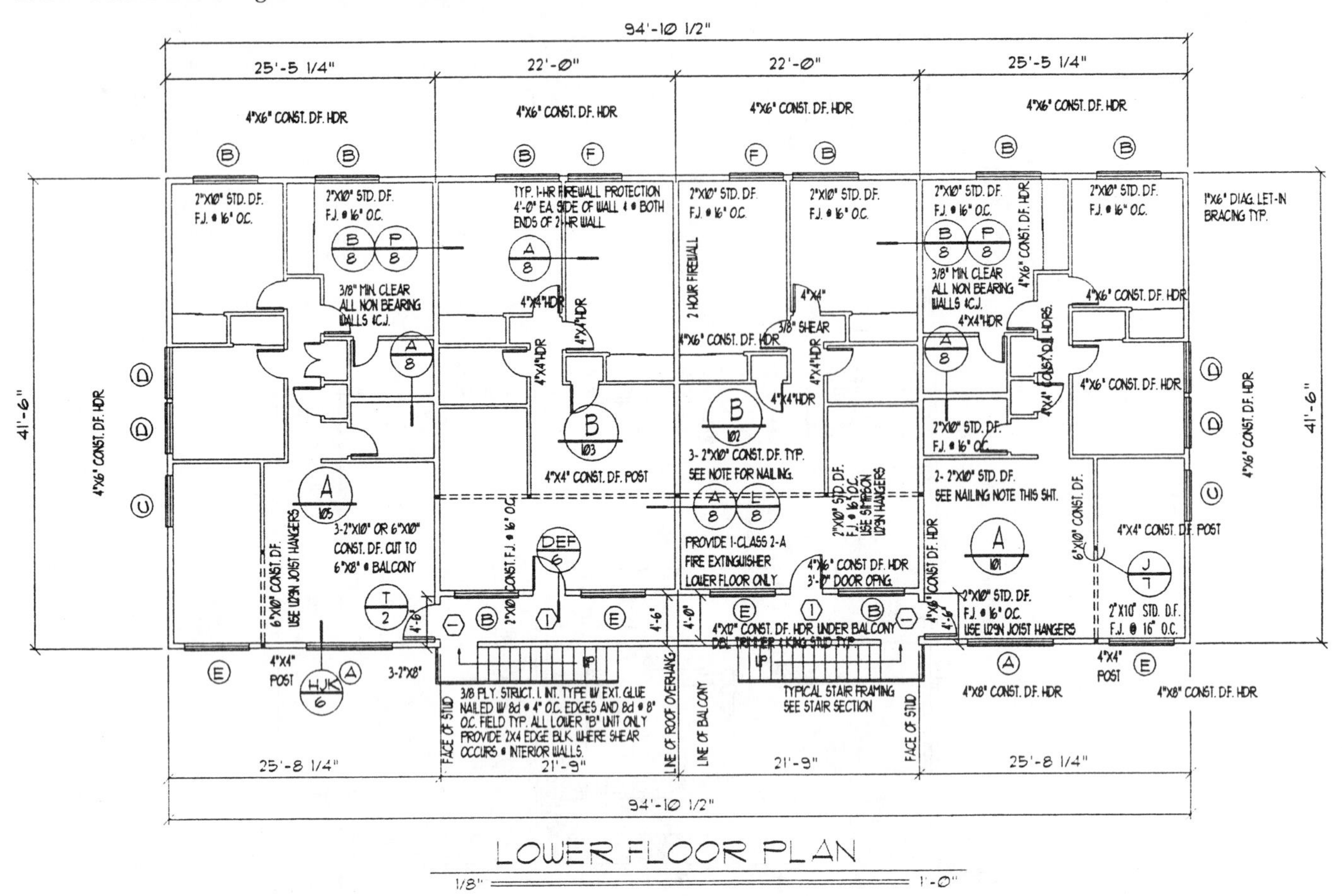

Figure 15–2 A floor plan can also be used to show how typical units will relate to each other. *Courtesy of Ken Smith, Structureform Masters, Inc.*

In commercial construction, the information is usually so complex that it will require its own plan. This would include framing, electrical, reflected ceiling, mechanical, and plumbing plans. The electrical and other HVAC drawings will be covered in Chapter 18.

The framing plan normally will contain all of the dimensions and information needed by the framing crew to form the basic shell of the structure. Architectural information related to the finishing material is placed on the floor plan. By dividing the material into separate plans, clarity is gained on the plans. An example of a framing plan can be seen in Figure 15–5. Not only is the information divided differently, the plans themselves are also different. Plans are typically grouped by architectural, structural, mechanical, and plumbing drawings, with each section numbered consecutively. Page numbers are preceeded by an A, S, M or P to represent each section of the job.

Foundation Plans

Foundations in commercial work are typically concrete slabs because of their durability and low labor cost. As seen in Figure 15–6, the foundation for a commercial project is similar to that of a residential project. The size of the footings will vary, but the same type of stresses affect a commercial project that affect a residence. They differ only in their magnitude. In commercial projects, it is common to have both a slab and a foundation plan. The slab, or slab-on-grade plan as it is sometimes called, is a plan view of the construction of the floor system and specifies the size and location of concrete pours. A slab-on-grade plan can be seen in Figure 15–7. The foundation plan is used to show below-grade concrete work. An example of a foundation plan can be seen in Figure 15–8.

Elevations

Elevations for commercial projects are very similar to the elevations required for a residential project. Figure 15–9 shows an example of an elevation for a restaurant. Notice that the elevation is used to show the same types of information as a residential elevation but uses much more detail. Because the size of many commercial structures is so large, commercial projects are sometimes drawn at a scale of ⅛" = 1'–0" or smaller and have very little detail added.

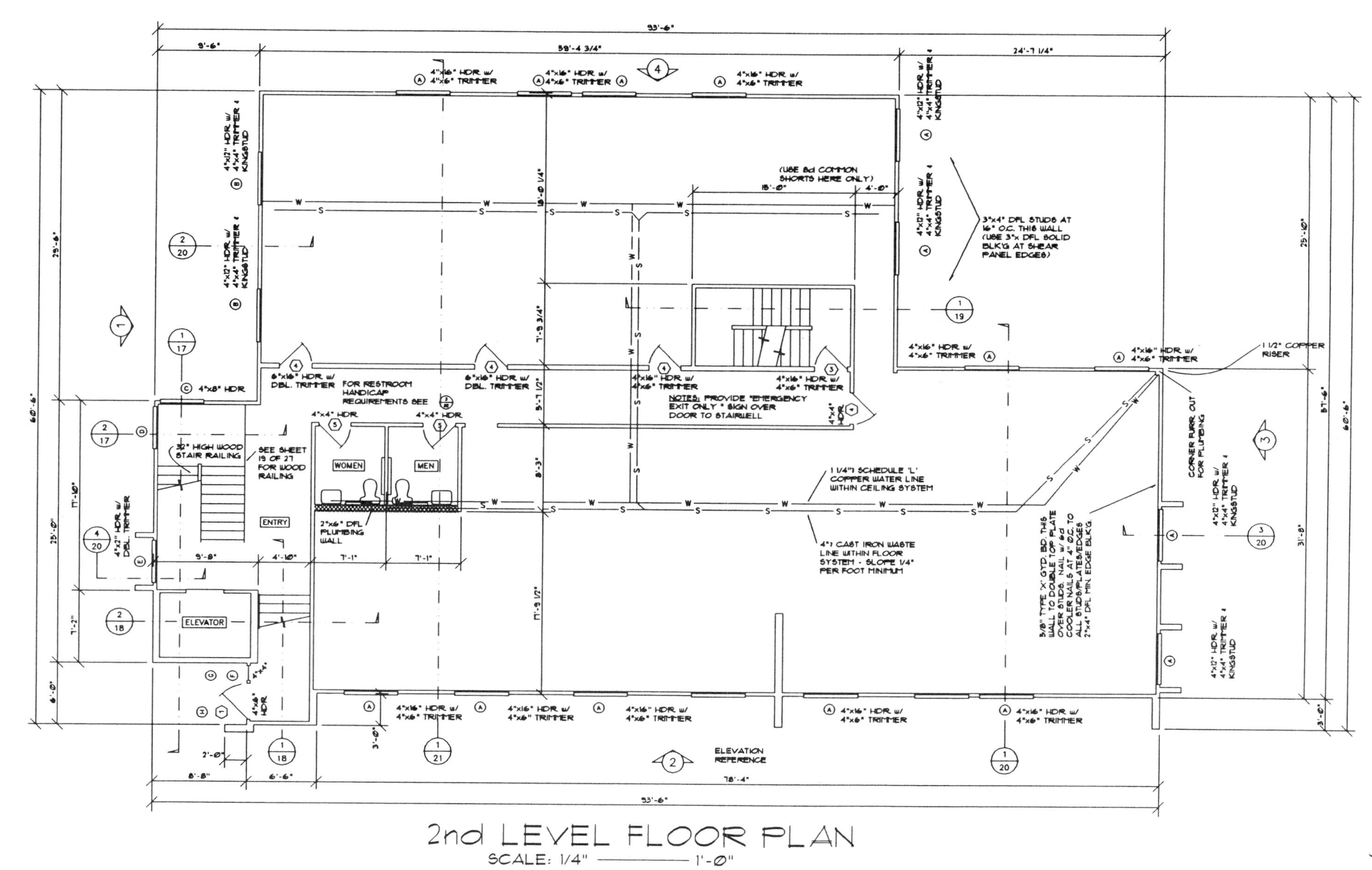

Figure 15–3 An office floor plan may only show the load bearing walls and allow each tenant to plan how their space will be partitioned. *Courtesy of Ken Smith, Structureform Masters, Inc.*

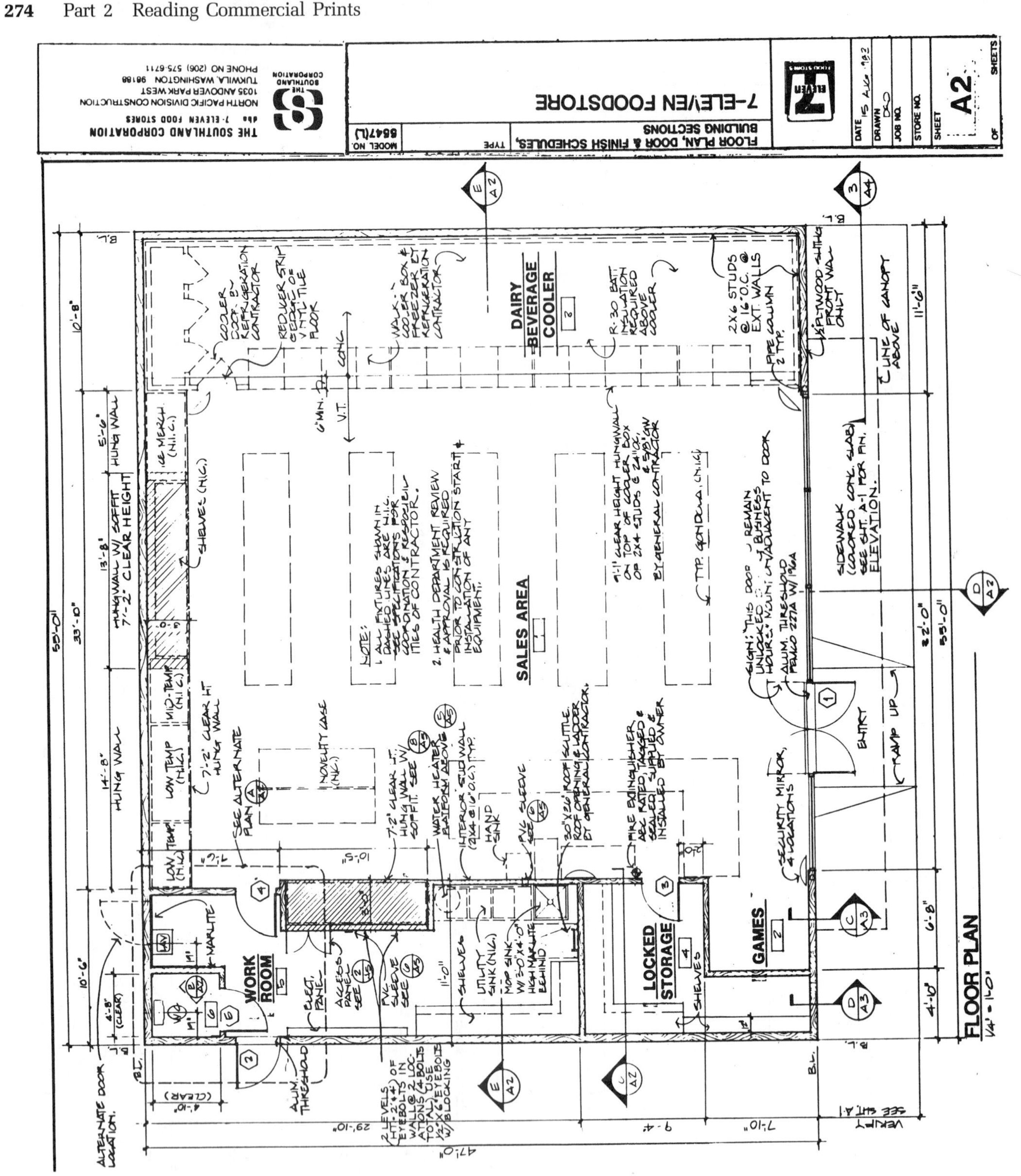

Figure 15–4 A floor plan for a retail sales outlet. *Courtesy of the Southland Corporation.*

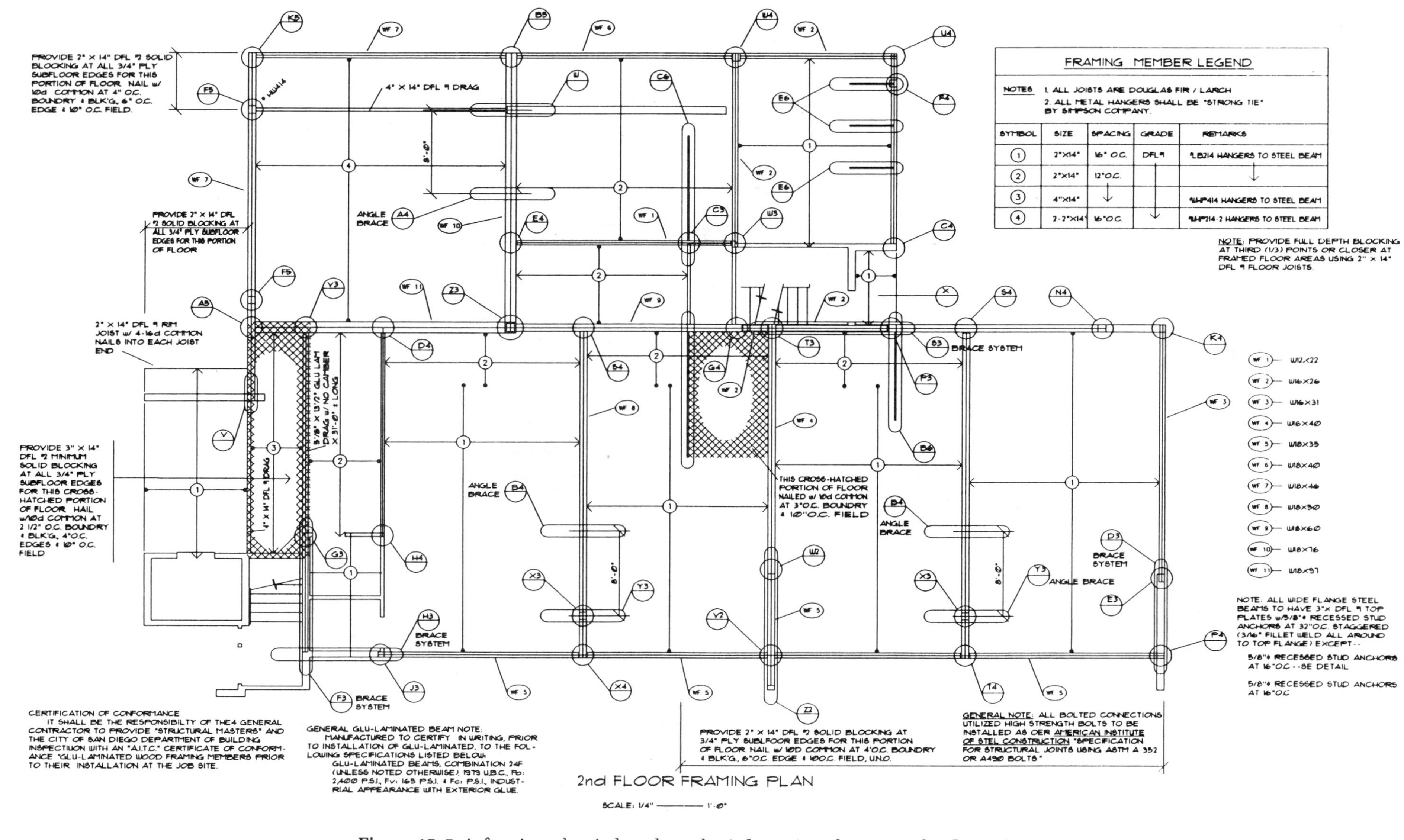

Figure 15–5 A framing plan is based on the information shown on the floor plan. The architectural information is removed and information related to the framing of that level is shown. Compare this plan with Figure 15–3. *Courtesy of Ken Smith, Structureform Masters, Inc.*

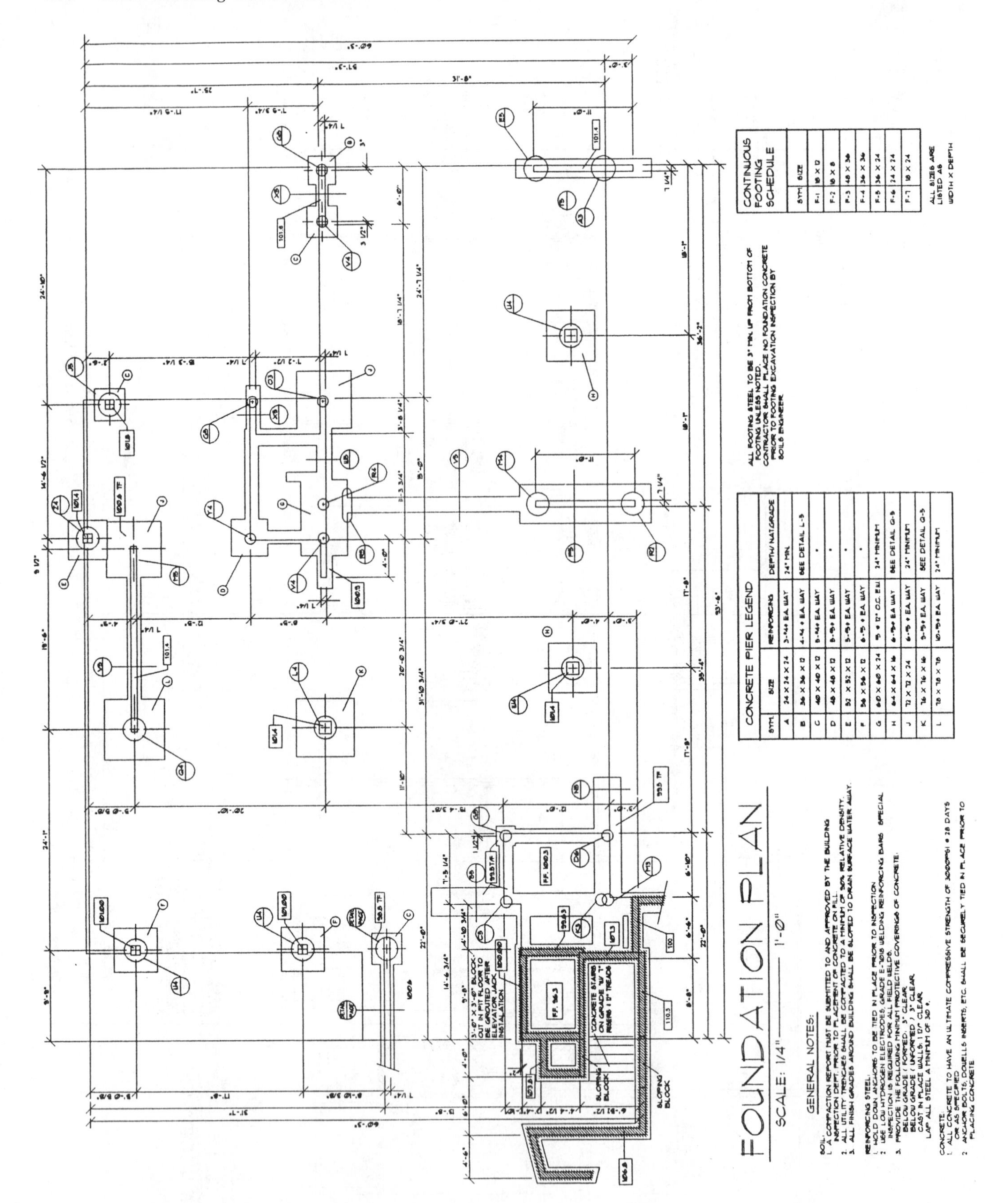

CONCRETE PIER LEGEND			
SYM.	SIZE	REINFORCING	DEPTH/ NAT.GRADE
A	24 X 24 X 24	3-#4 EA. WAY	24" MIN.
B	36 X 36 X 12	4-#4 EA. WAY	SEE DETAIL L-5
C	40 X 40 X 12	5-#4 EA. WAY	"
D	48 X 48 X 12	5-#5 EA. WAY	"
E	52 X 52 X 12	5-#5 EA. WAY	"
F	56 X 56 X 12	6-#5 EA. WAY	"
G	60 X 60 X 24	#5 @ 12" O.C. E.W.	24" MINIMUM
H	64 X 64 X 16	6-#5 EA. WAY	SEE DETAIL Q-5
J	72 X 72 X 24	6-#5 EA. WAY	24" MINIMUM
K	76 X 76 X 16	9-#5 EA. WAY	SEE DETAIL Q-5
L	78 X 78 X 78	10-#5 EA. WAY	24" MINIMUM

CONTINUOUS FOOTING SCHEDULE	
SYM.	SIZE
F-1	18 X 12
F-2	18 X 8
F-3	48 X 36
F-4	36 X 36
F-5	36 X 24
F-6	24 X 24
F-7	18 X 24

Figure 15–6 The foundation plan for commercial projects shows the location of concrete located below grade. *Courtesy of Ken Smith, Structureform Masters, Inc.*

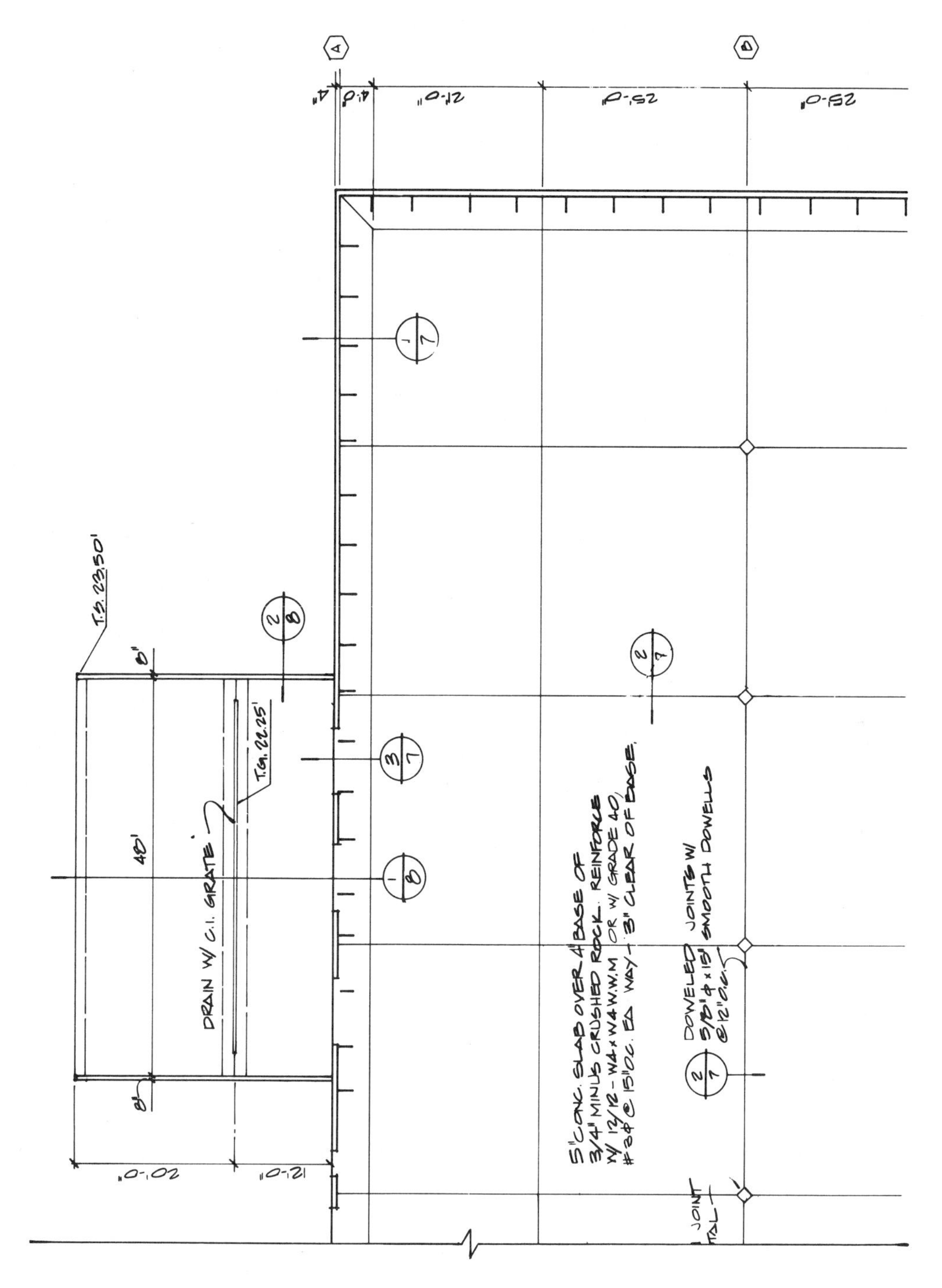

Figure 15–7 The Slab-on-Grade Plan shows the location of the floor slabs and all concrete located at grade level. *Courtesy of Ken Smith, Structureform Masters, Inc.*

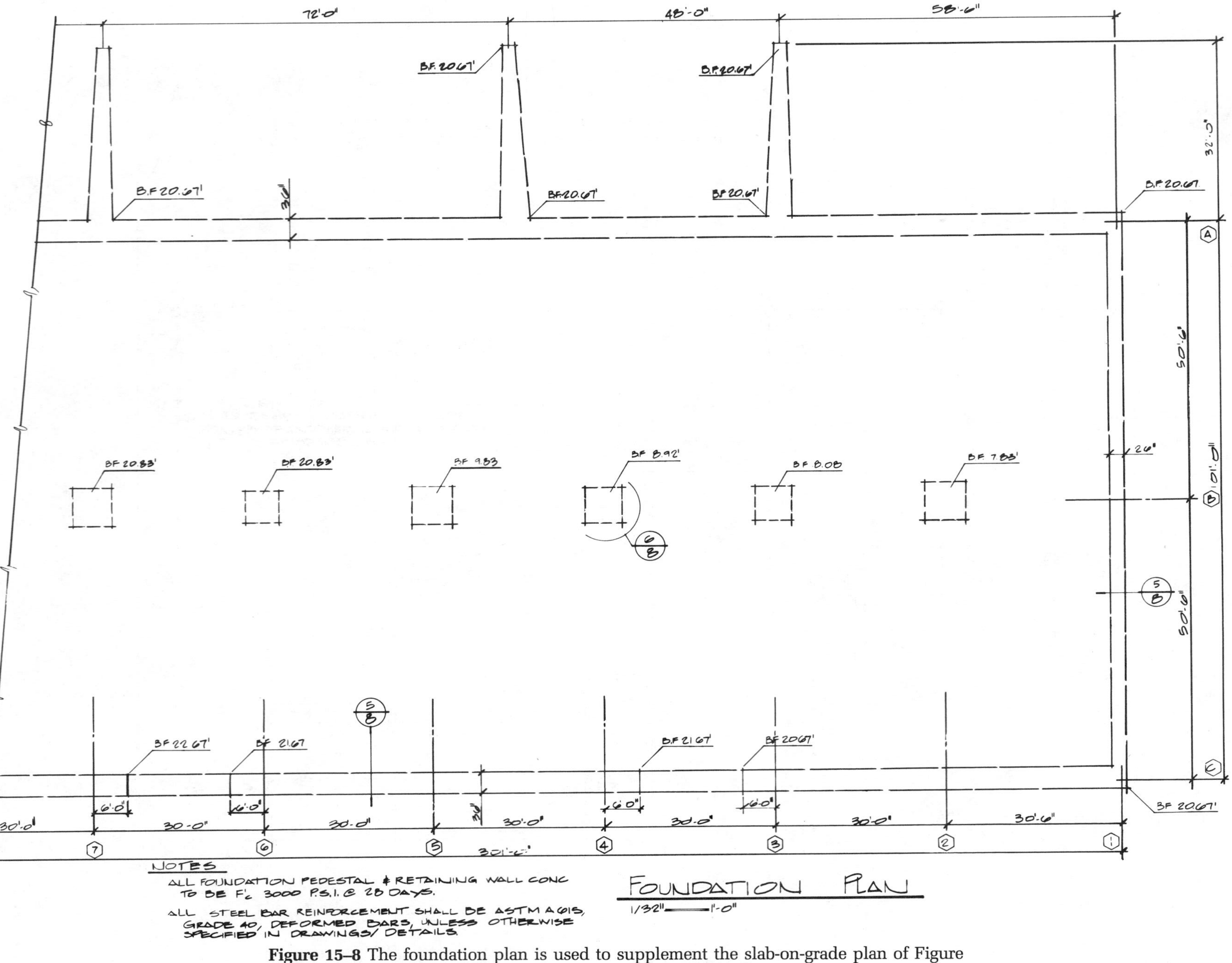

Figure 15–8 The foundation plan is used to supplement the slab-on-grade plan of Figure 15–8. *Courtesy of Ken Smith, Structureform Masters, Inc.*

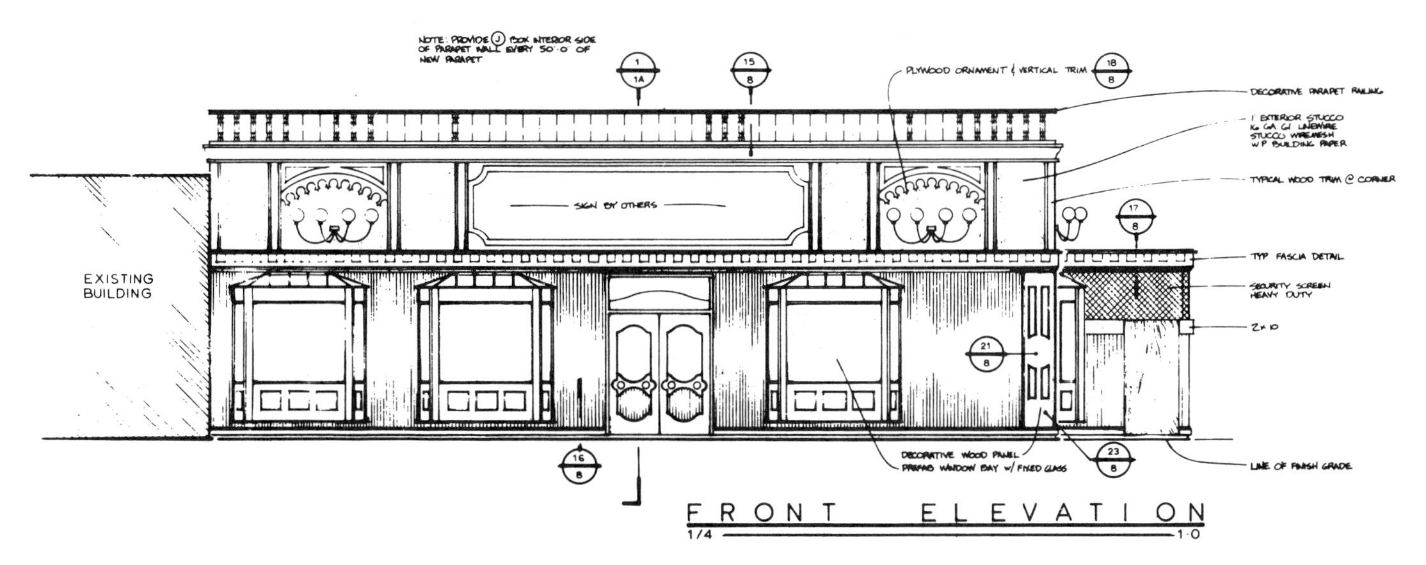

Figure 15–9 Elevations are used to show exterior finishes and reference for exterior finishing details. *Courtesy of Ken Smith, Structureform Masters, Inc.*

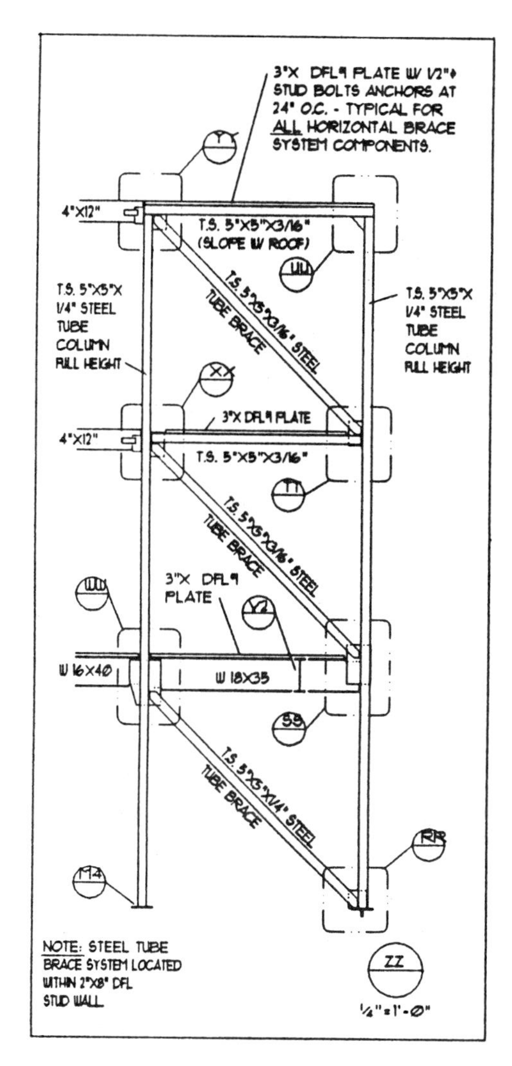

Figure 15–10 Sections are often used as a reference map to locate details or other sections. *Courtesy of Ken Smith, Structureform Masters, Inc.*

Sections

The sections for a commercial project are similar to those of a residential project, but the drawings may have a different use. On commercial plans the sections are used primarily as a reference map of the structure. Sections are usually drawn at a scale of 1/4" = 1'–0" and details of specific intersections are referenced to the sections. Figure 15–10 shows an example of this type of section. Another common use of section drawings for commercial projects is to use several partial sections.

Details

Because there are so many variables in construction techniques and materials, details are used to explain a specific area of construction. On residential plans details may be used to specify information in a complex area that cannot be easily seen in a section. On commercial drawings, details are an integral part of the plans. Details are typically grouped together depending on which construction crew will need them or which plan view they are supporting. These arrangements will typically place all roof details together, all wall details together, and all concrete details together. Figure 15–11 shows some of the details that were required for section 22 of Figure 15–10.

Roof Plans

Roof plans and roof drainage plans are typically used more in commercial projects than in residential

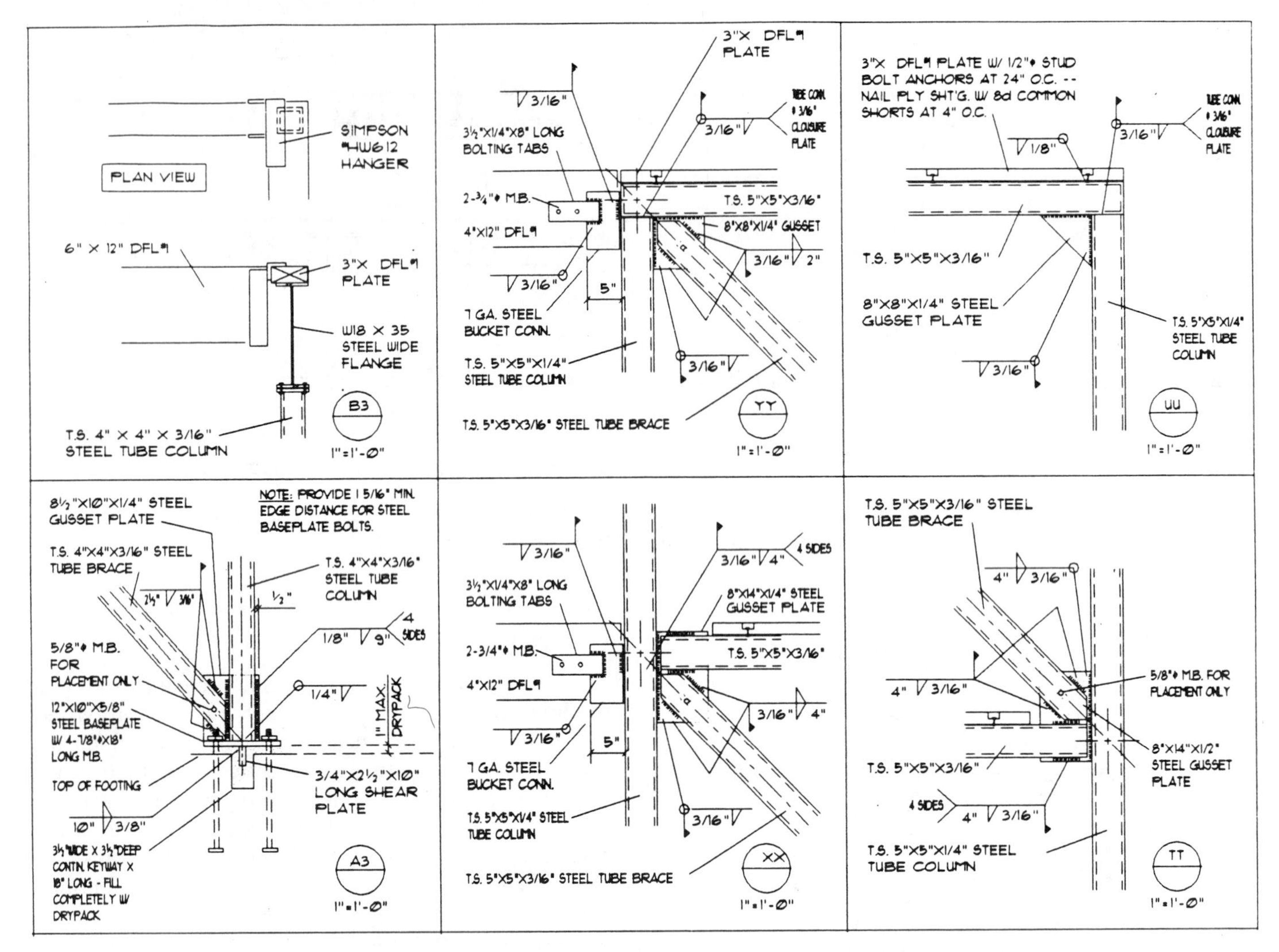

Figure 15–11 Details are often used to supplement a section or partial section. Notice detail DD is a reference drawing to help visualize how each detail relates to the project. *Courtesy of Ken Smith, Structureform Masters, Inc.*

work. Because the roof system is usually flat and contains concentrated loads from mechanical equipment, a detailed placement of beams and interior supports is usually required. Two of the most common types of roof systems used on commercial projects are the truss and the panelized systems. The use of trusses was discussed in Chapter 7. A panelized roof consists of large beams supporting smaller beams which in turn support 4' × 8' roofing panels. The system will be further discussed in Chapter 17. A roof drainage plan can be seen in Figure 15–12 and the roof framing plan can be seen in Figure 15–13.

Interior Elevations

The kind of interior elevations provided for a commercial project will depend on the type of project to be built. In an office or warehouse building, cabinets will be minimal and usually not drawn. For the office setting, an interior decorator may be used to coordinate and design interior spaces. On projects such as a restaurant, several pages of cabinet elevations and details may be required. These drawings are often grouped on pages labeled ss for stainless steel details. This requires coordinating the floor plan with the interior elevations.

Site-Related Elevations

The structure must be shown as it relates to the job site. This is typically done on the plot plan just as it would be done for a residential project. Commercial projects typically show not only the building, but also the parking and access facilities. Plot plans are typically supplemented by parking plans, grading plans, landscaping plans, and irrigation plans. Each can be seen in Chapter 20.

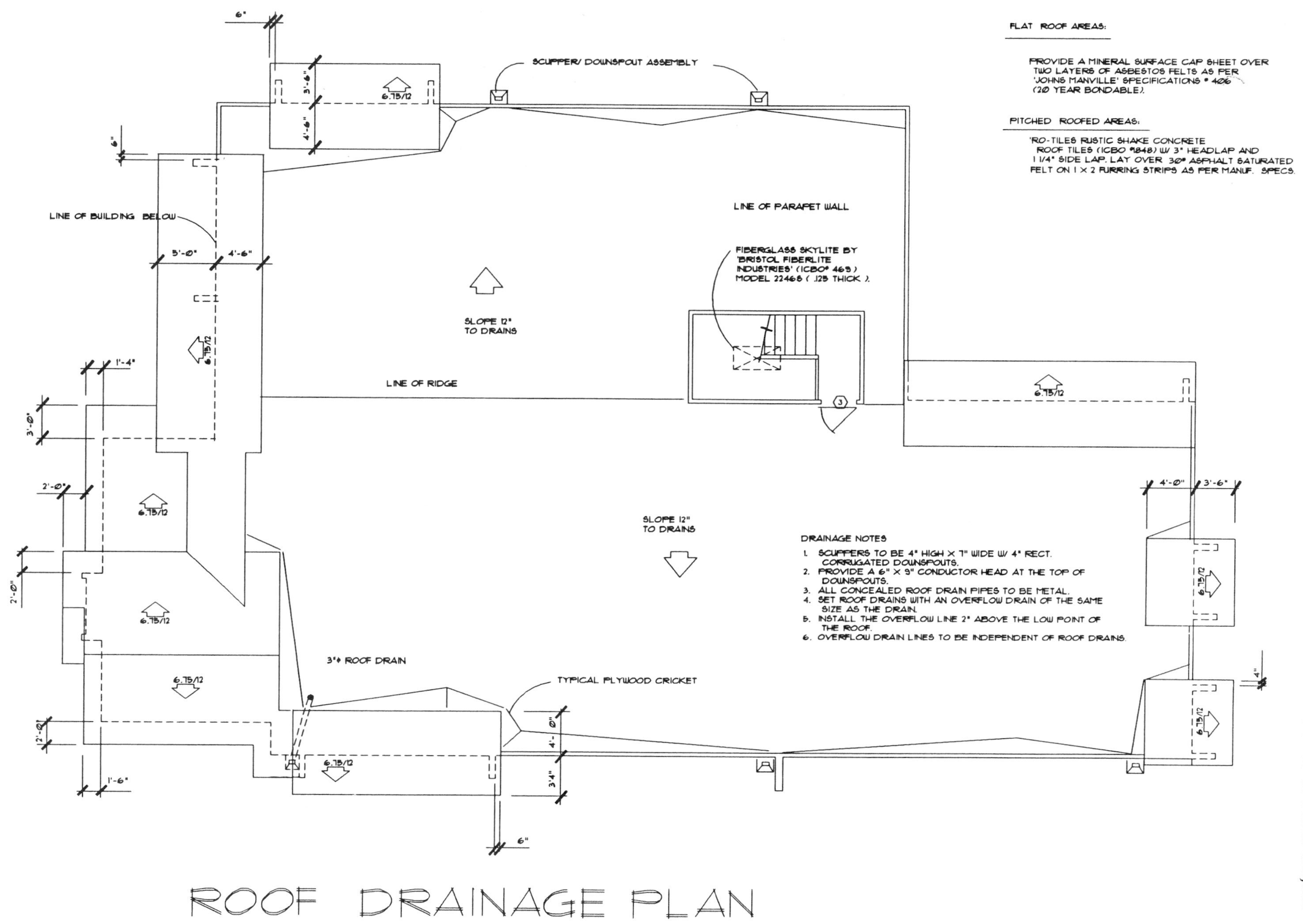

Figure 15–12 The roof drainage plan is used to show how water will be removed from the roof structure. *Courtesy of Ken Smith, Structureform Masters, Inc.*

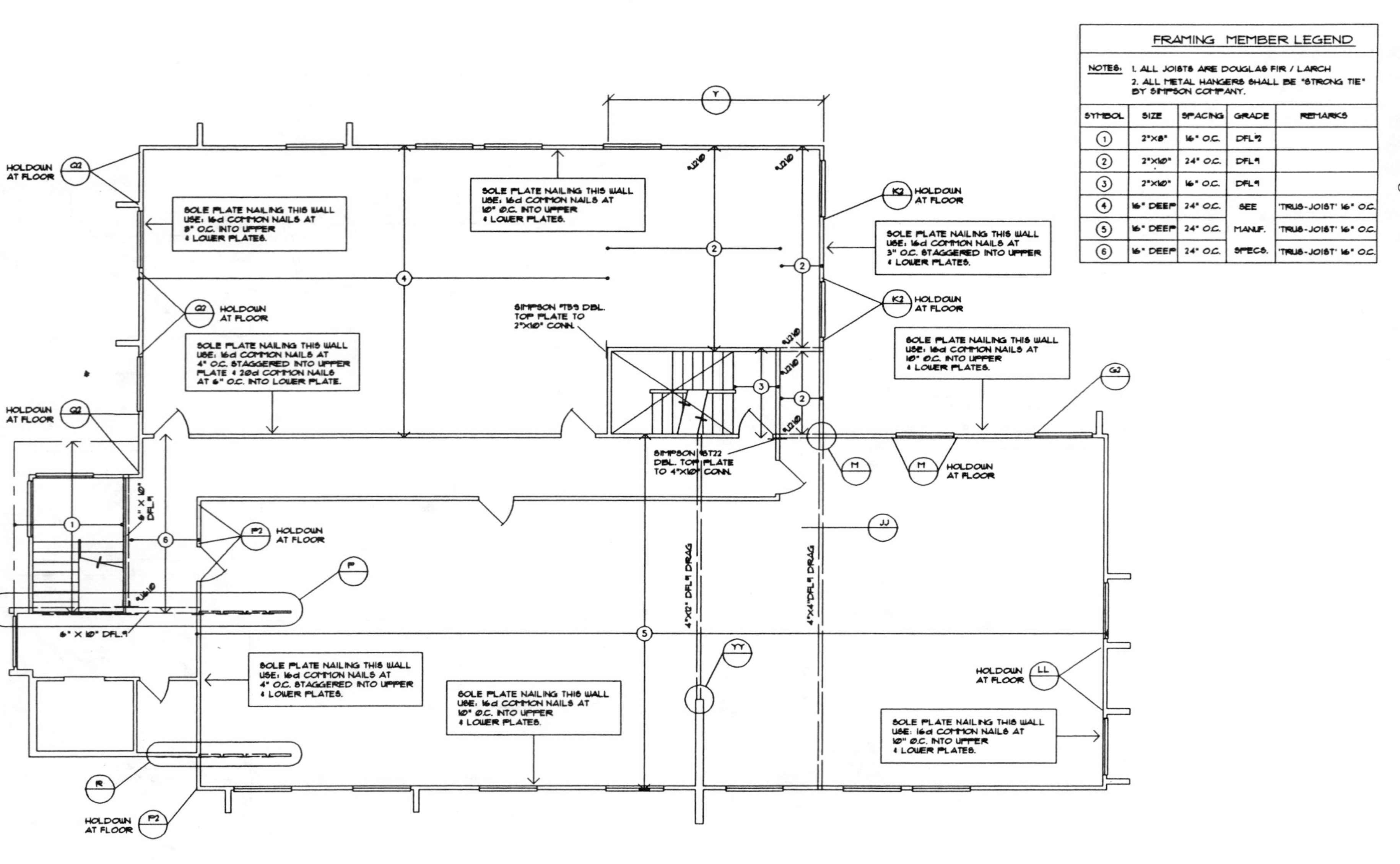

SYMBOL	SIZE	SPACING	GRADE	REMARKS
1	2"X8"	16" O.C.	DFL.2	
2	2"X10"	24" O.C.	DFL.1	
3	2"X10"	16" O.C.	DFL.1	
4	16" DEEP	24" O.C.	SEE	'TRUS-JOIST' 16" O.C.
5	16" DEEP	24" O.C.	MANUF.	'TRUS-JOIST' 16" O.C.
6	16" DEEP	24" O.C.	SPECS.	'TRUS-JOIST' 16" O.C.

Figure 15–13 The roof framing plan is similar to other framing plans showing the location, size, and direction of all framing members. Compare this plan with Figure 15–15. *Courtesy of Ken Smith, Structureform Masters, Inc.*

BUILDING CODES

Not only do you need to be familiar with differences in drawings, but you should also have an understanding of how building codes effect the projects on which you are working. As you work on commercial projects, the building codes will become much more influential. To be effective at the job site you must be able to understand the code that governs your area. Although there are many different codes in use throughout the country, most will be similar to the Uniform Building Code (UBC), Basic National Building Code (BOCA), or the Standard Building Code (SBC).

If you haven't done so already, check with the building department in your area and determine the code that will govern construction.

Purchase of a code book can greatly aid your understanding of construction drawings. Don't hesitate to mark or place tabs on certain pages to identify key passages. You won't need to memorize the book, but you will need to refer to certain areas often. Page tabs will help you quickly find needed tables or formulas.

DETERMINE THE CATEGORIES

To use a building code effectively, you must determine several classifications used to define a structure. These classifications will be used throughout the codes and include determining the occupancy group, location on the property, type of construction, floor area, height, and occupant load.

Occupancy Groups

The occupancy group specifies by whom or how the structure will be used. To protect the public adequately, buildings used for different purposes are designed to meet the hazards of that usage. Obviously, the safety requirements for the multileveled apartment with hundreds of occupants should be different than a single-family residence.

The code that you are using will affect how the occupancy is listed. The letter of the occupancy listing generally is the first letter of the word that it represents. Figure 15–14 shows a detailed listing of occupancy categories from the UBC.

Structures often have more than one occupancy group within the structure. Separation must be provided at the wall or floor dividing these different occupancies. Figure 15–15 shows a listing of required separations based on UBC requirements. If you are working on a structure that has a 10,000 sq ft office area and a parking garage, the two areas must be separated. By using the table in Figure 15–14 you can see that the office area is a B-2 occupancy and the garage area is a B-3 occupancy. By using the table in Figure 15–15, you can determine that a wall with a one-hour rating must be provided between these two areas. The hourly rating is given to construction materials specifying the length of time that the item must resist structural damage that would be caused by a fire. Various ratings are assigned by the codes for materials ranging from nonrated to a four-hour rating. Once the occupancy grouping is determined, you will need to turn to the section of the building code that introduces specifics on that occupancy grouping.

Building Location and Size

When the occupancy of the structure was determined in Figure 15–14, information on the location of the structure to the property lines was also given. The type of occupancy will affect the location of a building on the property. The location to the property lines will also affect the size and amount of openings that are allowed in a wall.

Type of Construction

Once the occupancy of the structure has been determined, the type of construction used to protect the occupants can be determined. The type of construction determines the kind of building materials that can or cannot be used in the construction of the building. The kind of material used in construction will determine the ability of the structure to resist fire. Five general types of construction are typically specified by building codes and are represented by the numbers 1 through 5.

Construction in building types 1 and 2 requires the structural elements such as walls, floors, roofs, and exits to be constructed or protected by approved noncombustible materials such as steel, iron, concrete, or masonry. Construction in types 3, 4, and 5 can be of either steel, iron, concrete, masonry, or wood. In addition to specifying the structural framework, the type of construction also dictates the material used for interior partitions, exit specifications, and a wide variety of other requirements for the building.

Building Area

Once the type of construction has been determined, the size of the structure can be determined. Figure 15–16 shows a listing of allowable floor area based on the type of construction to be used. These are basic sq ft sizes which may be altered depending on the different construction techniques that can be used. You will now need to turn to the areas of the code that cover these types of construction to determine types of materials that could be used.

TABLE NO. 5-A—WALL AND OPENING PROTECTION OF OCCUPANCIES BASED ON LOCATION ON PROPERTY TYPES II ONE-HOUR, II-N AND V CONSTRUCTION: For exterior wall and opening protection of Types II One-hour, II-N and V buildings, see table below and Sections 504, 709, 1903 and 2203.
This table does not apply to Types I, II-F.R., III and IV construction, see Sections 1803, 1903, 2003 and 2103.

	GROUP	DESCRIPTION OF OCCUPANCY	FIRE RESISTANCE OF EXTERIOR WALLS	OPENINGS IN EXTERIOR WALLS
ASSEMBLY	A See also Section 602	1—Any assembly building with a stage and an occupant load of 1000 or more in the building	Not applicable (See Sections 602 and 603)	
		2—Any building or portion of a building having an assembly room with an occupant load of less than 1000 and a stage 2.1—Any building or portion of a building having an assembly room with an occupant load of 300 or more without a stage, including such buildings used for educational purposes and not classed as a Group E or Group B, Division 2 Occupancy	2 hours less than 10 feet, 1 hour less than 40 feet	Not permitted less than 5 feet Protected less than 10 feet
		3—Any building or portion of a building having an assembly room with an occupant load of less than 300 without a stage, including such buildings used for educational purposes and not classed as a Group E or Group B, Division 2 Occupancy	2 hours less than 5 feet, 1 hour less than 40 feet	Not permitted less than 5 feet Protected less than 10 feet
		4—Stadiums, reviewing stands and amusement park structures not included within other Group A Occupancies	1 hour less than 10 feet	Protected less than 10 feet
BUSINESS	B See also Section 702	1—Gasoline service stations, garages where no repair work is done except exchange of parts and maintenance requiring no open flame, welding, or use of Class I, II or III-A liquids 2—Drinking and dining establishments having an occupant load of less than 50, wholesale and retail stores, office buildings, printing plants, municipal police and fire stations, factories and workshops using material not highly flammable or combustible, storage and sales rooms for combustible goods, paint stores without bulk handling Buildings or portions of buildings having rooms used for educational purposes, beyond the 12th grade, with less than 50 occupants in any room	1 hour less than 20 feet	Not permitted less than 5 feet Protected less than 10 feet
		3—Aircraft hangars where no repair work is done except exchange of parts and maintenance requiring no open flame, welding, or the use of Class I or II liquids Open parking garages (For requirements, See Section 709.) Heliports	1 hour less than 20 feet	Not permitted less than 5 feet Protected less than 20 feet
		4—Ice plants, power plants, pumping plants, cold storage and creameries Factories and workshops using noncombustible and nonexplosive materials Storage and sales rooms of noncombustible and nonexplosive materials	1 hour less than 5 feet	Not permitted less than 5 feet
EDUCATION	E See also Section 802	1—Any building used for educational purposes through the 12th grade by 50 or more persons for more than 12 hours per week or four hours in any one day 2—Any building used for educational purposes through the 12th grade by less than 50 persons for more than 12 hours per week or four hours in any one day 3—Any building used for day-care purposes for more than six children	2 hours less than 5 feet, 1 hour less than 10 feet[1]	Not permitted less than 5 feet Protected less than 10 feet[1]
HAZARDOUS	H See also Sections 902 and 903	1—Storage, handling, use or sale of hazardous and highly flammable or explosive materials other than Class I, II, or III-A liquids [See also Section 901 (a), Division 1.]	See Chapter 9 and the Fire Code	
		2—Storage, handling, use or sale of Classes I, II and III-A liquids; dry cleaning plants using Class I, II or III-A liquids; paint stores with bulk handling; paint shops and spray-painting rooms and shops [See also Section 901 (a), Division 2.] 3—Woodworking establishments, planing mills, box factories, buffing rooms for tire-rebuilding plants and picking rooms; shops, factories or warehouses where loose combustible fibers or dust are manufactured, processed, generated or stored; and pin-refinishing rooms	4 hours less than 5 feet, 2 hours less than 10 feet, 1 hour less than 20 feet	Not permitted less than 5 feet Protected less than 20 feet
		4—Repair garages not classified as a Group B, Division 1 Occupancy 5—Aircraft repair hangars	1 hour less than 60 feet	Protected less than 60 feet
		6—Semiconductor fabrication facilities and comparable research and development areas when the facilities in which hazardous production materials are used are designed and constructed in accordance with Section 911 and storage, handling and use of hazardous materials is in accordance with the Fire Code. [See also Section 901 (a), Division 6.]	4 hours less than 5 feet, 2 hours less than 10 feet, 1 hour less than 20 feet	Not permitted less than 5 feet, protected less than 20 feet
INSTITUTIONS	I See also Section 1002	1—Nurseries for the full-time care of children under the age of six (each accommodating more than five persons) Hospitals, sanitariums, nursing homes with nonambulatory patients and similar buildings (each accommodating more than five persons)	2 hours less than 5 feet, 1 hour elsewhere	Not permitted less than 5 feet Protected less than 10 feet
		2—Nursing homes for ambulatory patients, homes for children six years of age or over (each accommodating more than five persons)	1 hour	
		3—Mental hospitals, mental sanitariums, jails, prisons, reformatories and buildings where personal liberties of inmates are similarly restrained	2 hours less than 5 feet, 1 hour elsewhere	Not permitted less than 5 feet, protected less than 10 feet
	M[2]	1—Private garages, carports, sheds and agricultural buildings (See also Section 1101, Division 1.)	1 hour less than 3 feet (or may be protected on the exterior with materials approved for 1-hour fire-resistive construction)	Not permitted less than 3 feet
		2—Fences over 6 feet high, tanks and towers	Not regulated for fire resistance	
RESIDENTIAL	R See also Section 1202	1—Hotels and apartment houses Convents and monasteries (each accommodating more than 10 persons)	1 hour less than 5 feet	Not permitted less than 5 feet
		3—Dwellings and lodging houses	1 hour less than 3 feet	Not permitted less than 3 feet

[1]Group E, Divisions 2 and 3 Occupancies having an occupant load of not more than 20 may have exterior wall and opening protection as required for Group R, Division 3 Occupancies.

[2]For agricultural buildings, see Appendix Chapter 11.

NOTES: (1) See Section 504 for types of walls affected and requirements covering percentage of openings permitted in exterior walls.
(2) For additional restrictions, see chapters under Occupancy and Types of Construction.
(3) For walls facing yards and public ways, see Part IV.
(4) Openings shall be protected by a fire assembly having a three-fourths-hour fire-protection rating.

Figure 15–14 Structures are divided into different occupancy listings depending on how the building will be used and the danger that will be created by that use. *Reproduced from the 1991 edition of the* Uniform Building Code™, *copyright 1991, with permission of the publisher, the International Conference of Building Officials.*

TABLE NO. 5-B—REQUIRED SEPARATION IN BUILDINGS OF MIXED OCCUPANCY
(In Hours)

	A-1	A-2	A-2.1	A-3	A-4	B-1	B-2	B-3	B-4	E	H-1	H-2	H-3	H-4-5	H-6	I	M[2]	R-1	R-3
A-1		N	N	N	N	4	3	3	3	N	4	4	4	4	4	3	1	1	1
A-2	N		N	N	N	3	1	1	1	N	4	4	4	4	4	3	1	1	1
A-2.1	N	N		N	N	3	1	1	1	N	4	4	4	4	4	3	1	1	1
A-3	N	N	N		N	3	N	1	N	N	4	4	4	4	3	3	1	1	1
A-4	N	N	N	N		3	1	1	1	N	4	4	4	4	4	3	1	1	1
B-1	4	3	3	3	3		1	1	1	3	2	1	1	1	1	4	1	3[1]	1
B-2	3	1	1	N	1	1		1	1	1	2	1	1	1	1	2	1	1	N
B-3	3	1	1	1	1	1	1		1	1	2	1	1	1	1	4	1	1	N
B-4	3	1	1	N	1	1	1	1		1	2	1	1	1	1	4	N	1	N
E	N	N	N	N	N	3	1	1	1		4	4	4	4	3	1	1	1	1
H-1	4	4	4	4	4	2	2	2	2	4		1	1	1	2	4	1	4	4
H-2	4	4	4	4	4	1	1	1	1	4	1		1	1	1	4	1	3	3
H-3	4	4	4	4	4	1	1	1	1	4	1	1		1	1	4	1	3	3
H-4-5	4	4	4	4	4	1	1	1	1	4	1	1	1		1	4	1	3	3
H-6	4	4	4	3	4	1	1	1	1	3	2	1	1	1		4	3	4	4
I	3	3	3	3	3	4	2	4	4	1	4	4	4	4	4		1	1	1
M[2]	1	1	1	1	1	1	1	1	N	1	1	1	1	1	3	1		1	1
R-1	1	1	1	1	1	3[1]	1	1	1	1	4	3	3	3	4	1	1		N
R-3	1	1	1	1	1	1	N	N	N	1	4	3	3	3	4	1	1	N	

Note: For detailed requirements and exceptions, see Section 503.

[1]The three-hour separation may be reduced to two hours where the Group B, Division 1 Occupancy is limited to the storage of passenger motor vehicles having a capacity of not more than nine persons. This shall not apply where provisions of Section 702 (a) apply.

[2]For Agricultural buildings, see also Appendix Chapter 11.

Figure 15–15 If a structure contains more than one occupancy grouping, the areas may need to be separated and structural materials protected from fire. The office in Figure 15–3 is a B-2 occupancy built over a public parking area which is a B-3. By comparing B-2 with B-3, it can be seen that material providing a 1-hour separation is required. *Reproduced from the 1991 edition of the Uniform Building Code™, copyright 1991, with permission of the publisher, the International Conference of Building Officials.*

MOST PROTECTIVE — LEAST RESTRICTIVE

TABLE NO. 5-C—BASIC ALLOWABLE FLOOR AREA FOR BUILDINGS ONE STORY IN HEIGHT[1]
(In Square Feet)

	TYPES OF CONSTRUCTION								
	I	II			III		IV	V	
OCCUPANCY	F.R.	F.R.	ONE-HOUR	N	ONE-HOUR	N	H.T.	ONE-HOUR	N
A-1	Unlimited	29,900	Not Permitted						
A) 2-2.1	Unlimited	29,900	13,500	Not Permitted	13,500	Not Permitted	13,500	10,500	Not Permitted
A) 3-4[2]	Unlimited	29,900	13,500	9,100	13,500	9,100	13,500	10,500	6,000
B) 1-2-3[3]	Unlimited	39,900	18,000	12,000	18,000	12,000	18,000	14,000	8,000
B-4	Unlimited	59,900	27,000	18,000	27,000	18,000	27,000	21,000	12,000
E	Unlimited	45,200	20,200	13,500	20,200	13,500	20,200	15,700	9,100
H) 1-2[4]	15,000	12,400	5,600	3,700	5,600	3,700	5,600	4,400	2,500
H) 3-4-5	Unlimited	24,800	11,200	7,500	11,200	7,500	11,200	8,800	5,100
H-6	Unlimited	39,900	18,000	12,000	18,000	12,000	18,000	14,000	8,000
I) 1-2	Unlimited	15,100	6,800	Not Permitted	6,800	Not Permitted	6,800	5,200	Not Permitted
I-3	Unlimited	15,100	Not Permitted[5]						
M[6]	See Chapter 11								
R-1	Unlimited	29,900	13,500	9,100[7]	13,500	9,100[7]	13,500	10,500	6,000[7]
R-3	Unlimited								

[1]For multistory buildings, see Section 505 (b).
[2]For limitations and exceptions, see Section 602 (a).
[3]For open parking garages, see Section 709.
[4]See Section 903.
[5]See Section 1002 (b).
[6]For agricultural buildings, see also Appendix Chapter 11.
[7]For limitations and exceptions, see Section 1202 (b).

N—No requirements for fire resistance
F.R.—Fire Resistive
H.T.—Heavy Timber

Figure 15–16 Building codes restrict the size of a structure based on the type of construction to help ensure public safety. *Reproduced from the 1991 edition of the Uniform Building Code™, copyright 1991, with permission of the publisher, the International Conference of Building Officials.*

Determine the Height

The occupancy and type of construction will determine the maximum height of the structure. Figure 15–17 shows an example of a table from the UBC used to determine the allowable height of a structure. A 20,000 sq ft office building could be 12 stories (or 160 ft) high, if constructed of type 2 construction materials or of unlimited height if built with type 1 materials. This height is based on building requirements governing fire and public safety. Zoning regulations for a specific area may further limit the height of the structure.

Determine the Occupant Load

After it is determined how the structure will be used, it must be determined how many people may use the structure. Figure 15–18 is an example of the table used to compute the occupant load. In each occupancy, the intended size of the structure is divided by the occupant load factor to determine the occupant load. In a 20,000 sq ft office structure, the occupant load would be 200. This is found by dividing the size of 20,000 sq ft by the occupant factor of 100. The occupant load affects the number of exits, the size and locations of doors, and many other construction requirements.

USING THE CODES

The need to determine the six classifications of a building may seem senseless to you. Although these are procedures that the architect, or engineer will perform in the initial design stage, the contractor and construction crews should be aware of these classifications as basic skills are performed on the project. Many of the problems that you will need the code to solve will require knowledge of the six basic code limitations.

Once you feel at ease reading through the chapters dealing with general construction, work on the chapters of the code that deal with uses of different types of material. Start with the chapter that deals with the most common building material in your area. For most of you, this will be the section governing the use of wood. The use of wood is covered in Chapter 25 in the UBC, sections 1223 through 1227 of BOCA, and Chapter 17 of the SBC. Although you will be familiar with the use of wood in residential structures, the building codes will open many new areas that were not required in residential construction.

TABLE NO. 5-D—MAXIMUM HEIGHT OF BUILDINGS

OCCUPANCY	TYPES OF CONSTRUCTION								
	I	II			III		IV	V	
	F.R.	F.R.	ONE-HOUR	N	ONE-HOUR	N	H.T.	ONE-HOUR	N
	MAXIMUM HEIGHT IN FEET								
	Unlimited	160	65	55	65	55	65	50	40
	MAXIMUM HEIGHT IN STORIES								
A-1	Unlimited	4	Not Permitted						
A) 2-2.1	Unlimited	4	2	Not Permitted	2	Not Permitted	2	2	Not Permitted
A) 3-4[1]	Unlimited	12	2	1	2	1	2	2	1
B) 1-2-3[2]	Unlimited	12	4	2	4	2	4	3	2
B-4	Unlimited	12	4	2	4	2	4	3	2
E[3]	Unlimited	4	2	1	2	1	2	2	1
H-1	Unlimited	2	1	1	1	1	1	1	1
H) 2-3-4-5	Unlimited	5	2	1	2	1	2	2	1
H-6	3	3	3	2	3	2	3	3	1
I-1	Unlimited	3	1	Not Permitted	1	Not Permitted	1	1	Not Permitted
I-2	Unlimited	3	2	Not Permitted	2	Not Permitted	2	2	Not Permitted
I-3	Unlimited	2	Not Permitted[4]						
M[5]	See Chapter 11								
R-1	Unlimited	12	4	2[6]	4	2[6]	4	3	2[6]
R-3	Unlimited	3	3	3	3	3	3	3	3

[1]For limitations and exceptions, see Section 602 (a).
[2]For open parking garages, see Section 709.
[3]See Section 802 (c).
[4]See Section 1002 (b).
[5]For agricultural buildings, see also Appendix Chapter 11.
[6]For limitations and exceptions, see Section 1202 (b).

N—No requirements for fire resistance
F.R.—Fire Resistive
H.T.—Heavy Timber

Figure 15–17 The height of a structure is limited by codes based on the occupancy and type of construction used in the structure. *Reproduced from the 1991 edition of the Uniform Building Code™, copyright 1991, with permission of the publisher, the International Conference of Building Officials.*

TABLE NO. 33-A—MINIMUM EGRESS AND ACCESS REQUIREMENTS

USE[1]	MINIMUM OF TWO EXITS OTHER THAN ELEVATORS ARE REQUIRED WHERE NUMBER OF OCCUPANTS IS AT LEAST	OCCUPANT LOAD FACTOR[2] (Sq. Ft.)	ACCESS BY MEANS OF A RAMP OR AN ELEVATOR MUST BE PROVIDED FOR THE PHYSICALLY HANDICAPPED AS INDICATED[3]
1. Aircraft Hangars (no repair)	10	500	Yes
2. Auction Rooms	30	7	Yes
3. Assembly Areas, Concentrated Use (without fixed seats) Auditoriums Bowling Alleys (Assembly areas) Churches and Chapels Dance Floors Lobby Accessory to Assembly Occupancy Lodge Rooms Reviewing Stands Stadiums	50	7	Yes[4 5]
4. Assembly Areas, Less-concentrated Use Conference Rooms Dining Rooms Drinking Establishments Exhibit Rooms Gymnasiums Lounges Stages	50	15	Yes[4 6]
5. Children's Homes and Homes for the Aged	6	80	Yes[7]
6. Classrooms	50	20	Yes[8]
7. Dormitories	10	50	Yes[7]
8. Dwellings	10	300	No
9. Garage, Parking	30	200	Yes[9]
10. Hospitals and Sanitariums—Nursing Homes	6	80	Yes
11. Hotels and Apartments	10	200	Yes[10]
12. Kitchen—Commercial	30	200	No
13. Library Reading Room	50	50	Yes[4]
14. Locker Rooms	30	50	Yes
15. Malls (see Appendix Chapter 7)	—	—	—
16. Manufacturing Areas	30	200	Yes[7]
17. Mechanical Equipment Room	30	300	No
18. Nurseries for Children (Day-care)	7	35	Yes
19. Offices	30	100	Yes[7]
20. School Shops and Vocational Rooms	50	50	Yes
21. Skating Rinks	50	50 on the skating area; 15 on the deck	Yes[4]
22. Storage and Stock Rooms	30	300	No
23. Stores—Retail Sales Rooms Basement Ground Floor Upper Floors	 11 50 10	 20* 30 50	 Yes Yes Yes
24. Swimming Pools	50	50 for the pool area; 15 on the deck	Yes[4]
25. Warehouses	30	500	No
26. All others	50	100	

[1]For additional provisions on number of exits from Group H and I Occupancies and from rooms containing fuel-fired equipment or cellulose nitrate, see Sections 3320, 3321 and 3322, respectively.

[2]This table shall not be used to determine working space requirements per person.

[3]Elevators shall not be construed as providing a required exit.

[4]Access to secondary areas on balconies or mezzanines may be by stairs only, except when such secondary areas contain the only available toilet facilities.

[5]Reviewing stands, grandstands and bleachers need not comply.

[6]Access requirements for conference rooms, dining rooms, lounges and exhibit rooms that are part of an office use shall be the same as required for the office use.

[7]Access to floors other than that closest to grade may be by stairs only, except when the only available toilet facilities are on other levels.

[8]When the floor closest to the grade offers the same programs and activities available on other floors, access to the other floors may be by stairs only, except when the only available toilet facilities are on other levels.

[9]Access to floors other than that closest to grade and to garages used in connection with apartment houses may be by stairs only.

[10]See Section 1213 for access to buildings and facilities in hotels and apartments.

[11]See Section 3303 for basement exit requirements.

*i.e., for every 20 ◻, one person is allowed in the building.

Figure 15–18 The occupant load of a structure is determined by dividing the intended size of the structure by the occupant load factor. *Reproduced from the 1991 edition of the* Uniform Building Code™, *copyright 1991, with permission of the publisher, the International Conference of Building Officials.*

CHAPTER 15 TEST

Neatly print your answer in the space provided.

1. List five plans that use the floor plan as a base. ____.
2. What type of information is typically placed on a floor plan? ____.
3. What are the five basic drawings that form a set of plans? ____.
4. List four common types of commercial projects. ____.
5. What would be the occupancy rating of each of the four major types of commercial construction? ____.
6. List two different types of roof drawings and describe what each will show. ____.
7. Describe three methods of showing structural connections excluding a plan view. ____.
8. List two different drawings related to the foundation and describe what each will show. ____.
9. List five drawings that will describe the land the project will be constructed on. ____.
10. What is the occupancy of the room where your class meets. ____
11. What is the occupancy, the least restrictive type of construction, and height limitation for a 6,500 sq ft restaurant serving 360 persons? ____.
12. According to Table 33-a from the Uniform Building Code how many people could legally be allowed in a 6,500 sq ft restaurant? ____
13. What fire rating is required between a wall separating a B-2 from an I occupancy? ______________________
14. On what plan would you expect to find the floor joist for the second floor of a three level office structure? ____.
15. Would a building framed with type V be safer than a building constructed of type 3-N materials? Explain your answer. ____.

CHAPTER 15 EXERCISES

PROBLEM 15–1. Use Figure 15–1 on page 271 to answer the following questions.

1. Is this the upper or lower plan? Explain your answer. ______________________________
____
____.
2. What is the size of this unit in square feet? ______________________________
3. What size joist will be required? ______________________________
4. What size is the kitchen? ______________________________
5. Two 2 X members intersect two other 2 X's in the hallwall by the dining room. How will they be supported?
____
____.

PROBLEM 15–2. Use Figure 15–3 on page 273 to answer the following questions.

1. What page should you look on to find information about the elevator? ______________________________
2. What do the lines labeled 's' and 'w' represent? ______________________________
____
____.
3. What size pipe brings water up to this floor? ______________________________
4. On each side of the building there is a number in a circle with an arrow around it. What do these numbers represent? ______________________________
____
____.
5. Where can you find information about the stair railings? ______________________________
____
____.

PROBLEM 15–3. Use Figure 15–5 on page 275 to answer the following questions.

1. What does the symbol WF 5 represent? ______________________________
2. An area in the center of the building is shown with cross hatching. What does it represent? ____________
____
____.
3. What size joist are represented by #2? ______________________________
4. What standards are to be used in the production of the glu-lam beams? ______________________________
____
____
____.
5. How will the 2 × 14 floor joist be connected to steel beams? ______________________________
____
____
____.

PROBLEM 15–4. Use Figure 15–8 on page 278 to answer the following questions.

1. What scale was this plan drawn at? ______
2. What is the elevation of the bottom of the pier at grid B-6? ______
3. What is the spacing from grid 3 to grid 4? ______ From A to B? ______
4. What is the elevation of the footing 6 feet to the left of grid 6A? ______
5. Describe the steel to be used in the footings. ______

PROBLEM 15–5. Use Figure 15–11 on page 280 to answer the following questions.

1. What is ZZ describing? ______
2. How will the 3X top plate be connected to the steel beam? ______
3. What size columns will support the beams? ______
4. A steel plate will help connect a diagonal brace to the vertical column between the 2nd and 3rd floor level. What detail describes the connection? ______
   How big is the plate? ______
5. What members are connecting on the left side of the wall at the second floor? ______

PROBLEM 15–6. Use Figure 15–12 on page 281 to answer the following questions.

1. What is the roof pitch of the mansards on the left side of the roof? ______
2. How much do the flat roof areas slope? ______
3. What roofing will be used at the flat areas? ______
4. How will the overflow drains be installed? ______
5. What will the concrete tiles be installed over at the pitched areas? ______

Chapter 16

Common Commercial Construction Materials and Connectors

WOOD
CONCRETE BLOCK
POURED CONCRETE
STEEL CONSTRUCTION
COMMON CONNECTION METHODS
TEST

AS YOU work on commercial projects, you will be exposed to new types of structures and codes different from the ones you were using with residential drawing. You will also be using different types of materials than are typically used in residential construction. Most of the materials used in commercial construction can be used in residential construction, but they are not because of their cost and the associated cost of labor with each material. Wood is the exception. Common materials used in commercial construction include wood, concrete block, poured concrete, and steel.

WOOD

Wood is used in many types of commercial buildings in a manner similar to its use in residential construction. The western platform system, which was covered in Chapter 8, is a common framing method for multifamily and office buildings. Heavy wood timbers are also used for some commercial construction. You will need to be familiar with both types of construction.

Platform Construction

Platform construction methods in commercial projects are similar to residential methods. Wood is rarely used at the foundation level but is a common material for walls and intermediate and upper level floor systems. Trusses or truss joists are also common floor joist materials for commercial projects. Although each is used in residential construction, their primary usage is in commercial construction.

Walls. Wood is used to frame walls on many projects. Although 2 × 4 studs are used, 2 × 6 studs are more common. It is not uncommon to even find some support walls framed with 2 × 8 studs as a result of larger loads to be supported and greater stress. The biggest difference in wood wall construction is not in the framing method but in the covering materials. Depending on the type of occupancy and the type of construction required, wood framed walls may require a special finish to achieve a required fire protection.

In Chapter 16, you were introduced to five general categories of construction with type I very-resistive to fire and type V much less restrictive. Because of the larger occupant loads commercial structures must protect, wood framing members will need to be covered with fire-resistant material. This will typically mean the use of one layer of 5/8" type X gypsum board for 1-hour protection and two layers for 2-hour protection. Although there are other materials that can be substituted, if greater than 2-hour protection is required, a different framing material is needed. Figure 16–1 shows several different methods of finishing a wood wall to achieve various fire ratings.

Roofs. Wood is also used to frame the roof system of many commercial projects. Joists, trusses, and panelized systems are the most typically used framing systems. Both joist and truss systems have been discussed in Chapter 8. These systems usually allow the joists or trusses to be placed at 24 or 32 in. O.C. for commercial uses. An example of a truss system can be seen in Figure 16–2. The panelized roof system typically uses beams placed approximately 20 to 30 ft apart. Smaller beams called purlins are then placed between the main beams typically using an 8 ft spacing. Joists that are 2 or 3 in. wide are then placed between these purlins at 24 in. O.C. The roof

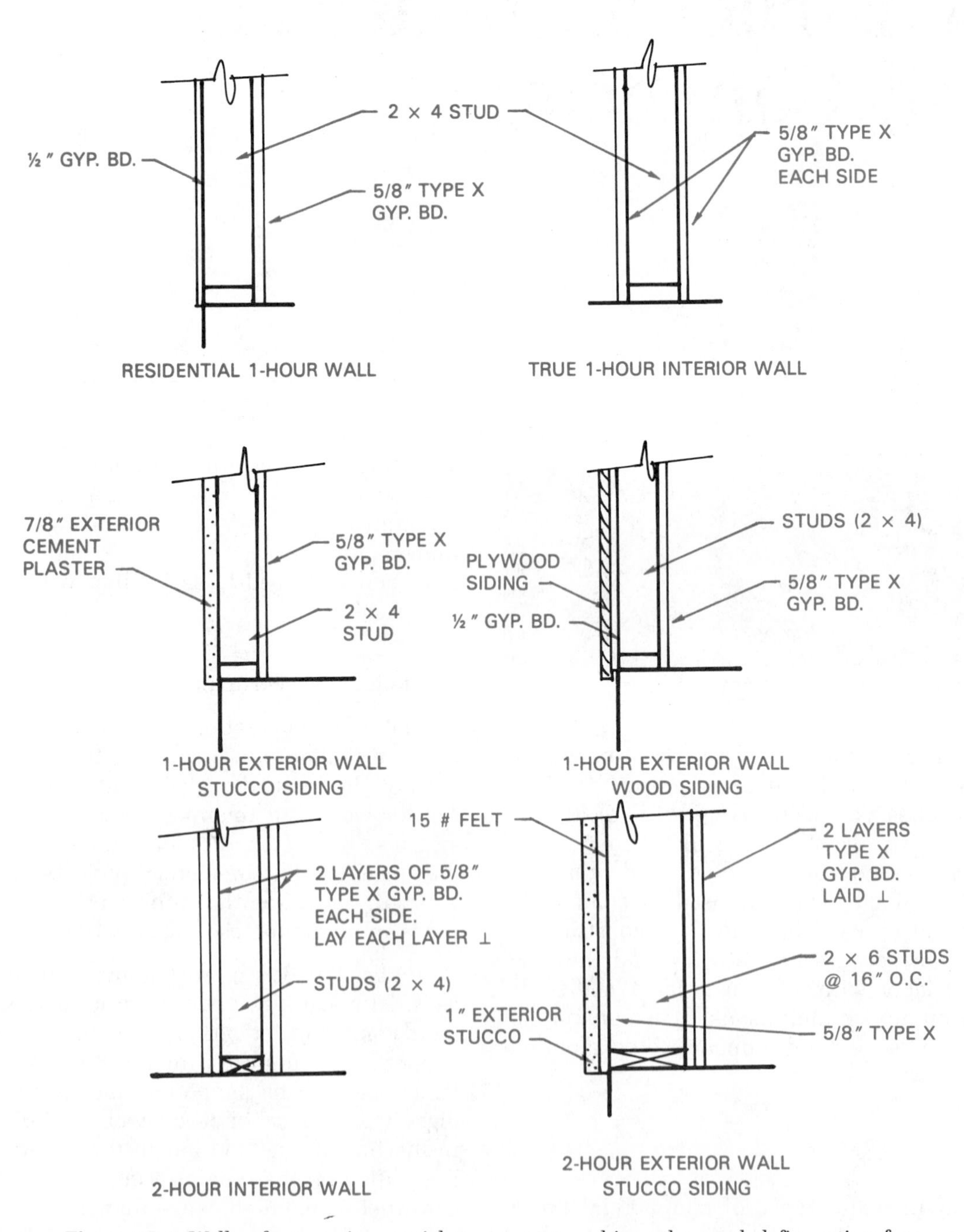

Figure 16–1 Walls often require special treatment to achieve the needed fire rating for types of construction.

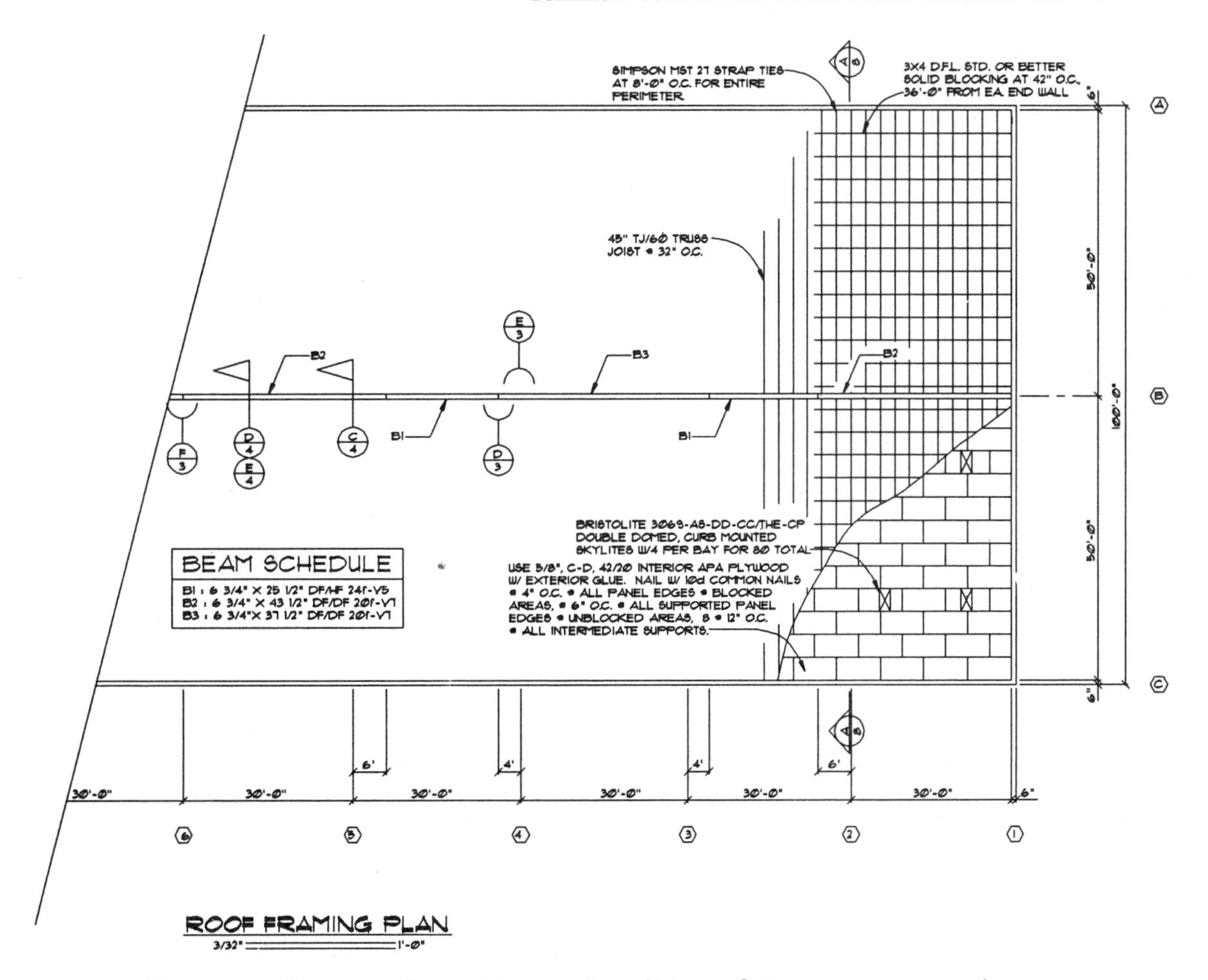

Figure 16–2 Trusses can be used to span long distances between supports creating open floor space.

is then covered with plywood sheathing. Figure 16–3 shows an example of a roof framing plan for a panelized roof system. The availability of materials, labor practices, and use of the building will determine which type of roof will be used.

No matter which system is used you will notice a difference in methods of beam support from residential to commercial construction. In residential construction, beams are typically supported at each end.

In commercial construction projects, as the loads and spans increase, the method of support is usually changed so that beams extend past its supporting column. This type of beam is called a cantilevered beam. By hanging a beam between two cantilevered beams, all three beam sizes can be decreased. Figure 16–4 shows how a cantilevered beam will react to its loads.

Heavy Timber Construction

In addition to standard uses of wood in the western platform system, large wood members are sometimes used for the structural framework of a building. This method of construction is typically used for both appearance and structural reasons. Heavy timbers have excellent structural and fire-retardant qualities. In a fire, heavy wood members will char on their exterior surfaces but will maintain their structural integrity long after an equal-sized steel beam will have failed.

Laminated Beams

Because of the difficulty of producing large beams in long lengths from solid wood, large beams are typically constructed from smaller members laminated together to form the larger beam. Laminated beams are a common material for buildings that require large amounts of open spaces. Three of the most common types of laminated beams are the single span, Tudor arch and the three-hinged arch beams. Examples of each can be seen in Figure 16–5.

The single span beam is often used in standard platform framing methods. Because of their increased structural qualities, a laminated beam can be used to replace a much larger sawed beam. Laminated beams will often have a curve, or camber, built into the beam. The camber is designed into the beam to help resist the loads to be carried.

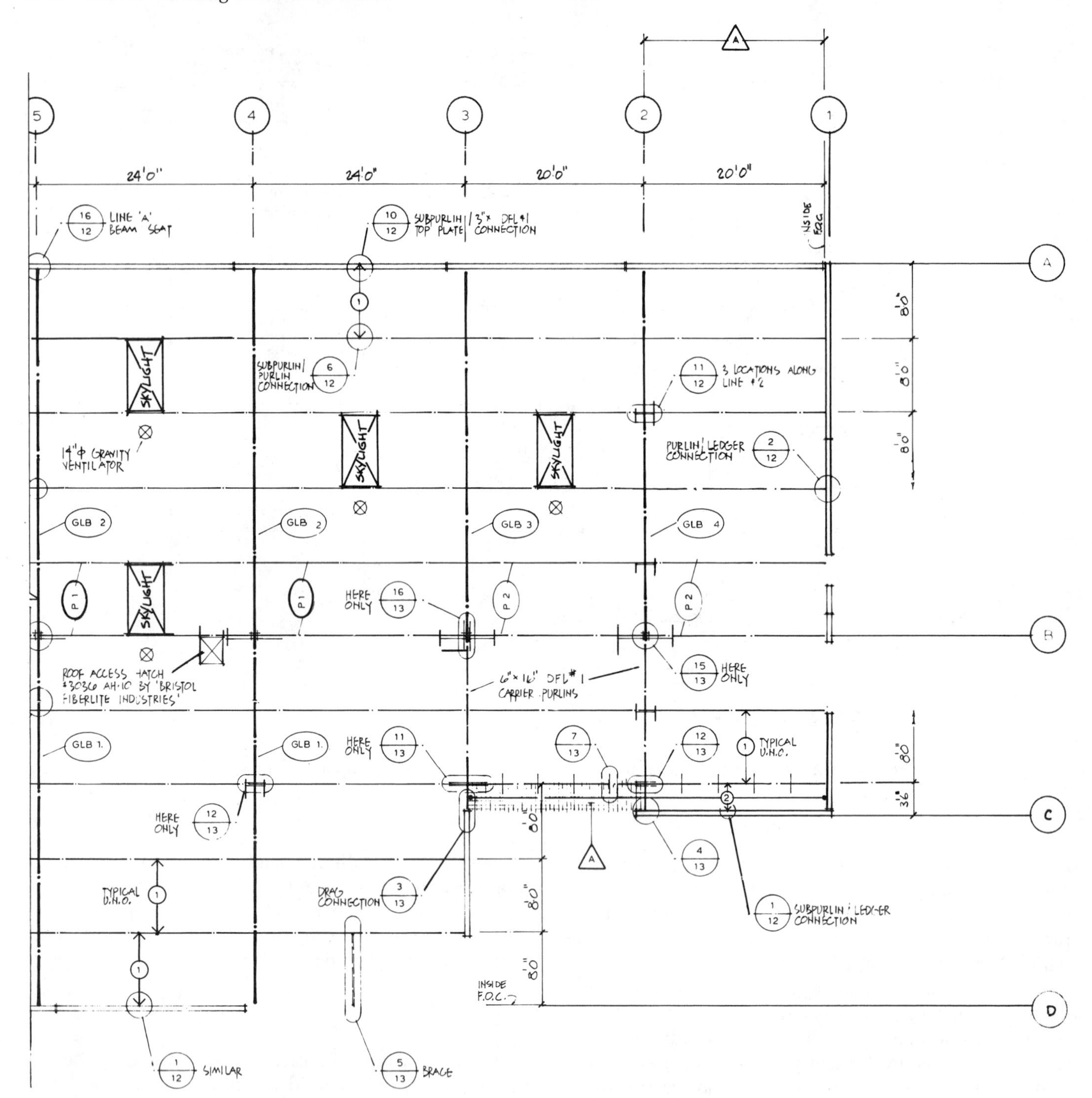

Figure 16–3 A panelized roof system uses beams placed at 20–30 foot intervals with smaller beams placed between them. *Courtesy of Structureform Masters.*

The Tudor and three-hinged arch members are a post-and-beam system combined into one member. These beams are specified on plans in a method similar to other beams. The main difference you will find when working with heavy timber is in how the units are specified. Typically a beam will be specified by a size and a type of wood such as 6 × 10 DFL #1. Because glu-lams are made of smaller, sawn lumber, which is glued together and then milled, you will notice that the sizes do not match sawn lumber sizes. Glu-lam beams come in widths of 3 1/8", 5 1/8", 6 3/4", 10 3/4" and 12 1/4". The depth ranges in size from 3" through 84" in 1 1/2" intervals. Typically, a glu-lam beam would be specified as: 6 3/4' × 16 1/2" f 22-V4 DF/DF.

After the size is given, the call out of a glu-lam will include a number followed by the letter *V* or *E* and another number. The first number is the fiber-bending stress that can be withstood. Fiber bending

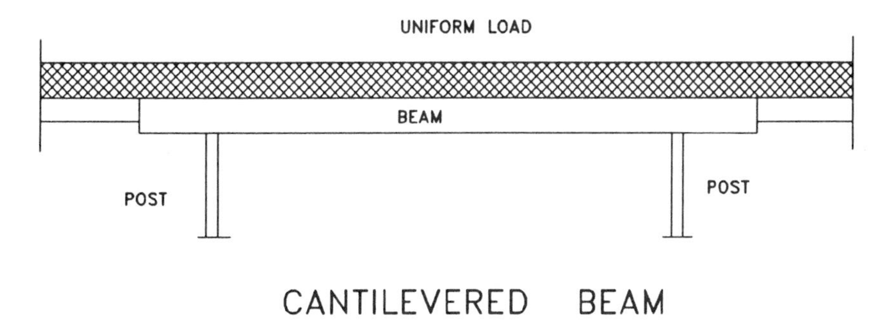

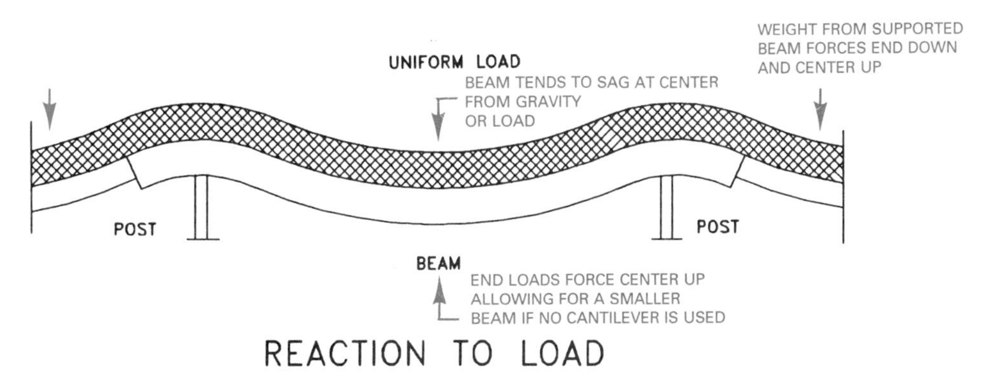

Figure 16–4 A cantilevered beam extends past its supports. A beam is hung from a cantilevered beam forces the ends of the cantilevered beam down and the center up.

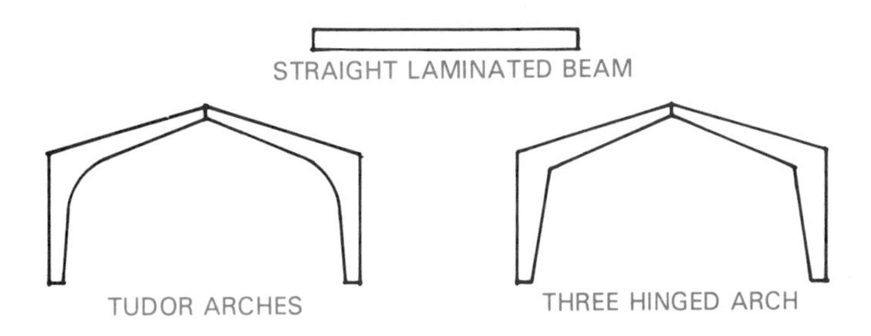

Figure 16–5 Common laminated beam shapes.

is the measurement of the action within the beam thar resists bending. A beam labeled 22F could resist 2200 units of stress in bending when the load is applied perpendicular to the wide face of the laminations. The bending stress is followed by the letter and number specification for inspection method. The *V* represents visually inspected lumber, and the *E* represents beams that have been tested by nondestructive methods. The number that follows the inspection methods relates to how and where lumber species may be combined.

The material the beam is to be made of is the final specification of the beam call out. Laminated beams can be made from a single specie such as Douglas Fir or from a combination of species. Common woods which might be used and their abbreviations include Douglas Fir (DF), hem-fir (HF), western woods or Canadian soft woods (WW), and southern pine (SP). It is important to realize when ordering material of the wide range of strengths, quality, and material that could be used to produce a specific beam.

CONCRETE BLOCK

In commercial construction, concrete blocks are used to form the wall system for many types of buildings. Blocks are typically manufactured in 8 × 8 × 16 modules. The sizes listed are width, height, then length. Other common sizes include 4 × 8 × 16, 6 × 8 × 16, and 12 × 8 × 16. The actual size of the block is smaller than the nominal size so that mortar joints can be included.

Concrete blocks are often reinforced with a wire mesh at every other course of blocks. Where the risk of seismic danger is great, concrete blocks are often required to have reinforcing steel called rebar placed within the wall. These bars are placed in a grid pattern throughout the wall to help tie the blocks together. The steel is placed in a block that has a channel running through it. This cell is then filled with grout to form a header or bond beam within the wall. Figure 16–6 shows an example of bond beam.

When the concrete blocks are required to support a load from a beam, a pilaster is often placed in the wall to help transfer beam loads down into the footing. Pilasters are also used to provide vertical support to the wall when the wall is required to span long distances. An example of a pilaster can be seen in Figure 16–7. Wood can be attached to concrete in several ways. Two of the most common methods are by the use of a seat as seen in Figure 16–8, or a metal connector as seen in Figure 16–9.

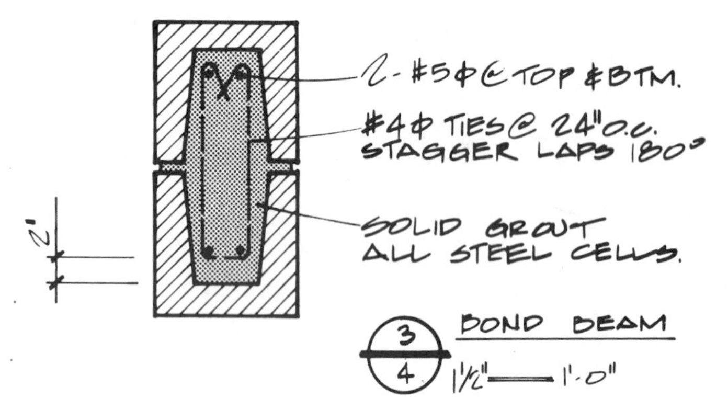

Figure 16–6 A bond beam is created in concrete block walls to allow for steel reinforcement placement.

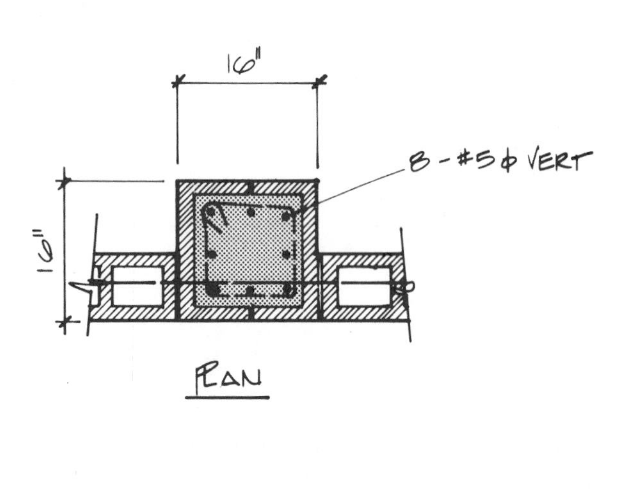

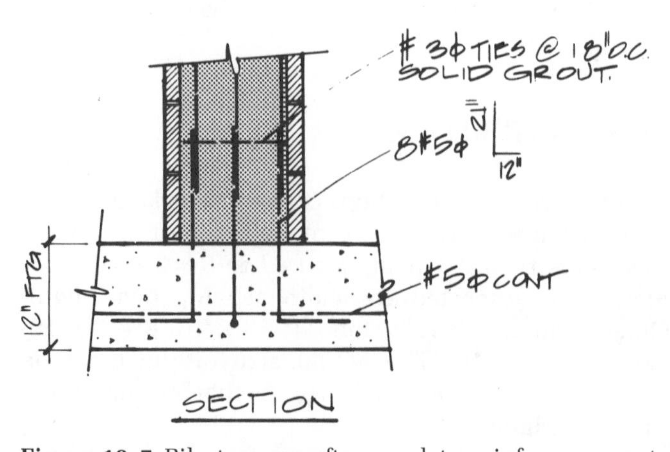

Figure 16–7 Pilasters are often used to reinforce concrete block walls and to provide support for roof beams.

POURED CONCRETE

Concrete is a common building material comprising sand and gravel bonded together with cement and water. One of the most common types of cement is portland cement which contains pulverized particles of limestone, cement rock, oyster shells, silica sand, shale, iron ore, and gypsum. The gypsum controls the time required for the cement to set.

Concrete can either be poured in place at the job site, formed off-site and delivered ready to be erected into place, or formed at the job site and lifted into place. Your area of the country, the office that you work in, and the type of structure to be built will

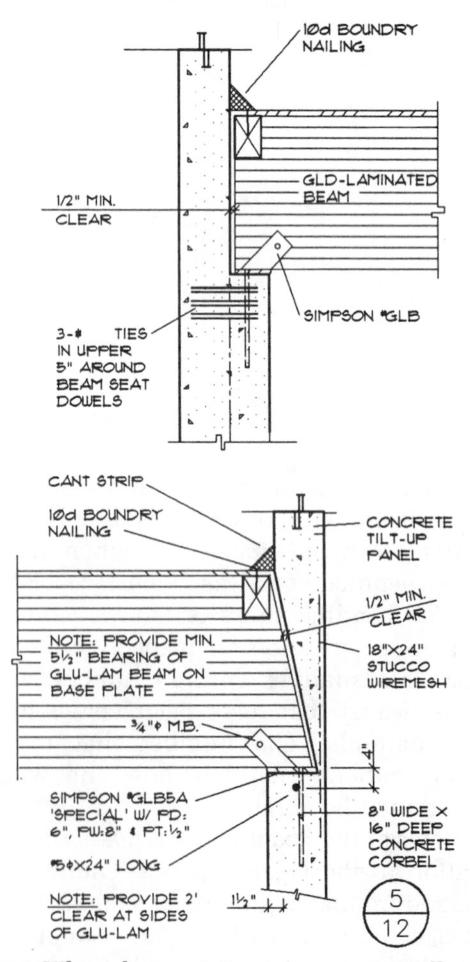

Figure 16–8 Where beams intersect concrete walls a pocket may be provided for the beam to rest on. *Courtesy of Structureform Masters.*

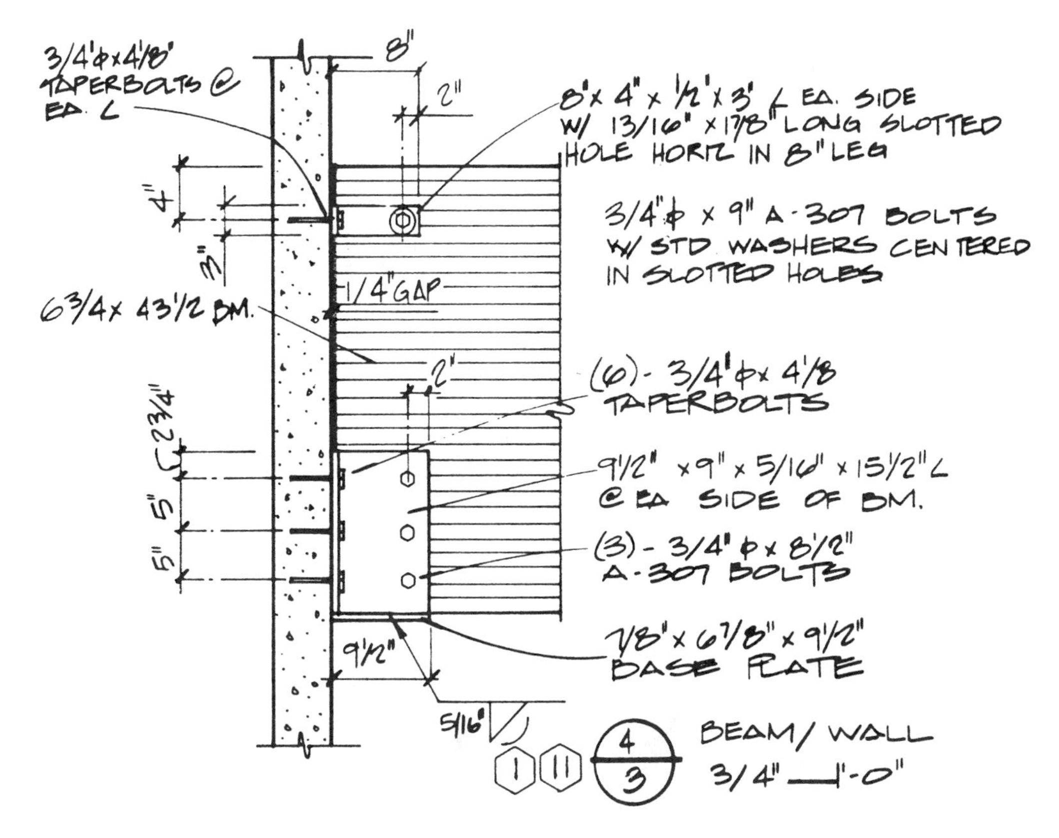

Figure 16–9 A metal hanger may be used in place of a pocket to provide support for a beam.

dictate which of these concrete construction methods will be used.

Cast in Place

Cast-in-place concrete is often used for walls, columns and floors above ground. Walls and columns are usually constructed by setting steel reinforcing in place and then surrounding it by wooden forms to contain the concrete. Once the concrete has been poured and allowed to set, the forms can be removed. It will be necessary to work with details showing not only sizes of the part to be constructed, but also steel placement within the wall or column. This will typically consist of drawing the vertical steel and the horizontal ties. Ties are wrapped around the vertical steel to keep the column from separating when placed under a load. Figure 16–10 shows two examples of column reinforcing methods. Figure 16–11 shows the drawings required to detail the construction of a rectangular concrete column. Depending on the complexity of the object to be formed, details for the column and for the forming system may be supplied.

Concrete is also used on commercial projects to form an above-grade floor. The floor slab either can be supported by a steel deck or be entirely self-supporting. The steel deck system is typically used

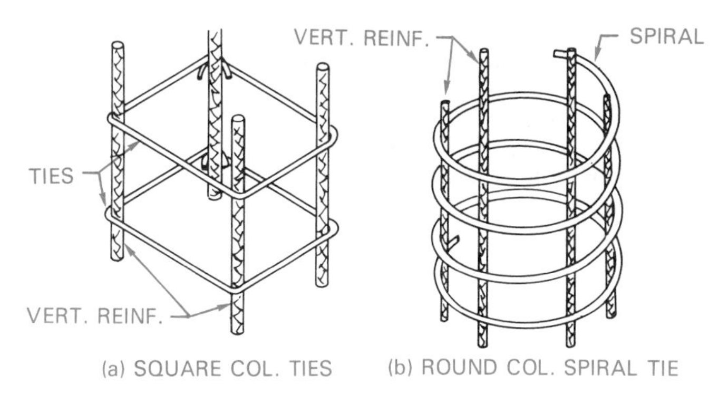

Figure 16–10 Two common methods of reinforcing concrete columns are with evenly spaced ties or with continuous spiral steel.

on structures constructed with a steel frame. Two of the most common poured-in-place concrete floor systems are the ribbed and waffle floor methods. Each can be seen in Figure 16–12.

The ribbed system is used in many office buildings. The ribs serve as floor joists to support the slab but are actually part of the slab. Spacing of the ribs will vary depending on the span and the reinforcing material. The waffle system is used to provide added support for the floor slab and is typically used in the floor system of parking garages.

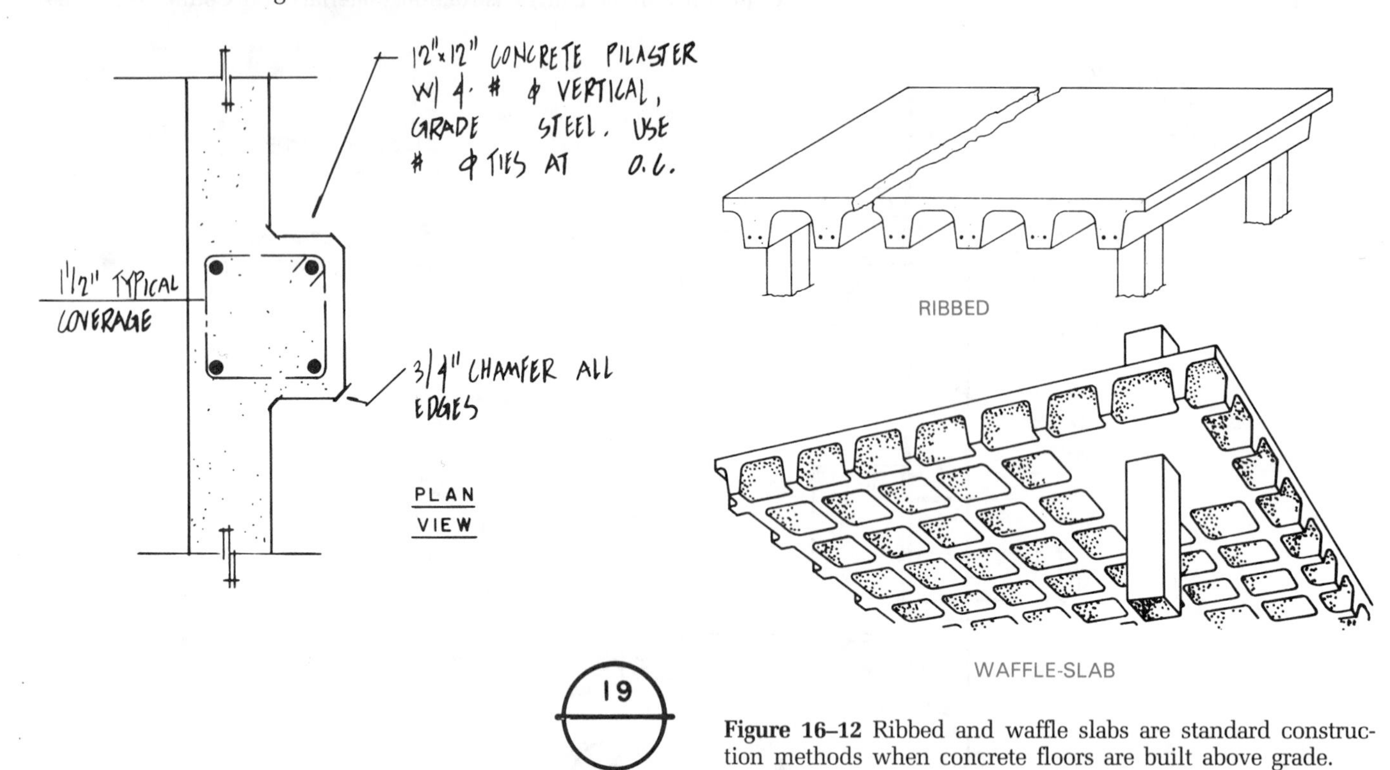

Figure 16–12 Ribbed and waffle slabs are standard construction methods when concrete floors are built above grade.

Figure 16–11 Details will be provided to show steel reinforcing location and spacing. *Courtesy of Structureform Masters.*

Precast Concrete

Precast concrete construction consists of forming walls or other components offsite and transporting the part to the job site. In addition to illustrating how precast members will be constructed, drawings may also include methods of transporting and lifting the part into place. Precast parts typically have an exposed metal flange so that the part can be connected to other parts. Figure 16–13 shows common details used for wall connections.

In addition to being precast, many concrete products are also prestressed. Concrete is prestressed by placing steel cables held in tension between the concrete forms while the concrete is poured around them. Once the concrete has hardened, the tension on the cables is released. As the cables attempt to regain their original shape, compression pressure is created within the concrete. The compression in the concrete helps prevent cracking and deflection and often allows the size of the member to be reduced. Figure 16–14 shows common shapes that are typically used in prestressed construction.

Tilt-up

Tilt-up construction is a method using preformed wall panels which are lifted into place. Panels may either be formed at the job or off site. Forms for a wall are constructed in a horizontal position and the required steel is placed in the form. Concrete is then poured around the steel and allowed to harden. Once the panel has reached its design strength, it can then be lifted into place. When using this type of construction, you will usually be working with a plan view to specify the panel locations as seen in Figure 16–15. Figure 16–16 shows an example of a typical steel placement drawing for a concrete panel.

STEEL CONSTRUCTION

Steel construction can be divided into the three categories of steel studs, prefabricated steel structures, and steel-framed structures.

Steel Studs

Prefabricated steel studs are used in many types of commercial structures to help meet the requirements of types 1, 2, and 3 construction methods. Steel studs offer lightweight, noncombustible, corrosion-resistant framing for interior walls, and load-bearing exterior walls up to four stories high. Steel members are available for use as studs or joists. Members are designed for rapid assembly and are predrilled for electrical and plumbing conduits. The standard 24 in. spacing reduces the number of studs required by about one-third when compared with

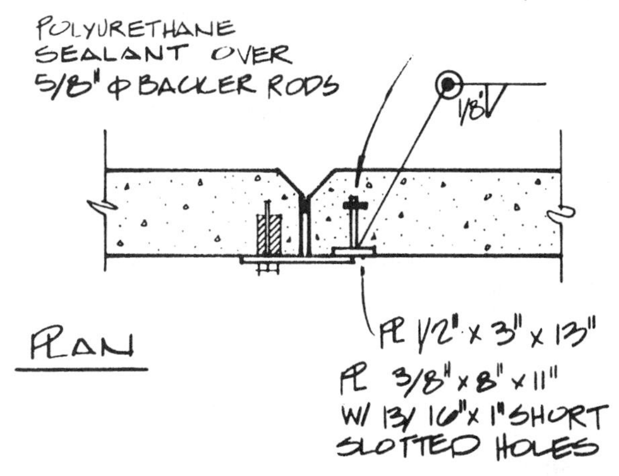

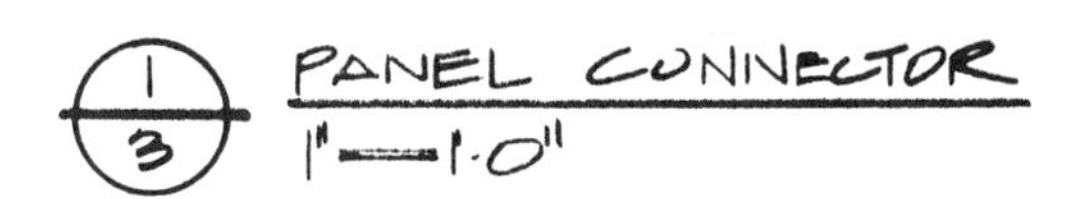

Figure 16–13 Wall details are provided to show how concrete panels are to be lifted, assembled and connected.

common studs spacing of 16 in. O.C. Widths of studs can range from 3⅝ to 10 in. but can be manufactured in any width. The material used to produce studs ranges from 12 to 20 gage steel, depending on the loads to be supported. Steel studs are mounted in a channel track at the top and bottom of the wall. This channel is similar to the top and bottom plates of a standard stud wall. Horizontal bridging is placed through the predrilled holes in the studs and then welded to the stud serving a function similar to solid blocking in a stud wall. Figure 16–17 shows components of steel stud framing.

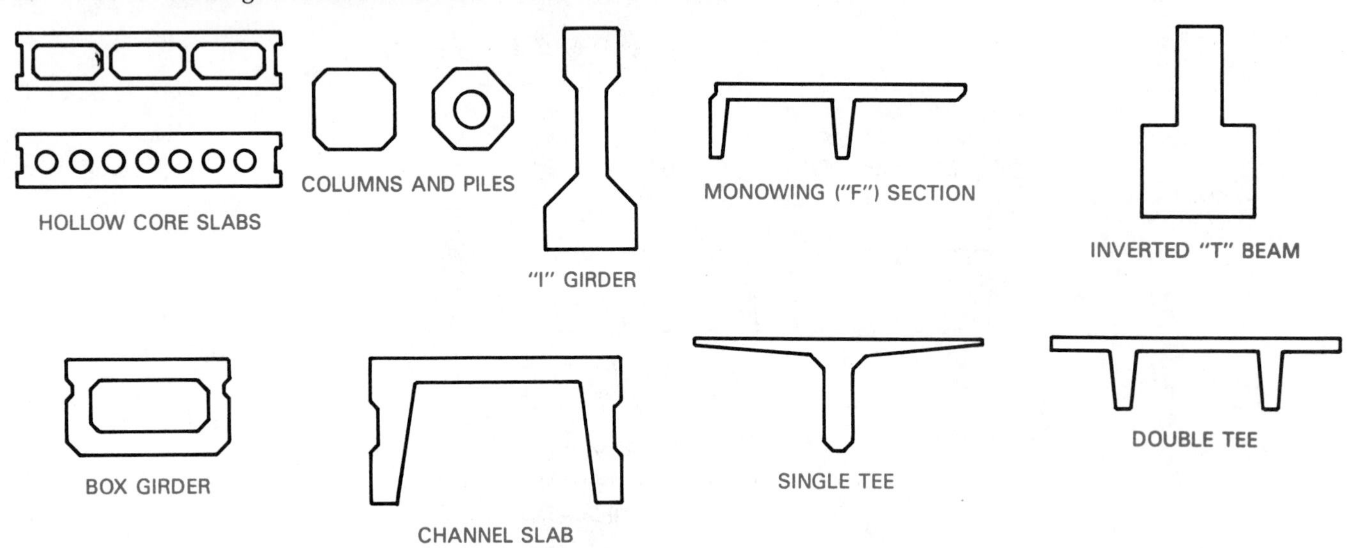

Figure 16–14 Common precast concrete shapes used in construction.

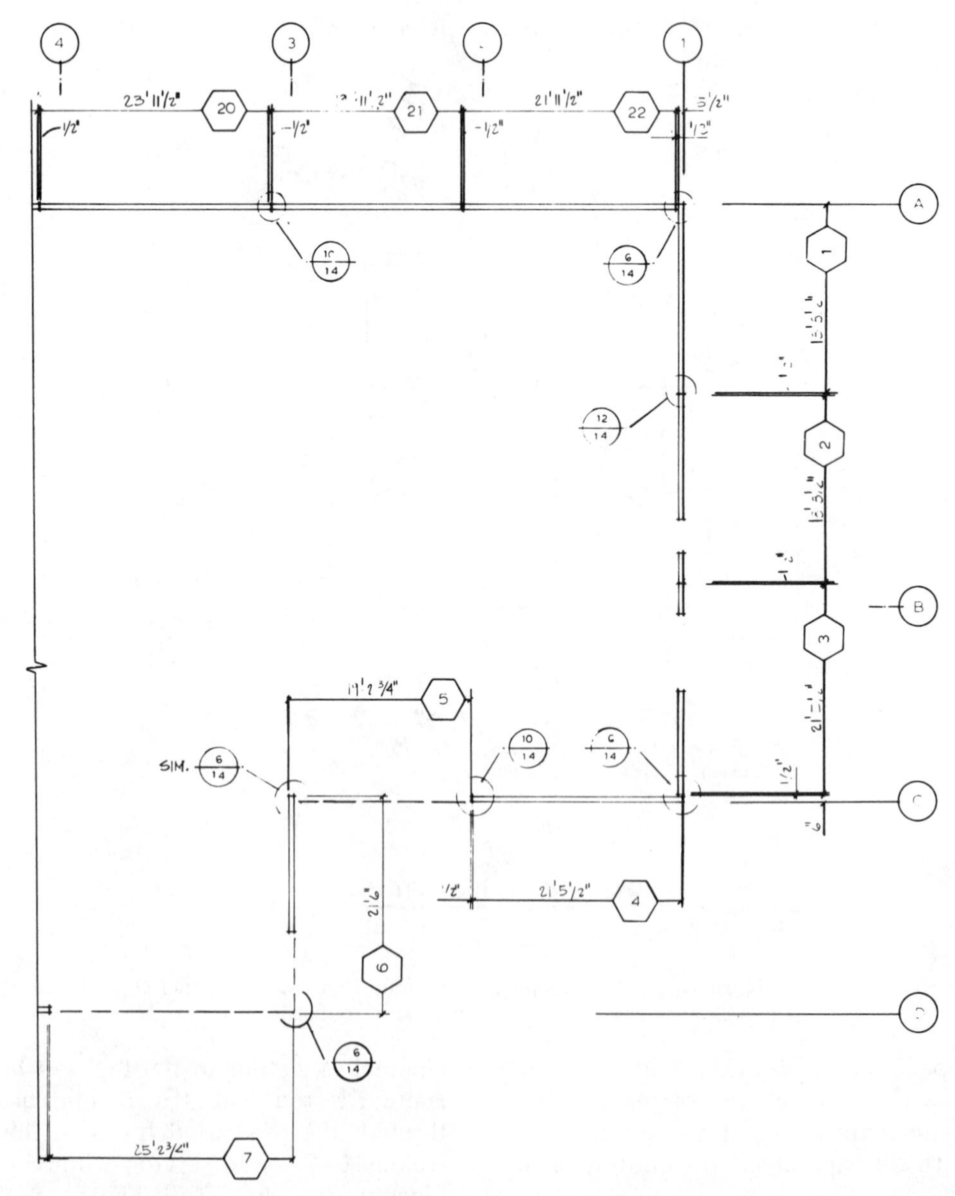

Figure 16–15 Looking similar to a floor plan, the panel plan is provided to show where each concrete panel will be placed. *Courtesy of Structureform Masters.*

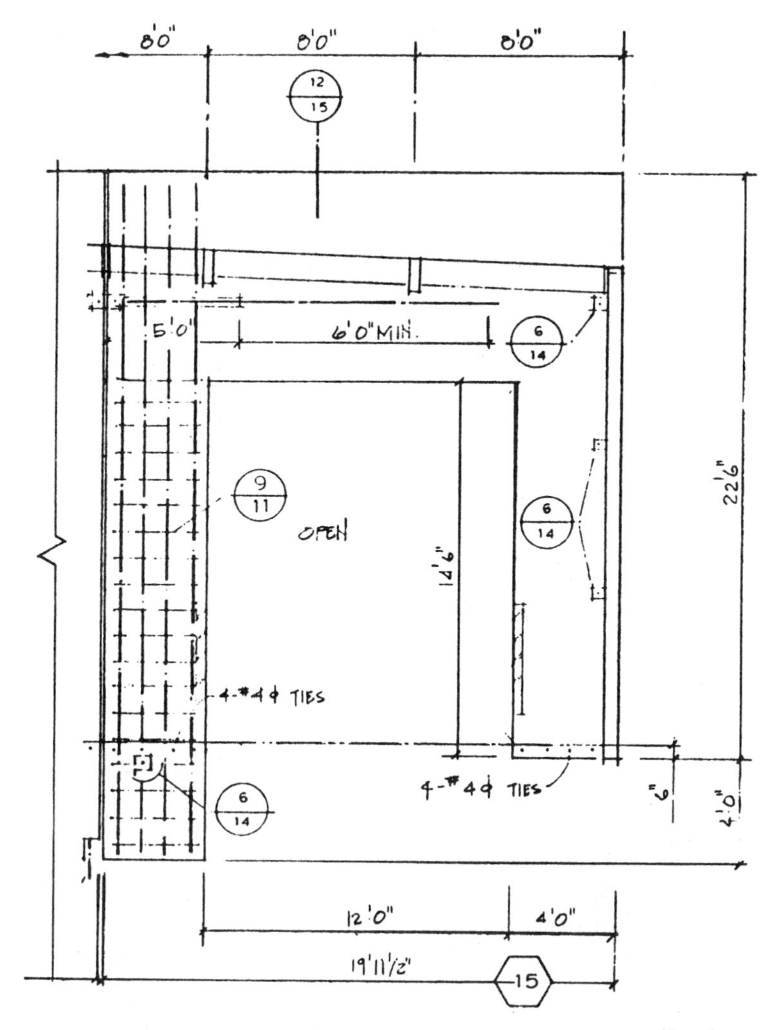

Figure 16–16 A panel elevation is provided to show opening locations and sizes and reinforcing steel for each panel. *Courtesy Structureform Masters.*

Prefabricated Steel Structures

Prefabricated or rigid frame buildings are common in many areas of the country because they provide fast erection time when compared with other types of construction. Standardized premanufactured steel buildings are sold as modular units with given spans, wall heights, and lengths in 12' or 20' increments.

The structural system is made up of the frame that supports the walls and roof. There are several different types of structural systems typically used, with the most typically used shown in Figure 16–18. The vertical wall member is bolted to what will become the inclined roof member to form one rigid member. This frame member is similar to a three hinged glu-lam arch.

The wall system is made of horizontal steel girts attached to the vertical frame. Usually the girts are a standard channel which is welded to the main frame. Metal siding is bolted to the girts to complete the walls. The roof system is steel purlins which span between the main frame. Steel roofing panels are bolted to the purlins. Figure 16–19 shows common methods of building with prefabricated structures.

Steel-Framed Buildings

Steel-framed buildings will require engineering and shop drawings similar to those used for concrete structures. You will most likely be working with engineering drawings similar to the one in Figure 16–20 and the shop drawings similar to the one in Figure 16–21.

Common Steel Products

Structural steel is typically identified as a plate, a bar, or by its shape. Plates are flat pieces of steel of various thickness used at the intersection of different members. Plates are typically specified on a drawing by giving the thickness, width, and length. The symbol ¶ is often used to specify plate material.

Bars are the smallest of structural steel products. Bars are either round, square, rectangular or hexagonal when seen in cross section. Bars are often used as supports or braces for other steel parts.

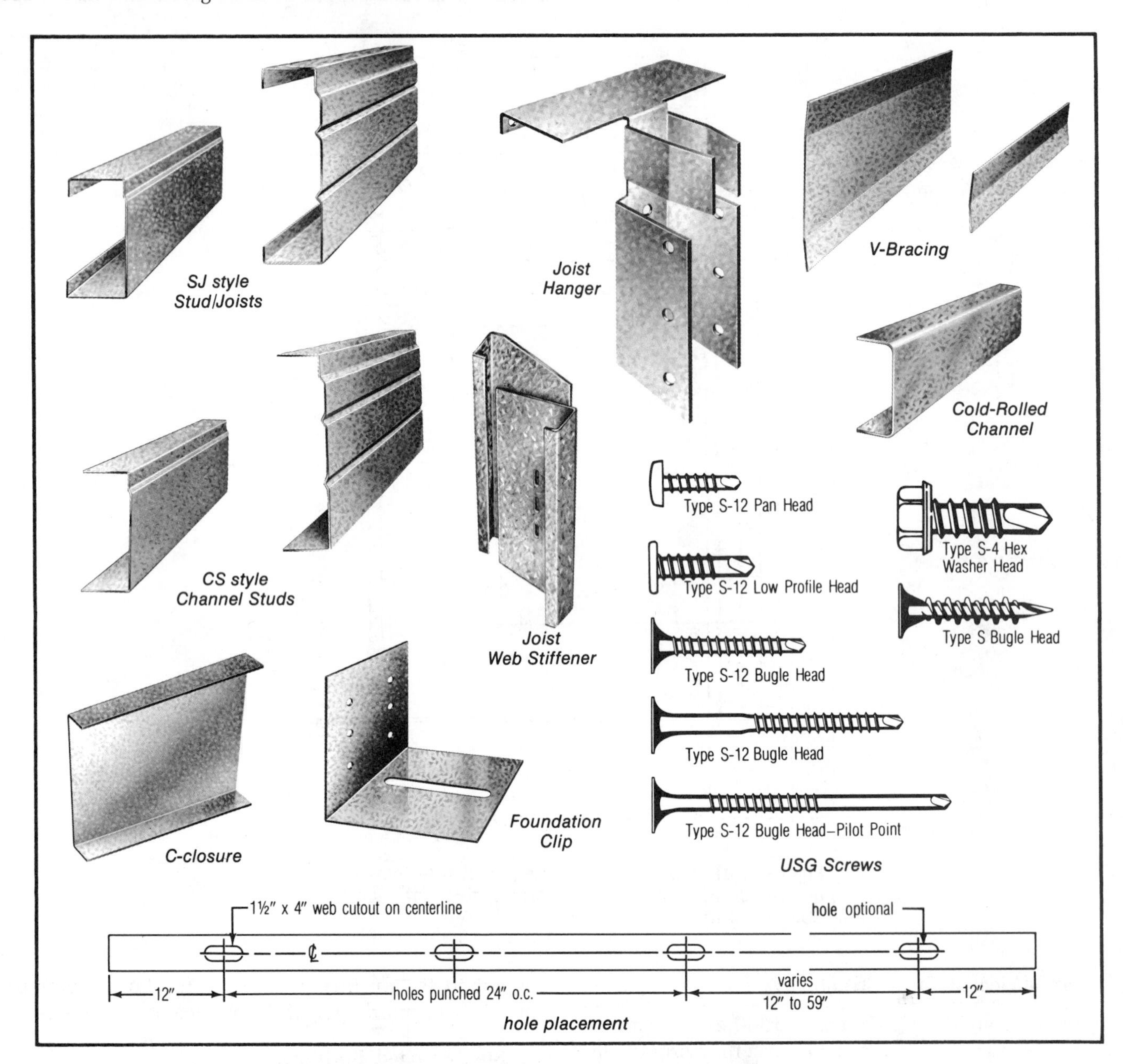

Figure 16–17 Typical components of steel stud construction. *Courtesy of United States Gypsum Corp.*

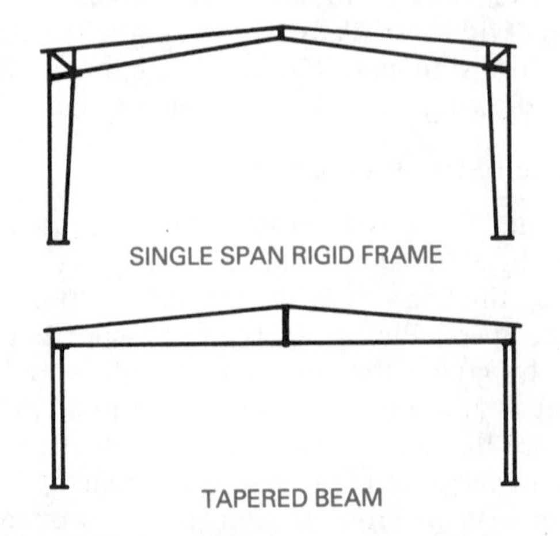

Figure 16–18 Common shapes of the vertical rigid frame of prefabricated steel structures.

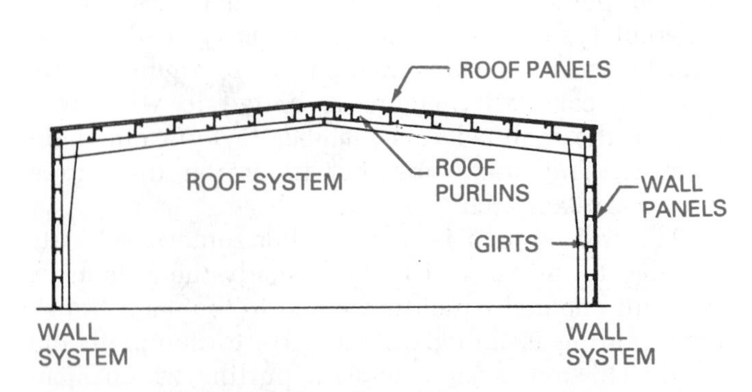

Figure 16–19 Horizontal steel purlins are placed between the rigid frame to support metal siding and roofing panels.

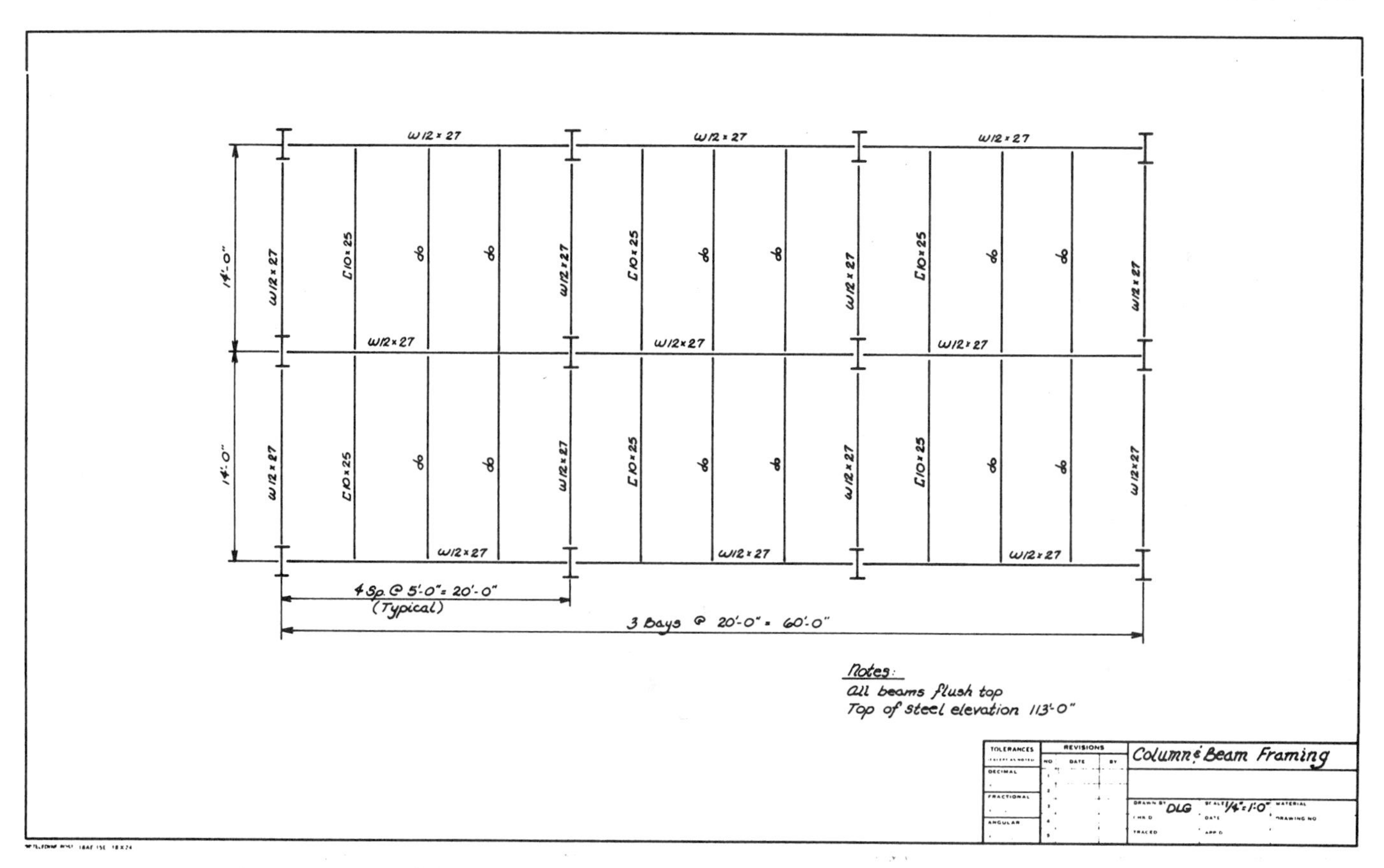

Figure 16–20 Steel frame construction requires drawings showing the location of each structural member. *From* Structural Drafting by Goetsch, *Delmar Publishers.*

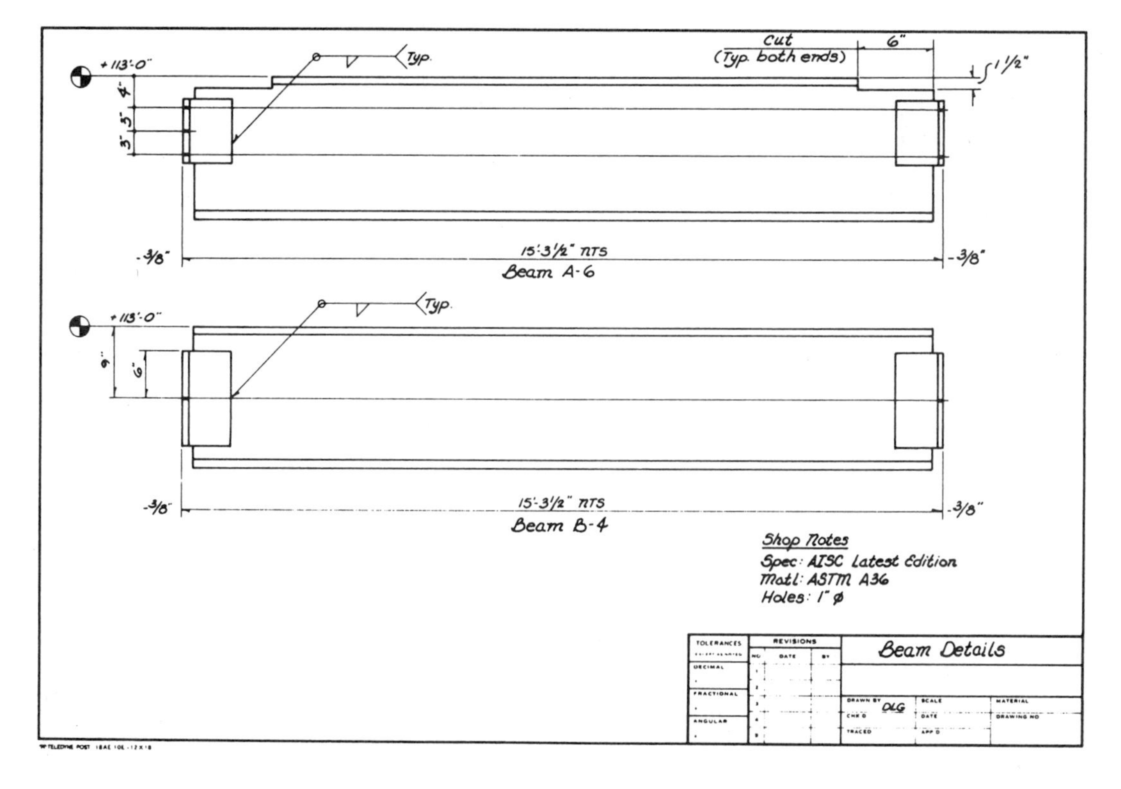

Figure 16–21 Fabrication drawings are needed for each steel member to show how it will be formed. *From* Structural Drafting by Goetsch, *Delmar Publishers.*

Structural steel is typically produced in the shapes that are seen in Figure 16–22. M, S, and W are the names given to steel shapes that have a cross-sectional area in the shape of the letter I. The three differ in the width of their flanges. The flange is the horizontal leg of the I shape and the vertical leg is the web. In addition to varied flange widths, the S shape flanges vary in depth.

Angles are structural steel components that have an L shape. The legs of the angle may be either equal or unequal in length but are usually equal in thickness. Channels have a squared C cross-sectional area and are represented by the letter C when specified in note form. Structural tees are cut from W, S, and M steel shapes by cutting the webs. Common designations include WT, ST, and MT.

Structural tubing is manufactured in square, rectangular, and round cross-sectional configurations. These members are used as columns to support loads from other members. Tubes are specified by the size of the outer wall followed by the thickness of the wall.

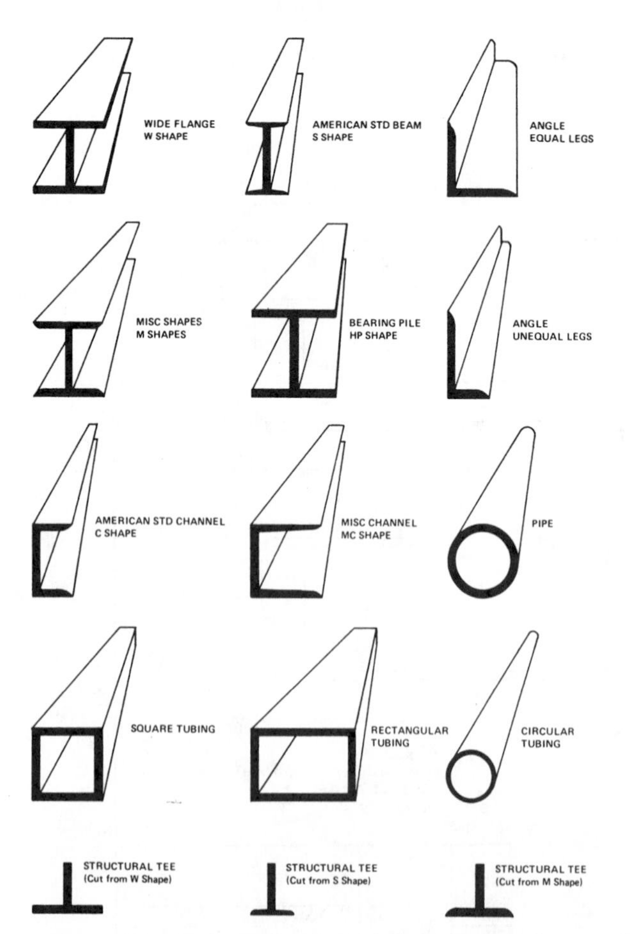

Figure 16–22 Standard structural steel shapes. *From* Structural Drafting by Goetsch, *Delmar Publishers.*

COMMON CONNECTION METHODS

Because different materials are being joined together, fast, economical, and safe methods must be used. The connection method selected by the engineer must be both suitable to the load to be held or transferred, and to the stress that will result at the connection. Nails, staples, bolts, and welds are the most common methods of connecting materials.

Nailing

Nails are often used for the fabrication of wood to wood members when their thickness is less than 1½". Thicker wood assemblies are normally bolted. Nailing patterns are usually described by the term penny and denoted by the letter d. Penny is the weight classification of nails determined by comparing the number of pounds of nails per 1000 nails. A thousand 16d nails would weigh 16 pounds. Penny is used to describe standard nails from 2d through 60d although nails larger than 20d are not common. Nails larger than 20d are typically specified by their diameter and may be referred to as spikes. Nails smaller than 20d are typically specified by their penny size and a spacing. If a pilot hole is required, the diameter is normally specified after the nail spacing. Figure 16–23 shows common sizes and types of nails.

An example of a typical nail callout might read:

16-2od @ 3" o.c. pre-drill w/ 5/32" holes.

Nailing locations are often specified by placement of edge, field or boundary. Edge nailing refers to nails placed at the edge of a sheet of plywood. Field nailing refers to the nails placed into the supports at the center of a piece of plywood. Boundary nailing refers to the location of nails placed around the edge of an entire area of plywood such as a shear wall or an area of a roof.

Typically a set of plans will include a nailing schedule based on the governing code for your area. This schedule will indicate nail placement and quantity for connections of structural members such as joist to plate, plate-to-studs, and studs-to-base plate.

Staples

Because of the speed involved, many construction crews have switched from nailing to staples. Staples are most often used for connecting roof, wall and floor sheathing to their respective support members.

Bolts

Bolts are used to provide a rigid connection between wood, steel, and concrete members. Bolts are typically used to connect wood to wood, wood

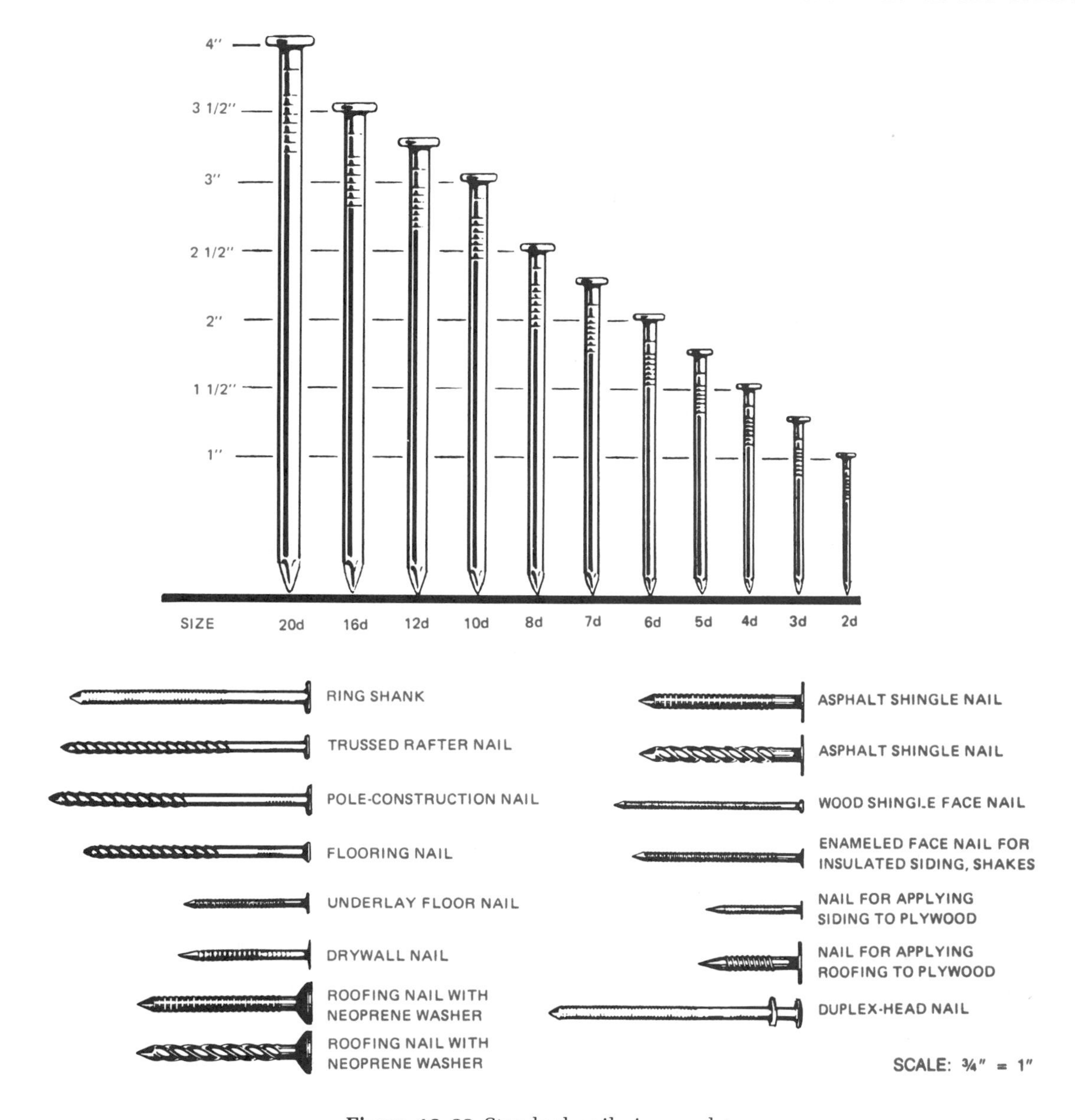

Figure 16–23 Standard nail sizes and types.

to steel, wood to concrete, and steel to concrete. Although bolts can be used to connect steel to steel, these connections are often made by welding.

When specified on plans, you should be able to find the diameter, length, and strength of the bolt. Another major concern when working with bolted connections is the location of the bolts. Many plans will contain a list of general specifications for bolts placed in concrete, steel or wood listing minimum area to surround the bolt. For instance bolts passing through wood may be required to be 1½" from an edge parallel to the grain and 3" minimum from the edge when perpendicular. Bolts in steel are often placed 1½" minimum from each edge, whereas 2" is preferred by many engineers for bolts from concrete. These general guidelines should be checked carefully on each plan.

Bolt strength is typically classified in accordance with the American Society for Testing Materials (ASTM) or American Institute of Steel Construction specifications. Some of the most common bolts specified in construction include:

A 307 Low carbon steel bolts
A 325 High-strength bolts
A 490 High-strength bolts
A 441 High-strength, low-alloy steel bolts
A 242 Corrosion-resistant high strength low alloy bolts

Washers or plates are also specified with bolt callouts. These pieces of hardware keep the bolt from pulling through the bolt hole. Washers are typically specified by a diameter, whereas plates are specified by a height, length, and thickness.

Welding

Welding is a method of providing a rigid connection between two or more pieces of steel. Through the process of welding, metal is heated to a temperature high enough to cause melting of the melting. The parts that are welded become one with the welded joint actually stronger than the original material. Welding offers better strength, better weight distribution of supported loads, and a greater resistance to shear or rotational forces than bolting can provide. There are a large number of welding processes available to industry with the most common welds in the construction field being shielded metal arc welding, gas tungsten arc welding, and gas metal arc welding.

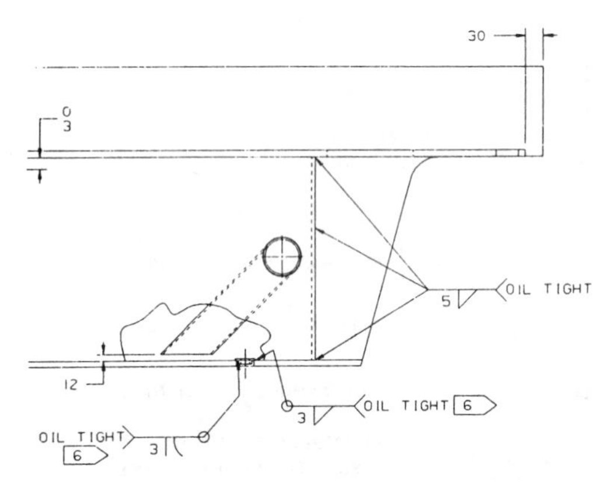

Figure 16–24 A reference line and leader line are used to show where a weld will take place. *Courtesy of Hyster-Yale Materials Handling, Inc.*

Welds are specified on a drawing by the use of a horizontal reference line connected to the parts to be welded by an inclined line with an arrow. The arrow touches the area to be connected. It is not uncommon to see the welding line bend to point into difficult to reach places or with more than one leader line extending from the reference line as seen in Figure 16–24. Information about the type of weld, the location of the weld, the welding process, and the size and length of the weld is all specified on or near the reference line. Figure 16–25 shows a welding symbol and the proper location of information.

Types of Welds

The type of weld is associated with the weld shape and/or the type of groove to which the weld is applied. Figure 16–26 shows the information that is associated with the types of welds.

Fillet Weld. A *fillet weld* is formed in the internal corner of the angle formed by two pieces of metal. The size of the fillet weld is shown on the same side of the reference line as the weld symbol and to the left of the symbol.

Square Groove Weld. A *square groove weld* is applied to a butt joint between two pieces of metal. The two pieces of metal will be spaced apart a given distance, known as the root opening. If the root opening distance is a standard in the company, this dimension is assumed. If the root opening is not standard,

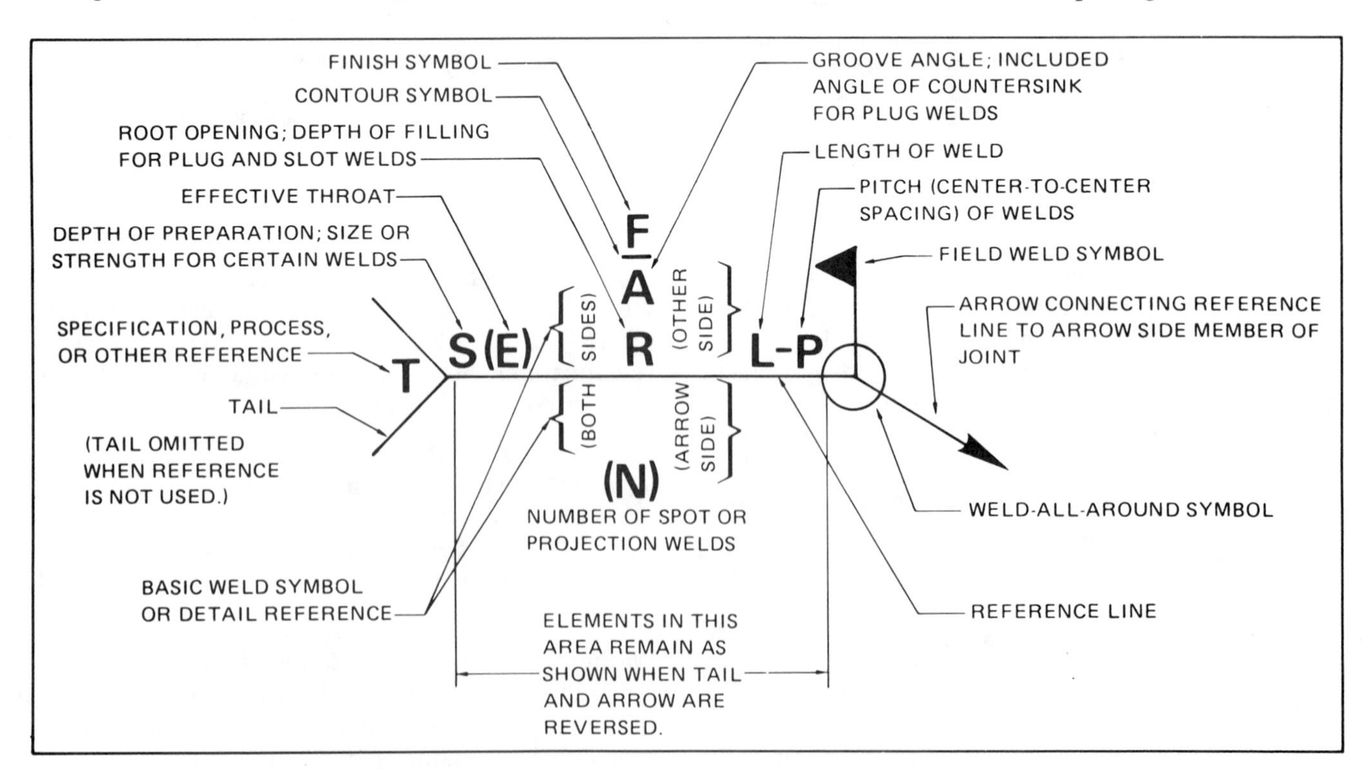

Figure 16–25 Standard location of elements of a welding symbol. *Courtesy of American Welding Society.*

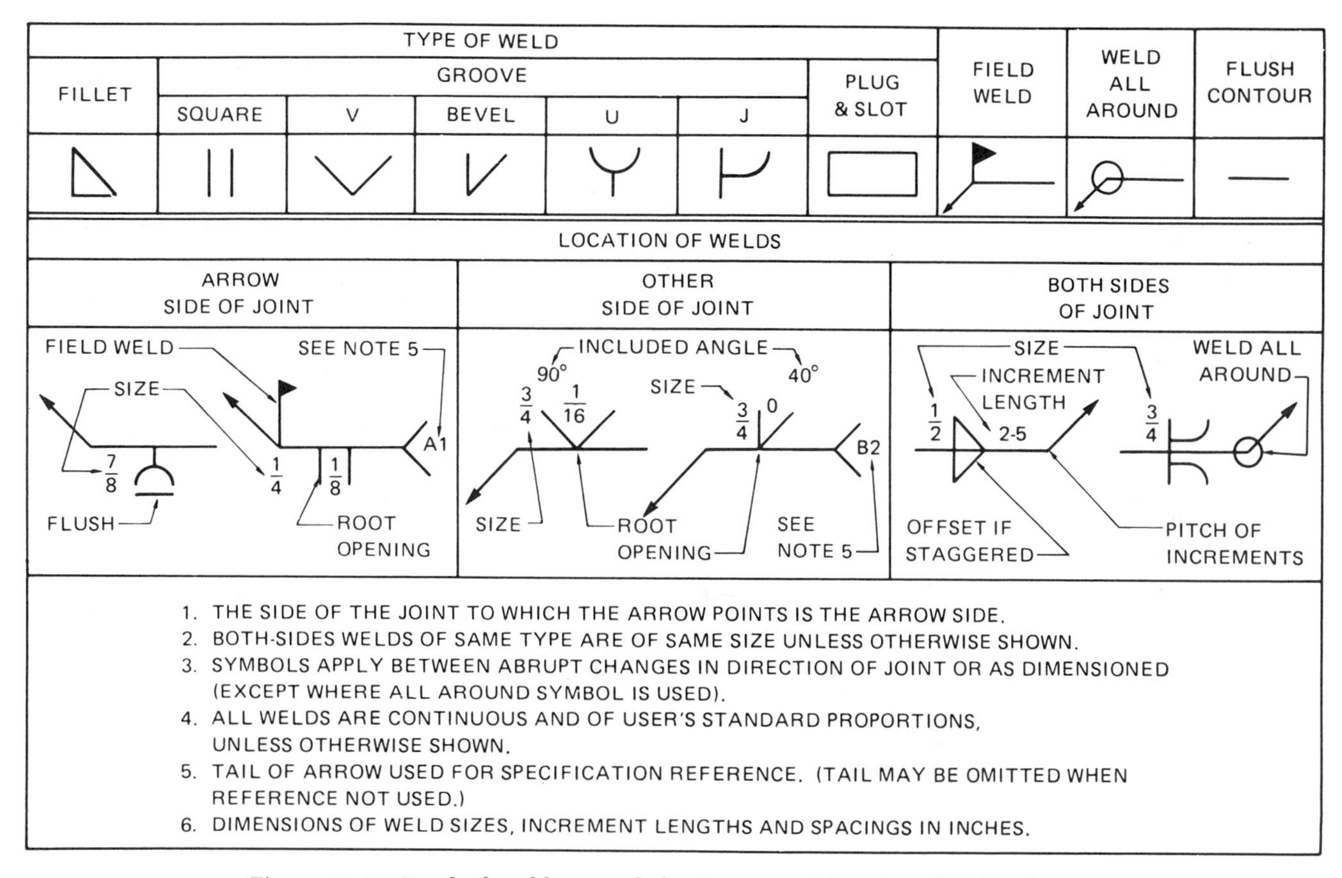

Figure 16–26 Standard welding symbols. *Courtesy of American Welding Society.*

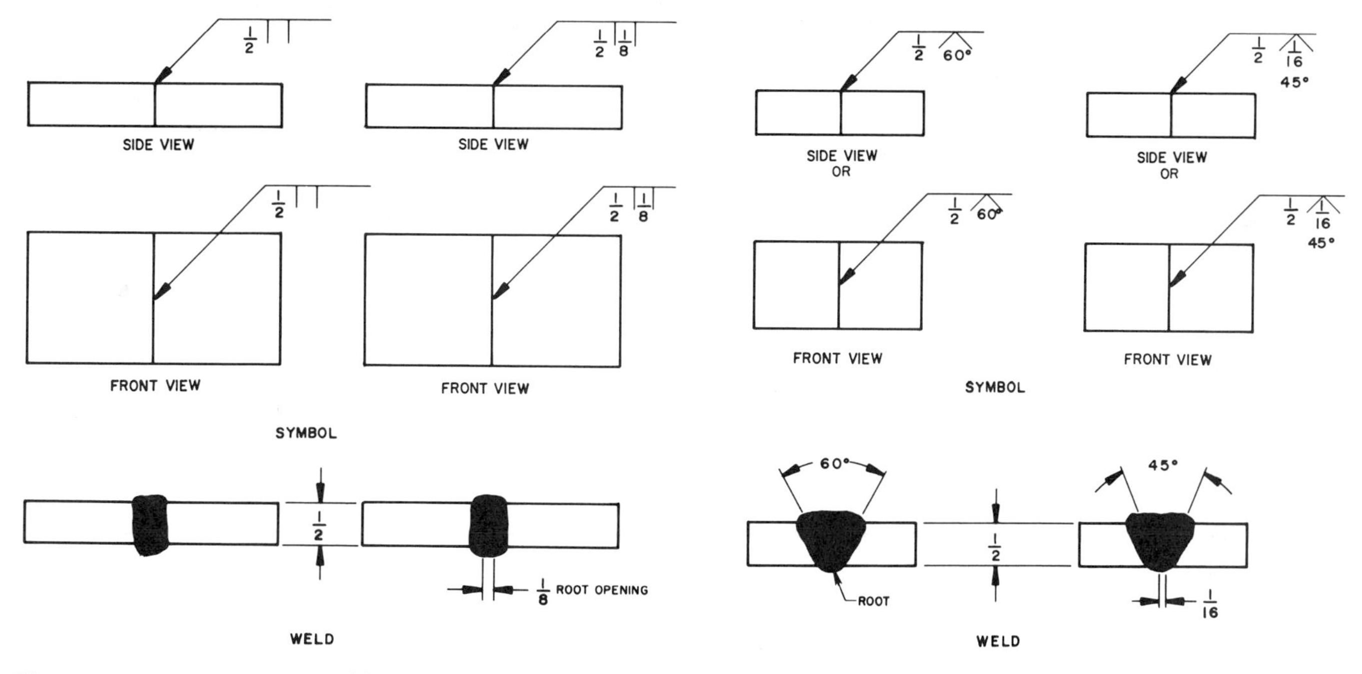

Figure 16–27 Square groove welds.

Figure 16–28 *V* groove welds.

the specified dimension is given to the left of the square groove symbol, as shown in Figure 16–27.

V-Groove Weld. A *V groove weld* is formed between two adjacent parts when the side of each part is beveled to form a groove between the parts in the shape of a V. The included angle of the V may be given with or without a root opening, as shown in Figure 16–28.

Bevel Groove Weld. The *Bevel groove weld* is created when one piece is square and the other piece has a beveled surface. The bevel weld may be given with a bevel angle and a root opening, as shown in Figure 16–29.

U Groove Weld. A *U groove weld* is created when the groove between two parts is in the form of a U. The angle formed by the sides of the U shape, the root, and the weld size are generally given. (See Figure 16–30.)

J Groove Weld. The *J groove weld* is necessary when one piece is a square cut and the other piece is in a J-shaped groove. The included angle, the root opening, and the weld size are given, as shown in Figure 16–31.

Field Weld. A *field weld* is a weld that will be performed in the field as opposed to in a fabrication shop. The field weld symbol is a flag attached to the reference line at the leader intersection, as shown in Figure 16–32.

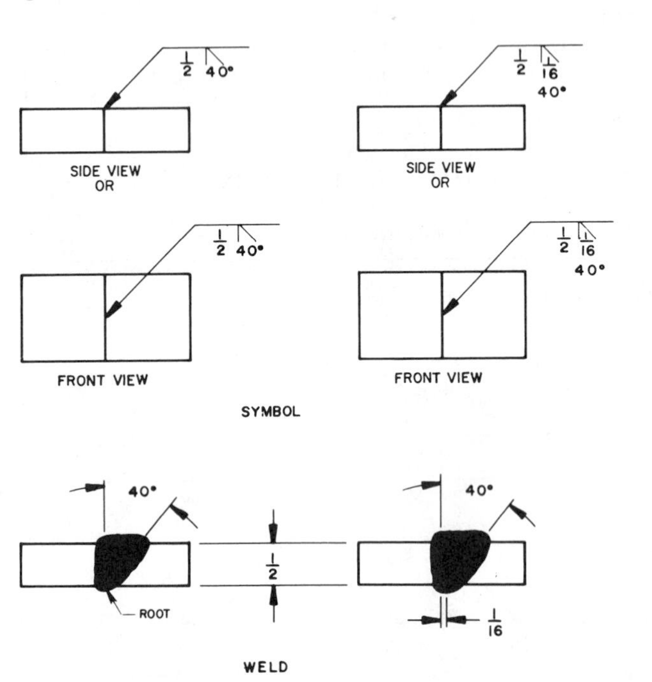

Figure 16–29 Bevel groove welds.

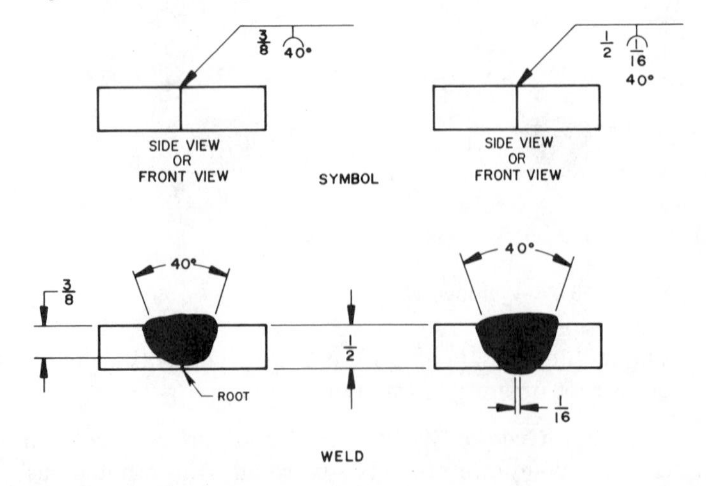

Figure 16–30 *U* groove welds.

Weld-all-around. Many offices still use the old symbol of a filled-in circle located at the bend of the leader line. When a welded connection must be performed all around a feature, the *weld-all-around* symbol is attached to the reference line at the junction of the leader. This will clarify that the weld surrounds the feature as opposed to a certain increment or length. (See Figure 16–25.)

Weld-Length and Increment. When a weld is not continuous along the length of a part, the weld length is given. In some situations, the weld along the length of a feature will be given in lengths spaced a given distance apart. The distance from one point on a weld length to the same corresponding point on the next weld is called the pitch; generally from center

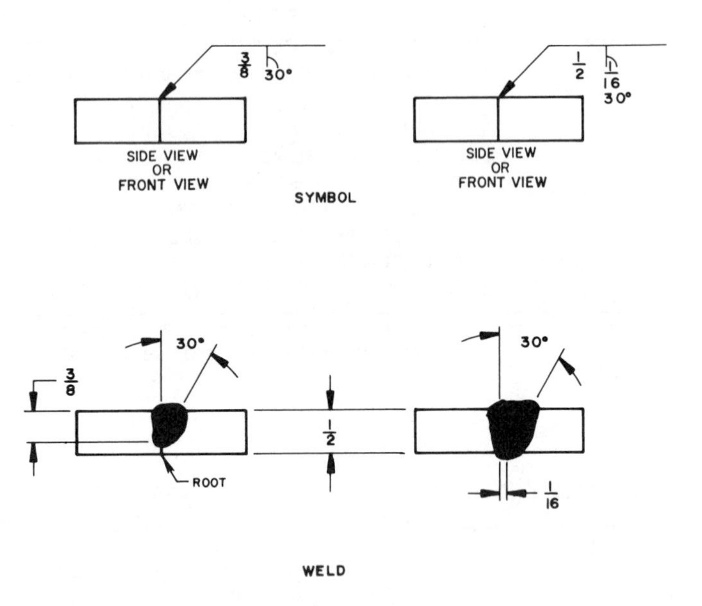

Figure 16–31 *J* groove welds.

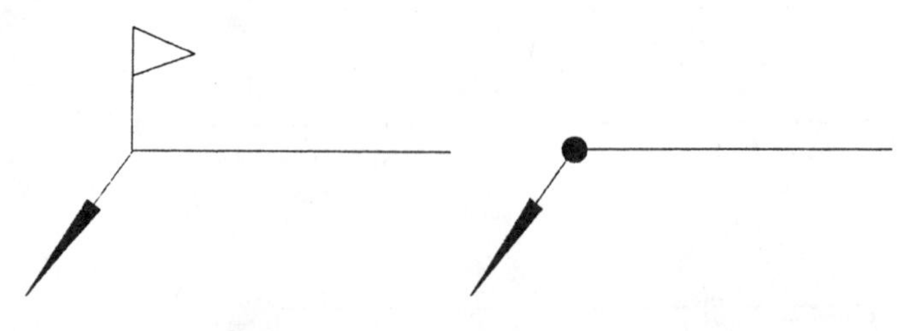

Figure 16–32 Two common methods of specifying field welds.

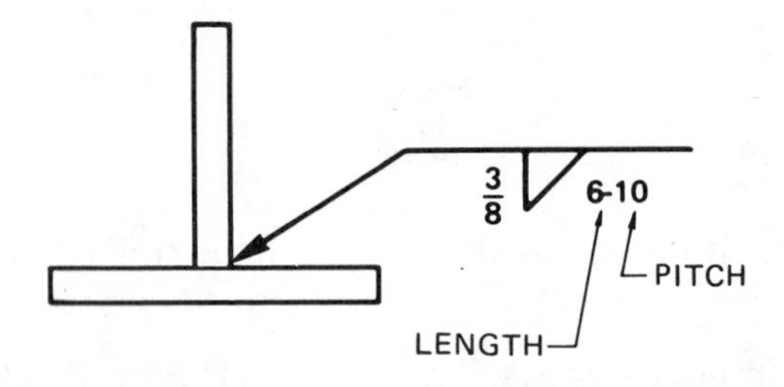

Figure 16–33 The weld length and spacing are shown to the right of the weld symbol.

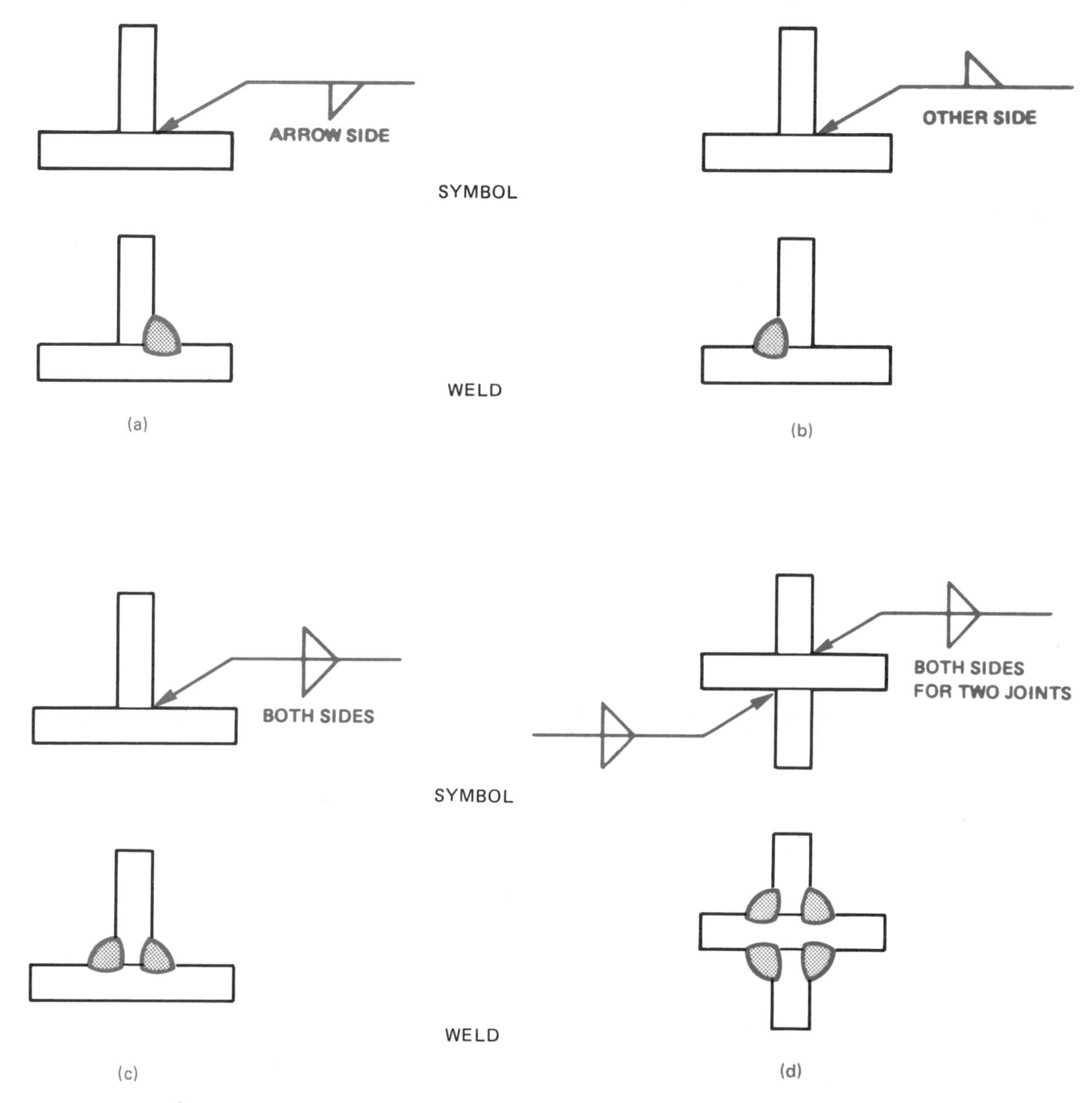

Figure 16–34 The placement of the symbol above or below the line specifies where the weld will take place. *Courtesy of American Welding Society.*

to center of welds. The weld length and increment are shown to the right of the weld symbol, as shown in Figure 16–33.

Weld Symbol Leader Arrow Related to Weld Location

Welding symbols are applied to the joint as the basic reference. All joints have an *arrow side* and an *other side.* When fillet and groove welds are used, the welding symbol leader arrows connect the symbol reference line to one side of the joint known as the arrow side. The side opposite the location of the arrow is called the other side. If the weld is to be deposited on the arrow side of the joint, the proper weld symbol is placed below the reference line, as shown in Figure 16–34*a*. If the weld is to be deposited on the side of the joint opposite the arrow, the weld symbol is placed above the reference line. (See Figure 16–34*b*.) When welds are to be deposited on both sides of the joint, then the same weld symbol is shown above and below the reference line, as shown in Figure 16–34*c* and *d*.

Weld Joints. The types of weld joints are often closely associated with the types of weld grooves already discussed. The weld grooves may be applied to any of the typical joint types. The weld joints used in most weldments are the butt, lap, tee, outside corner, and edge joints shown in Figure 16–35.

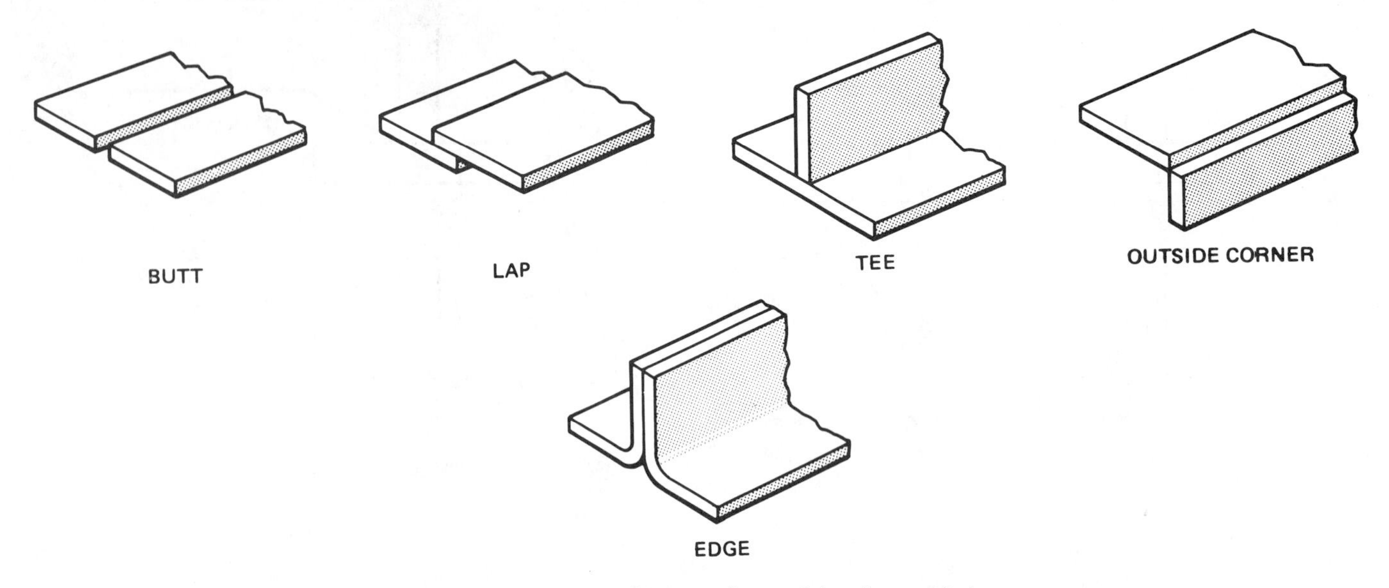

Figure 16–35 Types of joints of material to be welded.

CHAPTER 16 TEST

1. What is the most common type of wood construction used? ______________________________.
2. Why are stud sizes often larger in commercial than residential construction? ______________________________.
3. What is the typical method of achieving two-hour construction using wood construction? ______________________________.
4. Which chapter of the Uniform Building Code would give a complete listing of fire-proofing methods? ________
5. How would you achieve a one-hour wall if stucco were to be on the exterior side? ______________________________.
6. What qualities do laminated beams offer over sawn lumber. ______________________________.
7. What makes a simple beam? ______________________________.
8. What advantage does a cantilevered beam supporting another beam offer over simple beams. ______________________________.
9. List two different methods of framing roofs. ______________________________.
10. List three common shapes of glu-lam beams. ______________________________.
11. List five common glu-lam beam widths. ______________________________.
12. Glu-lam beams increase in depth by ____________ in. intervals.
13. List two methods or inspecting laminated beams and list the letter specification for each. ______________________________.

14. List four types of wood typically used to make laminated beams AND give the letter specification for each. ____

____

____.

15. List three sizes of concrete blocks other than concrete block. ____

____.

16. How can concrete block construction be protected from seismic damage? ______________________

17. List two types of drawings associated with concrete construction. ______________________

18. What are three methods of forming a concrete floor? ______________________________

____

____.

19. How is concrete prestressed? ____

____

____.

20. Steel studs are typically used to meet construction requirements of type ______________________.

21. What gauge is typically used for steel studs? ______________________________

22. What is the common spacing of the vertical members of rigid frame construction? ______________

23. List four methods of connecting material used in construction. ______________________

____

____.

24. How long is a 6d nail? ____

25. What term is used to describe nailing around an area of plywood? ______________________

26. How should ceiling joist be attached to the top plate? ______________________________

27. What type of bolt is an A-307? ____

28. Draw the symbol to represent the following types of welds.

______________	______________	______________
fillet	'V' bevel	weld all around

29. What type of weld is used when two pieces of steel butt together and each piece has square ends? ______

____.

30. Show each method of specifying that a weld is to be preformed in the field. ______________________

____

____.

Chapter 17
Reading Structural Drawings and Details

IN CHAPTER 17, you were introduced to the drawings that make up a set of working drawings. The core of the working drawings are the structural drawings. These include framing plans for each level of the structure, sections and the construction details that relate to each of these drawings. Understanding the major concepts contained on these drawings as well as how the drawings are integrated into the entire set of plans will greatly increase your ability to read plans.

STRUCTURAL DRAWINGS

The structural drawings contain the majority of information needed to erect the framework of a structure. The area of the country you are in, the type of building to be erected, and the occupancy of the structure will dictate what materials will be used. Common materials seen in structural drawing include information for steel frame, poured concrete, concrete block, steel studs, or wood studs. Typically several of these materials will be incorporated into the framework. The type of material to be used is often easily found on the framing plan.

FRAMING PLANS

Framing plans are normally drawn at a scale of ¼" = 1'–0". It is not uncommon to see framing plans drawn at a smaller scale because of the size of the structure. If a large complex structure is being represented, it may be divided into zones and placed on two or more sheets.

Framing plans are typically drawn for each level of the structure. For a three level structure this would require a roof framing, third, second, and first floor framing, and foundation plan. Notice that the plans are listed from the top of the structure down to the ground. By examining the plans from the top down, the location and need for supports and beams is often better understood.

The main goal of a framing plan is to represent the location and size of framing members such as beams, joist, post, and columns. These are the major components of the frame used to transfer weight to the soil through the foundation. Framing plans are also used to show materials used to resist stress from

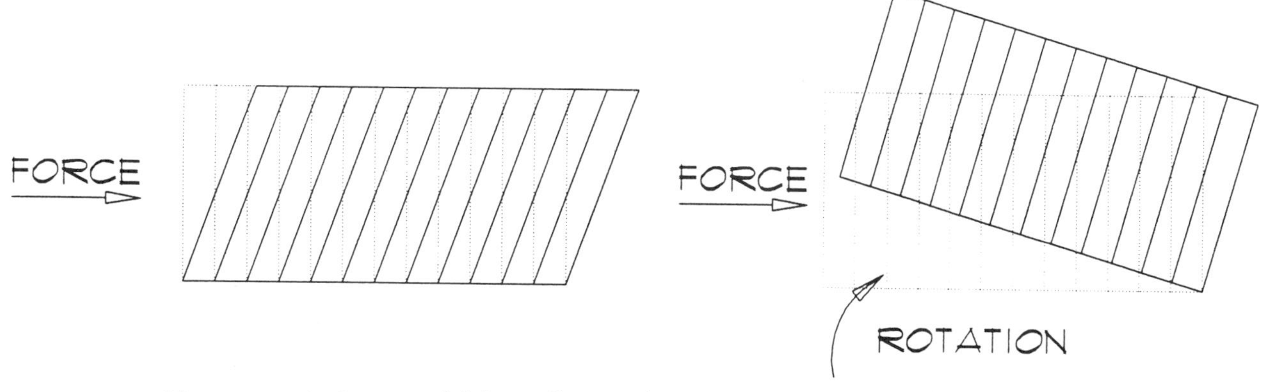

Figure 17–1 A shear panel (plywood) is used to resist motion that tends to turn a wall into a parallelogram. Once the wall is rigid it will tend to rotate as an entire unit.

forces caused by wind, flooding, and seismic activity. This would include information for specific areas regarding nailing, bolting, and welding to resist these stresses.

In thinking about how loads are supported, you would think that the walls hold up the roof. In many structures the walls do support the weight of the roof, but the roof actually keeps the walls erect and perpendicular. This situation can best be seen in a structure with the majority of the exterior support walls covered in glass. By making the roof stiff enough so that it will not twist or buckle, the roof will keep the wall columns from turning into parallelograms.

The rigid areas of a structure used to resist twisting are called diaphragms and may be found on a roof, floor, or wall. These areas can be recognized on framing plans by increased nailing or bolting and smaller spacing of framing members. Often a contractor will see a note requiring floor joist at 8" O.C. with plywood sheathing nailed at 2" O.C. and think the specification must be a mistake. This specification is typical of how a diaphragm would be represented on a framing plan to resist forces that come from outside the building. The second and third floor framing plans for the office at the end of the chapter show an example of a diaphragm on a floor framing plan for wood construction. If steel is used, cross-members placed on the diagonal can be used to form a triangular shape for better resistance. Framing plans using concrete walls will typically reflect added bolting where wood and concrete members intersect to resist lateral stress.

The foundation plan has been discussed in other chapters but should also be considered when thinking about the framing plans. The materials used to resist the loads and stresses of the upper levels must be transferred into the foundation. This will typically result in the use of special bolting patterns or metal connectors. Bolting patterns for a loadbearing wall may normally require 3/4" dia. bolts @ 48" O.C. If the wall has been designed to resist shear forces, the bolts may be placed as close as 12" O.C.

As loads and pressure attempt to turn a wall from a rectangle to a parallelogram, shear panels and diaphragms can be used to resist the stress. These reactions make a wall rigid and cause all of the individual studs to act as a unit, but the force will attempt to flatten the wall in one other way. Since the wall is a rigid unit, it will tend to rotate on one corner as seen in Figure 17–1. To keep the wall from rotating and resist forces of uplifting, a hold-down anchor is often used. These anchors may be used at the foundation level to attach the wall to the foundation or at the intersection of two walls to a floor. Examples of each can be seen in Figure 17–2.

Another major type of information found on framing plans is the location of cutting planes for sections. The cutting plane is used to locate the direction and location of each section. Reference bubbles may also be placed on the framing plans to tie details to a specific area.

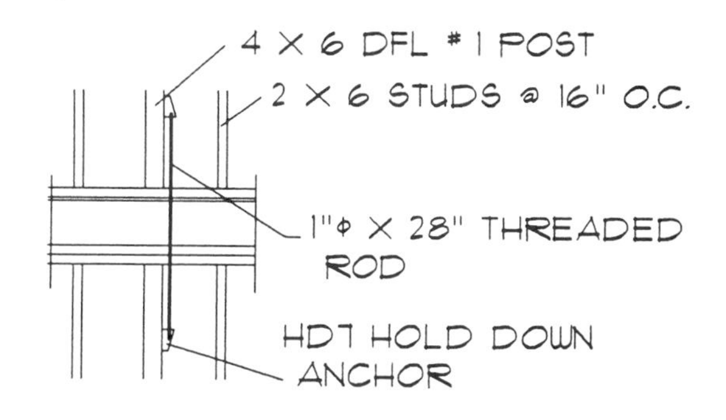

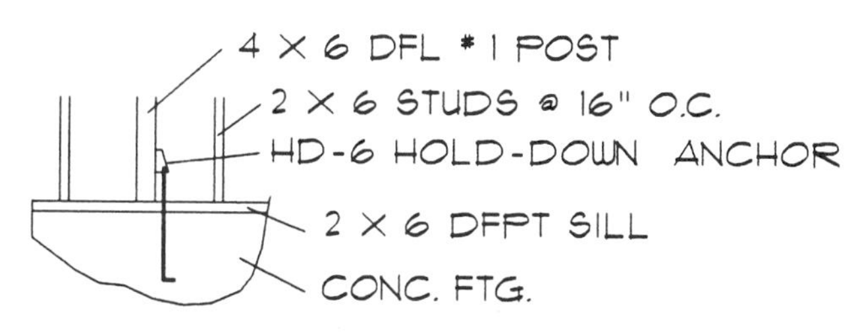

Figure 17–2 Hold-down anchors or tie straps are used to resist up-lift and rotation.

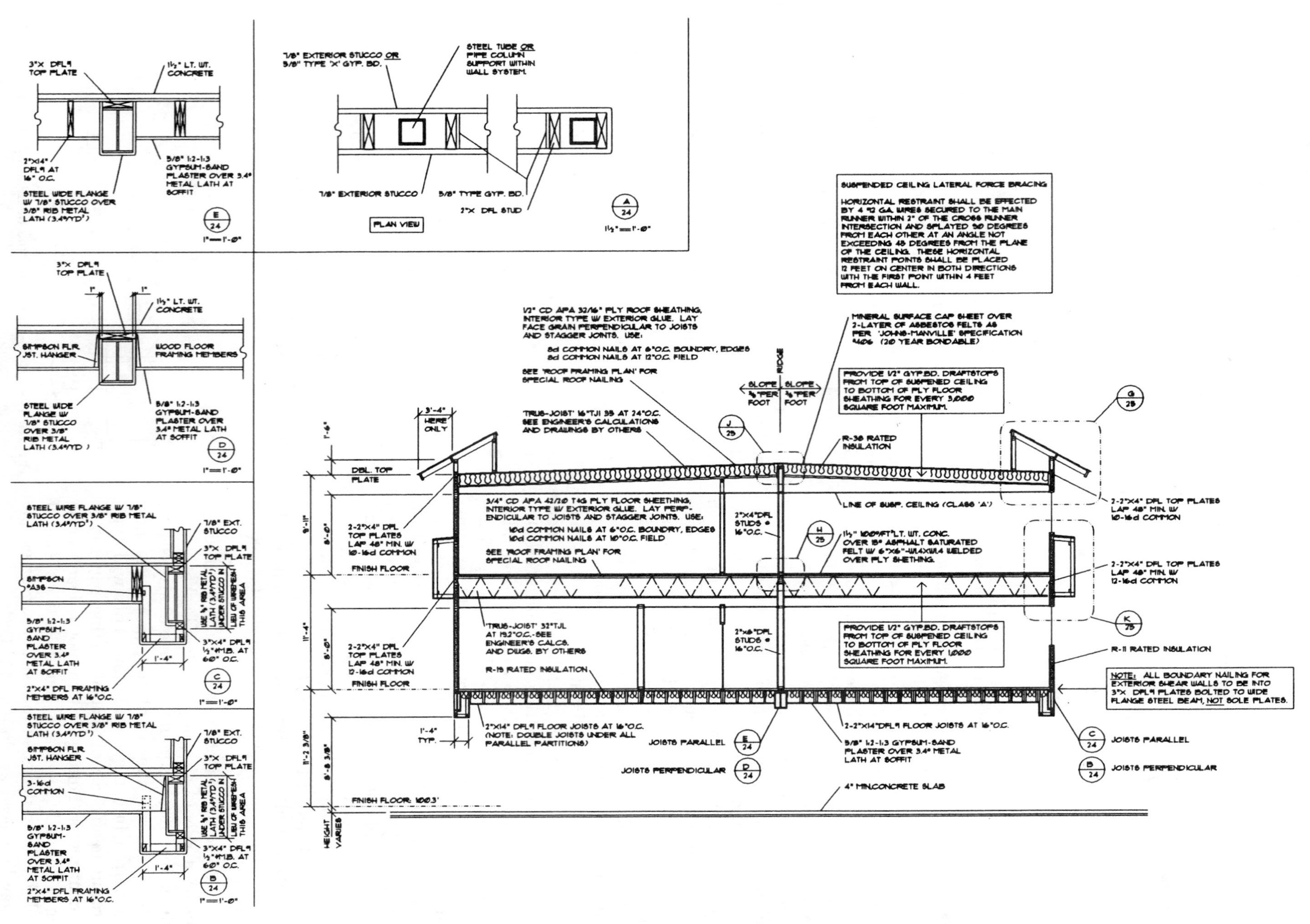

Figure 17–3 Reference bubbles are placed throughout drawings to show how details relate to other drawings. *Courtesy Structureform Masters.*

SECTIONS

Sections are used to show the vertical location of framing members. See chapters 8 and 15 for a complete review of sections and the importance in the set of plans. Sections are typically placed after the framing plans and before details in a set of plans. Most commercial plans use sections as a vertical reference map and try to show very little construction information. Reference bubbles are placed throughout the sections to show how details relate to specific construction methods, as seen in Figure 17–3.

DETAILS

Details are the key to construction. Although the framing plan may specify a shear panel, it is typically the details that will provide information to make all of the required connections. Sections may show how two beams intersect, but details will provide all of the needed dimensions for bolting of hangers or other connectors. When studying details, consider the scale and view orientation before looking for specific information. Details will vary greatly in scale. A scale of ¾" = 1'–0" through 1½" = 1'–0" is used to draw construction details. Scales of 1" = 1'–0" through 3" = 1'–0" are often used to represent details showing finishing material. Although you should never scale a blueprint, knowing the scale will help to interpret the size of the parts you are viewing.

Details may be a view of a member from the top or side of the object giving a view equal to what would be seen at the job site. Another common method of showing information in a detail would be to show the interior of a unit (such as wall) allowing all of the interior components to be seen. The method the detail is referenced on the framing plan or section will often tell the viewer what will be seen in the detail. Figure 17–4 shows examples of detail referencing for external and internal views.

FINDING DETAILS

Before you can read a detail you have to be able to find it. Sounds simple, but on a complex job some details tend to get lost. Details are often grouped together by the major material they are showing. Two common grouping methods are common on working drawings.

Details are typically grouped by the area they represent and placed after the sections. Roof details would be placed together, wall information in another area, and foundation information in another area. Each of these areas would in turn have details grouped together. Although the placement in the prints will vary from each office, grouping all truss or concrete or steel drawings together may allow only certain pages to be given to the roofing crew rather than an entire set of plans. Figure 17–5 shows a typical page of details grouped together by location in the structure.

The second method of locating details in a set of plans is to show the roof details with the roof framing plans, the wall details with the framing plan and the foundation details with the foundation plan. This

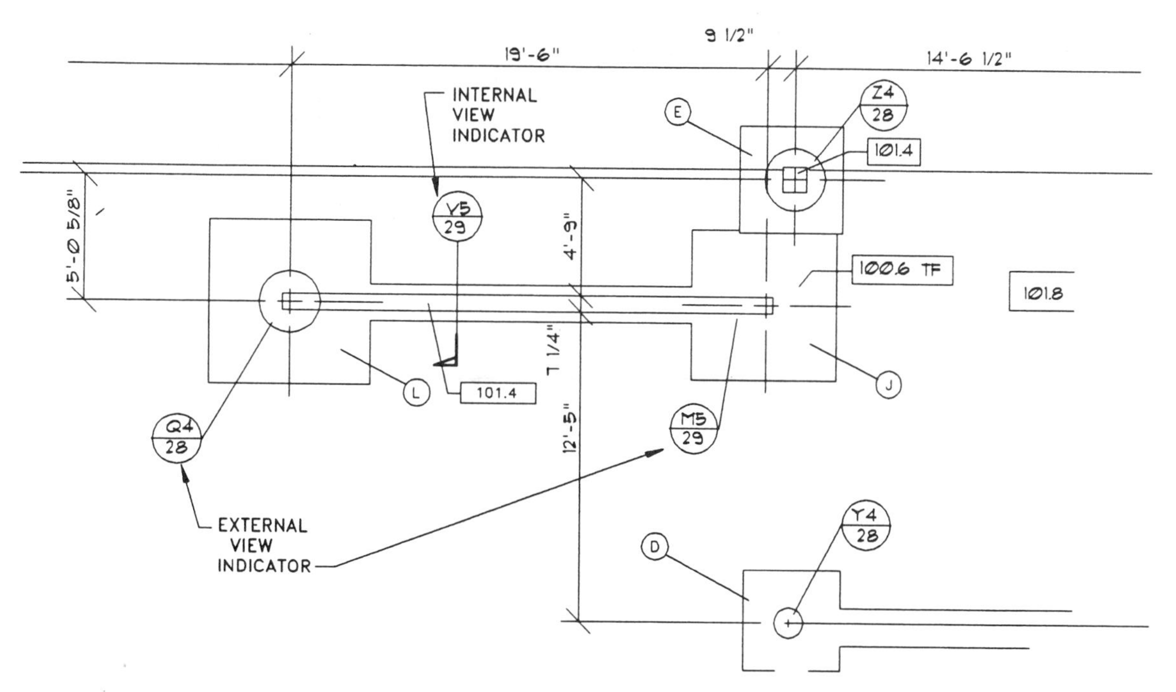

Figure 17–4 Reference bubbles can be used to show internal and external views. *Courtesy Structureform Masters.*

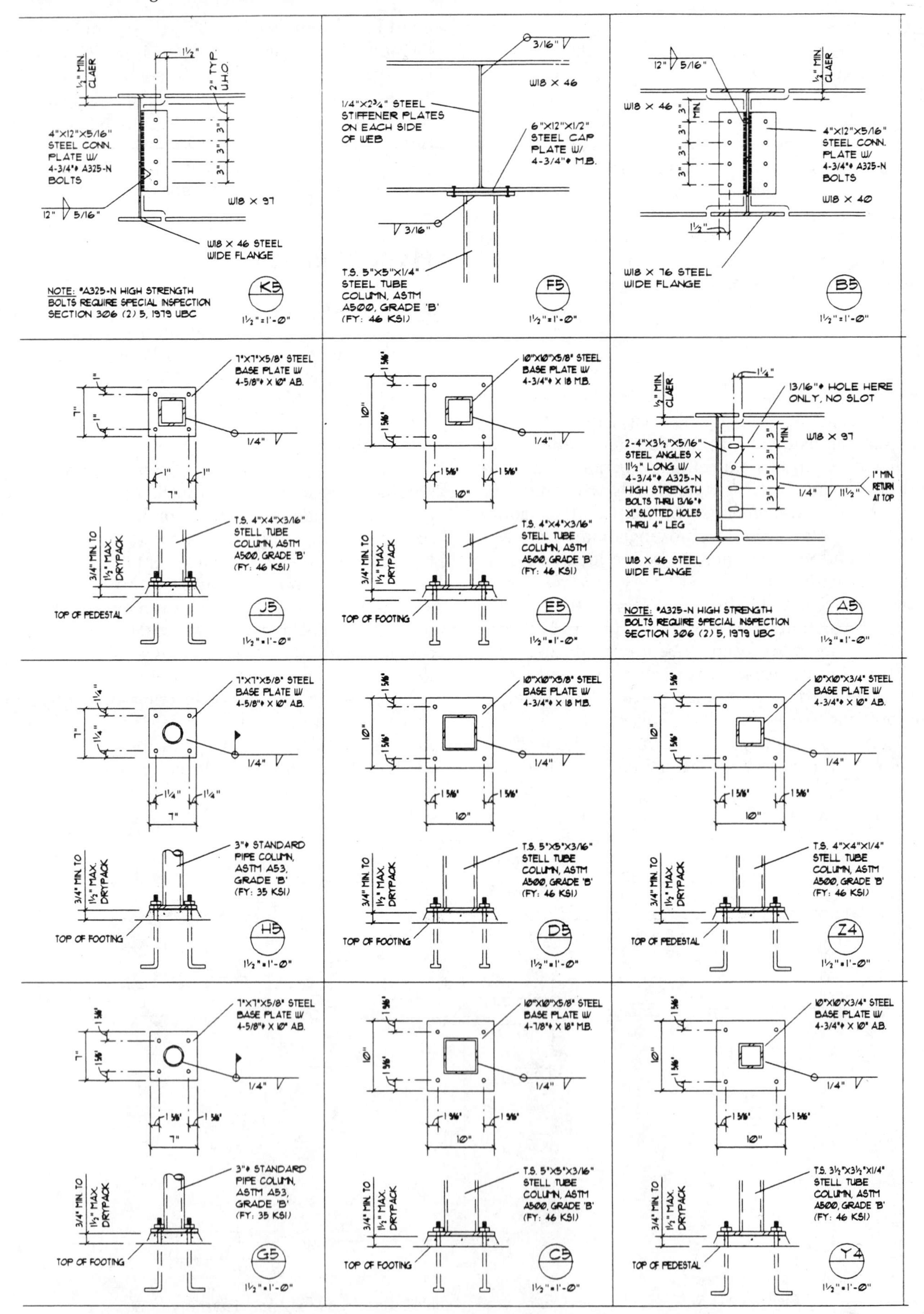

Figure 17–5 Details are often grouped together by types of information shown. *Courtesy Structureform Masters.*

method is often used on structures that are not very complex and will not require large amounts of details to explain construction methods.

No matter where the details are placed, they are typically referenced to other drawings by the use of bubbles containing a combination of letters or numbers. A bubble is placed by each detail. The page the detail is located on is placed in the bottom half of the bubble. The number or letter representing the details position on the page is placed in the upper half of the circle.

Some companies will use letters rather than numbers to identify details. The letter I, O, and Q are rarely used to avoid confusion with the numbers 1 and 0. If more than 23 details are used in the drawings, the details are assigned double or even triple letters. Figure 17–5 shows how the details might be numbered on a page.

Details are typically specified on each of the framing plans and sections. A specific detail will not be specified at every place that it occurs, but will typically be specified on each type of drawing that relates to that detail.

PAGE GROUPING

Because commercial drawing projects are so large they are often broken up into areas. Each sheet within a specific area is given a consecutive page number independent of other areas. 'S-5' would be the fifth page of the structural drawings. Areas would typically include:

A — architectural drawings: Plot, landscaping, sprinkler, irrigation, cabinet, millwork.
S — structural drawings: floor, framing, foundation plans, sections, and details.
E — electrical drawings: electrical plans, wiring diagrams, electrical load calculations.
M — mechanical drawings: HVAC plans, diagrams, and connection details.
P — plumbing drawings: floor and plot plans showing fresh and waste water lines.
EQ — equipment drawings: jobs with 'commercial kitchens' will typically have floor plans and elevations to show placement and type of equipment.

CHAPTER 17 TEST

1. What are the range of scales used to draw details showing finish work? ______
2. How many framing plans would be required for a five level office structure? ______
3. In what why is the roof used to hold up the walls? ______
______.
4. A rigid area used to resist twisting is called a ______.
5. What type of forces are hold-down anchors used to resist? ______
______
______.
6. Where would the connection of a beam to a wall typically be found in the structural drawings. ______
______
______.
7. What type of structural drawing would show the bolts used to hold a beam to a wall? ______
8. What symbols are typically never used when referring to details? ______
9. What is the typical scale used to represent framing plans? ______
______.
10. What shaped is often used to resist wracking? ______
11. Show a detail symbol representation for an internal view of a wall on a framing plan. ______
______
______
______.
12. Explain what the two symbols of a detail bubble represent. ______
______
______.

13. Show a detail symbol representation for an external view of a beam to wall connection on a framing plan. ____.

14. Give the common range of scales used to draw construction details. ______________________

15. What should be determined first when looking at a detail? ____.

CHAPTER 17 EXERCISES

PROBLEM 17–1. Use Figure 17–3 on page 314, 17–5 on page 316, and the following 14 framing plans and details shown on pages 320–333 to answer the following questions. Assume north to be at the top of the page. Provide the location (plan or detail number) and the answer in the space provided.

1. A stair is located on the west side of the structure. What size rafters will be used over the stairs to frame the roof? ______________________
2. How is the wall in the southwest corner of the structure tied to the roof? ______________________
3. What ties the Northeast wall to the Southeastern portion of the roof? ______________________
4. What type of metal connectors are to be used at the roof level? ____.
5. How will the plates of the Southeast wall (which runs north/south) be connected to floors and roof? ____.
6. What will support the floor of the upper office level on the south side of the structure? ____.

A shear wall is located to the southeast corner of the diaphragm on the 3rd floor framing plan. Answer the following questions based on that wall.

7. What size drag strut will connect to the wall? ____.
8. How will the wall plate on the west side of the wall be nailed? ____.
9. What size post will support the drag strut at the south end? ____.
10. What will connect the post to the floor? ____.

11. What size beam will support the post in the shear wall? ____
____
____.

12. Three beams intersect at the south end of the shear wall. What size steel plate and bolts will be used to connect them? ____
____
____.

13. Where can you find a view showing the entire shear wall? ____
____
____.

14. What size column will support the lowest level of the wall at the north end? Give the complete specification.
____
____.

15. What size footing will support the north column? ____
____
____.

16. What type of weld will connect the south lower column to the base plate? ___________________________
____
____.

17. What size footing is north of the footing supporting the shear wall? ______________________________
____.

18. What size brace will be used at the mid level of the shear panel between the north and south columns?
____.

19. How will the lower end of the diagonal brace at the upper level be held to the column? _____________
____
____.

20. A drag strut is shown at the upper level at the north end of the shear wall. How will it be connected to the column? ____
____.

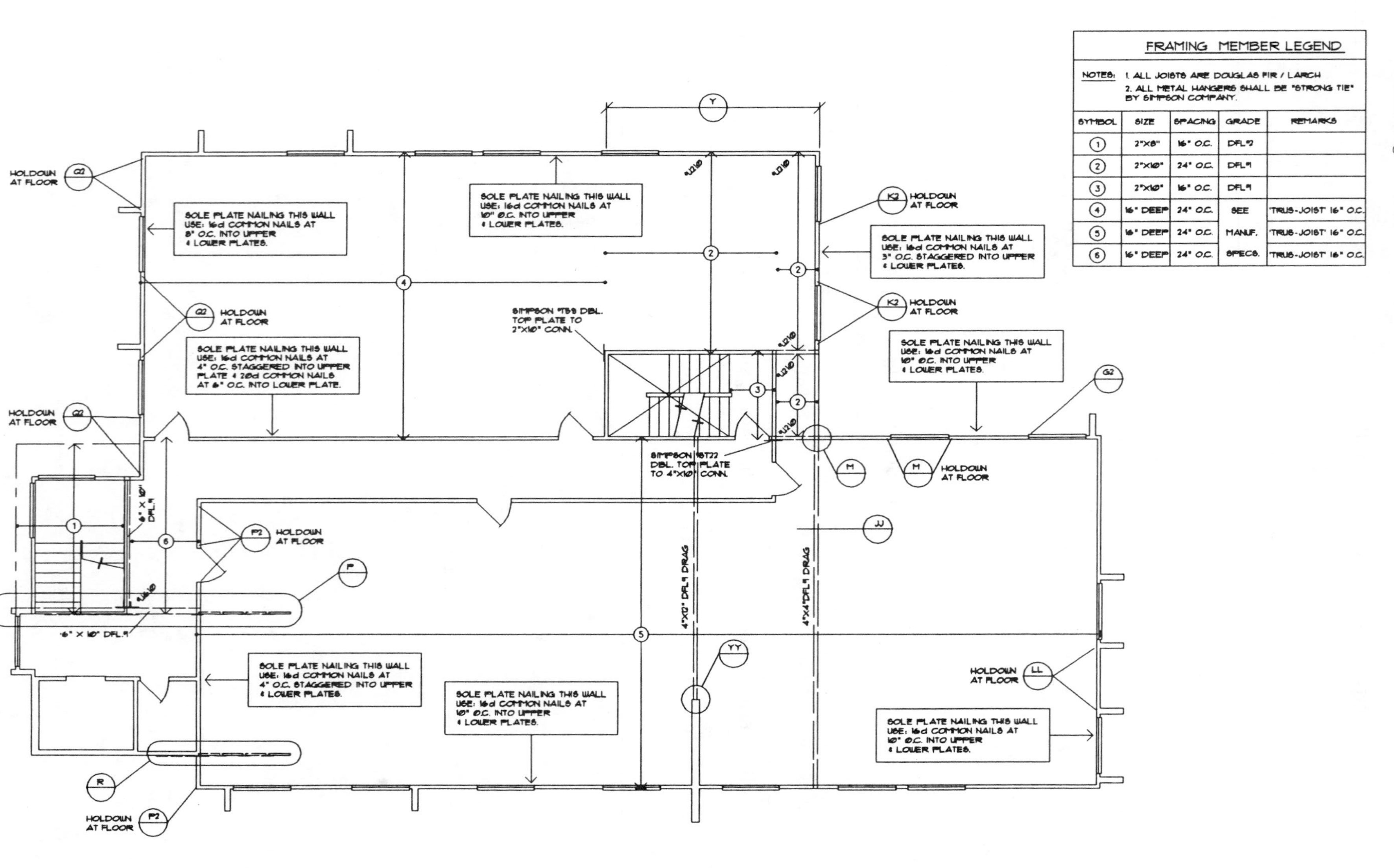

Problem 17–1. *Courtesy Structureform Masters.*

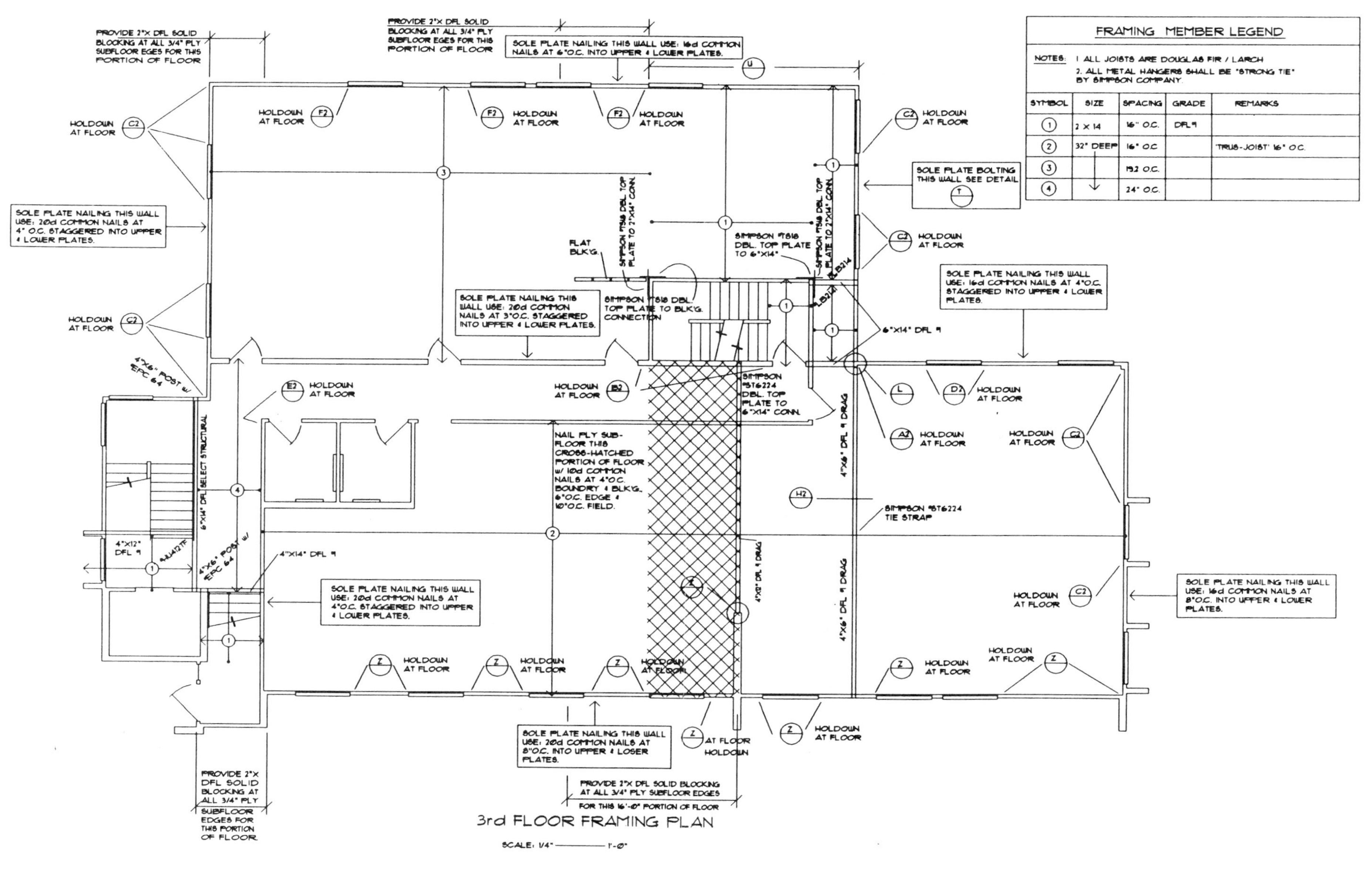

Problem 17–1. *Courtesy Structureform Masters.*

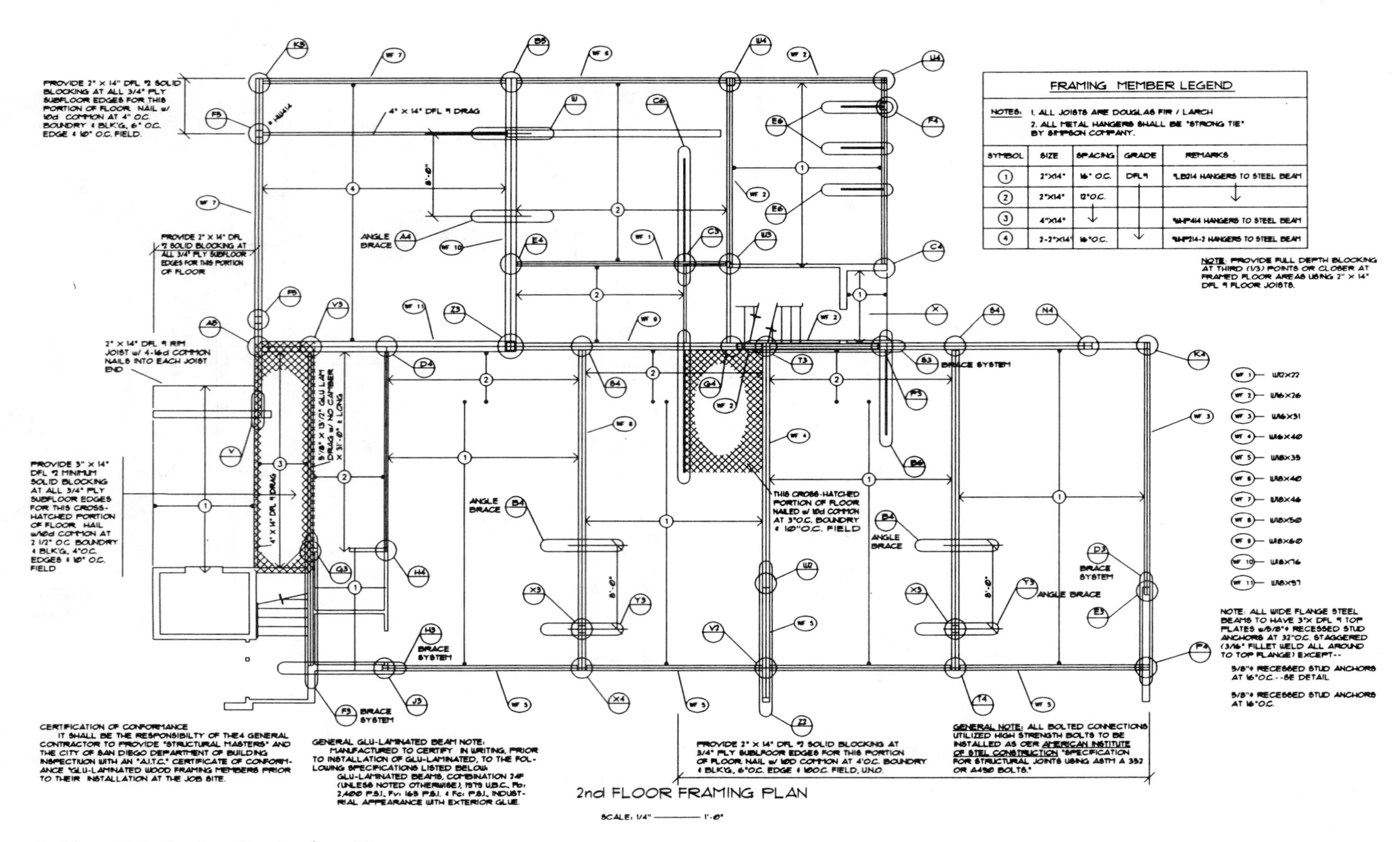

Problem 17–1. *Courtesy Structureform Masters.*

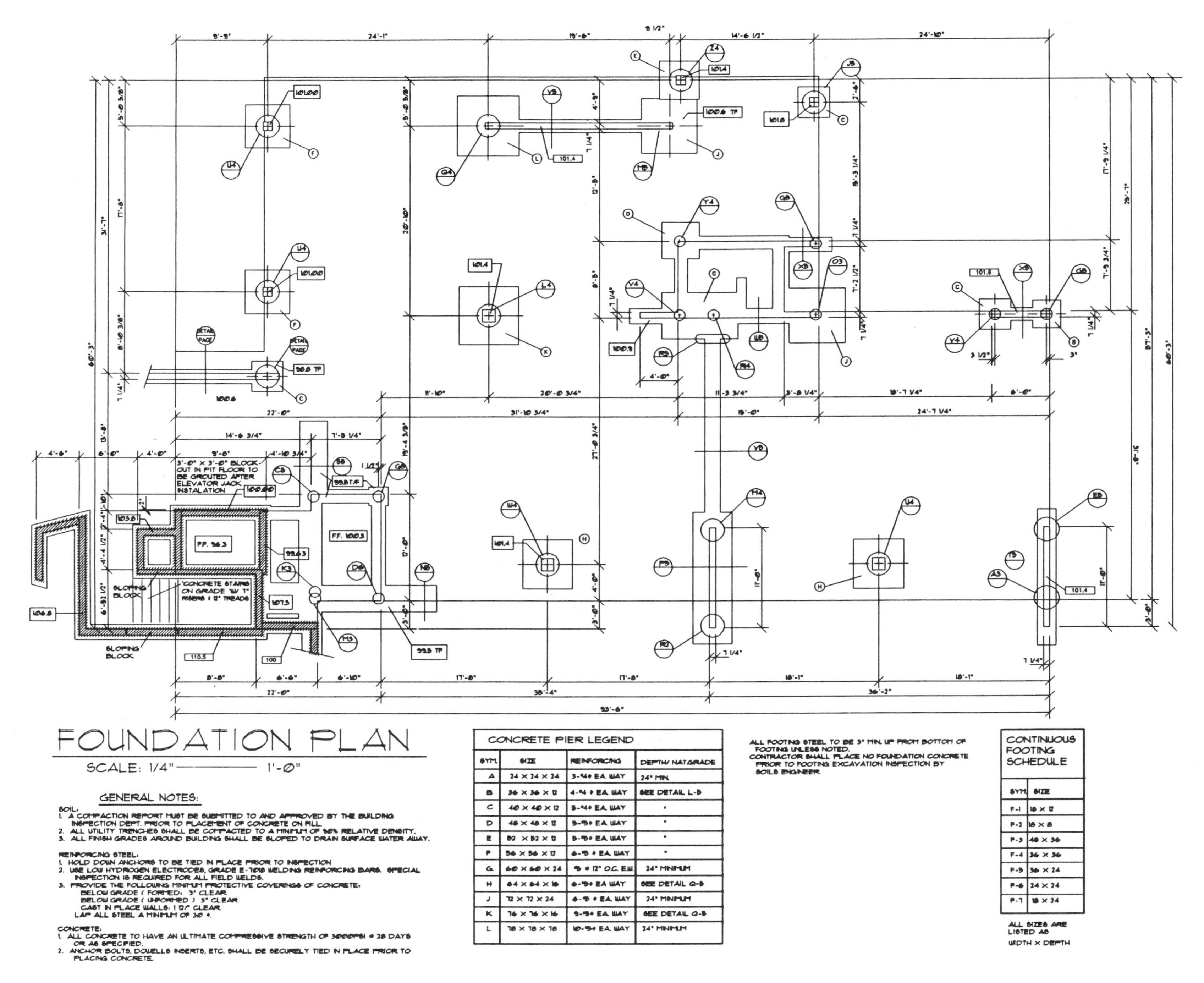

CONCRETE PIER LEGEND			
SYM.	SIZE	REINFORCING	DEPTH/ NAT.GRADE
A	24 X 24 X 24	3-#4ø EA. WAY	24" MIN.
B	36 X 36 X 12	4-#4 ø EA. WAY	SEE DETAIL L-B
C	40 X 40 X 12	5-#4ø EA. WAY	"
D	48 X 48 X 12	5-#5ø EA. WAY	"
E	82 X 82 X 12	8-#5ø EA. WAY	"
F	56 X 56 X 12	6-#5 ø EA. WAY	"
G	60 X 60 X 24	#5 @ 12" O.C. E.W.	24" MINIMUM
H	64 X 64 X 16	6-#5ø EA. WAY	SEE DETAIL Q-B
J	72 X 72 X 24	6-#5 ø EA. WAY	24" MINIMUM
K	76 X 76 X 16	9-#5ø EA. WAY	SEE DETAIL Q-B
L	78 X 78 X 78	10-#5ø EA. WAY	24" MINIMUM

CONTINUOUS FOOTING SCHEDULE	
SYM.	SIZE
F-1	18 X 12
F-2	18 X 8
F-3	48 X 36
F-4	36 X 36
F-5	36 X 24
F-6	24 X 24
F-7	18 X 24

Problem 17–1. *Courtesy Structureform Masters.*

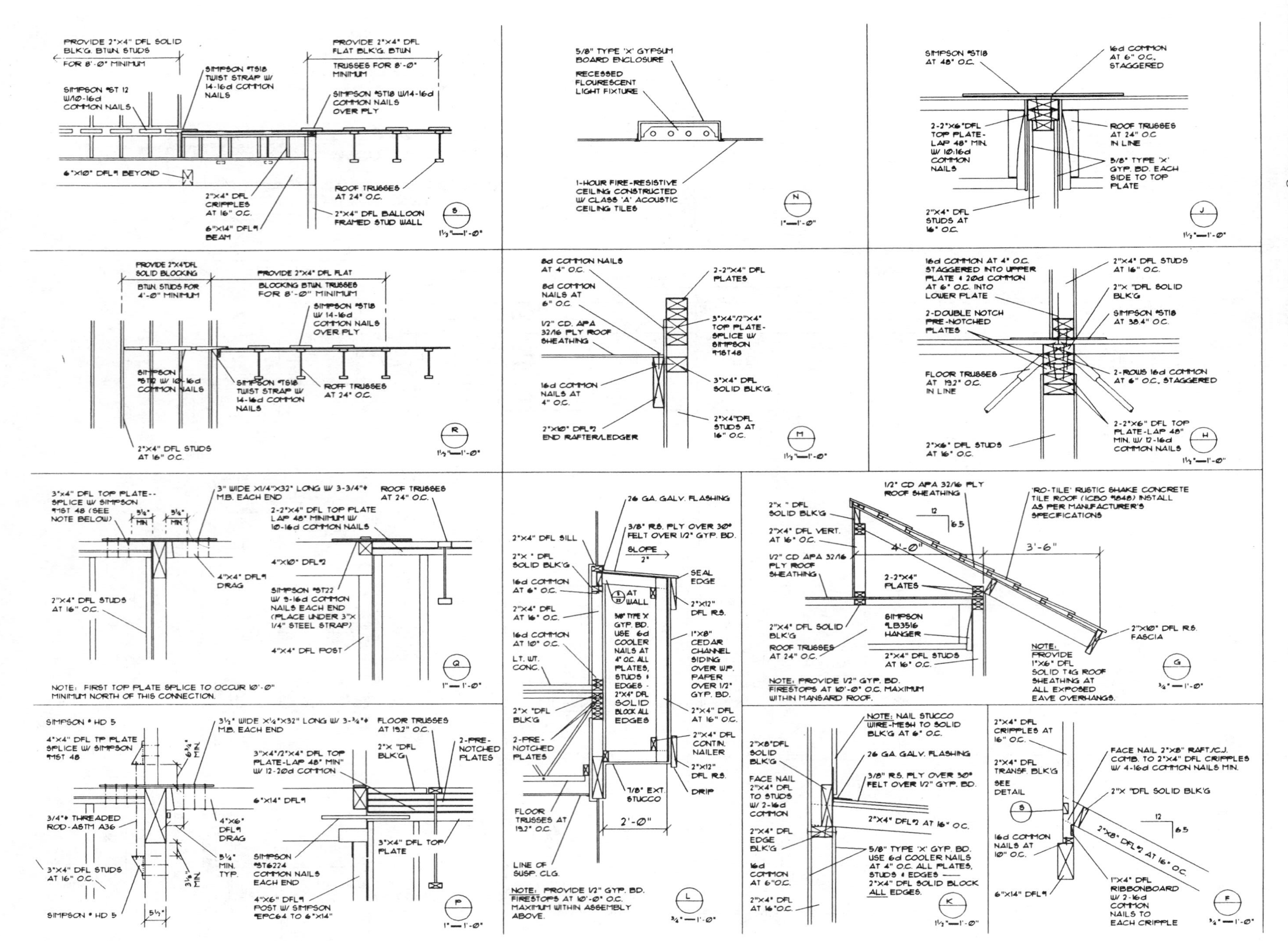

Problem 17–1. *Courtesy Structureform Masters.*

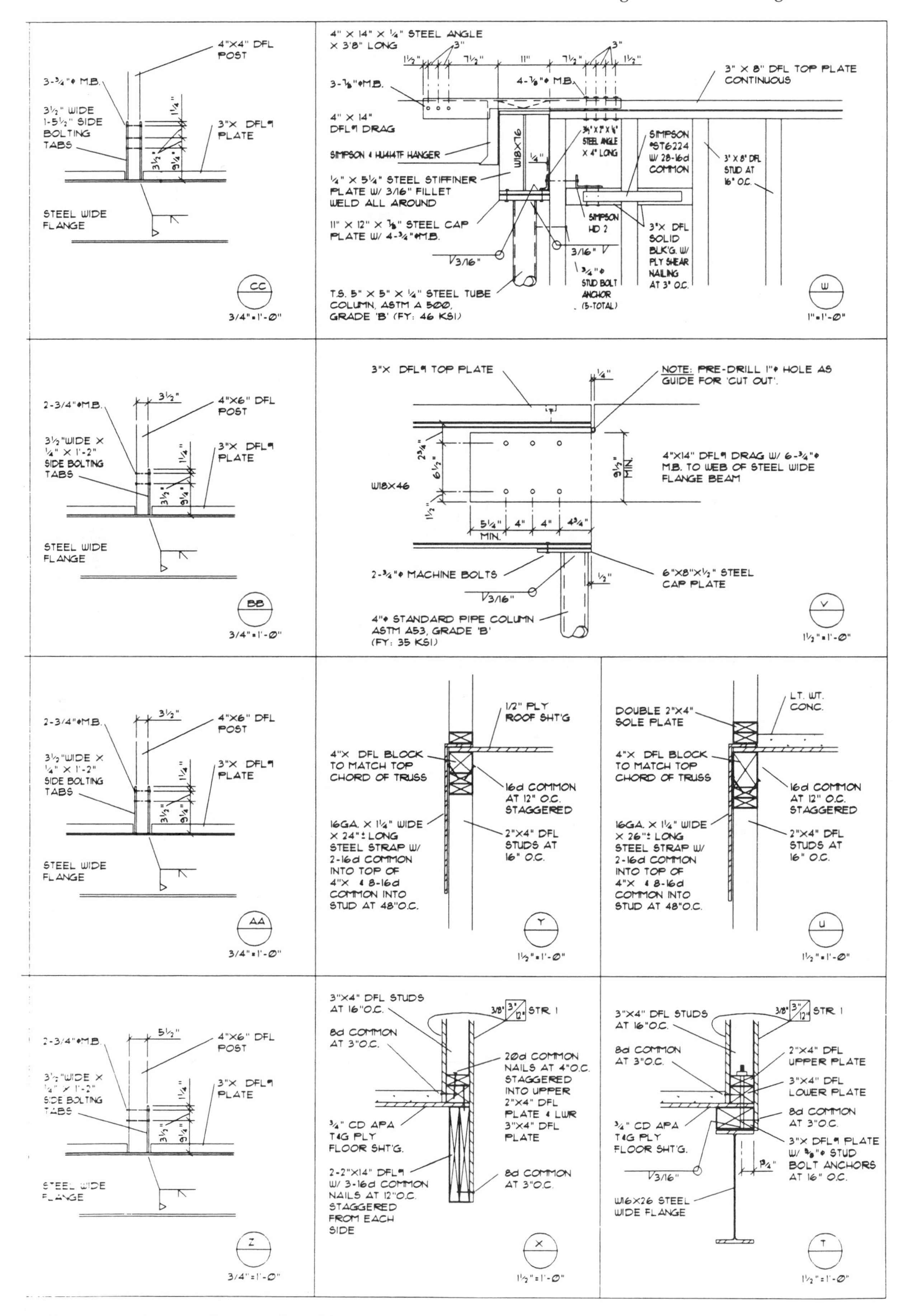

Problem 17–1. *Courtesy Structureform Masters.*

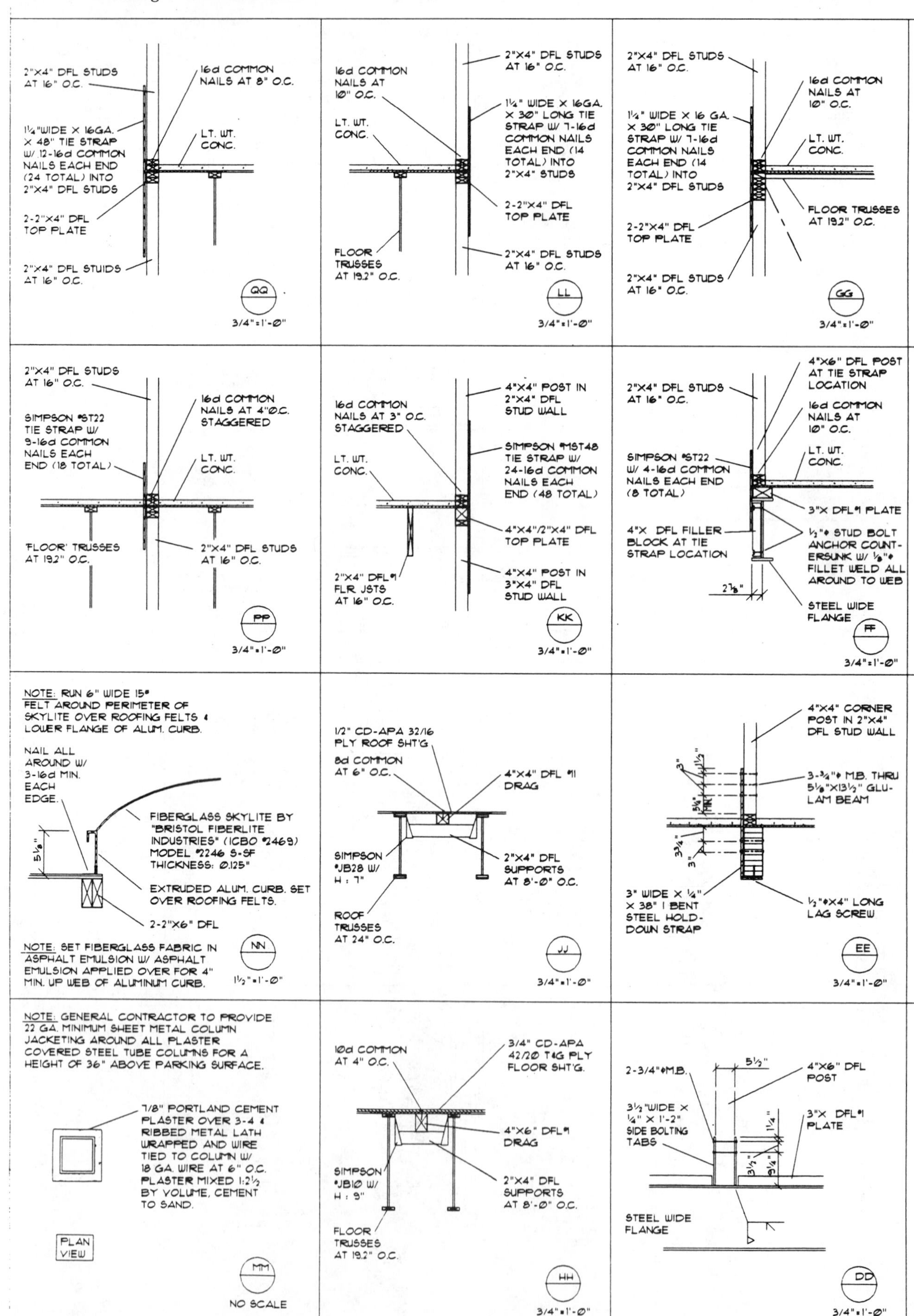

Problem 17–1. *Courtesy Structureform Masters.*

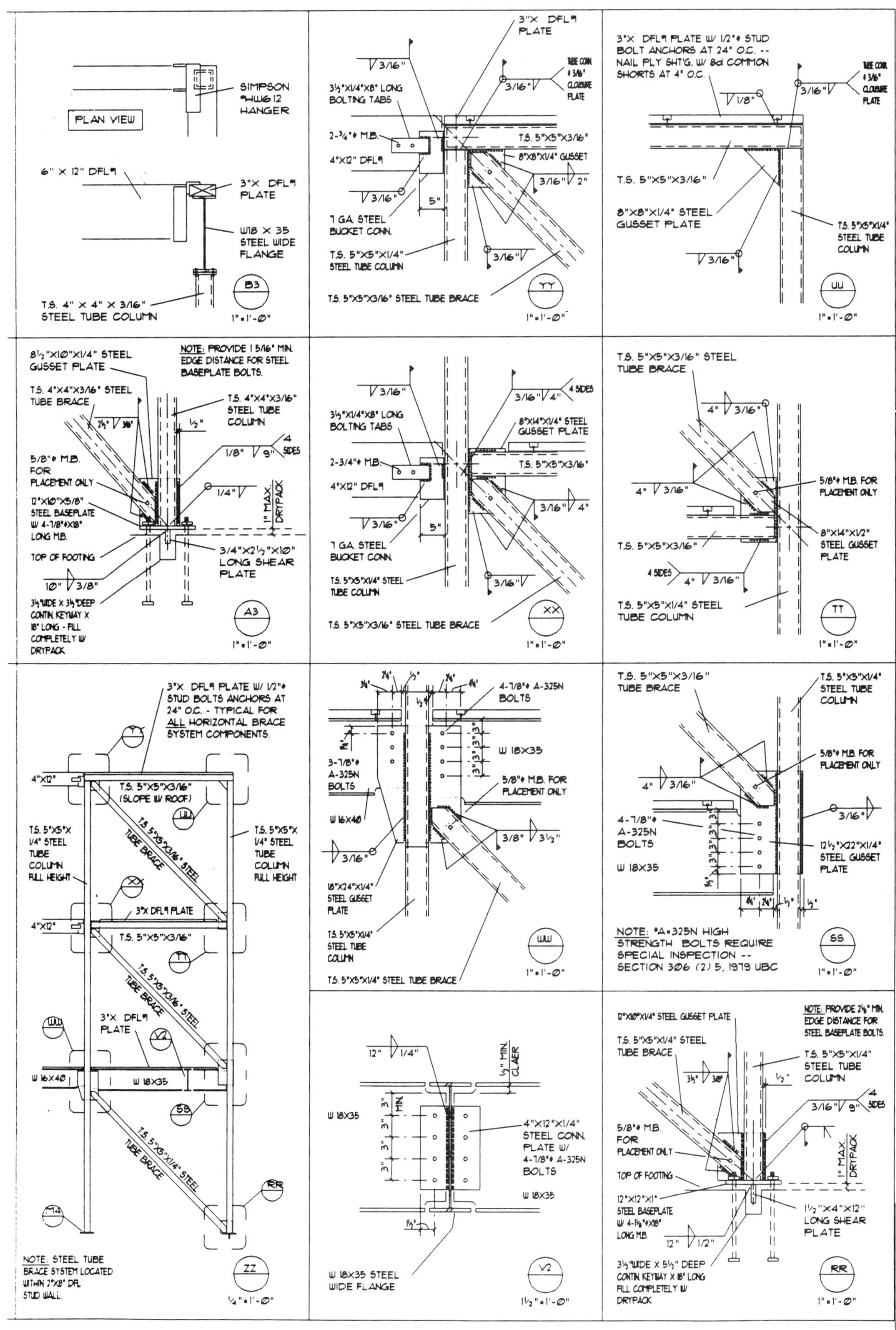

Problem 17–1. *Courtesy Structureform Masters.*

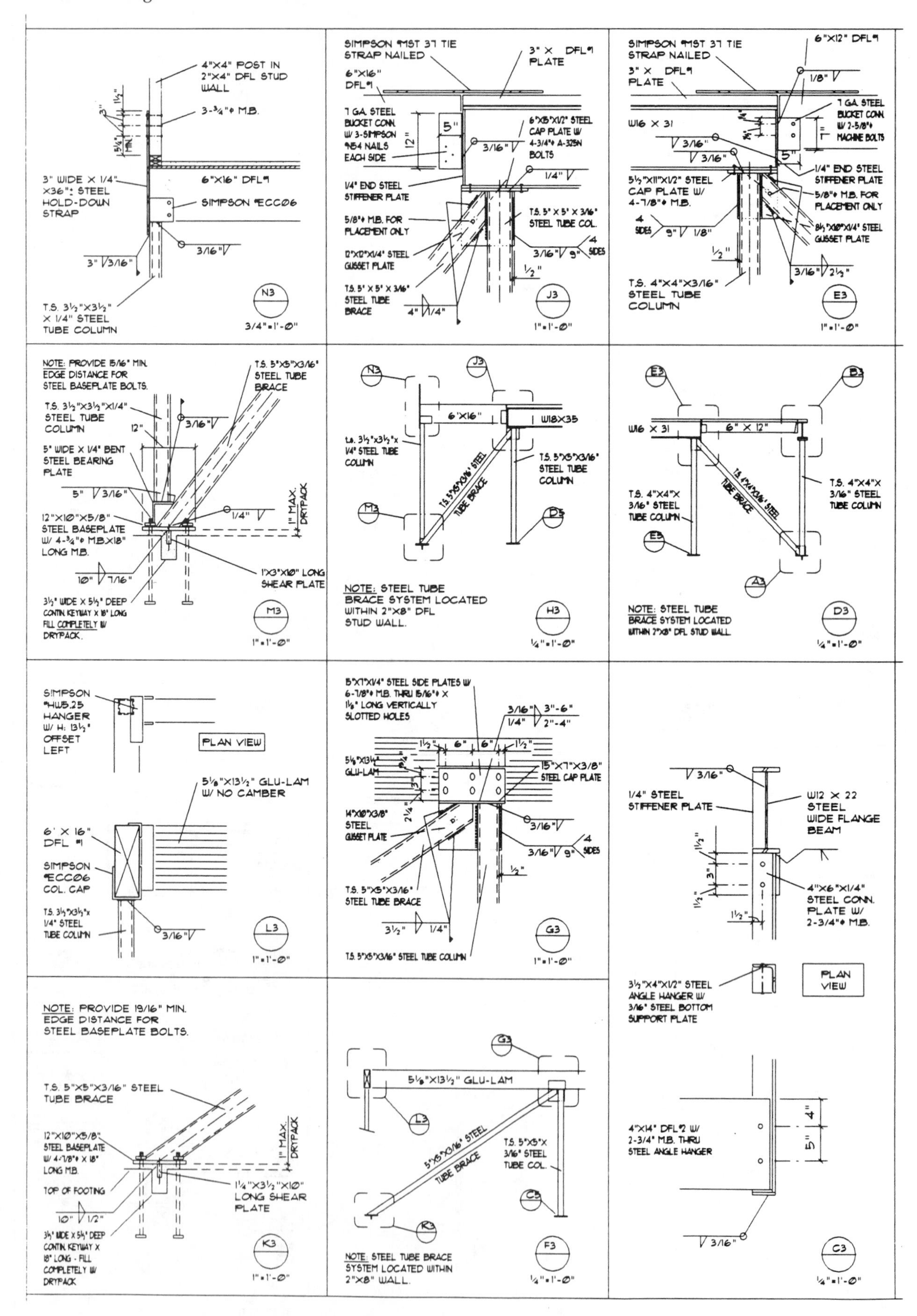

Problem 17–1. *Courtesy Structureform Masters.*

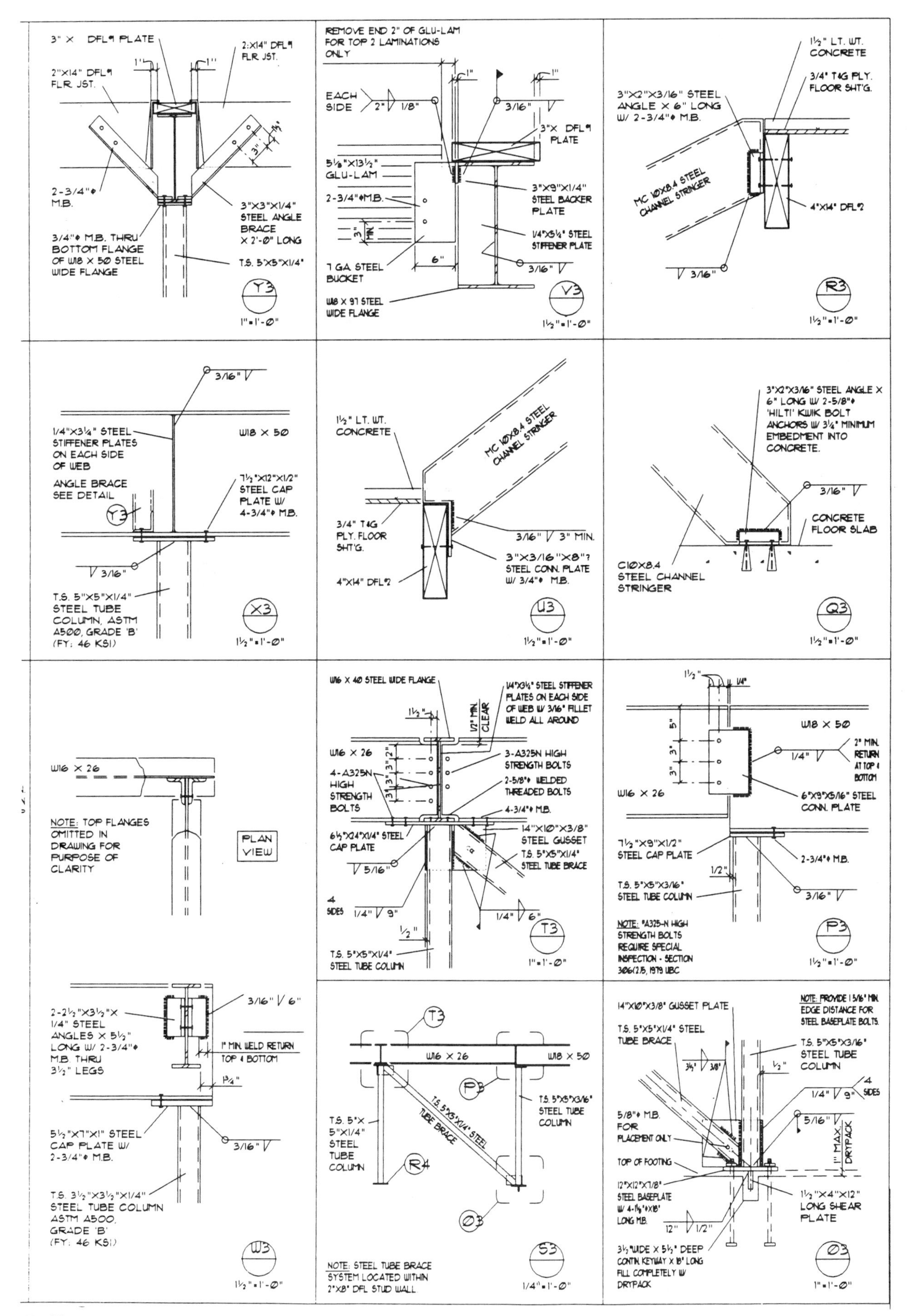

Problem 17–1. *Courtesy Structureform Masters.*

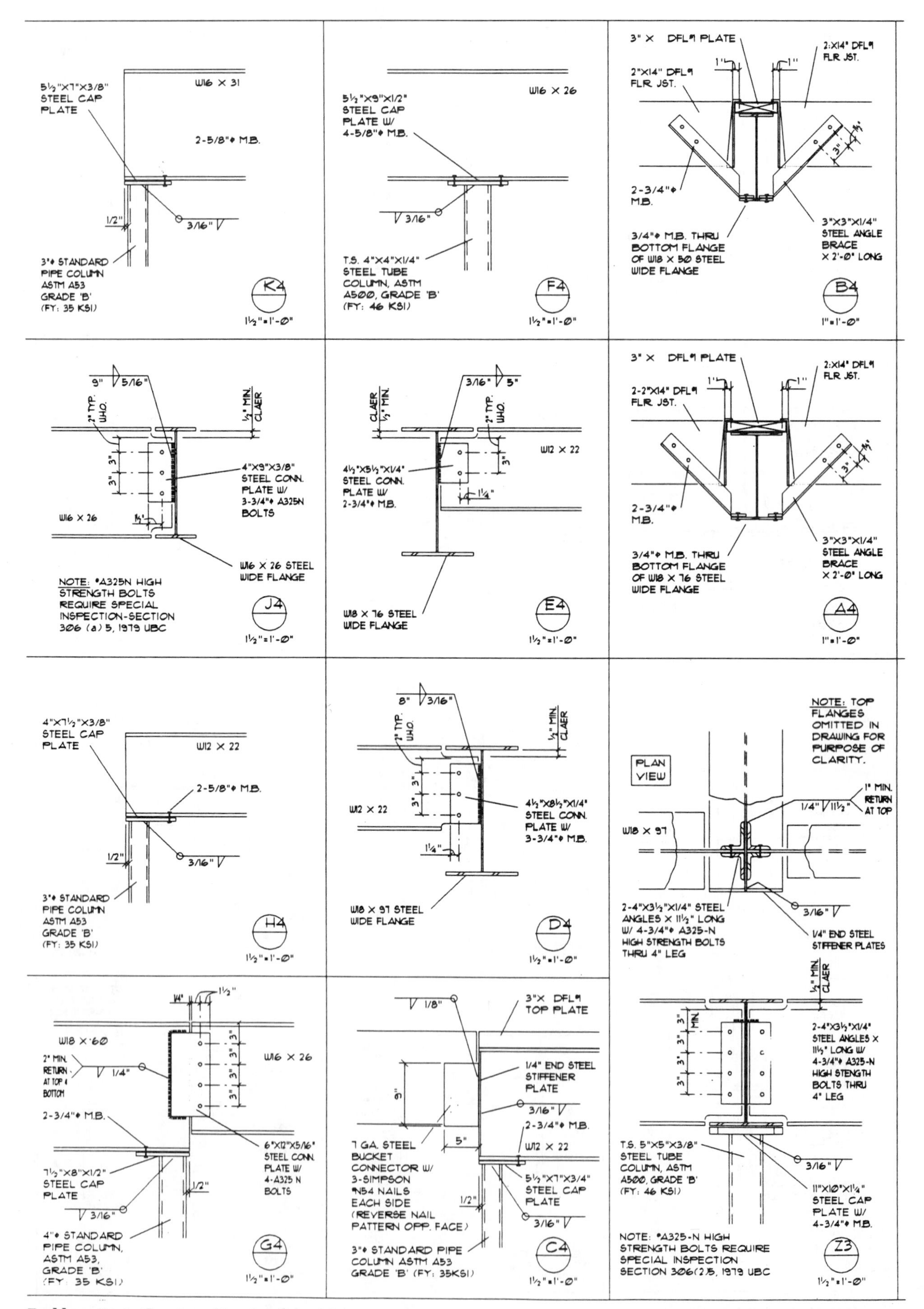

Problem 17–1. *Courtesy Structureform Masters.*

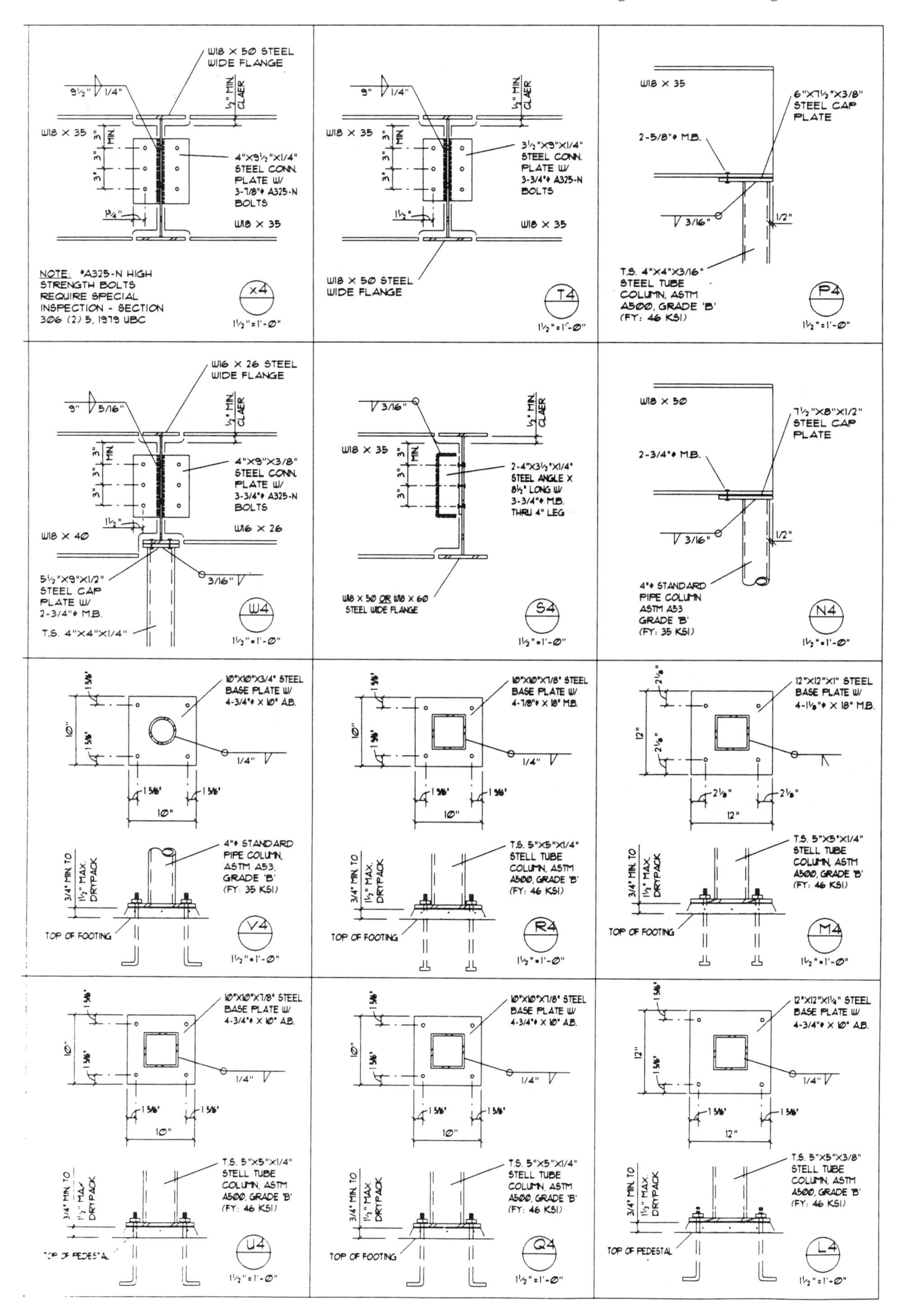

Problem 17–1. *Courtesy Structureform Masters.*

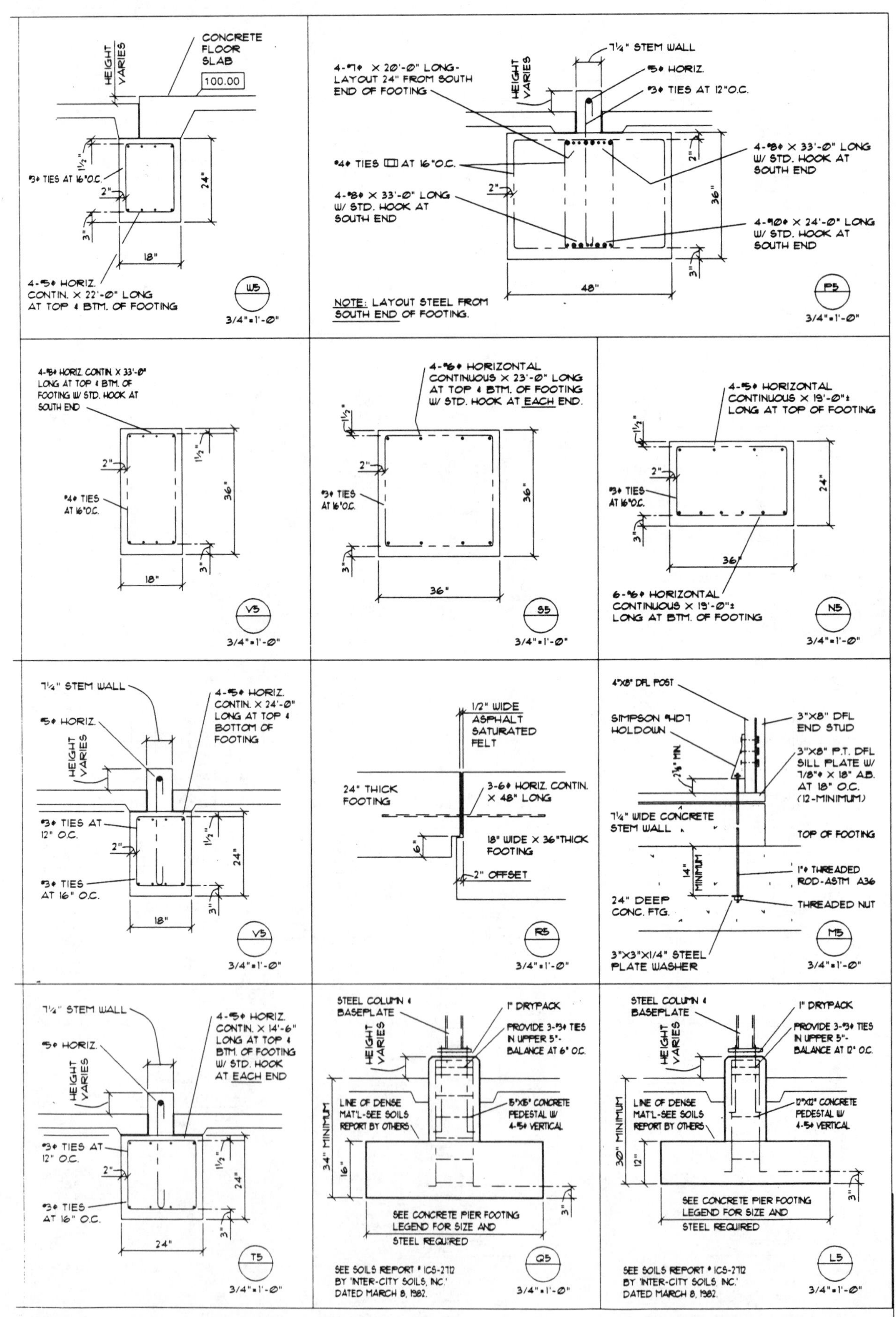

Problem 17–1. *Courtesy Structureform Masters.*

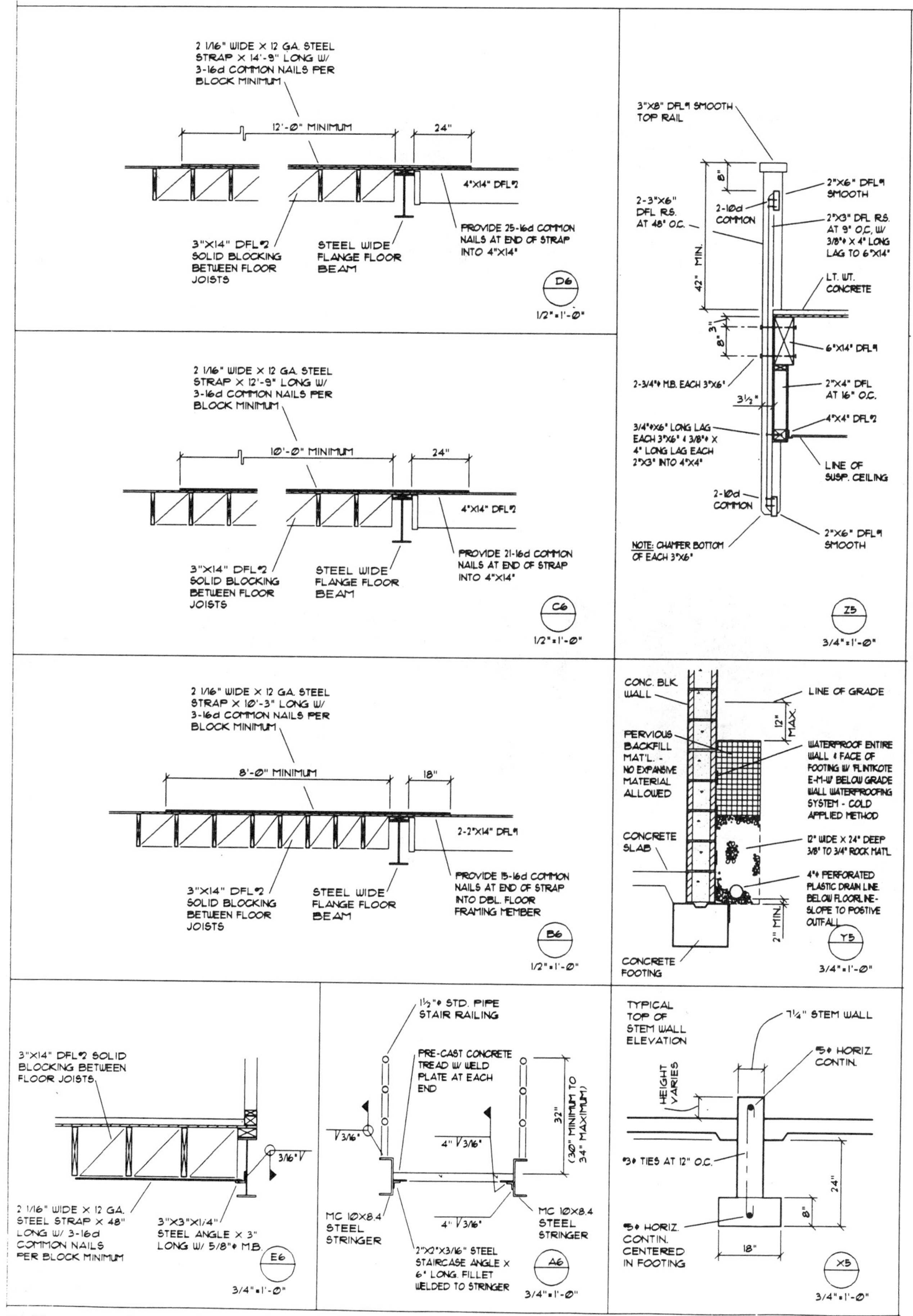

Problem 17–1. *Courtesy Structureform Masters.*

Chapter 18

Reading Commercial Electrical, HVAC, and Plumbing Plans

INTRODUCTION TO READING COMMERCIAL ELECTRICAL, PLUMBING, AND HVAC PLANS

A review of Chapter 4 "Reading Floor Plan Symbols for Electrical, Plumbing and HVAC," would be a good idea before starting this chapter. Many of the symbols and terms used in residential print reading are the same for commercial construction. This chapter covers differences that exist in reading prints for commercial electrical, plumbing, and HVAC construction. Symbols and techniques used in commercial drawings are often more complex and detailed than those used in residential prints.

READING COMMERCIAL ELECTRICAL PLANS

The drawing of the electrical layout is an important part of the total function and safety of a commercial or industrial facility. Local and national electrical codes provide specifications for installations and layout. Layout planning should play an important role in conjunction with code guidelines.

Electrical Symbols

In residential applications as discussed in Chapter 4. Commercial electrical plans are commonly separate from the floor plan and other architectural drawings and to other overlays. Common electrical symbols are shown in Figure 18–1.

Review Chapter 4 for residential electrical layouts.

When special characteristics are required, such as a specific size fixture, a location requirement, or any other specification, a local note may be applied next to the electrical symbol to briefly describe the situation. (See Figure 18–2.) Where a specification affects electrical installations on the entire layout, general notes may be used. (See Figure 18–3.)

Commercial Electrical Plan Examples

Commercial electrical plans often follow much more detailed installation guidelines than residential applications.

The electrical circuit switch legs for commercial applications are generally drawn as solid lines rather than dashed, as in residential electrical plans. The electrical circuit lines that continue from an installation to the service distribution panel are terminated

ELECTRICAL LEGEND	
SYMB	DESCRIPTION
	CIRCUIT CONDUCTOR IN CONDUIT
	CIRCUIT CONDUCTOR UNDERGROUND OR IN SLAB CONDUIT
	TELEPHONE CONDUIT
G	BRANCH CIRCUIT . SHORT LINES INDICATE THE NUMBER OF PHASE CONDUCTORS . LONG LINE INDICATES THE NEUTRAL . LONG LINE WITH "G" INDICATES GROUNDING WIRE
4V1 1 •12	PANEL DESIGNATIONS CIRCUIT BREAKER POLE NUMBER WIRE SIZE
	PANEL BOARD
T	TRANSFORMER
TEL	TELEPHONE CIRCUIT BOARD (TCB)
	DISCONNECT SWITCH (F=FUSED)
SD	FUEL ISLAND SAFETY DISCONNECT SWITCH
	GROUNDING DUPLEX RECEPTACLE . 15A . 120V
	GROUNDING, SINGLE RECEPTACLE, 30A, 120V
WP/B	RECEPTACLE WITH WEATHERPROOF BOX AND WEATHERPROOF COVER
GFI	RECEPTACLE OR CIRCUIT BREAKER WITH GROUND FAULT INTERRUPTER
RCR	RECESSED WALL CLOCK RECEPTACLE . 7 FT ABOVE FINISHED FLOOR
	480V, 40 AMP, 3 PHASE, RECEPTACLE & PLUG
	240V, 20A, ONE PHASE, W/NEUTRAL
	WALL MOUNTED TELEPHONE OUTLET
	FLOOR MOUNTED TELEPHONE OUTLET
	FLAT CONDUCTOR CABLE (FCC) TRANSITION BOX (TELEPHONE)
	FLAT CONDUCTOR CABLE (FCC) TRANSITION BOX (POWER)
	FLOOR MOUNTED ELECTRICAL DUPLEX OUTLET (FCC)
A1 F	480V . FUSED DISCONNECT W/80A FUSE
A2 F	240V . FUSED DISCONNECT W/200A FUSE

MECHANICAL EQUIPMENT LEGEND	
SYMB	DESCRIPTION
DO	DOOR OPENER
DF	DRINKING FOUNTAIN
HW	HOT WATER HEATER
ACOMP	AIR COMPRESSOR
W	WELDER
PG	PEDESTAL GRINDER
DP	DRILL PRESS
L	LATHE
GPX	GREASE PIT EXHAUST
BC	BRIDGE CRANE
T	THERMOSTAT
UH	UNIT HEATER
VE	VEHICLE EXHAUST
AL	AUTO LIFT PUMPING UNIT
FCP	FIRE ALARM PANEL
FAP	ANNUNCIATOR PANEL
SP	SECURITY PANEL
EF	EXHAUST FAN
ASU	FAN UNIT
EH	ELECTRIC HEAT
DM	DAMPER MOTOR
SP	FLUSH MOUNTED SPEAKER

LIGHTING LEGEND	
SYMB	DESCRIPTION
S	SINGLE POLE SWITCH
S2	DOUBLE POLE SWITCH
S3	THREE WAY SWITCH
S4	FOUR WAY SWITCH
SRC	SPDT SWITCH . MOMENTARY CONTACT . CENTER OFF REMOTE CONTROL STATION
SMC	SP3T SWITCH, ROTARY, W/INSCRIPTION " SECURITY LIGHTS" "AUTO OFF ON"
©	CONTACTOR

Figure 18–1 Commercial application symbols. *Courtesy Bonneville Power Administration.*

FIXTURE SIZE OUTLET HEIGHT

Figure 18–2 Special notes found on prints describing information about electrical fixtures.

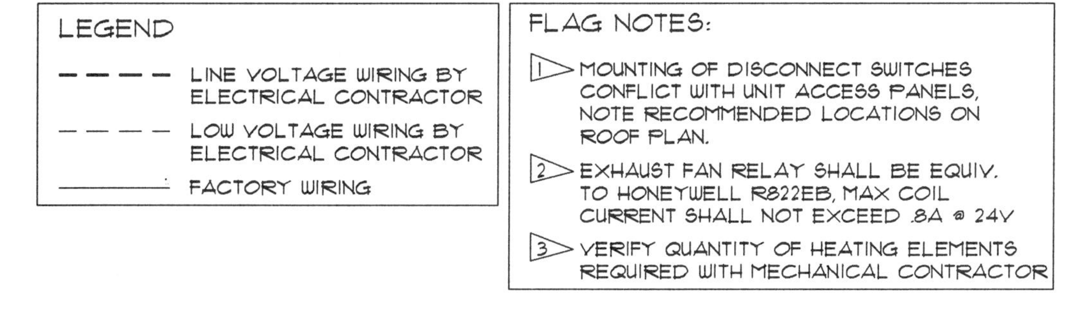

GENERAL NOTES:

1. POWER WIRING AND GROUNDING OF EQUIPMENT MUST COMPLY WITH LOCAL CODES.
2. ENSURE POWER SUPPLY AGREES WITH EQUIPMENT NAMEPLATE.
3. LOW VOLTAGE TO BE #18 AWG MINIMUM CONDUCTOR.
4. SEE HEAT PUMP & HEATER WIRING DIAGRAMS FOR POWER SUPP. CONN. DETAILS.
5. THESE FACTORY SUPPLIED POLARIZED PLUGS MUST BE FIELD-CONNECTED AS SHOWN. SOME HEATERS HAVE NO LEAD TO TERM. # 6. SEE HEATER WIRING DIAGRAM.
6. CONNECT COPPER CONDUCTORS ONLY IF AL. OR CU-CLAD-AL POWER WIRING IS USED, CONNECTORS WHICH MEET ALL APPLICABLE CODES AND ARE ACCEPTABLE TO THE INSPECTION AUTHORITY HAVING JURISDICTION SHALL BE USED.
7. THE ACCUSTAT HOOK-UP DIFFERS SIGNIFICANTLY FROM OTHER UNITS: FOLLOW THE HOOK-UP EXACTLY AS SHOWN.
8. SELECT SENSORS EXACTLY AS FOLLOWS: W1=70°F, W2=72°F, V1=76°F, V2=74°F.

Figure 18–3 General notes for electrical specifications.

next to the fixture and capped with an arrowhead, meaning that the circuit continues to the distribution panel. When multiple arrowheads are shown, this indicates the number of circuits in the electrical run. (See Figure 18–4.) For many installations where a number of circuit wires are used, the number of wires is indicated by slash marks placed in the circuit run. The number of slash marks equals the number of wires. (See Figure 18–5.)

There may be more than one commercial electrical overlay; for example, floor-plan lighting, electrical-plan power supplies, or reflected ceiling plan. The floor-plan lighting layout provides the location and identification of lighting fixtures and circuits. It is usually coordinated with a lighting fixture schedule as shown in Figure 18–6. In some applications a power supply plan is used to show all electrical outlets, junction boxes, and related circuits. (See Figure 18–7.) The reflected ceiling plan is used to show the layout for the suspended ceiling system as shown in Figure 18–8. Electrical plans for equipment installations may also be needed to supplement the power supply and lighting plans. Figure 18–9 shows the plan for roof installation of equipment. You may also find drawings to read both as schematic diagrams for specific electrical installations as shown in Figure 18–10.

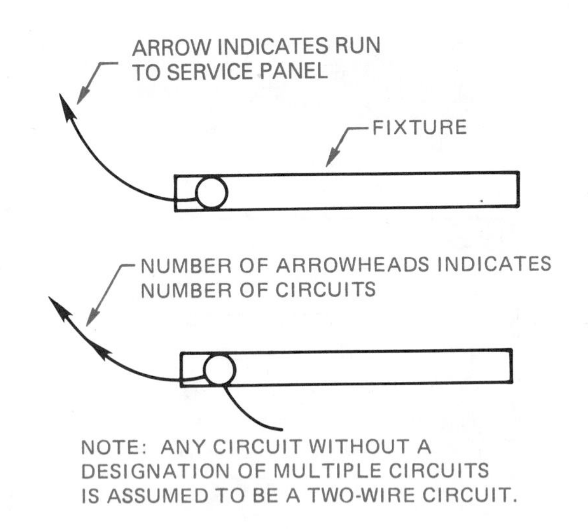

Figure 18–4 Typical electrical circuit designations.

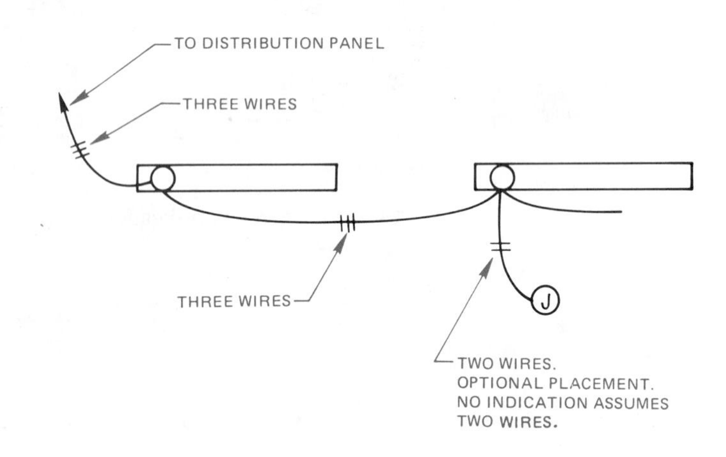

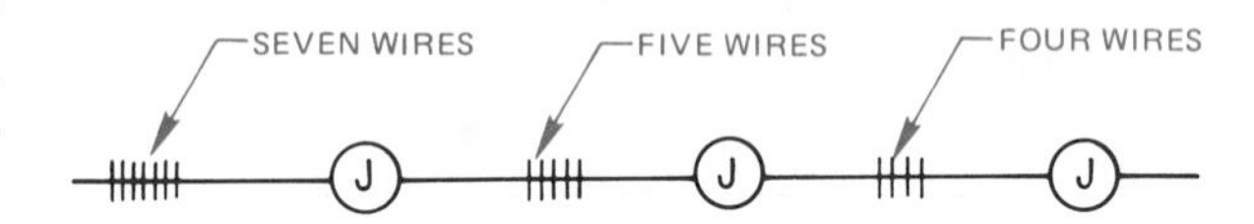

Figure 18–5 This is how the number of wires designated in a circuit run are shown on a print.

ELEMENTS OF COMMERCIAL HVAC PRINTS

Drawings for the HVAC system show the size and location of all equipment, duct work, and components with accurate symbols, specifications, notes, and schedules that form the basis of contract requirements for construction.

For commercial structures the HVAC plan may be prepared by an HVAC engineer also known as a mechanical engineer. The mechanical engineer is responsible for the HVAC design and installation. The engineer determines the placement of all equipment and the location of all duct runs and components. He or she also determines all of the specifications for unit and duct size based on calculations of structure volume, exterior surface areas and construction materials, rate of air flow, and pressure. An HVAC drawing is shown in Figure 18–11.

Several examples of duct system elements are shown in Figure 18–11.

Single- and Double-line HVAC Plans

HVAC plans are drawn over the outline of the floor plan. The HVAC plan is then drawn as an overlay with the floor plan. In most cases, when you are reading commercial HVAC prints you will see the floor plan with the door and window symbols displayed for reference and the HVAC plan shown as it would appear over the floor plan. HVAC drawings are generally shown separately from other plans such as the electrical because it would be too difficult to read the drawings were they combined. The HVAC plan shows the placement of equipment and duct work. The size (in inches) and shape (with symbols, Ø = round, □ = square or rectangular) of duct work and system component labeling is placed on the drawing or keyed to schedules. Drawings may be either single line or double line, depending on the needs of the client or how much detail must be shown. In many situations single line drawings are adequate to provide the equipment placement and duct routing as shown in Figure 18–12. Double-line drawings are often necessary when complex systems require more detail as shown in Figure 18–13.

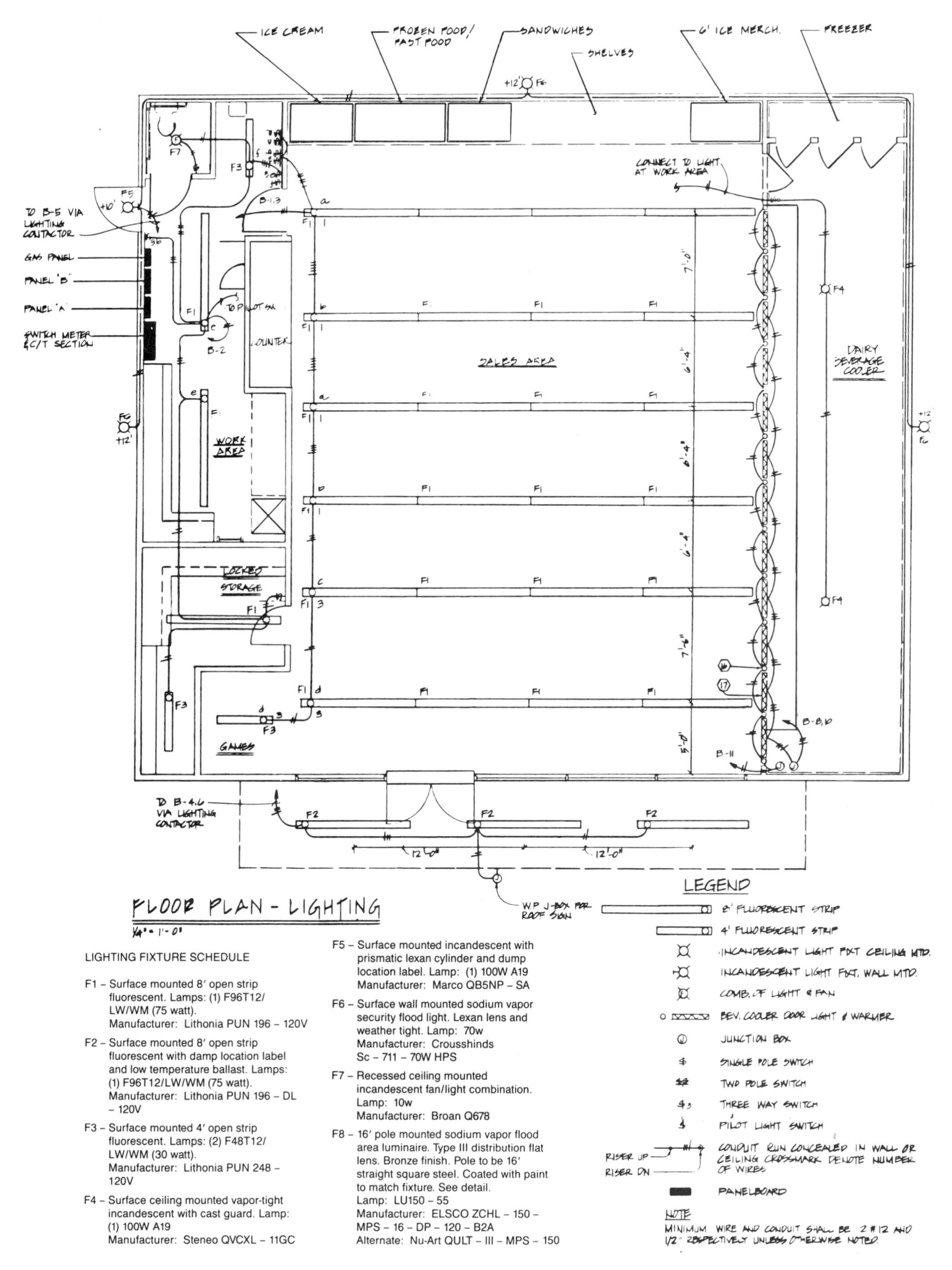

Figure 18–6 This print shows a typical Lighting-Floor Plan, Legend, and Lighting Fixture Schedule. *Courtesy The Southland Corporation.*

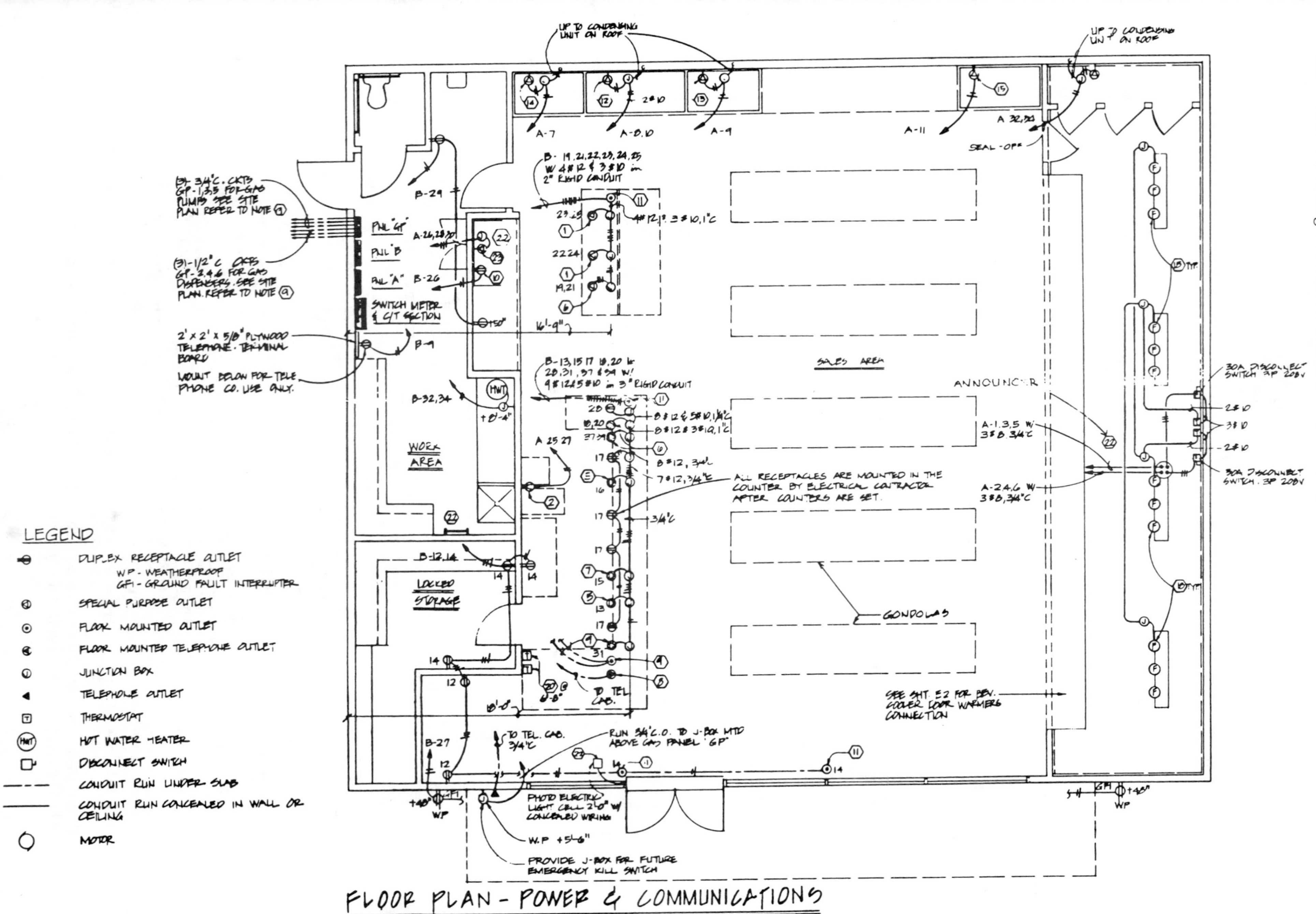

Figure 18–7 This print shows a typical Power Supply and Communications Floor Plan. *Courtesy The Southland Corporation.*

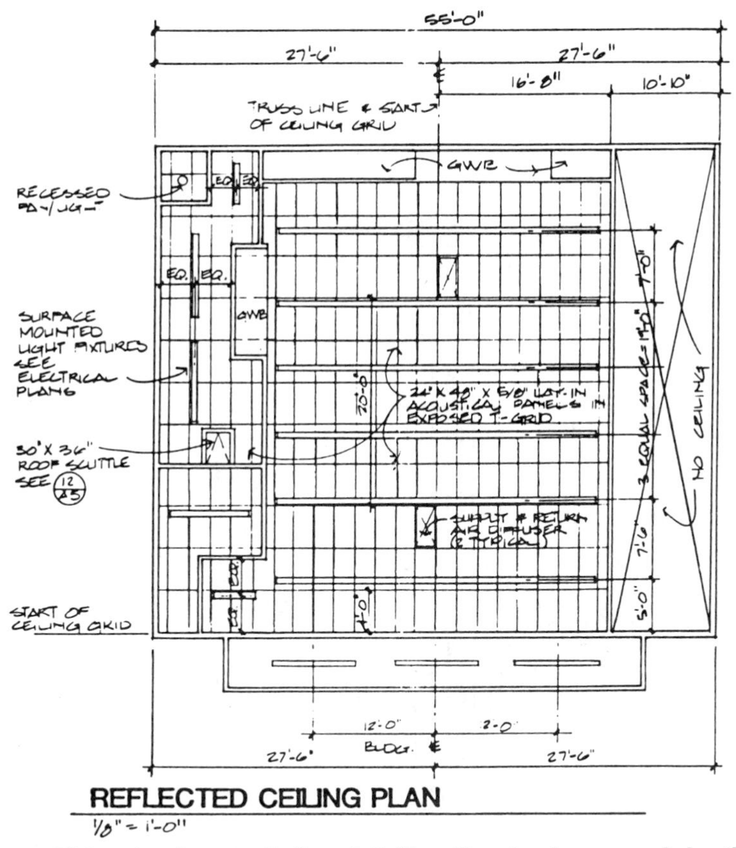

Figure 18–8 This print shows a Reflected Ceiling Plan for the suspended ceiling layout. *Courtesy The Southland Corporation.*

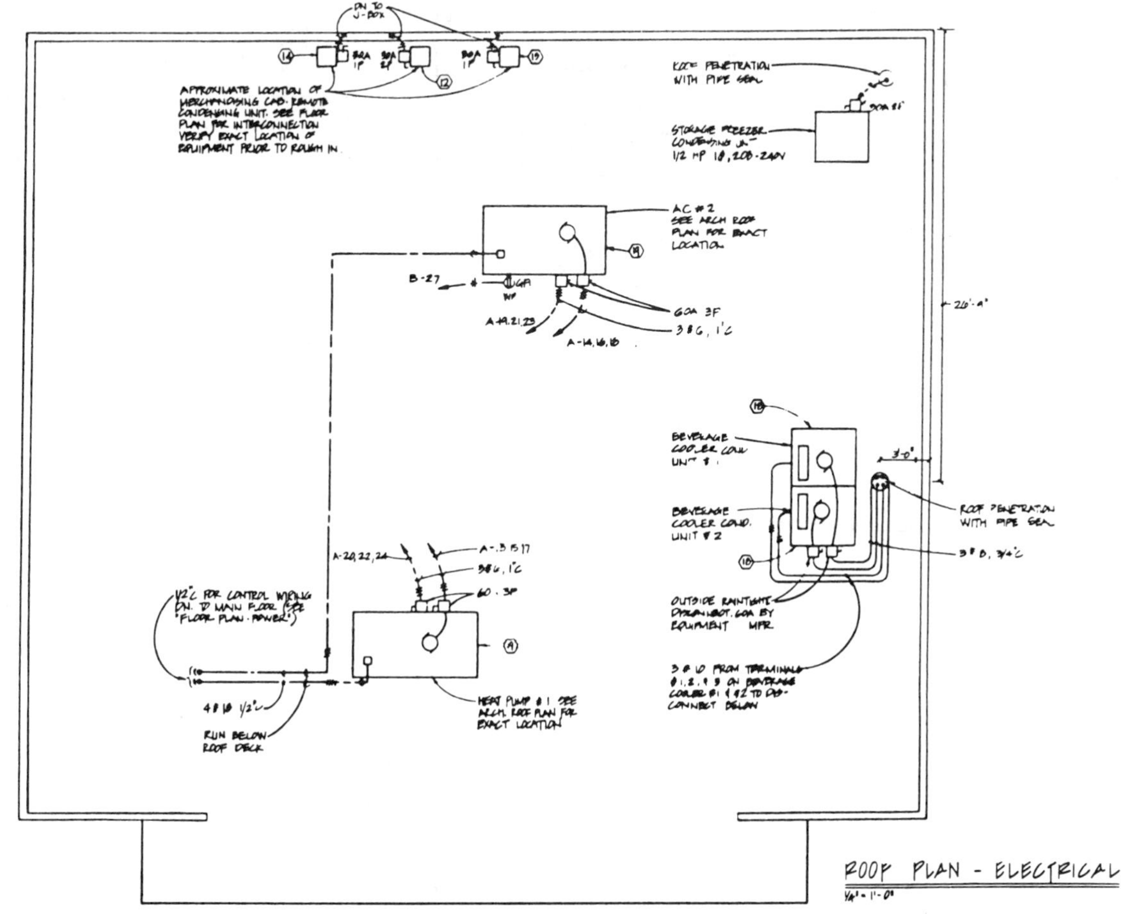

Figure 18–9 This print shows the electrical needs of equipment on the roof. *Courtesy The Southland Corporation.*

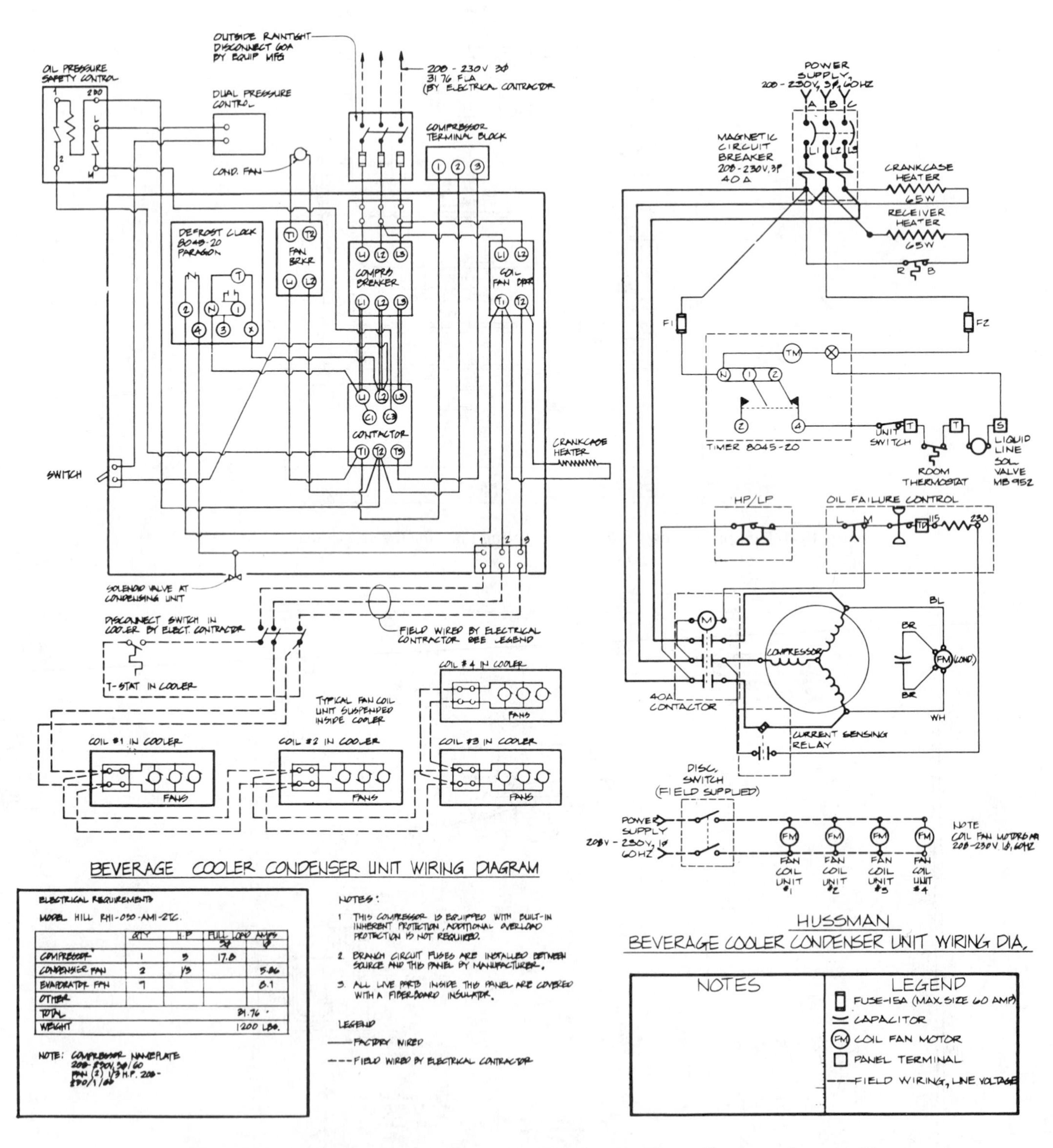

Figure 18–10 This print shows schematic wiring for specific applications. *Courtesy The Southland Corporation.*

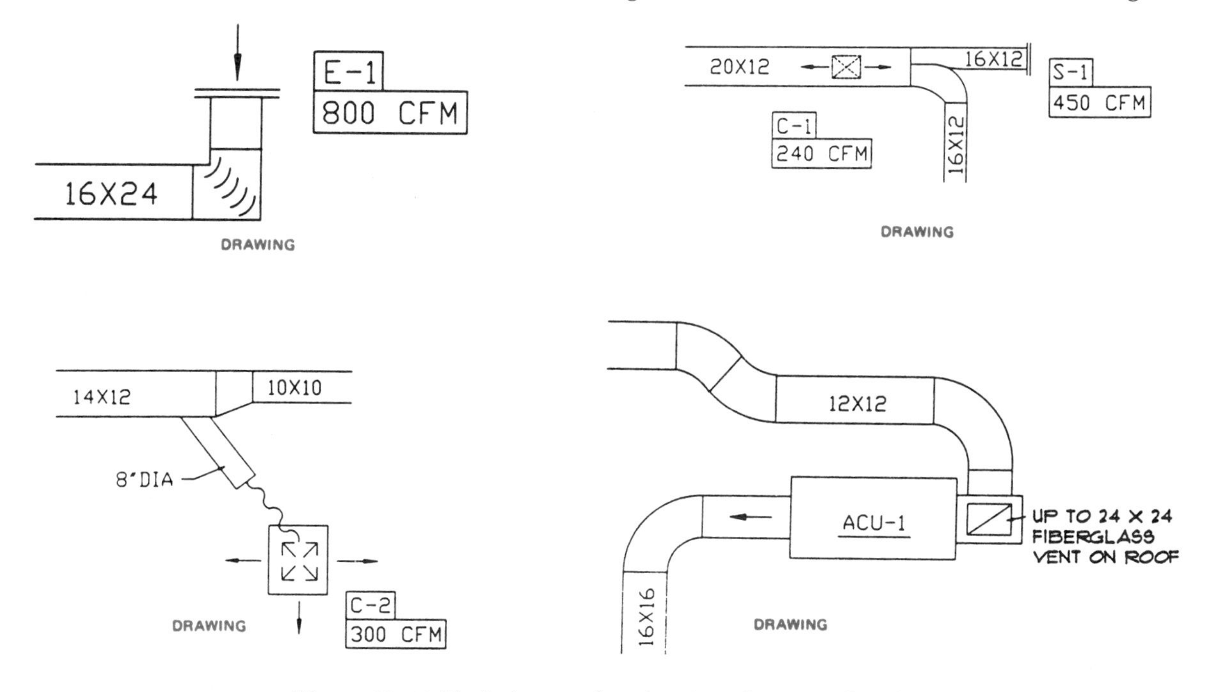

Figure 18–11 Typical examples showing elements of a plan.

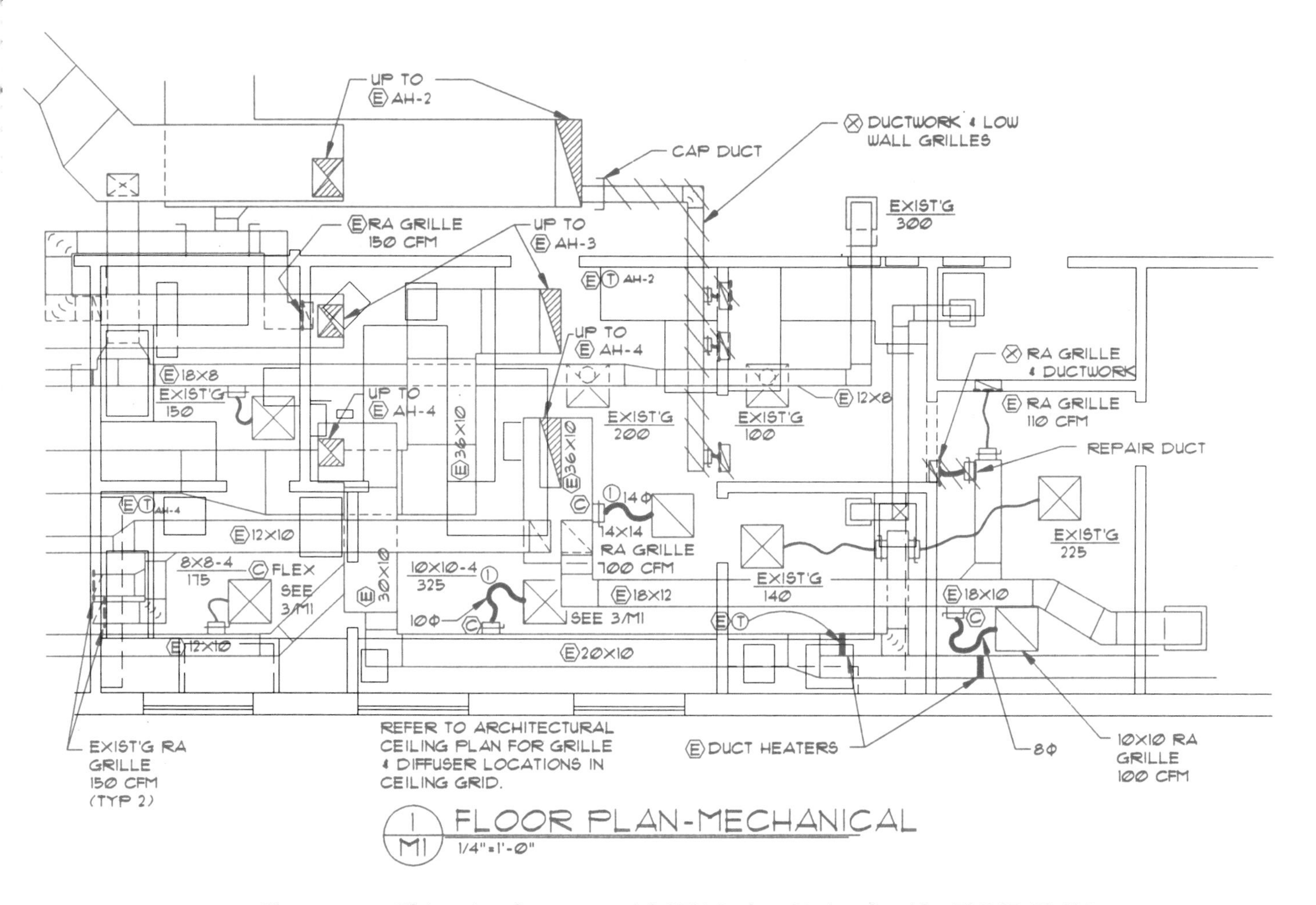

Figure 18–12 This print shows a partial HVAC plan. Notice the title: FLOOR PLAN-MECHANICAL. The term MECHANICAL refers to HVAC. *Courtesy System Design Consultants, Inc.*

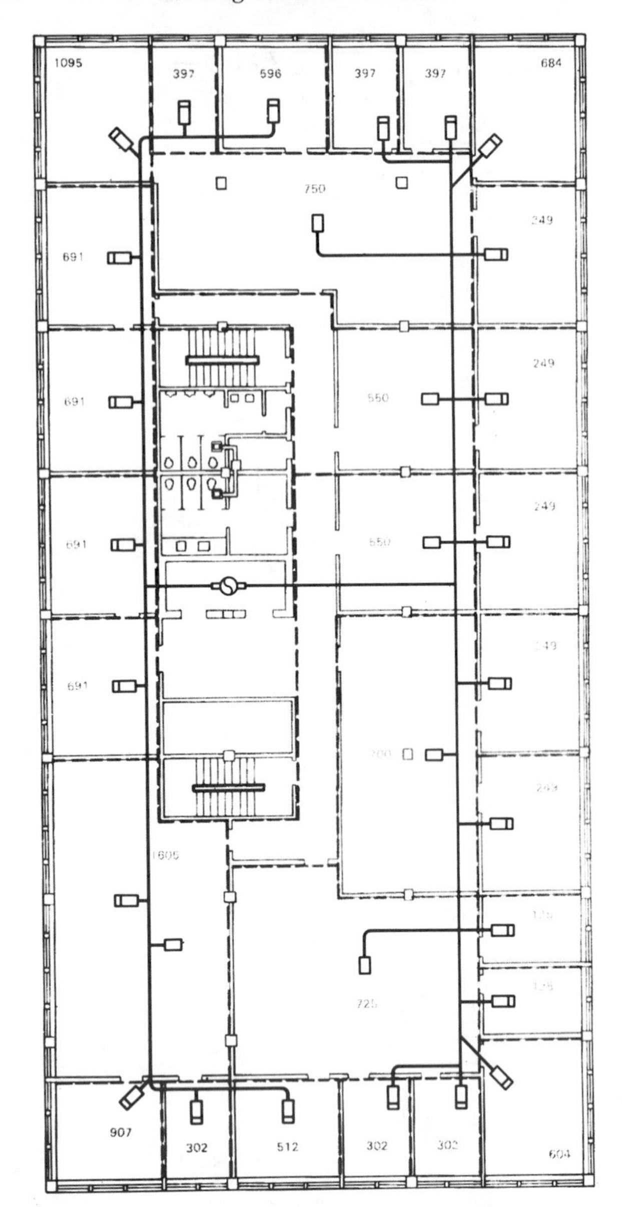

Figure 18–13 This print shows an example of a single-line ducted HVAC system with a layout of the proposed trunk and runout ductwork. *Reprinted by permission of The Trane Company, La Crosse, WI.*

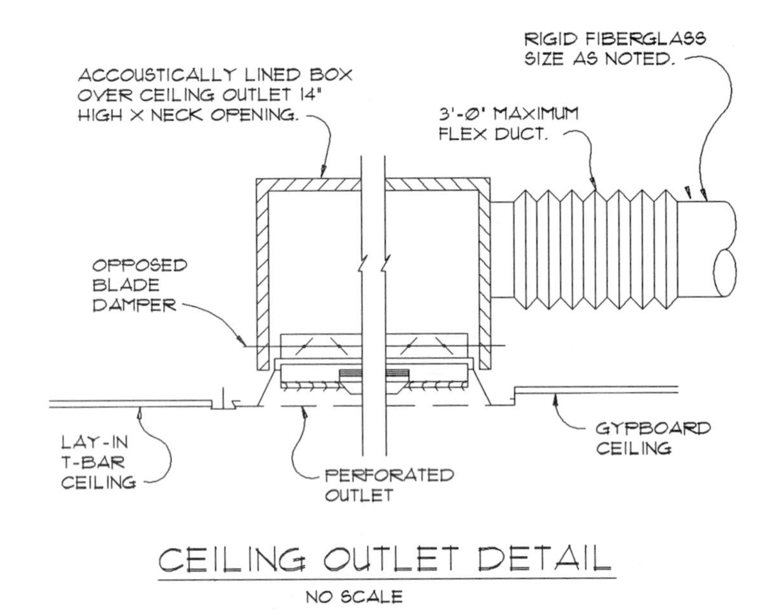

Figure 18–14 This is a typical detail drawing taken from a set of HVAC prints. *Courtesy W. Alan Gold Consulting Mechanical Engineer and Robert Evenson Associates AIA.*

Detail Drawings

Detail drawings are used to clarify specific features of the HVAC plan. Single- and double-line drawings are intended to establish the general arrangement of the system; they do not always provide enough information to fabricate specific components. When further clarification of features is required, detail drawings are made. A detail drawing is an enlarged view(s) of equipment, equipment installations, duct components, or any feature that is not defined on the plan. Detail drawings may be scaled or unscaled and provide adequate views and dimensions for sheet metal shops to prepare fabrication patterns as shown in Figure 18–14.

Section Drawings

Sections or sectional views are used to show and describe the interior portions of an object or structure that would otherwise be difficult to visualize. Section drawings may be used to provide a clear representation of construction details or a profile of the HVAC plan as taken through one or more locations in the building. There are two basic types of section drawings used in HVAC. One method is used to show the construction of the HVAC system in relationship to the structure. In this case, the building is sectioned and the duct system is shown unsectioned. This drawing provides a profile of the HVAC system. There may be one or more sections taken through the structure, depending on the complexity of the project. Figure 18–15 is a section through a plan showing the HVAC. The other sectioning method is used to show detail of equipment, or to show how parts of an assembly fit together. (See Figure 18–16.)

Schedules

Numbered symbols are used on the HVAC plan to key specific items to schedules. These schedules are used to describe items such as ceiling outlets, supply and exhaust grills, hardware, and equipment. Schedules include information regarding size, description, quantity used, capacity, location, vendor's specification, and any other information needed to con-

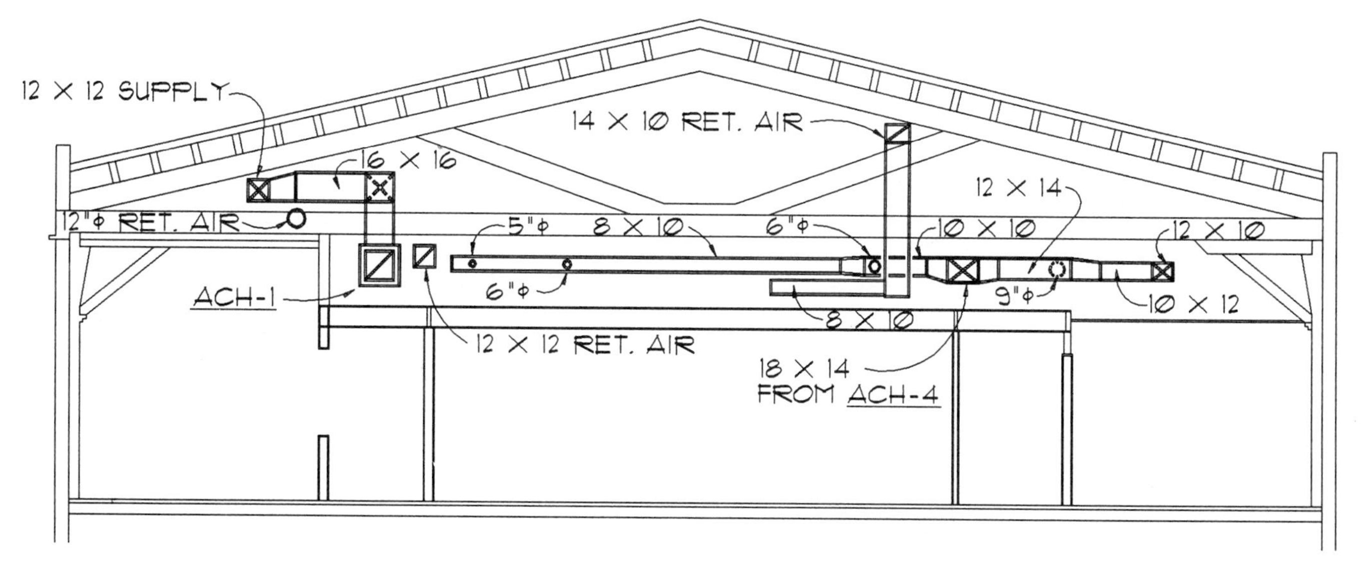

Figure 18–15 This is a typical section drawing taken from a set of HVAC prints. *Courtesy W. Alan Gold.*

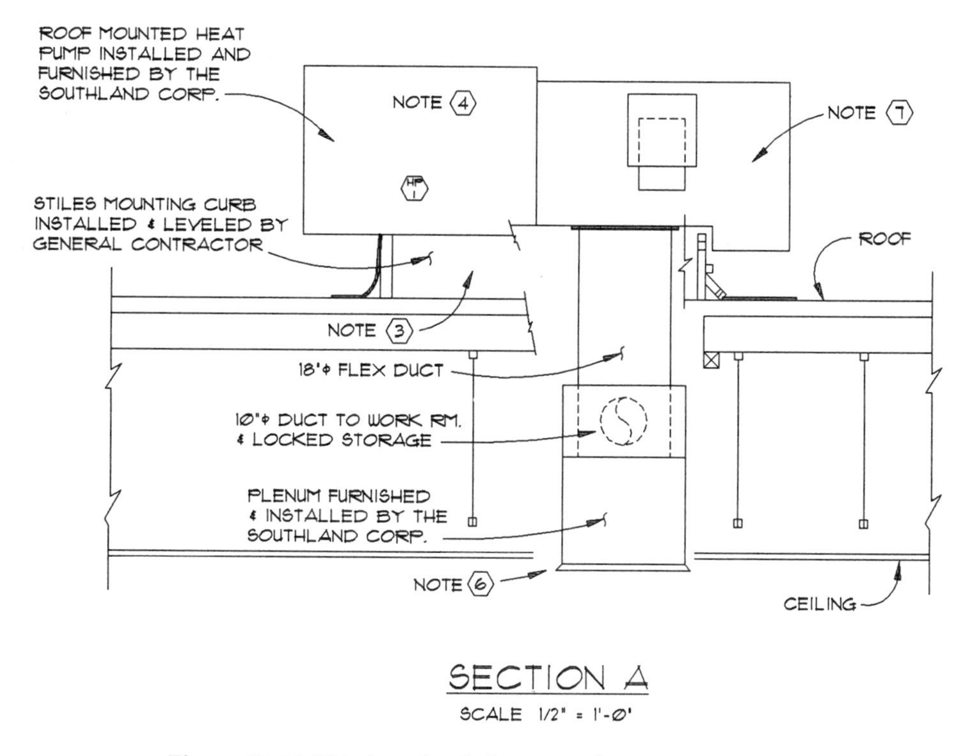

Figure 18–16 This is a detailed section showing HVAC equipment installation. *Courtesy The Southland Corporation.*

struct or finish the system. Items on the plan may be keyed to schedules by using a letter and number combination such as C-1 for CEILING OUTLET NO. 1, E-1 for EXHAUST GRILL NO. 1, or ACU-1 for AIR CONDITIONER UNIT NO. 1. An exhaust grill schedule for an HVAC plan is shown in Figure 18–17.

Pictorial Drawings

Pictorial drawings may be used as shown in Figure 18–18. They are used in HVAC for a number of applications, such as assisting in visualization of the duct system, and when the plan and sectional views are not adequate to show difficult duct routing.

EXHAUST GRILL SCHEDULE									
				DAMPER TYPE					REMARKS
SYMBOL	SIZE	CFM	LOCATION	FIRE DPR.	KEY OP. OPPBLD	KEY OP. EXTR	NO. DPR.	TYPE	
E-1	24x12	750	HIGH WALL	X	X			4	
E-2	18x18	720	CEILING	X	X			2	
E-3	10x10	240	CEILING		X			1	24x24 PANEL
E-4	10x10	350	CEILING		X			1	
E-5	12x12	280	CEILING		X			1	↓
E-6	10x10	500	CEILING	X	X			1	↓
E-7	6x6	350	↓		X			2	↓
E-8	12x6	50	↓		X			3	↓
E-9	12x6	200	↓		X			3	↓
E-10	12x8	150	↓		X			3	↓
E-11	10x10	290	↓		X			3	
E-12	9x4	160	↓		X			1	24x24 PANEL
E-13	9x4	75	HIGH WALL	X	X			4	

TYPE 1: KRUGER 1190 SERIES STEEL PERFORATED FRAME 23 FOR LAY-IN TILE

TYPE 2: KRUGER 1190 SERIES STEEL PERFORATED FRAME 22 FOR SURFACE MOUNT

TYPE 3: KRUGER EGC-5:1/2"X1/2"X1/2" ALUMINUM GRID.

TYPE 4: KRUGER S80H:35° HORIZ. BLADES 3/4"O.C.

Figure 18–17 This is a typical schedule taken from a set of HVAC prints.

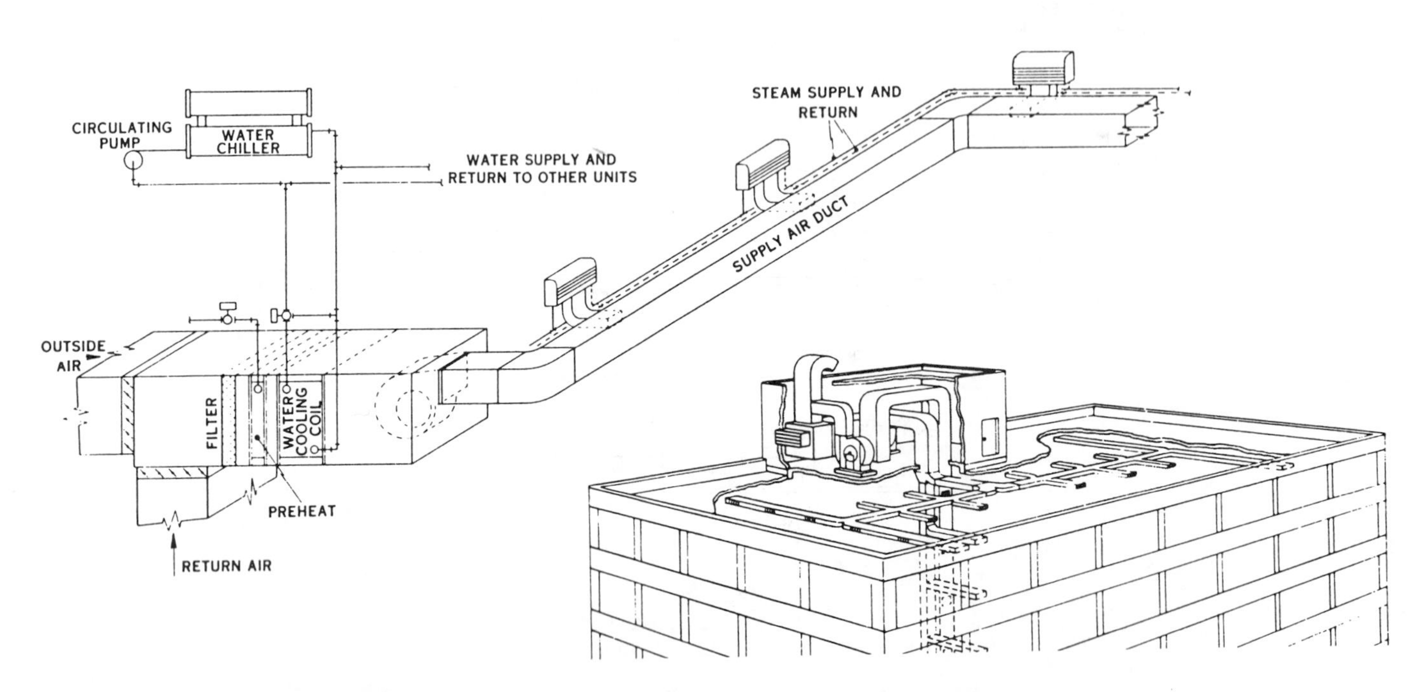

Figure 18–18 Sample pictorial drawings of HVAC systems or installations. *Reprinted by permission of The Trane Company, La Crosse, WI.*

READING COMMERCIAL PLUMBING PRINTS

In most cases, the plumbing drawings are found as an individual component of the complete set of plans for the commercial building. The architect or mechanical engineer prepares the plumbing drawings over an outline of the floor plan. This method keeps the drawing clear of any other unwanted information and makes it easier for the plumbing contractor to read the print. An example of this is shown in Figure 18–19. The plumbing plan shown in color is the only other information provided.

Common Commercial Plumbing Symbols

In residential plumbing drawings, you learned that displaying the fixture is the most important aspect

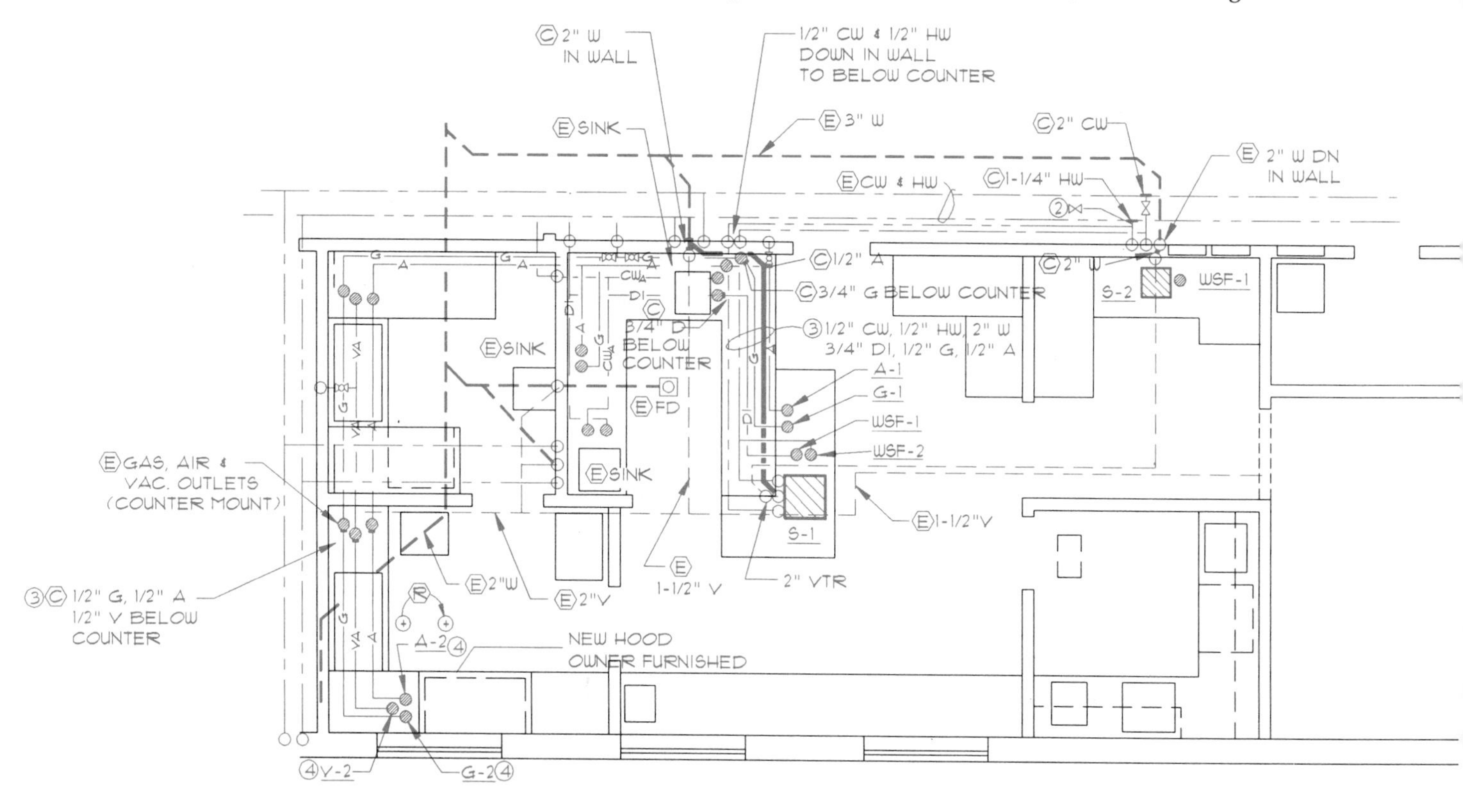

Figure 18–19 The floor plan is shown as an outline while the plumbing plan is detailed and labeled to clearly identify the installation for the person reading the print. *Courtesy System Design Consultants, Inc.*

SYMBOL	ABBREVIATION	SYMBOL NAME
(heavy solid line)	W	SANITARY WASTE ABOVE SLAB
(heavy dashed line)	W	SANITARY WASTE BELOW SLAB
——— — — ———	HW	DOMESTIC HOT WATER
— — — — — —	V	VENT
——— — ———	CW	DOMESTIC COLD WATER
——— G ———	G	GAS (NATURAL)
——— A ———	A	COMPRESSED AIR
———VA———	VA	LABORATORY VACUUM
———CW_A———	CW_A	COLD WATER (ASPIRATOR)
——— DI ———	DI	DEIONIZED WATER
(gate valve / globe valve symbols)		GATE VALVE/GLOBE VALVE
(pipe reducer symbol)		PIPE REDUCER
(sprinkler head symbol)		SPRINKLER HEAD

Figure 18–20 Common plumbing symbols found on a print. Any specific plumbing use may be applied to the pipe symbol by placing the proper abbreviation at a space provided in the pipe. *Courtesy System Design Consultants, Inc.*

of the drawing and usually the actual piping is left off the plan. Commercial plumbing drawings must clearly show the pipe lines and fittings as symbols with related information given in specific and general notes. Some of the more common plumbing symbols are shown in Figure 18–20. There are a variety of valve symbols that may be used in a plumbing print. A valve is a device used to regulate the flow of a gas or liquid. The symbols for common valves are shown in Figure 18–21. The direction the pipe is running is also important when reading plumbing prints. When the pipe is rising, the end of the pipe is shown as if it were cut through, and the inside is cross-hatched with section lines. When the pipe is turned down, or running away from you, the symbol appears as if you are looking at the outside of the pipe. Figure 18–22 shows several common symbols in side view, in a rising pipe view, and in a view with the pipe turning down.

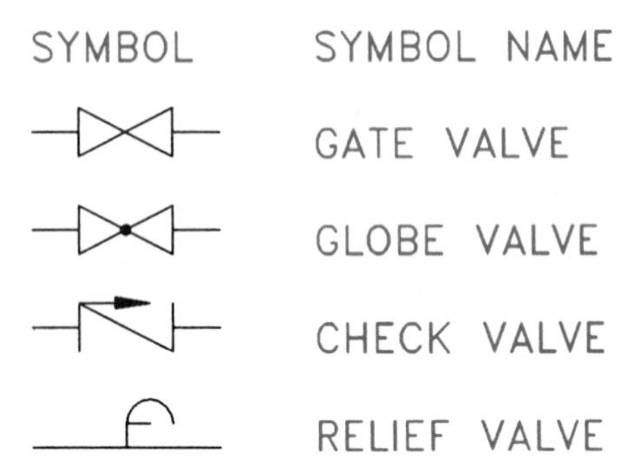

Figure 18–21 Common valve symbols found on a plumbing print. *Courtesy System Design Consultants, Inc.*

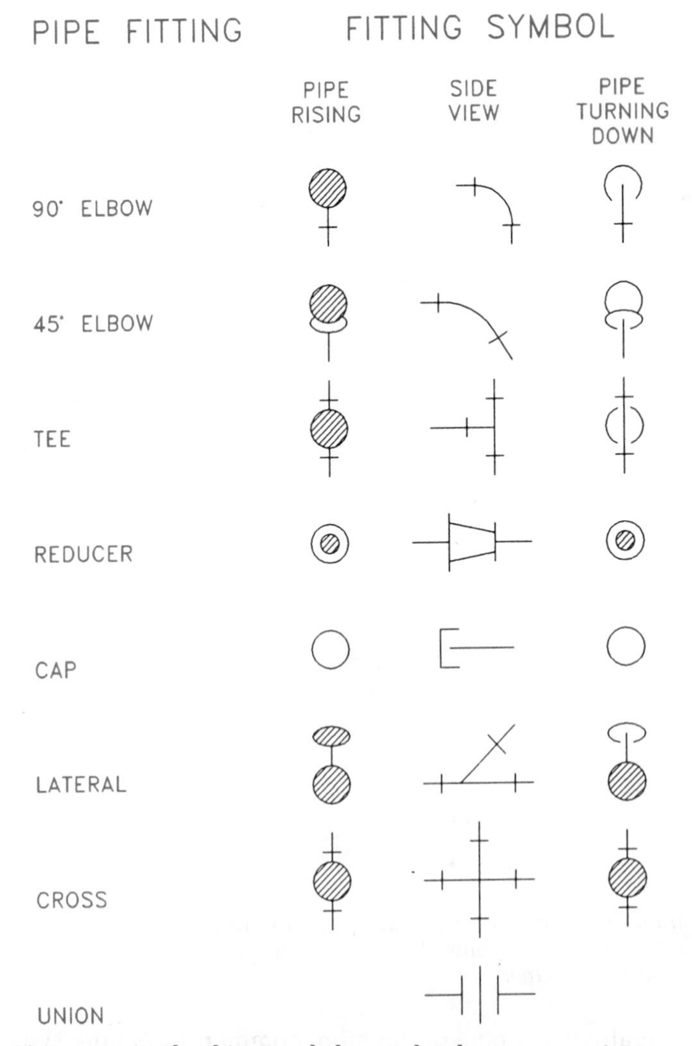

Figure 18–22 Plumbing symbols may be shown on a print as related to the direction you look at the fitting, either from the side, with the pipe rising, or with the pipe turned down. *Courtesy System Design Consultants, Inc.*

Pipe Size and Elevation Shown on the Print

The size of the pipe may be shown with a leader and a note or with an area of the pipe expanded to provide room for the size to be given as shown in Figure 18–23.

Some plumbing installations provide the elevation of a run of pipe or the elevation of a pipe fixture when a specific location is required. Elevation is the height of a feature from a known base usually given as 0 (zero elevation). The elevation of a pipe or fitting is given in feet and inches from the base. For example, EL 24'–6", where EL is the abbreviation of elevation, and 24'–6" is the height from the zero elevation base reference. The elevation may be noted on the print in relationship to a construction member such as TOC = Top of Concrete, or TOS = Top of Steel. The elevation may also be related to a location on the pipe such as BOP = Bottom of Pipe, or CL = Center Line of Pipe. Look at Figure 18–24.

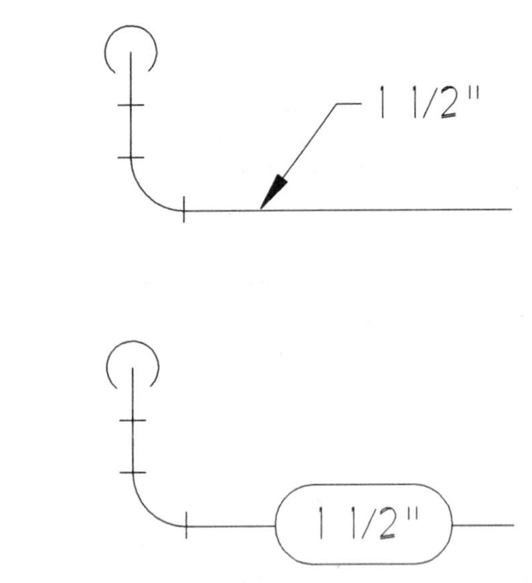

Figure 18–23 The size of plumbing pipe may be displayed on a print with a note or with a balloon placed on the pipe. *Courtesy System Design Consultants, Inc.*

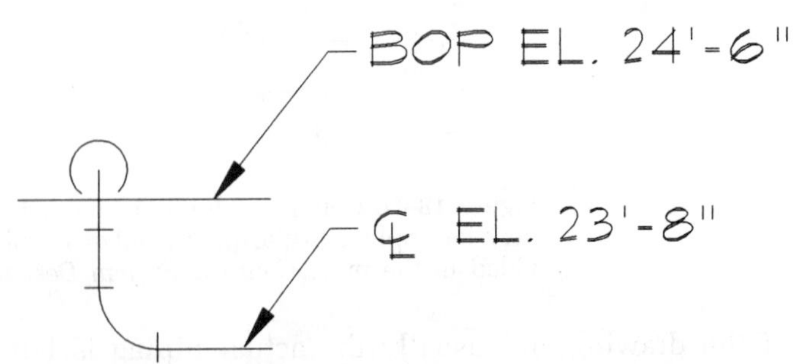

Figure 18–24 This example shows how the elevation of pipe and pipe fittings might be displayed on a print. *Courtesy System Design Consultants, Inc.*

Special Symbols, Schedules, and Notes on a Plumbing Print

As with any print, plumbing information may be provided in the form of special symbols, schedules, and general notes. Figure 18–25 shows special symbols that are used indicate specific procedures for pipes or fittings in the plumbing system. Figure 18–26 shows an example of a plumbing connection schedule and a general note.

In order to read plumbing prints properly it is important to become familiar with the symbols available. This chapter and Chapter 6 have introduced you to many of these symbols. Architectural and mechanical engineering offices often place a legend of symbols on the drawing for your reference. The legend displays symbols that are used on the print that you are reading. Always look at the title block, general notes, and legend to find general information about the print before you look at specific applications.

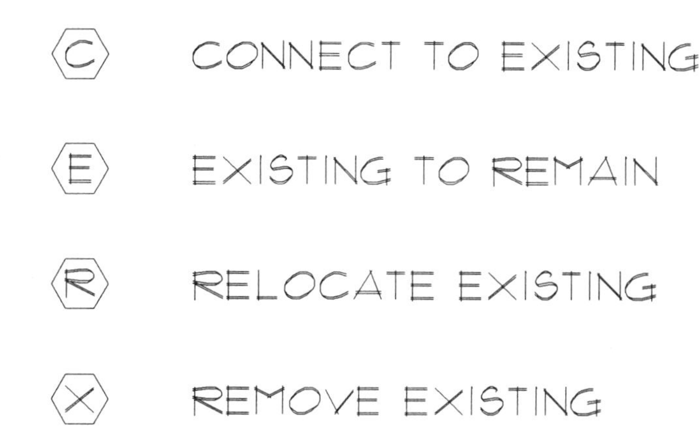

Figure 18–25 Sometimes special symbols are used on the print to designate specific applications. *Courtesy System Design Consultants, Inc.*

CONNECTION SCHEDULE (A)

SYMBOL	DESCRIPTION	CW	HW	W	V	REMARKS
S-1	SINK	1/2"	1/2"	2"	1-1/2"	STAINLESS
S-2	SINK	1/2"	1/2"	2"	1-1/2"	STAINLESS
WSF-1	WATER SUPPLY FITTING	1/2"	1/2"	—	—	ASPIRATOR
WSF-2	WATER SUPPLY FITTING	1/2"	1/2"	—	—	DEIONIZED WATER

(A) PLUMBING FIXTURES ONLY - G, VC AND A BRANCH SIZES NOTED ON 2/M1. VERIFY CONNECTION REQUIREMENTS WITH FIXTURES FURNISHED.

Figure 18–26 Schedules and general notes are used on plumbing prints to help keep information that is common to the entire print in one convenient location. *Courtesy System Design Consultants, Inc.*

CHAPTER 18 TEST

Fill in the blanks below with the proper word or short statement as needed to correctly complete the sentence or answer the question.

1. How is a special characteristic such as a specific size fixture, or a location requirement shown on an electrical print? ______________________________.

2. What do multiple arrowheads mean when you see them placed on the electrical circuit run? ______________________________.

3. What does it mean when you see a circuit run with three slash marks displayed? ______________________________.

4. Describe what is included in a floor plan lighting layout print. ______________________________.

5. What is another name for an HVAC engineer? ______________________________.

6. What do these symbols mean when found in an HVAC print?

   ∅ ____________.

   □ ____________.

7. Describe a detail drawing. ______________________________.

8. Numbered symbols that are used on the HVAC plan to key specific items to charts that are known as ______________________________.

9. How are residential and commercial plumbing drawings different? ______________________________.

10. Describe how a pipe or fitting looks on a print when it is rising. ______________________________.

11. Describe how a pipe or fitting looks on a print when it is turned down. ______________________________.

12. Explain two ways that the size of a pipe might be shown on a print. ______________________________.

13. Define elevation. ______________________________.

14. Explain the meaning of this note: BOP EL. 134'–0". ______________________________

CHAPTER 18 EXERCISES

PROBLEM 18–1. Given the following electrical circuit designations with leader lines pointing to different symbols, identify the symbol representations on the blanks provided for each.

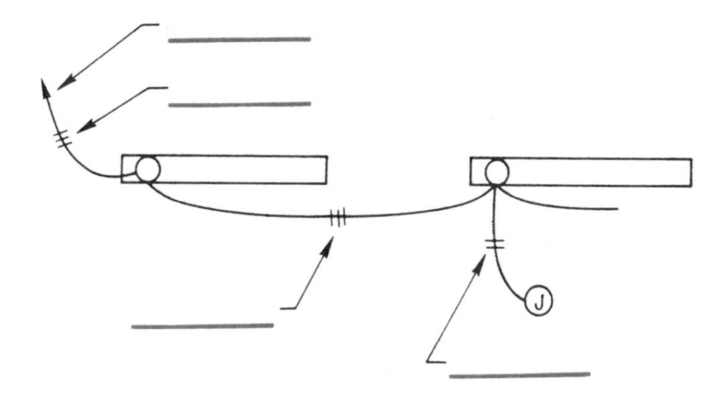

PROBLEM 18–2. Given the partial Power and Communication plan, legend, and related notes shown on pages 350–352 answer the following questions with short complete statements or words as needed: (Note: refer to all sheets of this problem when responding.)

1. How many duplex wall receptacles are there? __________
2. How many double duplex wall receptacles are there? __________
3. How many combination telephone/data wall outlets are there? __________
4. Locate the two panels shown in these partial plans and give the note associated with each. __________
5. There is an arrowhead at the end of some circuit runs and a note such as 2P2C-14, 16, 18. What does the arrowhead mean? __________

   What does the note 2P2C-14.16.18 mean? __________

6. The symbols D and P are generally placed together, what do they mean? __________
7. Refer to the symbols discussed in question number 6 and determine the number and size of wires in each circuit run from these symbols. __________
8. Count the number of junction boxes found in these partial plans. __________
9. What is the reference of the double asterisks (**). __________
10. What is the reference for this symbol: ③. __________

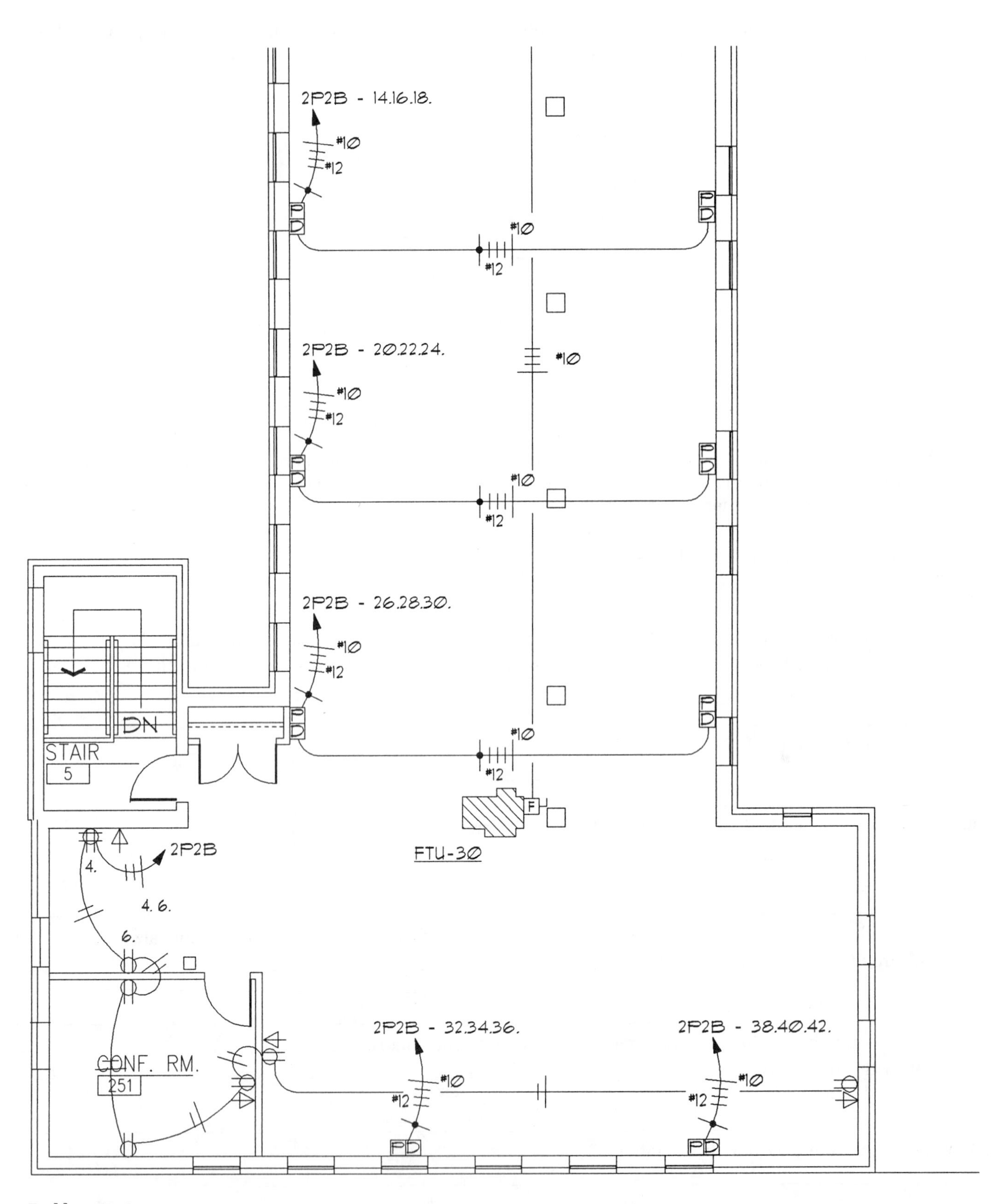

2P2B - 14.16.18.
#10
#12
2P2B - 20.22.24.
2P2B - 26.28.30.
STAIR
5
DN
FTU-30
F
2P2B
4.
4. 6.
6.
CONF. RM.
251
2P2B - 32.34.36.
2P2B - 38.40.42.
PD

Problem 18–2.

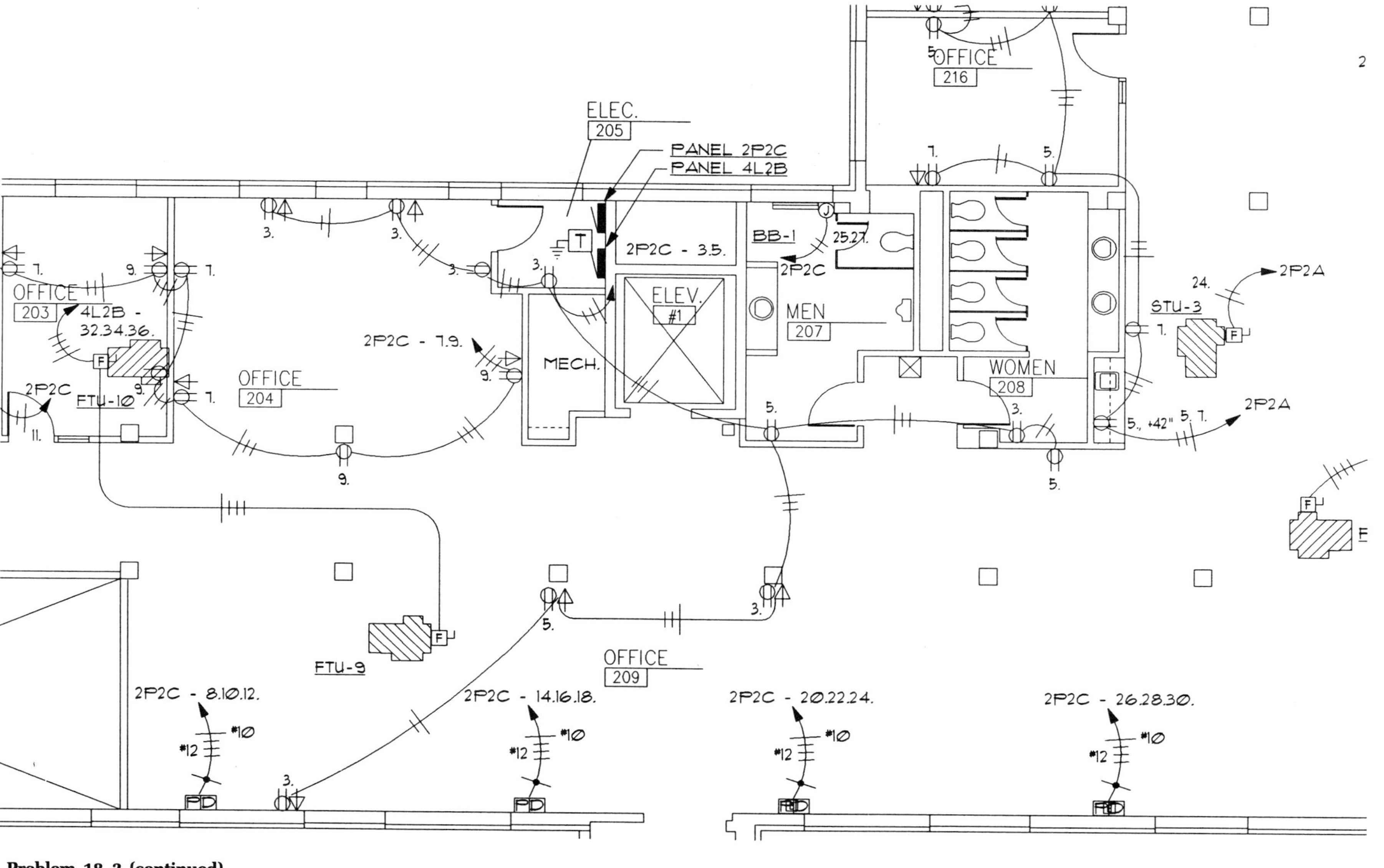

Problem 18–2 (continued).

GENERAL NOTES

1. VERIFY ELECTRIFIED PARTITION CONNECTION WITH PARTITION INSTALLER FOR EXACT WALL OR POKE THRU LOCATION PRIOR TO ROUGH-IN, TYPICAL ALL LOCATIONS.

SHEET NOTES

⟨1⟩ PROVIDE CONNECTIONS AS REQUIRED FOR OWNER FURNISHED EQUIPMENT. FIELD VERIFY NEMA CONFIGURATION WITH EXISTING EQUIPMENT.

⟨2⟩ PROVIDE 3/4" FIRERATED PLYWOOD FOR TELEPHONE TERMINAL BACKBOARD. PLYWOOD SHALL BE 8'-0" HEIGHTH AND ENTIRE LENGTH OF WALL.

⟨3⟩ PROVIDE WALL SWITCH FOR CONTROL OF EXHAUST FAN. MOUNT ADJACENT TO LIGHT SWITCHES, PROVIDE ENGRAVED NAMEPLATE " EXHAUST FAN ".

PARTIAL LEGEND (SEE SHEET E-13)

◀ WALL OUTLET: TELEPHONE *

◁ WALL OUTLET: DATA *

WALL OUTLET: COMBINATION TELEPHONE/DATA *

FIRE RATED FLUSH POKE THRU: OUTLETS AS SHOWN *

WALL RECEPTACLE: DUPLEX

WALL RECEPTACLE: DOUBLE DUPLEX

D FIRE RATED WALL POKE THRU: ELECTRIFIED PARTITION CONNECTION , WALL (DATA & TELEPHONE - ROUTE 1" C. WITH PULL STRING TO ACCESSIBLE CEILING SPACE) * *

E FIRE RATED POKE THRU: ELECTRIFIED PARTITION CONNECTION (DATA, TELEPHONE, POWER)

P FIRE RATED WALL POKE THRU: ELECTRIFIED PARTITION CONNECTION , WALL (POWER) * *

IG DENOTES ISOLATED GROUND

* NOTE: ROUTE 3/4" CONDUIT WITH PULL STRING TO ACCESSIBLE CEILING SPACE.

* * NOTE: MOUNT HORIZONTAL IN BASE

N

Problem 18–2 (continued).

PROBLEM 18–3. Given the Wiring Diagram shown on page 354, answer the following questions with short complete statements or words as needed:

1. What is the title of this wiring diagram? ______________________________ .
2. How many fuses are there in this wiring system? ______________ .
3. What is the standard fuse amperage recommendation? ______________ .
4. Give the magnetic circuit breaker specifications. ______________
5. How many coil fan motors are there? ______________ .
6. Count the number of capacitors. ______________ .
7. What are the power supply specifications? ______________________________ .
8. What are the timer specifications? ______________

PROBLEM 18–4. Given the HVAC Floor Plan, Legend, and notes shown on pages 355 and 356, answer the following questions with short complete statements or words as needed: (Note: refer to all sheets of this problem when responding.)

1. List the rectangular duct sizes given in this plan:

   ____
   ____
   ____
   ____
   ____
   ____
   ____

2. How many flexible ducts are there in this system? ______________ .
3. What do the following abbreviations mean?

   CFM ______________
   RA ______________
   SA ______________
   EA ______________
   TYP ______________

4. What do the following symbols mean?

   (T) ______________
   (C) ______________
   (E) ______________
   (R) ______________
   (X) ______________
   Ø ______________

5. How many thermostats are there in this system? ______________ .
6. What do these specifications mean?

   14Ø ______________
   18 × 12 ______________

7. What is the specification about removing and replacing the existing spin-in flex duct and diffuser? ______________________________ .
8. What is the formula for the size of the supply diffuser and return grille plenum? (Give the meaning of each element of the formula.) ______________________________ .

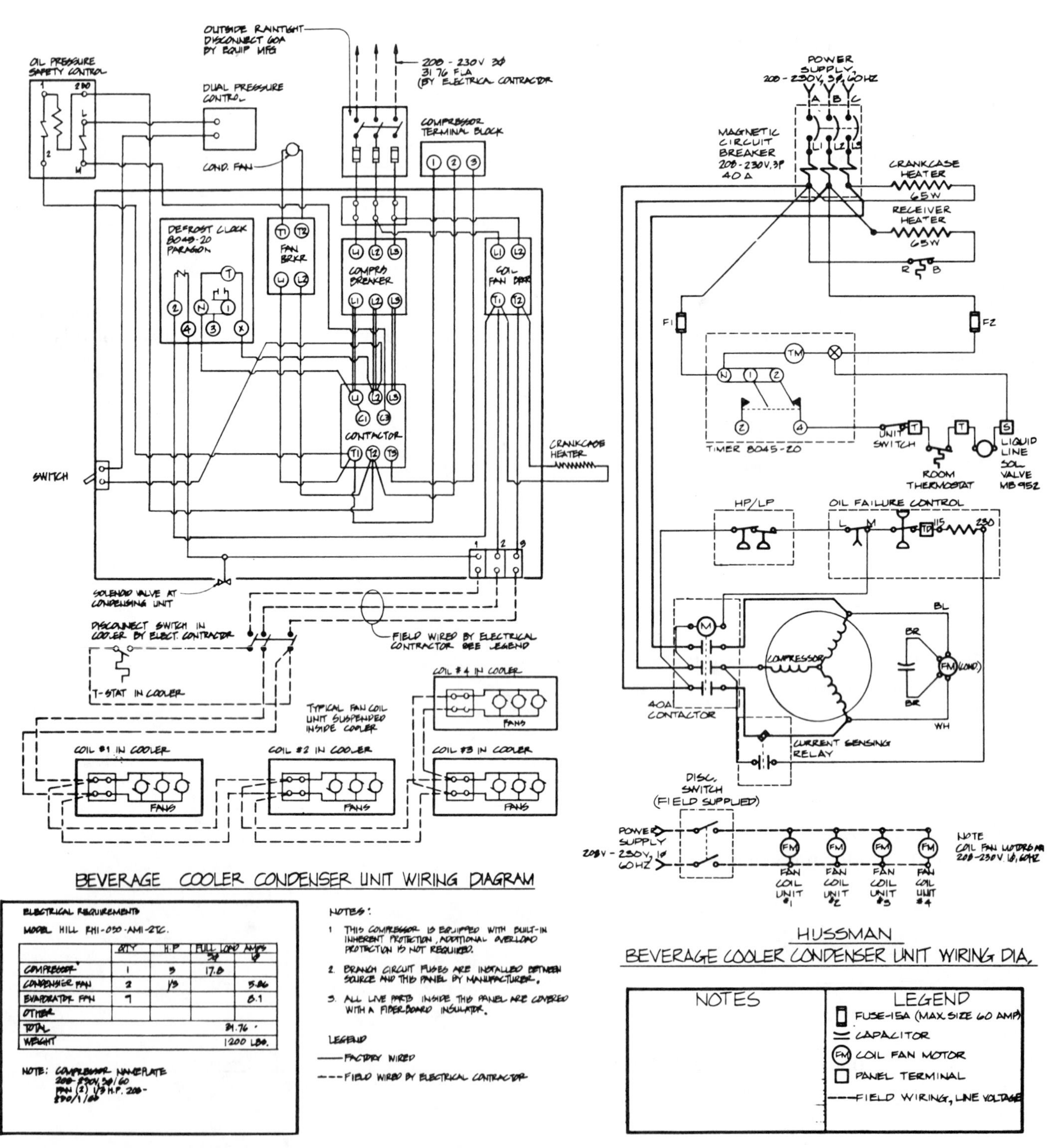

Problem 18–3. *Courtesy The Southland Corporation.*

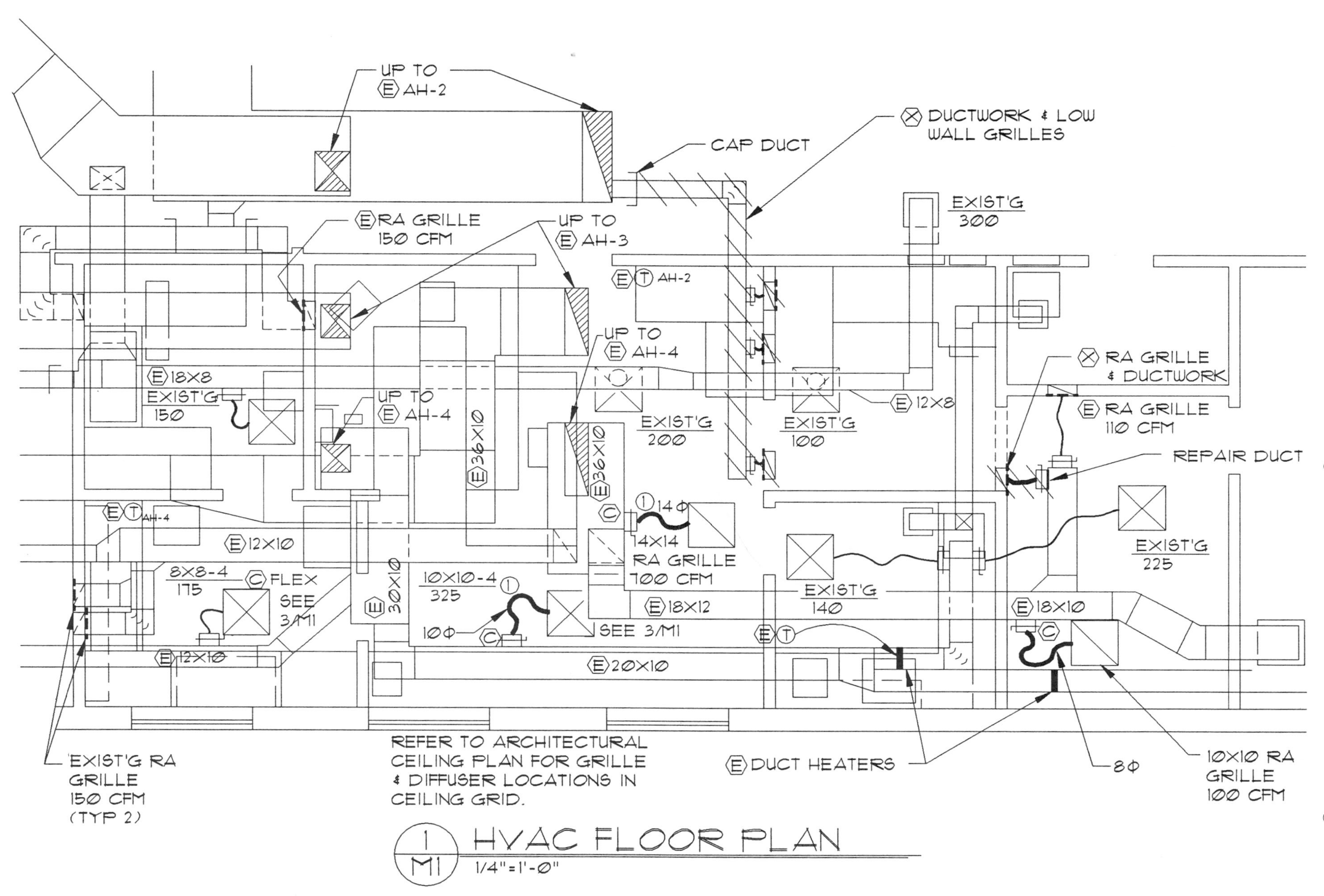

UP TO E AH-2
X DUCTWORK & LOW WALL GRILLES
CAP DUCT
EXIST'G 300
E RA GRILLE 150 CFM
UP TO E AH-3
E T AH-2
UP TO E AH-4
X RA GRILLE & DUCTWORK
E 18X8
EXIST'G 150
UP TO E AH-4
E 36X10
E 36X10
EXIST'G 200
EXIST'G 100
E 12X8
E RA GRILLE 110 CFM
REPAIR DUCT
E T AH-4
1 14Φ
C
14X14 RA GRILLE 700 CFM
E 12X10
EXIST'G 225
8X8-4 175
C FLEX SEE 3/M1
E 30X10
10X10-4 325
1
EXIST'G 140
E 18X12
E 18X10
10Φ
C
SEE 3/M1
E T
C
E 12X10
E 20X10
EXIST'G RA GRILLE 150 CFM (TYP 2)
REFER TO ARCHITECTURAL CEILING PLAN FOR GRILLE & DIFFUSER LOCATIONS IN CEILING GRID.
E DUCT HEATERS
8Φ
10X10 RA GRILLE 100 CFM
1
M1
HVAC FLOOR PLAN
1/4"=1'-0"

Problem 18–4.

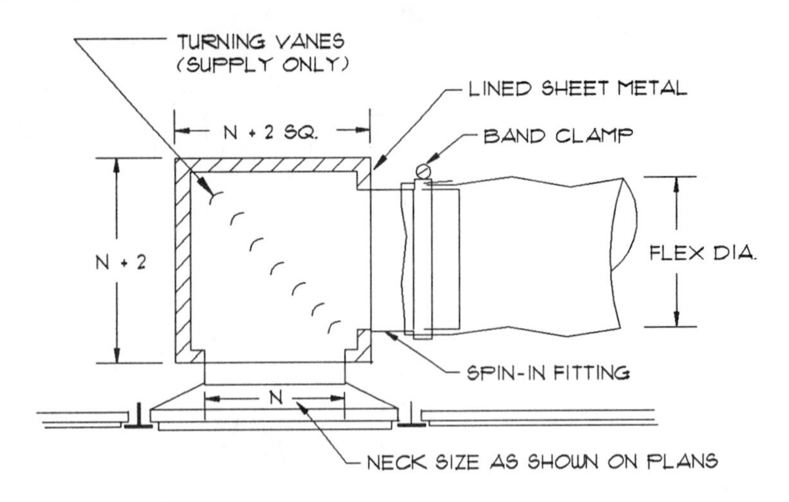

SUPPLY DIFFUSER/RETURN GRILLE PLENUM DETAIL

3 / M1

NO SCALE

LEGEND

8X8-2 / 2ØØ	NECK SIZE-THROWS/ CFM
	EXISTING (LIGHT)
	NEW (HEAVY)
	FLEX DUCT
	SPIN-IN W/VD
	DUCTWORK (DOUBLE LINE, SINGLE LINE)
	DIFFUSER, RA GRILLE
	DUCT RISER
SA	SUPPLY AIR
RA	RETURN AIR
EA	EXHAUST AIR
6ϕ	ROUND DUCT DIAMETER, INCHES
12x8	RECTANGULAR DUCT SIZE, INCHES
(T)	ROOM THERMOSTAT
(C)	CONNECT TO EXISTING
(E)	EXISTING TO REMAIN
(R)	RELOCATE EXISTING
(X)	REMOVE EXISTING

NOTES

(1) REMOVE AND REPLACE EXISTING SPIN-IN, FLEX DUCT AND DIFFUSER WITH SIZES AS NOTED.

Problem 18–4 (continued).

PROBLEM 18–5. Write the name of each symbol on the blanks provided below:

G

A

VA

CW_A

DI

PROBLEM 18–6. Write the name of each symbol on the blanks provided to the left and label the view, i.e. rising, side, or turning down, on the blank to the right of each symbol:

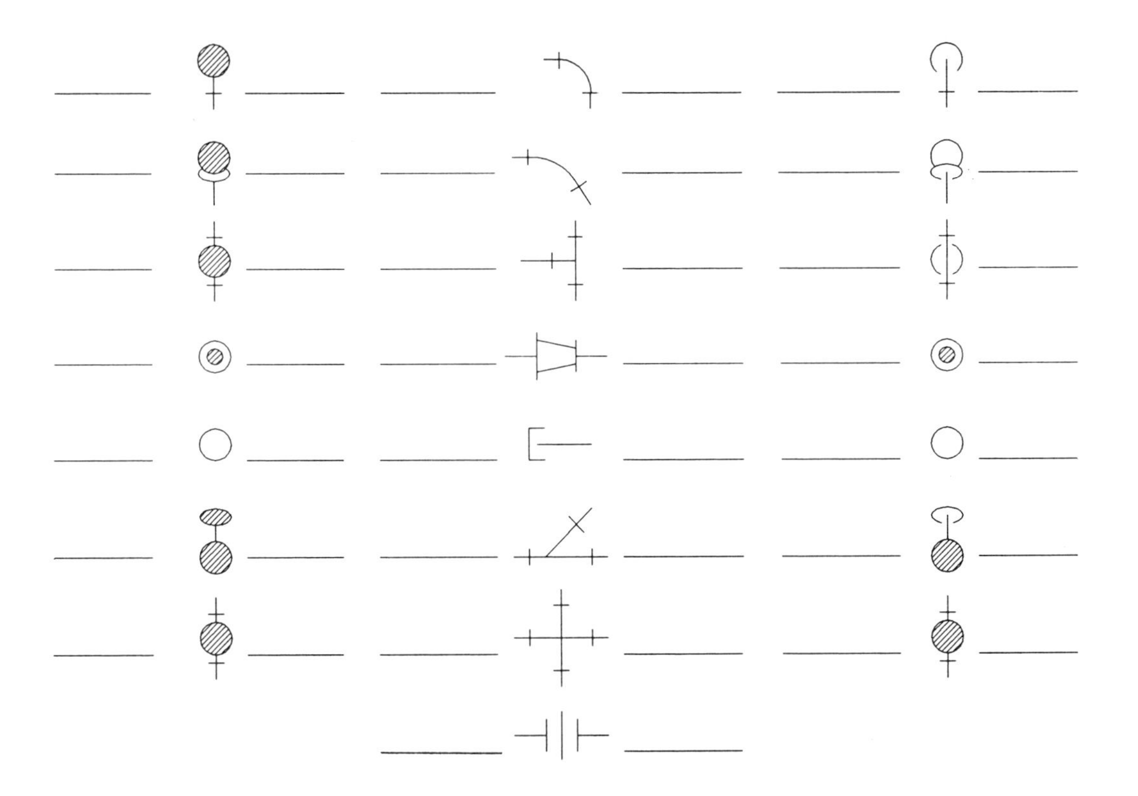

PROBLEM 18–7. Given the plumbing section shown on page 359, answer the following questions with short complete statements or words as needed:

1. What do the following abbreviations mean?
   EL. ____________
   TYP. ____________
   CL. ____________
   B.O. TANK. ____________
2. What is the elevation at the centerline of the pump? ____________
3. What is the elevation of the highest pipe? ____________.
4. What is the elevation of the 2"-S2B-12-10-LBH1503 pipe? ____________.
5. What is the difference in elevation between the bottom of the tank and the pipe described in question 4? ____________.
6. Name the type of valve found at the 2"-7B66T-C. ____________.
7. Describe the elbow connected to the 8"-S6M-12-10-LBH1541 pipe. ____________.
8. What is the difference in elevation between the bottom of the tank and the centerline of the pump? ____________.

PROBLEM 18–8. Given the HVAC Floor Plan, Legend, and notes shown on pages 360 and 361, answer the following questions with short complete statements or words as needed: (Note: refer to all sheets of this problem when responding.)

1. What do the following abbreviations mean?
   W ____________ A ____________
   HW ____________ VA ____________
   V ____________ CWA ____________
   CW ____________ DI ____________
   G ____________
2. What are the designated sizes of the below slab sanitary waste pipes? ____________.
3. What is the size of the above slab sanitary waste? ____________.
4. How many gate valves are shown? ____________.
5. Count the number of globe valves. ____________.
6. What do the following mean?
   S-1 ____________
   S-2 ____________
   WSF-1 ____________
   WSF-2 ____________
7. What is the material required for S-1? ____________.
8. Give the following pipe sizes for S-2:
   CW ____________
   HW ____________
   W ____________
   V ____________
9. How many sprinkler heads are there? ____________.
10. What are the specifications regarding placement of pipe in relation to the counter and the wall? ____________
11. How many floor drains are there? ____________.

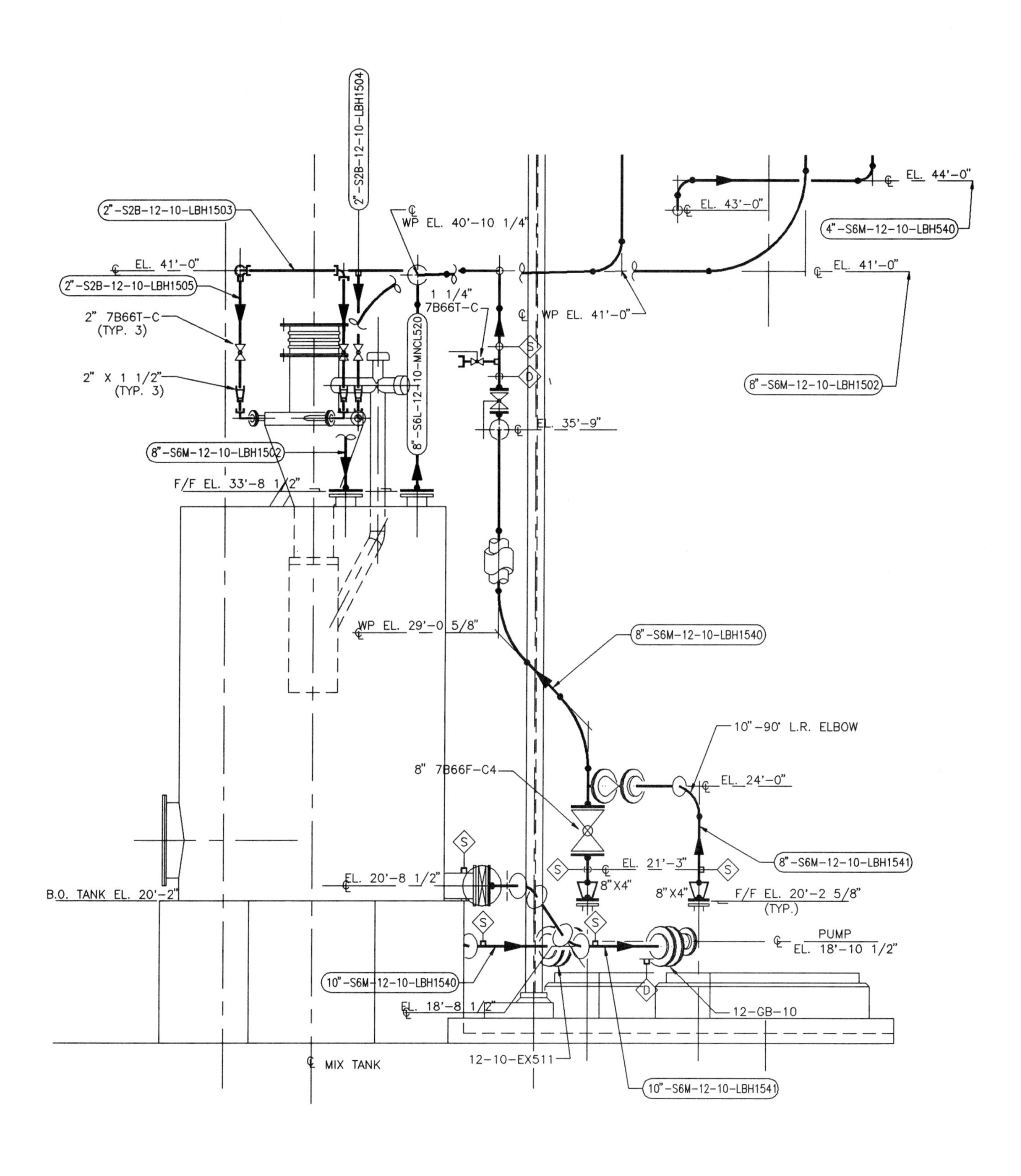

2"-S2B-12-10-LBH1503
2"-S2B-12-10-LBH1504
℄ EL. 41'-0"
2"-S2B-12-10-LBH1505
2" 7B66T-C (TYP. 3)
2" X 1 1/2" (TYP. 3)
8"-S6M-12-10-LBH1502
F/F EL. 33'-8 1/2"
℄ WP EL. 40'-10 1/4"
1 1/4" 7B66T-C
8"-S6L-12-10-MNCL520
℄ WP EL. 41'-0"
EL. 35'-9"
℄ EL. 43'-0"
℄ EL. 44'-0"
4"-S6M-12-10-LBH540
℄ EL. 41'-0"
8"-S6M-12-10-LBH1502
℄ WP EL. 29'-0 5/8"
8"-S6M-12-10-LBH1540
10"-90° L.R. ELBOW
8" 7B66F-C4
℄ EL. 24'-0"
8"-S6M-12-10-LBH1541
℄ EL. 20'-8 1/2"
B.O. TANK EL. 20'-2"
℄ EL. 21'-3"
8"X4"
8"X4"
F/F EL. 20'-2 5/8" (TYP.)
℄ PUMP EL. 18'-10 1/2"
10"-S6M-12-10-LBH1540
℄ EL. 18'-8 1/2"
12-GB-10
℄ MIX TANK
12-10-EX511
10"-S6M-12-10-LBH1541

SECTION
C
G201-E

Problem 18–7.

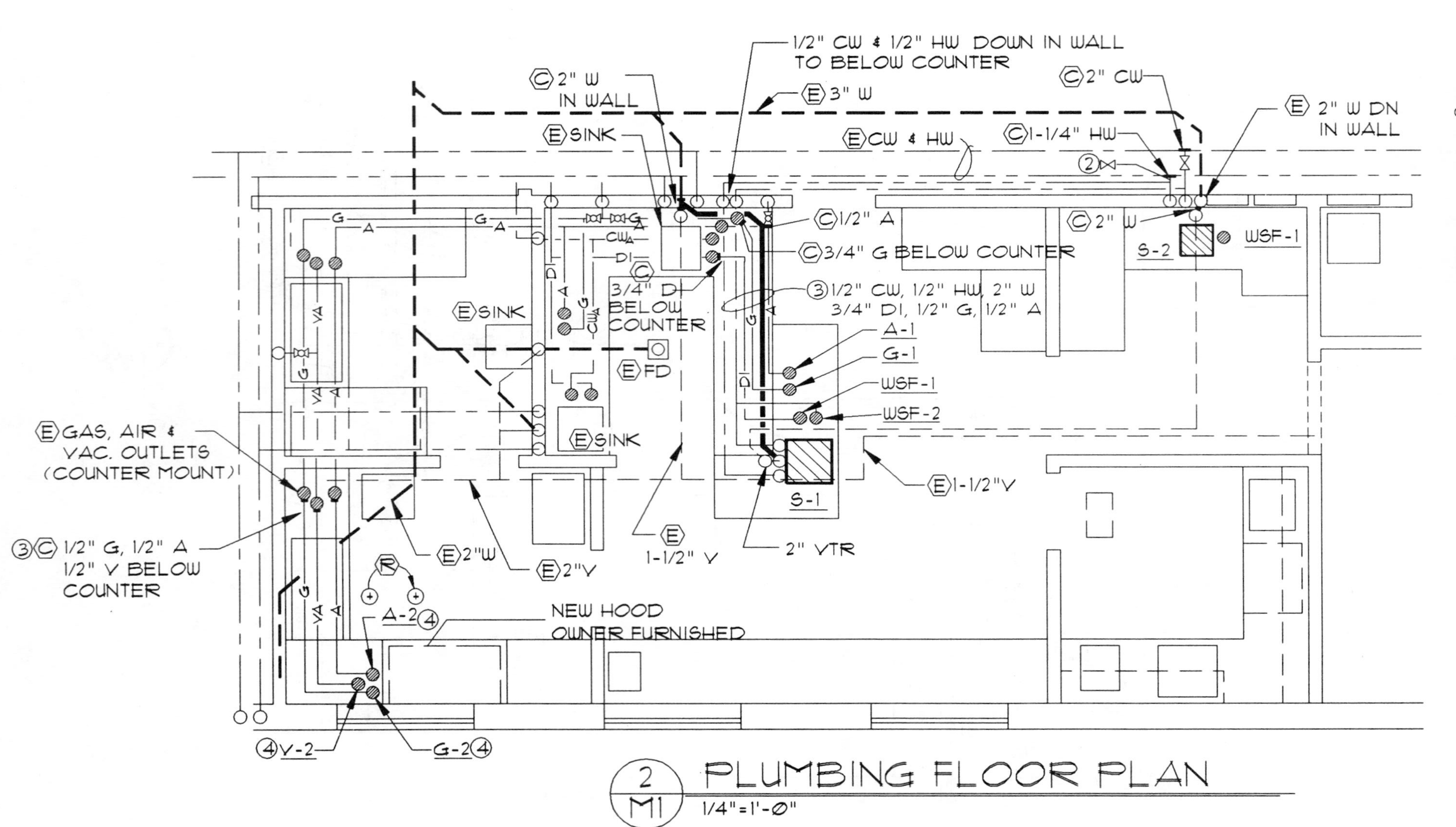
1/2" CW & 1/2" HW DOWN IN WALL TO BELOW COUNTER
2" W IN WALL
3" W
2" CW
2" W DN IN WALL
SINK
CW & HW
1-1/4" HW
1/2" A
3/4" G BELOW COUNTER
2" W
WSF-1
S-2
3/4" DI BELOW COUNTER
1/2" CW, 1/2" HW, 2" W 3/4" DI, 1/2" G, 1/2" A
A-1
G-1
WSF-1
WSF-2
SINK
FD
SINK
GAS, AIR & VAC. OUTLETS (COUNTER MOUNT)
1-1/2" V
S-1
1/2" G, 1/2" A 1/2" V BELOW COUNTER
2" W
2" V
1-1/2" V
2" VTR
NEW HOOD OWNER FURNISHED
A-2
V-2
G-2
PLUMBING FLOOR PLAN
2
M1
1/4"=1'-0"

Problem 18–8.

LEGEND

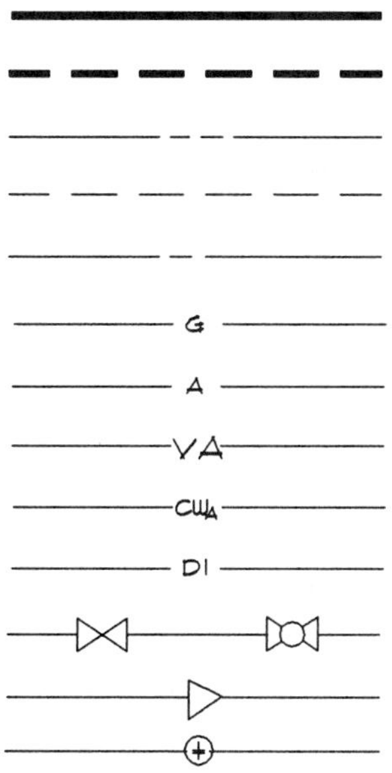

Line	Symbol	Description
	W	SANITARY WASTE ABOVE SLAB
	W	SANITARY WASTE BELOW SLAB
	HW	DOMESTIC HOT WATER
	V	VENT
	CW	DOMESTIC COLD WATER
G	G	GAS (NATURAL)
A	A	COMPRESSED AIR
VA	VA	LABORATORY VACUUM
CW_A	CW_A	COLD WATER (ASPIRATOR)
DI	DI	DEIONIZED WATER
		GATE VALVE/GLOBE VALVE
		PIPE REDUCER
		SPRINKLER HEAD
	C	CONNECT TO EXISTING
	E	EXISTING TO REMAIN
	R	RELOCATE EXISTING
	X	REMOVE EXISTING

CONNECTION SCHEDULE (A)

SYMBOL	DESCRIPTION	CW	HW	W	V	REMARKS
S-1	SINK	1/2"	1/2"	2"	1-1/2"	STAINLESS
S-2	SINK	1/2"	1/2"	2"	1-1/2"	STAINLESS
WSF-1	WATER SUPPLY FITTING	1/2"	1/2"	—	—	ASPIRATOR
WSF-2	WATER SUPPLY FITTING	1/2"	1/2"	—	—	DEIONIZED WATER

(A) PLUMBING FIXTURES ONLY - G, VC AND A BRANCH SIZES NOTED ON 2/M1. VERIFY CONNECTION REQUIREMENTS WITH FIXTURES FURNISHED.

NOTES

(2) PROVIDE GATE VALVE IN BRANCH BEFORE TAKEOFF.

(3) MOUNT ALL PIPING BELOW COUNTER TIGHT TO WALL

(4) PROVIDE HARD CONNECTION FROM GAS OUTLET TO 3/8" NPT HOOD GAS COCKS. PROVIDE MISC. FITTINGS AS REQUIRED. VERIFY EXACT CONNECTION LOCATIONS WITH ACTUAL HOOD FURNISHED BY OWNER.

Problem 18–8 (continued).

Chapter 19

Reading Interior Elevations and Details

INTRODUCTION TO COMMERCIAL INTERIOR ELEVATIONS AND DETAILS

MANY OF the same type of cabinet and millwork elements are used in commercial construction and residential buildings. Review Chapter 12 to refresh your memory on the basic cabinet components, types of mill work, and residential interior print reading techniques. Residential and commercial cabinet and mill work construction drawings can be very detailed and complex. While residential and commercial interior drawings techniques may be the same in many architectural offices, this chapter presents commercial interior drawings as a more advanced method of presenting interior construction drawings.

KEYING CABINET ELEVATIONS TO FLOOR PLANS

There are several methods that may be used to key the cabinet elevations to the floor plan.

Given the commercial drawing shown in Figure 19–1, notice the floor plan has identification numbers within a combined circle and arrow. The symbols, similar to that shown in Figure 19–2, are pointing to various areas that correlate to the same symbols labeled below the related elevations. A similar method may be used to relate cabinet construction details to the elevations. The detail identification symbol as shown in Figure 19–3 is used to correlate a detail to the location from which it originated on the elevation. With some plans, small areas such as the women's and men's rest rooms have interior elevations without specific orientation to the floor plan. The reason for this is that the relationship of floor plan to elevations is relatively obvious.

There may be various degrees of detailed representation needed for millwork depending on its complexity and the design specifications. The sheet from a set of architectural working drawings shown in Figure 19–4 and 19–5 are examples of how millwork drawings can provide very clear and precise construction techniques. Interior elevations and details often provide very complex construction information as shown in Figure 19–5.

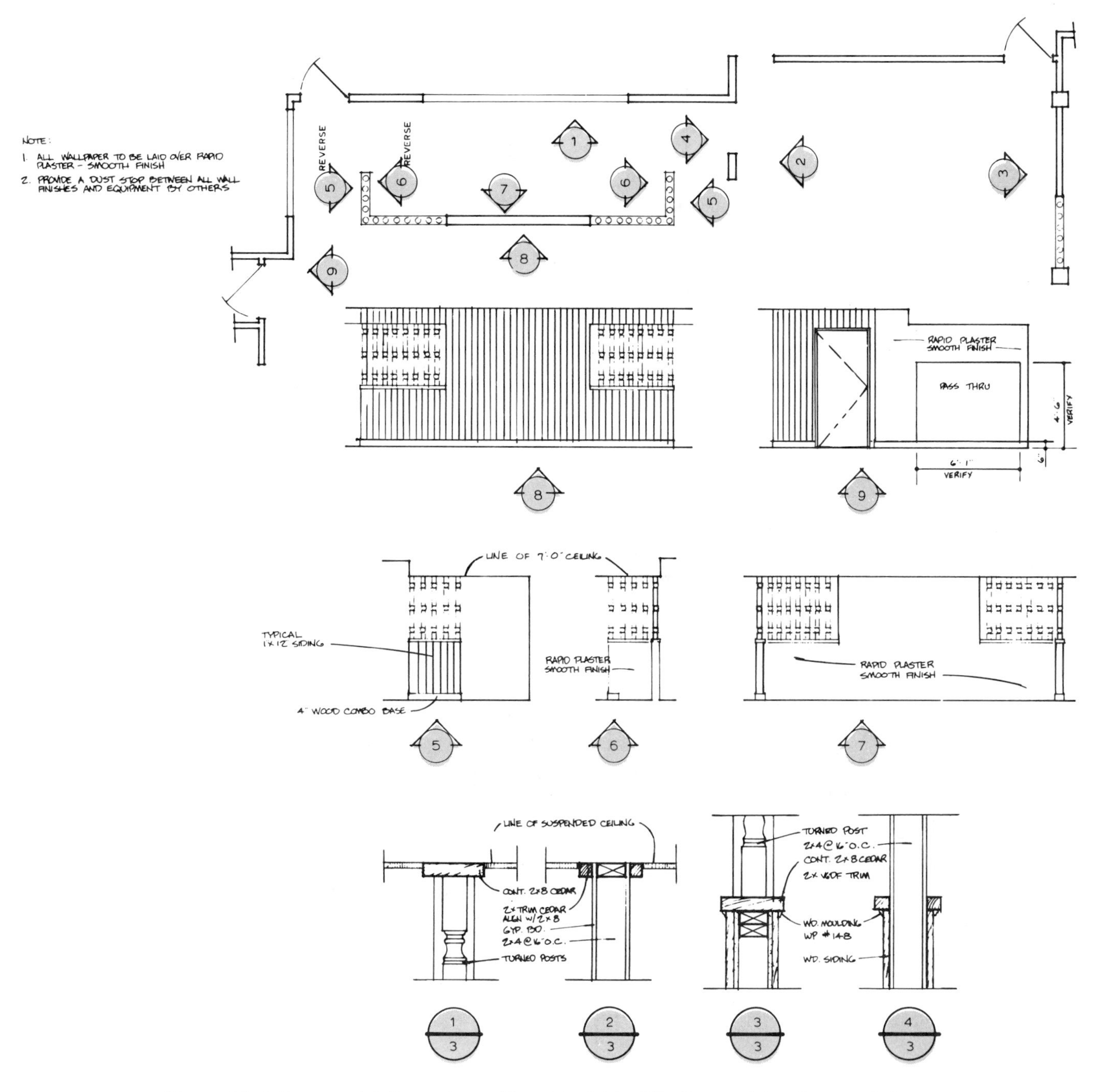

Figure 19–1 Commercial floor plan with cabinet identification and location symbols. *Courtesy Ken Smith Structureform Masters, Inc.*

Figure 19–2 Floor plan and elevation cabinet identification and location symbol.

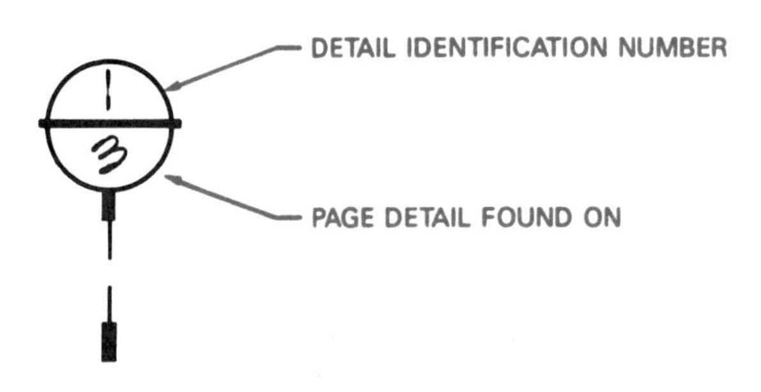

Figure 19–3 Detail identification symbol.

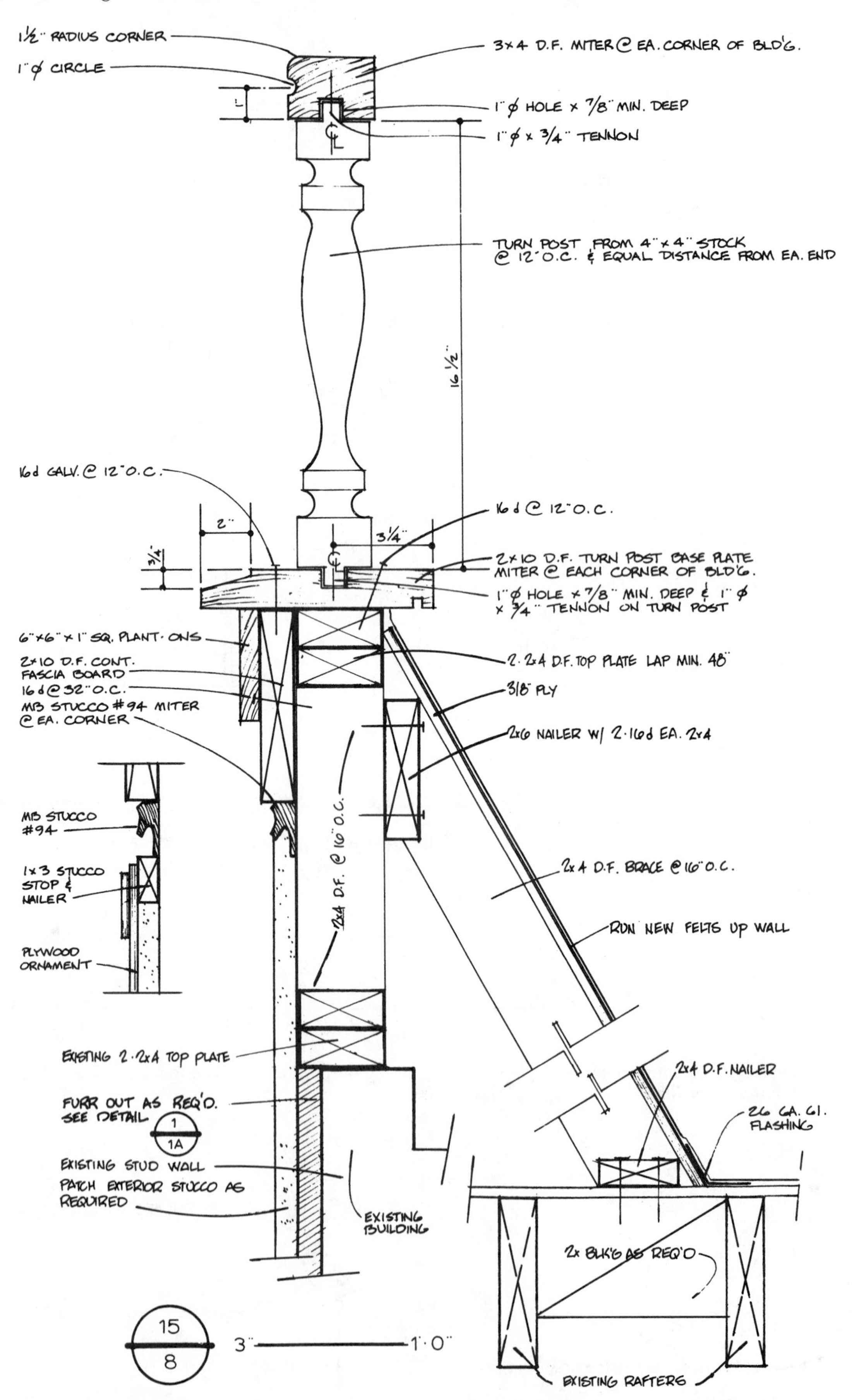

Figure 19–4 Millwork and cabinet construction details. *Courtesy Ken Smith Structure form Masters, Inc.*

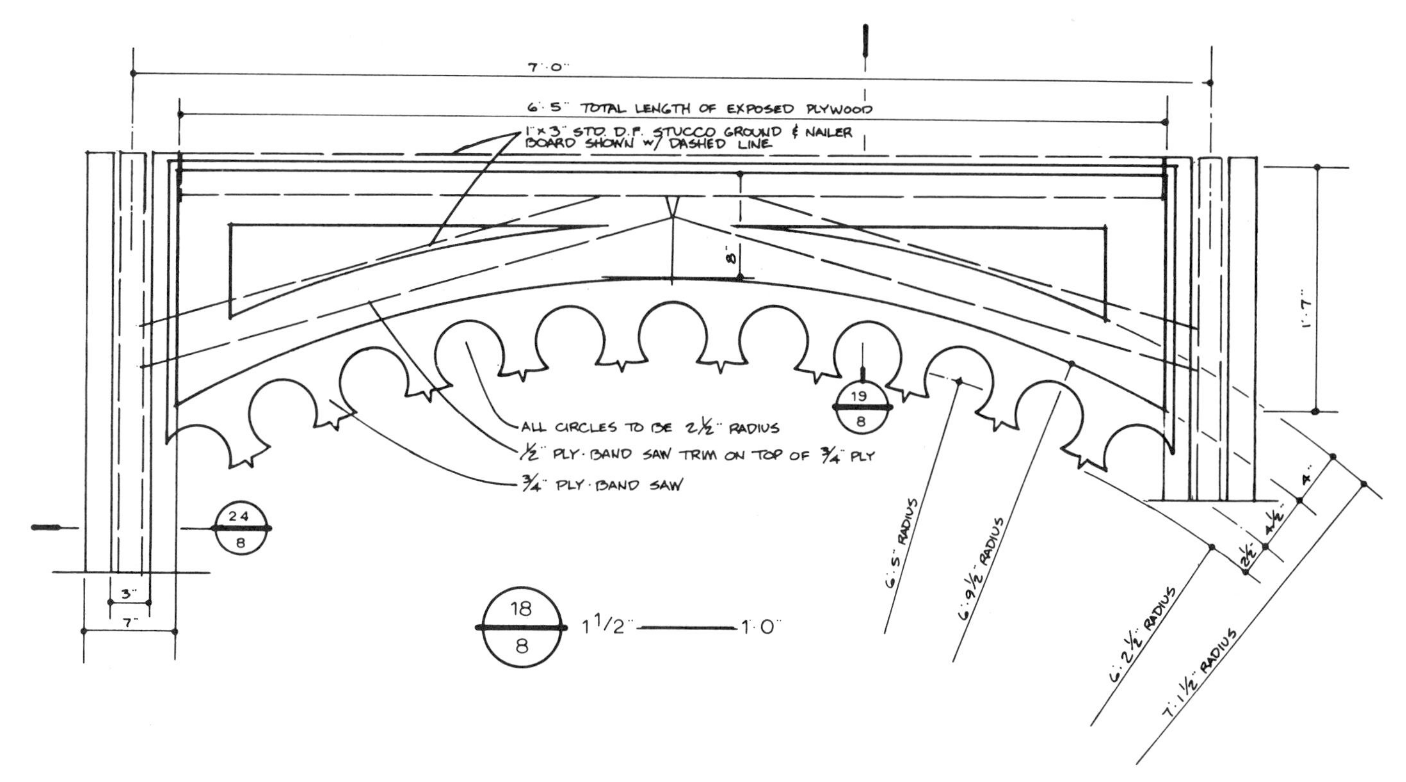

Figure 19–5 In addition to structural details, there may be interior elevations and details such as this finish detail for Farrell's restaurant. *Courtesy Structureform Masters, Inc.*

CHAPTER 19 TEST

Fill in the blanks below with the proper word or short statement as needed to correctly complete the sentence or answer the question.

1. Identify two ways that cabinet elevations may be correlated with the floor plan. ______________________
____
____.

2. Describe how the cabinet construction details may be correlated to the floor plan. ______________________
____
____
____.

CHAPTER 19 EXERCISES

PROBLEM 19–1. Given the commercial cabinet floor plan, elevations, and details shown on pages 367–371, answer the following questions with short complete statements, words, or dimensional information as needed: (Note: refer to all sheets of this problem when responding.)

1. What is the overall length, width, and height of the cabinet? ______________________
2. What are the dimensions of the cutting board? ______________________

3. How much longer is the space, where the cutting board fits, than the actual length of the cutting board? ______________________________.

4. How much wider is the space, where the cutting board fits, than the actual width of the cutting board? ______________________________.

5. How much higher is the space, where the cutting board fits, than the actual thickness of the cutting board? ______________________________.

6. What is the total number of 4" × 16" soffit vents used? ______________________________.

7. What are the C/O (cut out) dimensions for the WELLS MOD-300? ______________________________.

8. What are the C/O dimensions for the WELLS MOD-200 COUNTER TOP? ______________________________.

9. An electrical (abbreviated ELEC.) chase is a place to run electrical wires. What are the specifications for electrical chase in this cabinet? ______________________________

10. What are the specifications of the slots milled for the plexiglass carving shield? ______________________________.

11. Completely describe the structure supporting the plexiglass carving shield. ______________________________.

12. Give the complete dimensions and specifications of the plexiglass carving shield. ______________________________.

13. Give the dimensions of the 5" CONC. CURB. ______________________________.

14. What is the counter top material upon which the WELLS MOD-200 and WELLS MOD-300 units set? ______________________________.

15. What is the distance from the floor to the bottom of the glass hood? ______________________________

16. What does the symbol ℄ mean? ______________________________.

17. What does the symbol ∅ mean? ______________________________.

18. What does the abbreviation ID mean? ______________________________.

19. Give the note about the construction of the place where 3/8" ∅ ID TUBE supporting the carving shield is to be placed. ______________________________.

20. Describe the area where TILE is to be placed. ______________________________.

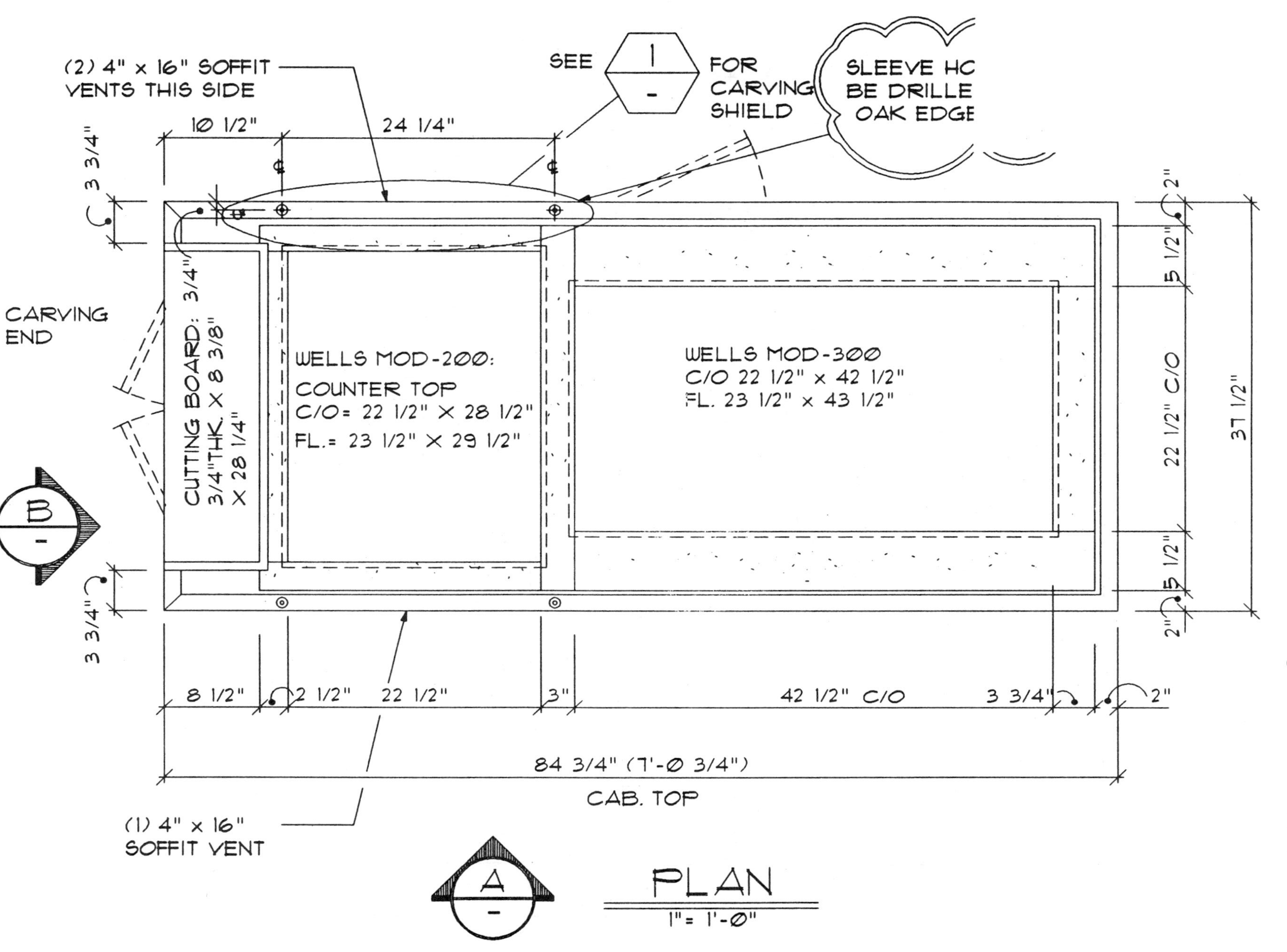

Problem 19–1. *Courtesy Alexander Manufacturing, Inc.*

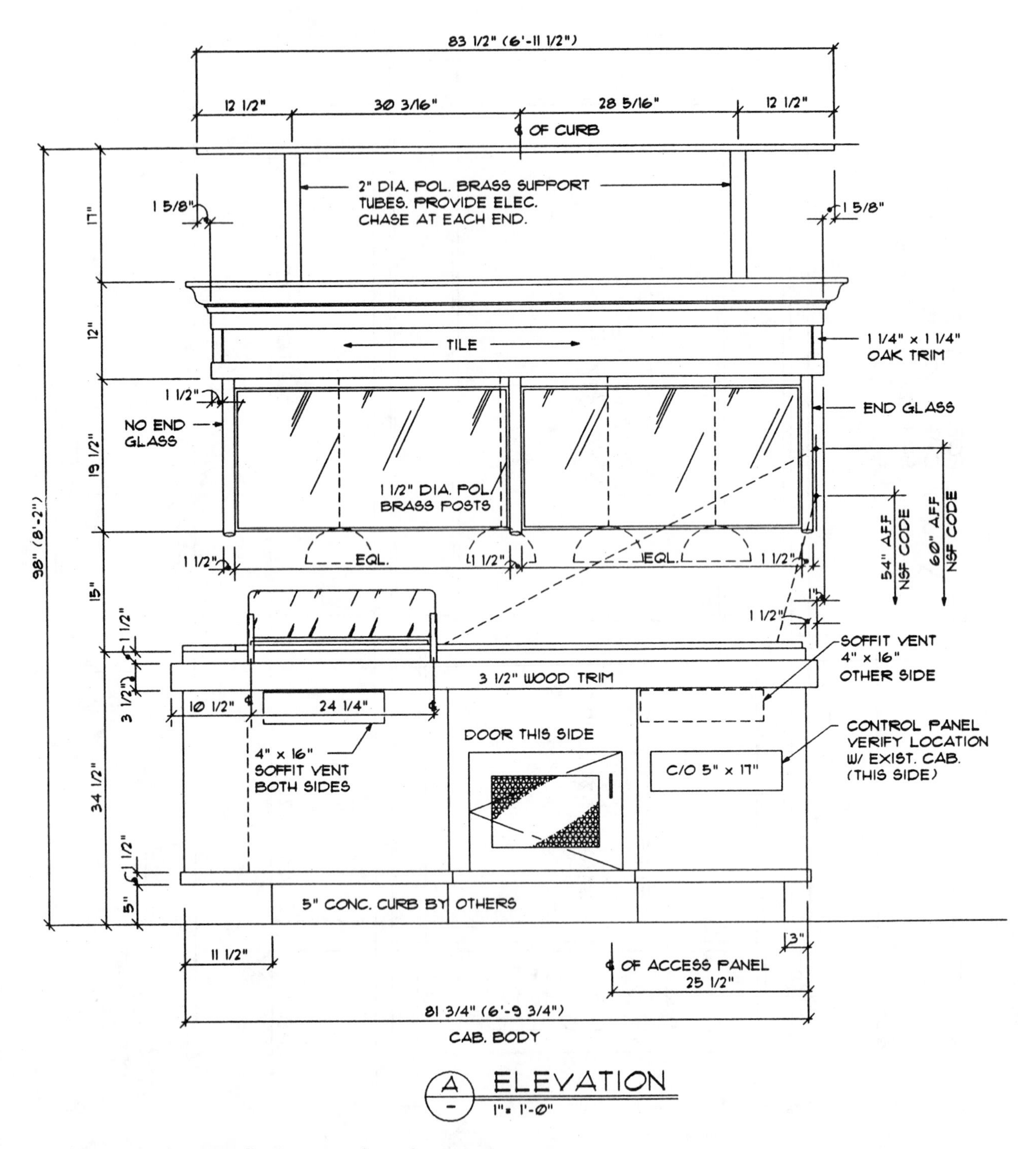

Problem 19–1 (continued). *Courtesy Alexander Manufacturing, Inc.*

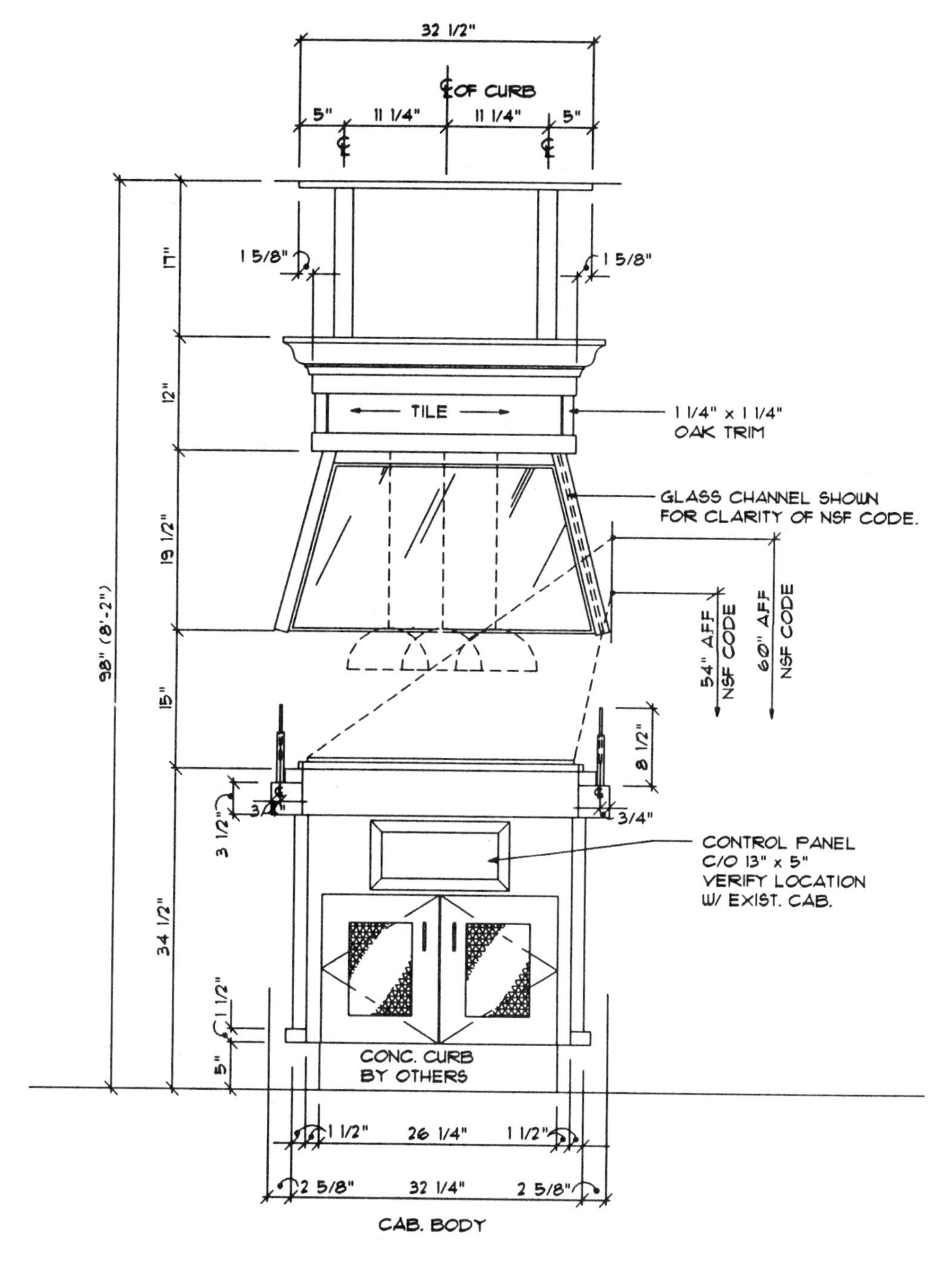

Problem 19–1 (continued). *Courtesy Alexander Manufacturing, Inc.*

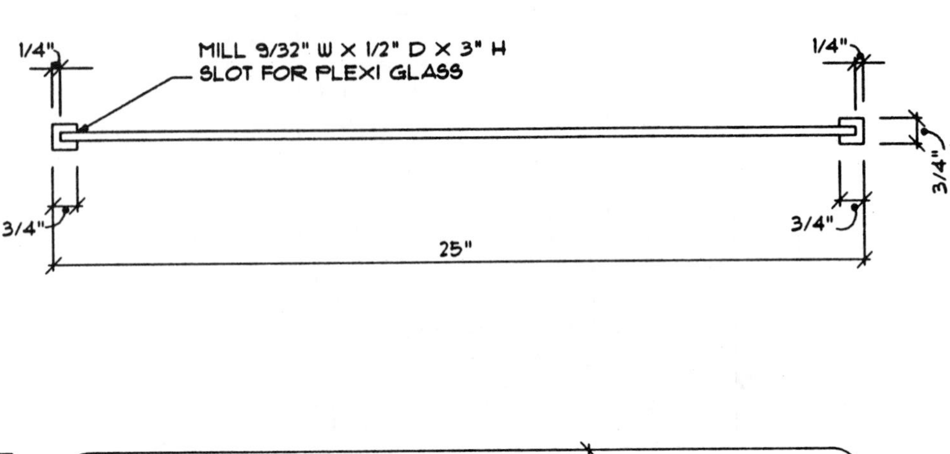

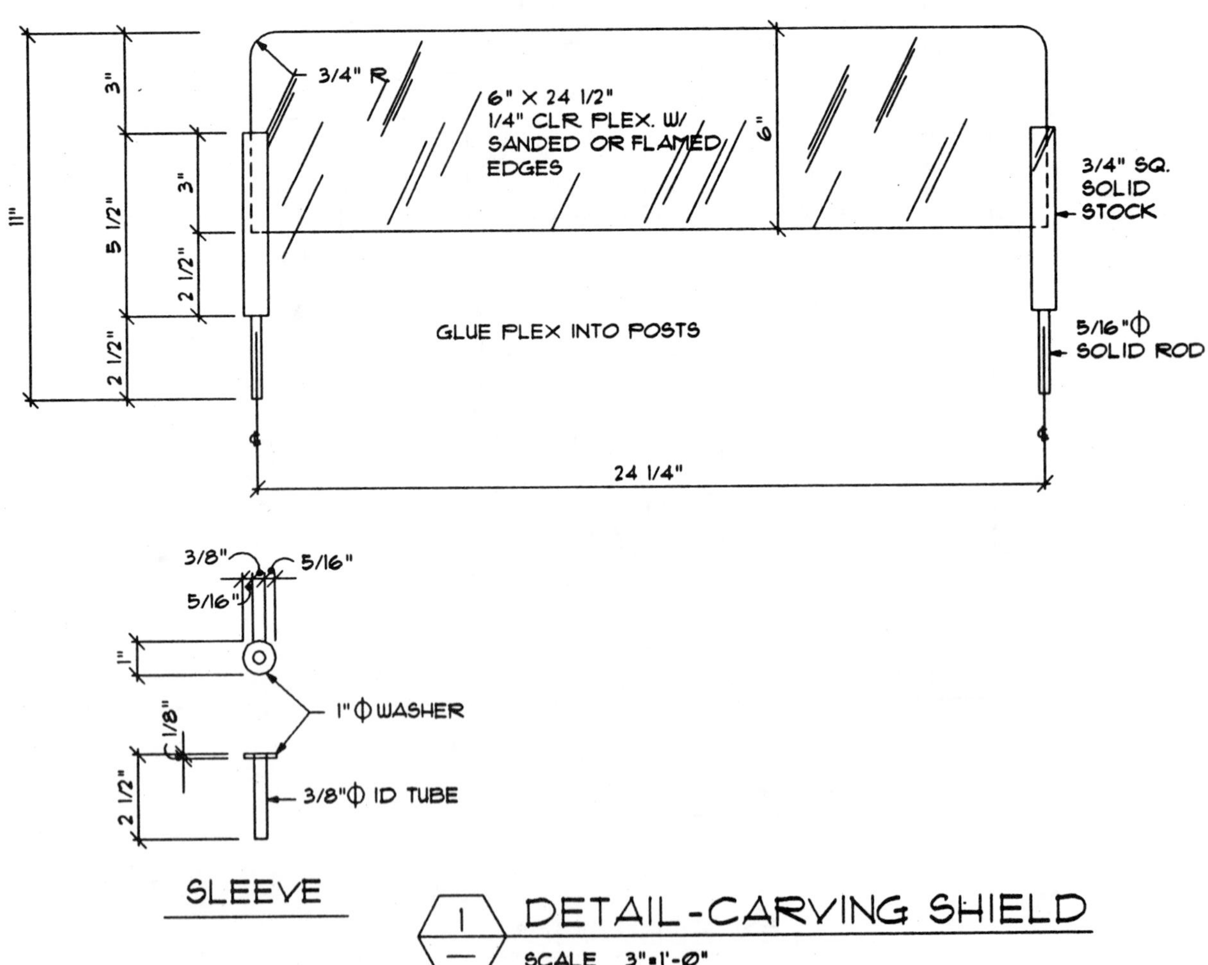

Problem 19–1 (continued). *Courtesy Alexander Manufacturing, Inc.*

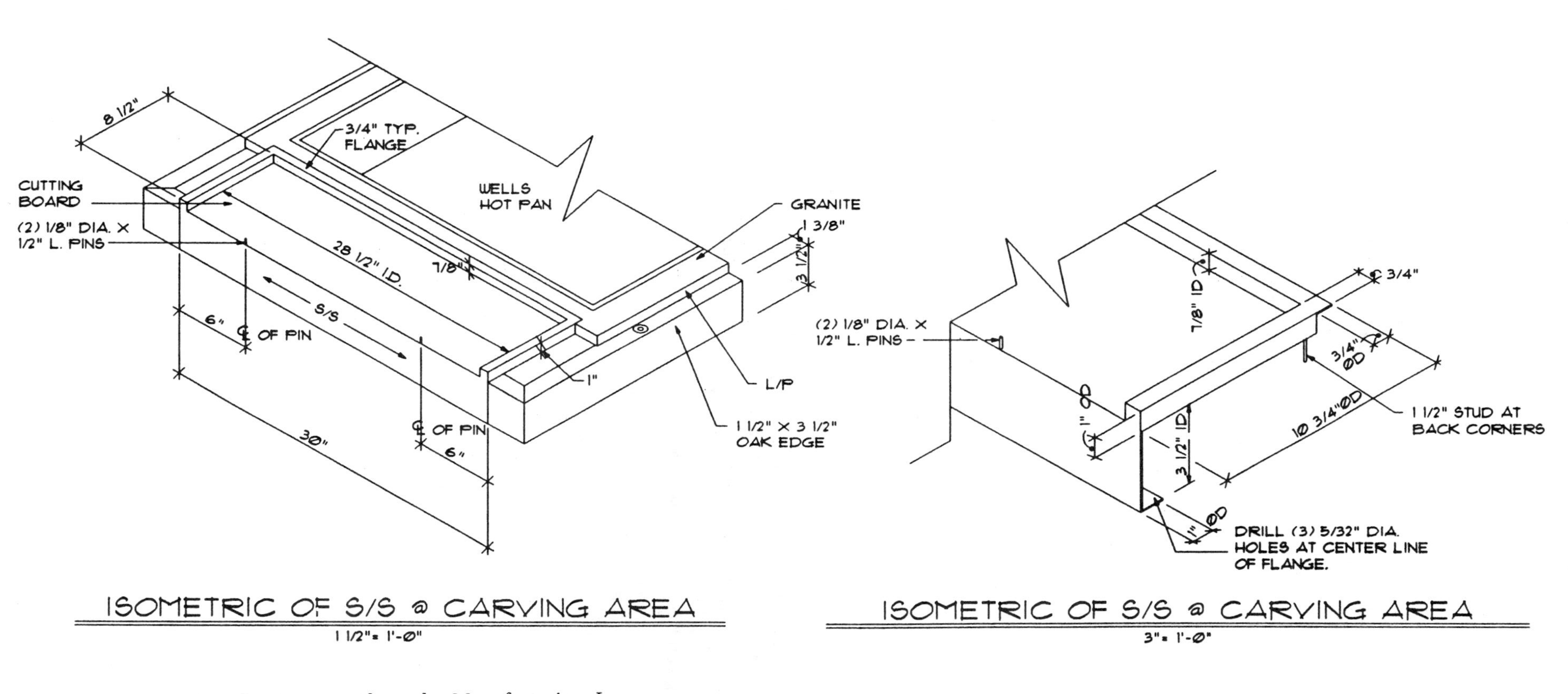

Problem 19–1 (continued). *Courtesy Alexander Manufacturing, Inc.*

Chapter 20
Reading Commercial Site Plans

INTRODUCTION TO READING COMMERCIAL SITE PLANS

MANY ASPECTS of reading commercial site plans are the same as reading residential plot plans. Much of the terminology is the same including legal descriptions. A review of Chapter 13 would be beneficial for you in remembering the basics of property legal descriptions, and the specific items that generally are included in a complete site plan. The terms "site plan" and "plot plan" are synonymous in this discussion. Some of the types of plans that are more commonly used in commercial construction include site analysis plans, planned unit development plans, parking plans, utility plans, and the site grading plan.

PLOT PLAN REQUIREMENTS

A *plot plan*, also known as a lot or site plan, is a map of a piece of land that may be used for any number of purposes. Plot plans may show a proposed construction site for a specific property. Plots may show topography with contour lines or the numerical value of land elevations may be given at certain locations. Plot plans are also used to show how a construction site will be excavated and are then known as a grading plan. See Chapter 13 for a complete review of plot plans.

Topography

Topography is a physical description of land surface showing its variation in elevation, known as relief, and locating other features. Surface relief can be shown with graphic symbols that use shading methods to accentuate the character of land or the differences in elevations can be shown with contour lines. Plot plans that require surface relief identification generally use *contour lines*. These lines connect points of equal elevation and help show the general lay of the land.

When the contour lines are far apart, the contour interval shows relatively flat or gently sloping land. When the contour lines are close together, then the contour interval shows a land that is much steeper.

Drawn contour lines are broken periodically and the numerical value of the contour elevation above sea level is inserted.

Site Analysis Plan

In areas where zoning and building permit applications require a design review, a site analysis plan may be required. The site analysis should provide the basis for the proper design relationship of the proposed development to the site and to adjacent properties. The degree of detail of the site analysis is generally appropriate to the scale of the proposed project. A site analysis plan, as shown in Figure 20–1, often includes the following:

- A vicinity map showing the location of the property in relationship to adjacent properties, roads, and utilities
- Site features such as existing structures and plants on the property and adjacent property
- The scale
- North direction
- Property boundaries
- Slope shown by contour lines, cross sections, or both
- Plan legend
- Traffic patterns
- Solar site information if solar application is intended
- Pedestrian patterns

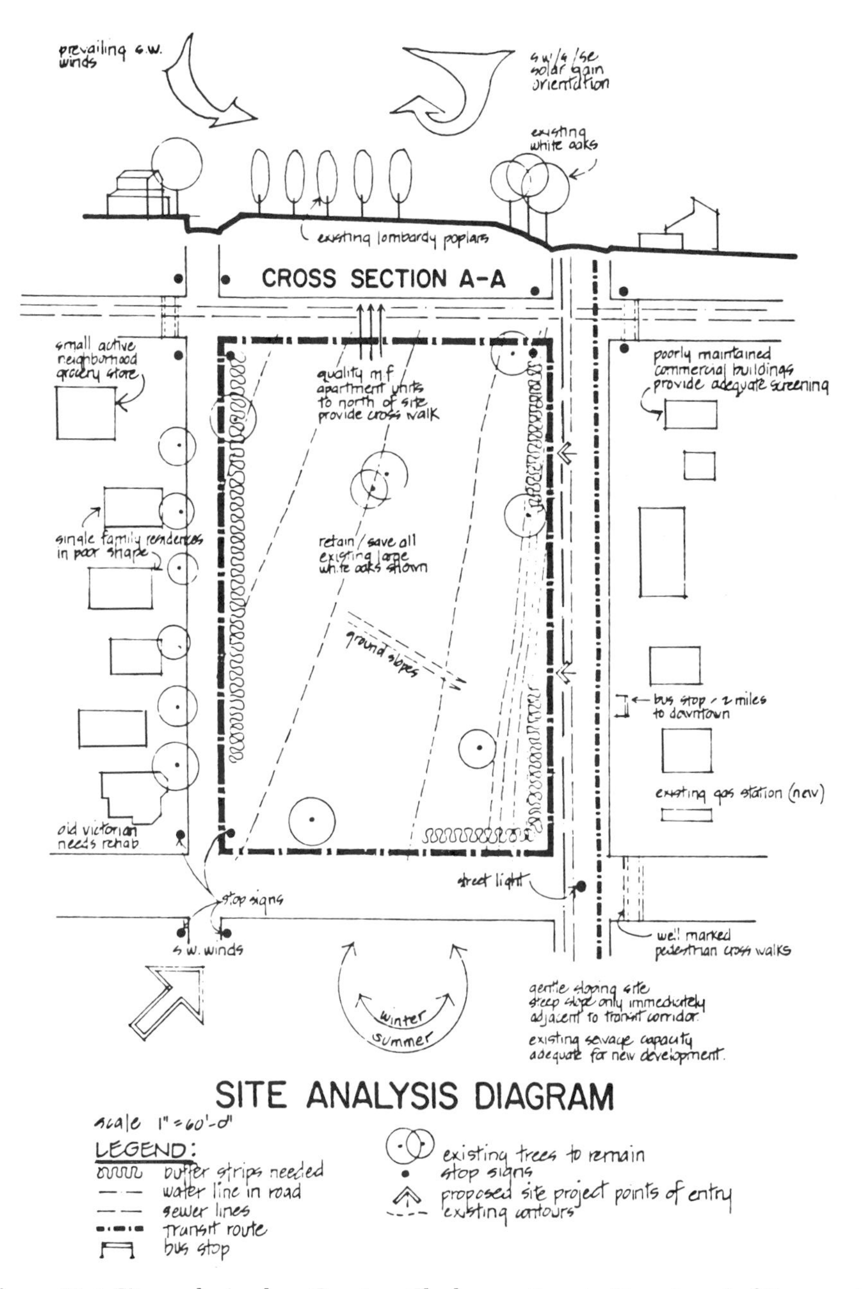

Figure 20–1 Site analysis plan. *Courtesy Clackamas County, Department of Transportation and Development.*

Planned Unit Development

A creative and flexible approach to land development is a planned unit development. Planned unit developments may include such uses as residential areas, recreational areas, open spaces, schools, libraries, churches, or convenient shopping facilities. Developers involved in these projects must pay particular attention to the impact on local existing developments. Generally the plats for these developments must include all of the same information shown on a subdivision plat, plus:

1. A detailed vicinity map, as shown in Figure 20–2.
2. Land use summary.
3. Symbol legend.
4. Special spaces such as recreational and open spaces, or other unique characteristics.

Figure 20–3 shows a typical planned unit development plan.

Plot Plan Parking Requirements

The site plan for a commercial project will resemble the plot plan for a residential project except for its size. One major difference is the need to plan and draw parking spaces, driveways, curbs, and walkways. In commercial projects, the plot plan is often drawn as part of the preliminary design study. Parking spaces are determined by the square footage of the building and the type of usage the building will receive. Parking requirements often dictate the layout of the floor plan. Figure 20–4 shows an example of a plot plan for an office building.

Grading Plan

Once basic information is placed on the plot plan to describe the structure, a Mylar print of the plan is usually given to a civil engineer or licensed surveyor so a grading plan can be prepared. A drafter working for a civil engineer then translates survey field notes into a grading plan as seen in Figure 20–5. Although titled a grading plan, the plan typically contains much more than just grading elevations. Drainage information is often placed on this plan as it relates to the structure, as well as walks and paving areas.

Utility Plan

Many commercial projects require extensive utility construction. Commercial site plans may include the layout for sanitary and storm sewers, water lines, gas lines, and electrical installations. Figure 20–6 shows a partial commercial utility plan.

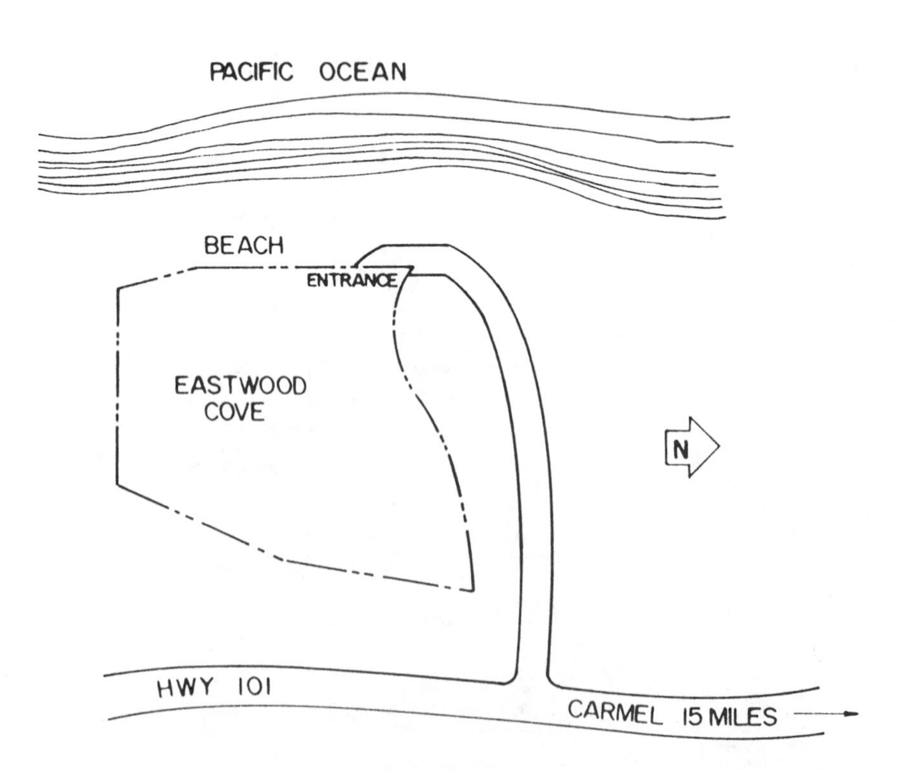

Figure 20–2 A vicinity map is often included with the commercial site plan to show where the construction project is located in relation to the surrounding area.

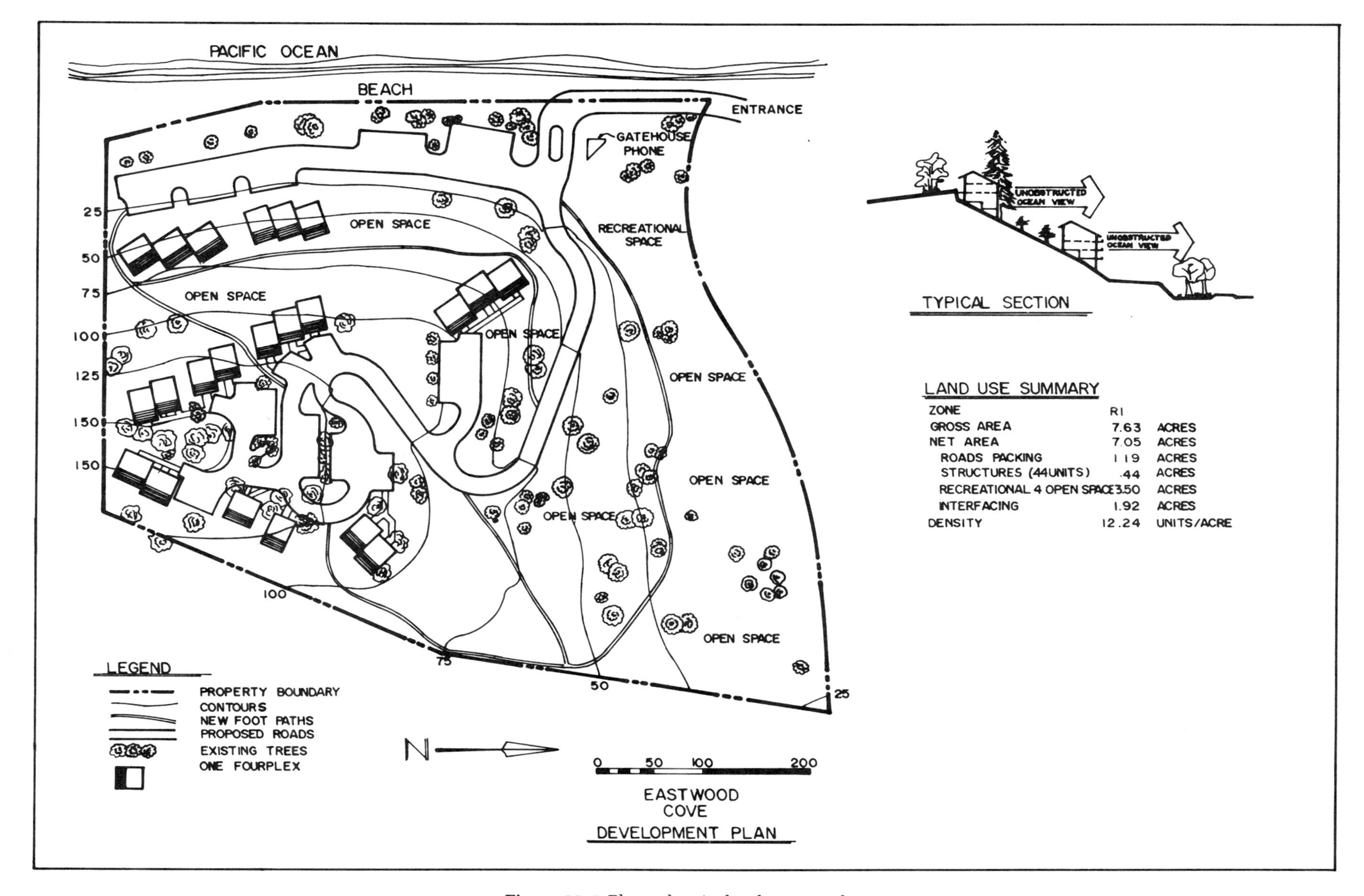

Figure 20–3 Planned unit development plan.

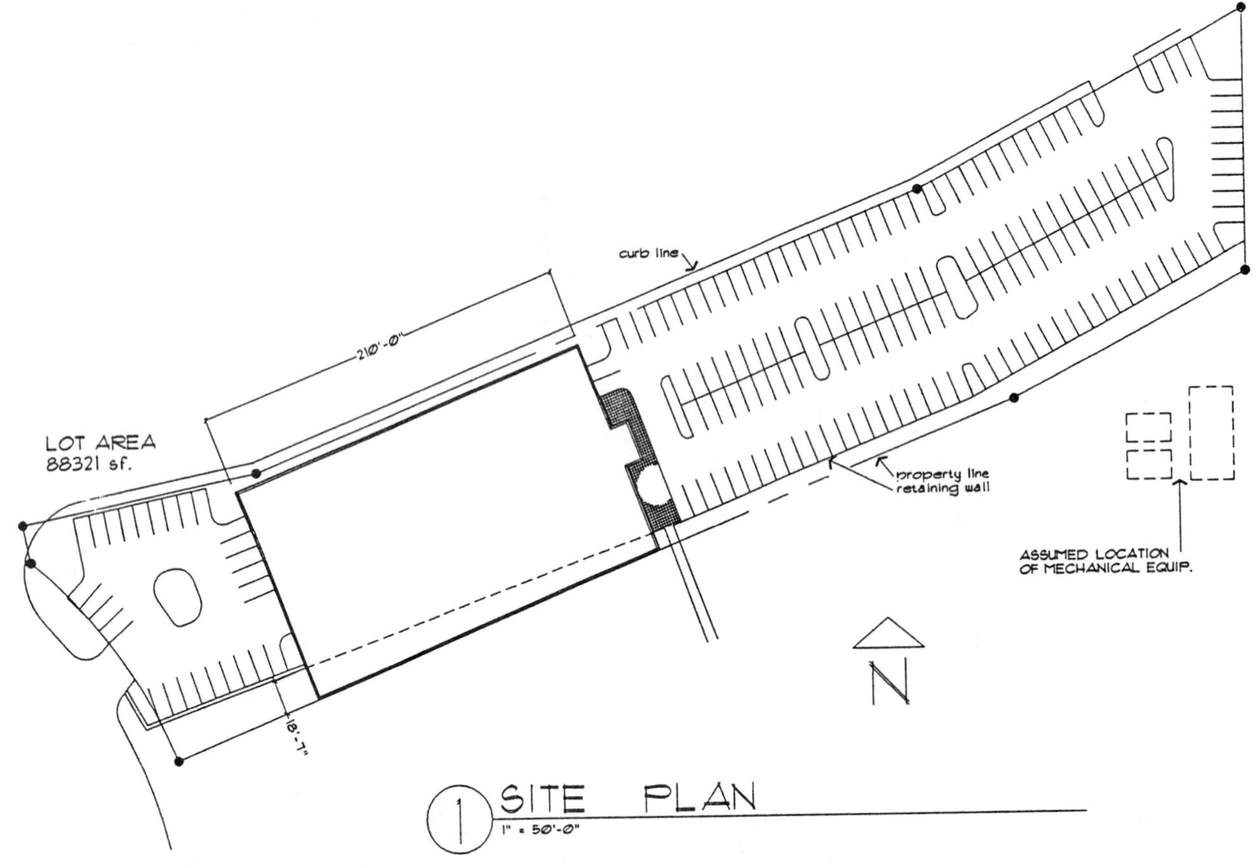

Figure 20–4 A plot, or site, plan for a commercial project shows the parking, curbs, sidewalks, and road ways. *Courtesy Soderstrom Architects.*

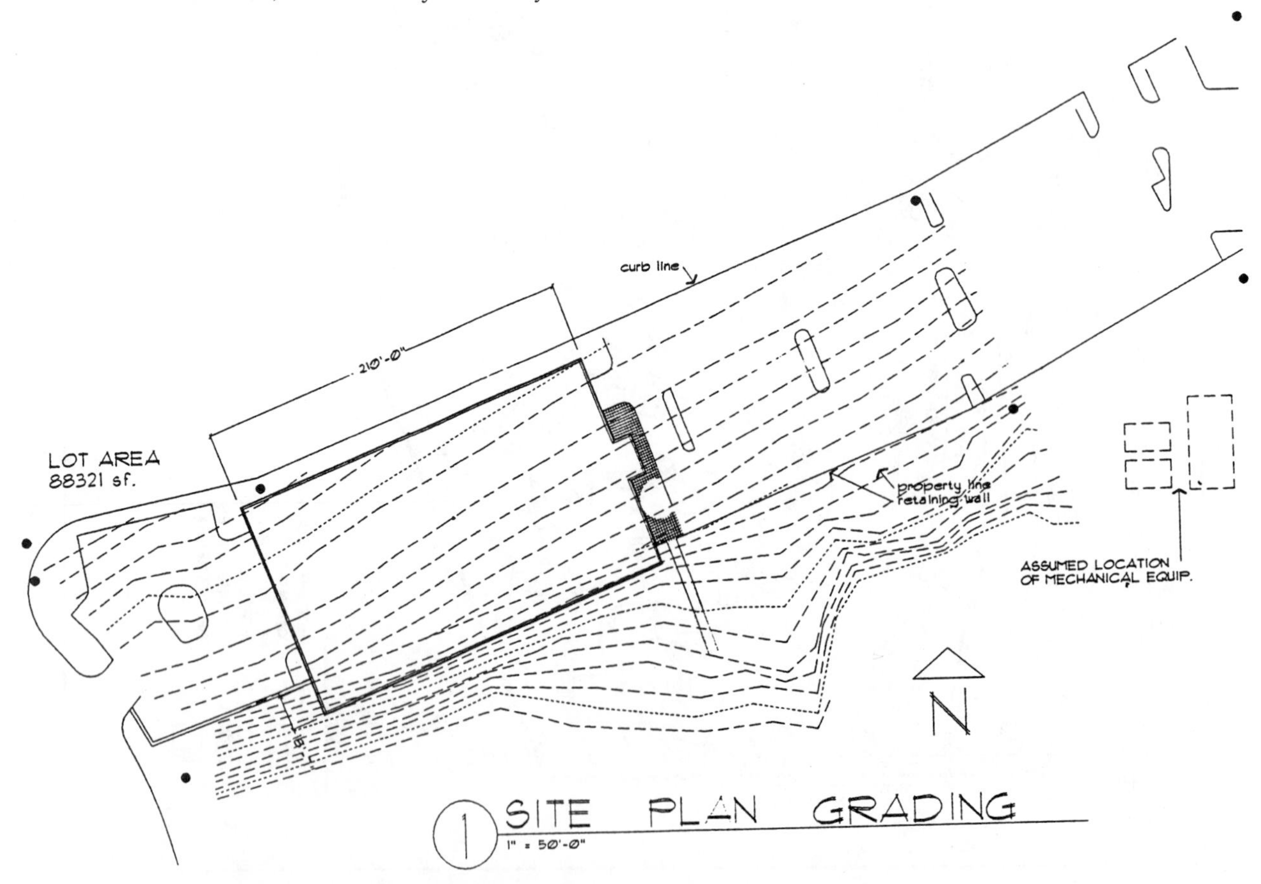

Figure 20–5 A commercial site plan often shows the grading information which includes contour lines, and cut and fill requirements. *Courtesy Soderstrom Architects.*

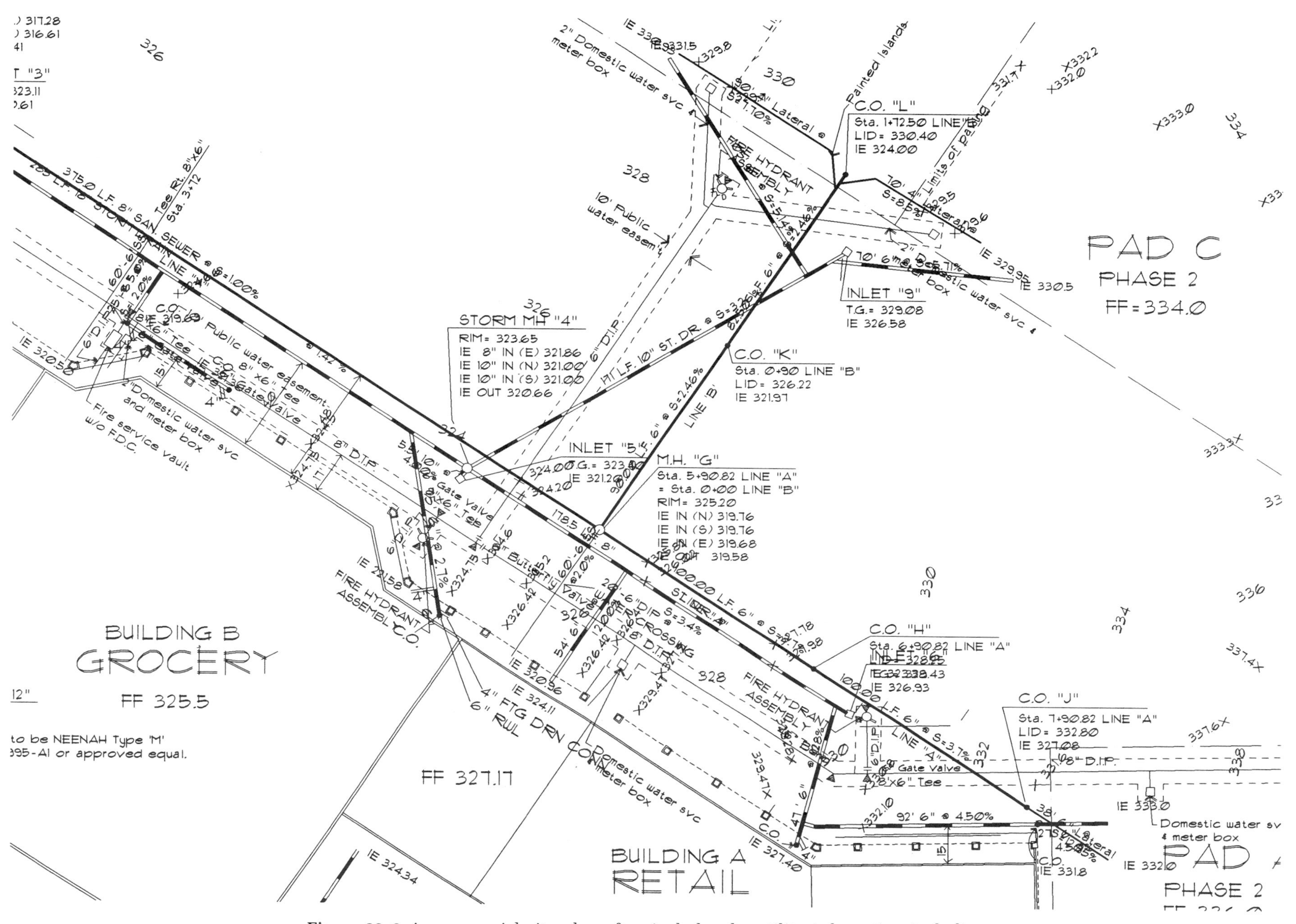

Figure 20–6 A commercial site plan often includes the utility information including sanitary and storm sewers, water lines and hook ups, and electrical requirements. *Courtesy Soderstrom Architects.*

Landscape Plan

The landscape plan will show suitable plants for the job site, and specify the plants by their proper Latin name on the plan. On typical commercial projects, a Mylar copy of the plot plan is given to a landscape architect who specifies the size, type, and location of plantings. A method of maintaining the plantings is also typically shown on the plan as seen in Figure 20–7.

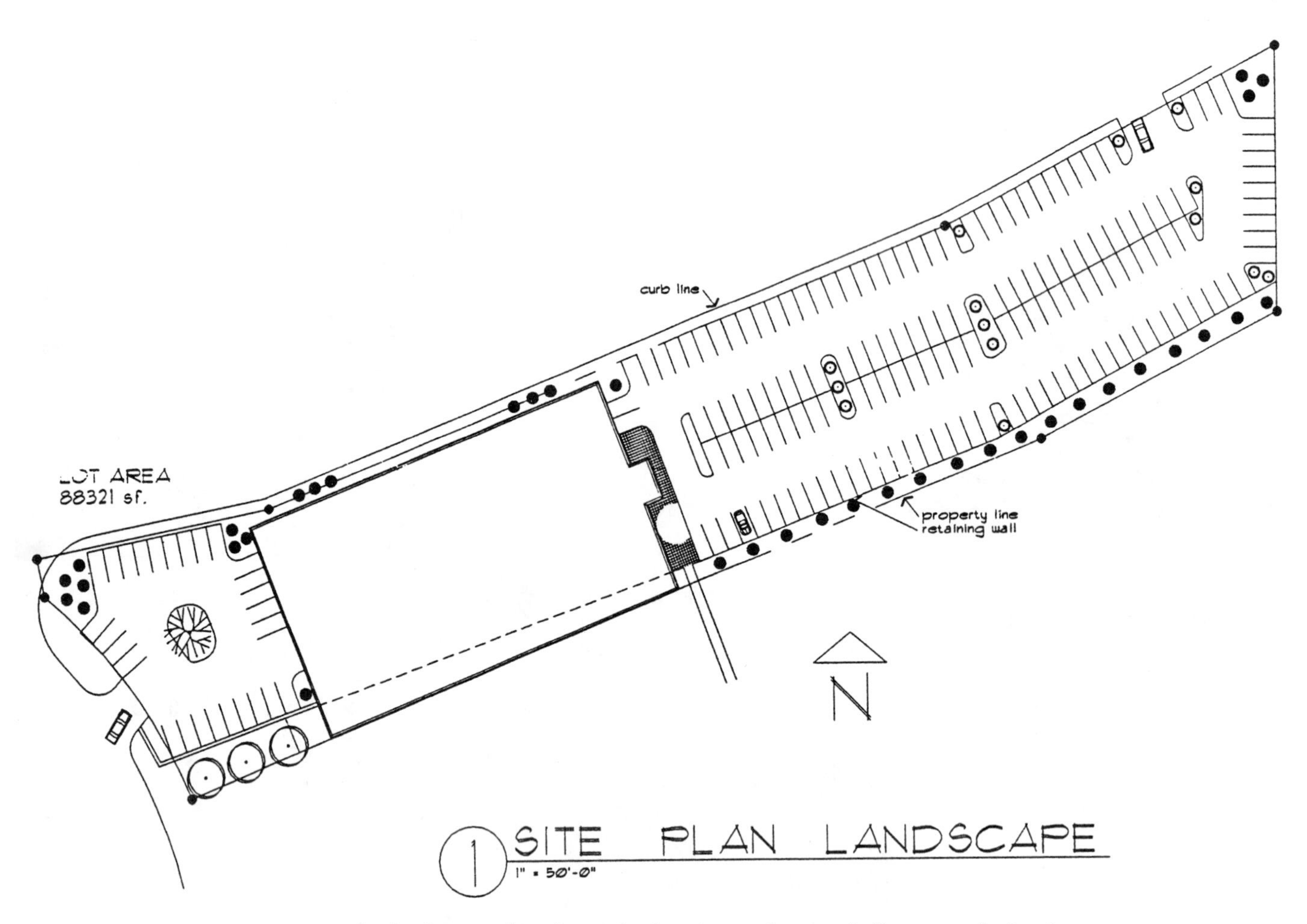

Figure 20–7 The landscape plan shows the location and type of all trees and plantings. *Courtesy Soderstrom Architects.*

CHAPTER 20 TEST

Fill in the blanks below with the proper word or short statement as needed to correctly complete the sentence or answer the question.

1. A plot plan is also known as a lot or site plan. Define plot plan. ____.

2. Name at least five items that are characteristic to most plot plans. ____.

3. Define topography. ____.

4. Define contour lines. ______
5. The vertical distance between contour lines is known as ______.
6. A plan that provides the basis for proper design relationship of the proposed development to the site and to adjacent properties is referred to as a ______.
7. Describe a planned unit development. ______
8. Identify at least two factors that determine parking requirements. ______
9. A plan that typically contains grading elevations, drainage information, walks, and paving areas is called a ______.
10. Name the type of plan where the size, type, location, and name of plantings are shown. ______
11. Commercial site plans often include the layout for sanitary and storm sewers, water lines, gas lines, and electrical installations on a ______.
12. In commercial construction is the plot or site plan usually part of the preliminary design study? ______.

CHAPTER 20 EXERCISES

PROBLEM 20–1. Given the Site Analysis Plan in Figure 20–1 on page 373 answer the following questions with short complete statements or words as needed:

1. What is indicated about the ground slope? ______.
2. What will happen to the white oaks? ______.
3. How far is it from the bus stop to downtown? ______.
4. What is required between the proposed site and the apartments to the North? ______
5. Where are the proposed site projects points-of-entry? ______.
6. Where is adequate screening required? ______.

PROBLEM 20–2. Given the commercial site plan shown on page 381, answer the following questions with short complete statements or words as needed:

1. What are the dimensions of the property? ______
2. Give the unit tabulation: ______.
3. What is the legal description? ______.
4. What is the name of the street in front of the proposed site? ______

5. Give the elevations at the following locations:

SE Property corner	______	East end face of curb	______
SW Property corner	______	West end face of curb	______
NE Property corner	______	East end top of slope	______
NW Property corner	______	West end top of slope	______

6. Describe the fence specifications: ______________________________ .
7. What is the number and dimensions of the parking spaces? ______________________________
8. Give the specifications for the construction of the parking area. ______________________________ .
9. What are the number and size of wheel stops used? ______________________________ .
10. How wide is the Alley? ______________________________
11. What is the dimension from the South property line to the proposed apartment building? ______________ .
12. What is the dimension from the East property line to the proposed apartment building? ______________ .
13. What is the dimension from the South property line to the Storage/Utility building? ______________ .
14. What is the dimension between the proposed Storage/Utility building and the proposed apartment building? ______________________________ .

PROBLEM 20–3. Given the Project Data and commercial site plan shown on pages 382 and 383, answer the following questions with short complete statements or words as needed:

1. What are the property metes and bounds?
   ____
   ____
   ____
   ____
2. What is the site area? ______________________________
3. What are the building areas?
   ____
   ____
   ____
4. What is the number and sizes of the proposed parking spaces? ______________________________
5. How wide is the proposed gate? ______________________________ .
6. What are the dimensions of the proposed building? ______________________________ .
7. What is the distance from the existing curb to the front of the proposed building? (show your calculations) ______________________________ .
8. Give the specifications of the existing drainage? ______________________________ .
9. How do you tell the difference between existing and proposed elevations? (explain or show an example) ______________________________ .
10. What is the property legal description? ______________________________ .
11. What is the finished floor elevation of the new warehouse? ______________________________ .
12. Describe the parking drive construction specifications. ______________________________ .
13. How many existing trees are there? ______________________________ .
14. What is the width of Whitaker Road? ______________________________ .

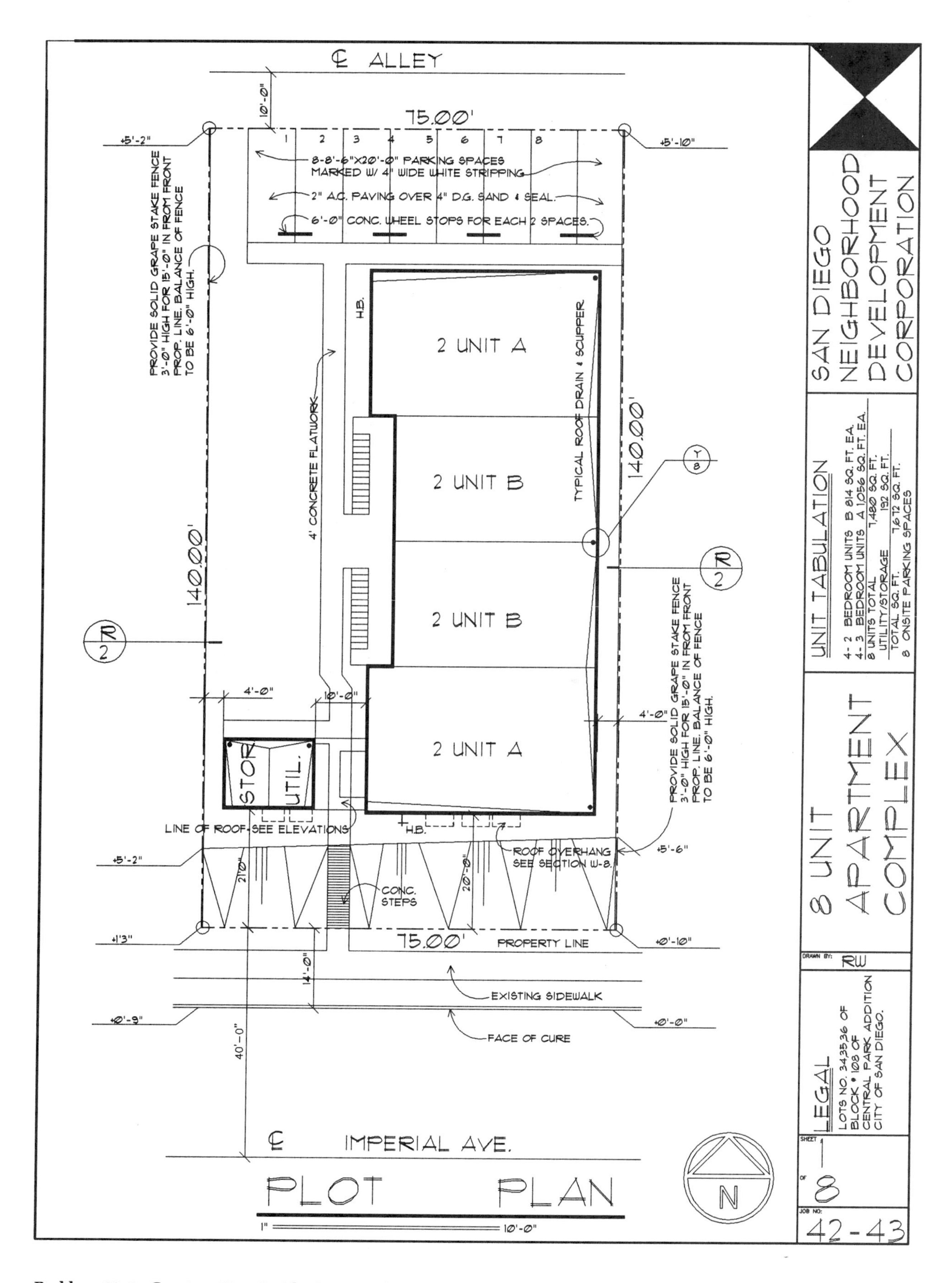

Problem 20–2. *Courtesy Ken Smith, Structureform Masters, Inc.*

PROJECT DATA

SITE AREA	22208 SQ FT .50 AC
BLDG. AREA	1600 SQ FT EXISTING 8100 SQ FT PROPOSED 9700 SQ FT TOTAL
OCCUPANCY	B-2
CONST. TYPE	UBC. TYPE
PARKG. REQD.	9 SPACES @ 9'x20'
LANDSCAPING REQD.	15% OF SITE 2220.8 3331 SQ FT REQD
LANDSCAPE PROVIDED	5,000 SQ FT
LEGAL DESCRIPTION	PARCEL A LOT 1 BLK 4 SPACE INDUSTRIAL PARK
ZONING	GI-2

98.30 EXISTING ELEVATION
(98.30). PROPOSED ELEVATION

Problem 20–3. *Courtesy Morrison Construction, Inc.*

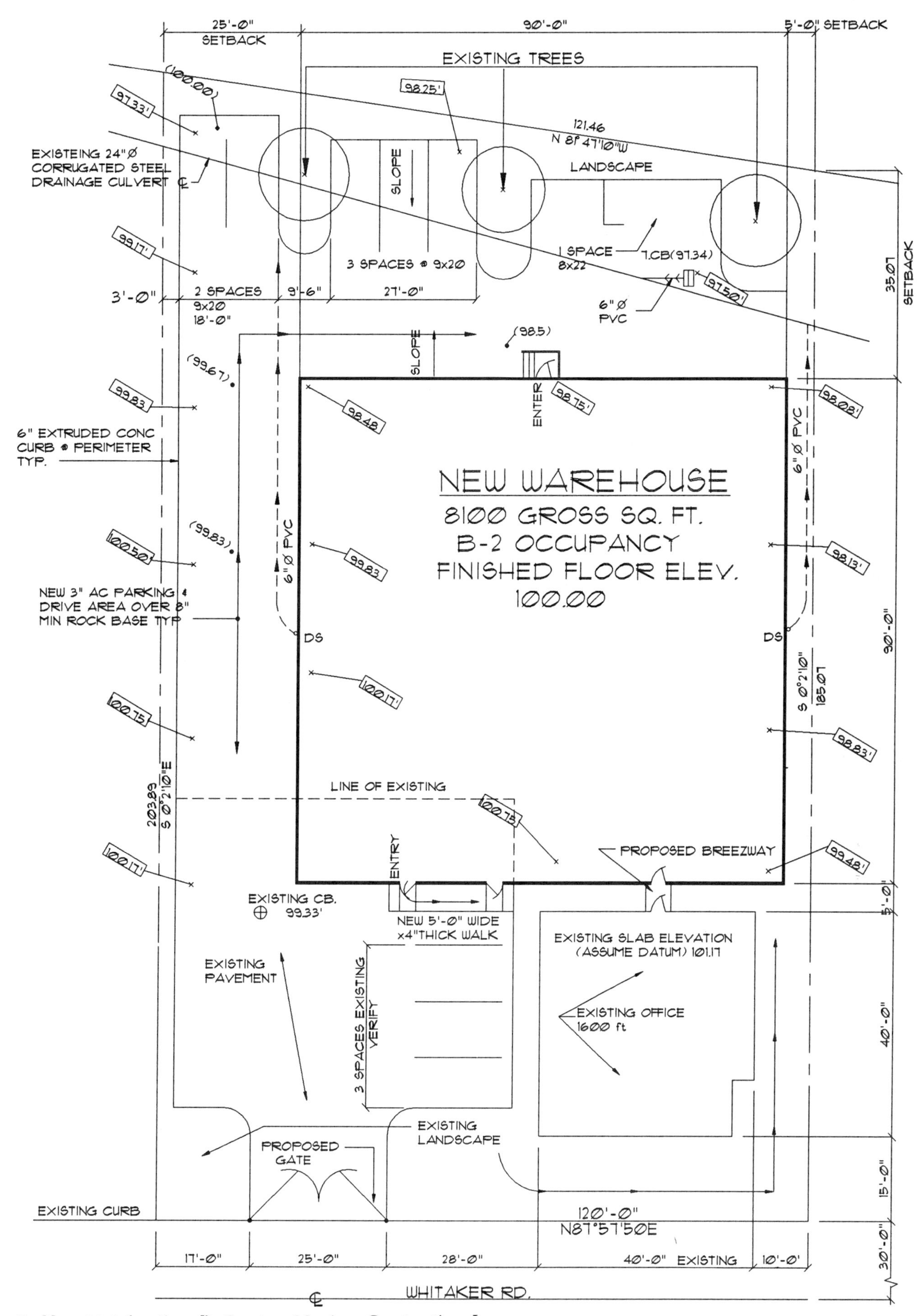

Problem 20–3 (continued). *Courtesy Morrison Construction, Inc.*

PROBLEM 20–4. Given the partial commercial site utility plan shown on page 385, answer the following questions with short complete statements or words as needed:

1. How many fire hydrant assemblies are there? ______________________________.
2. Refer to Man Hole (MH) number 4 and provide the following missing information:
   RIM = ________________
   IE (Inlet Elevation) 8" IN (E) ________________
   IE 10" IN (N) ________________
   IE 10" IN (S) ________________
   IE OUT ________________
3. What is the width of the public water easement? ______________________________.
4. Give the distance between the storm and sanitary sewers that run parallel. ________________.
5. Refer to CO (Clean Out) K and provide the following missing information:
   Sta. (Station) ________________ LINE B
   LID = ________________
   IE = ________________

PROBLEM 20–5. Given the partial commercial site contour survey shown on page 386, answer the following questions with short complete statements or words as needed:

1. How often are contour lines labeled? ______________________________
2. What is the contour interval? ______________________________
3. Notice how the contour lines form a "V" pattern, what is located where the contour lines meet at this "V"?
4. How are trees shown and labeled? ______________________________
   ____
   ______________________________.

PROBLEM 20–6. Given the commercial site landscape plan shown on page 387, answer the following questions with short complete statements or words as needed:

1. What is the name of the project? ______________________________.
2. Who is the architect? ______________________________
3. Name the streets next to this project and give their compass orientation to the project. ________________
   ____
   ______________________________.
4. Name the different groups of plant materials used in this landscaping plan. ________________
5. How many different types of trees are used? ______________________________.
6. How many handicapped parking spaces are there? ______________________________.
7. How are the plants on the drawing keyed to the plant list? ______________________________
   ____
   ______________________________.
8. How can you tell how many of each type of plant there are in the landscape plan? ________________
   ______________________________.
9. Identify the specific information listed in the Plant List. ______________________________
   ______________________________.
10. Give the size of the Thwa Plicata plants used. ______________________________
11. Give the size and recommended spacing of the Arctostaphylob Uva-Ursi. ________________
    ______________________________.
12. Give the size and remarks regarding the David Vibrunum. ______________________________
    ______________________________.

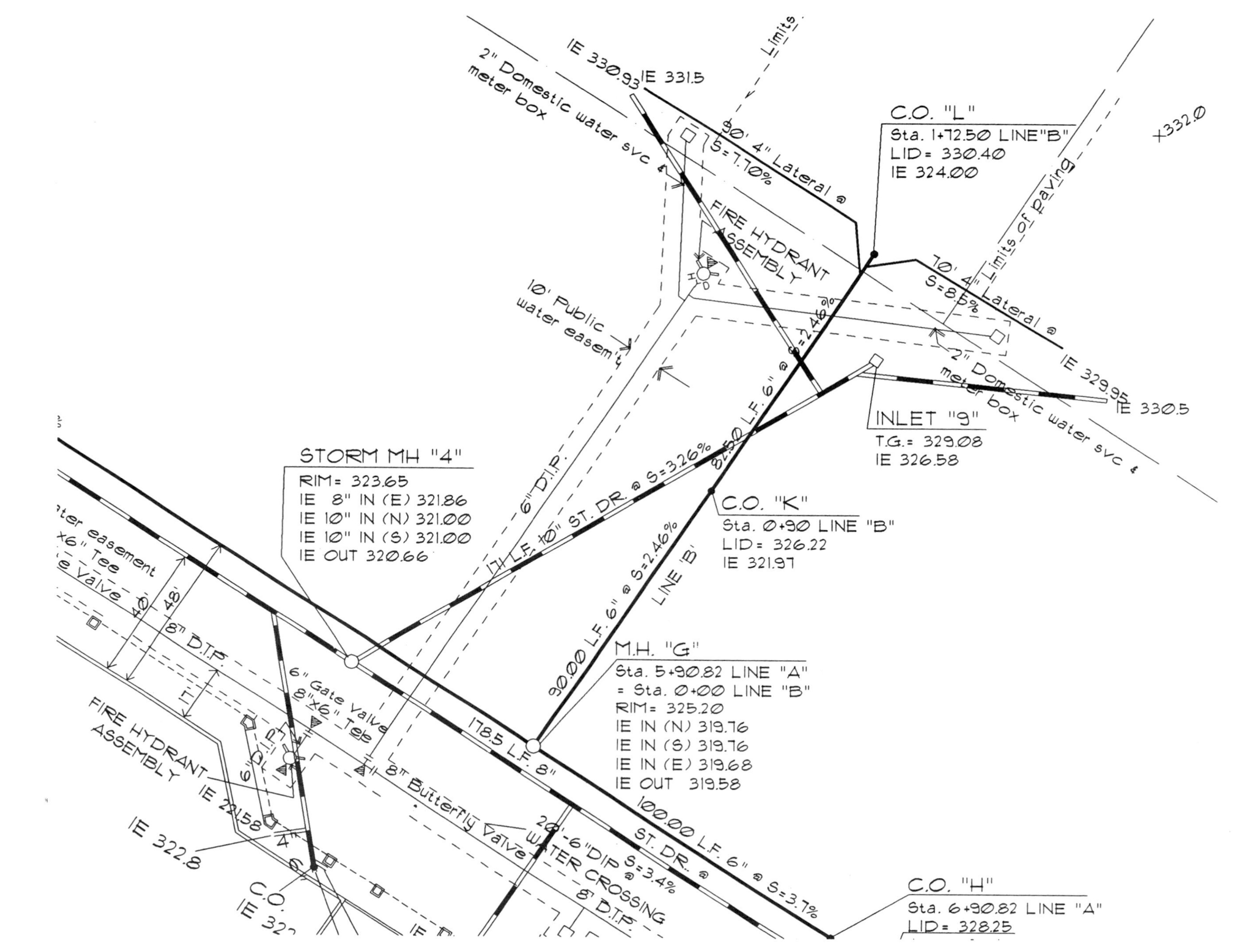

Problem 20–4. *Courtesy Soderstrom Architects.*

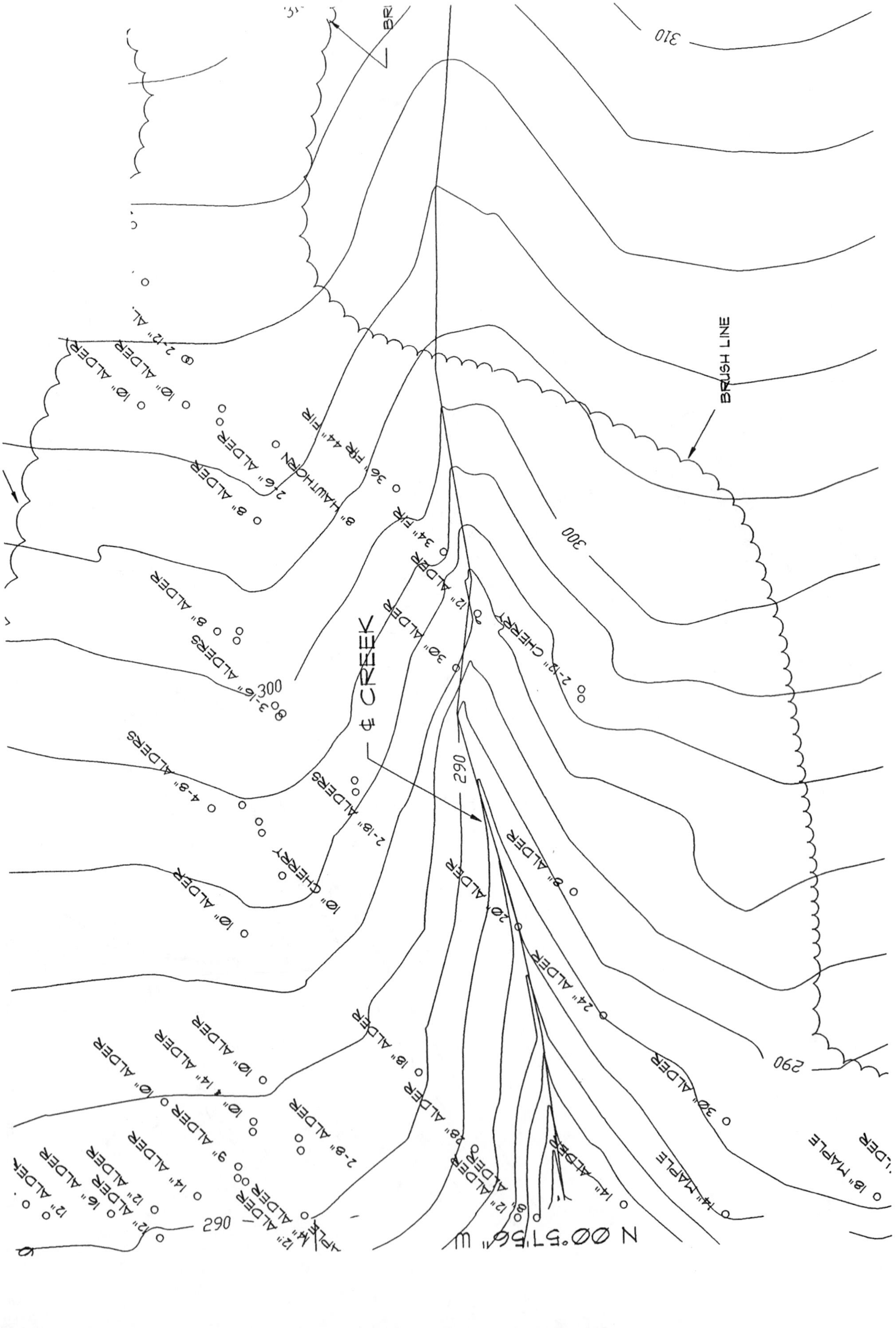

Problem 20–5. *Courtesy Soderstrom Architects.*

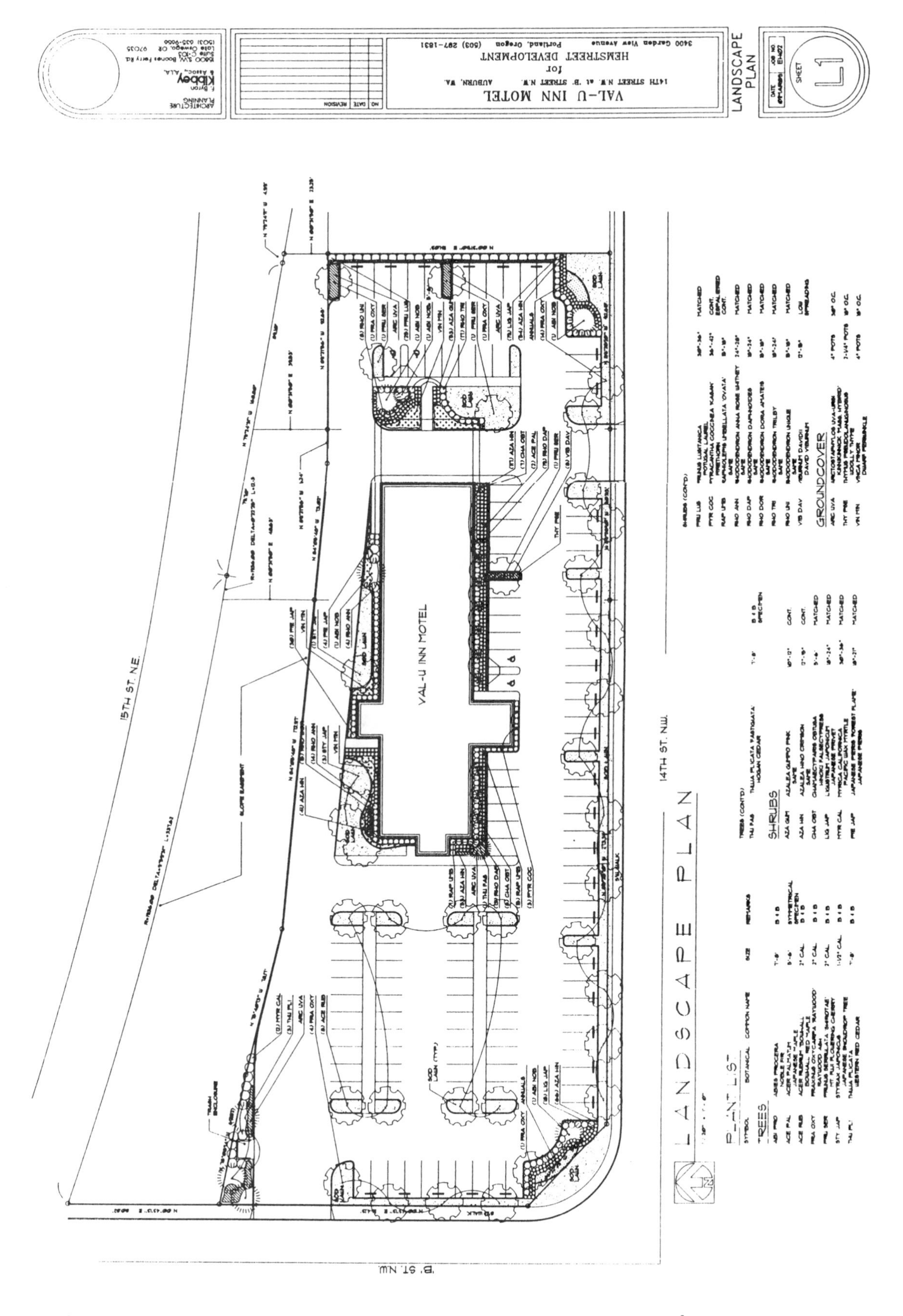

Problem 20–6. *Courtesy OTAK, Inc., for landscaping, and Kibbey & Associates, Architects, for the site plan.*

Courtesy American Plywood Association.

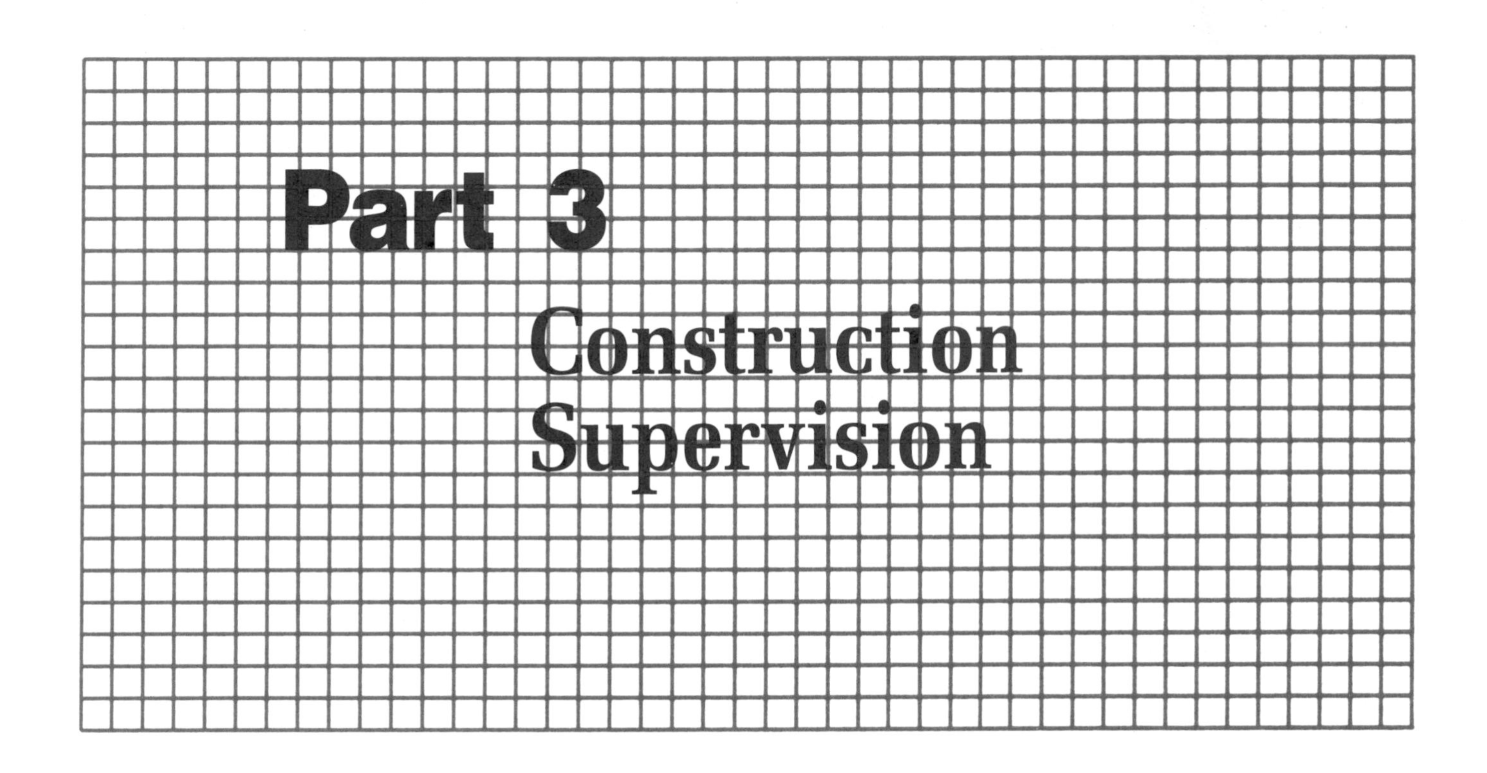

Part 3

Construction Supervision

Chapter 21
Construction Supervision Procedures

THERE ARE A few terms that you may already know including builder, contractor, and subcontractor that should be reviewed before you start this chapter. The terms, builder and contractor, mean the same thing. The contractor, often referred to as the general contractor, is the person or company that is responsible for the complete and proper construction of the structure. The contractor may or may not do all of the actual construction work. The contractor may do part of the hands-on construction and hire others work on parts of the construction. Subcontractors are hired by the contractor to complete specific elements of the construction project. The subcontractors are usually people who have certain skills that the contractor does not have. For example, subcontractors include such trades as electrical, plumbing, heating, excavation, masonry, cabinet construction, carpentry, finish carpentry, or carpet installation. Finish carpenters usually do final wood working such as door installation and millwork.

The building contractor may often be involved with several phases of preparation before construction of the project begins. Contractors are often involved with requirements including zone changes when necessary, specification preparation, building permit applications, bonding requirements, client financial statement, lender approval, and construction estimates.

LOAN APPLICATIONS

Loan applications vary depending on the requirements of the lender. Most applications for construction financing contain a variety of similar information. The FHA-proposed construction appraisal requirements are, in part, as follows:

1. Plot plan (three copies per lot).
2. Prints (three copies per plan). Prints should include the following:
   Four elevations including front, rear, right side, and left side.
   Floor plan.
   Foundation plan.
   Wall section

Roof plan.
Cross sections of exterior walls, stairs, etc.
Cabinet detail including a cross section.
Fireplace detail (manufacturer's detail if fireplace is prefabricated).
Truss detail (include name and address of supplier).
Heating plan. If forced air, show size and location of ducts and registers plus cubic feet per minute (CFM) at register.
Location of wall units with watts and CFM at register.

3. Specifications (one copy per set of prints).

CHANGE ORDERS

Any physical change in the plans or specifications should be submitted to the FHA on FHA Form 2577. It should be noted whether the described change will be an increase or decrease in value and by what dollar amount. It must be signed by the lender, the builder, and, if sold, the purchaser, who must also sign prior to submission to the FHA.

Most lending institutions' applications are not as comprehensive as the FHA while others may require additional information. The best practice is to always research the lender to identify the needed information. A well-prepared set of documents and drawings usually stands a better chance of funding.

BUILDING PERMITS

The responsibility of completing the building permit application may fall to the contractor. The builder should contact the local building official to determine the process to be followed. Generally, the building permit application is a basic form that identifies the major characteristics of the structure to be built, the legal description and location of the property, and information about the applicant. The application is usually accompanied by two sets of plans and up to five sets of plot plans. The fee for a building permit usually depends on the estimated cost of construction. The local building official determines the amount based on a standard schedule at a given cost per square foot. The fees are often divided into two parts: a plan-check fee paid upon application and a building-permit fee paid when the permit is received. There are other permits and fees that may or may not be paid at this time. In some cases, the mechanical, sewer, plumbing, electrical, and water permits are obtained by the general or subcontractors. Water and sewer permits may be expensive depending on the local assessments for these utilities.

CONTRACTS

Building contracts may be very complex documents for large commercial construction or short forms for residential projects. The main concern in the preparation of the contract is that all parties understand what will be specifically done, in what period of time and for what reimbursement. The contract becomes an agreement between the client, general contractor, and architect.

It is customary to specify the date by which the project is to be completed. On some large projects, dates for completion of the various stages of construction are specified. Usually, the contractor receives a percentage of the contract price for the completion of each stage of construction. Payments are typically made three or four times during construction. Another method is for the contractor to receive partial payment for the work done each month. Verify the method used with the lending agency.

The owner is usually responsible for having the property surveyed. The architect may be responsible for administering the contract. The contractor is responsible for the construction and security of the site during the construction period.

Certain kinds of insurance are required during construction. The contractor is required to have a license, bond, and liability insurance. This protects the contractor against being sued for accidents occurring on the site. The owner is required to have property insurance including fire, theft, and vandalism. Workman's compensation is another form of insurance that provides income for contractors' employees if they are injured at work.

The contract describes conditions under which the contract may be ended. Contracts may be terminated if one party fails to comply with the contract, when one of the parties is disabled or dies, and for several other reasons.

There are two kinds of contracts in use for most construction. One is the fix-sum and the other is the cost-plus contract. Each of these offers certain advantages and disadvantages.

Fix-sum (sometimes called lump-sum) contracts are used most often. With a fixed-sum contract, the contractor agrees to complete the project for a certain amount of money. The greatest advantage of this kind of contract is the owner knows in advance exactly what the cost will be. However, the contractor does not know what hidden problems may be encountered and so the contractor's price must be high enough to cover unforeseen circumstances such as excessive rock in the excavation or sudden increases in the cost of materials.

A cost-plus contract is one in which the contractor agrees to complete the work for the actual cost, plus a percentage of overhead and profit. The advantage of this type of contract is the contractor does not have to allow for unforeseen problems. A cost-plus contract is also useful when changes are to be made during the course of construction. The main disadvantage of this kind of contract is the owner does not know exactly what the cost will be until the project is completed.

COMPLETION NOTICE

The completion notice is a document that should be posted in a conspicuous place on or next to the structure. This legal document notifies all parties involved in the project that work has been substantially completed. There may be a very small part of work or cleanup to be done but the project, for all practical purposes, is complete. The completion notice must be recorded in the local jurisdiction. Completion notices serve several functions. Subcontractors and suppliers have a certain given period of time to file a claim or lien against the contractor or client to obtain reimbursement for labor or materials that have not been paid. Lending institutions often hold a percentage of funds for a given period of time after the completion notice has been posted. It is important for the contractor to have this document posted so the balance of payment can be received. The completion notice is often posted in conjunction with a final inspection that includes the local building officials and the client. Some lenders require that all building inspection reports be submitted before payment is given and may also require a private inspection by an agent of the lender.

BIDS

Construction bids are often obtained by the architect for the client. The purpose is to get the best price for the best work. Some projects require work be given to the lowest bidder, while other projects do not necessarily go to the lowest bid. In some situations, especially in the private sector, other factors are considered such as an evaluation of the builder's history based upon quality, ability to meet schedules, cooperation with all parties, financial stability, and license, bond, and insurance.

The bid becomes part of the legal documents for completion of the project. The legal documents may include plans, specifications, contracts, and bids. The following items are part of a total analysis of costs for residential building construction. The architect and client should clearly know what the bid includes.

1. Plans
2. Permits, fees, specifications
3. Roads and rod clearing
4. Excavation
5. Water connection (well and pump)
6. Sewer connection (septic)
7. Foundation, waterproofing
8. Framing including materials, trusses, and labor
9. Fireplace including masonry
10. Plumbing, both rough and finished
11. Wiring, both rough and finished
12. Windows
13. Roofing including sheetmetal and vents
14. Insulation
15. Drywall or plaster
16. Siding
17. Gutters, downspouts, sheet metal and rain drains
18. Concrete flatwork and gravel
19. Heating
20. Garage and exterior doors
21. Painting and decorating
22. Trim and finish interior doors including material and labor
23. Underlayment
24. Carpeting including the amount of carpet and padding, and the cost of labor
25. Vinyl floor covering amount and cost of labor
26. Formica
27. Fixtures and hardware
28. Cabinets
29. Appliances
30. Intercom/stereo system
31. Vacuum system
32. Burglar alarm
33. Weatherstripping and venting
34. Final grading, cleanup, and landscaping
35. Supervision, overhead, and profit
36. Subtotal of land costs
37. Financing costs

CONSTRUCTION INSPECTIONS

When the architect is responsible for the supervision of the construction project, then it is necessary to work closely with the building contractor to get the proper inspections at the necessary times. There are two types of inspections that most frequently occur. The regularly scheduled code inspections that are required during specific phases of construction. These inspections help ensure that the construction methods and materials meet local and national code requirements. The general intent of these inspections is to protect the safety of the occupants and the public. Another type of inspection is often conducted by the lender during certain phases of construction. The purpose of these inspections is to ensure the materials and methods described in the plans and specifications are being used. The lender has a valuable interest here. If the materials and methods are inferior or not the standard expected, then the value of the structure may not be what the lender had considered when a preliminary appraisal was made. Another reason for these inspections, and probably the reason that the builder likes best, is the dispersement inspection. These inspections may be requested at various times, such as monthly, or they may be related to a specific dispersement schedule of four times during construction, for example. The intended result of these inspections is the release of funds for payment of work completed.

When the architect supervises the total construction, then he or she must work closely with the contractor to ensure that the project is completed in a timely manner. When a building project remains idle, the overhead costs, such as construction interest, begin to add up quickly. A contractor that bids a job high but builds quickly may be able to save money in the final analysis. Some overhead costs go on daily even when work has stopped or slowed. The supervisor should also have a good knowledge of scheduling so inspections can be made at the proper time. If an inspection is requested when not ready, the building official may charge a fee for excess time spent. Always try to develop a good rapport with building officials so that encounter goes as smoothly as possible.

CHAPTER 21 TEST

Provide short complete answers or fill in the blanks:

1. List at least five items of information contained in most applications for construction financing: ________
____
____.

2. The completed building permit application is usually accompanied by ________________
____
____.

3. The building permit fee usually depends on the ________________
____.

4. What are the three main concerns of the contract? ________________
____
____.

5. The contract becomes an agreement between ________________
____.

6. Any physical changes in the plans of specifications should be submitted on a ________________.

7. The ________________ is a legal, recorded document that must be posted in a conspicuous place on the completed structure.

8. The completion notice serves several functions; list at least three: ________________
____
____
____.

CHAPTER 21 EXERCISES

Directions: All forms must be neatly hand-lettered or type, unless otherwise specified by your instructor.

PROBLEM 21–1. Complete the building permit application on page 395 based on this information:

- Project location address: 3456 Barrington Drive, your city and state
- Nearest cross street: Washington Street

- Subdivision name: Barrington Heights
- Township: 2S
- Range: 1E
- Section: 36
- Tax lot: 2400
- Lot size: 15000 sq ft.
- Building area: 2000 sq ft.
- Basement area: None
- Garage area: 576 sq ft.
- Stories: 1
- Bedrooms: 3
- Water source: Public
- Sewage disposal: Public
- Estimated cost of labor and materials: $88,500
- Plans and specifications made by: You
- Owner's name: Your teacher; address and phone may be fictitious
- Builder's name: You, your address and phone number
- You sign as applicant
- Homebuilder's registration no: Your social security number or another fictitious number
- Date: Today's date

BUILDING PERMIT APPLICATION

Amount Due ____________

TO BE FILLED IN BY APPLICANT

Project Location (Address) ____________

Nearest Cross Street ____________

Subdivision Name ____________ Lot ________ Block ________

Township ________ Range ________ Section ________ Tax Lot ________

Lot Size ________ (Sq. Ft.) Building Area ________ (Sq. Ft.) Basement Area ________ (Sq. Ft.) Garage Area ________ (Sq. Ft.)

Stories ________ Bedrooms ________ Water Source ________ Sewage Disposal ________

Estimated Cost of Labor and Material ____________

Plans and Specifications made by ____________ accompany this application.

Owner's Name ____________ Builder's Name ____________

Address ____________ Address ____________

City ________ State ________ City ________ State ________

Phone ________ Zip ________ Phone ________ Zip ________

I certify that I am registered under the provisions of ORS Chapter 701 and my registration is in full force and effect. I also agree to build according to the above description, accompanying plans and specifications, the State of Oregon Building Code, and to the conditions set forth below.

APPLICANT ____________ HOMEBUILDER'S REGISTRATION NO. ____________ DATE ________

I agree to build according to the above description, accompanying plans and specifications, the State of Oregon Building Code, and to the conditions set forth below.

APPLICANT ____________ DATE ________

Problem 21–1.

Glossary

ACOUSTICS The science of sound and sound control.

ADOBE A heavy clay soil used in many southwestern states to make sun-dried bricks.

AGGREGATE Stone, gravel, cinder, or slag used as one of the components of concrete.

AIR-DRIED LUMBER Lumber that has been stored in yards or sheds for a period of time after cutting. Building codes typically assume a 19 percent moisture content when determining joist and beams of air-dried lumber.

AIR DUCT A pipe, typically made of sheet metal, that carries air from a source such as a furnace or air conditioner to a room within a structure.

AIR TRAP A "U"-shaped piped placed in wastewater lines to prevent backflow of sewer gas.

ALCOVE A small room adjoining a larger room, often separated by an archway.

AMPERE (AMPS) A measure of electrical current.

ANCHOR A metal tie or strap used to tie building members to each other.

ANCHOR BOLT A threaded bolt used to fasten wooden structural members to masonry.

ANGLE IRON A structure piece of steel shaped to form a 90 degree angle.

APRON The inside trim board placed below a window sill. The term is also used to apply to a curb around a driveway or parking area.

AREAWAY A subsurface enclosure to admit light and air to a basement. Sometimes called a window well.

ASBESTOS A mineral that does not burn or conduct heat; it is usually used for roofing material.

ASHLAR MASONRY Squared masonry units laid with a horizontal bed joint.

ASH PIT An area in the bottom of the firebox of a fireplace to collect ash.

ASPHALT An insoluble material used for making floor tile and for waterproofing walls and roofs.

ASPHALTIC CONCRETE A mixture of asphalt and aggregate, which is used for driveways.

ASPHALT SHINGLE Roof shingles made of asphalt-saturated felt and covered with mineral granules.

ATRIUM An inside courtyard of a structure which may be either open at the top or covered with a roof.

AWNING WINDOW A window that is hinged along the top edge.

BACKFILL Earth, gravel, or sand placed in the trench around the footing and stem wall after the foundation has cured.

BAFFLE A shield, usually made of scrap material, to keep insulation from plugging eave vents. Also used to describe wind- or sound-deadening devices.

BALLOON FRAMING A building construction method that has vertical wall members that extend uninterrupted from the foundation to the roof.

BAND JOIST A joist set at the edge of the structure that runs parallel to the other joist. Also called a rim joist.

BANISTER A handrail beside a stairway

BASEBOARD The finish trim where the wall and floor intersect, or an electric wall heater that extends along the floor.

BASE COURSE The lowest course in brick or concrete masonry unit construction.

BASE LINE A reference line.

BASEMENT A level of a structure that is built either entirely below grade level (full basement) or partially below grade (daylite basement).

BATT A blanket insulation usually made of fiberglass to be used between framing members.

BATTEN A board used to hide the seams when other boards are joined together.

BATTER BOARD A horizontal board used to form footings.

BAY A division of space within a building, usually divided by beams or columns.

BEAM A horizontal structural member that is used to support roof or wall loads. Often called a header.

BEAMED CEILING A ceiling that has support beams that are exposed to view.

BEARING PLATE A support member, often a steel plate used to spread weight over a larger area.

BEARING WALL A wall that supports vertical loads in addition to its own weight.

BENCH MARK A reference point used by surveyors to establish grades and construction heights.

BEVELED SIDING Siding that has a tapered thickness.

BIBB An outdoor faucet which is threaded so that a hose may be attached.

BILL OF MATERIAL A part of a set of plans that lists all of the material needed to construct a structure.

BIRD BLOCK A block placed between rafters to maintain a uniform spacing and to keep animals out of the attic.

BIRD'S MOUTH A notch cut into a rafter to provide a bearing surface where the rafter intersects the top plate.

BLIND NAILING Driving nails in such a way that the heads are concealed from view.

BLOCKING Framing members, typically wood, placed between joist, rafters or studs to provide rigidity. Also called bridging.

BOARD AND BATTEN A type of siding using vertical boards with small wood strips (battens) used to cover the joints of the boards.

BOARD FOOT The amount of wood contained in a piece of lumber 1 in. thick by 12 in. wide by 12 in. long.

BOND The mortar joint between two masonry units, or a pattern in which masonry units are arranged.

BOND BEAM A reinforced concrete beam used to strengthen masonry walls.

BOTTOM CHORD The lower, usually horizontal, member of a truss.

BOX BEAM A hollow built-up structural unit.

BRIDGING Cross blocking between horizontal members used to add stiffness. Also called blocking.

BREEZEWAY A covered walkway with open sides between two different parts of a structure.

BROKER A representative of the seller in property sales.

BTU British thermal unit. A unit used to measure heat.

BUILDING CODE Legal requirements designed to protect the public by providing guidelines for structural, electrical, plumbing, and mechanical areas of a structure.

BUILDING LINE An imaginary line determined by zoning departments to specify on which area of a lot a structure may be built (also known as a setback).

BUILDING PAPER A waterproofed paper used to prevent the passage of air and water into a structure.

BUILDING PERMIT A permit to build a structure issued by a governmental agency after the plans for the structure have been examined and the structure is found to comply with all building code requirements.

BUILT-UP BEAM A beam built of smaller members that are bolted or nailed together.

BUILT-UP ROOF A roof composed of three or more layers of felt, asphalt, pitch, or coal tar.

BULLNOSE Rounded edges of cabinet trim.

BUTT JOINT The junction where two members meet in a square-cut joint; end to end, or edge to edge.

BUTTRESS A projection from a wall often located below roof beams to provide support to the roof loads and to keep long walls in the vertical position.

CABINET WORK The interior finish woodwork of a structure, especially cabinetry.

CANTILEVER Projected construction that is fastened at only one end.

CANT STRIP A small built-up area between two intersecting roof shapes to divert water.

CARRIAGE The horizontal part of a stair stringer that supports the tread.

CASEMENT WINDOW A hinged window that swings outward.

CASING The metal, plastic, or wood trim around a door or a window.

CATCH BASIN An underground reservoir for water drained from a roof before it flows to a storm drain.

CATHEDRAL WINDOW A window with an upper edge which is parallel to the roof pitch.

CAULKING A soft, waterproof material used to seal seams and cracks in construction.

CAVITY WALL A masonry wallformed with an air space between each exterior face.

CEILING JOIST The horizontal member of the roof which is used to resist the outward spread of the rafters and to provide a surface on which to mount the finished ceiling.

CEMENT A powder of alumina, silica, lime, iron oxide, and magnesia pulverized and used as an ingredient in mortar and concrete.

CENTRAL HEATING A heating system in which heat is distributed throughout a structure from a single source.

CESSPOOL An underground catch basin for the collection and dispersal of sewage.

CHAMFER A beveled edge formed by removing the sharp corner of a piece of material.

CHANNEL A standard form of structural steel with three sides at right angles to each other forming the letter C.

CHASE A recessed area of column formed between structural members for electrical, mechanical, or plumbing materials.

CHECK Lengthwise cracks in a board caused by natural drying.

CHECK VALVE A valve in a pipe that permits flow in only one direction.

CHIMNEY An upright structure connected to a fireplace or furnace that passes smoke and gases to outside air.

CHORD The upper and lower members of a truss which are supported by the web.

CINDER BLOCK A block made of cinder and cement used in construction.

CIRCUIT BREAKER A safety device which opens and closes an electrical circuit.

CLAPBOARD A tapered board used for siding that overlaps the board below it.

CLEARANCE A clear space between building materials to allow for air flow or access.

CLERESTORY A window or group of windows which are placed above the normal window height, often between two roof levels.

COLLAR TIES A horizontal tie between rafters near the ridge to help resist the tendency of the rafters to separate.

COLUMN A vertical structural support, usually round are made of steel.

COMMON WALL The partition that divides two different dwelling units.

COMPRESSION A force that crushes or compacts.

CONCRETE A building material made from cement, sand, gravel, and water.

CONCRETE BLOCKS Blocks of concrete that are precast. The standard size is 8 × 8 × 16.

CONDENSATION The formation of water on a surface when warm air comes in contact with a cold surface.

CONDUCTOR Any material that permits the flow of electricity. A drain pipe that diverts water from the roof (a downspout).

CONDUIT A bendable pipe or tubing used to encase electrical wiring.

CONTINUOUS BEAM A single beam that is supported by more than two supports.

CONTRACTOR The manager of a construction project, or one specific phase of it.

CONTROL JOINT An expansion joint in a masonry wall formed by raking mortar from the vertical joint.

CONVENIENCE OUTLET An electrical receptacle through which current is drawn for the electrical system of an appliance.

COPING A masonry cap placed on top of a block wall to protect it from water penetration.

CORBEL A ledge formed in a wall by building out successive courses of masonry.

CORNICE The part of the roof that extends out from the wall. Sometimes referred to as the eave.

COUNTERFLASH A metal flashing used under normal flashing to provide a waterproof seam.

COURSE A continuous row of building material such as shingles, stone or brick.

CRAWL SPACE The area between the floor joists and the ground.

CRICKET A diverter built to direct water away from an area of a roof where it would otherwise collect such behind a chimney.

CRIPPLE A wall stud that is cut at less than full length.

CROSS BRACING Boards fastened diagonally between structural members such as floor joists to provide rigidity.

CULVERT An underground passageway for water, usually part of a drainage system.

CUPOLA A small structure built above the main roof level to provide light or ventilation.

CURTAIN WALL An exterior wall which provides no structural support.

DAMPER A movable plate that controls the amount of draft for a woodstove, fireplace, or furnace.

DATUM A reference point for starting a survey.

DEADENING Material used to control the transmission of sound.

DEAD LOAD The weight of building materials or other unmoveable objects in a structure.

DECKING A wood material used to form the floor or roof, typically used in 1 and 2 in. thicknesses.

DESIGNER A person who designs buildings, but it not licensed, as is an architect.

DIVERTER A metal strip used to divert water.

DORMER A structure which projects from a sloping roof to form another roofed area. This new area is typically used to provide a surface to install a window.

DOUBLE HUNG A type of window in which the upper and lower halves slide past each other to provide an opening at the top and bottom of the window.

DOWNSPOUT A pipe which carries rain water from the gutters of the roof to the ground.

DRAIN A collector for a pipe that carries water.

DRESSED LUMBER Lumber that has been surfaced by a planing machine to give the wood a smooth finish.

DRY ROT A type of wood decay caused by fungi that leaves the wood a soft powder.

DRYWALL An interior wall covering installed in large sheets made from gypsum board.

DRY WELL A shallow well used to disperse water from the gutter system.

DUPLEX OUTLET A standard electrical convenience outlet with two receptacles.

DUTCH HIP A type of roof shape that combines features of a gable and a hip roof.

EASEMENT An area of land that cannot be built upon because it provides access to a structure or to utilities such as power or sewer lines.

EAVE The lower part of the roof that projects from the wall. See cornice.

EGRESS A term used in building codes to describe access.

ELBOW An L-shaped plumbing pipe.

ELEVATION The height of a specific point in relation to another point. The exterior views of a structure.

ELL An extension of the structure at a right angle to the main structure.

EMINENT DOMAIN The right of a government to condemn private property so that it may be obtained for public use.

ENAMEL A paint that produces a hard, glossy, smooth finish.

EQUITY The value of real estate in excess of the balance owed on the mortgage.

EXCAVATION The removal of soil for construction purposes.

EXPANSION JOINT A joint installed in concrete construction to reduce cracking and to provide workable areas.

FABRICATION Work done on a structure away from the job site.

FACADE The exterior covering of a structure.

FACE BRICK Brick that is used on the visible surface to cover other masonry products.

FACE GRAIN The pattern in the visible veneer of plywood.

FASCIA A horizontal board nailed to the end of rafters or trusses to conceal their ends.

FELT A tar-impregnated paper used for water protection under roofing and siding materials. Sometimes used under concrete slabs for moisture resistance.

FIBERBOARD Fibrous wood products that have been pressed into a sheet. Typically used for the interior construction of cabinets and for a covering for the subfloor.

FILLED INSULATION Insulation material that is blown or poured into place in attics and walls.

FILLET WELD A weld between two surfaces that butt at 90° to each other with the weld filling the inside corner.

FINISHED LUMBER Wood that has been milled with a smooth finish suitable for use as trim and other finish work.

FINISHED SIZE Sometimes called the dressed size, the finished size represents the actual size of lumber after all milling operations and is typically about 2 in. smaller than the nominal size, which is the size of lumber before planing.

NOMINAL SIZE (IN.)	FINISHED SIZE (IN.)
1	3/4
2	1 1/2
4	3 1/2
6	5 1/2
8	7 1/4
10	9 1/4
12	11 1/4
14	13 1/4

FIREBRICK A refractory brick capable of withstanding high temperatures and used for lining fireplaces and furnaces.

FIREBOX The combustion chamber of the fireplace where the fire occurs.

FIREBRICK A brick made of a refractory material that can withstand great amounts of heat and is used to line the visible face of the firebox.

FIRE CUT An angular cut on the end of a joist or rafter that is supported by masonry. The cut allows the wood member to fall away from the wall without damaging a masonry wall when the wood is damaged by fire.

FIRE DOOR A door used between different types of construction which has been rated as being able to withstand fire for a certain amount of time.

FIREPROOFING Any material that is used to cover structural materials to increase their fire rating.

FIRE RATED A rating given to building materials to specify the amount of time the material can resist damage caused by fire.

FIRE-STOP Blocking placed between studs or other structural members to resist the spread of fire.

FIRE WALL A wall constructed of materials resulting in a specified time that the wall can resist fire before structural damage will occur.

FLASHING Metal used to prevent water leaking through surface intersections.

FLOOR PLUG A 110 convenience outlet located in the floor.

FLUE LINER A terra-cotta pipe used to provide a smooth flue surface so that unburned materials will not cling to the flue.

FOOTING The lowest member of a foundation system used to spread the loads of a structure across supporting soil.

FOOTING FORM The wooden mold used to give concrete its shape as it cures.

FOUNDATION The system used to support a building's loads and made up of stem walls, footings, and piers. The term is used in many areas to refer to the footing.

FROST LINE The depth to which soil will freeze.

FURRING Wood strips attached to structural members that are used to provide a level surface for finishing materials when different-sized structural members are used.

GABLE A type of roof with two sloping surfaces that intersect at the ridge of the structure.

GABLE END WALL The triangular wall that is formed at each end of a gable roof between the top plate of the wall and the rafters.

GALVANIZED Steel products that have had zinc applied to the exterior surface to provide protection from rusting.

GAMBREL A type of roof formed with two planes on each side. The lower pitch is steeper than the upper portion of the roof.

GIRDER A horizontal support member at the foundation level.

GLUED-LAMINATED TIMBER (GLU-LAM) A structural member made up of layers of lumber that are glued together.

GRADE The designation of the quality of a manufactured piece of wood.

GRADING The moving of soil to effect the elevation of land at a construction site.

GRAVEL STOP A metal strip used to retain gravel at the edge of built-up roofs.

GREEN LUMBER Lumber that has not been kiln-dried and still contains moisture.

GROUND FAULT CIRCUIT INTERRUPTER (GFCI OR GFI) A 110 convenience outlet with a built-in circuit breaker. GFCI outlets are to be used within 5'–0" of any water source.

GROUT A mixture of cement, sand, and water used to fill joints in masonry and tile construction.

GUARDRAIL A horizontal protective railing used around stairwells, balconies, and changes of floor elevation greater than 30 in.

GUSSET A plate added to the side of intersecting structural members to help form a secure connection and to reduce stress.

HALF-TIMBER A frame construction method where spaces between wood members are filled with masonry.

HANGER A metal support bracket used to attach two structural members.

HARDBOARD Sheet material formed of compressed wood fibers used as an underlayment for flooring.

HEAD The upper portion of a door or window frame.

HEADER A horizontal structural member used to support other structural members over openings such as doors and windows.

HEADER COURSE A horizontal masonry course with the end of each masonry unit exposed.

HEADROOM The vertical clearance in a room over a stairway.

HEARTH The fire-resistant floor within and extending a minimum of 18 in. in front of the firebox.

HEARTWOOD The inner core of a tree trunk.

HIP ROOF A roof shape with four sloping sides.

HORIZONTAL SHEAR One of three major forces acting on a beam, it is the tendency of the fibers of a beam to slide past each other in a horizontal direction.

HOSE BIBB A water outlet that is threaded to receive a hose.

HUMIDIFIER A mechanical device that controls the amount of moisture inside of a structure.

INDIRECT LIGHTING Mechanical lighting that is reflected off a surface.

ISOMETRIC A drawing method which enables three surfaces of an object to be seen in one view, with the base of each surface drawn at 30° to the horizontal plane.

JACK RAFTER A rafter which is cut shorter than the other rafters to allow for an opening in the roof.

JACK STUD A wall member which is cut shorter than other studs to allow for an opening such as a window. Also called a cripple stud.

JAMB The vertical members of a door or window frame.

JOIST A horizontal structural member used in repetitive patterns to support floor and ceiling loads.

KILN DRIED A method of drying lumber in a kiln or oven. Kiln dried lumber has a reduced moisture content when compared to lumber that has been air dried.

KING STUD A full-length stud placed at the end of a header.

KNEE WALL A wall of less than full height.

LALLY COLUMN A vertical steel column that is used to support floor or foundation loads.

LAMINATED Several layers of material that have been glued together under pressure.

LATTICE A grille made by criss-crossing strips of material.

LEDGER A horizontal member which is attached to the side of wall members to provide support for rafters or joists.

LIEN A monetary claim on property.

LINTEL A horizontal steel member used to provide support for masonry over an opening.

LIVE LOAD The loads from all movable objects within a structure including loads from furniture and people. External loads from snow and wind are also considered live loads.

LOAD-BEARING WALL A support wall which holds floor or roof loads in addition to its own weight.

LOOKOUT A beam used to support eave loads.

MANSARD A sour-sided, steep-sloped roof.

MANTEL A decorative shelf above the opening of a fireplace.

MESH A metal reinforcing material placed in concrete slabs and masonry walls to help resist cracking.

METAL WALL TIES Corrugated metal strips used to bond brick veneer to its support wall.

MILLWORK Finished woodwork that has been manufactured in a milling plant. Examples are window and door frames, mantels, moldings, and stairway components.

MINERAL WOOL An insulating material made of fibrous foam.

MODULUS OF ELASTICITY (E) The degree of stiffness of a beam.

MOISTURE BARRIER Typically a plastic material used to restrict moisture vapor from penetrating into a structure.

MONOLITHIC Concrete construction created in one pouring.

MONUMENT A boundary marker set to mark property corners.

MORTAR A combination of cement, sand, and water used to bond masonry units together.

MUSDILL The horizontal wood member that rests on concrete to support other wood members.

MULLION A horizontal or vertical divider between sections of a window.

MUNTIN A horizontal or vertical divider within a section of a window.

NAILER A wood member bolted to concrete or steel members to provide a nailing surface for attaching other wood members.

NEWEL The end post of a stair railing.

NOMINAL SIZE An approximate size achieved by rounding the actual material size to the nearest larger whole number.

NONBEARING WALL A wall which supports no loads other than its own weight. Some building codes consider walls which support only ceiling loads as nonbearing.

NONFERROUS METAL Metal, such as copper or brass, that contains no iron.

NOSING The rounded front edge of a tread which extends past the riser.

ON CENTER A measurement taken from the center of one member to the center of another member.

OUTLET An electrical receptacle which allows for current to be drawn from the system.

OUTRIGGER A support for roof sheathing and the fascia which extends past the wall line perpendicular to the rafters.

OVERHAND The horizontal measurement of the distance the roof projects from a wall.

PARAPET A portion of wall that extends above the edge of the roof.

PARGING A thin coat of plaster used to smooth a masonry surface.

PARQUET FLOORING Wood flooring laid to form patterns.

PARTITION An interior wall.

PARTY WALL A wall dividing two adjoining spaces such as apartments or offices.

PILASTER A reinforcing column built into or against a masonry wall.

PILING A vertical foundation support driven into the ground to provide support on stable soil or rock.

PLANK Lumber which is 1½ to 3½ in. in thickness.

PLASTER A mix of sand, cement, and water, used to cover walls and ceilings.

PLAT A map of an area of land which shows the boundaries of individual lots.

PLATE A horizontal member at the top (top plate) or bottom (sole plate or sill) of walls used to connect the vertical wall members.

PLENUM An air space for transporting air from the HVAC system.

PLOT A parcel of land.

PLUMB True vertical.

PLYWOOD Wood composed of three or more layers, with the grain of each layer placed at 90° to each other and bounded with glue.

PORCH A covered entrance to a structure.

PORTICO A roof support by columns instead of walls.

PORTLAND CEMENT A hydraulic cement made of silica, lime, and aluminum that has become the most common cement used in the construction industry because of its strength.

POST A vertical wood structural member usually 4 × 4 or larger.

PRECAST A concrete component which has been cast in a location than the one in which it will be used.

PREFABRICATED Buildings or components that are built away from the job site and transported ready to be used.

PRESTRESSED A concrete component that is placed in compression as it is cast to help resist deflection.

PURLIN A horizontal roof member which is laid perpendicular to rafters to help limit deflection.

PURLIN BRACE A support member which extends from the purlin down to a load-bearing wall or header.

QUAD A courtyard surrounded by the walls of buildings.

QUARRY TILE An unglazed, machine-made tile.

QUARTER ROUND Wood molding that has the profile of one-quarter of a circle.

RABBET A rectangular groove cut on the edge of a board.

RAFTER The inclined structural member of a roof system designed to support roof loads.

RAFTER/CEILING JOINT An inclined structural member which supports both the ceiling and the roof materials.

RAKE JOINT A recessed mortar joint.

REBAR Reinforcing steel used to strengthen concrete.

REFERENCE BUBBLE A symbol used to designate the origin of details and sections.

REGISTER An opening in a duct for the supply of heated or cooled air.

REINFORCED CONCRETE Concrete that has steel rebar placed in it to resist tension.

RHEOSTAT An electrical control device used to regulate the current reaching a light fixture. A dimmer switch.

RELATIVE HUMIDITY The amount of water vapor in the atmosphere compared to the maximum possible amount at the same temperature.

RESTRAINING WALL A masonry wall supported only at the bottom by a footing that is designed to resist soil loads.

RETAINING WALL A masonry wall supported at the top and bottom, designed to resist soil loads.

R-FACTOR A unit of thermal resistance applied to the insulating value of a specific building material.

RIBBON A structural wood member framed into studs to support joists or rafters.

RIDGE BOARD A horizontal member that rafters are aligned against to resist their downward force.

RIDGE BRACE A support member used to transfer the weight from the ridge board to a bearing wall or beam. The brace is typically space at 48 in. O.C., and may not exceed a 45° angle from vertical.

RIM JOIST A joist at the perimeter of a structure that runs parallel to the other floor joist.

RISE The amount of vertical distance between one tread and another.

RISER The vertical member of stairs between the treads.

ROLL ROOFING Roofing material of fiber or asphalt that is shipped in rolls.

ROOF DRAIN A receptacle for removal of roof water.

ROUGH FLOOR The subfloor, usually hardboard, which serves as a base for the finished floor.

ROUGH HARDWARE Hardware used in construction, such as nails, bolts, and metal connectors, that will not be seen when the project is complete.

ROUGH IN To prepare a room for plumbing or electrical additions by running wires or piping for a future fixture.

ROUGH LUMBER Lumber that has not been surfaced but has been trimmed on all four sides.

ROUGH OPENING The unfinished opening between framing members allows for doors, windows, or other assemblies.

ROWLOCK A pattern for laying masonry units so that the end of the unit is exposed.

RUN The horizontal distance of a set of steps or the measurement describing the depth of one step.

SADDLE A small gable-shaped roof used to divert water from behind a chimney.

SASH An individual frame around a window.

SCAB A short member that overlaps the butt joint of two other members used to fasten those members.

SCRATCH COAT The first coat of stucco which is scratched to provide a good bond surface for the second coat.

SEASONING The process of removing moisture from green lumber by either air- (natural) or kiln-drying.

SEISMIC Earthquake-related forces.

SEPTIC TANK A tank in which sewage is decomposed by bacteria and dispersed by drain tiles.

SERVICE CONNECTION The wires that run to a structure from a power pole or transformer.

SHAKE A hand-split wooden roof shingle.

SHEAR The stress that occurs when two forces from opposite directions are acting on the same member. Shearing stress tends to cut a member just as scissors cut paper.

SHEAR PANEL A plywood panel applied to walls to resist wind and seismic forces by keeping the studs in a vertical position.

SHEATHING A covering material placed over walls, floors, and roofs which serves as a backing for finishing materials.

SHIM A piece of material used to fill a space between two surfaces.

SHIPLAP A siding pattern of overlapping rabbeted edges.

SILL A horizontal wood member placed at the bottom of walls and openings in walls.

SKYLIGHT An opening in the roof to allow light and ventilation that is usually covered with glass or plastic.

SLEEPERS Strips of wood placed over a concrete slab in order to attach other wood members.

SMOKE CHAMBER The portion of the chimney located directly over the firebox which acts as a funnel between the firebox and the chimney.

SMOKE SHELF A shelf located at the bottom of the smoke chamber to prevent down-drafts from the chimney from entering the firebox.

SOFFIT A lowered ceiling, typically found in kitchens, halls, and bathrooms to allow for recessed lighting or HVAC ducts.

SOIL STACK The main vertical waste-water pipe.

SOLDIER A masonry unit laid on end with its narrow surface exposed.

SOLE PLATE The plate placed at the bottom of a wall.

SPECIFICATIONS Written descriptions or requirements to specify how a structure is to be constructed.

SPLICE Two similar members that are jointed together in a straight line usually by nailing or bolting.

SPLIT-LEVEL A house that has two levels, one about a half a level above or below the other.

SQUARE An area of roofing covering 100 square feet.

STACK A vertical plumbing pipe.

STAIR WELL The opening in the floor where a stair will be framed.

STILE A vertical member of a cabinet, door, or decorative panel.

STIRRUP A U-shaped metal bracket used to support wood beams.

STOCK Common sizes of building material as they are sold.

STOP A wooden strip used to hold windows in place.

STRESSED-SKIN PANEL A hollow, built-up member typically used as a beam.

STRINGER The inclined support member of a stair that supports the risers and treads.

STUCCO A type of plaster made from Portland cement, sand, water, and a coloring agent that is applied to exterior walls.

STUD The vertical framing member of a wall which is usually 2 × 4 or 2 × 6 in size.

SUBFLOOR The flooring surface that is laid on the floor joist and serves as a base layer for the finished floor.

SUMP A recessed area in a basement floor to collect water so that it can be removed by a pump.

SURFACED LUMBER Lumber that has been smoothed on at least one side.

SWALE A recessed area formed in the ground to help divert ground water away from a structure.

TENSILE STRENGTH The resistance of a material or beam to the tendency of stretch.

TERMITE SHIELD A strip of sheet metal used at the intersection of concrete and wood surfaces near ground level to prevent termites from entering the wood.

TERRA-COTTA Hard-baked clay typically used as a liner for chimneys.

THERMAL CONDUCTOR A material suitable for transmitting heat.

THERMAL RESISTANCE Represented by the letter R, resistance measures the ability of a material to resist the flow of heat.

THERMOSTAT A mechanical device for controlling the output of HVAC units.

THRESHOLD The beveled member directly under a door.

THROAT The narrow opening to the chimney that is just above the firebox. The throat of a chimney is where the damper is placed.

TIMBER Lumber with a cross-sectional size of 4 × 6 in. or larger.

TOENAIL Nails driven into a member at an angle.

TONGUE AND GROOVE A joint where the edge of one member fits into a groove in the next member.

TRANSOM A window located over a door.

TRAP A U-shaped pipe below plumbing fixtures which holds water to prevent odor and sewer gas from entering the fixture.

TREAD The horizontal member of a stair on which the foot is placed.

TRIMMER Joist or rafters that are used to frame an opening in a floor, ceiling, or roof.

TRUSS A framework made in triangular-shaped segments used for spanning distances greater than is possible using standard components and methods.

VALLEY The internal corner formed between two intersecting roof surfaces.

VAPOR BARRIER Material that is used to block the flow of water vapor into a structure. Typically 6 mil (.006 in.) black plastic.

VENEER A thin outer covering or nonload bearing masonry face material.

VENTILATION The process of supplying and removing air from a structure.

VENT PIPE Pipes that provide air into the waste lines to allow drainage by connecting each plumbing fixture to the vent stack.

VENT STACK A vertical pipe of a plumbing system used to equalize pressure within the system and to vent sewer gases.

VESTIBULE A small entrance or lobby.

WAINSCOT A paneling applied to the lower portion of a wall.

WALLBOARD Large flat sheets of gypsum, typically ½ or ⅝ in. thick, used to finish interior walls.

WATERPROOF Material or a type of construction that prevents the absorption of water.

WATER STRIP A fabric or plastic material placed along the edges of doors, windows, and skylites to reduce air infiltration.

WEEP HOLE An opening normally in the bottom course of a masonry to allow for drainage.

WYTHE A single unit thickness of a masonry wall.

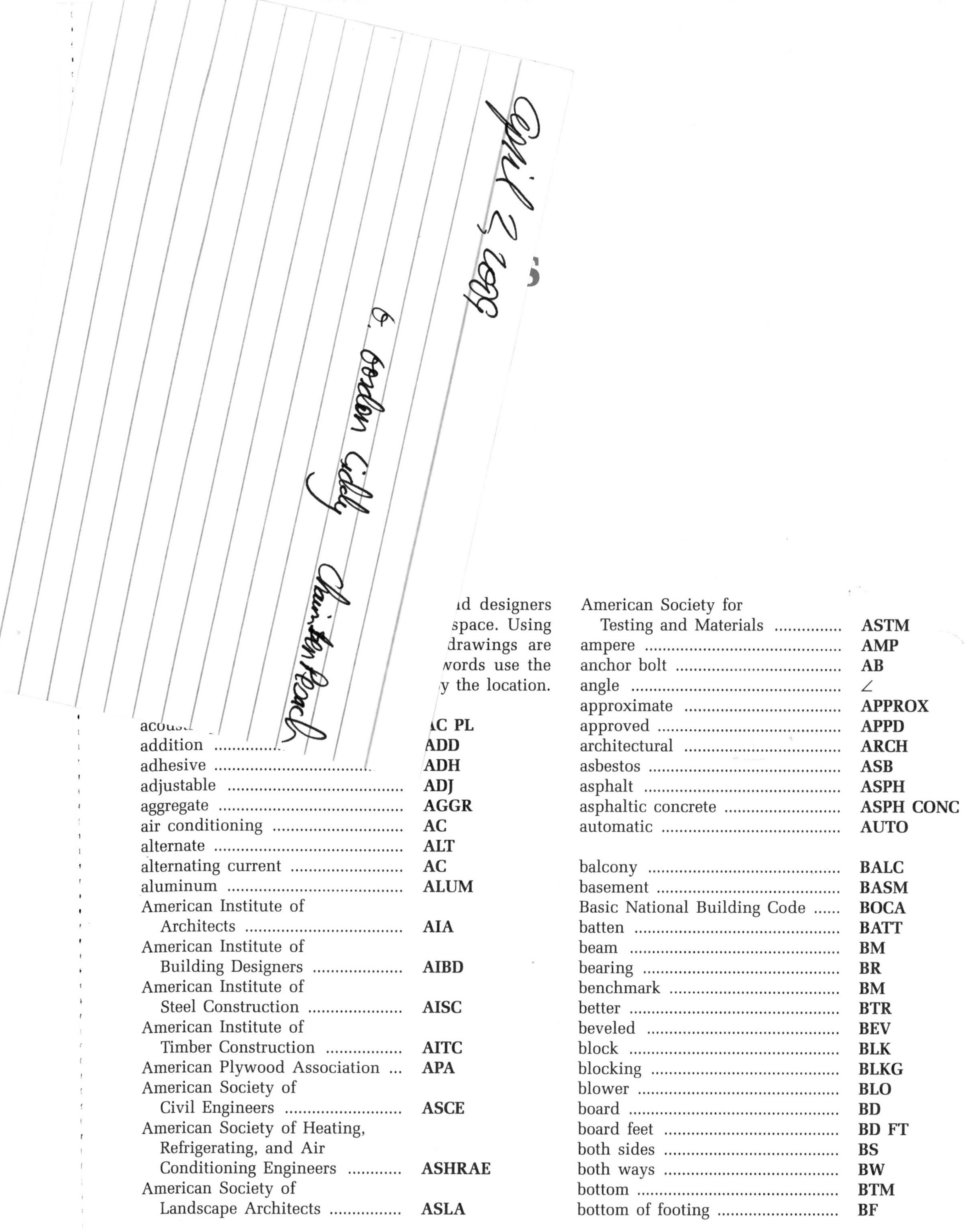

…d designers
…space. Using
…drawings are
…words use the
…y the location.

acou[illegible]	C PL
addition	ADD
adhesive	ADH
adjustable	ADJ
aggregate	AGGR
air conditioning	AC
alternate	ALT
alternating current	AC
aluminum	ALUM
American Institute of Architects	AIA
American Institute of Building Designers	AIBD
American Institute of Steel Construction	AISC
American Institute of Timber Construction	AITC
American Plywood Association	APA
American Society of Civil Engineers	ASCE
American Society of Heating, Refrigerating, and Air Conditioning Engineers	ASHRAE
American Society of Landscape Architects	ASLA
American Society for Testing and Materials	ASTM
ampere	AMP
anchor bolt	AB
angle	∠
approximate	APPROX
approved	APPD
architectural	ARCH
asbestos	ASB
asphalt	ASPH
asphaltic concrete	ASPH CONC
automatic	AUTO
balcony	BALC
basement	BASM
Basic National Building Code	BOCA
batten	BATT
beam	BM
bearing	BR
benchmark	BM
better	BTR
beveled	BEV
block	BLK
blocking	BLKG
blower	BLO
board	BD
board feet	BD FT
both sides	BS
both ways	BW
bottom	BTM
bottom of footing	BF

brass	BR
brick	BRK
British thermal unit	BTU
bronze	BRZ
building	BLDG
building line	BL
built-in	BLT-IN
buzzer	BUZ
by	×
cabinet	CAB
cast concrete	C CONC
cast iron	CI
catch basin	CB
caulking	CALK
ceiling	CLG
ceiling diffuser	CD
ceiling joist	CJ OR CEIL JST
cement	CEM
center	CTR
center to center	C-C
centerline	CL
centimeter	CM
ceramic	CER
chamfer	CHAM
channel	C
check	CHK
cinder block	CIN BLK
circuit	CIR
circuit breaker	CIR BKR
class	CL
cleanout	CO
clear	CLR
coated	CTD
cold water	CW
column	COL
combination	COMB
common	COM
composition	COMP
concrete	CONC
concrete masonry unit	CMU
conduit	CND
construction	CONST
Construction Standards Institute	CSI
contractor	CONTR
control joint	CJ
copper	COP
corridor	CORR
countersink	CSK
cubic	CU
cubic feet	CU FT
cubic feet per minute	CFM
cubic inch	CU IN
cubic yard	CU YD
damper	DPR
decibel	DB
decking	DK
degree	° or DEG
detail	DET
diagonal	DIAG
diameter	∅ or DIA
diffuser	DIF
dimension	DIMEN
dishwasher	D/W
disposal	DISP
double hung	DH
Douglas fir	DF
downspout	DS
drain	D
drinking fountain	DF
dryer	D
drywall	DW
each face	EF
each way	EW
elbow	EL
electrical	ELECT
elevation	ELEV
enamel	ENAM
engineer	ENGR
entrance	ENT
Environmental Protection Agency	EPA
estimate	EST
excavate	EXC
exhaust	EXH
existing	EXIST
expansion joint	EXP JT
exposed	EXPO
extension	EXTN
fabricate	FAB
face of studs	FOS
Fahrenheit	F
Federal Housing Administration	FHA
feet per minute	FPM
finished	FIN
finished floor	FIN FL
finished grade	FIN GR
finished opening	FO
firebrick	FBRK
fixture	FIX
flammable	FLAM
flange	FLG
flashing	FL
flexible	FLEX
floor	FLR
floor drain	FD
floor joist	FL JST
floor sink	FS
folding	FLDG
foot candle	FC
footing	FTG
foot pounds	FT LB

Term	Abbreviation
forced air unit	FAU
foundation	FND
furnace	FURN
furred ceiling	FC
galvanized	GALV
galvanized iron	GI
gage	GA
garage	GAR
girder	GIRD
glass	GL
glue laminated	GLU-LAM
grude	GR
grade beam	GR BM
gravel	GVL
grille	GR
ground	GND
ground fault circuit interrupter	GFCI
gypsum	GYP
gypsum board	GYP BD
hardboard	HDB
hardware	HDW
hardwood	HDWD
header	HDR
heater	HTR
heating	HTG
heating/ventilating/air conditioning	HVAC
height	HT
hemlock	HEM
hemlock-fir	HEM-FIR
hollow core	HC
horsepower	HP
hose bibb	HB
hot water	HW
hot water heater	HWH
hundred	C
illuminate	ILLUM
incandescent	INCAN
inch pounds	IN LB
incinerator	INCIN
inflammable	INFL
inside diameter	ID
inside face	IF
inspection	SP
install	INST
insulate	INS
insulation	INSUL
interior	INT
International Conference of Building Officials	ICBO
iron	I
jamb	JMB
joint	JT
joist	JST
junction	JCT
junction box	J-BOX
kiln dried	KD
kilowatt	KW
kilowatt hour	KWH
Kip (1,000 lb.)	K
knockout	KO
laboratory	LAB
landing	LDG
laundry	LAU
lavatory	LAV
level	LEV
light	LT
linear feet	LIN FT
linen closet	L CL
linoleum	LINO
living room	LIV
louver	LV
lumber	LUM
machine bolt	MB
manhole	MH
manufacturer	MANUF
marble	MRB
masonry	MAS
material	MAT
mechanical	MECH
medicine cabinet	MC
medium	MD
membrane	MEMB
metal	MTL
meter	M
mirror	MIRR
miscellaneous	MISC
mixture	MIX
model	MOD
modular	MOD
molding	MLDG
motor	MOT
mullion	MULL
National Association of Home Builders	NAHB
National Bureau of Standards	NBS
natural	NAT
natural grade	NAT GR
noise reduction coefficient	NRC
nominal	NOM
not applicable	NA
not in contract	NIC
not to scale	NTS
obscure	OBS
on center	OC

opening	**OPG**
opposite	**OPP**
ounce	**OZ**
outside diameter	**OD**
outside face	**OF**
overhead	**OVHD**
painted	**PTD**
panel	**PNL**
parallel	**// or PAR**
part	**PT**
partition	**PART**
pavement	**PVMT**
penny	**D**
per	**/**
perforate	**PERF**
perimeter	**PER**
permanent	**PERM**
perpendicular	**⊥ or PERP**
pi (3.1416)	**Π**
plaster	**PLS**
plasterboard	**PLS BD**
plastic	**PLAS**
plate	**P or PL**
platform	**PLAT**
plumbing	**PLMB**
plywood	**PLY**
polished	**POL**
polyethylene	**POLY**
polyvinyl chloride	**PVC**
position	**POS**
pound	**# or LB**
pounds per square foot	**PSF**
pounds per square inch	**PSI**
precast	**PRCST**
prefabricated	**PREFAB**
preferred	**PFD**
preliminary	**PRELIM**
pressure treated	**PT**
property	**PROP**
pull chain	**PC**
pushbutton	**PB**
quality	**QTY**
quantity	**QTY**
radiator	**RAD**
radius	**R or RAD**
random length and width	**R L & W**
receptacle	**RECP**
recessed	**REC**
redwood	**RDWD**
reference	**REF**
refrigerator	**REFR**
register	**REG**
reinforcing	**REINF**
reinforcing bar	**REBAR**
reproduce	**REPRO**
required	**REQD**
return	**RET**
revision	**REV**
ridge	**RDG**
riser	**RIS**
roof drain	**RD**
roofing	**RFG**
rough	**RGH**
rough opening	**RO**
round	**∅ or RD**
safety	**SAF**
sanitary	**SAN**
screen	**SCRN**
screw	**SCR**
second	**SEC**
section	**SECT**
select	**SEL**
select structural	**SEL ST**
self-closing	**SC**
service	**SERV**
sewer	**SEW**
sheathing	**SHTG**
siding	**SDG**
sill cock	**SC**
similar	**SIM**
single hung	**SH**
soil pipe	**SP**
solid block	**SOL BLK**
solid core	**SC**
Southern Building Code	**SBC**
Southern pine	**SP**
Southern Pine Inspection Bureau	**SPIB**
specifications	**SPECS**
spruce-pine-fir	**SPF**
square	**□ or SQ**
square feet	**or SQ FT**
square inch	**# or SQ IN**
stainless steel	**SST**
stand pipe	**ST P**
standard	**STD**
steel	**STL**
stirrup	**STIR**
stock	**STK**
storage	**STO**
storm drain	**SD**
structural	**STR**
structural clay tile	**SCT**
substitute	**SUB**
supply	**SUP**
surface	**SUR**
surface four sides	**S4S**
surface two sides	**S2S**
suspended ceiling	**SUSP CLG**
switch	**SW**

symbol .... SYM
symmetrical .... SYM
synthetic .... SYN

tangent .... TAN
tar and gravel .... T & G
telephone .... TEL
temperature .... TEMP
terra-cotta .... TC
terrazzo .... TZ
thermostat .... THRM
thickness .... THK
thousand .... M
thousand board feet .... MBF
threshold .... THR
through .... THRU
tongue and groove .... T & G
top of wall .... TW
tread .... TR
tubing .... TUB

Underwriters' Laboratories, Inc. ... UL
unfinished .... UNFIN
Uniform Building Code .... UBC
United States Department of Housing and Urban Development .... HUD
urinal .... UR
utility .... UTIL

valve V
vanity .... VAN
vapor barrier .... VB
vapor proof .... VAP PRF
ventilation .... VENT
vent pipe .... VP
vent stack .... VS
vent through roof .... VTR
vertical grain .... VERT GR
vinyl .... VIN
vinyl asbestos tile .... VAT
vinyl base .... VB
vinyl tile .... VT
vitreous .... VIT
vitreous clay tile .... VCT
volt V
volume .... VOL

wainscot .... WSCT
wall vent .... WV
washing machine .... WM
waste stack .... WS
water closet .... WC
water heater .... WH
waterproof .... WP
watt .... W
weather stripping .... WS
weatherproof .... WP
weep hole .... WH
weight .... WT
welded wire fabric .... WWF
white pine .... WP
wide flange .... W or WF
width .... W
window .... WDW
with .... W/
without .... W/O
wood .... WD
wrought iron .... WI

yellow pine .... YP

zinc .... ZN

Index

Note: Page numbers in **bold type** reference non-text material.